BIOLOGICAL OXIDATIONS

EDITED BY

THOMAS P. SINGER

Professor of Biochemistry, University of California School of Medicine and Chief, Molecular Biology Division, Veterans Administration Hospital, San Francisco, California

1968

INTERSCIENCE PUBLISHERS

A DIVISION OF JOHN WILEY & SONS, NEW YORK · LONDON · SYDNEY

Preface

Since the appearance of Green's *Mechanisms of Biological Oxidations* in 1940 no single volume dealing with this subject in a comprehensive manner has been published, although this has been one of the most widely investigated and productive areas of research in biochemistry. The contributors to this volume, most of whom have been actively engaged in teaching graduate students and post-doctoral fellows, have felt it desirable to publish a reasonably up-to-date treatise on biological oxidations, which students, fellows, and investigators in other branches of science might use to gain an overall acquaintance with the present state of knowledge in this field.

It has been the intention to emphasize reaction mechanisms, wherever possible, instead of the lexicographic presentation of data on individual enzymes (since such information is readily available in standard treatises, such as *Methods in Enzymology*), and to include not only topics where the information appears to be firmly established but to examine also the periphery of established knowledge and to dissect and discuss alternative hypotheses on controversial or unsettled problems, stating pros and cons in a logical and unbiased manner. It has been our hope that to the extent that we succeed along these lines, this volume might become more than a textbook and might be of value also to the practicing biochemist.

As is often the case in collective volumes, the original aims have been followed to varying extent depending on the nature of the subject matter, which in some cases precludes useful generalizations or demands different treatment for ease of presentation, as well as the predilections of the individual contributors. The differences in treatment are particularly apparent in dealing with areas where precise information is scanty: some authors have preferred to avoid these subjects, while others have dealt with these at length. Where interpretations of a given phenomenon or observation are at variance in different chapters, attention was called to alternative hypotheses wherever possible.

Regardless of differences in the method of presentation, however, a common thread which runs throughout this volume may be discerned: impressive though the advances have been in this field during the past decades, this remains a very lively, challenging, and productive area in research, and problems which remain for the future to resolve in many cases overshadow established knowledge.

This book is divided into two parts. The first part is concerned with overall processes, while the second deals with individual groups of enzymes classified according to prosthetic groups. In instances where the chemistry of the coenzymes is highly specialized or not part and parcel of the general biochemical background of the average reader, special chapters are included or the subject is briefly surveyed at the beginning of the individual chapter. Even in such instances those aspects of the chemistry are emphasized which bear directly on the catalytic mechanism.

Lastly it is well to point out, lest some of the contributors be accused of presenting outdated information, that progress in some of the areas covered in this volume has been uncommonly rapid in recent times and that there has been an unfortunate time lapse of almost two years between the receipt of the first and last manuscripts.

Thomas P. Singer

San Francisco
December, 1967

Authors

DANIEL I. ARNON, *Department of Cell Physiology, University of California, Berkeley, California.*

P. D. BOYER, *Molecular Biology Institute and Department of Chemistry, University of California, Los Angeles, California.*

FREDERICK L. CRANE, *Department of Biological Sciences, Purdue University, Lafayette, Indiana.*

ANDERS EHRENBERG, *Medicinska Nobelinstitutet, Biokemiska Institutionen, Stockholm, Sweden.*

QUENTIN H. GIBSON, *Division of Biological Sciences, Cornell University, Ithaca, New York.*

P. HANDLER, *Department of Biochemistry, Duke University Medical Center, Durham, North Carolina.*

OSAMU HAYAISHI, *Department of Medical Chemistry, Kyoto University Faculty of Medicine, Kyoto, Japan.*

PETER HEMMERICH, *Institut für anorganische Chemie, an der Universität Basel, Basel, Switzerland.*

F. M. HUENNEKENS, *Department of Biochemistry, Scripps Clinic and Research Foundation, La Jolla, California.*

MARTIN KLINGENBERG, *Physiologisch-Chemisches Institut der Philipps-Universität, Marburg, Germany.*

BO G. MALMSTRÖM, *Department of Biochemistry, University of Gothenburg, Gothenburg, Sweden.*

VINCENT MASSEY, *Department of Biological Chemistry and Biophysics Research Division, Institute of Science and Technology, The University of Michigan, Ann Arbor, Michigan.*

GRAHAM PALMER, *Biophysics Research Division, Institute of Science and Technology and Department of Biological Chemistry, The University of Michigan, Ann Arbor, Michigan.*

K. V. RAJAGOPALAN, *Department of Biochemistry, Duke University Medical Center, Durham, North Carolina.*

LARS RYDÉN, *Department of Biochemistry, University of Gothenburg, Gothenburg, Sweden.*

ANTHONY SAN PIETRO, *Charles F. Kettering Research Laboratory, Yellow Springs, Ohio.*

THOMAS P. SINGER, *Biochemistry Department, University of California School of Medicine and Molecular Biology Division, Veterans Administration Hospital, San Francisco, California.*

LUCILE SMITH, *Department of Biochemistry, Dartmouth Medical School, Hanover, New Hampshire.*

PHILIPP STRITTMATTER, *Department of Biological Chemistry, Washington University School of Medicine, St. Louis, Missouri.*

HORST SUND, *Universität Konstanz, Konstanz, Germany*

Contents

PART I: PROCESSES

The Respiratory Chain. By Martin Klingenberg 3
The Respiratory Chain System of Bacteria. By Lucile Smith . . . 55
Electron Transport and Phosphorylation in Photosynthesis by
Chloroplast. By Daniel I. Arnon 123
Microsomal Electron Transport. By Philipp Strittmatter 171
Oxidative Phosphorylation. By P. D. Boyer 193

PART II: THE OXIDIZING ENZYMES

Flavocoenzymes: Chemistry and Molecular Biology.
By Anders Ehrenberg and Peter Hemmerich 239
Mechanisms of Flavoprotein Catalysis. By Graham Palmer and
Vincent Massey 263
Metalloflavoproteins. By K. V. Rajagopalan and P. Handler . . 301
The Respiratory Chain-Linked Dehydrogenases. By Thomas P. Singer 339
Cytochromes. By Quentin H. Gibson 379
The Copper-Containing Oxidases. By Bo G. Malmström and
Lars Rydén 415
Folate and B_{12} Coenzymes. By F. M. Huennekens 439
Ferredoxin and Photosynthetic Pyridine Nucleotide Reductase.
By Anthony San Pietro 515
Quinones in Electron Transport. By Frederick L. Crane 533
Oxygenases (Oxygen-Transferring Enzymes). By Osamu Hayaishi . . 581
The Pyridine Nucleotide Coenzymes. By Horst Sund 603
The Pyridine Nucleotide-Dependent Dehydrogenases. By Horst Sund 641

Subject Index 707

PART I

PROCESSES

The Respiratory Chain

Martin Klingenberg, *Physiologisch-Chemisches Institut der Philipps-Universität, Marburg, Germany*

I.	Introduction	3
II.	The Principle of the Respiratory Chain	4
III.	Carriers of the Respiratory Chain	6
IV.	Composition of Respiratory Chain Preparations	8
V.	Submitochondrial Preparations and Fragments of the Respiratory Chain	12
VI.	The Structural and Molecular Organization of the Respiratory Chain	15
VII.	Mechanism of Interaction in the Respiratory Chain	18
VIII.	Inhibitors of Electron Transfer	20
IX.	Artificial Electron Donors and Acceptors of the Respiratory Chain	23
X.	The Controversial Role of Cytochrome b	25
XI.	The Steady States of the Respiratory Chain	27
	A. Control of Respiration	28
	B. The Redox Behavior of the Respiratory Chain in the Steady States	29
XII.	Reversibility of Electron Transfer	36
	A. Reversibility under Aerobic Conditions	36
	B. External Acceptors and Donors	39
	C. Pathways of Reversed Electron Transfer	41
	D. Reversed Electron Transfer under Anaerobiosis	42
XIII.	Kinetics of the Oxidation and Reduction of Respiratory Carriers	46
	References	49

I. Introduction

The respiratory chain is commonly regarded as the multienzyme complex which catalyzes the dehydrogenation of substrates and the transfer of reducing equivalents to oxygen. In a narrower sense, however, the respiratory chain comprises only the electron-transfer sequence linked to oxidative phosphorylation. Often the concept of the respiratory chain is

3

somewhere between these broad and narrow definitions, and, from the variety of dehydrogenases, only the structurally bound flavoproteins, such as succinate, DPNH, choline, and α-glycerophosphate dehydrogenase are selected to figure as members of the chain. However, only DPNH dehydrogenase is linked to the first phosphorylation step and can, according to the precise definition, be considered a member of the respiratory chain.

"The respiratory chain" remains an improvised term for a multienzyme system which does not necessarily reflect a multicomponent complex on a molecular basis such as pyruvate dehydrogenase or fatty acid synthetase in which a limited number of molecules form one supramolecular unit. Our present knowledge of the molecular organization of the respiratory chain allows no decision as to whether such a molecular unit or a more statistical arrangement exists. The chainlike sequence of hydrogen- and electron-transfer steps is operationally described by the term *respiratory chain*.

In this chapter an outline of our present understanding of the respiratory chain will be given. Our knowledge of the constituents and composition of the respiratory chain, and of the function, dynamics, and mechanism thereof, are presented. Only data on the respiratory chain from mammalian tissues are considered in this chapter. Since this is an introduction rather than a review, no complete reference to the literature is attempted. For more detailed presentation the reader is referred to a series of reviews and survey articles on the subject (1–12).

II. The Principle of the Respiratory Chain

A simple operational scheme gives the view on the respiratory chain represented in this chapter (Fig. 1).* The cytochromes, which early became the basis for the formulation of the respiratory chain (13,14), form its backbone: a uniform channel for the reducing equivalents coming from a variety of substrates, each of which is dehydrogenated by a specific enzyme. At one stage the dehydrogenases are flavoproteins and at another stage DPN-linked enzymes, with the exception that in the decarboxylating dehydrogenation of pyruvate and ketoglutarate, lipoic acid and flavoprotein precede the DPN system. The dehydrogenases are connected to the cytochromes by hydrogen-transferring coenzymes, DPN and ubiquinone. Thus, the pathways branch out toward the substrates and here the chain

---

* The reader is referred to the excellent survey by Nichols (8) for a detailed historical description of the development of the respiratory chain concept.

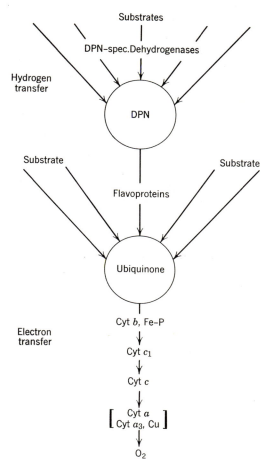

Fig. 1. Electron and hydrogen pathways in the respiratory chain.

concept is difficult to apply operationally. Furthermore, on the basis of the molecular organization, no chain-type arrangement can possibly be extended to the dehydrogenases. This will be discussed later.

The sequence of the components in the respiratory chain can be formulated by various methods which are combined to give the present picture. Thermodynamically, the reaction sequence may be determined by the redox potentials of the carriers which together form the total span of the redox potential between the substrate and oxygen. Kinetically, the reaction sequence can be analyzed directly by following the transfer of the reducing equivalents. Enzymically, the donor and acceptor specificity of the components can be studied after isolation of the enzymically active components, in some cases by reconstitution of the respiratory chain. These criteria for determining the respiratory chain sequence will be dealt with subsequently.

III. Carriers of the Respiratory Chain

A summary of the respiratory components, comprising the prosthetic group and the molecular weight estimated from the isolated enzymes, is given in Table I. Detailed descriptions of the isolated components are

TABLE I

Components of the Respiratory Chain

Component	Prosthetic group	Mol. wt[a]	Ref.
Cytochrome oxidase	5–6 heme a	530,000	15
"Complex I"; 3 cytochrome a,	2 Cu	6 × 72,000	
3 cytochrome a_3		(= 430,000)	
Cytochrome a; inactive monomer	1 heme a	72,000	16
Cytochrome c	1 heme c	12,900	17
Cytochrome c_1	1 heme c	38,000	18
Cytochrome b	1 protoheme	28,000	19
Fe-"complex III";	2 Fe	25,000	20
$g = 1.90 -$ Fe			
DPNH-DH			
"High mol wt form"[b]	FMN	ca. 550,000	21
"Low mol wt form"[b]	FMN	ca. 80,000	22–23a,b,110

[a] The molecular weights for the various cytochromes, except for c, are tentative, since homogeneous preparations have not been obtained.

[b] A detailed discussion and definition of the different preparations of DPNH dehydrogenase is given by Singer in Part II.

given in other chapters in this volume. In the isolated state these components are often not unequivocally defined molecules because of the great difficulties in isolating the mostly structurally bound components. The components vary in molecular weight and number of prosthetic groups according to the preparation, since they may still be impure, generally due to "structural protein," or they may be aggregated. This applies particularly to cytochrome oxidase and DPNH dehydrogenase. Table I also contains a non-heme iron protein isolated from complex III which is a redox active component but not necessarily a member of the main pathway.

The components contain various hemes and flavin nucleotides as electron transferring prosthetic groups. These groups are fixed to the protein either by a covalent bond such as in cytochrome c and succinate dehydrogenase or by intermolecular forces. In contrast, the hydrogen-transferring coenzymes, DPN and ubiquinone, can easily dissociate from the enzymes and can therefore generally be considered to be not bound to a protein.

A fundamental factor in formulating the sequence in the respiratory chain is the increase in the redox potentials of the carriers along the chain. The entire span of the redox potential difference between substrate and oxygen (about 1200 mV) may be divided up into smaller increments, with widely varying redox potential differences, among the respiratory components (Fig. 2). It is seen that the standard potentials closely follow the sequence of the electron-transfer components generally accepted on other

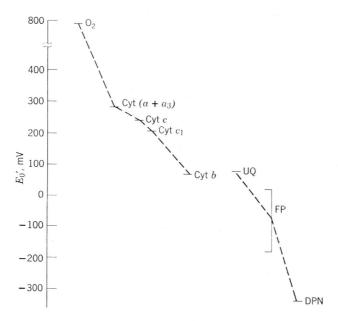

Fig. 2. The sequence of the respiratory carriers on the basis of the standard redox potentials. General references on redox potentials of respiratory components: 24–26; cytochrome b: 27,28; cytochrome a: 29,24; cytochrome c_1: 30; cytochrome c: 24,31; ubiquinone: 32,33; DPN: 34.

grounds (Fig. 1). The standard potentials determined for the isolated components, however, do not necessarily represent the redox potentials in the integrated respiratory chain, where the physicochemical properties of the components may be modified by the environment. Furthermore, the redox potentials may be changed by coupling to the energy transfer in oxidative phosphorylation, which will be discussed in Section XII. Nevertheless, the redox potentials have proven to be a fundamental guideline for assigning to a respiratory component a position in the chain.

IV. Composition of Respiratory Chain Preparations

The molar or stoichiometric composition of the respiratory chain is one basis for the understanding of its molecular construction. As a rule, only the content of the respiratory chain preparation per milligram of protein and the molar ratios can be given, since the existence of a supramolecular multienzyme complex of the respiratory chain and therefore its absolute composition remains unknown.

Of particular interest is the composition of the respiratory chain in preparations where the components can be considered to be fully retained such as in carefully isolated mitochondria (the respiratory chain can be depleted of easily dissociable components such as cytochrome c and DPN). Some data on the cytochrome content in the mitochondria from rat heart and liver and from beef heart are given in Table II. The table also contains data on the composition of fragments derived from beef heart mitochondria.

The major method for the assay of cytochrome content has been the spectrophotometric estimation which was first used on mitochondria by Chance and Williams (1). In subsequent estimations extended to mitochondria from muscles, a cytochrome content considerably higher than that in liver mitochondria was noted (39,40). The relative content of cytochromes a, c, and b in the mitochondria was found to approach a stoichiometric ratio of 1:1. Only cytochrome c_1 deviated largely from this relation, being contained in a considerably smaller ratio as determined from spectra at liquid air temperature ($-190°C$) (41) and extraction of cytochrome c (42).

A survey of the cytochrome content in mitochondria from several sources (42–44) demonstrated a variation of the cytochrome content in the different mitochondria. Furthermore, some variation in the stoichiometric ratio between the cytochromes was found. For example, in the mitochondria from liver a ratio of cytochrome c to cytochrome a of 1.1

TABLE II

The Composition of Mitochondria and Derived Fragmentary Respiratory Chain Preparations (in μ moles per g. protein)[a]

	Mitochondria					Derived from beef heart mitochondria								
	Rat heart	Rat liver	Beef heart			Keilin heart muscle	ETP_H	EP	Complexes				Succ.–cyt c red. complex	Cyt b–c particles
Components (6)	(6)	(6)	(35)	(36)	(21)	(35)	(21) (37)	(21) (37)	I	II (36)	III (36)	IV	(38)	(38)
Cyt a_3			1.13	0.63		1.05 (0.59)	1.6 (1.1)	2.9 (1.95)		0	0	8.9–9.4		
Cyt a	0.8	0.3	0.80	0.55	(0.89)	0.85 (0.57)	0.3	0.16		0	0	0		
Cyt c	0.65	0.28	0.65	0.66		0.56		0.48	trace	trace		0		
Cyt c_1	0.20	0.07									4.1	0	1.25	1.4
Cyt b	0.75	0.30	0.60	0.68		0.59	0.9	1.5	trace	4.8	8.5	0	1.97	2.4
UQ	4.0	2.4		3.4–4.0		4–6		5.3	4.5		2	1.8–3.0	0.86	0.8
DPN	7.5	4.0												
FAD			0.26 }0.46			0.09								
tryptic acid			0.11	0.2		0.12	0.28	0.43		4.6		0	0.40	
FMN			0.08		0.06	0.15	0.10 }0.38	0.084 }0.48						
Total flavin			0.45	0.66		0.36		0.91						
Fe			6.4	33		8.35	5.9	7.15	26	36	10		9.75	6.3
Cu			1.46	1.47		1.49	1.7	3.3				10		

[a] Values in parentheses for cytochrome a and a_3 are recalculated from the published values by using the extinction coefficients of van Gelder and Muijers (37a).

was measured and in mitochondria from kidney the ratio was 1.8. This variation of the cytochrome ratio was also noted in a comparison of liver with ascites tumor mitochondria (45).

Reliable results on the content of the hydrogen-transferring coenzymes DPN and ubiquinone, as well as the content of flavin, were possible with the introduction of sensitive enzymatic and chemical analysis. Both pyridine nucleotides, DPN and TPN, were found in comparatively high amounts in various mitochondria (46,47). The content of DPN reaches comparatively large excess to cytochrome a with an approximately constant molar ratio of 10 (43,47). In contrast, the molar ratio of TPN to cytochrome a can vary 30-fold among various mitochondria, from 20 to 0.6.

Ubiquinone is also contained in a considerably large molar excess to the cytochromes (48,49), and at a relatively constant molar ratio of ubiquinone to cytochrome a of 4–6. The high and constant content both of DPN and of ubiquinone to the cytochromes is in accordance with their analogous function as hydrogen carriers in the main pathway of the respiratory chain as indicated in Figure 1. In contrast, the large variation in TPN content is regarded to be in agreement with its position outside the main pathway of hydrogen transfer to the respiratory chain. Also, the comparison of the redox behavior of DPN and TPN in mitochondria shows that in the transitions between the steady states only DPN, and not TPN, follows closely the other respiratory components (50).

The determination of the content of various flavoproteins is still unsatisfactory. The spectrophotometric method does not give quantitative data since other components interfere and it cannot differentiate between the single flavoproteins. Only a few data based on chemical determination of the FAD and FMN content of the mitochondria are available (21,35,36). Flavins can in part be tentatively assigned to specific dehydrogenases: the peptide-bound portion of FAD represents the prosthetic group of succinate dehydrogenase and the FMN content the DPNH dehydrogenase, provided that no other as yet unknown flavoprotein contributes to these flavin portions.

The content of some DPN-specific dehydrogenases linked to the respiratory chain can be examined by measuring the enzymic activity in mitochondrial extracts (51). Malate dehydrogenase activity was found to follow closely the content of cytochrome a in various mitochondria and to be contained in a relatively high activity. DPN-specific isocitrate dehydrogenase was also found to approximately follow the cytochrome content, although at a much lower level (52). In contrast, the activities of TPN-specific isocitrate dehydrogenase and glutamate dehydrogenase were found to vary widely among different mitochondria (51).

The nearly constant proportion is thus found to extend from the cytochromes and coenzymes to some of the dehydrogenases (53). This indicates that these components act in close collaboration with the respiratory chain and therefore are parts of the main pathway of hydrogen transfer (51). Components with widely varying proportions can be regarded to be on a side path which is attributed to a special function of the specific mitochondria.

The molar composition of the respiratory chain, relative to cytochrome a content, is summarized for heart muscle mitochondria in Table III. The

TABLE III

Molar Composition of the Respiratory Chain in Mitochondria

Component	Beef heart[a]	Rat heart[b]
Cyt a_3	1.1	
Cyt a	1	$\equiv 1$
Cyt c	}1.2	0.8
Cyt c_1		0.3
Cyt b	1.1	0.95
Succ.-DH	0.2	
DPNH-DH[c]	0.14	
UQ	7	5
DPN		9.5
Non-heme Fe	1.2	
Cu	2.6	

[a] Calculated from data in Table II, using recalculated values in reference 37a.

[b] Data from reference 6.

[c] The DPNH dehydrogenase content, however, may be much lower than the FMN content (21).

cytochromes are contained in a ratio of approximately 1:1 with the exception of cytochrome c_1 (1) and the two flavin dehydrogenases, succinate and DPNH dehydrogenase, which reach only 1/5 or 1/7 of the cytochrome a content (6). It has been estimated (21) that only 31% of the FMN of beef heart mitochondria belongs to DPNH dehydrogenase. The two hydrogen-transport metabolites, ubiquinone and DPN, occur in a five- to ten-fold excess. Estimates on the content of the DPN-specific dehydrogenases show that malate dehydrogenase occurs in a molar ratio to cytochrome *a* of about 1/2, DPN- and TPN-specific isocitrate dehydrogenases as well as glutamate dehydrogenase in ratios of 1/10 to 1/30 (6).

V. Submitochondrial Preparations, Fragments, and Reconstitution of the Respiratory Chain

The classical preparations of mammalian origin for study of the respiratory chain can be regarded as consisting of submitochondrial particles. Here the respiratory chain is still functioning with the variety of structurally bound dehydrogenases which are not depleted from the respiratory chain by the preparations. Thus both succinate and DPNH are still actively oxidized in the Keilin–Hartree heart muscle preparation (54), which is extracted directly from whole heart muscle. Submitochondrial fragments have also widely been prepared starting from the isolated mitochondria. Differential centrifugation at pH 8 has been used to prepare the electron transport particle (55). After sonic fragmentation of mitochondria, particles with phosphorylative activity can be prepared (56). By further treatment with cholate under careful conditions an "elementary particle" preparation is obtained which still contains most structurally bound components of the respiratory chain (37). From liver mitochondria, submitochondrial preparations are obtained by digitonine treatment (57,58) and sonic treatment (59), both with phosphorylative activity.

Table II shows how the content of cytochromes is increased in the submitochondrial preparations in comparison to the intact mitochondria. The fragmentation removes soluble proteins, including cytochrome *c* and the DPN-specific dehydrogenases. The content of cytochrome *a* and cytochrome *b* is correspondingly increased. For the preparation of the "elementary particle," the removal of structurally bound protein by the cholate treatment appears to increase further the cytochrome content which is two- to threefold higher than in the intact mitochondria. The increase in the cytochrome content of liver mitochondria with digitonine treatment is comparatively small (58).

The submitochondrial fragments have been mostly used to isolate the structurally bound respiratory carriers (see other chapters in this volume). Efforts have also been directed to derive multicomponent portions by dividing and separating the respiratory chain starting from these fragments. The preparation of multicomponent particles is justified by the fact that the isolated single carriers are often in a state where they have not retained their enzymic activity. This applies in particular to structurally bound carriers such as cytochromes *a*, *b*, and c_1.

When, however, the carriers are retained in multicomponent particles, such as complex III or the cytochrome $b-c_1$ particle, they may retain the original conformation sufficiently to permit recombination with neighboring segments of the respiratory chain and to perform hydrogen or electron transport similar to the intact chain. These attempts to reconstruct the respiratory chain have been called "reconstitution"; they are intended to help the understanding of the constitution and function of the respiratory chain.

The multicomponent fragments of the respiratory chain are called complexes I–IV by Green and co-workers (60–63) and particles by King et al. (64–66). Isolation of these fragments requires that they be solubilized from the particulate respiratory chain preparation without losing their reconstitutive activity. For fragmentation of the chain and of the complexes, treatment with desoxycholate in the presence of KCl is employed. The isolation includes ammonium sulfate fractionation steps. Cholate is also employed by King's group of the isolation of the succinate–cytochrome c reductase complex (38). From this complex the "$b-c_1$ particle" is prepared by separating the succinate dehydrogenase by alkaline treatment (64). The preparation and properties of the flavoprotein dehydrogenases and of the corresponding particulate complexes I (DPNH, ubiquinone reductase) and II (succinate–ubiquinone reductase) are reported by Singer (see Part II).

A guideline for describing the various multicomponent particles is their content of respiratory components, which is given in Table II together with the composition of the intact respiratory chain preparation, all derived from beef heart as a common source. Complexes I and IV correspond largely to the purified components, DPNH dehydrogenase and cytochrome oxidase. Complexes II and III contain at least two components and have cytochrome b in common. Characteristic for complex III is the presence of cytochrome c_1 and an appreciable amount of ubiquinone; the latter is also present in complex I. The relative proportion of the components of the "$b-c_1$ particle" corresponds to that of complex III. The succinate–cytochrome c_1 reductase complex is a still larger unit and would correspond to the combination of complexes II and III, with a considerably lower content of the cytochromes.

Complexes I–IV may be recombined to give a preparation capable of transferring reducing equivalents from DPNH and succinate to oxygen (60). Figure 3 represents a working scheme for the reconstitution of the respiratory chain from the individual complexes (67). Of the carriers, only soluble cytochrome c must be added externally but combination of the components also requires the addition of phospholipids. It has been

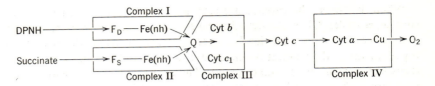

Fig. 3. Block scheme demonstrating the construction of the respiratory chain from the four complexes. The interaction of the complexes as catalyzed by diffusable components such as ubiquinone and cytochrome c. From Green and Oda (67).

implied that the reconstituted particle is equivalent to the parent "elementary particle." The reconstituted system responds to the specific inhibitors of electron transport such as amytal, antimycin A, rotenone, etc. Reconstitution occurs only at rather high concentration of the components but after recombination the complexes do not dissociate spontaneously (60,68). No analysis of its composition which could identify the reconstituted with the parent particle is given in these studies.

Reconstitution as studied by King et al. (38,64,66,69) concerns the recombination of soluble succinate dehydrogenase with respiratory chain preparations depleted of this enzyme or with the "b–c_1 particle" (38). As seen from the analysis in Table II it is not attempted to purify the multicomponents but to isolate them as functionally intact as possible. The components recombine in proportion to their concentration. The reconstitution was followed by the response of the redox state of the cytochromes, confirming the reconstitution of the respiratory chains linked to succinate dehydrogenase. In contrast to the reconstitution of complexes I and II, in this case no lipid is added and no preceding incubation in a concentrated mixture is required. The recombination is always reversible on dilution.

Reconstitution experiments raise the question as to the forces responsible for the structural organization of the respiratory chain so that the components are arranged in a manner allowing a specific sequence of electron transfer. Most probably, the specificity of the arrangement is based on the protein interaction. This would be in accordance with the current thinking that all biological organization is attributable to the primary structure of the proteins. Structural protein and lipid may form a rather unspecific matrix for this arrangement. Thus, in principle, only by mixing the isolated components of the respiratory chain could an association be achieved according to the thermodynamically most favorable organization of the respiratory chain.

VI. The Structural and Molecular Organization of the Respiratory Chain

The particle-bound nature of the respiratory chain has intrigued workers and inspired theories on the structural organization of the chain. The major conclusion has been that the structural arrangement of the respiratory components determines their mutual interaction by spatial accessibility (70,71). This theme is maintained in various contemporary theories of the respiratory chain organization and differentiated in several models (cf. survey by Nichols in ref. 8).

A great advance in the understanding of the organization of the respiratory chain was accomplished by the identification of mitochondria as carriers of the respiratory chain including the full complex of the dehydrogenases and the coenzymes. The intense penetration of the mitochondria by membrane structures as revealed by electron microscopy (61,73,74) gave a morphological explanation for the particle-bound nature of the respiratory components. The mitochondrial membrane could be assumed to be the binding structure. By comparing the mitochondria from various organs, in particular from liver and heart, an increase of the cytochrome content parallel to the density of intracristae membranes was observed (43,73).

Relevant data for the possible arrangement of the respiratory chain are furnished by the protein portion taken up by the respiratory components in the mitochondria. These can be estimated on the basis of the provisional molecular weights (Table I) and the contents of the components as referred to total protein (Table II). The protein share of the cytochromes can be calculated for beef heart mitochondria to be about 18% and for rat heart mitochondria about 16% (cf. also 44). In submitochondrial preparations, such as ETP, the share may increase to about 25–30% owing to the removal of soluble proteins from the mitochondria (75). In liver mitochondria a considerably smaller portion (6%) is estimated to be cytochrome protein, in accord with the much larger content of soluble proteins in these mitochondria. The cytochrome content of the mitochondrial membranes of liver mitochondria has been calculated to be 10–20% (76). In the "elementary particle" preparation, after removal of further structurally bound proteins, it may reach more than 50%. Thus, on the average about 20–25% of the protein layer in the mitochondrial membranes may consist of cytochromes. The protein portion taken up by the dehydrogenases is probably considerably smaller. A large part may be due to structural protein and to the specific enzymes involved in the energy transfer of the oxidative phosphorylation.

The density of packing on the membrane surface of respiratory components can be calculated from the average density of membranes in the mitochondria. It is assumed that the respiratory components occupy primarily one side of the membranes, e.g., the side towards the matrix. On this basis it can be calculated that the membrane surface area per cytochrome chain (consisting of one molecule each of cytochromes a_3, a, c, and b) is about 50,000 Å² (77,78). Assuming globular molecules for the cytochromes and using the provisional molecular weights in Table I, about 10% of the membrane surface may be calculated as taken up by cytochromes, in agreement with estimation by other means (44).

The molar and spatial relations are described in a scheme to give an impression of the principal spatial dimensions for the structural organization of the respiratory chain (Fig. 4). The special arrangement of the chain

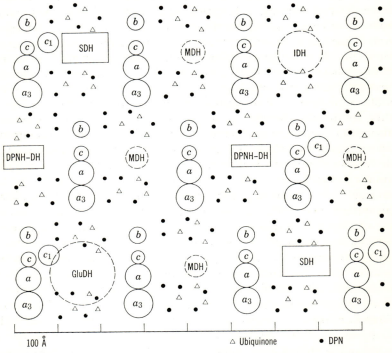

Fig. 4. The packing of the mitochondrial membrane surface with respiratory components, including several dehydrogenases. The dimensions of the molecules are estimated from the molecular weights, the density of surface packing from the content in heart muscle mitochondria of cytochromes and cristae membranes (77). Enzymes with dashed circles are in the hydrophilic layer.

assumed here is irrelevant to the question as to what per cent of the surface is occupied by the cytochromes. The low molar content of the dehydrogenases indicates a sparse distribution over the surface. Thus, there are relatively large distances between one dehydrogenase molecule and the 5–10 cytochrome chains which are to receive reducing equivalents from it. Therefore, hydrogen-transport metabolites must be required which can bridge these distances by diffusion. Ubiquinone is eminently suited for this role since it occurs in large excess over the cytochromes and is a small non-protein-bound molecule. Because of its lipophilic quality ubiquinone probably diffuses on the surface of the membrane between the lipid-bound components. In its transport function, ubiquinone may form a common pool for the hydrogen collected from the various flavoprotein dehydrogenases to be donated to the cytochromes. The action of ubiquinone as a diffusible component between the respiratory chain "complexes" was proposed (10). A hydrogen-distributing mechanism before the entrance to the cytochrome chain or the antimycin A block is also indicated by titration studies with antimycin A (79). Furthermore, hydrogen transfer between various flavoprotein dehydrogenases such as succinate and α-glycerophosphate dehydrogenases (80) is visualized here as requiring the existence of a diffusible hydrogen carrier.

Ubiquinone has in principle the same function as mitochondrial DPN for the DPN-linked dehydrogenases, which are also, with the exception of malate dehydrogenase, only sparsely distributed. Correspondingly, DPN is also contained in large excess over the cytochromes. It will perform its task more readily in the hydrophilic phase of the matrix as it is freely diffusible in water.

As indicated earlier, the stoichiometric relation of $1:1$ would allow the cytochromes to be organized spatially in single assemblies where they can react in sequence (1). The existence of three chains containing three molecules each of cytochromes a, b, and c to one molecule of cytochrome c_1 in one multienzyme complex has been concluded from analysis of the so-called "elementary particle" (61,37). At a molecular weight of 1.4×10^6 it was postulated to contain this number of cytochromes. The organized unit is supposed to be formed together with structural proteins and lipids as discussed before.

The preexistence of the "elementary particle" in the intact membrane has been contested (82) and it has been suggested that the "elementary particle" preparation actually represents fragments of multicomponent assemblies of the respiratory carriers. The identity of the "elementary particle" with a repeating, knoblike structure on the intermitochondrial membranes (83,62) could not be experimentally verified. The repeating unit, detected by electron microscopy with negative staining (84,85), was too small (80 Å diam) to be identified with the elementary particle (86). Furthermore, stripping the mitochondrial membrane of the repeating units did not remove the cytochromes (87).

VII. Mechanism of Interaction in the Respiratory Chain

In the discussion of the various authors on the organization of the cytochromes one might differentiate between the hypothesis of a chainlike interaction of the carriers as chemically emphasized in the "elementary particle" theory, and an arrangement corresponding to a multimolecular assembly where in principle all neighboring cytochromes can interact with each other. In this case, in principle, hydrogen coming from one dehydrogenase could also be transferred to neighboring cytochromes not directly connected to the dehydrogenase, and therefore all cytochromes could be reduced. The distances between the cytochromes would be overcome by some limited translational movement and rotations. In view of the spatial situation described in Figure 1 only an uneven distribution of the cytochromes in multichainlike bands or other associations would allow for such "lateral" interaction. This interaction among cytochromes, going for example from cytochrome b to various molecules of cytochrome c and from one cytochrome c to various molecules of cytochrome a, is implied in the oxysome theory (82). In a study of the competition for the respiratory chain between reducing equivalents originating from two dehydrogenases, a lateral interaction on the level of cytochrome c_1 was postulated (81). A branching mechanism (Fig. 5) has been discussed on the basis of

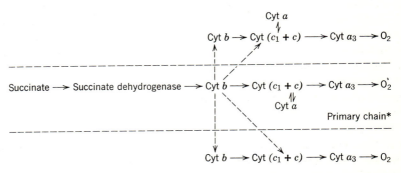

Fig. 5. Parallel and branch type interaction between the cytochrome chains as deduced from reconstitution studies with succinate dehydrogenase added to a deficient respiratory chain preparation. From Lee, Estabrook, and Chance (88).

titration of a succinate dehydrogenase-deficient preparation with succinate dehydrogenase (88) following earlier reconstitution experiments (89). The scheme does not take into consideration the function of ubiquinone, which in principle can distribute hydrogen from one dehydrogenase molecule to many cytochrome chains and thus abolishes the need for lateral interaction of cytochrome chains.

The electron transfer in the respiratory chain, and therefore the interaction of the respiratory components, is strongly temperature dependent, as are other enzyme reactions. Electrons can even be trapped in the cytochromes and prevented from further transfer at the temperature of liquid nitrogen (90). The temperature dependence excludes electron transfer by conduction bands in proteins in the respiratory chain. Thus, the interaction of the respiratory components can be assumed to include thermal movements such as translation which, however, is restricted by the binding. Rotation of the cytochromes has been assumed for the interaction of cytochromes in the chain (1); that is, the prosthetic groups would turn from the preceding to the following partner in the electron acceptor–donor process.

It has been attempted to support the thermal collision theory by the finding that electron transport can be inhibited on increasing the viscosity by addition of glycerol (90). This interpretation of the effect of glycerol may be questioned or interpreted differently by assuming that a hydrophilic environment is important for the electron-transfer carriers to react and it is disturbed by the glycerol. The inhibition of electron transport by replacement of H_2O with D_2O also supports this interpretation (91).

An important role in electron transfer has been attributed to mitochondrial lipids (92). Their role, however, can be considered to be primarily in the formation of mitochondrial membranes. There, the lipids form a base for binding the respiratory components together with a structural protein and thus provide correct spatial arrangements for the interaction. It appears today that the lipid membrane plays a more direct role in energy transfer than in electron transfer.

It must be stressed that in all considerations on the mechanisms of electron transfer it must be assumed that electron transfer is basically organized for performing energy transfer. Therefore, in all respiratory chain systems, even when they are uncoupled and fractionated, there may be present proteins and lipids which are essential for the sequence and organization of electron transfer but the function of which can be only understood by consideration of energy transfer (93).

VIII. Inhibitors of Electron Transfer

The classical inhibitors of the respiration of biological material, such as carbon monoxide, cyanide, and sulfide, all have been proved to act on the oxygen-accepting terminal of the respiratory chain (94,95). More recently, inhibitors have been identified which interfere with specific electron-transfer reactions in the respiratory chain. These inhibitors divide the respiratory chain into segments before and after the inhibition site. Therefore, inhibitors provide an important tool for elucidating the electron transfer sequence. Further, they permit studying the participation of respiratory components in electron transfer on either side of the point of inhibition in partial reactions of electron transport.

The favored view now is that the large variety of respiratory inhibitors act at only three sites of the sequence. These inhibition sites may be identical with the three energy-transfer sites (see Fig. 6). In support of this view, some interference of electron-transfer inhibitors with the energy transfer has been noted. One inhibition site (site I) has been identified in the dehydrogenase region at the level of DPNH dehydrogenase. The following class of compounds inhibits at this site: barbiturates, of which the most used derivative is amytal (96,97); the rotenoids, the most important representative of which is rotenone (98,99); and a number of steroid hormones, e.g., desoxycorticosterone, progesterone, and diethylsilbestrol (100,101). There is also a series of compounds such as methylene glycol, which, however, are rather heterogeneous as a group and not of great importance for electron-transfer studies.

It was first noted that these substances inhibit the respiration of DPN-linked substrates but not of flavin-linked substrates such as succinate or α-glycerolphosphate. Direct studies of the redox state of the respiratory components in isolated mitochondria indicated that amytal causes a reduction of DPN and oxidation of the flavoprotein when inhibiting the oxida-

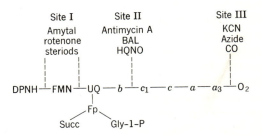

Fig. 6. Inhibition sites in the respiratory chain.

tion of a DPN-linked substrate (97). It does not inhibit the reduction of flavin by succinate (97) or α-glycerolphosphate (102). In contrast, in submitochondrial particles amytal causes reduction of flavoprotein while inhibiting the oxidation of DPNH (103). In purified DPNH dehydrogenase no inhibition of the DPNH dehydrogenation with various artificial electron acceptors was noted (104), leading to the conclusion that amytal acts after the DPNH dehydrogenase flavoprotein in the intact chain. An attempt has been made to explain these discrepancies by the assumption that the inhibition is shifted in going from the phosphorylating to the non-phosphorylating respiratory chain, due to some structural rearrangements (105). However, the evidence for the action of amytal between DPNH and flavin in intact mitochondria is based only on spectrophotometric measurements which do not allow observation of the DPNH dehydrogenase separate from other flavin dehydrogenases. Therefore, the data obtained on purified preparations of DPNH dehydrogenase putting the action of amytal and thus the location of inhibition site I after the flavin dehydrogenase may be valid also for the integrated system in mitochondria.

The action of amytal, the steroids, and other inhibitors of site I is also not clearly defined in other respects either. For example, at high concentration amytal can slightly inhibit the oxidation of succinate (106). Furthermore, some interference of amytal between the flavoprotein reduced by α-glycerolphosphate and the cytochromes has been noted (107). Amytal, as well as the steroids, appears to interfere with the energy transfer also, since uncouplers partially remove the effect of these inhibitors (108). Thus, the target point appears to be very close to the respiratory chain, where the inhibition site is common to both energy transfer and electron transfer. This supports the notion that the inhibition sites for electron transfer are located at the energy transfer sites.

Rotenone does not share the ambiguous properties of amytal in energy and electron transfer. Rotenone binds virtually irreversibly to mitochondria when exerting its inhibition. The titer of rotenone for liver mitochondria (25 mμ moles/g protein) for complete inhibition is very low, in view of the higher concentration of barbiturates required (K_i for amytal = 2 mM) (109). The low titer of rotenone corresponds to the comparatively low molar content of the DPNH dehydrogenase in mitochondria (6) (cf. Table II) without assuming an unknown carrier (109). This is substantiated by studies of the isolated DPNH-ubiquinone reductase (100). Rotenone could be shown to inhibit the reduction of the isolated DPNH-dehydrogenase (110), where it binds as a ratio of 1:1 to the content of FMN. However, objections have been raised against extrapolating the inhibition characteristics from this DPNH-dehydrogenase preparation to intact mitochondria (110a), which are to be discussed by Singer in Part II.

Inhibition site II of the respiratory chain is located at some point between ubiquinone and cytochrome c_1 (Fig. 6). The following inhibitors have been found to react at this region: naphthoquinone derivatives (111,112), British antilewisite (BAL) (113), antimycin A (114), and hydroxyquinoline-N-oxide derivatives (115). It was concluded that these inhibitors affect some compound between cytochrome b and cytochrome c, since cytochrome b became reduced and cytochromes c and a oxidized upon addition of these compounds to the respiratory chain (cf. detailed description in ref. 71).

A particularly interesting and valuable inhibitor at this site proved to be antimycin A since it is effective at a very low "titer." By detailed spectroscopic studies a point of action at the level of, or on the O_2 side of, cytochrome b was verified (116,117). The inhibition of antimycin A was found to depend on the ratio of inhibitor to enzyme preparation; i.e., a stoichiometric binding by the respiratory chain appears to occur (114,79). The inhibition starts to be effective only near the titration endpoint of full inhibition. These findings have been interpreted, as discussed above, to indicate an interaction between adjacent respiratory chains in the sense of the multiassembly theory of the cytochrome chain (79). Cytochrome b was thought to be a direct target of antimycin A, since about one molecule of antimycin A is required for both complete inhibition of respiration and full reduction of cytochrome b in heart muscle preparations (117). In liver, however, the antimycin A titer is only one-third of the cytochrome b content (118). For this and other reasons it has been concluded that a component after cytochrome b is the site of inhibition (118). This may be the non-hemin iron protein which is possibly located before cytochrome c_1 (20,119). The chelating properties of antimycin A would support this conclusion (120). Furthermore, in the presence of antimycin A cytochrome b can still donate electrons toward the substrate site in reversed electron transfer supported by ATP (121), but antimycin A is an effective inhibitor of reversed electron transfer from cytochromes c and a (122,123).

It is established that antimycin A inhibits on the O_2 side of ubiquinone (124,63,49,125,126). In all these cases ubiquinone is found to be reduced on the addition of antimycin A in the presence of a suitable substrate. Under certain circumstances ubiquinone is not reduced by succinate in the presence of antimycin A, e.g., when succinate dehydrogenase is inhibited by an uncoupler (49,121). This can be explained by the incomplete inhibition of electron transfer by antimycin A and does not suggest an antimycin A block before ubiquinone.

The close relation of site II inhibition to energy transfer is indicated by its stimulative effect on mitochondrial ATPase (127,128) and the inhibition of the $^{32}P_i$-ATP exchange (128,129). The inhibition at site II by hydro-oxyquinoline-N-oxide derivatives (115) was found to be partially removed on uncoupling by DNP (130). In contrast, inhibition by antimycin A was not influenced. This difference is reminiscent of the similar situation observed with amytal and rotenone at site I.

Possible interference with energy transfer by antimycin A is reflected in Chance's assumption that the antimycin A site is situated in the pathway of reverse electron transfer from succinate to DPN (123). As discussed below, direct interference with this reversed hydrogen transfer has not been observed in other laboratories. The non-interference by antimycin A is in agreement with the scheme of electron transport wherein hydrogen from succinate to DPN passes between the two dehydrogenases by mediation of the ubiquinone pool.

IX. Artificial Electron Donors and Acceptors of the Respiratory Chain

Artificial electron donors and acceptors have long been used in studies of the respiratory chain. In this way certain segments of the electron-transfer sequence can be isolated. This has been of great value for elucidating the sites of energy transfer and also of inhibitors of electron transport. The site of action of the artificial donors and acceptors, however, is often not specific and therefore makes interpretation of their action difficult. In principle, the site of action is determined by the redox potentials. There are certain artificial acceptors which exhibit a substrate-like specificity, since they resemble the natural donors or acceptors, and there are others which are completely unspecific. Other qualities such as the solubility to lipids and the ability to act as a one- or two-electron acceptor may determine the site of action.

Most of the artificial electron acceptors act on the flavoprotein-type dehydrogenase and they will be dealt with in the chapters describing the isolated dehydrogenases. Only those cases will be briefly considered which have been of interest in studying the integrated respiratory chain. Apparently in intact mitochondria the flavoprotein sites are not readily accessible to many acceptors. The main interest here has been to elucidate the function of quinones, in particular ubiquinone, by replacement with analogous quinones as acceptors or for reactivation of a ubiquinone-deficient chain. These aspects will be discussed in the chapter by Crane. The point of action of some important electron donors and acceptors is depicted in Figure 7.

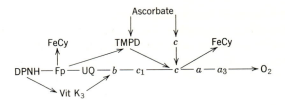

Fig. 7. External donors and acceptors interacting with the respiratory chain. (FeCy = ferricyanide, TMPD = tetramethylphenylenediamine, Vit K_3 = menadione.)

Before the role of ubiquinone became widely acknowledged, it had been proposed that phylloquinone functions at this site of the respiratory sequence (131). A DPNH-phylloquinone reductase has been isolated, which is inhibited by dicumarol (131a). Whereas evidence for the function of phylloquinone in the intact chain is lacking, an efficient "external" pathway can be established with the naphthoquinone derivative menadione (132,133). The flavoprotein involved is that responsible for the reduction of phylloquinone and shows a particularly high activity toward menadione (vitamin K_3). Because of its dual specificity for DPNH and TPNH the flavoprotein has been called "DT-diaphorase" (132). Thus, a bypass around the respiratory chain DPNH dehydrogenase can be formed bridging inhibition site I. The physiological role of this pathway may involve primarily events in the extramitochondrial space where most of the activity is located.

The one-electron acceptor ferricyanide has been very useful for studying various flavoprotein dehydrogenases. Ferricyanide also is an effective electron acceptor in respiratory chain preparations (134). In submitochondrial particles the reduction of ferricyanide is not inhibited by amytal or rotenone, indicating a possible direct interaction of ferricyanide with DPNH dehydrogenase before the inhibitor site (104). In intact liver mitochondria ferricyanide reduction was found to be sensitive to both amytal and antimycin A at low ferricyanide concentrations, which indicates an interaction of ferricyanide at the cytochrome c site (105). Thus, with ferricyanide as acceptor $P/2e = 1$ is obtained (135), and the rate of its reduction is controlled by the presence of phosphate acceptor (105). Ferricyanide is therefore useful in studying the electron-transfer sequence in a coupled system including the first two phosphorylation sites. At high concentrations ferricyanide can also interact directly with DPNH dehydrogenase in mitochondria and its reduction then becomes amytal insensitive (105,105a,135).

For studies of the isolated steps between cytochrome c and oxygen the addition of reduced cytochrome c was used formerly (cf. literature in ref. 136). However, electron-transport activity with cytochrome c, ferricyanide, silicomolybdate + ascorbate, etc. was low and coupled to a very low efficiency of oxidative phosphorylation (136). With the introduction of phenylenediamines, in particular N,N,N',N'-tetramethylparaphenylene-diamine (TMPD), as a mediator of electrons from ascorbate, an efficient electron donor became available for the cytochrome c stage with a high phosphorylation yield (133,138). As expected, respiration with ascorbate + TMPD is not inhibited by antimycin A.

Since a P/O ratio greater than 1 is found, two phosphorylation steps are assumed to be included in the oxidation of TMPD + ascorbate, both located after the antimycin A site (138). However, TMPD may also accept electrons, to some extent, from cytochrome b and thus circumvent the antimycin A block (139). This offers an explanation for the increased P/O ratio. The main entrance of TMPD has been shown to be located between the site of action of antimycin A and cytochrome c_1 (140).

The system ascorbate + TMPD has been extremely useful in studying reversed electron transfer both as electron donor below the cytochrome c level and for supplying energy generated in the cytochrome region for utilization in the other steps of the respiratory chain. This is discussed in the section on reversed electron transfer.

X. The Controversial Role of Cytochrome b

Whereas the cytochromes of types c and a are accepted to be on the main pathway of electron transport, the role of cytochrome b is not yet fully understood. Thus it is not clear whether cytochrome b joins the other cytochromes in the common channel for reducing equivalents from the various substrates, whether it is a member of a branch before the merging point, or whether it is on a special side path of its own. Cytochrome b is closest to the dehydrogenase region among the cytochromes. It has a considerably lower redox potential than the other cytochromes. Its close relation to the dehydrogenases is also reflected in fractionation studies of mitochondria, where it consistently follows the flavoprotein-containing fractions (cf. Table II).

The major arguments against the functioning of cytochrome b on the main pathway are kinetic. It was early indicated by Slater that cytochrome b does not participate in DPNH dehydrogenation, since in submito-chondrial Keilin-Hartree preparations cytochrome b is more slowly reduced than the other cytochromes (141). It was proposed that cytochrome b is on the particular pathway of succinate oxidation, which merges with DPNH dehydrogenase via an unidentified carrier (142). With refined spectrophotometric techniques, Chance (116) realized that cytochrome b is reduced much more slowly than the other cytochromes in Keilin-Hartree preparations by either DPNH or succinate. Thus, in these nonphosphorylating preparations cytochrome b is not on the direct path of oxidation of the two major substrates, DPNH and succinate. In tightly coupled liver mitochondria, however, cytochrome b was found to follow closely the kinetics of the other cytochromes (69) (cf. Table V), so that its position in the electron-transport chain before cytochrome c could be confirmed in accord with the classical sequence established from redox potentials. On this basis it may be visualized that cytochrome b is "dislocated" in nonphosphorylating particles in such a way that a new pathway around cytochrome b is established.

Another explanation of the difference would be that cytochrome b is always on a side path but is more tightly coupled to the respiratory chain in phosphorylating than in nonphosphorylating preparations. Thus, the equilibration of cytochrome b with the respiratory components, for example ubiquinone, is not rate limiting in phosphorylating preparations in contrast to the nonphosphorylating ones. A special role of cytochrome b in energy transfer was further supported by the finding of an absorption shift of the α band of cytochrome b to 566 mμ in an energy-linked steady state of mitochondria (B. Chance, personal communication).

The role of cytochrome b has been closely associated with the inhibition by antimycin A. A direct interaction appeared to be substantiated by a shift of the α band of cytochrome in the presence of antimycin A (140,144). This, however, would not agree with the fact that antimycin A also inhibits electron transport from all dehydrogenases, in cases where cytochrome b does not appear to be in the main pathway. It has been proposed that antimycin A may recouple cytochrome b to the respiratory chain (140). Whereas cytochrome b is only sluggishly reduced on inhibition of respiration—for example, by cyanide—it is very rapidly reduced after the addition of antimycin A. It has been speculated that cytochrome b may exist in two different forms in the mitochondria, the reduced form of which, under the influence of antimycin A, has a shifted absorption maximum (144). This influence is indirect, since the target of antimycin A appears to be not cytochrome b but a closely associated component (145).

In fragmentation studies the close association of cytochrome b with cytochrome c_1 is indicated by the isolation of complexes which contain a ratio of cytochromes $b/c_1 = 2$ (38). Cytochrome b was also found to be associated in the "complex II" containing succinate dehydrogenase (36). Therefore it has been proposed that cytochrome b occupies a bridge position between the dehydrogenases in complexes I and II (61), although cytochrome b was not reduced in these preparations. When, however, the submitochondrial particles are prepared under less severe conditions, a highly active reduction of cytochrome b is observed (27).

The function of cytochrome b has also been explored in studies of the reconstitution of succinate oxidase with soluble succinate dehydrogenase (88). In the respiratory fragment used, the rate and extent of reduction of cytochrome b increased on addition of successive increments of succinate dehydrogenase (88). In these particles cytochrome b is apparently less strongly dissociated through the fragmenting treatment than in Keilin-Hartree preparations. A branch-chain picture would agree with the statistical arrangement of the succinate dehydrogenase with respect to the cytochrome chains. These experiments were interpreted to suggest an intimate relation between cytochrome b and succinate dehydrogenase (88). It must be emphasized, however, that in intact mitochondria the molar ratio of succinate dehydrogenase to cytochrome b is only about 1:5 (6), whereas in the experiments quoted the full reactivation of cytochrome b with succinate dehydrogenase required a ratio of 1:1. Thus the notion that cytochrome b is directly linked to succinate dehydrogenase (141,88) is difficult to extend to the intact system.

The anomalies in the behavior of cytochrome b in reversed electron transfer complement the picture evolved from forward electron transfer. In anaerobic pigeon heart muscle mitochondria, upon treatment with ATP the fully reduced cytochromes a and c are largely oxidized, whereas the partially reduced cytochrome b becomes still more reduced (146). The same has been found in submitochondrial particles (140). This reduction of cytochrome b by ATP can be interpreted as an activation of the succinate dehydrogenase by ATP. In the redox patterns shown in Figure 11, cytochrome b also stands out in becoming more reduced on the addition of ATP whereas the other cytochromes and ubiquinone become oxidized. In contrast, in rat liver mitochondria cytochrome b follows closely the redox patterns of the other cytochromes on oxidation (122). Apparently in isolated heart muscle mitochondria cytochrome b is more dissociated than in the more intact liver mitochondria.

XI. The Steady States of the Respiratory Chain

The sequence of hydrogen- and electron-transfer carriers in the respiratory chain is particularly suited for studying the dynamics of a multienzyme system. The characteristic spectroscopic absorption shifts of electron carriers which accompany their redox changes lend themselves to the sensitive and rapid techniques of absorption measurements. These analytical advantages of the respiratory chain as a multienzyme system are shared with those of the electron-transfer system linked to photosynthesis.

A. Control of Respiration

Respiration, i.e., the overall electron flow in the respiratory chain, contributes valuable information on the steady state and dynamics of the respiratory chain. With a maximum supply of electrons and oxygen, the capacity for electron transfer is obtained, provided that there is no limitation imposed by oxidative phosphorylation, as discussed below. In this case, respiration may be limited by the internal electron transfer in the cytochrome chain. Generally it is difficult to saturate the cytochrome chain and respiration is limited rather by the velocity of hydrogen transfer from dehydrogenases. With an excess of a particular substrate the respiration may reflect the maximal activity of the respective dehydrogenase (which is not necessarily identical with the maximal velocity measurable with the isolated enzyme). It may be limited by the effective concentration of the hydrogen-accepting coenzymes DPN or ubiquinone or by a low substrate concentration.

The concentration of oxygen is rarely limiting to respiration because of the very low K_m value of the respiratory chain for oxygen. The approximate value of $K_m(O_2)$ for mitochondria is $3\mu M$ (148,149), which corresponds to about 1% of the oxygen concentration in a solution saturated with air at $25°C$. For this reason, under experimental conditions *in vitro*, a limitation by oxygen (except for anaerobiosis) is a rare case.

Of particular importance is the control of respiration by the reciprocal coupling of electron and energy transfer. This type of "respiratory control" was first observed in carefully isolated mitochondria. Respiration was found to be dependent on the substrates of energy transfer, ADP and P_i. Respiration is slow when mitochondria are supplied only with substrate and oxygen and is increased by the addition of ADP and P_i (150–152). Respiration is also stimulated, sometimes to a still higher extent, by the addition of an uncoupler of oxidative phosphorylation.

Respiratory control can be elegantly illustrated by a polarographic recording of oxygen concentration (Fig. 8) (153). Respiration is stimulated by ADP and again controlled after the phosphorylation of ADP to ATP. The degree of coupling can be described by the "respiratory control

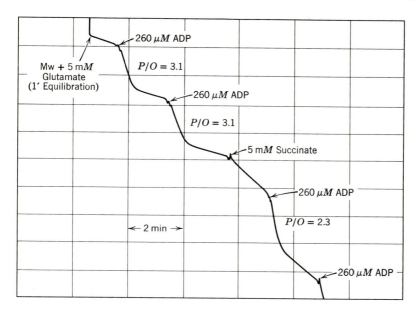

Fig. 8. Respiratory control as measured with the oxygen electrode in a suspension of pigeon heart mitochondria. From Chance and Hagihara (153).

ratio," (rate [+ ADP])/(rate [− ADP]). The ratio decreases on aging or destruction of the respiratory chain preparation and is a sensitive criterion for its "intactness" in isolated mitochondria. As shown in Figure 8, respiratory control may also be used to determine the phosphorylation efficiency (P/O ratio) (152).

The most important steady states of electron flow in the respiratory chain are established by respiratory control through energy transfer (152,154). These steady states have been called "controlled state" (lack of ADP and P_i), "active state" (excess ADP and P_i), and "uncoupled state" (with uncoupler) (50). In the nomenclature of Chance and Williams (154) the controlled state is called "state 4" and the active state "state 3." In the controlled state, respiration is limited inside the chain, in the active state at the terminal step, e.g., by the activity of the dehydrogenases.

B. The Redox Behavior of the Respiratory Chain in the Steady States

The function of a single carrier during the overall process of respiration can be followed by spectrophotometric and chemical means. Methodological progress in directly following the catalyst has greatly advanced our knowledge of the respiratory chain. The study of the steady-state behavior of the respiratory chain is of general interest for comprehension of the steady state in multicomponent systems.

The first quantitative measurements of steady state reduction of respiratory components were made by Chance on nonphosphorylating Keilin-Hartree preparations with the help of sensitive dual-beam spectrophotometry (155). In cytochromes a_3, a, and c, 13–35% reduction was observed. An evaluation of the second-order rate constants for electron-transfer reactions, from steady-state reduction data, was attempted by applying the mass action law to the system in a quasi-soluble state. Because of the unknown arrangement of the components, only effective concentrations could be assumed, so that the physical meaning of these rate constants remains obscure.

Chance and Williams (156,154) observed a close correlation in mitochondria between the steady-state reduction of respiratory carriers and the steady states of electron flow under the control of the energy-transfer system. A more complete picture became available when the redox state of the hydrogen-transfer coenzymes, DPN, TPN, and ubiquinone, could also be measured by enzymic chemical analysis (157,49,121).

These aspects are illustrated by a simultaneous recording of the respiration, the absorption of cytochrome b, and sensitive analysis of the reduction of ubiquinone and DPN in heart muscle mitochondria (121) (Fig. 9). On the first addition of succinate a relatively slow respiration is initiated in the controlled state and all three components become partially reduced. When respiration is stimulated in the active state by further addition of ADP and P_i, the carriers are oxidized and after the exhaustion of ADP they are again reduced. This transition from the active to the controlled state is slow and shows that the concentration of ADP becomes increasingly rate limiting. The redox changes appear to be closely synchronized for all components, indicating that the rate of transition between these steady states is controlled by the ADP concentration. On subsequent activation of respiration with an uncoupler the carriers become still more oxidized as with added ADP.

In a difference spectrum the difference of the reduction in the controlled and active states is demonstrated (1). Cytochromes b and c are more reduced and cytochrome a more oxidized on transition from the active to the controlled state. The steady-state reduction of the respiratory components, as evaluated from this type of observation, is summarized in the redox pattern, where per cent reduction is plotted according to the sequence of the respiratory components. In the controlled state the degree

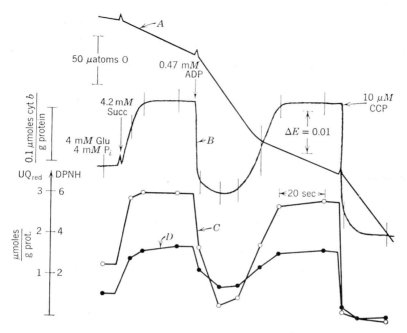

Fig. 9. The redox changes of various respiratory components during the transitions between various steady states. A simultaneous recording of respiration and the absorption of cytochrome b combined with simultaneous sampling for chemical analysis of the reduction of ubiquinone and DPN. (Glu = glutamate, Succ = succinate, CCP = carbonylcyanide phenylhydrazone). Experiment on a suspension of rat heart muscle mitochondria (121). (A) Respiration. (B) Absorption at 434–490 mμ. (C) DPNH and (D) reduced ubiquinone, in extracts.

of reduction decreases from the substrate to the oxygen side, except in the case of cytochrome b. With active substrates, such as succinate, glutamate plus malate, or α-glycerolphosphate, the components DPN, flavin, ubiquinone, and cytochrome b are reduced as a rule to more than 50% and cytochrome c, a, and a_3 to less than 50%. A break in the degree of reduction between cytochrome b and c is particularly pronounced in heart muscle mitochondria. In the active state the redox pattern is more flattened and in the uncoupled state the degree of reduction becomes equally low for all components.

The striking redox pattern in the controlled state reflects an accumulation of hydrogen and electrons in the respiratory chain, concomitant with the inhibition of respiration. This accumulation increases as the respective dehydrogenases become more active. A most pronounced decrease in the redox pattern is obtained with succinate in mammalian mitochondria, and with glycerolphosphate in insect flight muscle mitochondria (102), whereas DPN-linked substrates give a more flattened pattern. The continuous decrease of the redox state, concomitant with inhibition of respiration in the controlled state, reflects a multi-site inhibition, as suggested first by Chance and Williams (158). Respiratory control is the overall effect of inhibition at various electron-transfer steps. Correspondingly, the partial release of inhibition of respiration by ADP and P_i largely removes the buildup of reducing equivalents in the carriers as they become oxidized. From the redox pattern, it can be concluded, furthermore, that on un-coupling the inhibition is completely released.

As shown clearly in the difference spectrum (Fig. 10) and in the redox pattern (Fig. 11), the reduction of cytochrome a increases whereas that of all other components decreases, in the transition from the controlled to the active state. This crossover led Chance et al. (159) to formulate the crossover theorem by stating that the crossover indicates a site of inhibition and thus a site of energy transfer at the respiratory chain. A familiar case illustrating the theorem is the inhibition by antimycin A, which, when added in the steady state, leads to the reduction of the components below and to an oxidation above the inhibition site (cf. chapter on inhibitors). The multiple inhibition by the three energy-transfer coupling sites requires, however, a more detailed evaluation.

Besides the region between cytochrome a and c, a crossover can also be observed between cytochromes b and c and between DPN and flavoprotein. For example, in heart muscle mitochondria with glutamate + malate as substrate (instead of succinate as in Fig. 11) the crossover is found between cytochromes b and c (121). In flight muscle mitochondria the crossover may be with glycerolphosphate between cytochromes c and a and with pyruvate + malate between DPNH and flavoprotein (160). Furthermore, a slight inhibition of the respiration at the oxygen terminal may shift the crossover point further down to the substrate side. Thus with increasing concentration of azide the crossover can be moved from the oxygen side between cytochromes c and a to the substrate side between DPNH and flavoprotein (161).

In general, only at these three of the six reaction steps in the respiratory chain identified here has a crossover been observed in a wide variety of mitochondria and conditions (162). This would support identification of the crossover site with the energy-transfer site.

The examples show an obvious relation of the crossover with the activity of the terminal reactions, such as dehydrogenation and oxygen

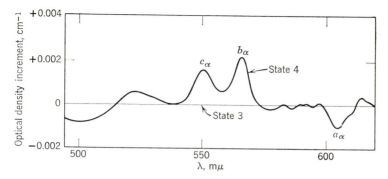

Fig. 10. Difference spectrum of a suspension of liver mitochondria between the controlled state (state 4) and active state (state 3), demonstrating the crossover between cytochromes a and c. From Chance and Williams (1).

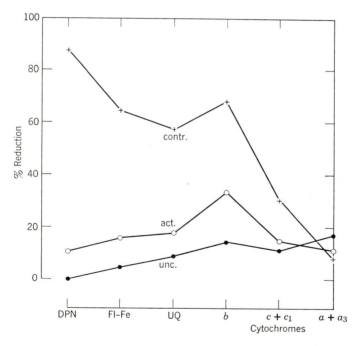

Fig. 11. Redox pattern of the respiratory chain in the controlled, active, and uncoupled states. Experiments on rat heart muscle mitochondria (121).

utilization, in an opposite sense. With increasing hydrogen pressure the crossover is shifted against the oxygen side, and vice versa.

The steady states controlled by the energy transfer have been analyzed in the search for a control mechanism in several ways. Relative rate constants for single electron-transfer reactions can be evaluated from the ratio of the oxidized to the reduced forms of the components under steady-state conditions (159), e.g.

$$A \xrightarrow{k_1} B \xrightarrow{k_2} C \xrightarrow{k_3}$$

$$k_1 A_{red} B_{ox} = k_2 B_{red} C_{ox}$$

The rate constants have been calculated to be 1/100 as large in the controlled state as in the active state. Thus, the observed inhibition of the overall rate would be a matter of control of the single-reaction velocities. A key assumption for these evaluations has been that DPN is reduced as much as 99% in the controlled state. However, DPN is found to be reduced only about 50% by the enzymic analysis (50). Furthermore, no corresponding difference of the single reaction rates could be observed directly (154).

In another approach the reduced carriers are assumed to exist largely in an inhibited "energy-rich" form (component $\sim$ I) (1) in the controlled state. The free carriers are released from the inhibited state in the energy transfer to another intermediate and finally to $ADP + P_i$:

$$C_{red} \sim I + X = C_{red} + X \sim I$$

$$X \sim I + ADP + P = ATP + X + I$$

With three of these interaction sites and a five-membered model of the respiratory chain, the rate equations were set up for analog computer solution (1,159). The computed reaction curves are shown in Figure 12 for the transient redox changes between the controlled and active states of the five components.

In the transition to the controlled state, cytochromes a_3, a, and c become more oxidized and cytochrome b more reduced, resulting in a crossover between cytochromes b and c. The analog computation also gives the shift of the crossover to the substrate side when the rate of interaction of oxygen with the respiratory chain is decreased.

These kinetic analyses can be regarded only as an analog representation in order to understand the steady-state behavior and kinetics of these systems. In principle, the observed crossovers can be explained either by changes in the rate constants or by assuming inhibited forms of the reduced or oxidized carriers. However, in both approaches a fully reduced state of

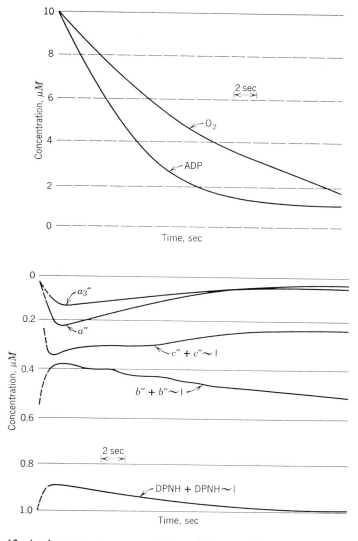

Fig. 12. Analog computer representation of the transition between the controlled and the active state under the assumption that the reduced carriers in the energy-rich state are inhibited for electron transfer (C∼I). The upper trace shows the kinetics of oxygen and ADP utilization after the addition of $10\mu M$ ADP, the lower traces the redox changes of the cytochromes and the DPN system. The computer solution shows a crossover between cytochromes c and b, i.e., the opposite direction of the redox changes, on the addition and exhaustion of ADP. From Chance and Williams (1).

DPN (99%)—as the first carrier in the chain—was a major assumption, a finding which could not be substantiated by enzymic analysis (47). As another explanation for the inhibition in the controlled state, the formation of an energy-dependent equilibrium in the respiratory chain will be discussed below.

XII. Reversibility of Electron Transfer

Reversibility of electron flow in the respiratory chain as part of the reversibility of oxidative phosphorylation is closely linked to steady-state reduction of respiratory components. The reversibility was first explicitly detected in the extensive reduction of mitochondrial DPN by flavin-linked substrates such as succinate (163) or α-glycerolphosphate (63,99). It was then noted as a general phenomenon in various mitochondria (47) and has since gained significance as a major controlling factor for electron transfer (164). The overall process linking reversed energy- and electron-transfer is part of the reversibility of oxidative phosphorylation. Only those aspects pertaining to electron transfer will be discussed here.

A. Reversibility under Aerobic Conditions

In the redox patterns (Fig. 11) the high degree of reduction of DPN in the presence of the flavin-linked substrate succinate was noted. Similar redox patterns can be observed with other flavin-linked substrates, e.g., with α-glycerolphosphate (100,102) and with fatty acids (165). In accordance with the sequence of electron transfer it is to be assumed that DPN is reduced by these substrates in a reversal of electron transfer. The pathway includes the first site of the oxidative phosphorylation from which it receives the energy required in reversed electron transfer. The overall equation may be written as follows:

$$SH_2 + DPN + ATP \longrightarrow S + DPNH + ADP + P_i$$

where SH_2 can be flavin-linked substrates.

In all mitochondria DPN was found to be much more extensively reduced by succinate than by DPN-linked substrates (47). It was assumed that the reduction of DPN includes a pathway, leading to a separate compartment containing most of the DPN, which is not necessarily identical with the hydrogen transfer path from DPN-linked dehydrogenases (166). On the basis of an enzymic study of the redox state of DPN (50,47,167), however, the pathways of forward and reversed electron transfer from succinate to DPNH dehydrogenase are assumed to be identical, so that the succinate-linked DPN reduction is the result of an energy-linked redox equilibrium. The postulate of an energy-rich, inhibited form of "DPNH~I" is then unnecessary (102). This equilibrium explains the high degree of reduction and concomitant inhibition of electron transfer.

The special position of DPN with respect to flavin-linked substrates (DPN is reduced by reversed electron transfer, the other components by forward electron transfer) is not apparent in the redox pattern (Fig. 10). DPN follows the general increase in the degree of reduction towards the substrate region. With DPN-linked substrates DPN is reduced to a smaller degree, similarly to the neighboring components.

This is best explained by extending the concept of the energy-rich redox equilibrium over the entire respiratory chain (122) (Fig. 14). In the equilibrium the principal characteristics of the redox patterns should be independent of the substrate. The ability of the flavin-linked substrates to maintain a highly reduced substrate region in the respiratory chain can be explained simply by the high activity of the dehydrogenases. DPN-linked substrates are usually less active in dehydrogenation, as measured by respiration in the active state, and therefore give smaller degrees of reduction.

The integration of the redox state of DPN into the redox pattern of the respiratory chain is demonstrated by the close correlation between the reduction of DPN and that of ubiquinone in the controlled state over a wide range of respiratory activity in the active state (Fig. 13). Respiration can be considered as a measure of the hydrogen pressure generated by the various substrates. The reduction of both coenzymes increases with the hydrogen pressure, each of them according to a common function for the various substrates. This is to be expected from the energy-linked redox equilibrium, since in equilibrium it is irrelevant for the redox state of the carrier whether hydrogen comes from the flavin or from the DPN side.

A scheme for the energy-linked redox equilibrium illustrates its mechanism (Fig. 14) (122). It can be assumed that three unspecified carriers linked to energy transfer and oxidative phosphorylation have changed their redox potential in forming an energy-rich intermediate and could thus eventually approach that of a preceding carrier. If the reduced carriers are assumed to form the intermediate, the redox potential would become more negative. The shift of the redox potential reflects the energy content of the intermediate, which can be equivalent to the phosphorylation potential ($= \log \{[ATP]/[ADP][P_i]\}$). The sequence of the redox and phosphorylation equilibria extends over the whole chain.

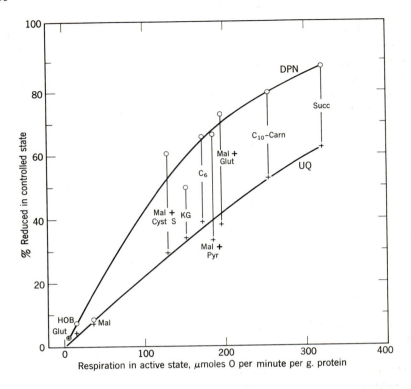

Fig. 13. The correlation of the redox states of DPN and ubiquinone in the controlled state when changing the degree of reduction by varying the substrates. The nearly parallel increase of the redox states of DPN and ubiquinone with dehydrogenase activity of the various substrates as reflected by the respiratory activity in the active state is shown. Experiments with rat heart muscle mitochondria (121). Succ = succinate. C_{10}-Carn = decanoyl carnitine. Mal = malate. Glut = glutamate. C_6 = caprylate. KG = α-ketoglutarate. Pyr = pyruvate. Cyst S = cysteine sulfinate. HOB = β-hydroxybutyrate.

On this basis the influence of the energy potential on the rate of electron transfer becomes understandable. In the absence of energy supply or in the uncoupled state, forward electron transfer is uninhibited. With increasing potential equilibrium is approached and the rate of forward electron transfer decreases until it is halted at equilibrium. When the energy potential rises beyond equilibrium the electron transfer will be reversed.

A competitive effect of ATP and ADP on the redox pattern is to be expected from the energy-linked redox equilibrium (Fig. 15) (164). When

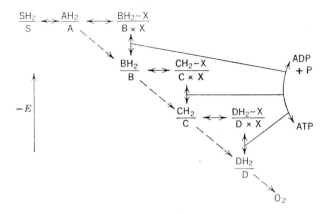

Fig. 14. The energy-linked redox equilibrium in the respiratory chain. The respiratory carriers are arbitrarily designated as A, B, C, and D (121). It is assumed that a reduced carrier forms energy-linked intermediates which would decrease its redox potential so that it can get into redox equilibrium with the preceding carrier. Thus, there are redox equilibria in the non-energy-linked and energy-linked forms with neighboring carriers and phosphorylation equilibria between the energy-rich and non-energy-rich forms of the same carrier. The sequence of the redox and phosphorylation equilibria forms an overall equilibrium extending over the chain.

ATP is added to the mitochondria in the active state, the reduction of the respiratory components increases, particularly toward the substrate region, and the redox pattern resembles more that of the controlled state. A crossover is induced by ATP between cytochromes c and a. Simultaneously the respiration in the active state is more inhibited. The ratio [ATP]/[ADP] could be shown to control respiratory activity. In the presence of ATP the K_M of ADP for respiration is greatly increased: with [ATP] = 0, K_M = 0.05 mM; with [ATP] = 5 mM, K_M = 0.3 mM (164). The control of respiration can be regarded as the result of the energy-linked equilibrium being regulated by the ratio {[ATP]/[ADP][P$_i$]}. The steady-state respiration will vary between the extremes of full equilibrium and uncoupling.

B. External Acceptors and Donors

The reversed hydrogen transfer from succinate is not only limited to the reduction of the endogenous DPN of the mitochondria but can also be extended to external acceptors. In this case the respiratory chain preparation is only a catalyst of an extensive overall process. In liver mitochondria DPNH-oxidizing substrates such as acetoacetate or ketoglutarate + NH_3

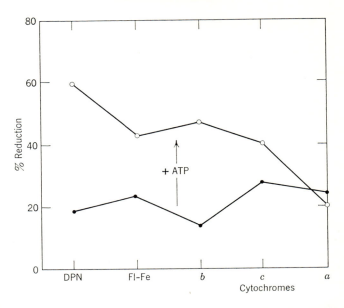

Fig. 15. The redox pattern of the aerobic respiratory chain under the influence of ATP, demonstrating the opposing control of electron transfer by ATP and ADP. Addition of excess ATP (5mM) to liver mitochondria in the active state (2mM Succ, 0.4mM ADP). The respiratory rate is simultaneously decreased (not shown) (cf. also 164).

can be reduced in large amounts by succinate (168–171). Energy may be supplied either by respiration or by ATP. The amounts of reduced substrates formed and energy equivalents utilized are several times larger than the amount of endogenous components of mitochondria. These results removed longstanding doubts about the existence of an energy-linked reversed electron transfer.

Of particular interest is the reversed electron transfer in submitochondrial particles, where the permeability barrier for pyridine nucleotides is removed and added DPN can be reduced by succinate in appreciable quantities (172–175). These reactions have been of great interest for elucidating the pathway of reversed electron and energy transfer. Closely related to the reversed electron transfer in the respiratory chain is the energy-linked transhydrogenation from DPNH to TPN (50,176,177). This reaction will not be dealt with in this context, since it is only loosely associated with the respiratory chain and its main interest resides in the elucidation of energy transfer (cf. 177 and review 11).

C. Pathways of Reversed Electron Transfer

It may follow from the definition of reversed electron transfer that the reversed and forward electron transfer have the same pathway. However, this may not be so *eo ipso* and other possible pathways have been considered (179). Today, evidence prevails in favor of the identity of the reversed and forward pathways. Thus even data on the reversed pathway can be used for elucidating forward pathways. A survey of the pathways of reversed electron transfer is given in Figure 16.

In the succinate-linked DPN reduction, according to the electron-transfer scheme in Figure 1, hydrogen is first transferred from succinate to the pool of ubiquinone and from there to the DPNH dehydrogenase. This includes the first step of oxidative phosphorylation and is in accordance with the stoichiometry of energy utilization, since about one energy equivalent is utilized per DPNH formed. Reliable measurements for this stoichiometry come from the energy-linked substrate → substrate hydrogen transfer reactions as sustained by energy derived from respiration (180, 181,181a) or from ATP-linked hydrogen transfer (170,171). The energy requirement had been overestimated in following only the reduction of endogenous pyridine nucleotides (182,183) because of the interference of the mitochondrial ATPase.

Also the inhibition by amytal (179,102) speaks for the reversal of the DPN → flavoprotein or flavoprotein → ubiquinone reaction—whichever is assumed to be the point of action of amytal—both being involved in the hydrogen transfer from succinate to DPN. A sensitivity to antimycin A has led to the assumption (123) that electrons are first transferred from succinate to cytochrome *b* and from there back to DPN. An

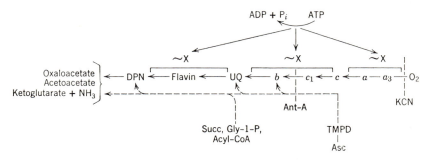

Fig. 16. Pathways of reversed hydrogen and electron transfer as observed on the respiratory components or with various external donors and acceptors.

interference of antimycin A with the succinate-linked DPN reduction was not observed by other authors (122,152,184,163,185,89). There is evidence that antimycin A interferes with succinate-linked DPN reduction at the branch of the energy transfer rather than at an electron transfer step (cf. below). The controversial discussion on the role of cytochrome b in the succinate-linked DPN reduction is related to its ambiguous role in forward electron transfer.

D. Reversed Electron Transfer under Anaerobiosis

Reversed electron transfer in the cytochrome region and thus in the whole respiratory chain can be demonstrated when the interaction of oxygen is eliminated under anaerobic conditions or by terminal respiratory inhibitors. Under these conditions the cytochromes are fully reduced. Now when energy is provided by supplying ATP, electrons are removed from the respiratory carriers including cytochrome a towards the substrate site (122,146). In the presence of an active, infinite hydrogen sink such as oxaloacetate, all carriers can be nearly fully oxidized in reversed electron transfer (Fig. 17).

The ATP-linked reversed electron transfer can be inhibited at various stages in the reversed pathway. The ATP-linked oxidation of cytochromes a and c is inhibited by antimycin A, by amytal, and by hydrogen-donating substrates which counteract the hydrogen-accepting substrates (122,179). This allows identification of the reversed pathway with forward electron transfer from a DPN reducing substrate.

The anaerobic reversed electron transfer can be coupled via the DPN pool to at least three different DPN-linked dehydrogenases—malate, glutamate, and hydroxybutyrate dehydrogenases—which are present in high concentrations in (rat) liver mitochondria. By suitable combinations of oxidizing and reducing substrates, different degrees of the redox ratio [DPNH]/[DPN] can be established. In the presence of ATP the redox ratio of cytochrome c (c_{red}/c_{ox}) is a linear function of DPNH/DPN over a wide range (Fig. 18). This is interpreted to show that under the influence of energy a redox equilibrium can extend over several carriers between DPN and cytochrome c (122). Also, a close correlation between the redox state of cytochrome c and the phosphorylation potential can be established under anaerobic conditions (186).

The redox equilibrium is more clearly demonstrated under anaerobic conditions since the steady-state forward electron transfer in the presence of oxygen is eliminated. At least part of the reducing equivalents which are removed from the cytochromes in ATP-linked reversed electron transfer

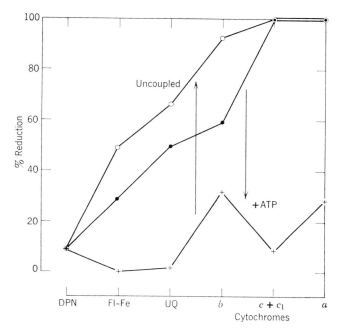

Fig. 17. The redox pattern of the anaerobic respiratory chain, as a result of reversed electron transfer after the addition of ATP. Experiments with rat heart muscle mitochondria (121). On the addition of ATP, the carriers become largely oxidized in reversed electron transfer to oxaloacetate. The addition of an uncoupler abolishes the effect of ATP and again causes reduction.

can be recovered by the mitochondrial DPN in a closed system without added hydrogen sinks (123). When this system is closed toward DPN by amytal, reducing equivalents from the cytochromes can in part be recovered in ubiquinone, as the cytochromes become oxidized and ubiquinone slightly reduced (121).

Electron transfer can also be reversed from the cytochrome region when the system is open with respect to the electron donors. From added reduced cytochrome c, electrons can be transferred to the mitochondrial DPN (187) driven by ATP under anaerobic conditions. Particularly interesting is reversed electron transfer from ascorbate + TMPD, which is assumed to interact at the site of cytochrome c (cf. Fig. 7). With these donors mitochondrial DPN and ubiquinone could be reduced (188,189). Ubiquinone was found to be reduced faster than DPN. This indicates

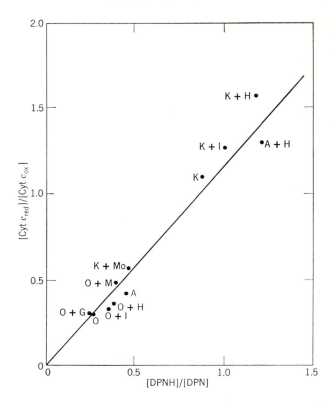

Fig. 18. The relation between the redox ratio of DPN and cytochrome c in the energy-linked redox equilibrium under anaerobic conditions. Experiments with liver mitochondria. (From ref. 122.) A = acetoacetate. G = glycerol-1-P. H = β-hydroxy-butyrate. I = isocitrate. K = α-ketoglutarate + NH_3. M = malate. Mo = malonate. O = oxaloacetate.

that reversed electron transfer from the cytochrome region to DPN involves ubiquinone.

Submitochondrial particles in combination with ascorbate and TMPD can be used to reduce added DPN in considerable amounts (190). With this reaction the energy requirements for reversed electron transfer from the cytochromes to DPN have been evaluated to be about two energy equivalents per molecule of DPNH. The inferred participation of two energy-transfer sites supports the identity of forward and reversed electron-transfer pathways.

The oxidation of ascorbate + TMPD by oxygen in the terminal segment of the cytochrome chain can be used to generate energy in the cytochrome region which is utilized in a succinate-linked DPN reduction. In these cases the electron transfer from TMPD ascorbate to DPN is eliminated by antimycin A. In this way the energy-transfer problem was studied in the reduction of DPN (191–193) and of exogenous hydrogen acceptors such as ketoglutarate + NH_3 (194) by succinate. These experiments demonstrate the transfer of energy equivalents generated in the terminal cytochrome region to the flavoprotein region.

A summary of the redox behavior of the respiratory chain under the variety of conditions described above,* is given in a simple descriptive survey of the redox patterns (Table IV). This is possible since the energy-

TABLE IV

Redox Patterns of the Respiratory Chain (176)

| Functional state | Terminal supply | | Pattern type |
	Substrate	Oxygen	
Controlled	High H pressure (flavin-linked substrate)	+	Strongly descending
Controlled	Medium H pressure, (DPN-linked substrate)	+	Descending
Active (only ADP)	High H pressure	+	Flat or slightly descending
	Medium H pressure	+	Flat
ATP/ADP-controlled	High H pressure	+	Descending
ATP-controlled	Endogenous	−	Descending (cytochrome largely oxidized)
	DPNH oxidizing	−	Flat (largely oxidized)
Uncoupled	High H pressure	+	Flat or ascending
	DPNH oxidizing	−	No gradient, fully reduced, ascending in dehydrogenase region.

linked redox equilibrium in the respiratory chain makes the redox pattern qualitatively inert to the various substrates so that qualitatively uniform types are observed. Controlling factors are the terminal conditions for hydrogen and oxygen "pressure" and the phosphorylation, i.e., energy potential.

* *Editor's Note:* See also chapters by San Pietro and Crane in Part II of this volume.

XIII. Kinetics of the Oxidation and Reduction of Respiratory Carriers

As described in the previous chapters, the steady-state reduction of the respiratory components can be used for evaluating kinetic data, such as the rate constants for the interaction between the respiratory components. However, in principle these rate constants are limited only to the type of the steady state, i.e., they may be different between a phosphorylating and an uncoupled respiratory chain. Furthermore, these constants are derived on the basis of the mass action law, of which the validity in the respiratory chain is difficult to judge. Direct measurements of the kinetic changes in the respiratory chain are therefore highly desirable. They are useful not only for collecting data of the single electron-transfer reactions, but also for evaluation of the position of a carrier in the respiratory sequence.

Measuring the kinetics of electron-transfer steps is only possible when the reaction is started fast enough to observe redox changes which are limited by the electron-transfer rate. This may not always be the case; in particular, transitions between different steady states may reflect comparatively slow quasi steady state changes. The coupling of electron and energy transfer may cause the rates of the redox changes to be limited by the energy transfer.

Oxidation of the respiratory components by addition of oxygen to an anaerobic respiratory chain appears to be the best method of studying electron-transfer rates. Under these conditions the respiratory components are first fully or largely reduced and then become rapidly oxidized. The electron or reducing equivalents stored in the respiratory components are transferred to oxygen in the order of their sequence from the oxygen terminal.

An example of measuring the rate of oxidation of cytochrome a by mixing the anaerobic suspension of mitochondria with oxygen in a rapid-flow apparatus is shown in Figure 19 (69). The rate constant is evaluated from the degree of oxidation during the few milliseconds of the steady-state flow. The first-order rate constant measured in this way on liver mitochondria at room temperature, as summarized in Table V (143,69), decreased strongly from cytochrome a_3 to cytochrome b and flavoprotein. This decreasing oxidation rate demonstrates kinetically and most directly the sequence of the electron flow in the respiratory chain. It confirms the sequence of the components established earlier on other grounds.

Rates of forward electron transport are also observed when the changes in the redox state are caused by interaction with the energy-transfer system. The transition from the controlled state to the active state on

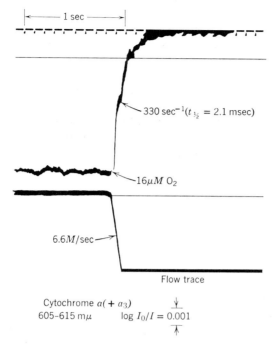

Fig. 19. The kinetics of oxidation of cytochrome a by the addition of oxygen to anaerobic rat liver mitochondria at 33°C. Experiment with a stopped flow apparatus. From Chance, Schöner, and de Vault (143). The rate constant was evaluated from the degree of reduction of cytochrome a during the brief period of steady-state flow.

TABLE V

Reaction Rates of the Oxidation of Respiratory
Components on Addition of Oxygen (143,69)

Reactions		k, sec^{-1}
$O_2 \rightarrow a_3$	25°	158
$\rightarrow a$		120
$\rightarrow c$		39
$\rightarrow b$		20
$\rightarrow$ fp		15
$O_2 \rightarrow a_3$	33°	490
$\rightarrow a$		330
$\rightarrow c$		210

addition of ADP reflects rates of the forward electron transport, since in the controlled state the components are largely reduced and electron flow is inhibited and in the active state electron flow is activated and the components are more oxidized.

In Figure 20 the kinetics of the oxidation of cytochrome b and DPNH during the transition from the controlled to the active state are compared. The large difference in the time needed for the transition (cytochrome b, 0.2 sec; DPNH, 12 sec) reflects the 20 times greater storage of reducing equivalents in DPNH. In fact, the initial transfer rates (as referred to

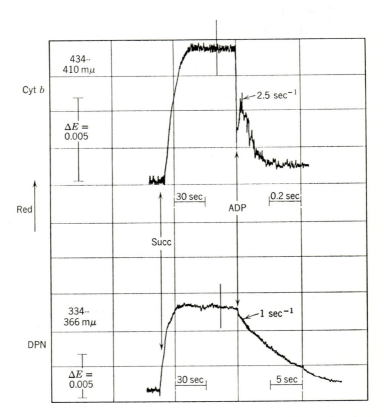

Fig. 20. Rate of oxidation of cytochrome b and of DPNH during the transition from the controlled state to the active state. Experiment on liver mitochondria at 11°C with the moving mixing chamber (195). The rate constants are referred to the electrons transferred during the oxidation. Upper traces: cytochrome b at 434–410 mμ. Lower traces: DPN at 334–366 mμ.

electrons) are only different by a factor of 2. It may indicate that the oxidation of DPNH is limited primarily by the diffusion of the DPNH molecules to DPNH dehydrogenase.

Rates of reversed electron transfer in the respiratory chain can be measured under anaerobic conditions by addition of ATP. Through a reversal of the oxidative phosphorylation the cytochromes are then oxidized from the fully reduced to the largely oxidized form, donating their electrons to the substrate site. These oxidation rates are only about 1/10 of those with oxygen (93). Equal rates are found for the oxidation of the various components in contrast to the oxidation rates with oxygen. It may be concluded that the reversed rate is determined primarily by the rate of the interaction of ATP with the respiratory chain. This rate appears to be the same for all three of the coupling sites since it probably employs a common step. The rate of the actual reversed electron transfer may be much higher.

References

1. B. Chance and G. R. Williams, in *Advances in Enzymology*, Vol. 17, F. F. Nord, Ed., Interscience, New York, 1956, p. 56.
2. E. C. Slater, in *Advances in Enzymology*, Vol. 20, F. F. Nord, Ed., Interscience, New York, 1958, p. 147.
3. D. E. Green, in *Advances in Enzymology*, Vol. 21, F. F. Nord, Ed., Interscience, New York, 1959, p. 73.
4. E. C. Slater, in *Structure and Function of Muscle*, Vol. II, New York, 1960, p. 105.
5. B. Chance, Ed., *Energy Linked Functions of Mitochondria*, Academic Press, New York, 1963.
6. M. Klingenberg, in *Ergebnisse der Physiologie*, Vol. 55, Springer-Verlag, Berlin, 1964, p. 132.
7. A. L. Lehninger, *The Mitochondrion*, W. A. Benjamin, New York, 1964.
8. P. Nicholls, "Cytochromes—A Survey," in *The Enzymes*, Vol. 8, London, 1963, p. 3.
9. E. Racker, in *Advances in Enzymology*, Vol. 23, F. F. Nord, Ed., Interscience, New York, 1961, p. 323.
10. Y. Hatefi, in *Advances in Enzymology*, Vol. 25, F. F. Nord, Ed., Interscience, New York, 1963, p. 275.
11. L. Ernster and C. P. Lee, *Ann. Rev. Biochem.*, **33**, 729 (1964).
12. D. R. Sanadi, *Ann. Rev. Biochem.*, **34**, 21 (1965).
13. O. Warburg, E. Negelein, and E. Haas, *Biochem. Z.*, **266**, 1 (1933).
14. D. Keilin, *Proc. Roy. Soc. (London)*, **A104**, 206 (1939).
15. S. Takemori, I. Sekuzu, and K. Okunuki, *Biochim. Biophys. Acta*, **51**, 464 (1961).
16. R. S. Criddle and R. M. Bock, *Biochem. Biophys. Res. Commun.*, **1**, 138 (1958).
17. A. Paleus, *Acta Chem. Scand.*, **9**, 335 (1955).

18. R. Bomstein, R. Goldberger, and H. Tisdale, *Biochim. Biophys. Acta*, **50**, 527 (1961).
19. R. Goldberger, A. L. Smith, H. Tisdale, and R. Bomstein, *J. Biol. Chem.*, **236**, 2788 (1961).
20. J. S. Rieske, R. E. Hansen, and W. S. Zaugg, *J. Biol. Chem.*, **239**, 3017 (1964).
21. T. Cremona and E. B. Kearney, *J. Biol. Chem.*, **239**, 2328 (1963).
22. T. E. King and R. L. Howard, *J. Biol. Chem.*, **237**, 1686 (1962).
23. H. R. Mahler, N. K. Sarkar, L. P. Vernon, and R. A. Alberty, *J. Biol. Chem.*, **199**, 585 (1952).
23a. B. Mackler, *Biochim. Biophys. Acta*, **50**, 141 (1961).
23b. H. Watari, E. B. Kearney, and T. P. Singer, *J. Biol. Chem.*, **238**, 4063 (1963).
24. E. G. Ball, *Biochem. Z.*, **295**, 262 (1938).
25. E. C. Slater, "Oxidation-Reduction Potentials and their Significance in Hydrogen Transfer," in *Handbuch der Pflanzenphysiologie*, Springer-Verlag, Berlin, 1960, p. 114.
26. W. M. Clark, *Oxidation-Reduction Potentials of Organic Systems*, Williams and Wilkins, Baltimore, 1960.
27. R. Goldberger, A. Pumphrey, and A. Smith, *Biochim. Biophys. Acta*, **58**, 307 (1962).
28. F. A. Holton and J. Colpa-Boonstra, *Biochem. J.*, **76**, 179 (1960).
29. W. W. Wainio, *J. Biol. Chem.*, **216**, 593 (1955).
30. D. E. Green, J. Järnefelt, and H. D. Risdale, *Biochim. Biophys. Acta*, **31**, 34 (1959).
31. F. L. Rodkey and E. G. Ball, *J. Biol. Chem.*, **182**, 17 (1950).
32. R. A. Morton, *Biochem. J.*, **73**, P2 (1959).
33. E. C. Slater, J. P. Colpa-Boonstra, and J. Links, "The Oxidation of Quinols by Mitochondrial Preparations," in *Ciba Foundation Symposium on Quinones in Electron Transport*, G. E. W. Wolstenholme and C. M. O'Connor, Ed., Little, Brown, Boston, 1961, p. 161.
34. F. L. Rodkey, *J. Biol. Chem.*, **213**, 777 (1955).
35. J. E. King, K. Nickel, and D. R. Jensen, *J. Biol. Chem.*, **239**, 1989 (1964).
36. D. E. Green and D. C. Wharton, *Biochem. Z.*, **338**, 335 (1963).
37. P. V. Blair, T. Oda, D. E. Green, and H. Fernandez-Moran, *Biochemistry*, **2**, 756 (1963).
37a. B. F. van Gelder and A. O. Muijers, *Biochim. Biophys. Acta*, **81**, 405 (1964).
38. S. Takemori and T. E. King, *J. Biol. Chem.*, **239**, 3546 (1964).
39. B. Chance and M. Baltscheffsky, *Biochem. J.*, **68**, 283 (1958).
40. M. Klingenberg and T. Bücher, *Biochem. Z.*, **331**, 312 (1959).
41. W. Estabrook, *J. Biol. Chem.*, **230**, 735 (1958).
42. P. Schollmeyer and M. Klingenberg, *Biochem. Z.*, **335**, 426 (1962).
43. M. Klingenberg, "Die funktionelle Biochemie der Mitochondrien," in GDNA Symposium in Rottach-Egern on *Funktionelle und Morphologische Organisation der Zelle*, P. Karlson, Ed., Springer-Verlag, Berlin, 1962, p. 69.
44. R. Estabrook and A. Holowinsky, *J. Biophys. Biochem. Cytol.*, **9**, 19 (1961).
45. B. Chance and B. Hess, *J. Biol. Chem.*, **234**, 2404 (1959).
46. K. B. Jacobson and N. O. Kaplan, *J. Biol. Chem.*, **226**, 603 (1957).
47. M. Klingenberg, W. Slenczka, and E. Ritt, *Biochem. Z.*, **332**, 47 (1959).
48. R. L. Lester and F. L. Crane, *J. Biol. Chem.*, **234**, 2169 (1959).
49. L. Szarkowska and M. Klingenberg, *Biochem. Z.*, **338**, 674 (1963).

50. M. Klingenberg and W. Slenczka, *Biochem. Z.*, **331**, 486 (1959).
51. D. Pette, M. Klingenberg, and T. Bücher, *Biochem. Biophys. Res. Commun.*, **7**, 425 (1962).
52. H. Goebell and M. Klingenberg, *Biochem. Biophys. Res. Commun.*, **13**, 212 (1963).
53. M. Klingenberg and D. Pette, *Biochem. Biophys. Res. Commun.*, **7**, 430 (1962).
54. D. Keilin and E. F. Hartree, *Biochem. J.*, **41**, 500 (1947).
55. F. L. Crane, J. L. Glenn, and D. E. Green, *Biochim. Biophys. Acta*, **22**, 475 (1956).
56. D. E. Green, A. W. Linnane, and D. M. Ziegler, *Biochim. Biophys. Acta*, **29**, 630 (1958).
57. C. Cooper and A. L. Lehninger, *J. Biol. Chem.*, **219**, 489 (1956).
58. T. M. Devlin, *J. Biol. Chem.*, **234**, 962 (1959).
59. C. T. Gregg, *Biochim. Biophys. Acta*, **74**, 573 (1963).
60. Y. Hatefi, A. G. Haavik, L. R. Fowler, and D. E. Griffith, *J. Biol. Chem.*, **237**, 2661 (1962).
61. G. E. Palade, in *Enzymes, Units of Structure and Function*, O. H. Gaebler, Ed., Academic Press, New York, 1956, p. 185.
62. D. E. Green, in GDNA Symposium in Rottach-Egern on *Funktionelle une Morphologische Organisation der Zella*, P. Karlson, Ed., Springer-Verlag, Berlin, 1962, p. 86.
63. Y. Hatefi, in *Advances in Enzymology*, Vol. 25, F. F. Nord, Ed., Interscience, New York, 1963, p. 275.
64. T. E. King and S. Takemori, *J. Biol. Chem.*, **239**, 3559 (1964).
65. T. E. King, *J. Biol. Chem.*, **238**, 4032 (1963).
66. T. E. King, *J. Biol. Chem.*, **236**, 2342 (1961).
67. D. E. Green and T. Oda, *J. Biochem.*, **49**, 742 (1961).
68. L. R. Fowler and S. H. Richardson, *J. Biol. Chem.*, **238**, 456 (1963).
69. B. Chance, *Discussions Faraday Soc.*, **20**, 205 (1955).
70. D. Keilin, *Ergeb. Enzymforsch.*, **2**, 239 (1933).
71. E. C. Slater, in *Advances in Enzymology*, Vol. 20, F. F. Nord, Ed., Interscience, New York, 1958, p. 147.
72. A. L. Lehninger, *Harvey Lectures*, **49**, 174 (1953–54).
73. G. E. Palade, *Anat. Record*, **114**, 427 (1952).
74. F. S. Sjöstrand and J. Rhodin, *J. Exptl. Cell Res.*, **4**, 426 (1953).
75. R. S. Criddle, R. M. Bock, D. E. Green, and H. Tisdale, *Biochemistry*, **1**, 827 (1962).
76. A. L. Lehninger, C. L. Wadkins, C. Cooper, T. M. Delvin, and G. L. Gamble, *Science*, **128**, 450 (1958).
77. M. Klingenberg and W. Vogell, unpublished.
78. M. Klingenberg, *Ber. Physik. Med. Ges. Würzburg*, **71**, 174 (1964).
79. M. B. Thorn, *Biochem. J.*, **63**, 420 (1956).
80. R. L. Ringler and T. P. Singer, *Arch. Biochim. Biophys.*, **77**, 229 (1958).
81. T. Kimura, T. P. Singer, and C. J. Lusty, *Biochim. Biophys. Acta*, **44**, 284 (1960).
82. B. Chance, R. W. Estabrook, and C. P. Lee, *Science*, **140**, 379 (1963).
83. D. E. Green, "Structure and Function of Subcellular Particles," in *Plenary Lecture, Proceedings of the 5th International Congress of Biochemistry in Moscow*, Pergamon Press, Oxford, 1963, p. 9.
84. H. Fernandez-Moran, *Circulation*, **26**, 1039 (1963).

85. H. Fernandez-Moran, *Science*, **140**, 381 (1963).
86. B. Chance and D. F. Parsons, *Science*, **142**, 1176 (1963).
87. B. Chance, D. F. Parsons, and G. R. Williams, personal communication.
88. C. Lee, R. W. Estabrook, and B. Chance, *Biochim. Biophys. Acta*, **99**, 32 (1965).
89. D. Keilin and T. E. King, *Proc. Roy. Soc.* (*London*), **B152**, 277 (1940).
90. B. Chance and E. Spencer, *Discussions Faraday Soc.*, **27**, 200 (1959).
91. H. Laser and E. C. Slater, *Nature*, **187**, 1115 (1960).
92. D. E. Green and R. L. Lester, *Federation Proc.*, **18**, 987 (1959).
93. M. Klingenberg, *Ber. Bunsen Ges. Physik. Chem.*, **68**, 747 (1964).
94. O. Warburg, *Schwermetalle als Wirkungsgruppen von Fermenten*, Verlag Dr. Werner Saenger, Berlin, 1946.
95. D. Keilin, *Proc. Roy. Soc.* (*London*), **B104**, 206 (1929).
96. L. Ernster, H. Löw, and O. Lindberg, *Exptl. Cell Res. Suppl.*, **3**, 124 (1955).
97. B. Chance, in *Enzymes, Units of Biological Structure and Function*, O. H. Gaebler, Ed., Academic Press, New York, 1956, p. 447.
98. P. E. Lindahl and K. E. Öberg, *Exptl. Cell Res.*, **23**, 228 (1961).
99. L. Ernster, G. Dallner, and G. F. Azzone, *Biochem. Biophys. Res. Commun.*, **10**, 23 (1963).
100. T. Bücher and M. Klingenberg, *Angew. Chem.*, **70**, 552 (1958).
101. K. L. Yielding, G. Tomkins, J. S. Munday, and I. J. Cowley, *J. Biol. Chem.*, **235**, 3413 (1960).
102. M. Klingenberg and T. Bücher, *Biochem. Z.*, **334**, 1 (1961).
103. R. W. Estabrook, *J. Biol. Chem.*, **227**, 1093 (1957).
104. S. Minakami, T. Cremona, R. L. Ringler, and T. P. Singer, *J. Biol. Chem.*, **238**, 1529 (1963).
105. R. W. Estabrook, *J. Biol. Chem.*, **236**, 3051 (1961).
105a. S. Minakami, et al., *J. Biol. Chem.*, **237**, 569 (1962).
106. A. M. Pumphrey and E. R. Redfearn, *Biochem. Biophys. Res. Commun.*, **8**, 92 (1962).
107. M. Klingenberg and P. Schollmeyer, *Biochem. Z.*, **333**, 335 (1960).
108. B. Chance and G. Hollunger, *J. Biol. Chem.*, **238**, 418 (1963).
109. L. Ernster, G. Dallner, and F. G. Azzone, *J. Biol. Chem.*, **238**, 1124 (1963).
110. R. L. Pharo and R. Sanadi, *Biochim. Biophys. Acta*, **85**, 346 (1964).
110a. J. Machinist and T. P. Singer, *Proc. Natl. Acad. Sci. U.S.*, **53**, 467 (1965).
111. E. G. Ball, C. B. Anfinsen, and V. Cooper, *J. Biol. Chem.*, **168**, 257 (1947).
112. E. M. Case and F. Dickens, *Biochem. J.*, **43**, 481 (1948).
113. E. C. Slater, *Nature*, **165**, 674 (1950).
114. R. van Potter and A. Reif, *J. Biol. Chem.*, **194**, 287 (1952).
115. J. W. Lightbown and F. L. Jackson, *Biochem. J.*, **63**, 130 (1956).
116. B. Chance, *Nature*, **169**, 169 (1952).
117. B. Chance, *J. Biol. Chem.*, **233**, 1223 (1958).
118. R. W. Estabrook, *Biochim. Biophys. Acta*, **60**, 236, 249 (1962).
119. J. S. Rieske, W. S. Zaugg, and R. E. Hansen, *J. Biol. Chem.*, **239**, 3023 (1964).
120. A. L. Tappel, *Biochem. Pharmacol.*, **3**, 289 (1960).
121. A. Kröger and M. Klingenberg, *Biochem. Z.*, **344**, 317 (1966).
122. M. Klingenberg, and P. Schollmeyer, *Biochem. Z.*, **335**, 243 (1961).
123. B. Chance and G. Hollunger, *J. Biol. Chem.*, **236**, 1562 (1961).
124. A. M. Pumphrey and E. R. Redfearn, *Biochem. J.*, **76**, 61 (1960).
125. E. R. Redfearn and A. M. Pumphrey, *Biochem. J.*, **73**, 3P (1959).

126. E. R. Redfearn and A. M. Pumphrey, *Biochem. Biophys. Res. Commun.*, **3**, 650 (1960).
127. D. K. Myers and E. C. Slater, *Nature*, **179**, 363 (1957).
128. H. Löw, P. Siekevitz, L. Ernster, and O. Lindberg, *Biochim. Biophys. Acta*, **29**, 392 (1958).
129. P. D. Boyer, W. W. Luchsinger, and A. B. Falcone, *J. Biol. Chem.*, **223**, 405 (1956).
130. J. L. Howland, *Biochim. Biophys. Acta.*, **73**, 665 (1963).
131. C. Martius, *Biochem. Z.*, **326**, 26 (1954).
131a. F. Maerki and C. Martius, *Biochem. Z.*, **333**, 111 (1960).
132. T. E. Conover and L. Ernster, *Biochim. Biophys. Acta*, **58**, 189 (1962).
133. T. E. Conover, L. Danielson, and L. Ernster, *Biochim. Biophys. Acta.*, **67**, 259 (1963).
134. J. H. Copenhaver and H. A. Lardy, *J. Biol. Chem.*, **195**, 225 (1952).
135. B. C. Pressman, *Biochim. Biophys. Acta.*, **17**, 274 (1955).
136. E. Jacobs and D. R. Sanadi, *Biochim. Biophys. Acta.*, **38**, 12 (1960).
137. E. Jacobs, *Biochem. Biophys. Res. Commun.*, **3**, 536 (1960).
138. J. L. Howland, *Biochim. Biophys. Acta*, **77**, 419 (1963).
139. C. H. Lee, K. Nordenbrand, and L. Ernster, *Proc. Intern. Symp.*, Amherst, Mass., 1964, in print.
140. D. Tyler, R. W. Estabrook, and R. Sanadi, *Biochem. Biophys. Res. Commun.*, **18**, 264 (1965).
141. E. C. Slater, *Biochem. J.*, **45**, 14 (1949).
142. E. C. Slater, *Biochem. J.*, **46**, 499 (1950).
143. B. Chance, B. Schöner, and D. Devault, *Proc. Intern. Symp.*, Amherst, Mass., 1964, in print.
144. A. M. Pumphrey, *J. Biol. Chem.*, **237**, 2384 (1962).
145. S. Takemori and T. E. King, *Science*, **144**, 852 (1964).
146. B. Chance, *J. Biol. Chem.*, **236**, 1544 (1961).
148. A. Beuder and M. Kiese, *Arch. Exptl. Pathol. Pharmakol.*, **224**, 312 (1955).
149. B. Chance, *Federation Proc.*, **16**, 671 (1957).
150. H. A. Lardy and H. Wellman, *J. Biol. Chem.*, **195**, 215 (1952).
151. H. A. Lardy, in *Proc. Intern. Congr. Biochem. 3rd, Brussels, 1955* (publ. 1956), p. 287.
152. B. Chance and G. R. Williams, *J. Biol. Chem.*, **217**, 383 (1955).
153. B. Chance and B. Hagihara, in *Proc. Intern. Congr. Biochem., 5th, Moscow, 1962* (publ. 1963), p. 3.
154. B. Chance and G. R. Williams, *J. Biol. Chem.*, **217**, 409 (1955).
155. B. Chance, *Nature*, **169**, 215 (1952).
156. B. Chance and G. R. Williams, *J. Biol. Chem.*, **217**, 395 (1955).
157. M. Klingenberg and W. Slenczka, *Biochem. Z.*, **31**, 334 (1959).
158. B. Chance and G. R. Williams, *J. Biol. Chem.*, **217**, 429 (1955).
159. B. Chance, G. R. Williams, W. H. Holmes, and J. Higgins, *J. Biol. Chem.*, **217**, 439 (1955).
160. M. Klingenberg, in *Proc. Intern. Congr. Biochem., 4th, Vienna, 1958* (publ. 1959), Vol. 15, p. 67.
161. B. Chance and G. R. Williams, *J. Biol. Chem.*, **221**, 477 (1956).
162. B. Chance, in *Haematin Enzymes*, Part 1, J. E. Falk, R. Lamberg, and R. K. Morton, Eds., Pergamon Press, London, 1961, p. 597.

163. J. M. Tager and E. C. Slater, *Biochim. Biophys. Acta*, **77**, 246 (1963).
164. M. Klingenberg, *Angew. Chem.*, **75**, 900 (1963).
165. M. Klingenberg and C. Bode, in *Recent Research on Carnitine*, G. Wolf, Ed., M.I.T. Press, Cambridge, Mass., 1965, p. 87.
166. B. Chance and G. Hollunger, *J. Biol. Chem.*, **236**, 1534 (1961).
167. M. Klingenberg, "Pyridinnukleotide und biologische Oxidation," *Zur Bedeutung der freien Nukleotide*, Springer-Verlag, Heidelberg, 1961, p. 82.
168. G. F. Azzone, L. Ernster, and E. C. Weinbach, *J. Biol. Chem.*, **238**, 1825 (1963).
169. J. M. Tager and E. C. Slater, *Biochim. Biophys. Acta*, **77**, 227 (1963).
170. M. Klingenberg and H. v. Häfen, *Biochem. Z.*, **337**, 120 (1963).
171. M. Klingenberg, G. Wenske, and H. v. Häfen, *Biochem. Z.*, **343**, 479 (1965).
172. H. Löw, H. Krueger, and D. M. Zeigler, *Biochem. Biophys. Res. Commun.*, **5**, 231 (1961).
173. H. Löw and I. Vallin, *Biochim. Biophys. Acta*, **69**, 361 (1963).
174. F. A. Hommes, *Biochim. Biophys. Acta*, **77**, 173 (1963).
175. H. Löw, I. Vallin, and B. Alm, in *Energy-Linked Functions of Mitochondria*, B. Chance, Ed., Academic Press, New York, 1963, p. 5.
176. M. Klingenberg and P. Schollmeyer, in *Proc. Intern. Congr. Biochem.*, *5th*, *Moscow*, *1962* (publ. 1963), p. 46.
177. L. Danielson and L. Ernster, *Biochem. Z.*, **338**, 188 (1963).
179. B. Chance and G. Hollunger, *J. Biol. Chem.*, **236**, 1562 (1961).
180. L. Ernster, in *Biological Structure and Function* (IUB/IUBS Symposium in Stockholm), Vol. II, Academic Press, New York, 1961, p. 139.
181. E. C. Slater and J. M. Tager, *Biochim. Biophys. Acta*, **77**, 276 (1963).
181a. L. Ernster, G. F. Azzone, L. Danielson, and E. C. Weinbach, *J. Biol. Chem.*, **238**, 1834 (1963).
182. B. Chance, *Biochem. Biophys. Res. Commun.*, **3**, 10 (1960).
183. B. Chance and T. T. Ito, *Nature*, **195**, 150 (1962).
184. D. R. Sanadi, T. E. Andreoli, R. L. Pharo, and S. R. Vyas, in *Energy-Linked Functions of Mitochondria*, B. Chance, Ed., Academic Press, New York, 1963, p. 26.
185. L. Ernster, *Proc. Intern. Congr. Biochem.*, *5th*, *Moscow*, *1962* (publ. 1963), p. 115.
186. M. Klingenberg, *Biochem. Z.*, **335**, 263 (1961).
187. B. Chance and U. Fugmann, *Biochem. Biophys. Res. Commun.*, **4**, 317 (1961).
188. H. S. Penefsky, *Biochim. Biophys. Acta*, **58**, 619 (1962).
189. H. S. Penefsky, in *Energy-Linked Functions of Mitochondria*, B. Chance, Ed., Academic Press, New York, 1963, p. 87.
190. H. Löw and I. Vallin, *Biochem. Biophys. Res. Commun.*, **9**, 307 (1962).
191. L. Packer, *Biochim. Biophys. Acta*, **74**, 127 (1963).
192. L. Packer, R. H. Marchant, and E. Corriden, *Biochim. Biophys. Acta*, **78**, 214 (1963).
193. J. M. Tager, G. L. Howland, and E. C. Slater, *Biochim. Biophys. Acta*, **58**, 616 (1962).
194. J. M. Tager, J. L. Howland, E. C. Slater, and A. M. Snoswell, *Biochim. Biophys. Acta*, **77**, 266 (1963).
195. M. Klingenberg, in *Rapid Mixing and Sampling Techniques in Biochemistry*, B. Chance, R. Eisenhardt, Q. H. Gibson, and K. K. Lonberg-Holm, Eds., Academic Press, New York, 1964, p. 61.

The Respiratory Chain System of Bacteria

LUCILE SMITH,* *Department of Biochemistry, Dartmouth Medical School, Hanover, New Hampshire*

	Introduction	56
I.	The Structure of the Bacterial Cell	57
II.	Studies on the Preparation and Fractionation of Broken-Cell Extracts of Bacteria and the Information They Give about the Structure of the Respiratory Chain System	60
	A. Preparation of Broken-Cell Extracts	60
	B. Localization of Respiratory Chain Enzymes in Cell-Free Extracts; The Nature of the Linkages of the Pigments to the Membrane	62
	C. Evidence of Changes in Structure on Rupture of the Membranes	67
	D. Composition and Properties of the Bacterial Membrane	68
III.	Properties of the Members of the Respiratory Chain	72
	A. Cytochromes	72
	1. Qualitative Studies of Cytochrome Complement	72
	2. Quantitative Measurements; Variations in Pigment Content with Changes in Growth Conditions	75
	a. Composition of the Growth Medium	75
	b. Oxygen Tension	75
	c. Age of Culture	76
	3. Oxidases	80
	4. b- and c-Type Cytochromes	84
	B. Dehydrogenases and Enzymes Reacting with Inorganic Substrates	86
	C. Quinones	89
	D. Reductases Reacting with Terminal Electron Acceptors other than O_2	92
	1. Nitrate	92
	2. Nitrite and Other Nitrogen Compounds	94
	3. Sulfate and Other Sulfur Compounds	95
	4. Organic Acids	95
	E. Non-Heme Iron	95
IV.	Sequence of Electron Transport and Mechanism of Action	96
V.	Transport of Substances into the Bacterial Cell	104
VI.	Respiratory Chain-Linked Phosphorylation	106
VII.	Photosynthetic Bacteria	112
VIII.	Summary	113
	References	113

* Career Research Development Awardee, United States Public Health Service.

Introduction*

Although there are different structural arrangements within the cells, the bacterial respiratory chain system has components and some properties similar to the system of mammalian mitochondria, described by Klingenberg and by Boyer (see p. 1 and p. 193, this volume). However, the differences which are apparent when the bacterial and mammalian systems are compared give another point of view when considering the nature of all such multienzyme complexes. Thus it is to these differences and the insight they give and the questions they raise regarding structural arrangements and mechanisms of interaction that this chapter will be largely directed.

Early work on the bacterial respiratory chain system has been covered rather thoroughly in a number of chapters and review articles (1–7). Emphasis will be given here to more recent work, not previously reviewed. The purpose of the chapter is not a cataloging of data on the bacterial system; but there will be an attempt to correlate some of the large amount of recently published work to the extent that this will be helpful in delineating the nature of these systems.

As a road map to follow, Figure 1 gives a summary of electron-transport

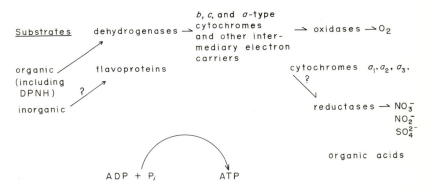

Fig. 1. Overall scheme of the respiratory chain system of different bacterial species. As in mammalian cells, the system functions to supply a utilizable energy source (usually ATP). The question marks indicate lack of knowledge of the site of interaction.

* Abbreviations used throughout: DPN, diphosphopyridine nucleotide; DPNH, reduced diphosphopyridine nucleotide; RNA, ribonucleic acid; DNA, deoxyribonucleic acid; ATP, adenosine triphosphate; ADP, adenosine diphosphate. The nomenclature of bacteria is that of the authors quoted.

and phosphorylating systems which should apply to a broad range of bacterial species. The enzymes are bound to membranous structures within the cells in such a way that electron- and energy-transfer reactions proceed rapidly.

Bacteria can respire rapidly, even with endogenous substrates. The latter can be decreased by "starvation" (aeration in the absence of substrate). Then on addition of a utilizable substrate rapid respiration follows immediately, except in a few instances where a permease must first be formed. During this rapid respiration, the turnover rates of the members of the respiratory chain system can be very high (4).

Some species of bacteria do not synthesize a typical respiratory chain system like that outlined in Figure 1. In these bacteria the energy-yielding reactions are largely fermentative, although some can respire via flavoprotein enzymes (7a). The latter will not be discussed in this chapter.

I. The Structure of the Bacterial Cell

Several books and reviews describe studies of the structure of the bacterial cell (8–13). For the purpose of this chapter, a brief summary should suffice.

Electron micrographs of thin sections of eubacteria fixed in a number of ways show no evidence of structures equivalent to mammalian mitochondria (6,8–10,14–18). Most of these micrographs reveal a cell wall with underlying cytoplasmic membrane, a concentration of "nuclear material" devoid of a membrane and small spherical structures, the ribosomes, populating the remainder of the cytoplasm. Figure 2 pictures a section of *Hemophilus parainfluenzae*, a bacterium which has been thoroughly studied with respect to its respiratory chain system.

Additional structures composed of layers of membranous material or clusters of vesicles are seen in micrographs of sections of a number of bacterial species. These may be membrane infoldings, often referred to as mesosomes, particularly on the inner rim of the developing transverse septa (19–26), suggestive of structures with a function in cell wall synthesis (19). However, other functions in cell replication have been suggested (13). Empty vesicles can be seen in the interior of empty envelopes of *Azotobacter agilis* which have been subjected to osmotic or temperature shock, so that the cell envelope is opened but not comminuted; these are occasionally attached to the peripheral membrane and may be largely removed from the ruptured cells by shaking with glass beads (27–29). Marr and co-workers point out that the internal structures could be either

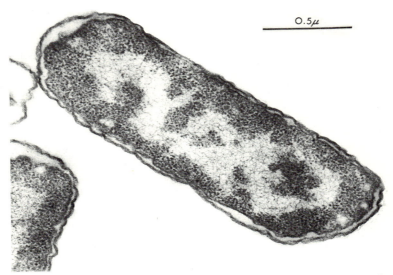

Fig. 2. Electron micrograph of thin section of *Hemophilus parainfluenzae*, obtained through the courtesy of Dr. S. F. Conti. The cells were fixed by a modification of the method of Ryter and Kellenberger (354), embedded in Epon 812, and the sections stained with uranyl acetate and lead.

preexisting extensions of the cell membrane or distortions of the membranes induced by treatments rupturing the envelopes. In other instances where internal membranes are visible within the cytoplasm, the possibility was not eliminated that these could be portions of the cytoplasmic membrane pulled away from the wall (14).

The membranes of bacteria visualized by electron microscopy resemble the membranes seen in other cells (14,18,22), the double-layered so-called "unit membrane" which is postulated to be a three-layered structure of phospholipid sandwiched between two protein layers (30). In addition, negative staining of an anaerobic bacterium (31) and of membrane fragments from *Bacillus stearothermophilus* and several mesophilic organisms (22) showed spherical units about 65–85 Å in diameter, apparently attached to the membranes by stalks and thus very reminiscent of the structures seen in negatively stained preparations of mammalian or yeast mitochondria (32–34) (see Klingenberg's chapter). In the preparation from *B. stearothermophilus* the spherical units were particularly visualized along the folded or broken edges of the membrane fragments (22). Since the spherical units are visualized even in anaerobic bacteria, there is no indication that they represent assemblies of respiratory chain enzymes, as postulated for the mitochondrial structures (see Klingenberg).

There is also evidence for the existence of a double envelope in bacteria from the production of plasmolysis under some conditions (6). The ability to remove the outer layer from a few Gram-positive bacteria by digestion with lysozyme (*Bacillus megaterium, Micrococcus lysodeikticus,* and *Sarcina lutea*) has shown clearly that the outer layer, the cell wall, serves only to give mechanical strength and defined form to these bacteria, enabling them to withstand the high pressures exerted by the intracellular solutes (6,12). The inner cytoplasmic membrane possesses the essential characteristic of semipermeability (6,35,36). The cell walls of these Gram-positive bacteria are composed predominantly of a mucopolymer which is degraded by lysozyme. If these bacteria are treated with lysozyme while suspended in a medium containing a nonpenetrating solute to balance the internal osmotic pressure of the bacteria, protoplasts (bacteria devoid of cell wall) remain (6,10,37). Within a certain range, the osmotic pressure of the medium has little influence on the volume of the resulting protoplast. However, if the concentration of the external solute is changed after the protoplasts are prepared, the protoplasts behave reversibly as osmometers (38) and finally lyse when the external osmotic pressure becomes too low. Substrates actively transported into the cell will also induce swelling and final lysis of the protoplasts (39). The presence of Mg^{++} ions appears to stabilize the protoplasts of *B. megaterium* (40), so that if they are exposed to low osmotic pressure in the presence of Mg^{++} ions ($0.005-0.01 M$), the cytoplasmic contents leak out and the "ghosts" remaining are spherical and of about the same diameter as the protoplasts. Under these conditions each protoplast appears to yield about one "ghost" (41), giving good evidence that the "ghosts" represent the plasma membranes. Osmotic rupture of protoplasts in the absence of Mg^{++} ions produces a heterogeneous mixture of membrane fragments.

Protoplasts suspended in a medium of appropriate osmotic pressure have permeability properties similar to those of intact cells and also similar synthetic and biochemical capacities (11,36,40), including the same respiratory capacity (37,42–44).

The cell walls of other Gram-positive and all Gram-negative bacteria contain the mucopolymer plus additional components; some have rather complex multilayered envelopes (11,45). The walls of some Gram-negative bacteria are predominantly lipoprotein. Since there is no enzyme which will selectively remove a layer of the envelope of the Gram-negative bacteria, the structural arrangement is not so clearly defined. In some Gram-negative bacteria an intimate association of the wall and membrane has been suggested (46). The mucopolymer of the envelope of *Pseudomonads* is in a layer where it is unavailable for digestion by lysozyme, but lysozyme will digest the mucopolymer of isolated cell wall-membrane preparations (46). Lysozyme can digest the mucopolymer of some bacteria which have cell walls containing other components, and removal of the mucopolymer leaves weakened cell walls so that the resulting "spheroplasts" are osmotically active (11,47). Deficient cell walls can also result from a number of other causes (11,36,48).

II. Studies on the Preparation and Fractionation of Broken-Cell Extracts of Bacteria and the Information They Give about the Structure of the Respiratory Chain System

A. Preparation of Broken-Cell Extracts

When protoplasts are lysed osmotically in the presence of Mg^{++}, the relatively unbroken membranes can be readily separated from the cytoplasmic constituents by centrifugation. Under these conditions a large number of polyribosomes seem to adhere to the membrane fraction (49). Isolated cytoplasmic membranes appear as approximately circular membranes in electron micrographs (50), and in some species there are dense granules associated with the membranes. However, there is evidence that these granules are free of enzymic activity and are not protein in nature (39,41,51). Electron micrographs of lysed protoplasts of *Micrococcus lysodeikticus* show structures giving the appearance of mesosomes in addition to the membrane fragments (23). Osmotic lysis of protoplasts in the absence of Mg^{++} ions yields a mixture of membrane fragments of varying sizes and, at least with *B. megaterium*, slightly more protein is released into the soluble fraction in the absence than in the presence of Mg^{++} ions (41). The membrane fraction separated from two strains of *B. megaterium* lysed in the absence of Mg^{++} ions (and treated with DNAase to decrease the viscosity before centrifugation) contained 13 and 15% of the protein of the total lysates (50). It is still not certain to what extent other substances or structures may adhere to the membrane fraction separated from lysates of bacterial protoplasts. Spheroplasts prepared in various ways can also be lysed osmotically; the membrane fractions obtained from these would contain some components of the cell wall in addition to the cytoplasmic membrane.

Protoplasts or spheroplasts or membrane fractions derived from them can be disrupted by exposure to sonic oscillation or by treatment with detergents (27,28,41,52); the extent of comminution of the membrane obtained varies according to the treatment. Thus particle fractions of varying ranges of sizes (and varying ease of sedimentation in the centrifuge) can be obtained.

Numerous mechanical methods for disrupting intact bacteria have been devised. Some of these, such as treatment with the Hughes press (46) or the French pressure cell (53), expel the cytoplasmic constituents leaving relatively intact cell wall-membranes which can be collected by centrifugation at low centrifugal forces. Treatments such as exposure to sonic oscillation or grinding with abrasives (alumina or powdered glass) result not only in opening of the cell, but also in comminution of the larger elements. The rates of release of structures and enzymic activities from intact cells of *Azotobacter vinelandii* and *Sarcina lutea* on continuous exposure to sonic oscillation have been measured (27,54,55). The small RNA-protein particles and some enzymes were released at the same rate as the rupture of the cells; thus these were present in the cytoplasm. The rates of release of some respiratory chain enzymes and phospholipid were slower and similar to the rate of decrease of size of the insoluble particles, and thus these were presumably derived from the breakdown of larger structures (54). The cytochromes and rhamnose (a cell wall constituent) were released at rates identical with the loss of weight from the sedimented particulate fraction, indicating that the cytochromes were contained in a structure associated with the envelope. This and considerable other evidence leaves little doubt that the respiratory chain enzymes are in the membrane of bacteria. There is cytochemical evidence for this (46,56–58), and "ghosts" derived from lysis of protoplasts or empty envelopes obtained in other ways have been shown to bear the cytochromes and some other respiratory chain enzymes (27,41). The particulate fraction from broken-cell extracts that bears this system is derived from the disrupted cytoplasmic membrane of Gram-positive bacteria and from some portion of the cell envelope of Gram-negative bacteria. After prolonged exposure to sonic oscillation and with a few other treatments of the cells, particles of very small size have been prepared which show respiratory activity (29,51,59–63); they are not sedimented by centrifugation at $140,000 \times g$ for one or more hours.

The internal vesicular structures seen in osmotically shocked *Azotobacter vinelandii*, described in the preceding section, can be removed from the readily sedimentable empty envelopes ("hulls") by shaking with glass beads, and this results in some loss of respiratory chain activity. From this it was concluded that the respiratory chain system in these bacteria is on both the external and internal membranes (29). This observation has only been made with *Azotobacter*.

The small insoluble fragments collected after rupturing intact cells or "hulls" or "ghosts" can have compositions and enzymic activities similar to those of the relatively intact membranes (27,29,37,46,56), but they require higher centrifugal forces for sedimentation. Extracts prepared from different species of bacteria broken in a number of different ways showed similar patterns in the ultracentrifuge, indicating a common distribution of macromolecular components (65). Also, particulate fractions derived from the same bacteria ruptured by different methods had similar composition and properties, although sometimes quantitative differences were seen (66–72), and particles of different sizes from the same broken-cell extract may have identical enzymic properties (1,73).

Electron micrographs of fixed sections of the particulate fractions from a number of kinds of bacteria show membranous structures irregular in size and shape, but sometimes approximately circular and with the same staining properties as the cell membrane (14,27,29,41,50,65,73–75). The closed vesicles present suggest that after rupture the membrane fragments may have been "healed" in a manner similar to that suggested for other membranes (76). The membrane pieces are less dense to the electron beam than the ribosomes, from which they can be separated by starch electrophoresis (29) or ultracentrifugation (77).

Figure 3 is an electron micrograph of a section of membrane fragments isolated from a cell-free extract of *Hemophilus parainfluenzae* prepared by grinding with alumina. There is no evidence for internal membranes in these cells, although obviously not all of the cytoplasmic material is removed from the fragments. The curls suggest that the membrane may have been under tension at the time of rupture.

B. *Localization of Respiratory Chain Enzymes in Cell-Free Extracts; The Nature of the Linkages of the Pigments to the Membrane*

The methods useful for studying the intact electron-transport chain or parts of it have been described in some detail (1,4,78–80). These include measurements of oxygen uptake and substrate oxidation by the entire system as well as the reactions of parts of the chain with appropriate acceptors or donors of electrons. The presence, amount, and oxidation-reduction reactions of pigments such as the cytochromes, flavoproteins, and pyridine nucleotides can be examined spectrophotometrically; the absorption spectra change on oxidation and reduction, and the properties of both forms of the pigments are known in many instances. Changes in absorption spectrum on

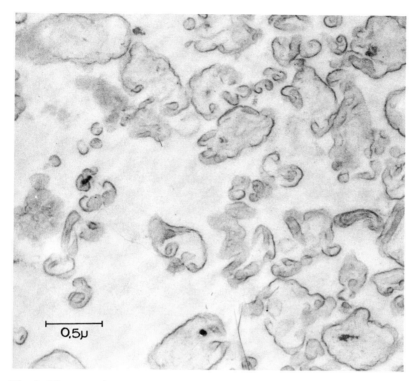

Fig. 3. Electron micrograph of thin section of particulate fraction of broken-cell suspension of *H. parainfluenzae*. The cells were ruptured by grinding with alumina and the preparation fixed and stained as described for Figure 2.

addition of known inhibitors or of other substances known to react with the various pigments may also be observed. Studies of turbid suspensions of bacteria or particles are often made most profitably by measuring difference spectra (for example, the difference in absorption spectrum between the reduced and oxidized forms) to minimize problems due to light scattering.

If the respiratory chain pigments are associated with the membrane, it would be expected that they would be found in the insoluble particulate fraction sedimented from the broken-cell extracts. However, when the membrane is extensively comminuted, small fragments produced may not be sedimented at the centrifugal forces employed in the separation and these will remain in the so-called "soluble fraction." Although the cytochromes and a few dehydrogenases are reported to be almost entirely localized in the membrane fraction (Table I) in many extracts, the results

TABLE I

Oxidative Activities of Particle Fractions Separated from Broken-Cell
Extracts of Different Bacterial Species[a]

Bacteria	Method of cell rupture	Activities found in membrane fraction	Ref.
Azotobacter vinelandii	Sonic oscillation, 5 min	DPNH, TPNH, succinate malate dehydrogenases, hydrogenase	29
	Sonic oscillation, 10 min, then treatment with 10% ethanol	DPNH, succinate, lactate, and malate oxidases	111
	Sonic oscillation, 25–30 min	DPNH oxidase; no succinate oxidase	186
	Sonic oscillation	DPNH, succinate, and malate oxidases	110
	Ground with powdered glass	DPNH, succinate, malate, and lactate oxidases	314
	Ground with alumina	DPNH, TPNH, succinate oxidases, oxidation of H_2 by O_2	310
Bacillus stearothermophilus	Treatment with lysozyme, then 50% acetone or ETOH–ether	DPNH, malate, and succinate oxidases	66
Proteus vulgaris	Shaken with glass beads	Succinate, lactate, and formate oxidases	63
	Sonic oscillation or Waring Blendor plus beads	DPNH and succinate oxidases	69
Bacillus cereus	Colloid mill, 10 min	DPNH and succinate oxidases	68
Aerobacter aerogenes	Shaken with glass beads	Succinate and lactate oxidases	63
Pseudomonas fluorescens	Sonic oscillator	DPNH, glucose, and gluconate oxidases; no TPNH oxidase	230
	Hughes press, sonic oscillation, or shaking with glass beads	Succinate and malate oxidases; nicotinic acid hydroxylase	46
Nitrobacter agilis	Sonic oscillation	Nitrite oxidase	110
Mycoplasma hominis	Sonic oscillation, 10 min	DPNH and succinate oxidases, CoA and short-chained acyl CoA derivatives oxidized	247

(continued)

TABLE I (*continued*)

Bacteria	Method of cell rupture	Activities found in membrane fraction	Ref.
Escherichia coli	Sonic oscillation, 40 min	DPNH, succinate, formate, α-glycero phosphate oxidases, hydrogenase	67
	Booth-Green mill	Succinate, lactate, formate oxidases	214
Mycobacterium phlei	Sonic oscillation	DPNH and succinate oxidases	106
Mycobacterium tuberculosis	Colloid mill	Only DPNH oxidase	75
Acetobacter peroxydans	Sonic oscillation or Hughes press or ground with alumina	Oxidases of D- and L-lactate, ethanol and primary alcohols, DPNH, acetaldehyde, malate	70
Acetobacter suboxydans	Sonic oscillation, 15 min	Oxidases of glucose, gluconate, galactose, arabinose, xylose, mannitol, inositol, and mesoerythrytol	44
	Sonic oscillation	Oxidases of glucose, manitol, sorbitol, erythrytol, glycerol, ethanol and propanol	89
Gluconobacter liquefaciens	Sonic oscillation or lysis or protoplasts	Oxidases of glucose, gluconate, 2-ketogluconate, lactate, ethanol, glycerol, arabinose, galactose, xylose	44
Serratia marcescens	Shaken with glass beads	Succinate, α-ketoglutarate and malate oxidases, lactate, formate, isocitrate dehydrogenases	88
Hemophilus parainfluenzae	Ground with alumina	DPNH, succinate, formate, D- and L-lactate oxidases	136

[a] Activities not listed were not measured.

of such localization studies are not always so clear-cut. A number of dehydrogenases are reported to be located in both the particulate and soluble fractions, even when the membrane is not subjected to extensive disruption (50,51,64). However, different experiments are not in agreement on the distribution of some enzymes, and good recovery of these enzymes is not always observed (51). Even more confusing, assays for the presence of dehydrogenases by measurements of dye reduction do not agree with measurements of oxygen uptake with the same substrates (51). Another problem apparent in these distribution studies is the extensive loss of some of the respiratory chain enzymes on preparation of the broken-cell extracts (60,81–85,87). Spheroplasts of *Bacillus subtilis* or protoplasts of *Bacillus megaterium* respire rapidly with endogenous substrate in a medium of proper osmotic pressure. After osmotic lysis, respiration of similar concentrations is too slow to measure (43,81). Added DPNH is rapidly oxidized, and with *B. megaterium* a moderate rate of respiration is observed on addition of malate or of glucose plus DPN. At a dilution where the respiration with DPNH, malate, or glucose plus DPN can be conveniently measured, respiration with all other substrates tested was too small to be measured, with or without added DPN. Although Storck and Wachsman reported respiration with other substrates in very concentrated lysates of *B. megaterium* (51), the respiration rates are very low compared with the respiration of the protoplasts with endogenous substrate (81).

Sometimes an increase of enzymic activity is observed on rupturing the bacterial membranes (43,50,81,86,87). This is usually explained as due to an increase in the "accessibility" of substrate to the proper site on an enzyme. The endogenous respiration of protoplasts of *B. megaterium* or spheroplasts of *B. subtilis* is not increased by addition of DPNH, but lysates or isolated membrane fragments have remarkably high DPNH oxidase activity (43,50,81). This seems to mean that the sites where DPNH reacts with the respiratory chain system are located on the inside of the membrane.

One thing that is clear from studies of broken-cell extracts is that the different respiratory chain pigments differ in firmness of attachment to the membrane. Although the succinic, lactic, and formic dehydrogenases of *Proteus vulgaris* and *Serratia marcescens* remained particle bound even after extensive shaking with glass beads (63,88), other dehydrogenases were increasingly removed from the particulate fraction of *Serratia marcescens* with increasing time of treatment (88). Increasing time of exposure to sonic oscillation also results in increasing loss of some dehydrogenases from the particulate fraction of extracts of several kinds of bacteria (48, 89–93). However, tests for recovery of the dehydrogenases in the supernatant fractions have not always been made. Two dehydrogenases which are very readily lost from the membrane are the D- and L-lactic dehydrogenases of *Hemophilus parainfluenzae*, which come off the membrane fragments just on standing in buffer in the cold for several hours (93), while other dehydrogenases remain firmly bound. The "diaphorase"

(DPNH-dye reductase) of *Mycobacterium tuberculosis* can be removed from the particles by freezing and thawing or by extraction of the lyophilized particles with alkaline buffer (92).

Most of the cytochromes remain particle-bound even after extensive disruption of the membranes. Cytochrome c is not removed from the small membrane fragments of a number of bacteria by treatment with salt or with trichloroacetic acid (29,94); in this respect the bacterial fragments resemble the small particles derived from heart muscle mitochondria. We have obtained preliminary evidence that cytochrome c can be removed from the unbroken cytoplasmic membrane of spheroplasts of *Micrococcus denitrificans* by treatment with salt (95). This indicates that the cytochrome c is at least partially bound by electrostatic linkages, as in mammalian mitochondria, where the cytochrome can be removed from swollen mitochondria, but not from membrane fragments, by suspension in salt (96). As with mitochondrial fragments, cytochrome c, but not the b- and a-type cytochromes, could be removed from the small particles of bacilli by treatment with lipase (97). There are a few reports of c-type cytochromes in the soluble fraction of a cell-free extract, or of instances where they are rather easily removed from the particles (86,98–100). Treatment of *Thiobacilli* with butanol removed only a part of the c-type cytochromes, leaving 30–40% membrane-bound (85); it is not known whether this means that different types of linkage were involved or that some of the particles had different structures. Cytochrome b_1 can be removed from *E. coli* particles by digestion with trypsin or by exposure to sonic oscillation (101,102). Quinones are firmly bound to the membrane (see Section III-C).

Section II-D describes the chemical composition of the membranes and Section IV some experiments on dissection of the particles; these give additional ideas of the kinds of forces holding the particles together. The variety of kinds of attachment possible in the different bacterial systems should make these systems excellent experimental material for further studies of the nature of the linkages of the pigments to the lipoprotein membrane.

C. Evidence of Changes in Structure on Rupture of the Membranes

Numerous pieces of evidence point to a change of the structure when intact membranes are fragmented. This is true for the mitochondrial membrane (103,104) as well as for bacterial membranes. Evidence of change in the membrane structure is seen when spheroplasts of *B. subtilis* are lysed osmotically (43). Immediately on lysis there is a large increase of DPNH oxidase activity, which decreases to a lower value after several minutes, then remains quite stable at the lower activity. Also the cytochromes may not be completely reduced enzymically in preparations of membrane fragments (52,105,106); thus the accessibility of the pigments for interaction must have changed. A puzzling observation is that sometimes recombination of fractions separated by centrifugation does not yield the same enzymic activities found in the original extract; some activities may be increased, others decreased. This kind of observation could be explained by assuming changes in the structure of the membrane-bound system during fractionation by centrifugation or by loss of enzymes from the membrane, or by both of these changes. The activity of an enzyme may change upon removal from a membrane (107,108), and some soluble pigments do not react rapidly with membrane-bound enzymes; the latter will be discussed later (Section IV). The indications of changes in structure when the intact membrane is fragmented urge caution in interpreting data on such preparations. They also make studies of distribution of enzymes in bacterial extracts difficult.

D. Composition and Properties of the Bacterial Membrane

The cytoplasmic membrane was reported to make up about 10% of the dry weight of *Micrococcus lysodeikticus* (109), 15–20% and 25% of *Bacillus megaterium* (41,49), 10–15% of *Staphylococcus aureus* (64). "Hulls" from *Azotobacter* (whole envelope) make up about 20% of the dry weight of the cells (29,110). Some of these values seem high and may mean that other substances or structures remain attached to the membrane during lysis.

Like the membranes of other cells, the most constant feature of the bacterial membranes is the high content of lipid. Table II shows the distribution of lipid and protein in the cytoplasmic membrane of several species of bacteria and in the cell wall-membrane of *Pseudomonas fluorescens* and the particulate fraction from the envelopes of *Azotobacter vinelandii*. The cell wall-membrane fraction of *P. fluorescens* is similar in composition to the protoplast membrane of the Gram-positive organisms (46) and there is obvious similarity of the bacterial membrane fractions to those derived from mammalian mitochondria (so-called ETP). Table III, taken from the data of Breummer et al. (111), illustrates this similarity.

The lipids of the bacterial membranes include a high content of phospholipids, but there are some notable differences in the lipid composition as compared with that of mammalian mitochondrial membrane. Although there are wide variations in the compositions of membranes from different bacterial species, some generalizations can be made: lecithin is rarely

TABLE II

Distribution of Lipid, Protein, and Carbohydrate in Membrane Fractions

Bacteria	Lipid, %	Protein, %	Carbohydrate, %	Ref.
Staphylococcus aureus	22.5	41		64
Bacillus megaterium	16–21	60–70		41
	15–20	65	1–10 (probably glycogen)	40
Micrococcus lysodeikticus	28	50	15–20 (largely polymer of mannose)	109
Azotobacter vinelandii	28.6			29
Pseudomonas fluorescens	16–18	50		46
Sarcina lutea	29	40	10 (hexosamine; possibly from cell wall)	36

encountered (41,112,113) and cholesterol is missing from nearly all species (41,112). There are significant quantities of phosphatidylethanolamine in some Gram-negative organisms (114), and the membranes of some Gram-positive bacteria contain highly polar phospholipids rich in phosphatidyl-glycerol-type lipids such as a complex polyphosphatidic acid (41,109,112, 115). Lipoamino acids, which are O-amino acid esters of phosphatidyl glycerol, are found in the lipid fraction of some bacterial membranes (112,115). There is carotenoid in the membrane of *Sarcina lutea* (55,116). The fatty acid components contain a high proportion of unsaturated fatty acids and of cyclopropane fatty acids (112,113). Kates has pointed out (112) that the presence of branched chain, cyclic, and unsaturated fatty acids would prevent close-packing and presumably give elasticity to the membrane.

In addition to lipid and protein, variable amounts of carbohydrate are associated with the membranes of some species. This may amount to up to 20% of the dry weight of the membrane fraction isolated from lysed protoplasts of *B. megaterium* and *M. lysodeikticus*. That from *M. lysodeikticus* is a polymer of mannose (109,117). The membrane fraction of *B. megaterium* can be separated into two parts, one containing mucoprotein and the other lipoprotein (118).

The disagreement in the various data concerning the RNA content of membrane fractions now appears to be explained by the association of polyribosomes with the membrane in the presence of the proper concentration of Mg^{++} ions (49). Under some preparative conditions (low Mg^{++} or treatment with detergent) the membrane fragments are free of RNA (29,109,119) or low in RNA (41,50,66). All data agree that there is no DNA in the membrane fraction (29,66,88,110).

TABLE III

Comparison of the Purified Particle from *Azotobacter vinelandii* and ETP (111)

| | Rate of oxidation (μmoles/min/mg protein) | | | | | | Hemes reduced by succinate or DPNH | Flavin ratio[a] |
| | DPNH | | | Succinate | | | | |
	O_2	Cytochrome c	Ferricyanide	O_2	Cytochrome c	Ferricyanide		
Purified particle from A. vinelandii	6.8	0.2[b]	3.5	2.4	0.3[b]	1.6	a_2, b_1, and c (c_4 and c_5)	1:5:16:2
ETP	5.7	0.2	5.0	2.0	0.1	2.2	a, b, c, and c_1	1:4:20:4[c]

[a] Atomic or molecular ratio of flavin to heme to non-heme iron to copper.
[b] Rate of oxidation with *A. vinelandii* cytochrome *c*.
[c] Based on revised value for flavin.

Since the membranes are largely composed of lipoprotein, it is not surprising that protoplasts of *B. megaterium* are lysed by treatment with lipase (120) or with detergents (119). Treatment with detergent also removes the polyribosomes from the membrane (49). In addition, enzymic activities associated with the membrane fractions are profoundly affected by such treatments, which have been shown to remove some of the enzymes from the membranes (75,121,122). This observation has led to the suggestion that the proteins of the membrane are associated with the lipid components and with each other by residual valences (6). The data discussed in Section II-B on the ease of removal of some enzymes from the membranes would be in agreement with this postulate for the binding of some of the pigments, but others may be bound with stronger bonds.

Protoplasts and spheroplasts behave as osmometers, so the membrane has the capacity to expand and contract. Since the electron-transport system is incorporated into the membrane, it might be expected that the ability of the pigments to interact would depend upon the osmotic pressure of the medium. However, suspensions of spheroplasts of *B. subtilis* in hypotonic sucrose, where they have been observed to swell, do not show decreased capacity for electron transport, as assessed by measurements of DPNH oxidase activity. Instead a large increase in activity is observed when the spheroplasts swell to the point of bursting (43). If the sucrose concentration is increased above $0.5M$ (the concentration in which the spheroplasts are prepared), some decrease in respiration rate is observed (30–70% inhibition in $0.8M$ sucrose). The explanation for this inhibition is not known; the respiration of cell-free extracts is relatively independent of the sucrose concentration. *Intact cells* of bacteria have been reported to show changes of turbidity or light scattering on increasing the salt concentration of the suspending medium (123) or on addition of utilizable substrates under some conditions (124). Salt concentrations which cause increases in turbidity (and give evidence of plasmolysis) in intact cells of *E. coli* are also observed to inhibit the respiration with glucose as substrate, whereas the respiration of broken-cell extracts is not affected (125). An explanation offered for the inhibition of respiration is that changes in the membrane by plasmolysis might modify the spatial arrangement of the membrane-bound respiratory chain system (125). However, effects of dehydration or possibly of pH change might also be involved.

Evidence has been cited to indicate that membrane fragments of sonic extracts of *M. phlei* behave as reversible osmotic systems in that they were observed to swell in hypotonic solution (91). However, changes in the osmotic pressure of the medium did not appear to affect the activities of the preparations (126). Although electron micrographs of membrane fragments of other bacteria show some vesicular structures, there are no other reports of swelling and shrinking accompanying changes in the osmotic pressure of the medium, and the respiratory activity of the fragments seems remarkably insensitive to such changes (81,127).

Membrane fractions separated from a cell-free extract at different centrifugal forces have sometimes been found to have similar compositions and activities and appear to differ only in size (1,75,86). However, occasionally there is evidence for differences in the composition of the fractions (7). The separation of two kinds of particles from *E. coli* extracts with different composition and activities has been reported (128). Different particle fractions from extracts of *Azotobacter vinelandii* show quite different specific activities of the DPNH and succinate oxidase systems in terms of protein content, and the more active particles have relatively intensified absorption peaks of reduced cytochromes compared with the intact cells (74), with a particularly high content of cytochrome *b* (62,110). A fraction from *Azotobacter* showed a relative intensification of the absorption band of cytochrome a_2 and a decreased content of *c*-type cytochromes (129). Weibull separated relatively light and heavy particulate fractions from a lysate of protoplasts of *B. megaterium* with different relative activities (41), and Godson et al. separated the membrane fraction of the same bacteria into a mucoprotein and a lipoprotein fraction (118). These observations give a suggestion of heterogeneity of the cytoplasmic membrane. This is one area of research where more work should surely be profitable.

III. Properties of the Members of the Respiratory Chain

Considerable knowledge has been gleaned about the individual members of the bacterial respiratory chain systems from studies with intact cells, with particulate preparations bearing dehydrogenases linked to the cytochrome systems and with degraded particulate fractions and purified enzymes.

A. Cytochromes

1. Qualitative Studies of Cytochrome Complement

The cytochromes and their heme derivatives have characteristic absorption spectra, which can be used in studying these pigments (1,4,130) (see Klingenberg, p. 1). Measurements of the difference in absorption spectrum between the reduced and oxidized forms in intact cells or particles give both qualitative and quantitative assays of these pigments. Also, measurements at the temperature of liquid nitrogen allow the visualization of overlapping bands because of the sharpening and intensification of the bands (131).

A few bacterial species have cytochromes with the same absorption spectra as those of mammalian mitochondria and yeast, but in most species the absorption peaks of the cytochromes are at different wavelengths (1,2,4). Figure 4 illustrates this difference with a comparison of the reduced minus oxidized difference spectra of mammalian mitochondria and intact cells of *B. subtilis* and *Azotobacter chroococcum*. A wide variety of combinations of *a*, *b*, *c*, and *o* cytochromes exists among the different bacterial species. (The different cytochrome types have different prosthetic groups and different types of linkage between heme and protein. This is summarized in reference 4 and discussed in the chapter by Gibson.) Table IV compiles data on the cytochrome components of a number

TABLE IV

Cytochrome Components of a Number of Species of Bacteria

Bacteria	Cytochromes	Ref.
Bacillus subtilis	a, a_3, b, c, c_1	133, 160, 161, 174
Staphylococcus aureus	a, o, b	173
Escherichia coli	a_1, a_2, o, b_1	214, 161
Aerobacter aerogenes	a_1, a_2, o, b_1	138, 161
Proteus vulgaris	a_1, a_2, o, b_1	293, 161
Azotobacter vinelandii	a_1, a_2, o, b, c, c_1	129, 161
Hemophilus parainfluenzae	a_1, a_2, o, b, c, c_1	162
Mycobacterium phlei	a, a_3, b, c, c_1	106
Acetobacter suboxydans	o, b, c, c_1	4, 161
Bacillus megaterium	a, a_3, o, b	81
Pseudomonas aeruginosa	a_1, a_2, b, c, c_1	199, 132
Micrococcus denitrificans	a, a_3, o, b, c	261, 353

of species examined at both room and low temperatures and in the presence of carbon monoxide, which combines with the oxidases (see Section III-A-3). These measurements give as complete a qualitative picture as possible. The data show that practically any possible combination can be found and that there is no essential requirement for any one type of cytochrome. The only requirement apparent is for a group of pigments with enough difference in redox potential to furnish energy to synthesize ATP from ADP (4).

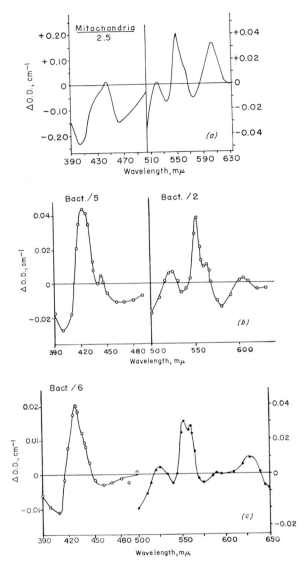

Fig. 4. Anaerobic minus aerobic difference spectra of (*a*) guinea pig liver mito-chondria (through courtesy of Dr. Britton Chance); (*b*) *Bacillus subtilis*; and (*c*) *Azotobacter chroococcum* (1,4).

2. Quantitative Measurements; Variations in Pigment Content with Changes in Growth Conditions

Quantitative estimates of the content of cytochromes in bacteria are difficult to judge, since the bacteria can greatly modify the total quantities and relative proportions of the different respiratory chain pigments with changes in culture conditions. Some of the factors shown to be influential follow.

a. Composition of the Growth Medium. Changes have been observed in the relative content of the different cytochromes with growth with different oxidizable substrates (132–135), which in one case seemed to be related to differences in the growth rates (133). However, changes in the pigments throughout the whole growth cycle were not followed in all cases. An unusual effect of carbon source is seen in *Hemophilus parainfluenzae*, where growth with glucose as substrate yields lactate respiration which is linked to the cytochrome system, but which is not cytochrome-linked when the cells are grown with gluconate or glucuronate (136). Even the pH of the medium may affect the yield of the respiratory chain pigments (137).

As would be expected, a deficiency of iron in the medium results in a loss or decrease of cytochromes, and some seem more sensitive to a lack of iron than others (138,139); there may also be a decrease in succinic dehydrogenase (138) and in quinone (140). However, the decreases in pigments do not always lead to the expected decrease in respiration with some substrates (138,139). The presence of respiratory chain inhibitors in the growth medium has also been observed to inhibit cytochrome synthesis (141,142).

b. Oxygen Tension. The growth of facultative anaerobes under anaerobic fermentative conditions depresses the synthesis of cytochromes in some bacteria (143–146), but the total amount of cytochromes may be greater in bacteria grown anaerobically using nitrate as a terminal electron acceptor in place of oxygen (see Section III-D-1) (147,149,151). The relative amounts of the different cytochromes may change with the presence or absence of oxygen as well as with changes in oxygen tension during growth (143,148, 151–153). When grown at varying measured oxygen concentrations, the maximal development of cytochrome a_2 of *Aerobacter aerogenes* was seen at $10^{-6}M$ oxygen; the content of cytochrome b_1 remained constant at concentrations between 10^{-6} and $10^{-3}M$ but decreased below $10^{-6}M$, while the concentration of cytochrome a_1 changed very little with changes in oxygen concentration (153). *Pseudomonas fluorescens* synthesizes a c-type cytochrome and a peroxidase only when the cells are grown at low concentrations of oxygen (154,155). There is an oxygen requirement for the synthesis of cytochromes by *Staphylococcus epidermidis* and also for the synthesis of a heme *a* from protoheme (156,157). The presence of oxygen during growth also affects the content of quinones in some bacteria (140,158), but not in all (140,146).

   c. Age of Culture. The cytochrome content and relative proportions of the cytochromes may change markedly as a bacterial culture ages, some cytochromes being synthesized largely during the later stages of growth (81,142,151,159–161). The content of different respiratory chain dehydrogenases also varies greatly during the growth cycle (93). Tables V–A and

TABLE V-A

Optical Density of DPNH-Reducible Cytochromes Formed under
Various Growth Conditions (151)[a]

Growth conditions	c_1	b_1	a_1	a_2	o	Fp
Log phase						
Aerobic	0.012[b]	0.012	0.001	0.001	0.003	0.008
Anaerobic	0.027	—[c]	0.010	0.001	0.003	0.008
Early stationary phase						
Aerobic	0.019	0.014[d]	0.011	0.035	0.006	0.027
Anaerobic	0.039	—[c]	0.049	0.001	0.009	0.032

   [a] The ΔOD for each cytochrome is given per 10 mg of bacterial protein/ml per cm light path.
   [b] Cytochrome c_1 maximum hidden by cytochrome b_1 maximum.
   [c] Indicates maximum hidden under an adjacent cytochrome maximum.
   [d] Seen in DPNH-reduced spectra of very old cells.

V–B summarize some representative data from quantitative studies of changes in the content of different respiratory chain pigments of *Hemophilus parainfluenzae* as a function of growth conditions (148,151), and Figures 5 and 6 illustrate the changes in cytochrome and dehydrogenase content with age of culture. As more data are accumulated, it should be possible to alter the composition of the respiratory chain system in defined ways and to test the results of these variations. This has been done to a limited extent with *Hemophilus parainfluenzae* (see Section IV).

   Under some conditions of growth *Hemophilus parainfluenzae* and *Thiobacillus denitrificans* synthesize relatively large quantities of *c*-type cytochromes; they can be 6% of the protein or 4% of the dry weight of the cells, respectively (162,163). The range of cytochrome content of *H. parainfluenzae* under various growth conditions is about 10^3–10^6 mole-

TABLE V-B

Maximum Cytochromes with Various Growth Conditions

Growth conditions	Generation time, min	Maximum turbidity	Maximum viable count, cells/ml	Maximum cytochrome formed				Terminal acceptor (concn)
				c_1 [a]	a_1	a_2	o	
Vigorous aeration	20	480	4×10^{10}	0.061	0.005	0.035	0.005	Oxygen (0.20mM)
Limited aeration	25	470	1.5×10^{10}	0.141	0.068	0.010	0.006	Oxygen, nitrate
Anaerobic	40	370	0.9×10^{10}	0.168	0.070	0.001	0.012	Nitrate (10mM)

[a] ΔOD of cytochromes measured after $Na_2S_2O_4$ reduction and expressed per 10 mg bacterial protein/ml per cm light path.

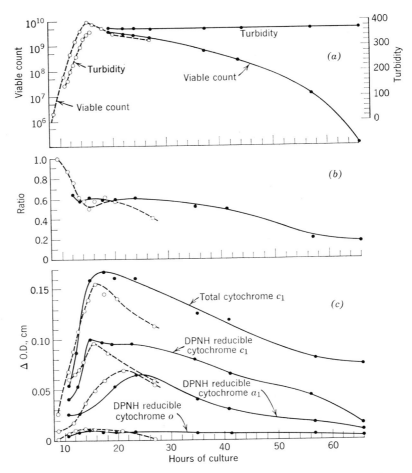

Fig. 5. (*a*) Viable count and turbidity. (*b*) Ratio of DPNH-reducible cytochrome *c* to $Na_2S_2O_4$-reducible cytochrome *c*. (*c*) ΔOD of reduced minus oxidized cytochromes (10 mg protein) cm^{-1}.

cules of cytochrome c_1, 10^3–10^4 oxidase molecules, and 10^3 molecules of cytochrome b_1 per bacterium (151). Table VI summarizes assorted data on the cytochrome content of isolated membrane fractions, where variations with growth conditions are not considered. The data show that the cytochrome content can be of the same order as that of the fragments of membrane from heart muscle mitochondria, although the relative ratios of the cytochromes may be quite different (see Section IV).

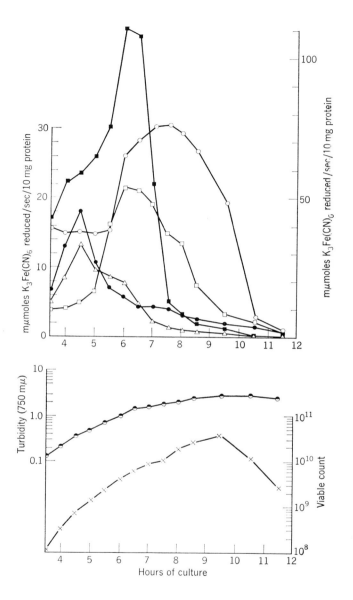

Fig. 6. Differential synthesis of the respiratory-chain-linked dehydrogenases during the growth cycle of *H. parainfluenzae*. The bacteria were grown in the presence of 20mM glucose with strong aeration. The activities were assayed by ferricyanide reduction with D-lactate (●); L-lactate (□); succinate (△); formate (○); and DPNH (■). The right-hand ordinate applies to the DPNH dehydrogenase activity. In the lower graph, (◓)represents turbidity measurements and (×) viable counts [93].

TABLE VI

Cytochrome Content of Particle Fractions of Bacteria

Preparation	Cytochrome content, mμmoles/mg protein				Total heme	Ref.
	$a + a_3$	b	$c + c_1$ (or $c + c_2$)	o		
M. phlei, particles from sonic extract	0.27	0.18	0.62			106
Dark grown Rhodospirillum rubrum, fragments from lysed protoplasts		0.3–0.46	0.1–0.15	0.2–0.28		86
Azotobacter agilis, purified "ETP"					5.0	111
Keilin-Hartree heart muscle particles	0.85	0.59	0.56			

3. Oxidases

The cytochromes which react rapidly with molecular oxygen are still little understood. None of the bacterial oxidases appears to be identical with that of mammalian tissues, where cytochrome a_3 reacts with oxygen, and a combination of cytochromes a and a_3 (cytochrome a does not react directly with oxygen) plus copper may oxidize cytochrome c with oxygen (see Gibson, p. 379). Like the mammalian system, many, but not all, bacteria and particle preparations can oxidize dyes such as p-phenylenediamine and hydroquinone (164), and this has been taken as evidence that similar oxidase systems are present. However, p-phenylenediamine does not react directly with cytochromes $a + a_3$ in the mammalian system, but does react with cytochrome c (165), and not all bacteria which can oxidize the dye contain cytochrome c.

The oxidation of p-phenylenediamine or of soluble cytochrome c by bacterial particulate preparations shows about the same sensitivity as the mammalian oxidase to inhibitors such as cyanide, azide, sulfide, or carbon monoxide (62,86,94,166–168), but the overall respiration of the bacteria or membrane fragments may be considerably less sensitive to inhibition by these substances and the sensitivity to cyanide may change on rupture of the bacteria (81). The precautions necessary in interpreting observations with inhibitors have been described previously (1). Inhibition by carbon monoxide is competitive with oxygen and is relieved by illumination with light absorbed by the carbon monoxide compound of the oxidase. Measurements of the action spectra of the relief of carbon monoxide inhibition by light have identified at least four different oxidases in bacteria (86,169–173). These differ in the relative affinities for O_2 and CO and in the relative sensitivities of the CO compounds to dissociation by light (171). *Bacillus subtilis* has cytochromes with absorption spectra similar to those of mammalian cytochromes a and a_3, but the relative contents of the two cytochromes differ (174). Cytochrome a_1 is another oxidase which may have the same heme (heme a) as mammalian cytochromes a and a_3 (175). The oxidase designated cytochrome a_2 has quite different absorption spectra from cytochromes a_1 and a_3 (170,171,176,177) and a very different prosthetic group, which is an iron complex of a hitherto undescribed chlorin (178). Another oxidase which is rather widespread among bacteria is called cytochrome o. The cytochrome o of two bacterial species appears to have protoheme as the prosthetic group (86,173), but this may not be true of cytochromes o in other species (171). At any rate, the data show that hemoproteins with very different heme compounds may function as oxidases. The fact that they react with similar substances (dyes and inhibitors) would point to the presence of similar groups with somewhat different reactivities and possibly different accessibilities.

Some bacteria synthesize more than one oxidase and may have as many as three (Table VII). When more than one oxidase is present, any one appears to be able to catalyze most of the respiration (171). And the loss of one of the oxidases from membrane fragments of *Aerobacter aerogenes* (84) or of cytochrome a from *Staphylococcus epidermidis* (156) resulted in no loss of respiratory ability. Thus the functional significance of the multiple oxidases is not understood. There is no evidence that a complex of two cytochromes (or two hemes) equivalent to cytochromes $a + a_3$ of the mammalian oxidase is present in the bacterial systems. Although the combination of cytochromes $a + a_3$ or $a + o$ is found in some species (174,181), this is not true of others. There is also no evidence of a heavy-metal component in the bacterial oxidases equivalent to the copper of the mammalian oxidase (86,179). A more extensive study of the different bacterial systems should give a better idea of the requirements for a functional oxidase.

The bacterial oxidases, like the mammalian one, show a high affinity for oxygen (86,180,181), and in the aerobic steady state the cytochromes remain oxidized down to low concentrations of oxygen (79). The lowest concentration of oxygen which supports maximal respiration of intact cells of *Hemophilus parainfluenzae* was observed to increase with decreasing electron flow to the oxidase. This would be expected, since it is the

TABLE VII

Oxidases of Bacteria

[Functional oxidases revealed by measurements of carbon monoxide action spectra (relief of carbon monoxide inhibition of respiration by light of different wavelengths)]

Bacteria	Oxidase(s)	Ref.
Acetobacter pasteurianum	a_1	170
Acetobacter peroxydans	a_1	161
Acetobacter suboxydans	o	170
Staphylococcus albus	o	170
Staphylococcus aureus	o	173
Rhodospirillum rubrum (dark grown)	o	350
Azotobacter vinelandii	o, a_1	161
Bacillus subtilis	a_3	161
Sarcina lutea	a_3	161
Escherichia coli, log phase	o	161
Escherichia coli, stationary phase	o, a_2	161
Proteus vulgaris, log phase	o	161
Proteus vulgaris, stationary phase	a_1, a_2, o	161
Aerobacter aerogenes, log phase	o	161
Aerobacter aerogenes, stationary phase	o, a_2	161

reduced oxidase with which oxygen reacts (182). Longmuir and Clarke observed that the pK_M (expressing the effect of pH on the K_M for oxygen) of a cell-free suspension of *Aerobacter aerogenes* (cytochromes a_1, a_2, and o) was constant between pH 5 and 8.5 and concluded that the oxidases of these bacteria react with oxygen without a change of charge in the combining group (183). There was a small variation of the pK_M with pH in a preparation from *Bacillus megaterium* (cytochromes $a + a_3$ and probably o).

The oxidases in cell-free suspensions of bacteria will oxidize reduced soluble c-type cytochromes isolated from the same species, but the rates are usually low compared with the respiration rates possible, even at relatively high concentrations of cytochrome c (62,86,98,129,163,184–189). The oxidases of a few bacterial species will oxidize soluble mammalian cytochrome c rapidly (98,168,187,190,191), but most do not oxidize it at all or do so at a very low rate. In some cases it was observed that oxidases react more readily with c-type cytochromes from the same or closely related species (192), but notable exceptions make any generalizations questionable. The oxidase of *Micrococcus denitrificans* can oxidize mammalian and its own cytochrome c equally rapidly, although the two cytochromes c have quite different properties (168,187,190). The reaction of the mammalian oxidase with soluble cytochrome c has been shown to involve many variables, and often the rate-limiting step appears to be that of getting the soluble cytochrome c to the proper reactive site in the proper orientation (see Gibson). This may explain the puzzling observations on the reactions of oxidases with soluble c-type cytochromes of different species. The effect of cytochrome c concentration on the oxidase kinetics observed with the bacterial oxidases is the same as that seen with the mammalian system (188,193). Incubating a particulate preparation from *Azotobacter vinelandii* at 30° for an hour (or treating it with calcium phosphate gel) increased the rate of oxidation of mammalian cytochrome c (188). Increase in the "exposure" of the oxidase to reaction with soluble cytochrome c has also been observed in suspensions of heart muscle particles following similar treatments (165).

There has been one report of a purified bacterial "oxidase" from *Pseudomonas aeruginosa*, which contains cytochrome a_2 and a c-type cytochrome (177,179). This soluble preparation can oxidize a soluble c-type cytochrome from these bacteria or p-phenylenediamine or hydroquinone with oxygen or nitrite with turnover rates of 2400 and 4000 min^{-1}, respectively, at 37° (194). Although the preparation has cytochrome oxidase activity, its function in the intact bacteria appears to be that of a nitrite reductase. The preparation can only be isolated from cells which have nitrite reductase after growth in nitrate (179), and it has all the nitrite reductase activity of the extracts of bacteria grown in this manner. Cells grown with air in the absence of nitrate seem to have another cytochrome oxidase (195). However, since the soluble preparation does have cytochrome oxidase activity, a number of the properties are of interest. When heme a_2 is separated by treatment with acetone–HCl (the c-type cytochrome is not split by this procedure), the protein containing the c-type cytochrome is insoluble, but it can recombine with heme a_2 at alkaline pH and the preparation is then soluble at pH 7. The recombined preparation has properties similar to, but not identical with, the original, and the cytochrome oxidase activity is partially restored (196,197). The protein free of heme a_2 would combine with heme a from beef heart, but this combination had very low cytochrome oxidase activity. However, the combination with protoheme showed 27% of the original activity in reducing nitrite (196).

A so-called "purified oxidase" from *Micrococcus denitrificans* was still particulate and showed absorption spectra of a c-type cytochrome but not of an a-type cytochrome (168). An interesting aspect of this report is the claim that the preparation was devoid of phospholipid, which has been thought to be essential for the mammalian oxidase.

4. *b*- and *c*-Type Cytochromes

The *b*- and *c*-type cytochromes that are members of the respiratory chain system can act as intermediary carriers between the oxidase and the dehydrogenase ends of the chain.

Cytochromes of the *c* type have been isolated from a large number of bacterial species and obtained in varying degrees of purity. As in the mammalian system, there are two *c*-type cytochromes in some species, the positions of the α peaks in the absorption spectra differing by several millimicrons. These thus appear to be equivalent to mammalian cytochromes *c* and c_1 (85,184,189,198–201). The bacterial *c*-type cytochromes have the same prosthetic group and kind of linkage between the prosthetic group and protein as mammalian cytochrome *c* (85,98,184,201,202) (see Gibson) and similar absorption spectra (66,97,135,184,187,189,192, 203,204). And like the mammalian cytochrome *c*, the bacterial pigments have low molecular weights (163,189,205), and are relatively heat stable (163,189,202).

Two bacterial cytochromes *c* have isoelectric points similar to that of the mammalian cytochrome *c*, around pH 10.05 (163,202), but most of the bacterial pigments have isoelectric points on the acid side of neutrality (98,187,189,201,204).

The properties of two highly purified cytochromes from *Pseudomonads* are listed in Table VIII (200,205–207) along with similar data for horse

TABLE VIII

Purified *c*-Type Cytochromes from *Pseudomonas aeruginosa* and
Pseudomonas fluorescens

	Ps. aeruginosa (205)	Ps. fluorescens (200)	Horse heart (352)
Molecular weight	7600–8100	ca. 9000	12,400
Isoelectric point		ca. 4.7	10.05
Peaks in absorption spectrum of reduced compound		550–551, 520, 413, 316	550, 520, 416, 315.5
Millimolar extinction coefficient of α-absorption peak of reduced compound	28.3	30	27.6
Ratio of α-absorption peak to protein peak	> 0.9	1.13–1.17	1.26
Redox potential	0.286 V at pH 6		0.260 V

heart cytochrome c. The bacterial cytochromes c showed a low content of arginine, cysteine, and histidine and less lysine and relatively more glutamic and aspartic acids than mammalian cytochrome c, accounting for the acid isoelectric point (200,206). The complete amino acid sequence of the pigment from *Pseudomonas fluorescens* has been determined (207). The sequence around the heme group is

gly cys val ala cys his ala

as compared with

lys (or arg) cys X Y cys his threo

for the same region of mammalian, yeast, and *Rhodospirillum rubrum* cytochrome c. In all, the heme is attached by thioether linkages with the two cysteines. Few similarities are seen in other parts of the sequence of the *Pseudomonas* pigment and the mammalian one except for the charge distribution over the N-terminal region which contains the heme group. There appears to be no clumping of amino acids with hydrophobic side chains in the bacterial cytochrome, but there are several concentrations of acidic residues. There can be only one coordinate linkage between the iron and histidine in the bacterial cytochrome, since only one histidine is present. A c-type cytochrome from an aerobic sulfate-reducing bacterium (see Section III-D-3) has an unusually high content of histidine and cysteine compared to other cytochromes c (206). So far it is not evident what the specific structural requirements of the protein part are for a functional c-type cytochrome.

A large proportion of the cytochromes c of the electron transport chain from substrate to oxygen have redox potentials close to that of mammalian cytochrome c, which is $+260$ mV (85,97,98,163,187,198,201,208, 209). The second c-type cytochrome in the bacteria that have two (the one with the α peak in the absorption spectrum at higher wavelengths) is reported to have a slightly lower (85,205,208) or slightly higher (189) redox potential. The c-type cytochromes involved in the electron-transport chain of photosynthetic bacteria or in sulfate reduction (see Sections VII and III-D-3) have very different redox potentials. Low-potential c-type cytochromes are synthesized by *E. coli* and other *Enterobacteriaceae* when they are grown under anaerobic conditions (143,210). These are in the soluble portion of the cell and not part of the respiratory chain system, but instead may be involved in fermentative reactions. Cytochromes with identical heme groups and linkages between the heme and protein can have very different redox potentials.

The interaction of the soluble cytochromes c with particulate oxidases and reductases is discussed in Sections III-A-3 and IV. Apparently the soluble cytochromes will interact with other soluble cytochromes having a proper redox potential (195,208,211).

Several b-type cytochromes have been isolated from bacteria, but most have not been highly purified. These all show hemochromogen-type absorption spectra similar to that of the mammalian cytochrome b, with slightly shifted positions of the absorption peaks in several (184,211). The prosthetic group is protoheme and it seems to be as easily split off as from the mammalian pigment (102,184,212). A soluble b-type cytochrome was released from the particulate fraction of aerobically grown $E.$ $coli$ by treatment with trypsin or exposure to sonic oscillation and was purified and crystallized (102). The redox potential of the purified pigment was -0.34 V, but was much higher than this in less pure preparations (0.01– 0.016). Addition of a "potential modifying protein," separated from the extract, to purified preparations lowered the potential of the crystalline preparation to 0.12 V. The absorption spectrum of the purified pigment was slightly different from that in the intact bacteria. Thus the purification procedure appears to have modified the cytochrome, and this may be true of other isolated b-type cytochromes, since all are autoxidizable (184,211–213), whereas they do not appear to be autoxidizable in the intact cells (214). Two purified soluble b-type cytochromes could be reduced by soluble enzymes from the corresponding bacteria, but additional soluble factors were required in each case (102,215).

Soluble b-type cytochromes other than the ones described above have been isolated from $E.$ $coli$ grown under different conditions, but the functions of these are unknown (216).

B. Dehydrogenases and Enzymes Reacting with Inorganic Substrates

An enormous number of compounds (organic and inorganic) can be oxidized by bacteria to furnish energy, and many microorganisms can manage with one compound as a substrate in the growth medium. This means that among the different bacterial species there exists a wide variety of dehydrogenases.

Some dehydrogenases are located in the cytoplasm; those that reduce pyridine nucleotide appear to be (5,7). The reduced DPN formed can then diffuse to the membrane to be oxidized by the DPNH dehydrogenase which is linked to the respiratory chain system (see below). The rate of reduction of DPN by the dehydrogenases in intact cells of $Hemophilus$ $parainfluenzae$ was found to be much lower than the rate of oxidation of the DPNH by the membrane-bound system (136). In these bacteria there appears to be one pool of pyridine nucleotide reducible by the various dehydrogenases, since the total amount of reduced pyridine nucleotide formed anaerobically in the presence of a number of substrates is the same, although the relative rates of reduction differ (136). These bacteria are freely permeable to pyridine nucleotide, and DPNH formed from the reduction of added DPN is oxidized at the same rate as added preformed DPNH (136).

Table I gives an idea of the different combinations of dehydrogenases that can be linked to the membrane-bound respiratory chain systems of different bacterial species; these dehydrogenases are bound in such a way that they can rapidly reduce the cytochrome systems of the membranes anaerobically in the presence of the appropriate substrates (59,217–219). The dehydrogenases are intimately associated with the cytochromes of the membrane, as shown by the isolation of a number of preparations of dehydrogenases containing b-type (59,70,220–222) or c-type cytochromes (223) which can be reduced in the presence of the substrate of the dehydrogenase.

DPNH dehydrogenase is almost universally present in the membrane fraction; dehydrogenases of succinate, malate, and formate are frequently present and occasionally those for D- and L-lactate. In addition, some membrane-bound dehydrogenases have been found to be characteristic of one kind of bacteria, such as the polyol dehydrogenase uniquely present in the particle fraction of *Acetobacter* (7,44,70,89). All the dehydrogenases appear to be distinct entities with different properties (93).

None of the membrane-bound dehydrogenases reduce pyridine nucleotide (58,70,73,93,136,217,218,224,225). There is evidence that those of DPNH, malate, succinate, lactate, and formate are flavoprotein in nature (58,93,106,218,222,226,227,227a). Only the dehydrogenases which remain attached to the membrane fragments in homogenates of *Hemophilus parainfluenzae* are active in electron transport (136), although in extracts of other bacteria some soluble dehydrogenases will interact with the particulate enzymes (see Section IV). The membrane-bound dehydrogenases can reduce a number of dyes and ferricyanide, and, at least in *H. parainfluenzae*, the reduction of ferricyanide seems to be a measure of the activities of the dehydrogenases themselves (93). As the dehydrogenases are removed from the membrane fragments into solution, the activity in reducing ferricyanide can be quantitatively recovered in the supernatant solution (93), except for formate dehydrogenase, which is less accessible for reaction with ferricyanide when it is membrane-bound (136).

Relatively few of the respiratory chain-linked dehydrogenases of bacteria have been obtained in a highly purified form. Bacterial succinate dehydrogenases from aerobic cells, but not from anaerobically cultured ones, are similar to the enzyme from mammalian tissues and yeast in overall properties and composition but show quantitative differences in reactivity (227a). (These are described in the chapter by Singer.) Several soluble preparations from bacterial extracts can oxidize DPNH with dyes or quinones or cytochrome c or have cyanide-insensitive DPNH oxidase activity (86,92,146,226,227). However, it is not known if these represent the solubilized DPNH dehydrogenases from the respiratory chain. More than one soluble DPNH-oxidizing enzyme can be obtained from some extracts; these have different reactivities and stability properties (92,227). Although all of the DPNH oxidase activity is associated with the particle fraction of some bacterial extracts (86,139,228), in others there is some in the supernatant fraction, usually small in amount. Soluble DPNH dehydrogenases with different properties are isolated from mitochondria using different solubilization and extraction procedures (see Singer). Thus the different soluble DPNH-oxidizing enzymes in the bacterial extracts could be artifacts of the preparative procedures, representing fragments of differing sizes and compositions and denaturation.

The soluble malate dehydrogenase isolated from extracts of *Mycobacteria* and *Acetobacter* is an example of a dehydrogenase which requires phospholipid for maximal activity when removed from the membranes (106,218).

Several species of bacteria synthesize a membrane-bound enzyme which is a dehydrogenase of glucose and other aldoses (221,229–233). There is no evidence for the participation of pyridine nucleotide and also no evidence that the enzyme is a flavoprotein (221,229–232,234). The enzyme contains a reducible group and in the oxidized form shows a broad band in the absorption spectrum centered at 347 mμ; the reduced form has a sharper absorption band with a peak at 337 mμ (235). A factor of unknown identity, dissociable from the protein at pH 3.2 at room temperature, appears to be the prosthetic group (235).

Dehydrogenase activity with a given substrate may be found in both the soluble and particulate fractions of a broken-cell extract. Sometimes the soluble enzyme is linked with pyridine nucleotide, while the membrane-bound one is not (58,106,234). This combination could serve a transport function between the membrane and the cytoplasm, as suggested in mammalian cells (see Klingenberg). In extracts of other species the enzymes in the soluble and particulate fractions have entirely similar properties and these are also similar to the solubilized particulate enzyme (92,219–221,234), although the soluble and particle-bound enzymes may show differences in affinity for substrates (231,232) or different re-activity with dyes (87,222) on removal from the membrane. It is not certain that the enzyme found in the soluble fraction represents enzyme which was removed from the membrane during preparation of the cell-free extract.

Some bacterial species contain enzymes that catalyze the oxidation of inorganic substances and that are linked to the membrane-bound respiratory chain system. A particulate preparation from *Ferrobacillus ferro-oxidans* can oxidize ferrous ions with oxygen over a pathway which seems to include at least two cytochromes (105,236). And intact cells of *thiobacilli* can reduce ferricyanide or cytochromes (endogenous or added soluble cytochromes) with thiosulfate or sulfite (237). The oxidation of inorganic sulfur compounds by these bacteria requires inorganic phosphate, and ATP is formed (238); the enzyme adenosine-5-phosphosulfate reductase is postulated to function in these oxidations (239). A particulate preparation from *Nitrobacter* can oxidize nitrite with the reduction of cytochromes, and this reaction appears to require ATP (240) (see Section III-D-1).

C. Quinones

The chemistry, distribution, and reactions of the quinones associated with the respiratory chain system are covered in the chapter by Crane.

Unlike mammalian cells, which contain appreciable quantities of ubiquinone and a relatively low content of vitamin K, different bacterial species may have either ubiquinones or vitamin K derivatives, or both, in substantial amounts. These quinones are not detectable in two organisms devoid of a cytochrome system (140), but one such species does synthesize quinones (241). In general, ubiquinones are found in some Gram-negative bacteria, vitamin K derivatives in some Gram-positive types, and both in *E. coli* and *Proteus vulgaris*. The quinones are found largely or entirely in the particulate fraction of the bacterial cell-free extracts (106,146,158, 242), where they are firmly bound (243) and where there may be an excess of quinone over the cytochrome pigments (158,243).

The assumption is usually made that the benzoquinones and naphthoquinones serve similar functions in the different organisms. Some of the evidence for the participation of the quinones in bacterial electron transport is the same as that suggesting a similar function in the mammalian system:

(*1*) If the cells or particulate fractions are extracted with certain organic solvents or irradiated with light of wavelength 360 mμ, the quinone content decreases and either treatment may lead to a decrease in overall electron transport with various substrates or a decrease in the rates of reduction of cytochrome c or dyes.These decreased activities can often be at least partially restored on addition of quinones to the extracts, particularly when they are added in combination with a lipid fraction from the bacteria. A puzzling aspect of this kind of experiment is the lack of specificity in the restorative effect of the quinones (244–246); often an entire range of quinones, including menadione, is effective, and in two instances ferricyanide is equally effective (347,348). Addition of a quinone may stimulate activities in the unextracted system, some being stimulated to greater extents than others (244,246,247,249,250), but the stimulated activities are sometimes not inhibited by substances which inhibit electron transport in the cytochrome chain (244,247). Also the reduction of the different cytochromes and the reaction with different substrates may not be equally restored by quinone addition (106,247,248). Finally, evidence has been presented suggesting that irradiation of *M. phlei* brings about the destruction of another substance in addition to the naphthoquinone (251). Thus the data are not as clear-cut as one would hope and in some instances suggest bypasses formed on addition of these active oxidation-reduction catalysts.

Irradiation with light of wavelength 360 mμ inhibits the growth of *E. coli* when the carbon source is a substance oxidized by the respiratory chain system (such as succinate), but growth is not inhibited in the presence of glucose, which can be fermented (252). The conclusion was drawn that the bacteria cannot obtain enough energy from oxidative metabolism when the quinones are reduced to a relatively low level (83% of the benzoquinone was destroyed). However, the irradiated cells were still able to oxidize succinate.

(2) The endogenous quinones of bacteria or of particulate preparations are found to be in the oxidized form under aerobic conditions (246) and to be partially reduced (25–40%) when anaerobic in the presence of a substrate (238,245,246,253), the extent of reduction varying with different substrates (128,246). The extent of reduction of the naphthoquinone of *H. parainfluenzae* with a given substrate increased as the related dehydrogenase increased in activity with changing growth conditions (246). The rates of oxidation and reduction of the endogenous quinones can be at least as rapid as the overall electron transport with the substrates tested, substantiating the possibility of a role in electron transport (246,253). Direct evidence for the oxidation and reduction of quinones in concert with known members of the bacterial respiratory chain system has not been seen by observations of changes in absorption spectrum of these

compounds during electron transport. This type of observaticn is difficult experimentally, since the changes in absorption spectrum are in the
ultraviolet region of the spectrum.

Cell-free extracts of a number of bacteria can reduce added ubiquinones
or naphthoquinones under anaerobic conditions in the presence of appropriate substrates (135,245,247,249,254,255), but the rate of reduction may
not be commensurate with the overall rate of electron transport with the
substrate used. The reduced quinones are oxidized on aeration of the
anaerobic suspension (247,249) or on addition of nitrate in extracts with
nitrate reductase activity (245). Some bacterial extracts have menadione
reductase activity (246,255–257). This may be localized in the particulate
fraction (52), and addition of menadione to the particles stimulates respiration. The menadione reductase activity is found in the soluble fraction
of broken-cell suspensions of other bacteria (146,255) and this has been
shown to be due to a flavoprotein which can also reduce acceptors such
as phenazine methosulfate or dichlorophenolindophenol (146,255). Since
flavoproteins are known to be relatively nonspecific with respect to hydrogen acceptors, the significance of menadione reductase activity as an expression of the biological role of quinones is questionable, particularly
since menadione can reduce cytochrome c non-enzymatically (256). Such
reactions may form artificial bypasses between parts of the electron-
transport chain. Like the DT diaphorase of mammalian cells, which can
be a part of such a bypass, the TPNH menadione reductase of *Mycobacterium tuberculosis* is inhibited by dicumarol (253).

(3) Addition of DPNH to a suspension of respiratory particles from *M.
phlei* resulted in the appearance of a free-radical signal in the ESR
spectrophotometer similar to one seen on addition of DPNH plus menadione to the particles (258). It was concluded that the hyperfine components of the signal at high pH would agree with the assumption that the
radicals responsible for the signal were naphthoquinones (259). Only a
small part of the total naphthoquinone appeared to be involved in the
reaction responsible for the signal (259), which disappeared on addition
of air, then reappeared more slowly (258). The appearance and disappearance of the free-radical signal was inhibited in the expected way by
inhibitors of the respiratory chain system, but there was considerable
disparity in the rate of appearance of the free radicals and the rate of
DPNH oxidase activity, and a rather high concentration of DPNH was
required for the maximal formation of free radicals.

The effects of inhibitors on the oxidation and reduction of quinones
in bacteria would indicate that they function in the same region of the
electron-transport chain as in the mammalian system, that is, between

b- and c-type cytochromes (86,246). As mentioned in Section III-A-2, the quinone content of some bacteria varies with growth conditions and during the growth cycle. The time course of change of quinone content of *H. parainfluenzae* was found to parallel that of cytochrome b_1 but not that of the other cytochromes or any of the flavoprotein dehydrogenases (243). Cytochrome b_1 and the quinone remained in a ratio of 1:14. This could indicate a close association of the cytochrome and the quinone, but the large excess of the latter and the different extents of reduction by the different dehydrogenases (246) are difficult to explain. Although a large increase in quinone content of *Staphylococcus albus* was accompanied by a similar increase in the respiration with lactate, similar increases in respiration of several other species were seen with little or no change in quinone content (140).

E. coli is an interesting bacterium from the point of view of studies of quinone function. It synthesizes both benzoquinones and naphthoquinones (297) and a "large-particle fraction," which contains only benzoquinones, can reduce menadione, but not vitamin K_2 with DPNH. A "small-particle fraction" contains both benzo- and naphthoquinones and can reduce both vitamin K_2 and menadione with DPNH (297).

Altogether, the experiments with the bacterial systems, and particularly with cell-free extracts, do not give a clear straightforward picture of the function of the different quinones in the electron-transport system. This may result in part from the problem mentioned above of creating artificial bypasses following the addition of active redox systems to preparations bearing respiratory chain enzymes. The quinones may well have a more interesting role than that of a member of a linear electron-transport sequence. The presence of both benzo- and naphthoquinones in some bacteria plus the possibility of wide variations in quinone content make the bacterial systems interesting experimental material.

In addition to its postulated role in electron transport, vitamin K has also been implicated in the coupled phosphorylation reactions leading to the formation of ATP in some bacteria. This will be discussed in Section VI.

D. Reductases Reacting with Terminal Electron Acceptors other than O_2

1. Nitrate

Several bacterial species can utilize nitrate as an alternative to oxygen as the ultimate electron acceptor in the anaerobic production of energy; the anaerobic reduction of nitrate can lead to production of ATP (see Section

VI). This process is often referred to as "nitrate respiration" or "dissimilatory nitrate reduction" to distinguish it from another system in some cells which reduces nitrate to ammonia for synthetic purposes (assimilatory nitrate reduction). There are species of bacteria which can carry out both processes (260,261) and the relative amounts of the two systems may depend upon growth conditions (262). An electron-transport system containing cytochromes is linked to the dissimilatory reduction of nitrate, as evidenced by the oxidation of cytochromes on addition of nitrate to intact bacteria under anaerobic conditions (162,184,237,261, 263,264). Particulate fractions can be isolated which anaerobically oxidize substrates such as DPNH, succinate, lactate, or formate with nitrate, and the endogenous cytochromes of the particles (as well as added soluble cytochromes) can be oxidized anaerobically with nitrate (100,136, 237,240,260,262,265–267). The cytochromes are also oxidized by oxygen. The reduction of nitrate by the appropriate substrates and the oxidation of the cytochromes by nitrate can be inhibited by substances which inhibit electron transport in the cytochrome chain involved in the reduction of oxygen (261–263,265,268), implying that similar pathways are involved in the reactions with nitrate and with oxygen. The two pathways may show different relative sensitivities to inhibitors (269); some cytochromes may be preferentially oxidized by nitrate (100,136,264,265). Nitrate oxidizes the same cytochromes of *H. parainfluenzae* as oxygen does, but with nitrate there is a relatively greater oxidation of cytochrome a_1 and flavoprotein, compared to the other cytochromes (162). Thus cytochrome a_1 either reacts preferentially with nitrate or it reacts rapidly with another enzyme which reacts directly with nitrate. The synthesis of cytochrome a_1 by these bacteria is stimulated by growth in the presence of nitrate (151). In other species the enzyme which reacts directly with nitrate does not seem to be identical with one which reacts directly with oxygen. In fact there is evidence that the nitrate reductase of some species is not a cytochrome, as shown by studies with a hemin-requiring mutant of *Staphylococcus albus* (268) and by isolation and purification of the nitrate reductase (100,262, 266,267,270), which can react directly with the reduced form of dyes such as methyl or benzyl viologen (262,265) or pyocyanine (268). Purification efforts suggest that the nitrate reductase of some species may be a flavoprotein (266,267,270) or a metalloflavoprotein (100,266). However, the purified enzyme from *E. coli* is reported to contain negligible flavin, but 40 atoms of iron and one of molybdenum per molecule (262). Nitrate reduction by *E. coli* cells or extracts is inhibited by cyanide and azide, but not by carbon monoxide (262,271). It is possible that the nitrate reductases of different species are quite different chemical entities.

If parts of the same electron-transport system are involved in the reduction of either nitrate or oxygen, a competition between the two processes would be expected; there are numerous examples of this. The oxygen uptake may be inhibited by nitrate and vice versa (148,264,272). In most species the affinity for oxygen is greater (148,260,266,273,274). One exception is *Nitrobacter agilis*; oxygen does not inhibit nitrate reduction by these bacteria, but nitrate inhibits oxygen uptake (100). Since the enzyme nitrate reductase of *Aerobacter aerogenes* appears to be actually inhibited by oxygen and the adaptive formation of the nitrate reductase may also be inhibited by oxygen (275), some caution must be exercised in judging the meaning of the competition between oxygen and nitrate.

The *Nitrobacter* can catalyze the reduction of nitrate to nitrite and the oxidation of nitrite to nitrate (100,240). A particle-bound preparation from *Nitrobacter winogradski* containing two cytochromes can reduce nitrate with DPNH with the synthesis of about two molecules of ATP per molecule of nitrate reduced (240). The oxidation of nitrite to nitrate appears to require ATP, since it is abolished by uncouplers of phosphorylation. The net result of the cyclic reduction of nitrate and oxidation of nitrite is the net production of either ATP or DPNH (240).

2. Nitrite and Other Nitrogen Compounds

Pseudomonas aeruginosa forms an adaptive nitrite reductase when grown in the presence of nitrate. A purified enzyme isolated from bacteria grown with nitrate accounted for all of the nitrite reductase activity. It shows absorption spectra characteristic of a_2- and c-type cytochromes and can oxidize a soluble cytochrome c from these bacteria with either nitrite or oxygen (177,179,194) (see Section III-A-3). A preparation of nitrite reductase from the same bacteria made by a different procedure had the same cytochromes but contained in addition FAD, iron, and copper (276). $FADH_2$, $FMNH_2$ or reduced pyocyanine, methylene blue, or menadione could serve as donors for nitrite reduction, and NO_2 and NH_2OH could replace nitrite and O_2 as acceptors in the oxidation of the soluble cytochrome c. In this preparative procedure there was some separation of the nitrite reductase and the cytochrome oxidase activity, suggesting that two different enzymes are involved.

Desulphovibrio desulphuricans reduce nitrite with hydrogen. A c-type cytochrome seems to be involved, since loss of cytochrome c by treatment of the cells with cetyl trimethyl ammonium bromide results in a decrease of activity which is restored on addition of soluble cytochrome c or a dye such as benzyl viologen (277,278). There is also some evidence that an electron-transport system containing cytochromes is involved in nitrite reduction in a halotolerant micrococcus (278a).

3. Sulfate and Other Sulfur Compounds

Sulfate-reducing bacteria such as *Desulphurvibrio desulphuricans* are strict anaerobes which can use sulfate and other sulfur compounds as terminal electron acceptors. A cytochrome chain is also involved in electron transport in these organisms; absorption bands of reduced cytochromes are seen anaerobically in the presence of hydrogen donors, and the cytochromes are oxidized on addition of sulfate, sulfite, thiosulfate, or tetrathionate (202,278). The cytochrome c_3 of these bacteria is autoxidizable, so it is also oxidized in the presence of oxygen (202,278). Cell-free extracts can oxidize soluble cytochrome c_3 with the same sulfur compounds anaerobically. Cells treated with cetyl trimethyl ammonium bromide lose their cytochrome c_3; the addition of the purified cytochrome restores the ability to reduce sulfite, thiosulfate, or tetrathionate, but not sulfate (202). ATP is required for the reduction of sulfate by the cell-free system in the presence of cytochrome c_3 (279). The form in which sulfate is reduced appears to be adenosine-5-phosphosulfate (APS) (239,254). The enzyme involved is APS reductase (239), and there may be an additional factor between the cytochrome c_3 and the APS reductase (254).

4. Organic Acids

Bacteroides rumincola is a rumen organism which grows in an anaerobic environment. The majority of strains require hemin for growth, and difference spectra obtained on addition of sodium dithionite show evidence of a *b*-type cytochrome, cytochrome *o*, and a flavoprotein. The cytochromes and flavoprotein are also reduced by adding substrates and then these can be oxidized by adding malate, oxaloacetate, fumarate, or CO_2 (280). The data suggest that cytochromes and flavoprotein are involved in electron transport coupling the oxidation of DPNH with the reduction of fumarate.

E. *Non-Heme Iron*

Evidence for the presence of non-heme iron has been cited for preparations of succinic dehydrogenase (200,227a) and of nitrate reductase (262,267,270) from bacteria. In one instance the reduction of nitrate with menadione was enhanced by the addition of ferrous ions. The purified preparation from *E. coli* which reduced nitrate with methyl viologen (see above section) contained considerable non-heme iron (262). Intact cells of several bacterial species in the reduced state (anaerobic plus substrate) showed a signal in the EPR spectrometer which had been attributed to an unusual state of iron (281). The properties of an iron protein isolated from *Azotobacter vinelandii* confirmed this suggestion (282). So far there is no evidence for the possible role of non-heme iron in the bacterial respiratory chain system.

IV. Sequence of Electron Transport and Mechanism of Action

As with the mammalian system, the sequence of interaction of the bacterial respiratory chain pigments has been established by (*1*) examining the kinetics of the oxidation-reduction reactions; (*2*) inhibitor studies; and (*3*) looking at partial reactions and attempts at fractionation of the system.

(*1*) Chance measured the rates of reduction of cytochromes a_2 and b_1 of *Aerobacter aerogenes* as the oxygen in solution was exhausted and showed that cytochrome a_2 is nearer to the end of the chain that reacts with oxygen (172). Tissières showed that the cytochrome a_2 is oxidized before cytochrome b_1 on addition of oxygen (138).

(*2*) Studies of inhibitors of the oxidase have been described in Section III-A-3, reaction with carbon monoxide being an indicator of the cytochromes which can react with oxygen. The lack of evidence for the reaction of cytochrome a with carbon monoxide or cyanide, but its reduction in the presence of these substances (174), would place cytochrome a next to the oxidase in those bacteria in which it occurs. Cyanide inhibits the reduction of the oxidase cytochrome a_2 of *Azotobacter vinelandii* (186) and a strain of *Achromobacter* (152), as it inhibits the reduction of cytochrome a_3 of the mammalian chain (283).

Electron transport in a few bacterial systems is blocked between b- and c-type cytochromes by addition of 2-alkyl-4-hydroxyquinoline-*N*-oxides (162,284,285), as in the mammalian system, showing that the b-type cytochromes are nearer the dehydrogenase end of the chain. Observations with other bacteria in the presence of this inhibitor are not as clear-cut (173,284,285); some studies with the inhibitor give evidence for the nonhomogeneity of the cytochrome b. This may be explained by the indications that the oxidase cytochrome o is also a b-type cytochrome. This was discussed in Section III-A-3. The respiration of some species is insensitive to inhibition by the alkyl hydroxyquinoline-*N*-oxides (59,105) or is inhibited only by relatively high concentrations of these substances (86,128). This is explained in a few organisms by a lack of penetration of the inhibitor into the intact cells (283). Antimycin A, which blocks the mammalian respiratory chain system at the same site as the alkyl hydroxy-quinoline-N-oxides, has no inhibitory effect on the respiration of most bacterial species (52,59,68,75,105,111,130,173,229,285), although a few species are inhibited by it at higher concentrations than are effective with the mammalian system (86,247,286). It must be remembered that anti-mycin A could be bound firmly by proteins at other sites in the bacteria or the cell-free extracts.

The bacterial respiratory chain system is inhibited by a number of sub-stances known to react with flavoproteins (105,106,136,247). However, the DPNH oxidase of some bacteria is completely or relatively insensi-tive to inhibition by sodium amytal (106,128) and rotenone (106,136), which inhibit DPNH oxidation in the mammalian system. The DPNH oxidase of one organism (*Mycoplasma hominis*) is inhibited by amytal (247), as are both the DPNH and succinate oxidases of dark-grown *Rhodospirillum rubrum* (86) and the non-DPN-linked malate oxidase of a *Pseudomonad* (58). The differences in the effects of inhibitors among the different species of bacteria and between some bacterial systems and the mammalian one are difficult to explain. As suggested for the effects of inhibitors on the oxidases (Section III-A-3), it would appear that all the systems contain pigments with similar groups, but that the groups may have different reactivities or different accessibilities to the reagents in-volved. Perhaps these differences could be profitably exploited in probing the nature of the reactions.

(3) A number of efforts to "purify" the bacterial respiratory chain system have resulted in the removal of some non-essential protein and yielded "electron-transport particles" of small size with increased specific activities with some substrates. But there has been little success in the dissection of the system into component parts (60,111,121,152,188,194, 229), even with methods that have been employed for dissecting the mam-malian system (see Klingenberg, this volume). Some attempts at purifica-tion yielded particles with decreased activities (66,265), altered reactivity (69), or loss or displacement of some dehydrogenases (121,229). Although the mammalian and bacterial membrane-bound systems have similar components, there must be some structural differences which make the dissection of the bacterial system more difficult.

A number of preparations have been made which contain a dehydro-genase and a *b*-type cytochrome, and in most of these the cytochrome can be reduced in the presence of the substrate of the dehydrogenase (59,70,

137,219–222,287), showing that in some bacteria the dehydrogenases are intimately linked to the b cytochromes. One preparation has been described containing a dehydrogenase and a c-type cytochrome (223). Some of these preparations contained other enzymes, not all functionally linked to the cytochrome.

Less information has been gleaned from studies of reactions of parts of the respiratory chain system with added soluble redox systems. Respiratory particles have been observed to reduce substances such as ferricyanide, dichlorophenolindophenol, menadione, vitamin K_2, and cytochrome c. However, often it has not been shown which member of the respiratory chain the added redox system is reacting with. Although some of the dehydrogenases of the system of *Hemophilus parainfluenzae* react equally well with ferricyanide when membrane bound or when removed from the membrane into solution, formate dehydrogenase appears to be less available for reaction when membrane-bound (93). Soluble mammalian cytochrome c can be reduced by many particulate preparations with various substrates, even by preparations from species which do not synthesize a c-type cytochrome. The soluble cytochrome c is obviously acting like other redox dyes. Soluble preparations of bacterial cytochrome c may be reduced more rapidly than the mammalian pigment, but even the reduction of its own cytochrome c by the bacterial particles appears to be slow compared to the interaction of the membrane-bound pigments (185,186). One c-type cytochrome isolated from a given bacterium may react with the membrane-bound system, while another does not; however, the two c-type cytochromes will interact in solution (185). This is additional evidence of the difficulty of getting the soluble cytochrome c to the proper reactive site, as discussed in Section III-A-3. In any case, many more bacterial particulate preparations can reduce soluble mammalian cytochrome c than can oxidize it. Apparently there are more specific requirements for the site where it can react with the oxidases [see Section III-A-3 and Gibson (this volume, p. 379)]. The soluble cytochrome c must react at different sites in the oxidase and reductase reactions, as it appears to with the system of heart muscle particles (103,288).

The observations reported to date are in agreement with a sequence of electron transport in bacterial systems outlined in Figure 1. The sequence may be determined by the spatial arrangements of the pigments on the membrane and this may change when the membranes are disrupted (see Section II-C).

There can be numerous dehydrogenases attached to the membranes in such a way that they can react rapidly with the cytochrome system. The overall rate of electron transport in *Hemophilus parainfluenzae* was shown to be limited by the reaction of the dehydrogenases with their substrates; changes in dehydrogenase activity give parallel changes in respiration (93,136). The proportion of the b- or c-type cytochromes that can be reduced anaerobically may vary in the presence of substrates of the different dehydrogenases (106,136,184,201). The greater the amount of a given dehydrogenase bound to the particles of $H.$ $parainfluenzae$ (as shown by the rate of ferricyanide reduction), the greater the proportion of b- or c-type cytochromes reduced under anaerobic conditions with the substrate of the dehydrogenase (93). When the dehydrogenases are removed from the membrane fragments, the respiration with the corresponding substrates decreases in proportion to the amount of the cytochromes which can be reduced anaerobically. The data of Table IX show that the respiration

TABLE IX

Comparison of Respiratory Rate of Bacteria and Aged Respiratory Particles of $H.$ $parainfluenzae$[a] (136)

Substrate	Oxygen utilization per 10 mg of protein, mμmoles/sec		Oxygen utilization per 0.05 absorbance of cytochrome b_1, mμmoles/sec	
	Bacteria	Particles	Bacteria	Particles
Formate	26.0	16.0	35.0	34.0
DPNH	4.16	3.12	7.9	7.9
D-Lactate	6.31	0.13	6.5	6.1
L-Lactate	7.40	0.25	7.1	7.0
Succinate	4.66	3.48	6.7	6.7

[a] Bacteria were harvested after 24 hours of incubation, centrifuged, washed with 50mM phosphate buffer (pH 7.6), ruptured with alumina, and resuspended in phosphate buffer. After centrifugation at 6000 × g for 10 min to separate intact bacteria and alumina, particles were collected by centrifugation at 12,000 × g and aged at 0° for 6 hr before use. Oxygen utilization is expressed as rate per 10 mg of protein or as rate per 0.05 absorbance unit of cytochrome b_1 reduced after anaerobiosis produced with that substrate.

rates of the particles, when expressed in terms of the cytochrome that can be reduced, are the same as those of the intact bacteria (136). Data of this kind indicate that the various flavoprotein dehydrogenases have different numbers of points of attachment to the respiratory chain system. The respiration rates of one strain of $H.$ $parainfluenzae$ are additive with a combination of some substrates but not with others. The rates of a mutant strain are less than additive with combinations of substrates, and mixtures of substrates do not give additive amounts of reduced b- and c-type cytochromes with either strain (136). A large proportion of the b- and c-type cytochromes can be reduced in the parent strain with DPNH and in the mutant in the presence of lactate or formate (93). The picture that emerges is one of dehydrogenases arranged in varying numbers around the different cytochrome chains in a three-dimensional array, with some overlapping of the flavoproteins with the different cytochrome assemblies. Additional evidence for the interaction of different dehydro-genases with the same cytochrome chain is seen in the inhibition of the oxidation of succinate by DPNH in particles from *M. phlei* (106) and the inhibition of DPNH oxidase of preparations of dark-grown *Rhodospiril-lum rubrum* by succinate (86).

Thus each of the flavoprotein dehydrogenases appears to be closely linked to some cytochrome chains. The situation is different with the membrane fragments from heart mitochondria, where the same number of cytochromes are reduced with either succinate or DPNH. However, there are not so many dehydrogenases bound to the mitochondrial membrane fragments as there are bound to the bacterial particles. The structure of the respiratory chain in the dehydrogenase–cytochrome region is not understood. Quinones have been postulated to function here (see Section III-C). An additional factor was required for the interaction of isolated soluble cytochrome b and flavoprotein from two bacterial species (102, 215), but the nature of the factor was not established. The addition of menadione could promote the interaction of formate dehydrogenase and cytochrome b_1 in a complex from *E. coli* which had lost the capacity to interact on standing in air (137). Since the original preparation contained no naphthoquinone, the added menadione was not replacing naphtho-quinone which had been destroyed on standing; it was not reported whether the preparation contained benzoquinone. Although the time course of synthesis of demethyl vitamin K_2 by a mutant of *H. parainfluenzae* paralleled that of cytochrome b_1 but not that of the other cytochromes or the flavoprotein dehydrogenases, the ratio of the quinone to cytochrome b_1 remained 14:1 during the growth cycle. Thus there is a large excess of the quinone (243). Taken all together, the data on the bacterial systems imply a close association of flavoprotein and cytochrome (usually b-type) with possibly an involvement of quinone. The suggestion that the quinone is functioning as a diffusible carrier is not supported by the fixed amount of cytochromes reduced by a given amount of flavoprotein.

Although the b- and c-type cytochromes are reduced to different extents anaerobically in the presence of substrates of the different membrane-linked dehydrogenases of *H. parainfluenzae*, the oxidases (cytochromes a_1, a_2, and o) are completely reduced with all substrates (162). This will be discussed below.

H. parainfluenzae synthesizes a large excess of cytochrome c_1 under some growth conditions (Section III-A-2). The cytochrome c_1 which can be reduced with DPNH can be 5–10 times the other cytochromes, and there is additional cytochrome c_1 in the cells which is not membrane bound. The latter can only be reduced with $Na_2S_2O_4$ and appears in solution on rupture of the cells (99). Some of the cytochrome c_1 on the membrane fragments can be removed by washing with buffer, but the respiration rate does not decrease until a considerable amount has been removed (99) (Table X). The soluble cytochrome c_1 cannot be oxidized or reduced by

TABLE X

Removal of Cytochrome c_1 from Respiratory Particles[a] (99)

Preparation	Cytochrome c_1 removed, $\Delta OD_{553m\mu}$	Respiration, μmolar O_2 sec^{-1}	
		+ Succinate	+ DPNH
Broken-cell extract		0.17	0.73–1.20
Particles washed once	0.087	0.21	1.90
Particles washed 2 times	0.014	0.19	1.19
Particles washed 3 times	0.004	0.067	0.19
Particles washed 4 times	0	0.065	0.20

[a] The broken-cell extract (8.2 mg of protein per ml) was centrifuged at 30,000 × g for 30 min; then the pellet was resuspended in 0.05M phosphate buffer, pH 7.6, to the volume of the original suspension. Subsequent washings were carried out in the same manner. The cytochrome c_1 removed by the washings was assayed by measuring the increase in optical density in the supernatant fluids on addition of $Na_2S_2O_4$.

the particulate enzymes, but it can be seen to undergo rapid oxidation and reduction as long as it is associated with the membrane-bound system (99,136).

Variations in the content of the different cytochromes in a number of bacteria showed no obvious relationship to the overall respiration rate with a given substrate (136,144,150,160). Quantitative studies have been made of the effect of changes in the content of the different respiratory-chain pigments of *H. parainfluenzae* on the overall respiratory activity (136). The differences in complement of cytochromes is exaggerated when the parent strain is compared with a mutant which synthesizes the same components at different relative rates. By selection of the proper growth conditions, bacteria can be harvested with a minimum of a 32-fold variation in the content of functional (membrane-bound) cytochrome c_1 and a nine-fold change in the content of functional cytochrome b_1. But the bacteria having these widely different proportions of the different cytochromes can have similar respiration rates with a number of substrates (136) (Table XI). Thus, wide variations in the relative proportions of the different cytochromes are compatible with rapid electron transport, the respiration rates in each case being determined by the content of the different dehydrogenases. These observations render untenable the concept of assemblies of respiratory pigments of fixed stoichiometry, as suggested for the mammalian system (32,90,289). The data from the bacterial systems can best be explained in terms of a three-dimensional array in which electrons can be transported through any number of pigments attached to the structure and in which loss of some of the pigments does not decrease the overall rate of electron transport. As pointed out above, some dehydrogenases have more points of attachment than others. Figure 7 is a two-dimensional representation of the kind of system indicated from the studies with bacteria.

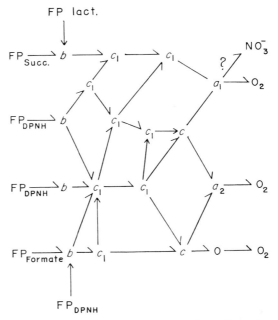

Fig. 7. Two-dimensional representation of membrane-bound respiratory chain system of *H. parainfluenzae.*

TABLE XI[a]

Comparison of Respiratory Rates Produced by Adding D-Lactate, L-Lactate, Succinate, DPNH, and Formate to *H. parainfluenzae* which Formed Principally Cytochrome b_1 with Bacteria which Formed Primarily Cytochrome c_1 (136)

Substrates	Mutant (cytochrome b_1 predominant)			Parental type (cytochrome c_1 predominant)		
	Rate of oxygen uptake	Rate expected if additive	Cytochrome b_1 reduced	Rate of oxygen uptake	Rate expected if additive	Cytochrome c_1 reduced
A. Added singly						
D-Lactate	2.23		0.028	0.67		0.048
L-Lactate	4.60		0.033	3.68		0.053
Succinate	1.48		0.026	2.12		0.064
DPNH	1.77		0.032	2.95		0.064
Formate	43.5		0.034	46.0		0.053
Na₂S₂O₄			0.034			0.102
B. Added sequentially						
Succinate + DPNH	2.95	3.25	0.033	5.10	5.07	0.062
Succinate + DPNH + L-lactate	6.75	7.85	0.031	8.90	8.75	0.063
Succinate + DPNH + formate	18.9	46.75	0.030	44.0	51.07	0.054
D- + L-Lactate	3.85	6.83	0.031	1.35	4.35	0.054

[a] Oxygen uptake was measured polarographically after addition of approximately 10 mg of bacterial protein and substrates at concentrations of 10 μM, except for DPNH (5 mM), at 30° in 50 mM phosphate buffer, pH 7.6, and is expressed as millimicromoles of O₂ per second per 10 mg of protein. Cytochrome b_1 was estimated as the difference in absorbance between 561 and 575 mμ in anaerobic (in the presence of the substrate) minus aerobic difference spectrum; cytochrome c_1 was estimated similarly at 553 and 575 mμ per 10 mg protein.

The mechanism of interaction of the membrane-bound pigments is still unknown. As discussed above, problems of getting soluble substances such as cytochrome c to the proper site for reaction with membrane-bound enzymes make these reactions poor models for studying the interaction between membrane-bound pigments. The oxidase of membrane fragments of heart mitochondria appears to react with added soluble cytochrome c only when the linkage between the endogenous cytochrome c and the oxidase is disrupted (165). The oxidase will not react with a large number of redox systems of appropriate redox potential, and this is unrelated to problems of charge (165,291) but may be a reflection of the complex nature of the oxidase (Section III-A-3). Soluble c- and b-type cytochromes will interact in solution even when some of these will not react with the membrane-bound system.

The extent of interaction of soluble enzymes with membrane-bound ones within the intact cell is even less understood. Broken-cell extracts of bacteria will respire without lag in the presence of a large number of substrates, including some not oxidized by the membrane-bound dehydrogenases. The rates may be low compared to those observed with the intact cells, even after addition of soluble cofactors (5,42). Some of the respiration must result from the oxidation of DPNH produced by the soluble dehydrogenases via the respiratory chain system. However, there are a number of reports of the interaction of soluble enzymes with the membrane-bound system which do not appear to involve soluble coenzymes (106, 292,293). The mechanism of this kind of interaction is not known. Some puzzling observations with broken-cell extracts are the synergistic effects sometimes seen on combining the soluble fraction with the particulate (51,296) fraction and, in other cases, the loss of respiratory activity on recombination, even in the presence of added coenzymes (51,88). The possibility of constituting a bypass of one or more endogenous electron-transport pigments with soluble redox systems is obvious; this has been demonstrated with soluble mammalian cytochrome c added to particles from heart mitochondria (103) and is evident in bacterial preparations in the presence of added menadione (297). Thus the extent of interaction of soluble substances (except for substrates and a few soluble coenzymes) with membrane-bound pigments in intact cells of bacteria would be difficult to judge from experiments with broken-cell extracts.

V. Transport of Substances into the Bacterial Cell

Regulation of the penetration of substances into the bacterial cell resides in the cytoplasmic membrane (6) (see Section I). The accumulation of substances within the cell against a concentration gradient would be an energy-requiring process, the energy being furnished by the metabolism of endogenous or exogenous substrates. Energy would also be required for the adaptive formation of "permeases" required for the penetration of specific substances, since this involves *de novo* protein synthesis (82, 294). Evidence of the requirement for energy is seen in the inhibition of the uptake of both organic and inorganic substances by inhibitors of respiration or substances which uncouple ATP synthesis (see below) (82,295,298) and in the supplementary O_2 uptake with endogenous substrates during accumulation (294) or the requirement for respiration for uptake by some cells (298). Changes in the light scattering of bacterial suspensions appear to accompany active transport and accumulation of compounds by the cells. Substrates (such as formate or glycerol) which enter *E. coli* by simple diffusion produce no change in the light-scattering properties of the cells (299), but the presence of formate as an energy source may increase the rate of change of the light scattering on addition of other substrates or inorganic salts to starved cells (300).

Mitchell and Moyle studied the properties of the transport systems in bacteria, which led to the interesting suggestion that the same membrane-bound enzymes (oxidation-reduction and group transfer) are involved in both metabolism and membrane transport (36). The transport process could be explained by anisotropic enzymes oriented in the membrane in such a way that some substances react from one side and others from the other side, the chemical groups passing across the membrane during the enzyme-catalyzed reactions. Or there could be sequential reactions where the product of the first reaction diffuses to the active center of a neighboring enzyme faster than it escapes into the medium (36). For example, studies of phosphate assimilation and phosphate exchange by *Staphylococcus aureus* showed that the system responsible for the initial step of phosphate assimilation had all the characteristics of a group-transferring system in the membrane (6,301). The phosphoryl groups, due to thermal movements, could become alternatively accessible to the medium or to the other side of the membrane, where an exchange might occur between the phosphoryl group and phosphate ions without a loss of bond energy of the carrier–phosphoryl compound. The formation of the R-phosphoryl compound may be coupled to the metabolism of glucose (6). The specific accumulation of succinate is explained similarly in terms of two enzyme systems located in the plasma membrane and in the cytoplasm (302,303). The system in the membrane is pictured as catalyzing the equilibrium between succinate and a succinyl compound (probably succinyl coenzyme A) in the presence of succinate, ATP, coenzyme A, and K^+ ions. The system in the cytoplasm does not require cofactors. The two systems could be involved in the transfer of succinyl groups across the membrane if the enzymes were arranged so that the active center of a phosphoryl-coenzyme A transferase were accessible to succinate only from the outside, but to succinyl coenzyme A and phosphoryl coenzyme A from the inside. Succinate could be transported as succinyl coenzyme A, then released by a deacylase. Other enzymes which react with the same substrate and which seem to be distributed between the membrane and the cytoplasm were described in Section III-B.

VI. Respiratory Chain-Linked Phosphorylation

Current concepts of the reactions involved in the phosphorylation of ADP to ATP coupled to the respiratory chain system are covered in the chapter by Boyer. Here all that remains is to describe the properties of the nonmitochondrial system of bacteria and the postulations that have emerged from studies of the bacterial systems.

Broken-cell suspensions of a number of bacterial species show phosphorylation of ADP to ATP accompanying the oxidation of various substrates by oxygen or by nitrate. Many of these species are obligate aerobes and cannot derive the energy for growth from fermentative reactions. With a few exceptions, the efficiency of the phosphorylation in the cell-free extracts is relatively low compared to that seen with mammalian mitochondria; P:O ratios (ratio of ATP formed to atoms of oxygen used) considerably less than 1 are observed in most extracts, although the rates of respiration and phosphorylation may be considerably higher. Attempts at assessing oxidative phosphorylation indirectly in *intact cells* have sometimes appeared to indicate low efficiency (304,305). However, P:O ratios measured directly (ATP content measured by the luciferin–luciferase method) with one species of intact bacteria approached those obtained with mammalian mitochondria: around 2.8 with β-hydroxybutyrate as substrate (306). The maximal P:O ratio obtained with cell-free extracts of the same bacteria is around 0.5. Thus, disruptive changes must accompany cell rupture. Even disrupted mammalian mitochondria, with which it would be more appropriate to compare the bacterial broken-cell extracts, can have P:O ratios only slightly lower than those of intact mitochondria. An explanation for the low ratios seen with the bacterial extracts may be the presence of soluble enzymes which can furnish nonphosphorylating bypasses in the oxidative reactions and phosphatases and phosphate-transferring enzymes which might lower the ATP content (126,128,307,308).

Respiration and phosphorylation are "obligately coupled" in suspensions of intact mitochondria; that is, respiration proceeds at a very low rate until the phosphate acceptor ADP is added. The inhibition of respiration in the absence of phosphate acceptor is usually referred to the inability of the high-energy compound of a respiratory chain pigment to interact in electron-transfer reactions (see chapter by Boyer). In most cell-free extracts of bacteria there is no effect of phosphate acceptor on the respiration. This "loose coupling" has been explained as resulting from splitting reactions of the high-energy intermediates. In *intact* bacteria there is some evidence of tight coupling of respiration to phosphorylation, to judge by a few examples of increase in respiration following procedures which would be expected to produce ADP or to uncouple phosphorylation. For example, a stimulation of respiration of *E. coli* is seen during the accumulation of a nonmetabolizable substrate, which would be an energy-requiring reaction (294,295). On the other hand, the transport of a nonmetabolizable amino acid into *B. megaterium* showed a direct relationship to the oxygen uptake. This respiration-controlled uptake did not show the "tight coupling" seen with *E. coli*; increased respiration stimulated amino acid uptake, but not vice versa (298). The observation that the respiration of numerous species of starved bacteria begins without lag and continues at the maximum rate on addition of an oxidizable substrate could be explained by the need for energy for transport of the substance into the cells. Formate is freely diffusible across the membrane of *E. coli* and respiration with this substrate is stimulated by a number of compounds which "uncouple" respiration from phosphorylation in mammalian mitochondria (299,309). A similar stimulation of respiration by an uncoupler was seen in *E. coli* to which potassium glutamate was added (300).

The respiratory chain-linked phosphorylating systems of mitochondria (or disrupted mitochondria) and of broken-cell suspensions of bacteria have some properties in common: (*1*) In both there is some dependence upon the structural integrity of the membrane-bound system; (*2*) they show the same specificity for ADP as phosphate acceptor (310) and require similar concentrations of Mg^{++} ions (52); (*3*) a number of inhibitors show similar effects on the two systems, although these effects vary widely with extracts from different bacterial species. The phosphorylating system of *Mycobacterium phlei* is uncoupled by 2,4-dinitrophenol, dicumarol, and thyroxine in concentrations rather similar to those which uncouple mitochondrial oxidative phosphorylation (308,311,312). But extracts from several other species require higher concentrations of DNP for uncoupling or are only partially uncoupled by it, and some systems are insensitive to uncoupling by DNP. Dinitrophenol can be bound by substances in the soluble fraction of *Azotobacter vinelandii* (311), and when the soluble fraction of the *Azotobacter* extract is replaced by that from *M. phlei* (which is sensitive to DNP) the *Azotobacter* system becomes sensitive to DNP (311). Thus in *Azotobacter* extracts the lack of uncoupling by DNP can be explained by the unavailability of the latter to the phosphorylating system. It is not known whether the lack of sensitivity (or low sensitivity) of extracts of other bacteria to uncouplers or inhibitors is a result of the binding of these compounds by other components of the cell-free extract. The respiration of two species of bacteria is inhibited by very low concentrations of substances which act as uncouplers of mitochondrial oxidative phosphorylation (136,313). The explanation of this puzzling observation is not known.

Phosphorylation linked to respiration is catalyzed by the insoluble fraction of some broken-cell extracts, even with particles of diameter approximately 1% of that of liver mitochondria. Sometimes particles of different size from the same extract show different $P:O$ ratios (62). However, in many cases respiration of the particulate fraction or phosphorylation or both is stimulated by the addition of the soluble supernatant fraction. The factors involved in this stimulatory effect seem to be removed from the membrane fragments into solution with relative ease. When the membrane fragments derived from sonication of protoplasts of *Micrococcus lysodeikticus* in low sucrose concentration are "shocked" osmotically by suspension in cold water, a factor required for phosphorylation goes into the soluble fraction and can be partially purified (52). The respiratory chain remains with the particulate fraction and is not affected by the soluble factor. The incubation of insoluble particles from *Azotobacter* extracts in media of low salt concentration resulted in the loss of a factor (or factors) into solution which, in the presence of Mg^{++} ions, could activate oxidative phosphorylation in particles so treated (314,315). The stimulating factors from several species appear to be heat labile and non-dialyzable (52,128,314); the one from *Azotobacter*, but not the one from *M. lysodeikticus*, could be replaced to some extent by serum albumin (314). The ease of dissociating factors which appear to be involved in phosphorylation reactions was one of the early promising aspects of working with the bacterial systems.

Extensive reconstitution studies have been made with soluble factors and particulate preparations from a few bacteria. Pinchot was able to separate cell-free extracts of *Alcaligenes faecalis* into three fractions

required for oxidative phosphorylation: (*1*) cytochrome-containing particles which catalyze electron transport without phosphorylation; (*2*) a
heat-labile factor; and (*3*) an RNA-like polymer, probably a tetranucleo-
tide. Reconstitution experiments with these fractions indicated that (*2*) is
bound to (*1*) in the presence of (*3*) and Mg^{++} ions (316–321). The interesting observation made with this system is that the soluble heat-labile
factor appears to be dissociated during electron transport in the presence
of DPNH in the form of a high-energy compound (postulated to be a
DPN-protein compound) which yields ATP when incubated with ADP
and inorganic phosphate (322). The heat-labile factor is then postulated
to reassociate with the particles when the high-energy bond is cleaved.
Labeled DPN was shown to be displaced from the protein factor in
amounts roughly equal to the ATP formed (323), and Mg^{++} ions were
required in the formation of ATP from the high-energy intermediate (324).
One difficulty in working with extracts of *A. faecalis* is the presence of
polynucleotide phosphorylase, which catalyzes a reversible reaction
between ADP (and other dinucleotides) and an RNA-like polymer and
inorganic phosphate. Although some evidence has been given to rule out
its involvement in oxidative phosphorylation (322), Pandit-Hovenkamp
found that most (90%) of the ^{32}P of the ATP formed in the presence of
^{32}P-labeled inorganic phosphate was not in the terminal phosphate and
was thus formed by an ADP–P_i exchange reaction, possibly by the polynucleotide phosphorylase (324a). The fractionated system from *M.
lysodeikticus* is essentially free of the polynucleotide phosphorylase, thus
eliminating this problem (52).

Extensive reconstitution studies have also been made with extracts
from *Mycobacterium phlei*, which yield higher P:O ratios than those
found with extracts from most other bacteria (325). A soluble fraction
contains factors which promote both respiration and phosphorylation
when added to the particulate fraction (91,311). So far these factors have
resisted separation, but some purification steps showed that a factor in
the heated supernatant fraction could be replaced by vitamin K_1 (the
bacteria actually contain vitamin K_9H). Further, irradiation of the
preparations to decompose naphthoquinones inactivated oxidative
phosphorylation, but it could be restored on addition of vitamin K_1
when DPNH, but not succinate, was the substrate being oxidized (326).
Fractions active in reconstituting oxidative phosphorylation were found
to have menadione reductase activity (311). This led to the conclusion
that the naphthoquinone is an "essential coenzyme for oxidation and
phosphorylation" (312) and to a scheme suggesting oxidation-reduction
reactions of vitamin K leading to an intermediate which could form a

high-energy phosphate compound (327). The scheme, which is similar to one proposed by Vilkas and Lederer (328), is largely based on the isolation of a naphthochromanol from a reaction mixture containing added vitamin K_1 and the subsequent reactions of this compound (327). It is discussed further in the chapter by Crane. Scott found little chemical evidence for this kind of mechanism in examinations of the chemical properties of similar compounds (329).

Quinones have also been implicated in oxidative phosphorylation in other species (128,135,146), on the basis of depletion and restoration experiments. The limitations of this type of observation have been discussed above, including the lack of specificity seen with some reactions stimulated by quinones. The "coupling factor" released from osmotically shocked fragments from *M. lysodeikticus* protoplasts, unlike the stimulatory factor from the *M. phlei* soluble fraction, contained no menadione reductase activity (52). The phosphorylating systems of *M. lysodeikticus* and *A. faecalis* are actually uncoupled on addition of menadione (52, 330). There are striking differences in the systems that are derived from extracts of different bacterial species; the significance of these differences is unknown.

ATPase and the P_i–ATP and ADP–ATP exchange reactions are believed to be expressions of reversible partial reactions coupling ATP synthesis to the high-energy intermediate formed in the oxidation-reduction reactions (see the chapter by Boyer). These activities are also found in bacterial extracts (52,74,117,126,310,319), although the distribution in the cell fractions is sometimes puzzling. The rates of the partial reactions of the particulate fraction from *M. lysodeikticus* were only a fraction (1/5 to 1/25) of the rate of overall oxidative phosphorylation. Then when the phosphorylating particles were subjected to osmotic shock, the enzymes catalyzing the partial reactions were lost from the particles into solution and the ATPase and the ADP–ATP exchange activities increased ten- to twenty-fold (52).

An interesting hypothesis concerning the mechanism of oxidative phosphorylation which is strikingly different from the more usual one assuming a series of high-energy intermediates evolved as an extension of Mitchell's studies on membrane structure and function in bacteria. Like his suggestions for transport of substances across the cell membrane (Section V), his mechanism for oxidative phosphorylation is also based on the concept of group translocation in membrane-bound enzyme systems with "supramolecular organization" (331,332). The concept of "chemiosmotic coupling" postulates an anisotropic reversible ATPase in the membrane accessible to OH^-, but not to H^+, from one side of the membrane and the reverse on the other side, the membrane being relatively impermeable to these ions. The membrane-bound electron-transfer system is postulated to be localized so that H^+ ions are generated on one side and OH^- ions on the other, creating a gradient of electrochemical activity of these ions. This would induce the specific translocation of OH^- and H^+ ions from the active center of the ATPase system, bringing about the dehydration of ADP and P_i to ATP. Mitchell's mechanism is diagrammed in Figures 1, 2, and 3 of reference 332. This mechanism offers an explanation for a number of observations on phosphorylating systems, such as the uncoupling accompanying changes in structure of the membranes and the existence of "loosely coupled" systems which lack respiratory control but have relatively normal P:O ratios. The mechanism requires that all phosphorylating preparations are closed membrane systems and is difficult to approach experimentally.

Recently, two other reactions characteristic of the mammalian oxidative phosphorylation system have been reported in bacterial preparations: reversed electron-transport (297,333) and an energy-dependent transhydrogenase (334).

In spite of the differences in cellular structure, the phosphorylating system of bacteria seems to have properties the same as those associated with the system of mammalian mitochondria. The apparent ease of dissociation of some of the factors involved in the phosphorylating reactions in bacteria has not so far led to much additional insight into the nature of the reactions involved. And the presence of other oxidative and phosphorylating enzymes in broken-cell extracts of bacteria have made experiments with these extracts difficult to interpret. Finally, the differences observed with extracts from different bacterial species are puzzling if similar mechanisms are assumed in all.

A number of bacterial species contain an enzyme which will catalyze the formation of polyphosphate from ATP. In some the reaction is readily reversible (335,336), but in others it is not (337). Bacteria containing the latter also synthesize a polyphosphatase which degrades polyphosphate to inorganic phosphate (338). Accumulation of polyphosphate is seen (localized in electron-opaque granules) when an ample supply of energy is available but growth is prevented by nitrogen starvation (337,339). The formation of the polyphosphate-containing granules is inhibited by concentrations of o-dinitrophenol which prevent formation of ATP in other organisms (337). Polyphosphate could act as a source of ATP, or as a store of phosphate in organisms where the polyphosphate does not break down to yield ATP (5,337).

VII. Photosynthetic Bacteria

Some photosynthetic bacteria can obtain energy for growth from either respiration or light-induced reactions; others can only grow with illumination. Cytochromes can be oxidized on illumination of either type (340–342), implicating the usual type of electron transport through a cytochrome chain as the energy-producing reaction. (Similar postulates have been made concerning photosynthesis in green plants; see Arnon, this volume.) Cytochromes have been isolated and purified from the photosynthetic bacteria (2,201,212,343), and the content of cytochromes may be greater when the cells are grown photosynthetically than when grown with oxygen in the dark (149). Cell-free extracts of photosynthetic bacteria catalyze rapid light-induced phosphorylation of ADP to ATP, the enzymes involved being localized in the "chromatophore fraction" probably derived from disintegration of intracellular membranes (334,345). A c-type cytochrome is observed to be oxidized when anaerobic broken-cell extracts of *Rhodospirillum rubrum* are illuminated under conditions where phosphorylation of ADP can proceed (346).

Broken-cell extracts of *R. rubrum*, an organism which can grow either anaerobically in light or aerobically in the dark, also synthesize ATP in the dark with oxygen, but at a much lower rate than the light-induced synthesis and with a low P:O ratio (306,346,347). The latter appears to result from disruptive changes during cell breakage (see Section VI). In these facultative heterotrophes there is evidence for two electron-transport chains, one being the respiratory chain system which can pass electrons to either oxygen or an oxidant produced on illumination, the other being activated only on illumination (346). This is seen in the oxidation of at least one more cytochrome on illumination than on oxygenation of anaerobic bacteria (348). Also, dark respiration and light-induced phosphorylation of ADP in cell-free extracts show different properties, such as different sensitivity to inhibitors (346). The two systems have sometimes been assumed to form parts of one large cyclic pathway (349). However, it was recently observed that additional ATP can be produced on illumination after the cells have been oxygenated, suggesting not only two pathways for ATP formation, but also different localization within the cells (306).

The dark respiratory chain system of *R. rubrum* contains b- and c-type cytochromes (348) and the oxidase appears to be an o-type cytochrome (86,350). When the bacteria are grown in the dark with aeration, a large part (86) of the c-type cytochrome, but not all (348), remains reduced in the presence of oxygen. At least part of the c cytochrome has the properties of the high-potential c cytochrome (cytochrome c_2) thought to be involved in light-induced reactions; it thus remains nonfunctional in the dark-grown cells (86). The localization of this cytochrome within the intact cell is not known.

The interesting suggestion has been made by Bose and Gest (351) that high-energy intermediates synthesized in light-induced reactions in *R. rubrum* drive the reduction of DPN by reversed electron transport (see Klingenberg's chapter). The evidence for this hypothesis is based, in part, on inhibition of light-dependent reductions by inhibitors of photophosphorylation.

VIII. Summary

This chapter has perhaps put undue emphasis on apparent ways in which the bacterial respiratory chain system differs from that of mammalian mitochondria. This is because some of the differences give a new point of view or a new experimental approach to an understanding of all such systems. One of the most promising characteristics of the bacterial systems is the possibility presented of systematically varying the relative proportions of the members. There is little need to reiterate the recurrent theme of the chapter, that nearly every aspect of work with the bacterial systems offers potentialities for obtaining new and different information about the functioning of these complex multicomponent systems. It was with the hope that some readers will be attracted to these experiments that this chapter was written.

References

1. L. Smith, *Bacteriol. Rev.*, **18**, 106 (1954).
2. W. M. Clark, N. O. Kaplan, and M. D. Kamen, *Bacteriol. Rev.*, **19**, 234 (1955).
3. M. I. Dolin, in *The Bacteria*, Vol. 2, I. C. Gunsalus and R. Y. Stanier, Eds., Academic Press, New York, 1961, p. 319.
4. L. Smith, in *The Bacteria*, Vol. 2, I. C. Gunsalus and R. Y. Stanier, Eds., Academic Press, New York, 1961, p. 365.
5. M. Alexander, *Bacteriol. Rev.*, **20**, 67 (1956).
6. P. Mitchell, *Ann. Rev. Microbiol.*, **13**, 407 (1959).
7. A. G. Marr, *Ann. Rev. Microbiol.*, **14**, 241 (1960).
7a. M. I. Dolin, in *The Bacteria*, Vol. 2, I. C. Gunsalus and R. Y. Stanier, Eds., Academic Press, New York, 1961, p. 425.

8. I. C. Gunsalus and R. Y. Stanier, Eds., *The Bacteria*, Vol. 1, Academic Press, New York, 1961, Chapters 2, 3, and 6.
9. E. M. Brieger, *Structure and Ultrastructure of Microorganisms*, Academic Press, New York, 1963.
10. "Bacterial Anatomy," *Symp. Soc. Gen. Microbiol.*, **6**, 1, 111, 150, 296, 318 (1956).
11. H. H. Martin, *J. Theoret. Biol.*, **5**, 1 (1963).
12. C. Weibull, *Ann. Rev. Microbiol.*, **12**, 1 (1958).
13. P. C. Fitz-James, *Symp. Soc. Gen. Microbiol.*, **15**, 369 (1965).
14. M. Beer, *J. Bacteriol.*, **80**, 659 (1960).
15. J. R. G. Bradfield, *Symp. Soc. Gen. Microbiol.*, **6**, 296 (1956).
16. G. B. Chapman and J. Hillier, *J. Bacteriol.*, **66**, 362 (1953).
17. E. Kellenberger and A. Ryter, *J. Biophys. Biochem. Cytol.*, **4**, 323 (1958).
18. S. F. Conti and M. E. Gettner, *J. Bacteriol.*, **83**, 544 (1962).
19. W. van Iterson, *J. Biophys. Biochem. Cytol.*, **9**, 183 (1961).
20. A. M. Glauert, E. M. Brieger, and J. M. Allen, *Exptl. Cell. Res.*, **22**, 73 (1961).
21. G. B. Chapman and A. J. Kroll, *J. Bacteriol.*, **73**, 63 (1957).
22. D. Abram, *J. Bacteriol.*, **89**, 855 (1965).
23. M. R. J. Salton and J. A. Chapman, *J. Ultrastruct. Res.*, **6**, 489 (1962).
24. C. Shinohara, K. Fukushi, and I. Suzuki, *J. Bacteriol.*, **74**, 413 (1957).
25. G. B. Chapman, J. H. Hanks, and J. H. Wallace, *J. Bacteriol.*, **77**, 205 (1959).
26. N. A. Lukoyanova, N. S. Gel'man, and V. I. Biryuzova, *Biokhimiya*, **26**, 786 (1961).
27. S. A. Robrish, and A. G. Marr, *J. Bacteriol.*, **83**, 158 (1962).
28. J. Pangborn, A. G. Marr, and S. A. Robrish, *J. Bacteriol.*, **84**, 669 (1962).
29. E. H. Cota-Robles, A. G. Marr, and E. Nilson, *J. Bacteriol.*, **75**, 243 (1958).
30. H. Davson and J. F. Danielli, *The Permeability of Natural Membranes*, Cambridge University Press, Cambridge, 1952.
31. H. A. Bladen, M. U. Nylen, and R. J. Fitzgerald, *J. Bacteriol.*, **88**, 763 (1964).
32. H. Fernandez-Moran, T. Oda, P. V. Blair, and D. E. Green, *J. Cell Biol.*, **22**, 63 (1964).
33. D. F. Parsons, *Science*, **140**, 985 (1963).
34. W. Stoeckenius, *J. Cell Biol.*, **17**, 443 (1963).
35. K. McQuillen, *Symp. Soc. Gen. Microbiol.*, **6**, 127 (1956).
36. P. Mitchell, and J. Moyle, *Symp. Soc. Gen. Microbiol.*, **6**, 150 (1956).
37. C. Weibull, *J. Bacteriol.*, **66**, 688, 696 (1953).
38. C. Weibull, *Exptl. Cell Res.*, **9**, 294 (1955).
39. J. T. Wachsman and R. Storck, *J. Bacteriol.*, **80**, 600 (1960).
40. C. Weibull, *Exptl. Cell Res.*, **10**, 214 (1956).
41. C. Weibull and L. Bergstrom, *Biochim. Biophys. Acta*, **30**, 340 (1958).
42. P. Mitchell and J. Moyle, *Biochem. J.*, **16**, 184 (1957).
43. L. Smith, *Biochim. Biophys. Acta*, **62**, 145 (1962).
44. J. de Ley and R. Dochy, *Biochim. Biophys. Acta*, **40**, 277 (1960).
45. M. R. J. Salton, in *The Bacteria*, Vol. 1, I. C. Gunsalus and R. Y. Stanier, Eds., Academic Press, New York, 1960, p. 115.
46. A. L. Hunt, A. Rodgers, and D. E. Hughes, *Biochim. Biophys. Acta*, **34**, 354 (1959).
47. W. R. Sistrom, *Biochim. Biophys. Acta*, **29**, 579 (1958).
48. S. E. Burrous and W. A. Wood, *J. Bacteriol.*, **84**, 364 (1962).

49. D. Schlessinger, *J. Mol. Biol.*, **7**, 569 (1963).
50. C. Weibull, H. Beckman, and L. Bergstrom, *J. Gen. Microbiol.*, **20**, 519 (1959).
51. R. Storck and J. T. Wachsman, *J. Bacteriol.*, **73**, 784 (1957).
52. S. Ishikawa and A. L. Lehninger, *J. Biol. Chem.*, **237**, 2401 (1962).
53. H. W. Milner, N. S. Lawrence, and C. S. French, *Science*, **111**, 633 (1950).
54. A. G. Marr and E. H. Cota-Robles, *J. Bacteriol.*, **74**, 79 (1957).
55. M. M. Mathews and W. R. Sistrom, *J. Bacteriol.*, **78**, 778 (1959).
56. E. Vanderwinkel and R. C. E. Murray, *J. Ultrastruct. Res.*, **1**, 185 (1962).
57. W. van Iterson and W. Leene, *J. Cell Biol.*, **20**, 361 (1964).
58. M. J. O. Francis, D. E. Hughes, H. L. Kornberg, and P. J. R. Phizackerly, *Biochem. J.*, **89**, 430 (1963).
59. N. B. Madsen, *Can. J. Biochem. Physiol.*, **38**, 481 (1960).
60. T. G. G. Wilson and P. W. Wilson, *J. Bacteriol.*, **70**, 30 (1955).
61. R. G. Eagon and A. K. Williams, *Arch. Biochem. Biophys.*, **79**, 401 (1959).
62. A. Tissières, H. G. Hovenkamp, and E. C. Slater, *Biochim. Biophys. Acta*, **25**, 336 (1957).
63. P. M. Nossal, D. B. Keech, and D. J. Morton, *Biochim. Biophys. Acta*, **22**, 412 (1956).
64. P. Mitchell, and J. Moyle, *Biochem. J.*, **64**, 19P (1956).
65. H. K. Schachman, A. B. Pardee, and R. Y. Stanier, *Arch. Biochem. Biophys.*, **38**, 245 (1952).
66. R. J. Downey, C. E. Georgi, and W. E. Militzer, *J. Bacteriol.*, **83**, 1140 (1962).
67. R. E. Asnis, V. G. Vely, and M. C. Glick, *J. Bacteriol.*, **72**, 314 (1956).
68. R. Doi and H. O. Halvorson, *J. Bacteriol.*, **81**, 51 (1960).
69. W. Feldman and D. J. O'Kane, *J. Bacteriol.*, **80**, 218 (1960).
70. J. de Ley and J. Schel, *Biochim. Biophys. Acta*, **35**, 154 (1959).
71. R. G. Eagon, *Can. J. Microbiol.*, **4**, 1 (1958).
72. R. Repaske, *J. Bacteriol.*, **68**, 555 (1954).
73. I. C. Gunsalus, C. F. Gunsalus, and R. Y. Stanier, *J. Bacteriol.*, **66**, 543 (1953).
74. P. E. Hartman, A. F. Brodie, and C. T. Gray, *J. Bacteriol.*, **74**, 319 (1957).
75. D. S. Goldman, M. J. Wagner, T. Oda, and A. L. Shug, *Biochim. Biophys. Acta*, **73**, 367 (1963).
76. P. Seikivitz and G. E. Palade, *J. Biochem. Biophys. Cytol.*, **4**, 309 (1958).
77. A. Tissières, in *Haematin Enzymes*, J. E. Falk, R. Lemberg, and R. K. Morton, Eds., Pergamon Press, London, 1961, p. 218.
78. T. P. Singer, in *The Enzymes*, Vol. 7, P. D. Boyer, H. Lardy, and K. Myrbäck, Eds., Academic Press, New York, 1963, p. 345.
79. B. Chance, *Science*, **120**, 767 (1954).
80. B. Chance, in *Methods of Enzymology*, Vol. 4, S. P. Colowick and N. O. Kaplan, Eds., Academic Press, New York, 1957, p. 273.
81. P. Broberg, unpublished experiments.
82. R. Repaske, J. Shroat, and D. Allman, *J. Bacteriol.*, **79**, 394 (1960).
83. N. S. Gel'man, M. A. Lukoyanova, and A. I. Oparin, *Biokhimiya*, **25**, 368 (1960).
84. A. Tissières, *Nature*, **169**, 880 (1952).
85. P. A. Trudinger, *Biochem. J.*, **78**, 673 (1961).
86. S. Taniguchi and M. D. Kamen, *Biochim. Biophys. Acta*, **96**, 395 (1965).
87. R. Hancock, *J. Gen. Microbiol.*, **25**, 429 (1961).
88. A. W. Linnane and J. L. Still, *Biochim. Biophys. Acta*, **16**, 305 (1955).
89. C. Widmer, T. E. King, and V. H. Cheldelin, *J. Bacteriol.*, **71**, 737 (1956).

90. M. M. Weber, A. F. Brodie, and J. E. Merselis, *Science*, **128**, 896 (1958).
91. A. F. Brodie and C. T. Gray, *Science*, **125**, 534 (1957).
92. W. Heinen, M. Kusunose, E. Kusunose, D. S. Goldman, and M. J. Wagner, *Arch. Biochem. Biophys.*, **104**, 448 (1964).
93. D. C. White, *J. Biol. Chem.*, **239**, 2055 (1964).
94. W. Militzer, T. B. Sonderegger, L. C. Tuttle, and C. E. Georgi, *Arch. Biochem.*, **26**, 299 (1950).
95. L. Smith, unpublished observations.
96. E. E. Jacobs and D. R. Sanadi, *J. Biol. Chem.*, **235**, 531 (1960).
97. L. P. Vernon and J. H. Mangum, *Arch. Biochem. Biophys.*, **90**, 103 (1960).
98. W. D. Butt and H. Lees, *Nature*, **182**, 732 (1958).
99. L. Smith and D. C. White, *J. Biol. Chem.*, **237**, 1337 (1962).
100. P. A. Straat and A. Nason, *J. Biol. Chem.*, **240**, 1412 (1965).
101. F. R. Williams and L. P. Hager, *Biochim. Biophys. Acta*, **38**, 566 (1960).
102. S. S. Deeb and L. P. Hager, *J. Biol. Chem.*, **239**, 1024 (1964).
103. P. W. Camerino and L. Smith, *J. Biol. Chem.*, **239**, 2345 (1964).
104. B. Chance, *J. Biol. Chem.*, **233**, 1223 (1958).
105. B. A. Blaylock and A. Nason, *J. Biol. Chem.*, **238**, 3453 (1963).
106. A. Asano and A. F. Brodie, *J. Biol. Chem.*, **239**, 4280 (1964).
107. M. Somlo, *Biochim. Biophys. Acta*, **65**, 333 (1962).
108. H. Watari, E. B. Kearney, T. P. Singer, D. Basunski, J. Hauber, and C. J. Lusty, *J. Biol. Chem.*, **237**, PC 1731 (1962).
109. A. R. Gilbey and A. V. Few, *Biochim. Biophys. Acta*, **29**, 21 (1958).
110. M. Alexander and P. W. Wilson, *Proc. Natl. Acad. Sci. U.S.*, **41**, 843 (1955).
111. J. H. Bruemmer, P. W. Wilson, J. L. Glenn, and F. L. Crane, *J. Bacteriol.*, **73**, 113 (1957).
112. M. Kates, *Advan. Lipid Res.*, **2**, 17 (1964).
113. M. D. Yudkin, *Biochem. J.*, **82**, 40P (1962).
114. T. Kaneshiro and A. G. Marr, *J. Lipid Res.*, **3**, 184 (1962).
115. C. K. Huston, P. W. Albro, and G. B. Grindley, *J. Bacteriol.*, **89**, 768 (1965).
116. J. W. Brown, *Biochim. Biophys. Acta*, **52**, 368 (1961).
117. D. Hughes, *J. Gen. Physiol.*, **29**, 39 (1962).
118. G. N. Godson, G. D. Hunter, and J. A. V. Butler, *Biochem. J.*, **81**, 59 (1961).
119. J. A. V. Butler and G. N. Godson, *Biochem. J.*, **88**, 176 (1963).
120. S. Spiegelman, A. I. Aronson, and P. C. Fitz-James, *J. Bacteriol.*, **75**, 102 (1958).
121. A. L. Hunt, *Biochem. J.*, **69**, 2P (1958).
122. C. Widmer, T. E. King, and V. H. Cheldelin, *J. Bacteriol.*, **71**, 737 (1956).
123. D. Henneman and W. W. Umbreit, *J. Bacteriol.*, **87**, 1266 (1964).
124. L. Packer and M. Perry, *Arch. Biochem. Biophys.*, **95**, 379 (1961).
125. D. Henneman and W. W. Umbreit, *J. Bacteriol.*, **87**, 1274 (1964).
126. A. F. Brodie and C. T. Gray, *J. Biol. Chem.*, **219**, 853 (1956).
127. H. G. Hovenkamp, *Biochim. Biophys. Acta*, **34**, 485 (1959).
128. E. R. Kashket and A. F. Brodie, *Biochim. Biophys. Acta*, **78**, 52 (1963).
129. A. Tissières, *Biochem. J.*, **64**, 582 (1956).
130. L. Smith, *Arch. Biochem. Biophys.*, **50**, 299 (1954).
131. R. W. Estabrook, in *Haematin Enzymes*, J. E. Falk, R. Lemberg, and R. K. Morton, Eds., Pergamon Press, London, 1961, p. 436.
132. E. Azoulay, *Biochim. Biophys. Acta*, **92**, 458 (1964).
133. P. Chaix and J. F. Petit, *Biochim. Biophys. Acta*, **25**, 481 (1957).

134. J. Jensen and E. Thofern, *Z. Naturforsch.*, **8b**, 599 (1953).
135. M. Richmond and N. O. Kjeldgaard, *Acta Chem. Scand.*, **15**, 266 (1961).
136. D. C. White and L. Smith, *J. Biol. Chem.*, **239**, 2956 (1964).
137. A. W. Linnane and C. W. Wrigley, *Biochim. Biophys. Acta*, **77**, 408 (1963).
138. A. Tissières, *Biochem. J.*, **50**, 279 (1951).
139. W. S. Waring and C. H. Werkman, *Arch. Biochem.*, **4**, 75 (1944).
140. D. H. L. Bishop, K. P. Panda, and H. K. King, *Biochem. J.*, **83**, 606 (1962).
141. J. H. Fellman and R. C. Mills, *J. Bacteriol.*, **79**, 800 (1960).
142. D. C. White, *J. Bacteriol.*, **85**, 84 (1963).
143. C. T. Gray, J. W. T. Wimpenny, and D. E. Hughes, *Biochim. Biophys. Acta*, **67**, 157 (1963).
144. P. Schaffer, *Biochim. Biophys. Acta*, **9**, 261 (1952).
145. E. Englesberg, J. B. Levy, and A. Gibor, *J. Bacteriol.*, **68**, 178 (1954).
146. R. J. Downey, *J. Bacteriol.*, **88**, 904 (1964).
147. L. P. Vernon, *J. Biol. Chem.*, **222**, 1045 (1956).
148. D. C. White, *J. Biol. Chem.*, **238**, 3757 (1963).
149. R. J. Porra and J. Lascelles, *Biochem. J.*, **94**, 120 (1965).
150. F. Moss, *Australian J. Exptl. Biol. Med. Sci.*, **30**, 531 (1952).
151. D. C. White, *J. Bacteriol.*, **83**, 851 (1962).
152. S. Mitzushima and K. Arima, *J. Biochem. Tokyo*, **47**, 837 (1960).
153. F. Moss, *Australian J. Exptl. Biol. Med. Sci.*, **34**, 395 (1956).
154. H. Lenhoff and N. O. Kaplan, *Nature*, **172**, 730 (1953).
155. H. Lenhoff, D. J. D. Nicholas, and N. O. Kaplan, *J. Biol. Chem.*, **220**, 967 (1956).
156. N. J. Jacobs and S. F. Conti, *J. Bacteriol.*, **89**, 675 (1965).
157. N. J. Jacobs, R. E. Heady, and J. M. Jacobs, *J. Bacteriol.*, **87**, 1406 (1964).
158. E. R. Kashket and A. F. Bodie, *Biochim. Biophys. Acta*, **40**, 550 (1960).
159. P. Chaix and G. Roncoli, *Biochim. Biophys. Acta*, **6**, 268 (1950).
160. P. Chaix and J. F. Petit, *Biochim. Biophys. Acta*, **22**, 66 (1956).
161. L. N. Castor, and B. Chance, *J. Biol. Chem.*, **234**, 1587 (1959).
162. D. C. White and L. Smith, *J. Biol. Chem.*, **237**, 1332 (1962).
163. J. P. Aubert, J. Millet, and G. Milhaud, *Ann. Inst. Pasteur*, **96**, 559 (1959).
164. S. Yamagutchi, *Acta Phytochim. (Japan)*, **8**, 263 (1935).
165. L. Smith and P. W. Camerino, *Biochemistry*, **2**, 1432 (1963).
166. M. Richardson, *J. Bacteriol.*, **74**, 699 (1957).
167. T. Yamanaka and K. Okunuki, *Biochim. Biophys. Acta*, **67**, 379 (1963).
168. L. P. Vernon and F. G. White, *Biochim. Biophys. Acta*, **25**, 321 (1957).
169. B. Chance, L. Smith, and L. N. Castor, *Biochim. Biophys. Acta*, **12**, 289 (1953).
170. L. N. Castor and B. Chance, *J. Biol. Chem.*, **217**, 453 (1955).
171. L. N. Castor and B. Chance, *J. Biol. Chem.*, **234**, 1587 (1959).
172. B. Chance, *J. Biol. Chem.*, **202**, 383, 397 (1953).
173. H. W. Taber and M. Morrison, *Arch. Biochem. Biophys.*, **105**, 367 (1964).
174. L. Smith, *J. Biol. Chem.*, **215**, 847 (1955).
175. W. A. Rawlinson and J. H. Hale, *Biochem. J.*, **45**, 247 (1949).
176. H. F. Holden, and R. Lemberg, *Australian J. Expl. Biol. Med. Sci.*, **17**, 133 (1939).
177. T. Yamanaka and K. Okunuki, *Biochim. Biophys. Acta*, **67**, 394 (1963).
178. J. Barrett, *Biochim. J.*, **64**, 626 (1956).
179. T. Yamanaka, S. Kijimoto, K. Okunuki, and K. Kusai, *Nature*, **194**, 759 (1962).

180. I. S. Longmuir, *Biochem. J.*, **57**, 81 (1954).
181. G. D. Ludwig and S. A. Kuby, *Federation Proc.*, **14**, 247 (1955).
182. D. C. White, *J. Biol. Chem.*, **238**, 3757 (1963).
183. I. S. Longmuir and B. J. Clarke, *Biochem. J.*, **63**, 57 (1956).
184. L. P. Vernon, *J. Biol. Chem.*, **222**, 1035 (1956).
185. P. A. Trudinger, *Biochem. J.*, **78**, 680 (1961).
186. R. Repaske and J. J. Josten, *J. Biol. Chem.*, **233**, 466 (1958).
187. M. D. Kamen and L. P. Vernon, *Biochim. Biophys. Acta*, **17**, 10 (1955).
188. E. C. Layne and A. Nason, *J. Biol. Chem.*, **231**, 889 (1958).
189. A. Tissières and R. H. Burris, *Biochim. Biophys. Acta*, **20**, 436 (1956).
190. N. Newton, unpublished experiments.
191. R. S. Pokallus and D. Pramer, *Arch. Biochem. Biophys.*, **105**, 208 (1964).
192. T. Yamanaka and K. Okunuki, *J. Biol. Chem.*, **239**, 1813 (1964).
193. T. Horio, T. Higashi, T. Yamanaka, M. Matsubara, and K. Okunuki, *J. Biol. Chem.*, **236**, 944 (1961).
194. T. Yamanaka, *Nature*, **204**, 253 (1964).
195. T. Yamanaka, S. Kijimoto, and K. Okunuki, *J. Biochem. Tokyo*, **53**, 416 (1963).
196. T. Yamanaka and K. Okunuki, *Biochem. Z.*, **338**, 62 (1963).
197. T. Yamanaka and K. Okunuki, *Biochim. Biophys. Acta*, **59**, 755 (1962).
198. I. W. Sutherland, *Biochim. Biophys. Acta*, **73**, 162 (1963).
199. T. Horio, *J. Biochem. Tokyo*, **45**, 195 (1958).
200. R. P. Ambler, *Biochem. J.*, **89**, 341 (1963).
201. L. P. Vernon, *J. Biol. Chem.*, **222**, 1045 (1956).
202. J. Postgate, *J. Gen. Microbiol.*, **14**, 545 (1956).
203. H. Lenhoff, and N. O. Kaplan, *J. Biol. Chem.*, **220**, 967 (1956).
204. T. Horio, T. Higashi, M. Sasagawa, K. Kusai, M. Nakai, and K. Okunuki, *Biochem. J.*, **77**, 194 (1960).
205. T. Horio, T. Higashi, M. Nakai, K. Kusai, and K. Okunuki, *Nature*, **182**, 1307 (1958).
206. M. L. Koval, T. Horio, and M. D. Kamen, *Biochim. Biophys. Acta*, **52**, 241 (1961).
207. R. P. Ambler, *Biochem. J.*, **89**, 349 (1963).
208. T. Horio, *J. Biochem. Tokyo*, **45**, 267 (1958).
209. T. Yamanaka, K. Miki, and K. Okunuki, *Biochim. Biophys. Acta*, **77**, 654 (1963).
210. T. Fujita and R. Sato, *Biochim. Biophys. Acta*, **77**, 690 (1963).
211. Y. Birk, W. S. Silver, and A. H. Heim, *Biochim. Biophys. Acta*, **25**, 227 (1957).
212. J. A. Orlando and T. Horio, *Biochim. Biophys. Acta*, **50**, 367 (1961).
213. F. L. Jackson and V. O. Lawton, *Biochim. Biophys. Acta*, **35**, 76 (1959).
214. D. Keilin and C. H. Harpley, *Biochem. J.*, **35**, 688 (1941).
215. J. G. Hauge, *Arch. Biochem. Biophys.*, **94**, 308 (1961).
216. E. Itagaki and L. P. Hager, *Federation Proc.*, **24**, 545 (1965).
217. H. Walker and R. G. Eagon, *J. Bacteriol.*, **88**, 25 (1964).
218. M. Benziman and Y. Galanter, *J. Bacteriol.*, **88**, 1010 (1964).
219. J. G. Hauge and E. H. Murer, *Biochim. Biophys. Acta*, **81**, 251 (1964).
220. F. J. S. Lara, *Biochim. Biophys. Acta*, **33**, 565 (1959).
221. J. G. Hauge, *Biochim. Biophys. Acta*, **45**, 250 (1960).
222. E. Itagaki, T. Fujita, and R. Sato, *J. Biochem. Tokyo*, **52**, 131 (1962).
223. T. Nakayama, *J. Biochem. Tokyo*, **49**, 240 (1961).
224. I. C. Gunsalus, C. F. Gunsalus, and R. Y. Stanier, *J. Bacteriol.*, **66**, 538 (1953).

225. T. Kimura and J. Tobari, *Biochim. Biophys. Acta*, **73**, 399 (1963).
226. M. Kogut and J. W. Lightbown, *Biochem. J.*, **73**, 15P (1959).
227. J. W. Lightbown and M. Kogut, *Biochem. J.*, **73**, 14P (1959).
227a. T. P. Singer, in *Oxidases and Related Redox Systems*, Vol. 1, T. E. King, H. S. Mason, and M. Morrison, Eds., Wiley, New York, 1965, p. 448.
228. R. W. Wheat, J. Rust, and S. J. Ajl, *J. Cellular Comp. Physiol.*, **47**, 317 (1956).
229. T. E. King and V. H. Cheldelin, *J. Biol. Chem.*, **224**, 579 (1957).
230. W. A. Wood and R. F. Schwerdt, *J. Biol. Chem.*, **201**, 501 (1953).
231. J. G. Hauge, *Biochim. Biophys. Acta*, **45**, 263 (1960).
232. J. G. Hauge and E. H. Mürer, *Biochim. Biophys. Acta*, **81**, 244 (1964).
233. M. Szymona and M. Doudoroff, *J. Gen. Microbiol.*, **22**, 167 (1960).
234. J. G. Hauge, *J. Bacteriol.*, **82**, 609 (1961).
235. J. G. Hauge, *J. Biol. Chem.*, **239**, 3630 (1964).
236. P. R. Dugan and D. C. Lundegren, *J. Bacteriol.*, **89**, 825 (1965).
237. G. Milhaud, J. P. Aubert, and J. Millet, *Comp. Rend.*, **246**, 1766 (1958).
238. H. D. Peck and E. Fisher, *J. Biol. Chem.*, **237**, 190 (1962).
239. H. D. Peck, T. E. Deacon, and J. T. Davidson, *Biochim. Biophys. Acta*, **96**, 429 (1965).
240. L. Kiesow, *Proc. Natl. Acad. Sci. U.S.*, **52**, 980 (1964).
241. M. I. Dolin, *J. Biol. Chem.*, **238**, PC4109 (1963).
242. M. M. Weber, G. Rosso, and H. Noll, *Biochim. Biophys. Acta*, **71**, 355 (1963).
243. D. C. White, *J. Bacteriol.*, **89**, 299 (1965).
244. A. Temperli and P. W. Wilson, *Nature*, **193**, 171 (1962).
245. E. Itagaki, *J. Biochem. Tokyo*, **55**, 432 (1964).
246. D. C. White, *J. Biol. Chem.*, **240**, 1387 (1965).
247. P. J. Van Demark and P. F. Smith, *J. Bacteriol.*, **88**, 122 (1964).
248. M. M. Weber, A. F. Brodie, and J. E. Merselis, *Science*, **128**, 896 (1958).
249. M. Benziman and L. Perez, *Biochem. Biophys. Res. Commun.*, **19**, 127 (1965).
250. P. D. Bragg, *J. Bacteriol.*, **88**, 1019 (1964).
251. M. M. Weber and G. Rosso, *Proc. Natl. Acad. Sci. U.S.*, **50**, 710 (1963).
252. E. R. Kashket and A. F. Brodie, *J. Bacteriol.*, **83**, 1094 (1962).
253. D. S. Goldman, M. J. Wagner, T. Oda, and A. L. Shug, *Biochim. Biophys. Acta*, **73**, 391 (1963).
254. H. D. Peck and E. Fisher, *J. Biol. Chem.*, **237**, 198 (1962).
255. A. Asano, T. Kaneshiro, and A. F. Brodie, *J. Biol. Chem.*, **240**, 895 (1965).
256. M. J. Cormier and J. R. Totter, *J. Am. Chem. Soc.*, **76**, 4744 (1954).
257. P. D. Bragg, *Biochim. Biophys. Acta*, **96**, 263 (1965).
258. M. M. Weber, T. Hollocher, and G. Rosso, *J. Biol. Chem.*, **240**, 1776 (1965).
259. M. M. Weber, T. Hollocher, and G. Rosso, *J. Biol. Chem.*, **240**, 1783 (1965).
260. S. Taniguchi, R. Sato, and F. Egami, *McCollum-Pratt Symposium on Inorganic Nitrogen Metabolism*, W. D. McElroy and B. Glass, Eds., Johns Hopkins Press, Baltimore, 1956, p. 87.
261. R. Sato, *McCollum-Pratt Symposium on Inorganic Nitrogen Metabolism*, W. D. McElroy and B. Glass, Eds., Johns Hopkins Press, Baltimore, 1956, p. 163.
262. S. Taniguchi and E. Itagaki, *Biochim. Biophys. Acta*, **44**, 263 (1960).
263. E. Itagaki, T. Fujita, and R. Sato, *Biochim. Biophys. Acta*, **51**, 390 (1961).
264. T. Onishi, *J. Biochem. Tokyo*, **53**, 71 (1963).
265. K. Hori, *J. Biochem. Tokyo*, **53**, 354 (1963).
266. C. A. Fewson and D. J. D. Nicholas, *Biochim. Biophys.*, *Acta*, **49**, 335 (1961).

267. J. C. Sadana and W. D. McElroy, *Arch. Biochem. Biophys.*, **67**, 16 (1957).
268. J. P. Chang and J. Lascelles, *Biochem. J.*, **89**, 503 (1963).
269. A. Ota, T. Yamanaka, and K. Okunuki, *J. Biochem. Tokyo*, **55**, 131 (1964).
270. S. D. Wainwright, *Biochim. Biophys. Acta*, **18**, 583 (1955).
271. W. Joklik, *Australian J. Sci. Res.*, **B3**, 28 (1950).
272. C. M. Gilmour, R. P. Bhatt, and J. V. Mayeux, *Nature*, **203**, 55 (1964).
273. V. B. D. Skerman and I. C. MacRae, *Can. J. Microbiol.*, **3**, 215 (1957).
274. V. B. D. Skerman and I. C. MacRae, *Can. J. Microbiol.*, **3**, 505 (1957).
275. F. Pinchinoty and L. D'Ornana, *Nature*, **191**, 879 (1961).
276. G. C. Walker and D. J. D. Nicholas, *Biochim. Biophys. Acta*, **49**, 350 (1961).
277. F. Pinchinoty and J. C. Senez, *Comp. Rend. Soc. Biol.*, **150**, 744 (1956).
278. M. Ishimoto, J. Koyama, and Y. Nagai, *J. Biochem. Tokyo*, **41**, 763 (1954).
278a. A. Asano, *J. Biochem. Tokyo*, **46**, 781 (1959).
279. F. Egami, M. Ishimoto, and S. Taniguchi, in *Haematin Enzymes*, J. E. Falk, R. Lemberg, and R. K. Morton, Eds., Pergamon Press, London, 1961, p. 392.
280. D. C. White, M. P. Bryant, and D. R. Caldwell, *J. Bacteriol.*, **84**, 822 (1962).
281. D. J. D. Nicholas, P. W. Wilson, W. Heinen, G. Palmer, and H. Beinert, *Nature*, **196**, 433 (1962).
282. Y. I. Shethna, P. W. Wilson, R. E. Hansen, and H. Beinert, *Proc. Natl. Acad. Sci. U.S.*, **52**, 1263 (1964).
283. D. Keilin and E. F. Hartree, *Proc. Roy. Soc. (London)*, **B127**, 167 (1939).
284. J. W. Lightbown and F. L. Jackson, *Biochem. J.*, **63**, 130 (1956).
285. F. L. Jackson and J. W. Lightbown, *Biochem. J.*, **69**, 63 (1958).
286. L. Packer, *Arch. Biochem. Biophys.*, **78**, 54 (1958).
287. A. M. Pappenheimer and E. D. Hendee, *J. Biol. Chem.*, **180**, 597 (1949).
288. L. Smith and K. Minnaert, *Biochim. Biophys. Acta*, **105**, 1 (1965).
289. P. V. Blair, T. Oda, D. E. Green, and H. Fernandez-Moran, *Biochemistry*, **2**, 756 (1963).
290. D. E. Green, D. C. Wharton, A. Tzagoloff, J. S. Rieske, and G. B. Brierley, in *Oxidases and Related Redox Systems*, T. E. King, H. S. Mason, and M. Morrison, Eds., Wiley, New York, 1965.
291. L. Smith and N. Newton, in *Oxidases and Related Redox Systems*, T. E. King, H. S. Mason, and M. Morrison, Eds., Wiley, New York, 1965.
292. A. Asano and A. F. Brodie, *Biochem. Biophys. Res. Commun.*, **13**, 416 (1963).
293. H. S. Moyed and D. J. O'Kane, *J. Biol. Chem.*, **218**, 831 (1956).
294. A. Kepes and J. Monod, *Comp. Rend.*, **244**, 1550 (1957).
295. A. Kepes, *Biochim. Biophys. Acta*, **40**, 70 (1960).
296. H. E. Swim and H. Gest, *J. Bacteriol.*, **68**, 755 (1954).
297. E. R. Kashket and A. F. Brodie, *J. Biol. Chem.*, **238**, 2564 (1963).
298. R. E. Marquis and P. Gerhardt, *J. Biol. Chem.*, **239**, 3361 (1964).
299. C. R. Bovell and L. Packer, *Biochem. Biophys. Res. Commun.*, **13**, 435 (1963).
300. C. R. Bovell, L. Packer, and R. Helgerson, *Biochim. Biophys. Acta*, **75**, 257 (1963).
301. P. Mitchell, *Nature*, **180**, 134 (1957).
302. P. Mitchell and J. Moyle, *Biochem. J.*, **72**, 21P (1959).
303. P. Mitchell and J. Moyle, *J. Gen. Microbiol.*, **21**, iii (1959).
304. L. Klungsöyr, T. E. King, and V. H. Cheldelin, *J. Biol. Chem.*, **227**, 135 (1957).
305. A. H. Stouthamer, *Biochim. Biophys. Acta*, **56**, 19 (1962).
306. L. Smith and J. Ramirez, *Federation Proc.*, **24**, 609 (1965).

307. A. Asano and A. F. Brodie, *Biochem. Biophys. Res. Commun.*, **19**, 121 (1965).
308. A. F. Brodie, M. M. Weber, and C. T. Gray, *Biochim. Biophys. Acta*, **25**, 448 (1957).
309. C. R. Bovell, L. Packer, and G. R. Schonbaum, *Arch. Biochem. Biophys.*, **104**, 458 (1964).
310. I. A. Rose and S. Ochoa, *J. Biol. Chem.*, **220**, 307 (1956).
311. A. F. Brodie, *J. Biol. Chem.*, **234**, 398 (1959).
312. A. F. Brodie and J. Ballantine, *J. Biol. Chem.*, **235**, 226 (1960).
313. Y. Avi-Dor, *Acta. Chem. Scand.*, **17**, 144 (1963).
314. H. G. Hovenkamp, *Biochim. Biophys. Acta*, **34**, 485 (1959).
315. H. G. Pandit-Hovenkamp, *Abstr. 6th Intern. Congr. Biochem., New York, 1964*, X, p. 785.
316. G. B. Pinchot, *Biochem. Biophys. Res. Commun.*, **1**, 17 (1959).
317. S. Shibko and G. B. Pinchot, *Arch. Biochem. Biophys.*, **93**, 140 (1961).
318. S. Shibko and G. B. Pinchot, *Arch. Biochem. Biophys.*, **94**, 257 (1961).
319. G. B. Pinchot, *J. Biol. Chem.*, **205**, 65 (1953).
320. G. P. Pinchot, *J. Am. Chem. Soc.*, **77**, 5763 (1955).
321. G. B. Pinchot, *J. Biol. Chem.*, **229**, 1 (1957).
322. G. B. Pinchot, *Proc. Natl. Acad. Sci. U.S.*, **46**, 929 (1960).
323. G. B. Pinchot and M. Hormanski, *Proc. Natl. Acad. Sci. U.S.*, **48**, 1970 (1962).
324. J. J. Scocca and G. B. Pinchot, *Biochim. Biophys. Acta*, **71**, 193 (1963).
324a. H. G. Pandit-Hovenkamp, *Biochim. Biophys. Acta*, **99**, 552 (1965).
325. A. F. Brodie, and C. T. Gray, *Biochim. Biophys. Acta*, **17**, 146 (1955).
326. A. F. Brodie and J. Ballantine, *J. Biol. Chem.*, **235**, 226 (1960).
327. P. J. Russell and A. F. Brodie, *Biochim. Biophys. Acta*, **50**, 76 (1961).
328. M. Vilkas and E. Lederer, *Experientia*, **18**, 546 (1962).
329. P. M. Scott, *J. Biol. Chem.*, **240**, 1374 (1965).
330. G. B. Pinchot, *Biochim. Biophys. Acta*, **23**, 660 (1957).
331. P. Mitchell, *Biochem. J.*, **79**, 23P (1961).
332. P. Mitchell, *Nature*, **191**, 144 (1961).
333. M. I. H. Aleem, H. Lees, and D. J. D. Nicholas, *Nature*, **200**, 759 (1963).
334. S. Murthy and A. F. Brodie, *J. Biol. Chem.*, **239**, 4292 (1964).
335. J. A. Cole and D. E. Hughes, *J. Gen. Microbiol.*, **38**, 65 (1965).
336. A. Kornberg, S. R. Kornberg, and E. S. Sims, *Biochim. Biophys. Acta*, **20**, 15 (1956).
337. D. E. Hughes and A. Muhammed, *Colloq. Intern. Centre Natl. Rech. Sci. (Paris)*, **106**, 591 (1962).
338. A. Muhammed, S. Rodgers, and D. E. Hughes, *J. Gen. Microbiol.*, **20**, 482 (1959).
339. D. E. Hughes, S. F. Conti, and R. C. Fuller, *J. Bacteriol.*, **85**, 577 (1963).
340. L. N. M. Duysens, *Nature*, **173**, 692 (1954).
341. B. Chance and L. Smith, *Nature*, **175**, 803 (1955).
342. J. M. Olson and B. Chance, *Arch. Biochem. Biophys.*, **88**, 26 (1960).
343. J. W. Newton and M. D. Kamen, *Biochim. Biophys. Acta*, **21**, 71 (1956).
344. G. Cohen-Bazire, in *Bacterial Photosynthesis*, H. Gest, A. San Pietro, and L. P. Vernon, Eds., Antioch Press, Yellow Springs, Ohio, 1963, p. 89.
345. R. C. Fuller, S. F. Conti, and D. B. Mellin, in *Bacterial Photosynthesis*, H. Gest, A. San Pietro, and L. P. Vernon, Eds., Antioch Press, Yellow Springs, Ohio, 1963, p. 71.

346. L. Smith and M. Baltscheffsky, *J. Biol. Chem.*, **234**, 1575 (1959).
347. D. M. Geller, in *Bacterial Photosynthesis*, H. Gest, A. San Pietro, and L. P. Vernon, Eds., Antioch Press, Yellow Springs, Ohio, 1963, p. 161.
348. L. Smith and J. Ramirez, *Arch. Biochem. Biophys.*, **79**, 233 (1959).
349. M. Nishimura and B. Chance, *Biochim. Biophys. Acta*, **66**, 1 (1963).
350. T. Horio and C. P. S. Taylor, *J. Biol. Chem.*, **240**, 1772 (1965).
351. S. K. Bose and H. Gest, *Proc. Natl. Acad. Sci. U.S.*, **49**, 337 (1963).
352. E. Margoliash and A. Schejter, *Advan. Protein Res.*, **21**, 113 (1966).
353. P. B. Scholes, unpublished experiments.
354. A. Ryter and D. Kellenberger, *Z. Naturforsch.*, **13b**, 597 (1958).

Electron Transport and Phosphorylation in Photosynthesis by Chloroplasts*

DANIEL I. ARNON, *Department of Cell Physiology, University of California, Berkeley, California*

	Introduction	124
I.	Introduction	124
II.	Energy Requirements for Carbon Assimilation	125
III.	Photosynthetic Capacity of Isolated Chloroplasts	128
	A. Separation of Enzymes of Carbon Assimilation from Chlorophyll Pigments	130
	B. Separation of Light and Dark Phases of Photosynthesis in Chloroplasts	130
IV.	Chloroplasts and the Early Reactions of Photosynthesis	131
V.	Discovery of Photophosphorylation	132
VI.	Concept of Cyclic Photophosphorylation	134
VII.	Noncyclic Photophosphorylation	136
VIII.	Ferredoxins in Bacteria and Green Plants	137
IX.	Definition of Ferredoxin	138
X.	Spectral Characteristics and Oxidation-Reduction Potentials	139
XI.	Some Chemical Properties of Ferredoxin	142
XII.	Ferredoxin and NADP Reduction	143
	A. Mechanism of NADP Reduction of Chloroplasts	143
XIII.	Photoreduction of Ferredoxin Coupled with Photoproduction of Oxygen	145
XIV.	Noncyclic Photophosphorylation with Ferredoxin	146
XV.	Cyclic Photophosphorylation with Ferredoxin	147
XVI.	Pseudocyclic Photophosphorylation with Ferredoxin	149
XVII.	Two Light Reactions in Photosynthesis: Observations on Whole Cells	150
XVIII.	One or Two Light Reactions in Noncyclic Photophosphorylation?	152
	A. Noncyclic Photophosphorylation without Oxygen Evolution	153

* This chapter is based on an article, "The Photosynthetic Activity of Isolated Chloroplasts," in *Physiological Reviews* (in press, 1966). The investigations from the writer's laboratory that are discussed herein have been aided by grants from the National Institutes of Health, the Office of Naval Research, and the Charles F. Kettering Foundation.

XIX.   Mechanisms of Cyclic and Noncyclic Photophosphorylation   156
XX.   The Hypothesis of Cyclic and Noncyclic Photophosphorylation
       as the Two Light Reactions in Chloroplasts   159
XXI.   Quenching of Chloroplast Fluorescence by Cyclic and Non-
       cyclic Photophosphorylation   163
XXII. A Copper Protein and Quinones in the Photosynthetic
       Apparatus   165
XXIII. Concluding Remarks.   165
       References   166

*Are not gross Bodies and Light convertible into one another,
and may not Bodies receive much of their activity from the
Particles of Light which enter their Composition? . . . The
Changing of Bodies into Light, and Light into Bodies, is
very conformable to the Course of Nature, which seems
delighted with Transmutations.*

Sir Isaac Newton, *Opticks* (1721)

I. Introduction

Photosynthesis may be broadly defined as the utilization of solar energy
by green plants and certain green and purple bacteria for the synthesis of
organic carbon compounds. The dominant form of photosynthesis on our
planet is plant photosynthesis, encountered in unicellular and multi-
cellular green plants, in which carbon dioxide is the source of carbon and
water is the reductant that provides the electrons and protons that constitute the hydrogen atoms needed for biosynthesis. The main products of
plant photosynthesis are carbohydrates with oxygen gas as the important
by-product.

Green and purple bacteria, and a few species of algae (1), carry on a
"bacterial" type of photosynthesis, which differs from plant photosynthesis in that oxygen is never produced. In bacterial photosynthesis
the source of carbon may be carbon dioxide, but often it is a simple
organic compound, acetate for example. The reductant in bacterial
photosynthesis is not water (hence no oxygen evolution) but reduced
sulfur compounds, exogenous hydrogen gas, or hydrogen derived from
decomposition of organic compounds. The main products of bacterial
photosynthesis are not carbohydrates but amino acids, proteins and
polymers of certain fatty acids.

The central problem of photosynthesis is the identification of those
photochemical reactions that provide the energy required for the endergonic assimilation of carbon dioxide or other carbon donors into cellular

constituents. Our knowledge of this phase of photosynthesis, with which we will be concerned in this article, came mainly from investigations of plant photosynthesis at the subcellular level. The literature pertaining to this subject has grown so large that no comprehensive survey of it will be undertaken here. In this article the author will supplement a selective discussion of the literature with a unified treatment of the work from his laboratory on the photochemical reactions of chloroplasts. In addition, a new hypothesis that is now being actively tested will be presented as an alternative to the older hypotheses on the nature of the photochemical reactions of chloroplasts. Before dealing with reactions of isolated chloroplasts, we will consider briefly their relevance to photosynthesis at the cellular level.

II. Energy Requirements for Carbon Assimilation

In the 1950s it became evident that the speculative proposals of Thimann (2), Lipmann (3), and Ruben (4) about CO_2 assimilation in photosynthesis had a basis in fact. Their suggestions that CO_2 reduction in photosynthesis is a reversal of the well-known oxidation reaction of glycolysis—namely, the reduction of 1,3-diphosphoglyceric acid to glyceraldehyde-3-phosphate —received experimental support from the work of Calvin and his associates (5), who identified 3-phosphoglyceric acid and other intermediate products of glycolysis among the early products of photosynthesis. Calvin et al. (5) proposed a photosynthetic carbon cycle—a cyclic sequence of reactions for assimilation of CO_2 in which the only reductive step is the reduction of 1,3-diphosphoglyceric acid to glyceraldehyde-3-phosphate.

The uniqueness of the photosynthetic carbon cycle lay not in its reductive step, which, as already mentioned, was a reversal of a well-known reaction of glycolysis, but in the manner in which 3-phosphoglyceric acid is formed as the first stable product of CO_2 fixation. Work in the laboratories of Calvin (5), Horecker et al. (6), Ochoa et al. (7), and Racker (8) established the presence in photosynthetic tissues of two new enzymes, phosphoribulokinase and ribulose diphosphate carboxylase (carboxydismutase), which jointly accounted for the entry of CO_2 into the metabolism of photosynthetic cells through a reaction with a five-carbon phosphorylated sugar, ribulose 1,5-diphosphate. Phosphoribulokinase catalyzes the phosphorylation of ribulose-5-phosphate into ribulose 1,5-diphosphate:

$$\text{Ribulose-5-P} + \text{ATP} \xrightarrow{\text{Mg}^{++}} \text{ribulose-di-P} + \text{ADP} \qquad (1)$$

Ribulose diphosphate carboxylase catalyzes the reaction in which ribulose 1,5-diphosphate, on combining with CO_2, is split to give two molecules of 3-phosphoglyceric acid (PGA):

$$\text{Ribulose-di-P} + CO_2 \longrightarrow 2PGA \tag{2}$$

However, even these two seemingly unique features of carbon assimilation in photosynthesis were soon found in nonphotosynthetic bacteria as well (see review, 9). For example, Trudinger (10) and Aubert et al. (11) found the entire photosynthetic or, as it has been called by Racker (8), the reductive pentose phosphate cycle, in the nonphotosynthetic sulfur bacterium *Thiobacillus denitrificans*. It thus became clear that CO_2 assimilation by way of the reductive pentose phosphate cycle is only indirectly dependent on the photochemical phase of photosynthesis, since all the reactions of the photosynthetic carbon cycle occur also in nonchlorophyllous cells.

An entirely different concept of photosynthesis has been put forward by Warburg (12), who questions the characterization of 3-phosphoglyceric acid as the first stable product of CO_2 assimilation. Warburg considers that the reductive step in CO_2 assimilation is not the reduction of 1,3-diphosphoglycerate by $NADPH_2$ but the deoxygenation by light of an "activated carbon dioxide" that is bound to chlorophyll. The activated carbon dioxide "$(H_2CO_3)^*$" is reduced, in a photochemical reaction that requires only one quantum of light, to a carbohydrate, "(H_2CO)" (Equation (3)). Ordinary carbon dioxide, in a hydrated form, is activated in the dark (Equation (5)) at the expense of energy that is liberated by another dark reaction in which two-thirds of the carbohydrate, formed by the light reaction, is oxidized by molecular oxygen (Equation (4)). Warburg implicates NADP and a respiratory chain of hydrogen carriers (13) in the oxidation of the newly formed carbohydrate (Equation (4)) and phosphorylations and glutamic acid in the activation of carbon dioxide (Equation (5)). The overall reaction (Equation (6) multiplied by 3) results in the assimilation of 1 mole of carbon dioxide at the expense of 3 light quanta:*

$$(H_2CO_3)^* + 1N_0h\nu = (H_2CO) + O_2 \tag{3}$$

$$\tfrac{2}{3}(H_2CO) + \tfrac{2}{3}O_2 = \tfrac{2}{3}H_2CO_3 + 76 \text{ kcal} \tag{4}$$

$$H_2CO_3 + 76 \text{ kcal} = (H_2CO_3)^* \tag{5}$$

$$\cdot \;\; \cdot \;\; \cdot \;\; \cdot \;\; \cdot \;\; \cdot \;\; \cdot \;\; \cdot \;\; \cdot \;\; \cdot \;\; \cdot \;\; \cdot$$

$$\text{Sum} \qquad \tfrac{1}{3}H_2CO_3 + 1N_0h\nu = \tfrac{1}{3}(H_2CO) + \tfrac{1}{3}O_2 \tag{6}$$

* The extensive literature on quantum efficiency in photosynthesis, which is beyond the scope of this article, remains divided into two categories. One supports the concept that the quantum requirement is close to the theoretical limit of three (142); the other supports the concept that it approaches eight (E. Rabinowitch, *Photosynthesis and Related Processes*, Vol. II, Part 1, Interscience, New York, 1951).

Despite the great contributions of Warburg to many branches of biochemistry, including photosynthesis, his concept of the mechanism of carbon dioxide assimilation has not gained wide acceptance among other investigators in this field. One reason is the lack of isolation and chemical characterization of the "active carbon dioxide" (H_2CO_3)* and of the first carbohydrate (H_2CO) formed. On the other hand, neither can the prevailing concept of carbon dioxide assimilation in photosynthesis, the reductive pentose phosphate cycle, be considered as well established in all respects. To consider only the enzymic apparatus involved, a serious difficulty is the low affinity of ribulose diphosphate carboxylase for carbon dioxide (8)—a difficulty that is particularly noteworthy because photosynthesis in nature operates at low partial pressures of carbon dioxide. Another problem is that the activities of several enzymes of the reductive pentose phosphate cycle, ribulose diphosphate carboxylase, fructose diphosphatase, sedoheptulose diphosphatase, and transaldolase, "are not high enough to satisfy the rate of overall CO_2 fixation observed in intact cells" (14). Despite these and other objections (15), the reductive pentose phosphate cycle is widely considered to account for carbon assimilation in photosynthesis because it provides for a cyclic regeneration of a CO_2 acceptor and includes some known enzyme systems and chemical intermediates that also precede the formation of carbohydrates in nonphotosynthetic cells.

The energy requirement (assimilatory power) for the reductive pentose phosphate cycle is well defined. First, as shown in Equation (1), the formation of ribulose 1,5-diphosphate from ribulose-5-phosphate requires ATP; and second, the reduction of 3-phosphoglycerate to form the first carbohydrate (glyceraldehyde-3-phosphate) requires both ATP and reduced pyridine nucleotide (Equations (7) and (8)).

$$\text{3-PGA + ATP} \xrightarrow{\text{phosphoglycerate kinase}} \text{1,3-diPGA} \qquad (7)$$

$$\text{1,3-diPGA + NADPH}_2 \xrightleftharpoons{\text{glyceraldehyde-3-phosphate dehydrogenase}} \text{glyceraldehyde-3-P} + \text{NADP} + \text{H}_3\text{PO}_4 \qquad (8)$$

Thus, the distinction between carbon assimilation in green plants and carbon assimilation in nonphotosynthetic cells would lie in the manner in which they generate the assimilatory power, which consists of ATP and reduced pyridine nucleotide. Nonphotosynthetic cells form ATP and reduced pyridine nucleotide at the expense of chemical energy, whereas plant cells form them at the expense of radiant energy.

The process by which green plants use radiant energy to generate ATP by several different reactions is now known as photosynthetic phosphorylation. Photosynthetic phosphorylation (photophosphorylation) is subdivided into cyclic photophosphorylation, a reaction which produces only ATP, and noncyclic photophosphorylation, in which the production of ATP is stoichiometrically coupled with electron transfer that results in oxygen evolution and the reduction of ferredoxin, an iron-containing protein native to chloroplasts. Reduced ferredoxin in turn leads to the reduction of NADP. The discovery and characterization of these partial reactions of photosynthesis came not from investigations with intact cells but with isolated chloroplasts. It will be useful, therefore, to begin our discussion of photophosphorylation with a discussion of the overall photosynthetic capacity of isolated chloroplasts.

III. Photosynthetic Capacity of Isolated Chloroplasts

Biochemical research on photosynthesis by isolated chloroplasts rests on the premise that in photosynthesis, as was the case earlier in fermentation and respiration, the elucidation of the constituent reactions and their mechanisms would most likely come when the process is reconstructed outside the intact cell. Since, within photosynthetic cells, chloroplasts contain all the photosynthetic pigments and were observed, almost a century ago (16–18), to produce starch and oxygen on illumination, it was thought for many years that photosynthesis in green plants begins and ends in chloroplasts.

It is often difficult for the student of photosynthesis today to realize that this view was never supported by critical experimental evidence and was largely abandoned after Hill (19,20) showed that isolated chloroplasts could evolve oxygen but could not assimilate carbon dioxide. In this reaction, which became known as the Hill reaction, isolated chloroplasts evolved oxygen when carbon dioxide was replaced by ferric oxalate or, as found later by Warburg (21), by other nonphysiological electron acceptors, benzoquinone and ferricyanide. The ability of isolated chloroplasts to assimilate CO_2 was reinvestigated when the very sensitive $C^{14}O_2$ technique became available, but here again the results were negative (22,23). Such $C^{14}O_2$ fixation as was observed was limited in scope. Thus, Fager (24,25) found no fixation of $C^{14}O_2$ by chloroplasts but by a protein preparation ("enzyme") from spinach leaves. The fixation of CO_2 by the "enzyme" was enhanced in the presence of the illuminated chloroplast preparation but did not proceed beyond phosphoglycerate. There was no evidence for a reductive assimilation of CO_2 to the level of carbohydrate.

Without proof that isolated chloroplasts were the site of total photoassimilation of carbon dioxide, photosynthesis in the early 1950s came to be regarded, like fermentation in the days of Pasteur, as a process that cannot be separated from the structural and functional complexity of whole cells.* Nevertheless, the possibility remained that the observed restricted photosynthetic capacity of isolated chloroplasts was merely a consequence of inappropriate experimental methods that were used in different laboratories, including that of the writer. This proved to be the case. By changing their experimental methods, Arnon, Allen, and Whatley found in 1954 that isolated spinach chloroplasts, unaided by other cellular particles or enzyme systems, reduced CO_2 to the level of carbohydrates, including starch, with a simultaneous evolution of oxygen at physiological temperatures and with no energy supply except visible light (28–30). By using the new experimental methods, or modifications thereof, the conversion of $C^{14}O_2$ to phosphorylated sugars and starch by isolated chloroplasts was confirmed and extended in other laboratories (31–35) (compare also 36–38).

Like most cellular processes that are reconstructed *in vitro*,† the rates of this extracellular CO_2 assimilation were low. Nevertheless, they provided reproducible biochemical evidence which has finally documented the frequently asserted, but never before proved, thesis that chloroplasts are the cytoplasmic structures in which the complete photosynthetic process takes place.

---

* To illustrate: In 1953, Rabinowitch (26) wrote that "the task of separating it [photosynthesis] from other life processes in the cell and analyzing it into its essential chemical reactions has proved to be more difficult than was anticipated. The photosynthetic process, like certain other groups of reactions in living cells, seems to be bound to the structure of the cell; it cannot be repeated outside that structure." In a review in 1954, Lumry et al. (27) summarized the many investigations with isolated chloroplasts as pointing to the conclusion that the chloroplast was "a system much simpler than that required for photosynthesis," and was the site of only "the light-absorbing and water-splitting reactions of the over-all photosynthetic process."

† It may be interesting to recall the earliest and the latest event of this kind in biochemistry. The historic significance of Eduard Buchner's demonstration of fermentation by a yeast extract (*Les Prix Nobel en 1907*), an event which ushered in the era of modern enzymology, was minimized on the grounds that his cell-free fermentation constituted only 1.6–4.6% of the fermentation activity of a corresponding quantity of intact yeast cells. (See, for example, M. Rubner, *Die Ernährungsphysiologie der Hefezelle bei al Koholischer Gärung*, Veit, Leipzig, 1913, p. 59.) In recent times, the achievement of cell-free synthesis of DNA was thus described by Arthur Kornberg (*Les Prix Nobel en 1959*): "The first positive results represented the conversion of only a very small fraction of the acid-soluble substrate into an acid-insoluble fraction (50 or so counts out of a million added)."

A. Separation of Enzymes of Carbon Assimilation from Chlorophyll Pigments

The ability of isolated chloroplasts to fix CO_2 was almost completely lost when they were treated with a dilute salt solution or with water (39). This treatment disrupted the whole chloroplasts and gave "broken chloroplasts" or "grana"—the water-insoluble fraction containing the chlorophyll pigments and capable of carrying out oxygen evolution—and a straw-colored chloroplast extract (CE) which contained the water-soluble CO_2-fixing enzymes that were subsequently separated and characterized (40,41).

Grana alone lacked the requisite enzymes for CO_2 assimilation (39). The capacity for CO_2 assimilation was restored by recombining grana with the chloroplast extract. The ease with which the soluble CO_2-fixing enzymes and, presumably, other soluble components are lost into an aqueous phase during the isolation of chloroplasts explained their low rates of CO_2 assimilation and also many of the previous failures to observe any CO_2 fixation at all by isolated chloroplasts.

B. Separation of Light and Dark Phases of Photosynthesis in Chloroplasts

The reconstituted chloroplast system (grana plus CE) required ATP and $NADPH_2$ for CO_2 assimilation (39). It appeared likely that in isolated chloroplasts the light phase of photosynthesis was limited to the formation of ATP and $NADPH_2$ and that these two substances were the first stable, chemically identifiable intermediates that were formed at the expense of radiant energy and were then used to drive the dark enzymic reactions of CO_2 assimilation.

Direct experimental support for this view was obtained by separating, in time and space, the light and dark phases of photosynthesis in chloroplasts (42). The light phase was completed first without adding CO_2 to a reaction mixture, which contained substrate amounts of NADP, ADP, and inorganic phosphate; illuminated chloroplasts converted these into substrate amounts of $NADPH_2$ and ATP (with an accompanying evolution of oxygen). The chloroplasts were then fractionated and the green portion (grana) discarded. Next, CO_2 was supplied to the remaining water-soluble chloroplast extract in the dark, was assimilated at the expense of previously formed ATP and $NADPH_2$, and yielded the same phosphorylated sugars and intermediate products in the dark as in the light (42). The same CO_2 assimilation by the enzymes in the chloroplast extract also occurred at the expense of exogenous ATP and $NADPH_2$ of nonphotochemical origin (42).

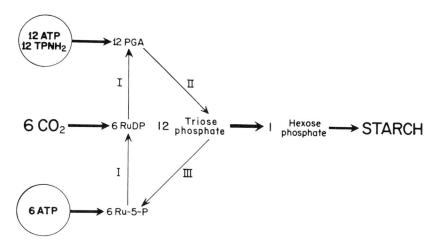

Fig. 1.   Condensed diagram of the reductive pentose phosphate cycle. The cycle consists of three phases. In the carboxylative phase (I), ribulose-5-phosphate (Ru-5-P) is phosphorylated to ribulose diphosphate (RuDP) which then accepts a molecule of CO_2 and is cleaved to 2 molecules of phosphoglyceric acid (PGA); in the reductive phase (II) PGA is reduced to triose phosphate; in the regenerative phase (III) triose phosphate is converted partly into Ru-5-P and partly into hexose phosphate and starch. The reactions of the carboxylative and reductive phases are driven by ATP and $NADPH_2$ ($TPNH_2$) formed via cyclic and noncyclic photophosphorylation, but otherwise all the reactions of the cycle are independent of light. One complete turn of the cycle results in the assimilation of 1 mole of CO_2 at the expense of 3 moles of ATP and 2 moles of $NADPH_2$.

The individual steps and enzymes of CO_2 assimilation by isolated chloroplasts were next identified and isolated (40,41) and found to be in good agreement with the work of other investigators on CO_2 assimilation by photosynthetic and nonphotosynthetic cells (5,9). A general scheme for CO_2 assimilation in whole cells and by isolated chloroplasts is diagrammatically summarized in Figure 1.

IV. Chloroplasts and the Early Reactions of Photosynthesis

Once the complete photosynthetic capacity of isolated chloroplasts was experimentally established, it was possible to concentrate on them rather than on whole cells in the search for those photochemical reactions which generate the first chemically defined, energy-rich products that are formed *prior* to the conversion of CO_2 into organic compounds. It was clear that these products must include ATP and a reductant with a reducing power at least equal to that of $NADPH_2$. The advantages of isolated chloroplasts for investigations of this aspect of photosynthesis are substantial. Chloroplasts cannot respire; they lack the terminal respiration enzyme, cytochrome oxidase (43–45). This feature insures that the early products of photosynthesis in chloroplasts would not be confused with intermediate products (including ATP) of respiration—a possibility that cannot be excluded with certainty in intact cells and which, therefore, has been the subject of much controversy in research on photosynthesis. Furthermore, isolated and fragmented chloroplasts, unlike whole cells, do not have permeability barriers to the entry of such key intermediates as ADP, NADP, and, as will be shown later, small protein molecules (ferredoxin). Thus, working with isolated chloroplasts, it is possible to supply these normally catalytic substances in substrate amounts and thereby follow, on a fairly large scale, their conversion, under the influence of light, to energy-rich products. This feature, together with the use of radioactive phosphate, proved to be a great experimental advantage in the discovery of photosynthetic phosphorylation.

V. Discovery of Photophosphorylation

The first experiments with the sensitive P^{32} technique to test the ability of isolated chloroplasts to form, on illumination, ATP gave negative results (46). The most plausible model for ATP formation in photosynthesis became one that envisaged a collaboration between chloroplasts and mitochondria. Chloroplasts would, in that scheme, reduce NAD photochemically, and mitochondria would reoxidize it with oxygen and form ATP via oxidative phosphorylation (47). This model posed a serious physiological problem. Photosynthesis in saturating light can proceed at a rate almost thirty times as great as the rate of respiration. It was difficult to see, therefore, how the respiratory mechanisms of mitochondria could cope with the ATP requirement in photosynthesis.

In 1954, Arnon, Allen, and Whatley (28) discovered a light-induced ATP formation by isolated spinach chloroplasts unaided by mitochondria. This process, which they named photosynthetic phosphorylation (photophosphorylation) to distinguish it from the respiratory (oxidative) phosphorylation by mitochondria, was independent of CO_2 assimilation. ATP was formed under conditions when no CO_2 was supplied to the reaction mixture and the reaction vessels contained KOH in the center well. The possibility cannot be excluded that, even under these conditions, residual CO_2 may have had a catalytic function on the photochemical reaction, as observed by Warburg et al. (48) and Stern and Vennesland (49) for the photoproduction of oxygen by chloroplasts. What can be excluded is that substrate amounts of carbon compound(s) were first synthesized in the light from exogenous CO_2 and were then used as electron donors for the formation of ATP.

Several unique features distinguished this type of photophosphorylation (28,50) from ATP formation in fermentation (substrate level phosphorylation) or in oxidative phosphorylation: (1) ATP was formed only in chlorophyll-containing structures and was independent of any other organelles or enzyme systems; (2) no oxygen was consumed or produced; (3) no energy-rich chemical substrate was consumed, the only source of energy being that of the absorbed photons; (4) ATP formation was not accompanied by a coupled oxidation-reduction involving an external electron donor and acceptor (Equation (9)).

$$n\text{ADP} + n\text{P}_i \xrightarrow{h\nu} n\text{ATP} \qquad (9)$$

The discovery of photosynthetic phosphorylation in chloroplasts was soon followed by evidence of a similar phenomenon in photosynthetic bacteria. Frenkel (51) observed a light-induced formation of ATP in cell-free preparations of *Rhodospirillum rubrum*. The occurrence of photosynthetic phosphorylation was confirmed and extended to cell-free preparations of all types of photosynthetic organisms: in photosynthetic bacteria by Williams (52), Geller and Gregory (53), Kamen and Newton (54), and Anderson and Fuller (55); in algae by Thomas and Haans (56) and Petrack and Lipmann (57); and in isolated chloroplasts by Avron and Jagendorf (58), Wessels (59), and many others. It was thus established that in all photosynthetic cells a major phosphorylation site is always associated with the chlorophyll pigments and supplies, independently of respiration or fermentation, the ATP needed in photosynthesis.

Soon after the demonstration of photosynthetic phosphorylation in isolated chloroplasts, attempts were made to compare its rate with that of CO_2 assimilated by illuminated whole cells. Since, as with most newly discovered cell-free reactions, the rates of photosynthetic phosphorylation were rather low, there was little inclination at first to accord this process quantitative importance (60) as a mechanism for converting light into chemical energy.

With further improvement in experimental methods, Arnon and his colleagues obtained rates of photosynthetic phosphorylation up to 170 times as high (61) as those they originally described (50), and even these high rates were exceeded by Jagendorf and Avron (62). The improved rates of photosynthetic phosphorylation were equal to or greater than the maximum known rates of carbon assimilation in intact leaves. It appeared, therefore, that isolated chloroplasts retain, without substantial loss, the enzymic apparatus for photosynthetic phosphorylation—a conclusion which is in harmony with evidence that the phosphorylating system is tightly bound in the water-insoluble grana portion of the chloroplasts.

When photophosphorylation was first extended from chloroplasts to photosynthetic bacteria, a question arose whether these two processes were fundamentally similar in not requiring a chemical substrate. Frenkel (51) found that photophosphorylation in a cell-free preparation from *Rhodospirillum rubrum* became dependent on a substrate (α-ketoglutarate) when the chlorophyll-containing particles were washed. However, in later experiments, Frenkel (63) and other investigators (54,55,64) found that the role of α-ketoglutarate and other organic acids in the bacterial system was regulatory and not that of a substrate. When this basic point was clarified, the fundamental similarity of photophosphorylation in chloroplasts and in bacterial systems was no longer in doubt.

VI. Concept of Cyclic Photophosphorylation

Once the main facts of photophosphorylation were firmly established, the next objective was to explain its mechanism. All known cellular phosphorylations occur at the expense of free energy liberated during electron transport from a high-energy electron donor to an electron acceptor, but there was no evidence for this in photophosphorylation. Thus, to explain the source of energy for photophosphorylation it was necessary to choose between two alternatives: a special mechanism, unrelated to electron transport and without counterpart in cellular physiology, or a novel type of electron transport that is induced by light but is hidden in the structure of chloroplasts.

After abandoning early attempts to link photophosphorylation with photochemical splitting of water (28,29), Arnon adopted the second alternative (65,66). His hypothesis was that a chlorophyll molecule, on absorbing a quantum of light, becomes excited and promotes an electron to an outer orbital with a higher energy level. This high-energy electron is then transferred to an adjacent electron acceptor molecule with a strongly electronegative oxidation-reduction potential. The transfer of an electron from excited chlorophyll to an adjacent electron acceptor molecule, present in chloroplasts, is the energy conversion step proper. By transforming a flow of photons into a flow of electrons it constitutes a mechanism for generating a strongly electronegative reductant at the expense of the excitation energy of chlorophyll.

Once the strongly electronegative reductant is formed, no further input of energy is needed. Subsequent electron transfers within the chloroplast liberate energy, since they constitute an electron flow from the electronegative reductant to electron acceptors with more electropositive redox potentials (65,66). In the end, the electron originally emitted by the excited chlorophyll molecule returns to the electron-deficient chlorophyll molecule and the quantum absorption process is repeated. Because of the envisaged cyclic pathway traversed by the emitted electron, the process was named *cyclic photophosphorylation* (65,66).

The number of exergonic electron transfer steps that are coupled with phosphorylation is still under investigation, but in theory the electron "cascade" that is now envisaged as operating in cyclic photophosphorylation can liberate free energy that is sufficient for more than one phosphorylation.

The chemical identity of the chloroplast constituent which first accepts an electron from an excited chlorophyll molecule was unknown. This substance was apparently lost during the isolation of chloroplasts since their cyclic photophosphorylation depended on the addition of an exogenous catalyst. What was puzzling was that the function of a catalyst was fulfilled by one of several different substances of a physiological or nonphysiological character. An example of the former is menadione (vitamin K_3) (67) and of the latter, phenazine methosulfate (62). However, as discussed later, recent evidence points to ferredoxin (mentioned already in connection with noncyclic photophosphorylation) as being also the key endogenous electron acceptor in cyclic photophosphorylation by chloroplasts. All the other catalysts now appear to act as substitutes for ferredoxin.

A cyclic electron flow that is driven by light and that liberates chemical energy, used for the synthesis of the pyrophosphate bonds of ATP, is unique to photosynthetic cells. The idea has been discussed elsewhere (65,66) that cyclic photophosphorylation rather than photolysis of water (68) is the common denominator of plant and bacterial photosynthesis and represents, from the point of view of evolution, the most primitive photosynthetic mechanism.

Light-induced electron flow mechanisms in photosynthesis have been postulated at various times [see, for example, Katz (69) and Levitt (70)] but they had little influence on contemporary research. Without valid biochemical evidence for the formation of early photosynthetic products such as ATP that are directly coupled with electron transport, the theoretical proposals for electron flow mechanisms in photosynthesis could not be adequately defended against the theoretical arguments leveled against them. Cyclic photophosphorylation supplied the evidence that photosynthesis includes an energy conversion process related to phosphorus turnover in which free energy released by a light-induced electron flow is trapped in the synthesis of the pyrophosphate bonds to ATP.

VII. Noncyclic Photophosphorylation

In 1957, Arnon, Whatley, and Allen (71) discovered a second type of photophosphorylation which provided the first direct experimental evidence for a coupling between light-induced electron transport and the synthesis of ATP. Here, in contrast to cyclic photophosphorylation, ATP formation was stoichiometrically coupled with a light-driven transfer of electrons from water to NADP (or to a nonphysiological electron acceptor such as ferricyanide) and a concomitant evolution of oxygen. Moreover, ATP formation in this coupled system greatly increased the rate of electron transfer from water to ferricyanide (71–73) or to NADP (74) and the rate of the concomitant oxygen evolution. It thus became apparent that the electron-transport system of chloroplasts functions more effectively when it is coupled, as it would be under physiological conditions, to the synthesis of ATP. The conventional Hill reaction (19,20) could thus be viewed as an electron-transport system that is uncoupled from photophosphorylation (71).

In extending the electron flow concept to this new reaction, Arnon (65,66) envisaged that a chlorophyll molecule excited by a captured photon transfers an electron to NADP (or to ferricyanide). Electrons thus removed from chlorophyll are replaced by electrons from water (OH^- at pH 7) with a resultant evolution of oxygen. In this manner, light induces an electron flow from OH^- to NADP and a coupled phosphorylation. Because of the unidirectional or noncyclic nature of this electron flow, this process was named *noncyclic photophosphorylation* (71,65).

More recent evidence has established that in noncyclic photophosphorylation illuminated chloroplasts do not react directly with NADP but react with ferredoxin (75). Reduced ferredoxin in turn reduces NADP by a mechanism that is independent of light.

Ferredoxin has thus emerged as a key substance in both cyclic and noncyclic photophosphorylation. Since the properties of ferredoxin and its role in the energy conversion process of photosynthesis have only recently been recognized, the pertinent evidence will now be reviewed in some detail.

VIII. Ferredoxins in Bacteria and Green Plants*

Prior to 1961 there was no evidence to challenge the view that chloroplasts, and only chloroplasts, contain a protein factor or an enzyme that catalyzes the photochemical reduction of NADP. But in that year K. Tagawa and M. Nozaki (unpublished data from this laboratory) and Losada et al. (76) isolated a "pyridine nucleotide reductase" from an organism devoid of chloroplasts, the photosynthetic bacterium *Chromatium*. The bacterial protein was able to replace the native chloroplast protein in mediating the photoreduction of NADP and the evolution of oxygen by chloroplasts, although *Chromatium* cells, from which this protein was isolated, are incapable of evolving oxygen in light. This finding indicated that proteins similar to those functioning in the NADP-reducing apparatus of chloroplasts were also present in photosynthetic bacteria devoid of chloroplasts, but this observation acquired a new significance a year later when Tagawa and Arnon (75) obtained the same effect with a protein, ferredoxin, from a nonphotosynthetic organism.

Ferredoxin is the name given by Mortenson et al. (77) to a protein containing iron which is neither a heme protein nor a flavoprotein. Mortenson et al. (77) isolated this protein from *Clostridium pasteurianum*, a nonphotosynthetic anaerobic bacterium which normally lives in the soil without any exposure to light. In this, and in other nonphotosynthetic, obligately anaerobic bacteria where ferredoxin was found, it appeared to

---

* For further discussion, see chapter by San Pietro.

function as a link between the enzyme hydrogenase and different electron donors and acceptors (77,78). Thus, the distribution of ferredoxin seemed likely to be limited to those obligately anaerobic, nonphotosynthetic bacteria that contain an active hydrogenase system. There was nothing to indicate that bacterial ferredoxin was in any way linked with photosynthesis.

It soon became clear, however, that ferredoxin-like proteins are present in all photosynthetic cells and play a key role in the energy transfer mechanisms of photosynthesis (Tagawa and Arnon (75)). It was recognized that, between 1952 and 1960, proteins which we now call ferredoxins had been isolated from chloroplasts of several species of green plants and had been assigned various functions under such different names as "methaemoglobin-reducing factor" (79), "TPN-reducing factor" (80), "photosynthetic pyridine nucleotide reductase" (PPNR) (81), and the "haem-reducing factor" (82). All these terms are now known to be synonymous and have been replaced by the term ferredoxin. It also became apparent that the "red enzyme" isolated in 1962 in Warburg's laboratory is analogous to ferredoxin (83,84). A detailed review of the earlier work and of the evidence which led to the adoption of the ferredoxin nomenclature is given elsewhere (94).

IX. Definition of Ferredoxin

Tagawa and Arnon (75) crystallized ferredoxin from the nonphotosynthetic bacterium *C. pasteurianum* and found that it was also able to replace the native chloroplast protein in the photoreduction of NADP. The same investigation led to other findings: (*a*) The chloroplast protein, like ferredoxin of *C. pasteurianum*, contained iron and was reversibly oxidized and reduced with characteristic changes in its absorption spectrum [the presence of iron in the chloroplast protein ("PPNR") was also independently observed by Fry and San Pietro (85), Horio and Yamashita (86), Katoh and Takamiya (87), and Gewitz and Voelker (84)]. (*b*) Crystalline ferredoxin from *C. pasteurianum* has a remarkably low oxidation-reduction potential ($E_0' = -417$ mV, at pH 7.55), close to the potential of hydrogen gas and about 100 mV more electronegative than the oxidation-reduction potential of pyridine nucleotides. (*c*) The oxidation-reduction potential of the spinach chloroplast protein was also strongly electronegative ($E_0' = -432$ mV, at pH 7.55).

These similarities led Tagawa and Arnon (75) to extend the name ferredoxin to the chloroplast protein and to other iron-containing proteins of photosynthetic cells and anaerobic bacteria that have an oxidation-reduction potential close to that of hydrogen gas and are, at least in part, functionally interchangeable in the photoreduction of NADP by isolated chloroplasts. In the new terminology, the family of ferredoxins would include those non-heme, non-flavin iron proteins that transfer to appropriate enzyme systems some of the most "reducing" electrons in cellular metabolism—electrons released by the photochemical apparatus of photosynthesis or by the H_2–hydrogenase system. Ability to catalyze the photoreduction of NADP by washed chloroplasts was included provisionally in the definition of ferredoxins because, in the experience of our laboratory, all the ferredoxins that had been tested so far exhibit this property. By contrast, the replaceability of different ferredoxins in other enzymic reactions is less consistent. Another feature, to be discussed later, of all ferredoxins which have been isolated so far is their labile sulfur that is liberated, on acidification, as hydrogen sulfide.

It is to be noted that this provisional definition allows for dissimilarities of some properties among different ferredoxins. For example, the absorption spectra of ferredoxins from bacterial cells, whether photosynthetic or nonphotosynthetic, resemble each other but differ significantly from the type of spectrum common to ferredoxins from algae and from chloroplasts of higher plants. In fact, we now distinguish, on the basis of spectral characteristics, two types of ferredoxins: the bacterial type and the chloroplast type.

A definitive characterization of ferredoxins as a group of electron carriers must await the isolation of a common prosthetic group in ferredoxins of different species. Pending the isolation of a common prosthetic group, it seems useful to retain the tentative definition of ferredoxins as iron-containing proteins which function as electron carriers on the "hydrogen side" of pyridine nucleotides. This tentative definition stresses the present distinction between ferredoxins and all the heme or non-heme iron proteins (including flavoproteins) with more electropositive oxidation-reduction potentials that serve as electron carriers on the "oxygen side" of pyridine nucleotides.

X. Spectral Characteristics and Oxidation-Reduction Potentials

Unlike cytochromes, which exhibit well-defined absorption peaks in the reduced state, ferredoxins have distinctive absorption peaks in the oxidized state. On reduction of ferredoxins, the absorption peaks in the longer wavelengths disappear.

The first bacterial ferredoxin to be crystallized, that of C. pasteurianum, exhibited in its oxidized state a distinctive spectrum with peaks in the visible and ultraviolet at 390, 300, and 280 mμ (75). The crystalline preparation gave an absorption ratio of 390 mμ/280 mμ = 0.79. These spectral characteristics of ferredoxin from *C. pasteurianum* were confirmed and extended by Buchanan et al. (88) and Lovenberg et al. (89) to ferredoxins of other species of *Clostridium*, which they prepared in crystalline form.

Figure 2 shows that the absorption spectrum of ferredoxin of the photosynthetic bacterium *Chromatium* closely resembles that of ferredoxin from the nonphotosynthetic *Clostridium* species. In the oxidized state, the absorption spectrum of *Chromatium* ferredoxin exhibits a flat peak at 385 mμ, a shoulder at 300 mμ, and a peak at 280 mμ. In the purest preparation the ratio of optical density, 385 mμ/280 mμ, was 0.74 (90).

As shown in Figure 2, *Chromatium* ferredoxin was reduced by three

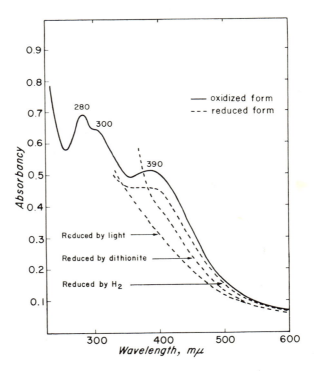

Fig. 2.   Reduction of *Chromatium* ferredoxin by H_2, sodium dithionite, and illuminated spinach chloroplasts (90).

methods: (*a*) H_2 gas in the dark, in the presence of a hydrogenase preparation from *C. pasteurianum*; (*b*) sodium dithionite, in the dark; and (*c*) photochemically, using a heated preparation of spinach chloroplasts and reduced dichlorophenol indophenol as the electron donor. Complete reduction of *Chromatium* ferredoxin was obtained only photochemically.

Figure 3 shows that the absorption spectrum of ferredoxin (oxidized state) from the blue-green alga *Nostoc* is of the chloroplast type (91). It resembles closely the absorption spectrum of ferredoxin from spinach chloroplasts and is different from the absorption spectrum of bacterial ferredoxins. The absorption peaks of *Nostoc* ferredoxin in the visible and

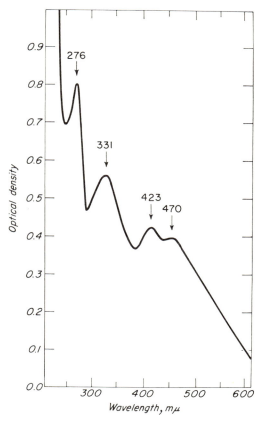

Fig. 3.   Absorption spectrum of *Nostoc* ferredoxin in the oxidized state (91).

in the ultraviolet are 470, 423, 331, and 276 mμ, as compared with 463, 420, 325, and 274 mμ for spinach ferredoxin. The purest preparation of *Nostoc* ferredoxin (91) had a ratio of optical density, 423 mμ/276 mμ, of 0.57. A preliminary determination of the oxidation-reduction potential of *Nostoc* ferredoxin gave a value of $E_0' = -405$ mV, at pH 7.55 (91).

The similarity between the ferredoxin of *Nostoc* and spinach is interesting from an evolutionary point of view. Blue-green algae are considered to be the most primitive algae, not too distant on the evolutionary scale from photosynthetic bacteria. They reproduce vegetatively and, like photosynthetic bacteria, do not have their photosynthetic pigments localized in chloroplasts but distributed throughout the outer part of the cell. However, unlike photosynthetic bacteria, the photosynthesis of blue-green algae is accompanied by the evolution of oxygen. It is an interesting question whether the occurrence of the chloroplast type rather than of the bacterial type of ferredoxin in blue-green algae is related to the type of the photosynthetic pigment system and oxygen evolution that distinguish algal photosynthesis from bacterial photosynthesis.

XI. Some Chemical Properties of Ferredoxin

Apart from iron, ferredoxins of chloroplasts and bacteria are noted for containing "labile sulfur," i.e., an inorganic sulfur group which is equimolar with iron. This was first observed in spinach ferredoxin by Fry and San Pietro (85) and independently in Warburg's laboratory in the "red enzyme" (or ferredoxin) of *Chlorella* (83,84). Buchanan et al. (88) and Lovenberg et al. (89) found inorganic sulfur in bacterial ferredoxins. The inorganic sulfur in ferredoxin is liberated as hydrogen sulfide upon acidification. Both iron and inorganic sulfur are loosely bound to the protein and the removal of one is accompanied by the removal of the other. Upon the loss of iron or labile sulfur, ferredoxin loses its spectral characteristics and also its biochemical activity.

Ferredoxins are small protein molecules. The bacterial ferredoxin, first estimated to have a molecular weight of around 12,000 (75), is now known to have a molecular weight of around 6000 (89). The chloroplast ferredoxin is estimated to have a molecular weight of approximately 13,000 (92). The iron content of bacterial and chloroplast ferredoxin varies. Thus the chloroplast ferredoxin of spinach has, on a molar basis, two atoms of iron per molecule whereas the bacterial ferredoxin of *Chromatium* has three and that of *Clostridium* has seven. A summary of some chemical properties of several ferredoxins is given in Table I.

TABLE I

Some Properties of Bacterial and Spinach Ferredoxins

	C. pasteurianum	Chromatium	Spinach
Iron (atoms/molecule protein)	7	3	2
Inorganic sulfur (moles/mole protein)	7	3	2
Molecular weight (approx.)	6000	6000	13,000
Redox potential (mV at pH 7.55)	−417	−490 (approx.)	−432

XII. Ferredoxin and NADP Reduction

To test the effectiveness of different ferredoxins in catalyzing the photoreduction of NADP, several ferredoxins were crystallized from organisms other than *C. pasteurianum*. Crystalline ferredoxins from such diverse sources as spinach chloroplasts (93), the blue-green alga *Nostoc* (91), and the photosynthetic bacterium *Chromatium* (90), were found to be effective substitutes for the native spinach chloroplast ferredoxin in catalyzing the reduction of NADP by illuminated spinach chloroplasts. What remained to be clarified was the mechanism of this reaction.

A. Mechanism of NADP Reduction of Chloroplasts

The elucidation of the nature of ferredoxin as an electron carrier led to the resolution of the mechanism of NADP reduction by chloroplasts into three steps: (*1*) a photochemical reduction of ferredoxin; (*2*) reoxidation of ferredoxin by a flavoprotein enzyme, ferredoxin-NADP reductase; and (*3*) reoxidation of the reduced ferredoxin-NADP reductase by NADP (94).

Shin et al. (95) isolated and crystallized ferredoxin-NADP reductase from spinach chloroplasts. Its absorption spectrum showed a typical flavoprotein absorption spectrum with peaks at 275, 385, and 456 mμ and minima at 321 and 410 mμ. Shin and Arnon (94) found that the oxidation-reduction of the flavin component of ferredoxin-NADP reductase was an intermediate step in the transfer of electrons from reduced ferredoxin to NADP.

Ferredoxin-NADP reductase catalyzed directly the reduction of either NADP or NAD but its affinity for NADP was very much greater (94). The Michaelis constant for NAD was found to be $3.75 \times 10^{-3} M$, which was about 400 times greater than the K_m found for NADP ($9.78 \times 10^{-6} M$).

The great difference between the affinities of ferredoxin-NADP reductase for NADP and NAD and the approximately equal concentrations of NAD and NADP in the cell (96) insure the specificity of the enzyme for NADP under physiological conditions.

The reduction of NAD and NADP by the reduced ferredoxin-NADP reductase is reversible. This reversibility of the action of the enzyme accounts for its apparent secondary function as diaphorase and as a transhydrogenase (97–99) which had been reported before its primary function as a NADP reductase was recognized. The reported specificity of its diaphorase and transhydrogenase action for $NADPH_2$ can now be explained by the low affinity of the enzyme for $NADH_2$.

The experiments which have shown that reduced ferredoxin does not transfer electrons directly to NADP have yielded no evidence for the existence of a "bound" NADP (99). Our evidence indicates that the electron transfer in the reduction of NADP by chloroplasts is not by a transhydrogenation reaction from a reduced "bound" NADP but by direct reduction of free NADP by reduced ferredoxin-NADP reductase. This is discussed in greater detail elsewhere (94).

Under physiological conditions (solid lines in the scheme below), the electron flow in the reduction of pyridine nucleotides by chloroplasts can now be summarized as follows:

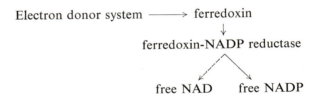

Since ferredoxin and not NADP is the electron acceptor in the photochemical reactions of chloroplasts, the experimentally established reducing potential that is generated by chloroplasts in the course of photophosphorylation is extended by over 100 mV beyond the redox potential of NADP. Moreover, as will be discussed later, the photoreduction of 2 moles of ferredoxin in noncyclic photophosphorylation is accompanied by the formation of one pyrophosphate bond of ATP, which is equivalent to a potential span of about 220 mV (compare 100). Thus, when water is the electron donor [$\frac{1}{2}H_2O = \frac{1}{4}O_2 + H^+ + e^-$, $E_0' = +0.82$ (pH 7)], the photoreduction of ferredoxin [$E_0' = -0.43$ V (pH 7)] and its accompanying photophosphorylation constitute a conversion of radiant energy into chemical energy that is equivalent to a potential span of about 1.25 V.

It should be pointed out that the emphasis here is not on a theoretical reducing potential that can be generated by chloroplasts but on the experimental isolation and characterization of a reductant, native to photosynthetic cells, that is formed by the photochemical acts of photosynthesis. As will be shown later, even on the assumption of a one-quantum process, it would still be theoretically possible for illuminated chloroplasts to generate in noncyclic photophosphorylation somewhat stronger reductants than reduced ferredoxin. [Table III in Section XX shows that in a one-quantum process the energy input of one mole-equivalent (einstein) of quanta of light at 680 mμ would be used up to the extent of 81%.] However, all such possibilities must remain speculative without evidence that photosynthetic cells contain reductants stronger than ferredoxin.*

XIII. Photoreduction of Ferredoxin Coupled with Photoproduction of Oxygen

The key role assigned to ferredoxin in the photosynthetic electron transport is subject to a rigid test. The evolution of oxygen by chloroplasts is uniquely dependent on light, and it occurs only in the presence of a proper electron acceptor. Thus, photoproduction of oxygen by chloroplasts should accompany the photoreduction of ferredoxin. Such direct demonstration, however, was technically difficult because reduced ferredoxin is readily oxidized by oxygen. However, when the rapid back reaction between reduced ferredoxin and evolved oxygen was impeded, the stoichiometry between the photoreduction of ferredoxin and photoproduction of oxygen became measurable.

The techniques used involved measuring oxygen evolution polarigraphically and determining the photoreduction of ferredoxin by the decrease in optical density at 420 mμ (101). Traces of oxygen were rigidly excluded prior to turning on the light. When the light was turned on, ferredoxin was reduced; when it was turned off, ferredoxin was reoxidized.

---

* Thus, the observations of Zweig and Avron (*Biochem. Biophys. Res. Commun.*, **19**, 397 (1965)), Kok et al. (*Biochim. Biophys. Acta*, **109**, 347 (1965)) and Black (*ibid.*, **120**, 332 (1966)) on the photoreduction by chloroplasts of nonphysiological dyes, some of which have polarigraphically measured oxidation-reduction potentials that are more negative than that of ferredoxin, provide no evidence that chloroplasts contain reductants more negative than ferredoxin. Moreover, there is often a lack of correspondence between polarographic measurements and true redox potentials. For example, NADP [E_0' (pH 7) $= -0.320$ V] was found polarographically to have a half-wave potential of -1.01 V (H. Rembold and H. Metzger, *Z. Physiol. Chemie* **348**, 194 (1967)).

The amount of ferredoxin reduced in the light was equal to the amount of ferredoxin reoxidized in the dark. The photoreduction of ferredoxin was accompanied by the evolution of oxygen, the reoxidation of ferredoxin by the consumption of oxygen. No oxygen was evolved without the addition of ferredoxin (101).

Of special interest was the stoichiometry between the ferredoxin added and oxygen produced. Table II shows that the stoichiometry between ferredoxin added and oxygen produced was 4:1 and remained the same with different amounts of added ferredoxin. This substantiates the conclusion, based on the stoichiometry of NADP reduction (92,102), that the photoreduction of ferredoxin involves a transfer of one electron.

TABLE II

Stoichiometry between Oxygen Evolution and Photoreduction of Ferredoxin by Isolated Chloroplasts (101)

(mμmoles)

Expt.	Test No.	Ferredoxin added	O_2 produced	$\dfrac{\text{Ferredoxin added}}{O_2 \text{ produced}}$
A	1	110	26	4.2
	2	110	27	4.1
	3	110	27	4.1
B	1	128	32	4.0
	2	128	33	3.9
	3	128	32	4.0
C	1	154	37	4.2
	2	154	42	3.7
	3	154	39	3.9

XIV. Noncyclic Photophosphorylation with Ferredoxin

A requirement for ferredoxin [then called "TPN-reducing factor" (80)] in noncyclic photophosphorylation was already observed when the process was first discovered (71) but only in recent experiments was the photoreduction of ferredoxin with its accompanying oxygen evolution linked with a stoichiometric ATP formation (101). The amount of ATP formed was proportional to the amount of ferredoxin added in a molar ratio of approximately 1 ATP to 2 ferredoxins (P:$2e$ = 1). This ratio is consistent with the other evidence that the oxidation-reduction of ferredoxin involves a transfer of one electron. Thus, noncyclic photophosphorylation can now be summarized by the equation

$$4Fd_{ox} + 2ADP + 2P_i + 2H_2O \xrightarrow{h\nu} 4Fd_{red} + 2ATP + O_2 + 4H^+ \qquad (10)$$

In experiments with isolated chloroplasts, it is usually more convenient to measure noncyclic photophosphorylation by using catalytic amounts of ferredoxin and stoichiometric amounts of NADP, which, unlike chloroplast ferredoxin, is commercially available and relatively stable to oxygen. However, this is merely an operational convenience which must not obscure the key role of ferredoxin in this type of photophosphorylation.

XV. Cyclic Photophosphorylation with Ferredoxin

As already mentioned, apart from noncyclic photophosphorylation, ferredoxin was also found to catalyze cyclic photophosphorylation.* Evidence for a ferredoxin-catalyzed cyclic photophosphorylation which proceeds anaerobically without the addition of other cofactors was obtained after the experimental conditions for this type of photophosphorylation had been established (103). These conditions include the use of an effective inhibitor of the electron flow from OH^- which results in oxygen evolution. It thus became clear that ferredoxin-catalyzed cyclic photophosphorylation and noncyclic photophosphorylation are mutually exclusive. Cyclic photophosphorylation catalyzed by ferredoxin can be unmasked only when noncyclic photophosphorylation is stopped.

Another way to demonstrate this mutually exclusive relation between cyclic and noncyclic photophosphorylation is to use monochromatic light above 700 mμ (103). Chloroplasts illuminated in this region of far-red light cannot produce oxygen; that is, they cannot remove electrons from water but can still carry on cyclic photophosphorylation catalyzed by ferredoxin. Here no inhibitor of photoproduction of oxygen is necessary since the far-red light serves as a physical equivalent of a chemical

* The question is sometimes raised whether cyclic photophosphorylation exists *in vivo*. The question cannot be settled by experiments with isolated chloroplasts. Such experiments, like those on oxidative phosphorylation by isolated mitochondria, measure only the ability of subcellular particles to perform certain reactions under the admittedly nonphysiological conditions of an artificial, extracellular environment. However, evidence has recently been accumulating from experiments with whole cells that is consistent with the operation of cyclic photophosphorylation *in vivo* [see Forti and Parisi, *Biochim. Biophys. Acta*, **71**, 1 (1963) and the three articles by Nultsch, Simonis, and Tanner, and Loos and Kandler in *Currents in Photosynthesis*, J. B. Thomas and J. C. Goedheer, Eds., Donker, Rotterdam, 1966].

inhibitor: it allows cyclic photophosphorylation to proceed and makes photoproduction of oxygen impossible (104).

Cyclic and noncyclic photophosphorylation are also sharply distinguished by their differential sensitivity to several inhibitors. Low concentrations of antimycin A, 2,4-dinitrophenol, and desaspidin, a chlorobutyrophenone derivative, inhibit cyclic but do not inhibit noncyclic photophosphorylation (103–108). Figure 4 illustrates a differential sensitivity to inhibition by antimycin A of cyclic photophosphorylation catalyzed by ferredoxin, menadione, or phenazine methosulfate; it also illustrates that, under the very low light intensity under which these experiments were carried out, ferredoxin is a much more effective catalyst for the conversion of radiant energy into pyrophosphate bond energy than either menadione or phenazine methosulfate. This result is also consistent with the physiological nature which we ascribe to the ferredoxin-catalyzed cyclic photophosphorylation.*

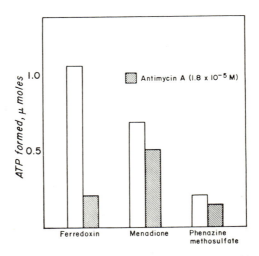

Fig. 4. Differential effect of antimycin A on cyclic photophosphorylation catalyzed by ferredoxin, menadione, and phenazine methosulfate. Illumination: monochromatic light, 714 mμ. (D. I. Arnon, *Science*, **40**, 1060 (1965)).

* Pertinent to this conclusion are very recent experiments which give rates of ferredoxin-catalyzed photophosphorylation that are up to 34 times higher than those reported earlier and are comparable with maximum rates of photosynthesis by green leaves of land plants (D. I. Arnon, H. Y. Tsujimoto, and B. D. McSwain, *Nature*, in press, 1967).

XVI. Pseudocyclic Photophosphorylation with Ferredoxin

When photophosphorylation was first discovered there was no net consumption or evolution of oxygen, but the presence of oxygen was required for ATP formation (28). This seemingly catalytic function of oxygen was found in experiments with whole chloroplasts to which no exogenous cofactors of photophosphorylation were added. When such cofactors were added, the dependence on oxygen ceased and ATP formation proceeded under anaerobic conditions (50,67,109) in what we now call cyclic photophosphorylation.

In later experiments (110) the nature of the oxygen-dependent photophosphorylation was clarified as a special case of noncyclic photophosphorylation in which molecular oxygen served as the terminal electron acceptor. No net oxygen consumption occurred because, in the oxygen-dependent photophosphorylation [renamed pseudocyclic photophosphorylation (110)] the consumption of oxygen at the terminal end of electron transfer is balanced by the production of oxygen at the site of electron donation by water (OH$^-$ at pH 7).

What remained obscure about pseudocyclic photophosphorylation was whether a catalyst was present in chloroplasts to mediate the transfer of electrons to oxygen. Evidence was obtained some years ago (80) for a requirement of ferredoxin ("TPN-reducing factor") in pseudocyclic photophosphorylation but the possible significance of these earlier observations became apparent only recently when the properties of ferredoxin were elucidated (75,94). Under aerobic conditions, reduced ferredoxin is preferentially reoxidized by NADP (via ferredoxin-NADP reductase), but when no NADP (in the oxidized form) is available, reduced ferredoxin will be reoxidized by molecular oxygen. In this manner, ferredoxin would act as a catalyst in pseudocyclic photophosphorylation. The photoreduction of ferredoxin with water as the electron donor and the oxidation of reduced ferredoxin by molecular oxygen would thus constitute a mechanism for generating ATP under aerobic conditions when reduced ferredoxin is not reoxidized by NADP.

Recent experiments have yielded direct evidence for this interpretation but have also revealed an interesting variation (108). Ferredoxin catalyzed a mixed type of photophosphorylation in air. Despite the presence of oxygen, total photophosphorylation at low concentrations of ferredoxin included an appreciable cyclic (anaerobic) component. At high concentrations of ferredoxin, phosphorylation was predominantly of the pseudocyclic type (108). This occurred in yellow light (108) that could support oxygen evolution (104). In far-red monochromatic light (> 708 mμ) that cannot support oxygen evolution, photophosphorylation under aerobic conditions was always of the cyclic type (104).

These results suggest that *in vivo*, when oxygen is present in the vicinity of chloroplasts, its reactivity with reduced ferredoxin does not constitute an electron "leak" which dissipates the chemical energy that was trapped in the photoreduction of ferredoxin. Such dissipation of chemical energy under aerobic conditions is prevented, first, by the greater affinity of ferredoxin for the enzyme system that brings about the reduction of NADP (94) than for oxygen and, second, by the ability of ferredoxin to catalyze ATP formation in the presence of oxygen by pseudocyclic photophosphorylation and even, under certain conditions, by cyclic photophosphorylation.

XVII. Two Light Reactions in Photosynthesis: Observations on Whole Cells

One of the widely held ideas today is that the photochemistry of photosynthesis involves the collaboration of two light reactions. Before considering this idea in relation to cyclic and noncyclic photophosphorylation by isolated chloroplasts, it will be useful to review its application to photosynthesis in intact cells.

The subject, in its modern phase, had its origin in measurements of quantum efficiency of photosynthesis in intact cells. Warburg et al. (111) observed that the quantum efficiency of photosynthesis in *Chlorella* cells illuminated by red light (645 mμ) was enhanced by the addition of catalytic amounts of blue light. The maximum effectiveness of the catalytic light was at 460 mμ. Warburg attributed this effect to the photoactivation of a photosynthetic enzyme ("Luminoferment") that had a maximum absorption at that wavelength. Emerson et al. (112), also measuring the quantum efficiency of *Chlorella* cells, observed a significant enhancement by supplementary light only when the main illumination was at wavelengths longer than 680 mμ. They reported that, to be effective, the intensity of the supplementary light must not be catalytic but be of sufficient intensity to give "measurable photosynthesis" by itself. Not only blue light but "all regions of the visible spectrum were effective, at least to 644 mμ" (112).

These differences led Emerson et al. (112) to conclude that their enhancement effect was fundamentally different from that observed by Warburg et al. Nevertheless, in a recent review of the extensive literature that now surrounds this subject, Smith and French (113) concluded that, in the light of present knowledge, these differences were no longer substantive and that Emerson et al. (112) were observing different manifestations of the same enhancement phenomenon that was reported by Warburg et al. (111).

Emerson and Chalmers (114) suggested from observations on *Chlorella* and other algae that, at the long wavelengths of light that are absorbed mainly by chlorophyll *a*, full efficiency of photosynthesis depends on the absorption of light of shorter wavelength by an "accessory" photosynthetic pigment: chlorophyll *b* in *Chlorella* and the phycobilins in other algae. Cells illuminated simultaneously by a long and a short wavelength light beam exhibited an "enhancement" effect; that is, a rate of photosynthesis higher than the sum of the rates obtained for each beam separately.

The hypothesis of two light reactions also received support from the experiments of Blinks on chromatic transients, defined as "temporary alterations of photosynthetic rate (10–15%) when the wavelength of incident light is changed, even though the intensities have been adjusted to give equal steady state rates" (115). Myers and French (116) found the same action spectra for enhancement of photosynthesis (measured at 700 mμ) and for chromatic transients when illumination was changed from 700 mμ to other wavelengths. They also reported (117) the same enhancement effect whether illumination by the two beams of light was given simultaneously or alternated in a period of a few seconds.

Further support for the hypothesis of two photochemical reactions came from the interpretation which Duysens et al. (118–120) placed on his observations of optical density changes at 420 mμ in the red alga, *Porphyridium cruentum*, which they attributed to the oxidation and reduction of cytochrome *f*. They observed that *Porphyridium* cells illuminated by long-wavelength radiation (680 mμ) oxidized cytochrome *f*. Reduction of cytochrome *f* occurred either by turning off the actinic beam of 680 mμ or by adding a second beam of shorter wavelength radiation at 560 mμ. The reducing effect of the short-wavelength beam was abolished when the cells were poisoned by an inhibitor of oxygen evolution.

Duysens et al. (118–120) interpreted these observations as decisive evidence for two photochemical reactions that operate in series in photosynthesis: (i) a short-wavelength reaction ("system 2") which reduces cytochrome *f* and depends mainly on light absorbed by the accessory pigment, phycoerythrin, and (ii) a long-wavelength reaction ("system 1") which oxidizes cytochrome *f*. Witt et al. (121) similarly interpreted their observations of spectral changes resulting from flash illumination. As for the significance of these reactions to the mechanism of photosynthesis, Duysens and Amesz suggested that "the simplest hypothesis is that the oxidation of cytochrome by system 1 is accompanied by a reduction of CO_2, probably via phosphopyridine nucleotide, and that the photochemical reduction of cytochrome by system 2 is accompanied by a simultaneous oxidation of water to oxygen" (120).

XVIII. One or Two Light Reactions in Noncyclic Photophosphorylation?

The hypothesis of two photochemical reactions in photosynthesis was based on observations with whole cells that had no direct bearing on the biochemical mechanisms of energy conversion in photosynthetic phosphorylation investigated with isolated chloroplasts. Here the main problem was how to interpret two different manifestations of the photochemical activity of chloroplasts: cyclic photophosphorylation that yields ATP as the sole product and noncyclic photophosphorylation (including its pseudocyclic variant) where the formation of ATP is linked with the photoreduction of ferredoxin and the photooxidation of water with a resultant evolution of oxygen.

Prior to the recognition of the role of ferredoxin, the electron flow hypothesis (65,66) attempted to explain the facts of cyclic and noncyclic photophosphorylation in terms of one light reaction: excitation of chlorophyll by photon capture followed by electron transfer. The divergence between the cyclic and noncyclic pathways was envisaged to lie in the fate of the ejected electron and in the source of its replacement. In noncyclic photophosphorylation the electron from excited chlorophyll was transferred to NADP and then used in the reduction of carbon dioxide, whereas in cyclic photophosphorylation the electron was transferred to a "catalyst," then to a cytochrome, and back to chlorophyll. Thus, in cyclic photophosphorylation no electrons were removed from the photosynthetic apparatus and hence none had to be replenished, whereas in noncyclic photophosphorylation the electrons transferred from excited chlorophyll to NADP were replaced by water, through the intermediary of a postulated cytochrome.

The inclusion of cytochromes as electron carriers in cyclic photophosphorylation presented no thermodynamic difficulties but the suggestion that cytochromes may mediate the electron transfer from water was not in agreement with the known properties of chloroplast cytochromes. The most oxidizing cytochrome component of chloroplasts, cytochrome f, has (at pH 7) a redox potential of $+0.365$ V (122), whereas the oxidation of water would require a cytochrome with a redox potential more electropositive than $+0.82$ V (see Section XII-A). Thus, the hypothesis that was put forward had to assume that our knowledge of the oxidation-reduction potentials of chloroplast cytochromes was incomplete (65,66).

In the intervening years no evidence for such strongly oxidizing chloroplast cytochromes has come to light, although Kamen (123) speculated that cytochrome f could exist in a ferryl state (Fe^{4+}) that would be sufficiently electropositive for the photooxidation of water.

Hill and Bendall (124) attempted to explain differently the experimental findings of noncyclic photophosphorylation. Their scheme was not based on any new observations and was not concerned with cyclic photophosphorylation. They postulated two light reactions in noncyclic photophosphorylation: one related to the reduction of cytochrome b_6 ($E_0' = -0.06$ V) and the oxidation of water ($E_0' = 0.82$ V) with an accompanying oxygen evolution, and a second light reaction that is associated with oxidation of cytochrome f ($E_0' = +0.365$ V) and reduction of NADP ($E_0' = -0.32$ V). The formation of ATP in their scheme was coupled, by analogy with oxidative phosphorylation, with electron transfer from reduced cytochrome b_6 to oxidized cytochrome f (124).

The hypothesis of Hill and Bendall (124) had the advantage of invoking known chloroplast cytochromes to account for the formation of ATP but it did not remove the difficulty in relating the properties of cytochromes b_6 and f to the least understood component of noncyclic photophosphorylation, the photooxidation of water. Arnon (65,66) symbolized the ignorance in this area by postulating the existence of some yet unrecognized chloroplast cytochrome or unrecognized properties of known chloroplast cytochromes, whereas Hill and Bendall (124) symbolized it by postulating two hypothetical entities, X and Y. Their scheme envisaged (124) that in one light reaction water was split, X was reduced, and cytochrome f oxidized; whereas in the second light reaction Y was oxidized, cytochrome b_6 reduced, and oxygen evolved. Another difference was that the scheme of Hill and Bendall (124) extended the photooxidation of water to photosynthetic bacteria, whereas an essential point in Arnon's hypothesis was that photooxidation of water is peculiar to plant photosynthesis (65,66).

Without new experimental evidence it was difficult to decide which of these two hypotheses had greater merit in explaining the nature of noncyclic photophosphorylation. However, in 1961, Losada, Whatley, and Arnon (76) reported new experiments which gave support to the idea that noncyclic photophosphorylation is indeed the sum of two component light reactions: (1) a photooxidation of water that yields oxygen and reduces an electron carrier (at an intermediate redox potential) which serves as an electron donor for (2) the photoreduction of NADP that is coupled with the formation of ATP.

A. Noncyclic Photophosphorylation without Oxygen Evolution

Losada et al. (76) based this conclusion on experiments with chloroplasts in which experimental conditions were so arranged that in one case illumination of chloroplasts gave an evolution of oxygen coupled to the reduction of an indophenol dye but gave no ATP formation. In another case, illumination of chloroplasts gave ATP formation that was coupled with a noncyclic electron flow from the reduced indophenol dye to NADP (125). Losada et al. (76) characterized this second reaction as a "bacterial type" of noncyclic photophosphorylation in accordance with the concept that photosynthetic bacteria cannot use water as an electron donor and hence produce no oxygen (65) but are able to maintain a noncyclic electron flow from other, less oxidized, electron donors to NAD (126,127).

Losada et al. (76) considered that in their experiments the artificial electron carrier, 2,6-dichlorophenol indophenol, joined the two light reactions by serving as the electron acceptor in the photooxidation of water and as the electron donor in the reduction of NADP. Losada et al. (76) thought that the function of the electron carrier (see *A* in Fig. 5) could be performed *in vivo* either by a cytochrome (124) or by a quinone (128). This scheme for noncyclic photophosphorylation in chloroplasts was represented (76) by a zigzag diagram (Fig. 5) which, with minor alterations, has been widely used since to depict the concept of two light reactions joined by a middle dark reaction (see, for example, Fig. 4 in ref. 101).

The scheme of Losada et al. (76) seemed to receive additional support when Nozaki et al. (129) found, in the chlorophyll-containing particles of photosynthetic bacteria, a noncyclic photophosphorylation which appeared to differ from that in chloroplasts by lacking only the photooxidation of

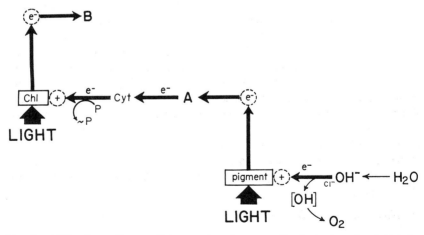

Fig. 5.   1961 scheme for two light reactions in noncyclic photophosphorylation by chloroplasts (76).

water component. Nevertheless, the validity of this interpretation became open to some question after Trebst and Eck (130) and other investigators (131–134) found that 2,6-dichlorophenol indophenol in the reduced form can catalyze cyclic photophosphorylation. The matter remained complicated because, under the conditions of the experiments of Losada et al. (76) with chloroplasts and of Nozaki et al. (129,135) with bacterial particles (strict anaerobicity and a great excess of ascorbate), 2,6-dichlorophenol indophenol failed to catalyze any phosphorylation unless a unidirectional, noncyclic electron flow from the reduced indophenol dye to pyridine nucleotide was established and maintained.

Thus, the question remained unsettled whether the noncyclic photophosphorylation that accompanied the photoreduction of NADP when water was the electron donor was the same photophosphorylation that accompanied the photoreduction of NADP when reduced indophenol dye was the electron donor. New evidence that these two photophosphorylations were indeed different came from recent experiments with desaspidin (108 and earlier references cited therein). When reduced indophenol dye was the electron donor for NADP reduction, the coupled photophosphorylation showed the same sensitivity to low concentrations of desaspidin as was observed for all types of cyclic photophosphorylation, including that catalyzed by ferredoxin (108). By contrast, when water was the electron donor for NAD reduction, the coupled photophosphorylation was resistant to inhibition by low concentrations of desaspidin* (108).

These results with desaspidin led Arnon, Tsujimoto, and McSwain (108) to abandon the earlier interpretation of Losada, Whatley, and Arnon (76) and its later elaborations (101,136) and to conclude that the photophosphorylation which accompanies the photoreduction of NADP by reduced indophenol dye occurs at a "cyclic" site that is not involved in the noncyclic photophosphorylation that accompanies the photoreduction of NADP by water. The new interpretation is consistent with the recent evidence that ferredoxin-dependent cyclic and noncyclic photophosphorylation in chloroplasts are two different photochemical reactions that are experimentally distinguished by their differential sensitivity to inhibitors (antimycin A, dinitrophenol, and desaspidin) and by their differential response to monochromatic light above 700 mμ. Arnon et al. (108) then

---

* Several papers dealing with the effect of desaspidin on photophosphorylation appeared after this article was completed. The arguments of Hind (*Nature*, **210**, 703 (1966); *Plant Physiol.*, **41**, 1237 (1966)) and Gromet-Elhanan and Avron (*Plant Physiol.*, **41**, 1231 (1966)) against the above interpretation of the effect of desaspidin were dealt with by Tsujimoto, McSwain, and Arnon (*Plant Physiol.*, **41**, 1376 (1966)).

proposed a new hypothesis which accounts for the differences between cyclic and noncyclic photophosphorylation and reinterprets the earlier observations on the relation between electron flow and photophosphorylation.

XIX. Mechanisms of Cyclic and Noncyclic Photophosphorylation

The present hypothesis (108) envisages that cyclic and noncyclic photophosphorylation in chloroplasts are parallel mechanisms for ATP formation in photosynthesis and do not share a common phosphorylating site. Cyclic photophosphorylation in chloroplasts is considered to include phosphorylations that are coupled with a flow of electrons from excited chlorophyll to ferredoxin, and then from reduced ferredoxin to cyclochrome b_6, thence to cytochrome f and back to chlorophyll (Fig. 6). The inclusion of cytochromes b_6 and f in the cyclic pathway only is based on the sensitivity to antimycin A inhibition of the ferredoxin-catalyzed cyclic photophosphorylation and the insensitivity of noncyclic photophosphorylation to that inhibitor.

Antimycin A is a well-known inhibitor of oxidative phosphorylation by mitochondria where the direct participation of the corresponding cytochromes, b and c, is well documented and where antimycin A inhibition is considered to be indicative of electron transport between cytochromes b and c (137,138). It may also be pertinent that ferredoxin-catalyzed cyclic photophosphorylation, but again not the noncyclic type, is sensitive to inhibition by 2,4-dinitrophenol—a well-known inhibitor of oxidative phosphorylation in mitochondria. Although these analogies between oxidative phosphorylation in mitochondria and cyclic photophosphorylation in chloroplasts are strongly suggestive, they cannot be considered as established without further evidence that is now being actively sought.

The span between the redox potentials of cytochromes b_6 and f (-0.06 and 0.365 V, respectively) is large enough to accommodate a phosphorylation. Likewise, the span between the redox potentials of ferredoxin (-0.43 V) and cytochrome b_6 is large enough to accommodate another phosphorylation. Two phosphorylation sites are specified tentatively in the cyclic electron-transport chain, but additional phosphorylation sites are thermodynamically possible and are not excluded.

The ferredoxin-catalyzed cyclic photophosphorylation appears to be the physiological one (103). However, as already discussed, cyclic photophosphorylation *in vitro* proceeds readily without ferredoxin when catalyzed by one of several dyes or other artificial cofactors. Since such

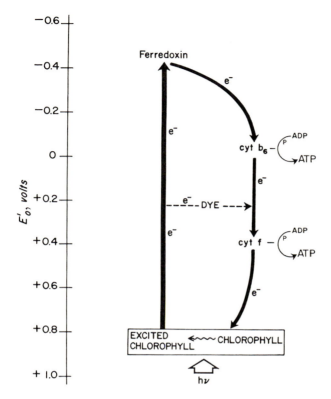

Fig. 6. Present scheme for cyclic photophosphorylation in chloroplasts (108).

cyclic photophosphorylations are resistant to inhibition by antimycin A, it seems reasonable to conclude that they bypass the cytochrome b_6 site (see dotted line in Fig. 6).

The present concept of noncyclic photophosphorylation in plants is illustrated by Figure 7 (108). The physiological electron acceptor is ferredoxin but it may be replaced by nonphysiological electron acceptors (Hill reagents) with an attendant drop in the light-generated reducing potential. Figure 7 (dotted lines) illustrates this for ferricyanide and benzoquinone (BQ).

ATP formation in noncyclic photophosphorylation is envisaged as being coupled to the photooxidation of OH^-—a coupling that would account for the consistent stoichiometry, $P/2e = 1$, between oxygen evolution and ATP formation. The existence of this coupling is also

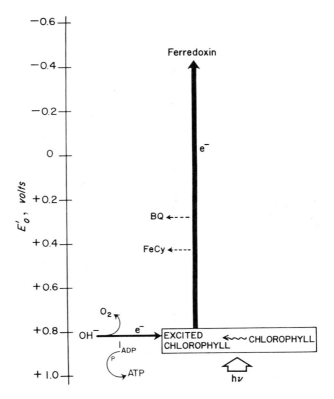

Fig. 7.   Present scheme for noncyclic photophosphorylation in chloroplasts (108).

strongly supported by the already cited evidence that the rate of electron flow from OH^- is greatly increased when accompanied by phosphorylation (71,73,74).

Thus, the photochemical event that is considered to be common to cyclic and noncyclic photophosphorylation is the photoreduction of ferredoxin by an electron donated by excited chlorophyll. What distinguishes noncyclic from cyclic photophosphorylation is the dependence on water as the ultimate electron donor for the reduction of ferredoxin and the resultant evolution of oxygen as a by-product. In cyclic photophosphorylation there is no consumption of an exogenous electron donor but a recycling of electrons derived from the constituents of the photosynthetic apparatus.

How can ferredoxin participate in both cyclic and noncyclic photophosphorylation? The fact has been already stressed that in isolated chloroplasts cyclic photophosphorylation occurs only when noncyclic photophosphorylation (and the related NADP reduction) is stopped. The regulatory mechanism(s) in the cell for switching from noncyclic to cyclic photophosphorylation is unknown, but one possibility has experimental support: the availability of NADP in the oxidized form (103). As long as oxidized NADP is available, electrons will flow from water to ferredoxin and thence [via ferredoxin-NADP reductase (94)] to NADP. However, when CO_2 assimilation temporarily ceases for lack of ATP, it is envisaged (103) that NADP accumulates in the reduced state and electrons from reduced ferredoxin begin to "cycle" within the chloroplasts, giving rise to cyclic photophosphorylation. The additional ATP thus generated would reestablish CO_2 assimilation, which, in turn, would restore $NADPH_2$ to its oxidized form and thereby reestablish noncyclic photophosphorylation.

XX. The Hypothesis of Cyclic and Noncyclic Photophosphorylation as the Two Light Reactions in Chloroplasts

If the reductive pentose phosphate cycle is accepted as the proven mechanism of carbon dioxide assimilation, then the function of the light phase of photosynthesis ceases to be a matter of speculation and becomes identified with the generation of ATP and $NADPH_2$ by way of cyclic and noncyclic photophosphorylation. The assimilation of 1 mole of CO_2 to the level of glucose by the reductive pentose cycle requires ATP and $NADPH_2$ in a ratio of 3:2. Since noncyclic photophosphorylation supplies $NADPH_2$ and ATP in a ratio of 1:1 (see Sections II and VII), the additional ATP must come from cyclic photophosphorylation. Cyclic photophosphorylation would be necessary even apart from the ATP requirement for glucose synthesis. After glucose is formed, more ATP would be required to form starch, the main product of photosynthesis in leaves. [In spinach chloroplasts ATP is expended in the formation of ADP-glucose from which the glucosyl moiety is transferred to a starch primer (139,140).] Moreover, as was pointed out elsewhere (66,71), cyclic photophosphorylation may be an important mechanism for providing the large supplies of ATP that are needed for protein synthesis and for other endergonic processes in the cell that are different from carbohydrate synthesis.

The present hypothesis (108) identifies cyclic and noncyclic photo-phosphorylation as two parallel and complementary pathways of energy conversion which jointly generate the ATP and $NADPH_2$ that are required for carbon assimilation in green plants. Both pathways are operative in white light, but in monochromatic light cyclic photophosphorylation can occur at wavelengths longer than 700 mμ that will not support noncyclic photophosphorylation. Thus, cyclic photophosphorylation may be viewed as a "long-wavelength" light reaction and noncyclic photophosphorylation as a "short-wavelength" light reaction.

The present hypothesis envisages that, in noncyclic photophosphorylation, the transfer of an electron from OH^- via chlorophyll to ferredoxin involves a single photochemical act. The evidence for the widely held view (shared until recently by this writer) that this electron transfer requires the collaboration of two photochemical reactions working in series seems no longer compelling for two reasons. First, the results of the chloroplast experiments with dichlorophenol indophenol which were used as evidence for the separation of noncyclic photophosphorylation into two component photochemical reactions (76) can now be differently interpreted in the light of new evidence (108) (see also Section XVIII). Second, as will be discussed below, the observations with whole cells, which were used to support the idea of two photochemical reactions as components of noncyclic photophosphorylation are also consistent with the idea that the two photochemical reactions may be cyclic and noncyclic photophosphorylation.

The idea of a single photochemical act in noncyclic photophosphorylation by chloroplasts is also compatible with the recent findings by Warburg of high quantum efficiency in leaves: a requirement of 5.5 quanta per CO_2 (141) as contrasted with his best earlier measurements in leaves of about 15 quanta per CO_2 (142). As far as the overall energy balance is concerned, Table III shows that 1 quantum of light at 680 mμ—the longest wavelength

TABLE III

Energy Balance of Noncyclic Photophosphorylation of the Plant Type

---

Energy input of 1 einstein, $\lambda = 680$ mμ: 42 kcal

$Ferredoxin_{ox} = Ferredoxin_{red} + e^-$; $E_0' = -0.43$ V (pH 7)

$\frac{1}{2}H_2O = \frac{1}{4}O_2 + H^+ + e^-$; $E_0' = +0.82$ V (pH 7)

Potential span between O_2 and $ferredoxin_{red}$: 1.25 eV = 1.25 × 23.06 = 28.8 kcal

Energy requirement (per electron) to form 1 ATP: 10 kcal/2 = 5 kcal

Excess of energy input over output: 42 − (28.8 + 5) = 8.2 kcal

---

that can still support a maximum rate of plant photosynthesis (143)—contains more than sufficient energy for the transfer of an electron from OH^- to ferredoxin and for the coupled noncyclic phosphorylation. Nevertheless, it must be stressed again that the electron pathway from OH^- to chlorophyll remains the least understood part in the mechanism of photosynthesis and its elucidation must be left to future research. Only a few of the cofactors and catalysts involved therein are now known: chloride ions (21,144), manganese (145,146), and plastoquinone (128,147,148) [plastoquinone is also involved in cyclic photophosphorylation in chloroplasts (147,149)].

The present hypothesis of cyclic and noncyclic photophosphorylation as two parallel and complementary photochemical reactions in photosynthesis would explain differently some of the observations on whole cells that are now used to support the concept of two light reactions working in series (see reviews 113, 143, and 150). "Red drop" would mean that at the longer wavelengths of light noncyclic photophosphorylation becomes limiting, thereby reducing the availability of $NADPH_2$ and decreasing the overall efficiency of the process. "Enhancement" would mean that the addition of light of shorter wavelength to far-red light restores noncyclic photophosphorylation, removes the shortage of $NADPH_2$, and thereby gives a synergistic rather than an additive increase in the overall efficiency of photosynthesis.

Among the other observations that would be explained by the present hypothesis is the oxidation of chloroplast cytochromes in longer wavelengths of light and their reduction in shorter wavelengths (120).

The present hypothesis envisages that, at the shorter wavelengths of light, ferredoxin, when reduced by electrons from water, reduces in turn, by an electron "spillover" effect, other constituents of chloroplasts, including cytochromes that may not be involved in noncyclic photophosphorylation. The cytochromes would remain in the reduced state as long as the electron flow from water to ferredoxin is maintained. Cytochromes would become oxidized during cyclic photophosphorylation when the flow of electrons from water is stopped and reduced cytochromes become the internal electron donors for the reduction of ferredoxin. Under laboratory conditions, this cyclic electron flow from cytochromes to ferredoxin would be established either by long-wavelength illumination or by short-wavelength illumination in the presence of inhibitors of photo-oxidation of water.

The "chlorophyll" in the diagrams of cyclic and noncyclic photophosphorylation (Figs. 6 and 7) represents the complex of chlorophyll *a* and b pigments and includes those accessory pigments that are involved in light absorption. In the past the term "accessory pigments" was applied to photosynthetically active pigments other than chlorophyll. Currently, the term "accessory pigments" in green algae and plants has come to denote all photosynthetic pigments other than chlorophyll a. Included among accessory pigments are now chlorophyll b and even certain variants of chlorophyll a (113). The main photochemical activity is thought to be confined to such forms of chlorophyll a as Ca683, Ca695 (113) and P700 (151). The existence of these varied forms of chlorophyll a has been inferred from spectroscopic observations, but in no case have they been extracted from either algae or chloroplasts. Hence, it would seem to be premature to consider them as established chemical entities. It is possible that they reflect different states of aggregation of association of chlorophyll a with other molecules. It also seems premature to consider the P700 form of chlorophyll a as the light collector or energy sink (151) for all the photochemical reactions of chloroplasts. For example, observations in this laboratory indicate that in monochromatic light cyclic photophosphorylation will proceed up to 730 mμ; that is, at wavelengths longer than the effective energy sink of P700.

Since far-red illumination is beyond the known absorption limits of chlorophyll b, it seems reasonable to conclude that cyclic photophosphorylation can proceed without light that is absorbed by this pigment. But there is no way at present to measure photochemical reactions by isolated leaf chloroplasts at wavelengths which are absorbed solely by chlorophyll b to the exclusion of chlorophyll a.

An efficient transfer of excitation energy from the accessory pigments to chlorophyll a is assumed, in accordance with the earlier work of Duysens (152). It is envisaged that, upon illumination by white light, an energetically heterogenous population of excited chlorophyll molecules is formed, each with its own vibrational substates. As already mentioned, experiments with spinach chloroplasts suggest that only when chlorophyll molecules are excited by wavelengths shorter than 700 mμ can they receive sufficient energy to give rise to noncyclic photophosphorylation and the coupled oxygen evolution. It is an attractive possibility that the threshold wavelength for maximum efficiency of noncyclic photophosphorylation is 685 mμ—the wavelength which coincides with the fluorescence peak of chloroplasts (Fig. 8). For the energetically less difficult cyclic photophosphorylation, in which not water ($E_0' = 0.820$ V) but cytochrome f ($E_0' = 0.365$ V) is the presumed electron donor, wavelengths longer than 685 mμ (up to the absorption limit of chlorophyll a) would be equally effective.

XXI. Quenching of Chloroplast Fluorescence by Cyclic and Noncyclic Photophosphorylation

Fluorescence of chloroplasts represents that portion of absorbed radiant energy which is not converted into chemical energy (or heat) but is reemitted as radiation. If our thesis is correct that cyclic and noncyclic photophosphorylation constitute the energy conversion process of photosynthesis, then they should act as quenchers of chloroplast fluorescence.

Figure 8 shows a marked quenching of chloroplast fluorescence by the addition of ferredoxin (153). This result is consistent with the hypothesis that ferredoxin establishes a cyclic electron flow: the degradation of the energy of a molecule excited by photon capture can occur by electron transfer to an appropriate electron acceptor molecule. An additional quenching of chloroplast fluorescence was observed upon adding ADP and inorganic phosphate (Fig. 8)—an observation which supports the idea

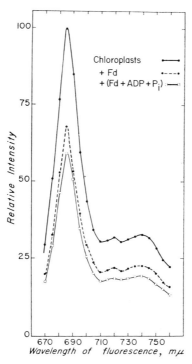

Fig. 8. Quenching of chloroplast fluorescence by ferredoxin, ADP, and inorganic phosphate (153).

that the cyclic electron flow catalyzed by ferredoxin is accelerated by concomitant phosphorylation.

Quenching of chloroplast fluorescence by noncyclic photophosphorylation is shown in Figure 9. The addition of ferredoxin and NADP gave a pronounced quenching effect. The addition of ADP and inorganic phosphate to the ferredoxin–NADP system gave a marked additional quenching effect. These results are consistent with earlier findings that the rate of noncyclic electron flow is markedly increased by a concomitant phosphorylation.

The quenching of chloroplast fluorescence by cyclic and noncyclic photophosphorylation is consistent with their characterization as the energy conversion reactions in photosynthesis. The energy of captured photons may be dissipated as fluorescence or may generate an electron flow which yields the chemical energy stored in the pyrophosphate bonds of ATP and in the reducing potential of ferredoxin.

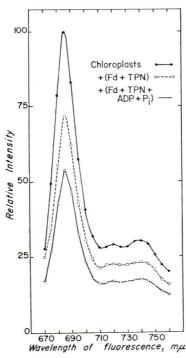

Fig. 9.   Quenching of chloroplast fluorescence by NADP (TPN), ferredoxin, ADP, and inorganic phosphate (153).

XXII. A Copper Protein and Quinones in the Photosynthetic Apparatus

Apart from cytochromes, the photosynthetic apparatus is now known to contain quinones and a copper protein that might serve as electron carriers in photochemical reactions. Katoh et al. (154,155) have discovered a copper-containing protein, plastocyanin, in *Chlorella* and in leaf tissue. Plastocyanin is blue when oxidized and colorless when reduced. The redox potential of plastocyanin at pH 7 is 0.40 V. Oxidized plastocyanin is reduced by illuminated chloroplasts and, once reduced, is not readily reoxidized by oxygen.

Our knowledge of the occurrence of quinones in chloroplasts began with the localization in them of vitamin K_1 by Dam et al. (156)—a finding that was first documented by bioassay and was recently confirmed by direct chemical methods (157). Apart from the naphthoquinone, vitamin K_1, the early work of Koffler and the more recent work of Crane and co-workers has led to the recognition as normal constituents of chloroplasts of seven benzoquinones: plastoquinone A, B, C, and D and α-, β-, and γ-tocopherol-quinones (see review 158).

The literature on the possible role of plastoquinones and plastocyanin as electron carriers has recently been reviewed by Vernon and Avron (150) from the point of view of two photochemical reactions working in series. Experiments are now under way in this laboratory to test the adequacy of the new hypothesis presented herein in explaining the role of plastocyanin and plastoquinone as electron carriers.

XXIII. Concluding Remarks

When Emil Fischer (159) discussed photosynthesis at the beginning of this century, he predicted that the elucidation of "the precise nature of the assimilation (photosynthesis) process ... will only be accomplished when biological research, aided by improved analytic methods, has succeeded in following the changes which take place in the actual chlorophyll granules (chloroplasts)."

This review has attempted to present some of the recent progress toward fulfilling Fischer's prophecy. The "chlorophyll granules" used in most of the research described herein were spinach leaf chloroplasts. The "improved analytic methods" that led to the discovery of photosynthetic phosphorylation and the synthesis of carbohydrates by isolated chloroplasts were, apart from the special biochemical techniques for isolation and fractionation of functioning chloroplasts, the labeled isotope technique introduced into biology by von Hevesy (160) and the development of paper partition chromatography by Martin and Synge (161).

The concept of photosynthesis that emerged from the work with isolated chloroplasts differs from the conventional view of photosynthesis as a process of CO_2 assimilation. Photosynthesis appears to be first and foremost a process for converting the radiant energy of sunlight into chemical energy. This conversion is directly linked to cyclic and noncyclic photophosphorylation which involve the esterification of inorganic phosphate and the reduction of the iron-protein, ferredoxin. The first chemically defined products of the energy conversion process in photosynthesis proved to be not intermediates of carbon assimilation but ATP and reduced ferredoxin. The basic facts of cyclic and noncyclic photophosphorylation are now well established, but the hypotheses invoked to explain their mechanisms are subject to change as further refinements of experimental techniques bring new facts to light.

References

1. E. G. Pringsheim and W. Wiessner, *Nature*, **188**, 919 (1960).
2. K. V. Thimann, *Science*, **88**, 506 (1938).
3. F. Lipmann, in *Advances in Enzymology*, Vol. 1, F. F. Nord, Ed., Interscience, New York, 1941, p. 99.
4. S. Ruben, *J. Am. Chem. Soc.*, **65**, 279 (1943).
5. M. Calvin, in *Proc. Intern. Congr. Biochem. 3rd Brussels, 1955* (publ. 1956), p. 211.
6. A. Weissbach, B. L. Horecker, and J. Hurwitz, *J. Biol. Chem.*, **218**, 795 (1956).
7. E. B. Jakoby, D. O. Brummond, and S. Ochoa, *J. Biol. Chem.*, **218**, 811 (1956).
8. E. Racker, *Arch. Biochem. Biophys.*, **69**, 300 (1957).
9. J. R. Quayle, *Ann. Rev. Microbiol.*, **15**, 119 (1961).
10. P. A. Trudinger, *Biochem. J.*, **64**, 274 (1956).
11. J. P. Aubert, G. Milhaud, and J. Millet, *Ann. Inst. Pasteur*, **92**, 515 (1957).
12. O. Warburg and G. Krippahl, *Z. Physiol. Chem.*, **332**, 225 (1963).
13. H. S. Gewitz and W. Voelker, *Z. Physiol. Chem.*, **330**, 124 (1962).
14. A. Peterkofsky and E. Racker, *Plant Physiol.*, **36**, 409 (1961).
15. M. Gibbs and O. Kandler, *Proc. Natl. Acad. Sci. U.S.*, **43**, 446 (1957).
16. J. Sachs, *Lectures on the Physiology of Plants*, Clarendon Press, Oxford, 1887, p. 299.
17. T. W. Engelmann, *Arch. Ges. Physiol. (Pflüger's)*, **30**, 95 (1883).
18. T. W. Engelmann, *Botan. Ztg.*, **46**, 661 ff., 1888.
19. R. Hill, *Proc. Roy. Soc. (London)*, **B127**, 192 (1939).
20. R. Hill, *Symp. Soc. Exptl. Biol.*, **5**, 223 (1951).
21. O. Warburg, *Heavy Metal Prosthetic Groups and Enzyme Action*, Clarendon Press, Oxford, 1949, p. 213.

22. A. H. Brown and J. Franck, *Arch. Biochem.*, **16**, 55 (1948).
23. A. A. Benson and M. Calvin, *Ann. Rev. Plant Physiol.*, **1**, 25 (1950).
24. E. W. Fager, *Arch. Biochem. Biophys.*, **41**, 383 (1952).
25. E. W. Fager, *Biochem. J.*, **57**, 264 (1954).
26. E. Rabinowitch, *Sci. Am.*, Nov. 1953, p. 80.
27. R. Lumry, J. D. Spikes, and H. Eyring, *Ann. Rev. Plant Physiol.*, **5**, 271 (1954).
28. D. I. Arnon, M. B. Allen, and F. R. Whatley, *Nature*, **174**, 394 (1954).
29. D. I. Arnon, Paper presented at the Cell Symposium, AAAS, Berkeley Meeting, 1954; *Science*, **122**, 9 (1955).
30. M. B. Allen, D. I. Arnon, J. B. Capindale, F. R. Whatley, and J. L. Durham, *J. Am. Chem. Soc.*, **77**, 4149 (1955).
31. M. Gibbs and M. A. Cynkin, *Nature*, **182**, 1241 (1958).
32. N. E. Tolbert, *Brookhaven Symp. Biol.*, **11**, 271 (1958).
33. M. Gibbs and N. Calo, *Plant Physiol.*, **34**, 318 (1959).
34. R. M. Smillie and R. C. Fuller, *Plant Physiol.*, **34**, 651 (1959).
35. R. M. Smillie and G. Krotkov, *Can. J. Botany*, **37**, 1217 (1959).
36. R. Ueda, *Botan. Mag.*, **62**, 731 (1949).
37. L. R. Irmak, *Rev. Fac. Sci. Univ. Istanbul*, **B20**, 237 (1955).
38. J. B. Thomas, A. J. M. Haans, and A. A. Van Der Leun, *Biochim. Biophys. Acta*, **25**, 453 (1957).
39. F. R. Whatley, M. B. Allen, L. L. Rosenberg, J. B. Capindale, and D. I. Arnon, *Biochim. Biophys. Acta*, **20**, 462 (1956).
40. M. Losada, A. V. Trebst, and D. I. Arnon, *J. Biol. Chem.*, **235**, 832 (1960).
41. A. V. Trebst, M. Losada, and D. I. Arnon, *J. Biol. Chem.*, **235**, 840 (1960).
42. A. V. Trebst, H. Y. Tsujimoto, and D. I. Arnon, *Nature*, **182**, 351 (1958).
43. R. Hill, in *Proc. Intern. Congr. Biochem. 3rd, Brussels, 1955* (publ. 1956), p. 225.
44. W. O. James and V. S. R. Das, *New Phytologist*, **56**, 325 (1957).
45. H. Lundegårdh, *Nature*, **192**, 243 (1961).
46. S. Aronoff and M. Calvin, *Plant Physiol.*, **23**, 351 (1948).
47. W. Vishniac and S. Ochoa, *J. Biol. Chem.*, **198**, 501 (1952).
48. O. Warburg, G. Krippahl, H. Gewitz, and W. Völker, *Z. Naturforsch.*, **14b**, 712 (1959).
49. B. K. Stern and B. Vennesland, *J. Biol. Chem.*, **235**, PC51 (1960).
50. D. I. Arnon, F. R. Whatley, and M. B. Allen, *J. Am. Chem. Soc.*, **76**, 6324 (1954).
51. A. W. Frenkel, *J. Am. Chem. Soc.*, **76**, 5568 (1954).
52. A. M. Williams, *Biochim. Biophys. Acta*, **19**, 570 (1956).
53. D. M. Geller and J. S. Gregory, *Federation Proc.*, **15**, 260 (1956).
54. M. Kamen and J. W. Newton, *Biochim. Biophys. Acta*, **25**, 462 (1957).
55. I. C. Anderson and R. C. Fuller, *Arch. Biochem. Biophys.*, **76**, 168 (1958).
56. J. B. Thomas and A. J. M. Haans, *Biochim. Biophys. Acta*, **18**, 286 (1955).
57. B. Petrack and F. Lipmann, in *Light and Life*, W. D. McElroy and B. Glass, Eds., Johns Hopkins Press, Baltimore, 1961, p. 621.
58. M. Avron and A. T. Jagendorf, *Nature*, **179**, 428 (1957).
59. J. S. C. Wessels, *Biochim. Biophys. Acta*, **25**, 97 (1957).
60. E. Rabinowitch, in *Research in Photosynthesis*, H. Gaffron, Ed., Interscience, New York, 1957, p. 345.

61. M. B. Allen, F. R. Whatley, and D. I. Arnon, *Biochim. Biophys. Acta*, **27**, 16 (1958).
62. A. T. Jagendorf and M. Avron, *J. Biol. Chem.*, **231**, 277 (1958).
63. A. W. Frenkel, *J. Biol. Chem.*, **222**, 823 (1956).
64. D. Geller and F. Lipmann, *J. Biol. Chem.*, **235**, 2478 (1960).
65. D. I. Arnon, *Nature*, **184**, 10 (1959).
66. D. I. Arnon, in *Light and Life*, W. D. McElroy and B. Glass, Eds., Johns Hopkins Press, Baltimore, 1961, p. 489.
67. D. I. Arnon, F. R. Whatley, and M. B. Allen, *Biochim. Biophys. Acta*, **16**, 607 (1955).
68. C. B. Van Niel, in *Advances in Enzymology*, Vol. 1, F. F. Nord, Ed., Interscience, New York, 1941, p. 263.
69. E. Katz, in *Photosynthesis in Plants*, J. Franck and W. E. Loomis, Eds., Iowa Press, Ames, Iowa, 1949, p. 287.
70. L. S. Levitt, *Science*, **118**, 696 (1953); **120**, 33 (1954); *Experientia*, **15**, 16 (1959).
71. D. I. Arnon, F. R. Whatley, and M. B. Allen, *Science*, **127**, 1026 (1958).
72. D. I. Arnon, F. R. Whatley, and M. B. Allen, *Biochim. Biophys. Acta*, **32**, 47 (1959).
73. M. Avron, D. W. Krogmann, and A. T. Jagendorf, *Biochim. Biophys. Acta*, **30**, 144 (1958).
74. H. E. Davenport, *Biochem. J.*, **77**, 471 (1960).
75. K. Tagawa and D. I. Arnon, *Nature*, **195**, 537 (1962).
76. M. Losada, F. R. Whatley, and D. I. Arnon, *Nature*, **190**, 606 (1961).
77. L. E. Mortenson, R. C. Valentine, and J. E. Carnahan, *Biochem. Biophys. Res. Commun.*, **7**, 448 (1962).
78. R. C. Valentine, R. L. Jackson, and R. S. Wolf, *Biochem. Biophys. Res. Commun*, **7**, 453 (1962).
79. H. E. Davenport, R. Hill, and F. R. Whatley, *Proc. Roy. Soc.* (*London*), **B139**, 346 (1952).
80. D. I. Arnon, F. R. Whatley, and M. B. Allen, *Nature*, **180**, 182, 1325 (1957).
81. A. San Pietro and H. M. Lang, *J. Biol. Chem.*, **231**, 211 (1958).
82. H. E. Davenport and R. Hill, *Biochem. J.*, **74**, 493 (1960).
83. O. Warburg, Remarks in discussion of a paper by D. I. Arnon in *La Photosynthese*, Colloques Internationaux du Centre National de la Recherche Scientifiques, No. 119, Paris, 1963, p. 540.
84. H. S. Gewitz and W. Voelker, *Z. Physiol. Chem.*, **330**, 124 (1962).
85. K. T. Fry and A. San Pietro, *Biochem. Biophys. Res. Commun.*, **9**, 218 (1962).
86. T. Horio and T. Yamashita, *Biochem. Biophys. Res. Commun.*, **9**, 142 (1962).
87. S. Katoh and A. Takamiya, *Biochem. Biophys. Res. Commun.*, **8**, 310 (1962).
88. B. B. Buchanan, W. Lovenberg, and J. C. Rabinowitz, *Proc. Natl. Acad. Sci. U.S.*, **49**, 345 (1963).
89. W. Lovenberg, B. B. Buchanan, and J. C. Rabinowitz, *J. Biol. Chem.*, **238**, 3899 (1963).
90. R. Bachofen and D. I. Arnon, *Biochim. Biophys. Acta*, **120**, 259 (1966).
91. A. Mitsui and D. I. Arnon, in preparation.
92. F. R. Whatley, K. Tagawa, and D. I. Arnon, *Proc. Natl. Acad. Sci. U.S.*, **49**, 266 (1963).
93. K. Tagawa and D. I. Arnon, unpublished data, 1963.
94. M. Shin and D. I. Arnon, *J. Biol. Chem.*, **240**, 1405 (1965).

95. M. Shin, K. Tagawa, and D. I. Arnon, *Biochem. Z.*, **338**, 84 (1963).
96. D. G. Anderson and B. Vennesland, *J. Biol. Chem.*, **207**, 613 (1954).
97. M. Avron and A. T. Jagendorf, *Arch. Biochem. Biophys.*, **65**, 475 (1956).
98. M. Avron and A. T. Jagendorf, *Nature*, **179**, 428 (1957).
99. D. L. Keister, A. San Pietro, and F. E. Stolzenbach, *Arch. Biochem. Biophys.*, **98**, 235 (1962).
100. F. Lipmann, in *Currents in Biochemical Research*, D. E. Green, Ed., Interscience, New York, 1946, p. 137.
101. D. I. Arnon, H. Y. Tsujimoto, and B. D. McSwain, *Proc. Natl. Acad. Sci. U.S.*, **51**, 1274 (1964).
102. T. Horio and A. San Pietro, *Proc. Natl. Acad. Sci. U.S.*, **51**, 1226 (1964).
103. K. Tagawa, H. Y. Tsujimoto, and D. I. Arnon, *Proc. Natl. Acad. Sci. U.S.*, **49**, 567 (1963).
104. K. Tagawa, H. Y. Tsujimoto, and D. I. Arnon, *Proc. Natl. Acad. Sci. U.S.*, **50**, 544 (1963).
105. H. Baltscheffsky and D. Y. De Kiewiet, *Acta. Chem. Scand.*, **18**, 2406 (1964).
106. Z. Gromet-Elhanan and D. I. Arnon, *Plant Physiol.*, **40**, 1060 (1965).
107. M. Losada and D. I. Arnon, in *Metabolic Inhibitors*, R. M. Hochster and J. M. Quastel, Eds., Academic Press, New York, 1963, p. 559.
108. D. I. Arnon, H. Y. Tsujimoto, and B. D. McSwain, *Nature*, **207**, 1367 (1965).
109. F. R. Whatley, M. B. Allen, and D. I. Arnon, *Biochim. Biophys. Acta*, **16**, 605 (1955).
110. D. I. Arnon, M. Losada, F. R. Whatley, H. Y. Tsujimoto, D. O. Hall, and A. A. Horton, *Proc. Natl. Acad. Sci. U.S.*, **47**, 1314 (1961).
111. O. Warburg, G. Krippahl, and W. Schröder, *Z. Naturforsch.*, **10b**, 631 (1955).
112. R. Emerson, R. Chalmers, and C. Cederstrand, *Proc. Natl. Acad. Sci. U.S.*, **43**, 133 (1957).
113. J. H. C. Smith and C. S. French, *Ann. Rev. Plant Physiol.*, **14**, 181 (1963).
114. R. Emerson and R. V. Chalmers, *Phycol. Soc. Am. News Bull.*, **11**, 51 (1958).
115. L. R. Blinks, in *Comparative Biochemistry of Photoreactive Systems*, M. B. Allen, Ed., Academic Press, New York, 1960, p. 367.
116. J. Myers and C. S. French, *J. Gen. Physiol.*, **43**, 723 (1960).
117. J. Myers and C. S. French, *Plant Physiol.*, **35**, 963 (1960).
118. L. N. M. Duysens, in *Progress in Photobiology*, B. C. Christensen and B. Buchmann, Eds., Elsevier, Amsterdam, 1961, p. 135.
119. L. N. M. Duysens, J. Amesz, and B. M. Kamp, *Nature*, **190**, 510 (1961).
120. L. N. M. Duysens and J. Amesz, *Biochim. Biophys. Acta*, **64**, 243 (1962).
121. H. T. Witt, A. Müller, and B. Rumberg, *Nature*, **191**, 194 (1961).
122. H. E. Davenport and R. Hill, *Proc. Roy. Soc. (London)*, **B139**, 327 (1952).
123. M. D. Kamen, in *Comparative Biochemistry of Photoreactive Systems*, M. B. Allen, Ed., Academic Press, New York, 1960, p. 323.
124. R. Hill and F. Bendall, *Nature*, **186**, 136 (1960).
125. L. P. Vernon and W. S. Zaugg, *J. Biol. Chem.*, **235**, 2728 (1960).
126. A. W. Frenkel, *Brookhaven Symp. Biol.*, **11**, 276 (1958).
127. L. P. Vernon and O. K. Ash, *J. Biol. Chem.*, **234**, 1878 (1959).
128. N. I. Bishop, *Proc. Natl. Acad. Sci. U.S.*, **45**, 1696 (1959).
129. M. Nozaki, K. Tagawa, and D. I. Arnon, *Proc. Natl. Acad. Sci. U.S.*, **47**, 1334 (1961).
130. A. Trebst and H. Eck, *Z. Naturforsch.*, **16b**, 455 (1961).

131. Z. Gromet-Elhanan and M. Avron, *Biochem. Biophys. Res. Commun.*, **10**, 215 (1963).
132. D. L. Keister, *J. Biol. Chem.*, **238**, PC2590 (1963).
133. J. S. C. Wessels, *Biochim. Biophys., Acta*, **79**, 640 (1964).
134. M. Avron, *Biochem. Biophys. Res. Commun.*, **17**, 430 (1964).
135. M. Nozaki, K. Tagawa, and D. I. Arnon, in *Bacterial Photosynthesis*, H. Gest, A. San Pietro, and L. P. Vernon, Eds., Antioch Press, Yellow Springs, Ohio, 1963, p. 175.
136. K. Tagawa, H. Y. Tsujimoto, and D. I. Arnon, *Nature*, **199**, 1247 (1963).
137. B. Chance and C. R. Williams, in *Advances in Enzymology*, Vol. 17, F. F. Nord, Ed., Interscience, New York, 1956, p. 65.
138. E. Racker, in *Advances in Enzymology*, Vol. 23, F. F. Nord, Ed., Interscience, New York, 1961, p. 323.
139. E. Recondo and L. F. Leloir, *Biochem. Biophys. Res. Commun.*, **6**, 85 (1961).
140. H. P. Gosh and J. Preiss, *Biochem.*, **4**, 1354 (1965).
141. O. H. Warburg and P. Z. Ostendorf, *Z. Naturforsch.*, **18b**, 933 (1963).
142. O. H. Warburg, *New Methods of Cell Physiology*, Interscience, New York, 1962.
143. L. N. M. Duysens, *Progr. Biophys.*, **14**, 1 (1964).
144. J. M. Bové, C. Bové, F. R. Whatley, and D. I. Arnon, *Z. Naturforsch.*, **18b**, 683 (1963).
145. A. Pirson, *Z. Botan.*, **31**, 193 (1937).
146. E. Kessler, *Arch. Biochem. Biophys.*, **59**, 527 (1955).
147. D. W. Krogmann, *Biochem. Biophys. Res. Commun.*, **4**, 275 (1961).
148. D. I. Arnon and A. A. Horton, *Acta Chem. Scand.*, **17**, S135 (1963).
149. F. R. Whatley and A. A. Horton, *Acta Chem. Scand.*, **17**, S140 (1963).
150. L. P. Vernon and M. Avron, *Ann. Rev. Biochem.*, **34**, 269 (1965).
151. B. Kok, *Biochim. Biophys. Acta*, **48**, 527 (1961).
152. L. N. M. Duysens, Doctoral thesis, University of Utrecht, Holland, 1952.
153. D. I. Arnon, H. Y. Tsujimoto, and B. D. McSwain, *Proc. Natl. Acad. Sci. U.S.*, **51**, 927 (1965).
154. S. Katoh, *Nature*, **186**, 533 (1960).
155. S. Katoh, I. Suga, I. Shiratori, and A. Takamiya, *Arch. Biochem. Biophys.*, **94**, 136 (1961).
156. H. Dam., E. Hjorth, and I. Kruse, *Physiologia Plantarum*, **1**, 379 (1948).
157. L. P. Kegel and F. L. Crane, *Nature*, **194**, 1282 (1962).
158. D. I. Arnon and F. L. Crane, in *Biochemistry of Quinones*, R. A. Morton, Ed., Academic Press, New York, 1965, p. 433.
159. E. Fischer, *J. Chem. Soc.*, **91**, 1749 (1907).
160. G. Von Hevesy, *Les Prix Nobel en 1944*, Nobel Foundation, Stockholm, 1945.
161. A. J. P. Martin, *Les Prix Nobel en 1952*, Nobel Foundation, Stockholm, 1953; R. L. M. Synge, *ibid.*

Microsomal Electron Transport

Philipp Strittmatter, *Department of Biological Chemistry, Washington University School of Medicine, St. Louis, Missouri*

I.	Introduction.	171
II.	Physical and Chemical Properties of Microsomes	172
III.	Oxidative Reactions of Microsomes	173
IV.	Chemical Properties of Oxidative Components of Microsomes	174
	A. Cytochrome b_5	174
	B. The Carbon Monoxide Binding Pigment	176
	C. TPNH Cytochrome c Reductase	177
	D. DPNH Cytochrome b_5 Reductase	177
V.	The Mechanism of Microsomal Hydroxylation Reactions	179
VI.	The Mechanisms of Isolated Microsomal Flavoproteins	181
	A. TPNH Cytochrome c Reductase	182
	B. DPNH Cytochrome b_5 Reductase	184
VII.	The Overall Problem of Microsomal Electron Transport	188
	References	189

I. Introduction

The problems involved in understanding microsomal electron transport are emphasized by the marked contrasts in these processes with the oxidative events that occur within the mitochondrion. The highly organized structural elements of cells, the mitochondria, are geared mainly to a coupled reduction of oxygen by a variety of substrates and the production of ATP or its chemical equivalent. The microsomes, however, are derived submicroscopic particles produced by the fragmentation of intracellular membranes, which contain a spectrum of enzymic activities concerned with general as well as specialized tissue functions. These reactions include a number of enzymes that fall in the class of mixed function oxidases (1–3). In addition, oxidative sequences involving reduced pyridine nucleotides as substrates were identified in some of the earliest biochemical characterizations of microsomes (4,5). Since then a heme protein and two flavoproteins have been isolated from the membrane fragments, and more recently, evidence has been obtained that a carbon monoxide-binding pigment in microsomes is also a heme protein. A discussion of microsomal electron transfer must therefore attempt to define the reductive steps in several biosynthetic pathways as well as other electron transport sequences in terms of identified oxidative components of this cell fraction. In the first case, the problem is one of limited information concerning the mechanisms of known metabolic events. The second is more frustrating because, even though the isolation of highly active enzymes which initiate reduced pyridine nucleotide oxidation has been achieved, the identity of the terminal electron acceptors, and therefore their biological significance, remains obscure.

II. Physical and Chemical Properties of Microsomes

Tissue homogenization techniques, which yield cell-free preparations, also result in the conversion of the endoplasmic reticulum into microsomes. Originally these particles were isolated by differential centrifugation of liver homogenates in a single density medium (6,7). Consequently, similar particles have been obtained from most mammalian tissues including kidney, adrenal cortex, testes, brain, mammary, pancreas, and thyroid. In reviewing the properties of the endoplasmic reticulum and microsomes, Siekevitz (8,9) has emphasized the physical and functional heterogeneity of this macromolecular structure. By both gradient differential and isopycnic centrifugation, it has been possible to separate heavy (rough) and light (smooth) membrane fragments in which separation depends upon the dense ribosome particles bound to the membrane in the heavy component. There is also a spectrum of membrane fragment sizes in microsomal fractions depending upon the particular physical manipulation involved in the tissue homogenization procedure. The enzymes that participate in electron transport are concentrated in the lipoprotein membrane fragments, particularly in the smooth fraction. The lipid of the membranes, which represents 30–50% of the dry weight of the particles, is largely phospholipid and there is also a significant but smaller amount of cholesterol and cholesterol esters. The concentrations of both flavin and heme are relatively high in this lipoprotein environment, approximately 0.2 μmole of flavin (personal communication from Dr. Helen B. Burch) and 2.5 μmoles of heme per gram of microsomal protein (10,11).

III. Oxidative Reactions of Microsomes

Electron transport in microsomes involves mainly three types of oxidative reactions. A considerable number of reactions in steroid and drug metabolism have in common mixed function oxidases that utilize both TPNH and molecular oxygen. Most ubiquitous are the hydroxylases which are represented schematically by Reaction (1):

$$R + TPNH + H^+ + O_2 \rightleftharpoons ROH + TPN^+ + H_2O \qquad (1)$$

The TPNH is usually the specific hydrogen donor in microsomal hydroxylations in which an oxygen atom from molecular oxygen appears in the product, the typical mixed function oxidase mechanism discovered by Mason et al. (1). Reviews by Hayaishi (12; see also chapter in this volume) deal with the mechanism of this and other classes of oxygenases. Related microsomal reactions of drug metabolism may be considered hydroxylations and have been reviewed by Brodie (2) and by Shuster (3). They show a similar requirement for TPNH and oxygen and result in sulfoxidation of thioethers, O-dealkylation, N-dealkylation, and S-demethylation. In the presence of iron pyrophosphate, a TPNH- and oxygen-dependent peroxidation of lipids also occurs (13).

Two groups of investigators (14,15) have obtained evidence that microsomal preparations from liver, kidney, adrenal cortex, and retina catalyze an DPNH-dependent reduction of semidehydroascorbate to ascorbate (Reaction (2)):

$$2 \text{ Semidehydroascorbate} + DPNH \rightleftharpoons 2 \text{ Ascorbate} + DPN^+ \qquad (2)$$

They propose that this may provide a route for the formation of ascorbate in reductive reactions that appear to require this electron donor in these tissues, or in ion transport. More recently, ascorbate has also been implicated in the hydroxylation of peptide-bound proline in colagen biosynthesis by microsomes (16).

The third general type of oxidative sequence is shown in Reaction (3):

$$\text{Reduced pyridine nucleotide} + 2 \text{ Cyt } c \text{ (Fe}^{3+}) \rightleftharpoons$$
$$\text{Oxidized pyridine nucleotide} + 2 \text{ Cyt } c \text{ (Fe}^{2+}) \qquad (3)$$

The reaction which utilizes DPNH as electron donor was one of the first enzymic activities detected in microsomal particles (4). There is a similar TPNH-cytochrome c reductase activity (5) which is identified with the flavoprotein TPNH-cytochrome c reductase (17,18). A large part of the DPNH-cytochrome c reductase activity, however, involves a specific

DPNH-cytochrome b_5 reductase (19,20) (Reaction (4)) and the rapid reduction of cytochrome c by the reduced microsomal heme protein (Reaction (5)).

$$DPNH + 2 \text{ Cyt } b_5 (Fe^{3+}) \rightleftharpoons NAD^+ + H^+ + 2 \text{ Cyt } b_5 (Fe^{2+}) \qquad (4)$$

$$\text{Cyt } b_5 (Fe^{2+}) + \text{Cyt } c (Fe^{3+}) \rightleftharpoons \text{Cyt } b_5 (Fe^{3+}) + \text{Cyt } c (Fe^{2+}) \qquad (5)$$

The overall reaction (Reactions (4) + (5)) not only represents one of the most active catalytic sequences of microsomes, but accounts for almost 50% of the DPNH cytochrome c reductase activities of whole homogenates of several tissues.

The following discussion* will focus on three main aspects of these oxidative reactions. Thus, the chemical properties of several oxidative components which can either be identified within microsomes or isolated in homogeneous forms will form the basis for a subsequent description of the recent advances in the general functional role of microsomes in a large class of hydroxylation reactions. Finally, the two flavoproteins which have been isolated from microsomes will be reviewed as objects for detailed studies on the mechanisms of flavin catalysis. In stressing these three areas, which have yielded most readily to experimental examination, the remaining problems of microsomal electron transport should be defined more clearly.

IV. Chemical Properties of Oxidative Components of Microsomes

A. Cytochrome b_5

The absorption bands of reduced cytochrome b_5 have been observed at intervals since the initial spectroscopic experiments of MacMunn (21,22). However, it was not until 1952 that it was recognized, by C. F. Strittmatter and Ball (11,23), that the marked red color of the microsomal pellet is largely dependent upon the presence of this heme protein. The cytochrome is released from microsomes by incubation with purified pancreatic lipase (24) and can be isolated as a homogeneous protein from rabbit and calf liver with a molecular weight in the latter case (25) of 13,000. This protein contains one mole of heme. Similar preparations have been obtained from a variety of tissues and species (26–28) and one report of the crystallization of the cytochrome appeared in a brief communication (29).

---

* Many of the experimental details were presented recently in a symposium on microsomal electron transport (9,53,58,60,61).

In the visible region the oxidized form of cytochrome b_5 has an absorption maximum at 413 mμ (Fig. 1, curve 1) (24,25) and in the reduced form the maxima appear at 423, 526, and 556 mμ, in good agreement with the maxima reported for the cytochrome in microsomal suspensions (11). Below 400 mμ there are broad absorption bands for the oxidized form at 355–370 mμ and for the reduced heme protein at 320–340 mμ. The broad and low-intensity absorption band of the oxidized form from 250 to 285 mμ (curve 2) results from light absorption by both the heme group and the 3 tyrosyl and 1 tryptophan residues in the peptide chain (25). The isolated heme protein is relatively stable at neutral and slightly alkaline pH values at 25° and can be stored indefinitely at −15°. It is reduced by a variety of reducing agents, e.g., sodium hydrosulfite, cysteine, and DPNH containing catalytic amounts of DPNH cytochrome b_5 reductase. It is in turn oxidized rapidly by cytochrome c, potassium ferricyanide, ferric chloride, and various dyes of appropriate potential, but very slowly by oxygen. Spectrophotometric oxidation-reduction titrations (24) indicate that the isolated cytochrome has an E'_0 at pH 7 and 25° of + 0.02 V. This

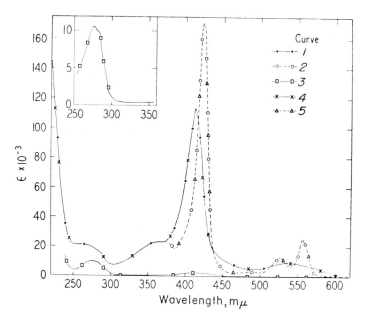

Fig. 1. Absorption spectra of: (1) oxidized cytochrome b_5; (2) reduced cytochrome b_5; (3) apocytochrome b_5; (4) apocytochrome b_5 plus heme; (5) apocytochrome b_5 plus heme and sodium hydrosulfite. From reference 25.

is 0.1 V higher than the estimated E_0' for this cytochrome when it is bound to microsomal particles (11), and may reflect the alteration in the equilibrium of this oxidation-reduction system as a result of the loss of heme protein–lipoprotein interactions. It is significant for the following discussion that neither the oxidized nor the reduced cytochrome will react with carbon monoxide or cyanide.

Completely resolved apocytochrome b_5 preparations can be obtained (Fig. 1, curve 3) which will recombine with heme to yield the original cytochrome spectrum (curve 4). Chemical or enzymic reduction (curve 5) also yields an unaltered reduced heme protein spectrum. The heme binding involves, as one ligand, a specific histidyl residue, since the formation of one mole of mono-azo imidazole group per mole of an apocytochrome derivative results in complete inhibition of recombination. The heme binding also appears to involve other interactions of the metalloporphyrin and changes in the protein conformation in which the iron in the heme group plays a crucial role (30).

B. The Carbon Monoxide Binding Pigment

Klingenberg (31) and Garfinkle (32) first detected a CO-binding pigment in mammalian liver microsomal preparations by the difference spectrum resulting from the intense absorption band which appears at 450 mμ when dithionate-reduced microsomes are equilibrated in a CO atmosphere. This pigment has been examined in microsomes largely as the CO complex (33,34); however, Hashimoto, Yamano, and Mason (35) observed an electron spin resonance signal in microsomal preparations which they attributed to an Fe_x compound. They suggested that it might arise from a heme protein containing the microsomal heme in excess of that which is bound to cytochrome b_5. Because the CO-binding pigment, or P-450, will also form a reversible ethyl isocyanide complex, a reaction characteristic of a hemochromogen with a dissociation constant similar to CO, Omura and Sato (34) also concluded that the CO-binding pigment is a heme protein. It is reduced slowly by DPNH and TPNH and oxidized by molecular oxygen (34). Indirectly, Estabrook, Cooper, and Rosenthal (36), in experiments described in more detail below, showed that the CO complex is photodissociable.

Omura and Sato (37) obtained additional evidence that P-450 is a heme protein when they attempted to purify this pigment. All solubilization treatments altered the spectrum of the CO pigment and yielded partially purified preparations of a heme protein, termed P-420 again for the absorption maximum of its CO complex. This derivative of P-450 has a

Soret peak at 414 mμ in the oxidized form and the dithionate reduced pigment has α, β, and Soret peaks at 559, 530, and 426 mμ, respectively. The oxidation-reduction potential, E'_0, is −0.02 V at pH 7.0 and 20°. Most significant is the fact that this preparation clearly contains heme, one to two moles of non-heme iron, and no flavin. In the crude, solubilized microsomal digest, the heme protein is reduced by DPNH and TPNH, but this activity is lost during further purification. The reduced P-420 is readily oxidized by oxygen.

The two heme proteins, cytochrome b_5 and P-450 are present in approximately equal amounts in microsomes and account for all of the iron protoporphyrin in this cell fraction (34,36). In rabbit liver microsomes (34) the values are 1.12 and 1.55 mμmoles per milligram of microsomal protein for cytochrome b_5 and P-450, respectively. A similar value for cytochrome b_5 in rat liver microsomes has been reported (11) and this corresponds to three times the cytochrome c content of rat heart muscle (11).

C. TPNH Cytochrome c Reductase

The flavoprotein TPNH cytochrome c reductase has been extracted from microsomes with pancreactic lipase by Williams and Kamin (17) and with trypsin by Phillips and Langdon (18). The properties of the isolated enzyme indicate that it is probably identical with the liver TPNH cytochrome c reductase isolated from whole-liver acetone powder in 1950 by Horecker (38). The highly purified enzyme (39) contains 2 moles of FAD per mole of protein, no appreciable iron and has a molecular weight of 60,000–90,000. In the near ultraviolet and visible region the enzyme has a typical flavoprotein spectrum, in which the flavin absorption bands are altered by interaction with the protein, yielding a pronounced shoulder at 485 mμ and a molar extinction of 11.3 × 10^3 at 450 mμ. The turnover number of the isolated enzyme with TPNH and cytochrome c as substrates is approximately 1200–1400/min. TPNH oxidation also occurs with a number of one- and two-electron acceptors including dichlorophenol indophenol, neotetrazolium, and potassium ferricyanide. The reaction with oxygen is extremely slow, less than two per minute.

D. DPNH Cytochrome b_5 Reductase

The isolation of cytochrome b_5 reductase involves an initial extraction of the enzyme from microsomes with snake venom (20). The molecular weight of the purified protein, both by minimum molecular weight determinations based on FAD content and by physical measurements

(20,40), is 38,000–41,000. The enzyme therefore contains 1 mole of flavin per mole and appears to be homogeneous not only by this criterion but also on the basis of electrophoresis at several pH values and stoichiometric titrations with substrates. Iron and a number of other transition elements are not present in the flavoprotein.

The resolution of cytochrome b_5 reductase by a brief acid ammonium sulfate treatment (40) yields apoprotein preparations (Fig. 2, curve 2) which will bind one equivalent of either FAD or FMN (curves 3 and 4). Their spectra are almost identical with that of the original enzyme (curve 1). The fluorescence of the isoalloxazine ring is completely quenched in either case and the position and intensity of the absorption bands are altered most conspicuously by the appearance of a distinct shoulder at 485–490 mμ. This coenzyme–protein interaction appears to involve a specific tyrosyl residue, since iodination of one phenolic group on the apoenzyme completely blocks flavin interaction. The results of more indirect experiments suggest that a specific protein conformation is also required for flavin binding. The apoenzyme is extremely labile to incubation at temperatures above 10° or with trypsin at 0°, procedures which do not affect the holoenzyme. Furthermore, the ability of apoprotein prepara-

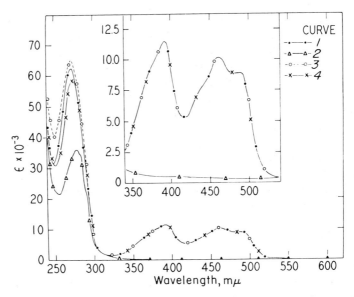

Fig. 2. Absorption spectra of: (1) cytochrome b_5 reductase; (2) aporeductase; (3) aporeductase plus FAD; (4) aporeductase plus FMN. From reference 40.

tions to recombine with flavin at a neutral pH is only slowly recovered after resolution at pH 1.8. The absence of the AMP moiety of the coenzyme and then the second phosphate group yields a progressive increase in the apparent dissociation constant of the flavin from less than $10^{-9}M$ for FAD to $5 \times 10^{-5}M$ for riboflavin, and conversely a decrease in the enzymic activity of the flavin–protein complexes to 15% of the maximum value (40). A preliminary examination of the interaction of free flavins with the protein indicates that the nitrogen in the 3 position of the iso-alloxazine ring must be free, since the fluorescence of the 3-methyl analog of riboflavin is unaffected by the apoenzyme.

Although the addition of reduced pyridine nucleotides to the aporeductase does not result in appreciable spectral changes above 300 mμ, an interaction between nucleotides and apoenzyme was readily detected by a marked enhancement of the reduced pyridine nucleotide fluorescence (42). One mole of NADH or its analogs is bound per mole of apoprotein, and the apparent dissociation constants are approximately $2 \times 10^{-7}M$. This interaction apparently involves one of the three reactive sulfhydryl groups on the apoprotein, because the pyridine nucleotide is completely displaced by one equivalent of p-mercuribenzoate. Both DPN and adenosinedi-phosphoribose abolish the fluorescence enhancement and therefore also displace the reduced pyridine nucleotides from the apoenzyme at nucleotide concentrations corresponding to dissociation constants for the DPN–apoenzyme complex of 10^{-5} and $10^{-6}M$ for the complex with adenosinedi-phosphoribose. This indicates not only that the adenosinediphosphoribose moiety is involved in the binding, but also that the oxidation state of the pyridine ring markedly affects the stability of the complexes. In the reduced form the pyridine nucleotide is bound much more tightly than in the oxidized form.

V. The Mechanism of Microsomal Hydroxylation Reactions

The hydroxylation of various drugs and steroids is representative of one of the clearly defined and general biological functions of microsomes. Initially, considerable effort was involved in documenting the stoichiometry shown in Reaction (1) and required by the mechanism of this type of reaction (cf. chapter by Hayaishi) for a number of substrates (2,3). More recently, several details concerning the mechanism of these microsomal reactions have been revealed by experiments which implicate two of the oxidative components of microsomes which were described above. Ryan and Engel (43) first demonstrated that the C_{21} hydroxylation of 17-hydroxy-progesterone by adrenocortical microsomes is inhibited by CO, and this type of inhibition has been confirmed subsequently for the C_{21} hydroxylation and extended as well to other steroid and drug hydroxylation reactions (44,45). Even more significant was the observation of Ryan and Engel (43) that the inhibition by CO is reversed by light.

By 1963, the CO-binding pigment, P-450, had been detected in liver microsomes by Klingenberg (31) and Garfinkle (32). Estabrook, Cooper, and Rosenthal (36) first identified this pigment in adrenal microsomes and then utilized the elegant action spectrum technique of Warburg (46) to obtain direct evidence that P-450 is the CO-sensitive component of the C_{21}–hydroxylase system. The photochemical action spectrum for the reversal of CO inhibition of the adrenocortical hydroxylation reaction with a maximum at 452 mμ coincided with the absorption maximum of P-450. The wide distribution of P-450 in various tissue microsomes and additional data on the CO sensitivity of a number of hydroxylation reactions (31,32,36,43–45) suggest that this CO-binding pigment is a common reactant in this class of microsomal reactions.

In a recent series of studies, Ernster and his colleagues (47–53) have utilized the phenomenon of drug-induced synthesis of liver microsomal hydroxylating enzyme systems in the rat to extend the description of these enzymic components. The initial *in vivo* demonstration of drug stimulation of hydroxylating capacity by Brown, Miller, and Miller (54) had already been extended to *in vitro* systems (55–57) and the results suggested that an increased synthesis of drug-metabolizing enzymes actually occurred. Ernster et al. (50,53) examined rat liver microsomes after phenobarbital treatment of rats for changes in (*1*) hydroxylation and oxidative demethylase activity using aminopyrine as substrate; (*2*) TPNH cytochrome *c*, neotetrazolium, and dichlorophenol indophenol reductase activities; (*3*) DPNH cytochrome *c* reductase activity; and (*4*) the content of P-450 and cytochrome b_5.

Only the TPNH-specific reductase activities and the P-450 showed a similar fivefold and concomitant increase with the oxidative demethylation or hydroxylation activity. These increases appear to represent new protein synthesis, since all are sensitive to actinomycin D and puromycin. The TPNH cytochrome *c* reductase therefore also appears to be a component of hydroxylation systems, and the involvement of P-450 is confirmed by an entirely different experimental design. Ernster and Orrenius (53) consequently included the following microsomal components in the scheme for hydroxylation:

$$\text{TPNH} \longrightarrow \text{Flavoprotein} \overset{?}{\dashrightarrow} \text{P-450} \longrightarrow \text{ROH}_3 + O_2 \tag{6}$$

with Cyt c below Flavoprotein and RH + HCHO below ROH$_3$.

Additional support for this mechanism (52,53) is the observation that the decay of activity for the hydroxylation reaction following drug induction is paralleled by decreases in both oxidative components utilized in this mechanism.

Further examination of this mechanism by Ernster and Orrenius (53), in which the binding of ^{14}C-labeled drugs to microsomes was studied, have shown that 0.3–0.5 mμmoles of ^{14}C-labeled aniline can be bound per milligram of microsomal protein. Remarkably, this binding of aniline to microsomes from livers of rats, previously induced with phenobarbital, was increased to 1.3 mμmoles per milligram of protein. Furthermore, the binding is inhibited 40–90% by CO, and the amount of P-450 is roughly equal to the total ^{14}C-labeled aniline bound in the absence of CO. Ernster and Orrenius conclude that these data do not support the idea of the existence of specific hydroxylases related to individual drugs, at least for the substrates which they have examined. These results present the unusual situation in which there is a direct stoichiometry and apparently an interaction between a specific substrate and an enzyme which appears to be the common catalyst in reactions involving molecular oxygen and a variety of drugs and perhaps steroids.

Recent studies of Omura et al. (58), on a related hydroxylation reaction catalyzed by adrenocortical mitochondria, the steroid-11-β-hydroxylase, implicate a ferridoxin-like nonheme iron protein as the electron carrier between a TPNH-specific flavoprotein and adrenocortical mitochondrial P-450. Whether or not a similar catalyst is present in microsomes must still be determined.

VI. The Mechanisms of Isolated Microsomal Flavoproteins

The summary of the oxidative enzyme activities of microsomes (cf. Section III) emphasizes the fact that reduced pyridine nucleotides are the major electron donors for microsomal oxidation-reduction reactions. Two flavoproteins, one specific for TPNH and the other for DPNH, have been isolated in soluble forms (cf. Section IV) and therefore could be subjected to more detailed studies. TPNH cytochrome c reductase is of particular interest because there is now evidence that it may be the initial enzymic step in microsomal hydroxylations. The DPNH cytochrome b_5 reductase, on the other hand, provides a direct pathway for the reduction of the only other heme protein in microsomes, and represents the most active catalytic system of the oxidative enzyme complement of this cell fraction.

A. TPNH Cytochrome c Reductase

The studies on the mechanism of the TPNH cytochrome c reductase have been concerned with the detection of possible intermediates in the catalytic mechanism. Kinetic studies with catalytic amounts of enzyme (39) and analysis by the method of Alberty (59) are consistent with a mechanism in which TPNH first reacts with the enzyme to yield TPN and reduced enzyme which in turn reacts with electron acceptors, and are clearly distinguished from the kinetics expected for the alternate mechanism involving the formation of a ternary complex.

The initial interaction of TPNH with the reductase was observed directly as a partial reaction with stoichiometric amounts of enzyme and substrate (Fig. 3) (39,60). Under anaerobic conditions, the addition of 1 mole of TPNH per mole of protein, thus 0.5 moles per mole of enzyme flavin, results in a partial bleaching of the flavin absorption band in the 400–500 mμ region and the appearance of a broad absorption band from 500–600 mμ. Masters, Kamin, Gibson, and Williams (39) suggest that these spectral changes represent the formation of a half-reduced enzyme

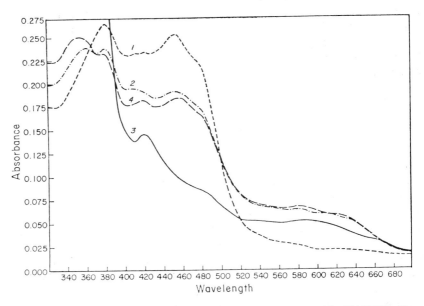

Fig. 3. Anaerobic titration of cytochrome c reductase with NADPH. From Masters, Kamin, Gibson, and Williams (39).

in which each of the two flavins is similarly in the half-reduced state. Excess TPNH will then yield a further bleaching of the flavin absorption band at 450 mμ and a partial decrease in the absorption band at 500 mμ to the fully reduced enzyme in which all flavin is present as $FADH_2$. When air is admitted to the reduced enzyme, the excess TPNH is initially oxidized and the enzyme flavin is subsequently oxidized, but only to the same half-reduced level obtained with 0.5 mole of substrate in the initial anaerobic experiment. This partially reduced form of the enzyme is stable for hours.

Stopped-flow experiments (39,60) were used to detect the slow and fast steps in the enzyme reduction and oxidation by various electron acceptors. All of the reductive sequences, i.e., oxidized to half reduced, half reduced to fully reduced, and oxidized to fully reduced, are sufficiently rapid to meet the kinetic requirements of the catalytic system. The oxidation of fully reduced enzyme by cytochrome c, ferricyanide, dichlorophenol indophenol, or menadione involves the rapid formation of half-reduced enzyme, again at rates comparable to the maximum enzyme turnover. However, the oxidation of the half-reduced enzyme, either by one- or two-electron acceptors invariably proceeds at a slow rate. Kamin et al. (60) have summarized these experiments in the mechanism shown in Figure 4. The TPNH cytochrome c reductase, according to this scheme, utilizes a shuttle between the fully reduced and half-reduced enzyme in the rapid reduction of both one- and two-electron acceptors in solution. In this hitherto undescribed form of flavin catalysis, both the electron donor and electron acceptor interactions with the enzyme must result in electron transfers involving two molecules of flavin, and on the oxidation side this occurs with either one- or two-electron acceptors. This mechanism obviously presents intriguing questions concerning the extent of flavin–flavin interaction and the nature of the electron donor and acceptor interactions with the enzyme that will accommodate this sequence of catalytic events and result in the unusually stable half-reduced enzyme.

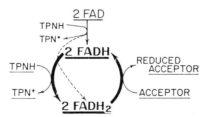

Fig. 4. Proposed mechanism for microsomal NADPH cytochrome c reductase. From Kamin, Masters, Gibson, and Williams (60).

B. DPNH Cytochrome b_5 Reductase

In addition to the rapid reduction of cytochrome b_5, the flavoprotein DPNH cytochrome b_5 reductase also catalyzes a slow hydrogen transfer from DPNH to analogs of DPN, and the rapid reduction of a number of chemical oxidants (Table I) (61). The enzyme activity is extremely low

TABLE I

Substrate Specificity of Cytochrome b_5 Reductase

Electron donor	Electron acceptor	Turnover[a]
DPNH	Potassium ferricyanide	29,000
	Cytochrome b_5	29,000
	Dichlorophenol indophenol	5,200
	Oxygen	1.4
	Cytochrome c	0
AcPyADH	Potassium ferricyanide	3,500
PyAlDH	Potassium ferricyanide	400
TPNH	Potassium ferricyanide	2.8
DPNH	AcPyADH	3.0

[a] Moles substrate oxidized per mole enzyme per minute at pH 8.1 and 25°. From reference 61.

with either oxygen or cytochrome c as electron acceptor, or with TPNH as electron donor. Most of the studies on the intermediates in these reactions evolved from the initial observation that DPNH protects the enzyme from inhibition of all of these activities by sulfhydryl group reagents (62). This led to the detection of a stable reduced enzyme–DPN complex (2,63), which was identified by the difference spectrum obtained by the addition of one equivalent of DPNH to the enzyme under anaerobic conditions (Fig. 5, curve 1). The flavin absorption bands are reduced and the absorption band at 317 mμ was shown to arise from DPN interaction with the free reduced enzyme. Analogs of DPNH yield a similar complex (curves 2 and 3), but TPNH only slowly reduces the flavin and TPN is not bound to the reduced enzyme (curve 4). The DPN binding involves a single sulfhydryl group, the same one essential to enzyme activity (62) and DPNH interaction with the apoenzyme (42) (cf. Section IV).

On the basis of these data, a reduced flavoprotein–oxidized pyridine nucleotide complex was assumed to be an intermediate in the enzyme reaction. To involve this complex as an intermediate in the slow hydrogen

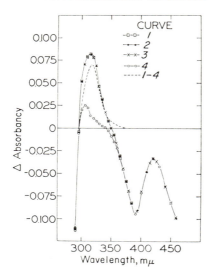

Fig. 5. Difference spectra of reduced cytochrome b_5 reductase complexes (63). Difference spectra were plotted for the absorbance change resulting from the addition of the following nucleotides to the reductase: (*1*) NADH; (*2*) desamino analog of NADH; (*3*) AcPyADH; (*4*) NADPH.

transfer reaction from DPNH to its oxidized analogs, the replacement of DPN on the reduced enzyme by an analog followed by a reversal of reduced enzyme formation was assumed. In testing this hypothesis (64) the hydrogen transfer was found to be flavin dependent and to involve direct and stereospecific hydrogen transfer from the 4A position of the nicotinamide ring to the 4A position of the analog. These results would be expected in such a mechanism of hydrogen transfer and also imply that direct hydrogen transfer to the flavoprotein may occur. Stoichiometric, rapid mixing experiments (65) showed that flavin reduction by DPNH was equal to the turnover of the enzyme and was therefore the rate-limiting step in the enzyme reaction. Both the flavin reduction and the overall catalytic reaction show a marked deuterium rate effect when DPNH or its analogs contain deuterium in the 4A position of the dihydropyridine ring. The formation of the reduced enzyme–pyridine nucleotide complex therefore is stereospecific and involves the same hydrogen atom as the hydrogen transfer reactions from DPNH to its analogs.

A second stable intermediate in the enzyme reaction was detected by the unusual properties of the PyAlADH interactions with the reductase. The existence of such an intermediate had already been suggested by the observation that the enzyme, which is largely in the oxidized form during turnover, is still protected from sulfhydryl group reagents by substrate (65). Fortunately, PyAlADH reacts with cytochrome b_5 reductase to yield an apparent intramolecular equilibrium in which the flavin is 65% reduced and the PyAlADH 35% reduced (66). The equilibrium mixture was thus assumed to include 65% reduced flavoprotein–oxidized pyridine nucleotide, the usual stable complex formed with DPNH or AcPyADH, and 35% oxidized flavoprotein–PyAlADH. This latter complex, which presumably precedes flavin reduction, is characterized by complete quenching of the pyridine nucleotide fluorescence and protection of the essential sulfhydryl group. Since the pyridinealdehyde analog is completely displaced from the enzyme by one equivalent of DPNH, the reactions leading to the formation of both the oxidized and reduced enzyme complexes appear to be reversible. This was confirmed by using high concentrations of borate buffer to recover AcPyADH from a reduced enzyme–AcPyAD complex (67). Furthermore, deuterium introduced as AcPyADH(D_A) was again recovered in the 4A position, indicating that the reversal involves the exact reaction sequence involved in complex formation.

Micro stopped-flow experiments (68) showed that these enzyme complexes satisfy both kinetic requirements for intermediates of the enzyme reaction, rapid formation and oxidation. The oxidized enzyme–reduced pyridine nucleotide complexes are formed too rapidly to measure. However, both flavin reduction and the appearance of the 317 mμ absorption band, the two distinguishing features of the reduced enzyme–oxidized pyridine nucleotide complex, proceed by first-order reactions with rate constants equal to enzyme turnover. Complete oxidation of reduced enzyme–pyridine nucleotide complexes and free reduced enzyme, by either cytochrome b_5 or potassium ferricyanide, is always more rapid than flavin reduction by the reduced nucleotide used as electron donor in each case. The oxidative sequence, either with ferricyanide and free reduced enzyme or with cytochrome b_5 and reduced enzyme–pyridine nucleotide complexes, can be clearly separated into an initial rapid one-electron transfer and a subsequent measurable one-electron oxidation. In the former situation, the transient spectral band often attributed to the flavin semi-quinone form in the 500–600 mμ region (69,70) could be distinguished.

The reaction sequence in the cytochrome b_5 reductase system (Fig. 6) (68) has been described in terms of the enzyme–coenzyme and substrate complexes which these kinetic and coenzyme binding studies (cf. Section IV) implicate as reaction intermediates. An initial rapid reaction which precedes flavin reduction (*1*) is followed by the rate-limiting step of direct

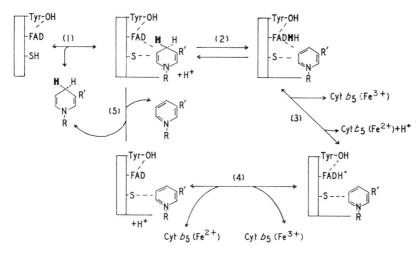

Fig. 6. The bar represents the protein: R, adenosinediphosphoribose; R', an amide, acetyl, or aldehyde group ; the hydrogen to the left at position 4 of the dihydropyridine ring, the hydrogen in position A. From reference 68.

hydrogen transfer (2), and oxidation involves a rapid one-electron reaction (3) and the final, slower second one-electron transfer (4). The two proposed intermediates have been detected as stable complexes. In the first case the reduced nucleotide and the flavin interact with at least two protein sites and possibly the nucleotides are sufficiently near to one another to permit fluorescence quenching of the reduced pyridine ring by energy transfer to the flavin. The carbon–hydrogen bond cleavage, which initiates the oxidation reactions, is sensitive to both the R' group and to substitution of deuterium in the 4A position of the dihydropyridine ring, and the rate of this reaction varies almost 3 orders of magnitude.

The hydrogen atom involved in the direct transfer does not participate in a rate-limiting step in the final oxidation (Reaction (4)). However, both the pyridine nucleotide bound to the reduced enzyme and the metalloporphyrin in the cytochrome b_5 do affect the rate of this reaction (68). This implies that the pyridine nucleotide remains bound to the enzyme during the entire oxidative cycle and that an interaction between pyridine nucleotides and half-oxidized enzyme affects the reactivity of the flavin with electron acceptors. The mechanism involves the minimum number of intermediates and each reaction may be more complex. In particular, the oxidation reactions obviously must involve flavoprotein–cytochrome interactions. Two other features of the enzyme mechanism also deserve comment. First, the exclusion of appreciable exchange of the proton involved in the direct hydrogen transfer requires that this proton either is confined in a hydrophobic area which excludes water, or participates in bonds which do not permit appreciable exchange with those of the medium. Second, the overall catalytic cycle includes complete reduction of the flavoprotein and then complete oxidation by two one-electron oxidations. These events would normally be obscured during catalysis but have emerged from the stopped-flow experiments and studies on partial reactions.

VII. The Overall Problem of Microsomal Electron Transport

The fact that this discussion has been largely limited to a consideration of the properties of the oxidation-reduction components of microsomes and the mechanisms of the reactions catalyzed by two flavoproteins, which have been successfully removed from the lipoprotein environment of the endoplasmic reticulum, emphasizes the major gap in an understanding of microsomal electron transport. The hydroxylation reactions as a class remain as the only well-documented electron-transfer reactions which fulfill an obvious physiological function. Two of the electron-transfer components of microsomes are now implicated in these reactions (cf. Reaction (6)). The unsettled questions concerning this mechanism center on the details of the pathway from the flavoprotein to P-450 and on a basis for the substrate specificity of these reactions. Progress here may depend upon the success in separating and purifying unaltered partial reaction sequences. It may then be possible to determine the relationship between the Fe_x component of Mason et al. and P-450, for example, and the extent to which the mechanism of TPNH cytochrome c reductase in solution applies to the overall hydroxylation reaction.*

The problem of electron transfer as a whole is much more perplexing. The microsomal particles have a capacity for carrying out a rapid oxidation of reduced pyridine nucleotides with a variety of electron acceptors. The rates of these reactions far exceed the ability of microsomes to carry out hydroxylations. The information that relates directly to a possible function for this oxidative capacity is that the most active electron donor is DPNH rather than TPNH and that the microsomal cytochrome b_5 is specifically

* There is now more recent spectral and electron spin resonance evidence (73–77) that substrate binding to P-450 may in fact occur at or near the heme, and these data also implicate Fe_x in these reactions.

reduced in an exceedingly rapid reaction that can at least in part be identified with a specific flavoprotein catalyst. In the search for a cellular electron acceptor (or acceptors) for reduced cytochrome b_5, its participation in reductive synthesis (71) and a direct interaction with the terminal electron transport chain of mitochondria have been considered (71,72). Although not conclusive, the experiments of Chance (72) gave no evidence of the latter route for cytochrome b_5 oxidation, nor is this heme protein autooxidizable at an appreciable rate (63). Reactions have also been considered which involve this electron-transport system in ascorbate formation (14,15) and more recently by Siekevitz in ion transfer or as an energy source for mechanical movement of cellular membranes (9). A critical and informative evaluation of these possibilities, however, awaits decisive experimental designs.

References

1. H. Mason, W. C. Fowlks, and E. W. Peterson, *J. Am. Chem. Soc.*, **17**, 2914 (1955).
2. B. B. Brodie, "Drug Metabolism-Subcellular Mechanisms," in *Enzymes and Drug Metabolism (Ciba Foundation Symposium)*, L. J. Mongar and A. V. S. deReuck, Eds., Little, Brown, Boston, 1962, p. 317.
3. L. Shuster, *Ann. Rev. Biochem.*, **33**, 571 (1964).
4. G. H. Hogeboom, *J. Biol. Chem.*, **177**, 847 (1949).
5. G. H. Hogeboom and W. C. Schneider, *J. Biol. Chem.*, **186**, 417 (1950).
6. A. Claude, *Biol. Symp.*, **10**, 111 (1943).
7. G. H. Hogeboom, A. Claude, and R. D. Hotchkiss, *J. Biol. Chem.*, **165**, 615 (1946).
8. P. Siekevitz, *Ann. Rev. Physiol.*, **25**, 15 (1963).
9. P. Siekevitz, *Federation Proc.*, **24**, 1153 (1965).
10. T. Omura and R. Sato, *J. Biol. Chem.*, **239**, 2370 (1965).
11. C. F. Strittmatter and E. G. Ball, *Proc. Natl. Acad. Sci. U.S.*, **38**, 191 (1952).
12. O. Hayaishi, Ed., *Oxygenases*, Academic Press, New York, 1962, p. 1.
13. P. Hochstein and L. Ernster, *Biochem. Biophys. Res. Commun.*, **12**, 388 (1963).
14. H. Kersten, W. Kersten, and H. Staudinger, *Biochim. Biophys. Acta*, **27**, 598 (1958).
15. H. Heath and R. Fiddick, *Biochem. J.*, **94**, 114 (1965).
16. B. Peterkofsky and S. Udenfreund, *Proc. Natl. Acad. Sci. U.S.*, **53**, 335 (1965).
17. C. H. Williams, Jr., and H. Kamin, *J. Biol. Chem.*, **237**, 587 (1962).
18. A. H. Phillips and R. G. Langdon, *J. Biol. Chem.*, **237**, 2652 (1962).
19. P. Strittmatter and S. F. Velick, *J. Biol. Chem.*, **221**, 277 (1956).
20. P. Strittmatter and S. F. Velick, *J. Biol. Chem.*, **228**, 785 (1957).
21. C. A. MacMunn, *J. Physiol.*, **5**, Proc. xxiv (1884).
22. C. A. MacMunn, *Phil. Trans. Roy. Soc.*, **177**, 267 (1886).
23. C. F. Strittmatter, Thesis, Harvard University, 1952.
24. P. Strittmatter and S. F. Velick, *J. Biol. Chem.*, **221**, 253 (1956).

25. P. Strittmatter, *J. Biol. Chem.*, **235**, 2492 (1960).
26. D. Garfinkle, *Biochim. Biophys. Acta*, **21**, 119 (1956).
27. K. Krisch and H. Staudinger, *Biochem. J.*, **331**, 37 (1958).
28. I. Raw, R. Molinari, D. F. do Amaral, and H. R. Mahler, *J. Biol. Chem.*, **233**, 225 (1958).
29. I. Raw and W. Colli, *Nature*, **184**, 1798 (1959).
30. J. Ozols and P. Strittmatter, *J. Biol. Chem.*, **239**, 1018 (1964).
31. M. Klingenberg, *Arch. Biochem. Biophys.*, **75**, 376 (1958).
32. D. Garfinkle, *Arch. Biochem. Biophys.*, **77**, 493 (1958).
33. T. Omura and R. Sato, *J. Biol. Chem.*, **237**, PC1375 (1962).
34. T. Omura and R. Sato, *J. Biol. Chem.*, **239**, 2370 (1964).
35. Y. Hashimoto, T. Yamano, and H. S. Mason, *J. Biol. Chem.*, **237**, PC3843 (1962).
36. R. W. Estabrook, D. Y. Cooper, and O. Rosenthal, *Biochem. Z.*, **338**, 741 (1963).
37. T. Omura and R. Sato, *J. Biol. Chem.*, **239**, 2379 (1964).
38. B. L. Horecker, *J. Biol. Chem.*, **183**, 593 (1950).
39. B. S. L. Masters, H. Kamin, Q. H. Gibson, and C. H. Williams, Jr., *J. Biol. Chem.*, **240**, 921 (1965).
40. P. Strittmatter, *J. Biol. Chem.*, **234**, 2661 (1959).
41. P. Strittmatter, *J. Biol. Chem.*, **236**, 2329 (1961).
42. P. Strittmatter, *J. Biol. Chem.*, **236**, 2336 (1961).
43. K. Ryan and L. Engel, *J. Biol. Chem.*, **225**, 103 (1957).
44. S. Orrenius, G. Dallner, and L. Ernster, *Biochem. Biophys. Res. Commun.*, **14**, 329 (1964).
45. D. Y. Cooper, S. Levin, S. Narasimhulu, O. Rosenthal, and R. W. Estabrook, *Science*, **147**, 400 (1965).
46. O. Warburg, *Biochem. Z.*, **152**, 479 (1924).
47. S. Orrenius and L. Ernster, *Biochem. Biophys. Res. Commun*, **16**, 60 (1964).
48. S. Orrenius and L. Ernster, *Biochem. J.*, **92**, 37P (1964).
49. R. Nilsson, S. Orrenius, and L. Ernster, *Biochem. Biophys. Res. Commun.*, **17**, 303 (1964).
50. S. Orrenius, J. L. E. Ericcson, and L. Ernster, *J. Cell. Biol.*, in press.
51. S. Orrenius, *J. Cell. Biol.*, in press.
52. S. Orrenius, *Abstr.*, *2nd Meeting Federation European Biochem. Soc.*, Wien, 1965.
53. L. Ernster and S. Orrenius, *Federation Proc.*, **24**, 1190 (1965).
54. R. R. Brown, J. A. Miller, and E. C. Miller, *J. Biol. Chem.*, **209**, 211 (1954).
55. A. H. Conney, E. C. Miller, and J. A. Miller, *Cancer Res.*, **16**, 450 (1956).
56. A. H. Conney, R. R. Brown, J. A. Miller, and E. C. Miller, *Cancer Res.*, **17**, 628 (1957).
57. A. H. Conney, E. C. Miller, and J. A. Miller, *J. Biol. Chem.*, **228**, 753 (1957).
58. T. Omura, R. Sato, D. Cooper, O. Rosenthal, and R. W. Estabrook, *Federation Proc.*, **24**, 1181 (1965).
59. R. A. Alberty, *J. Am. Chem. Soc.*, **75**, 1928 (1953).
60. H. Kamin, B. S. S. Masters, Q. H. Gibson, and C. H. Williams, Jr., *Federation Proc.*, **24**, 1164 (1965).
61. P. Strittmatter, *Federation Proc.*, **24**, 1156 (1965).
62. P. Strittmatter, *J. Biol. Chem.*, **233**, 748 (1958).
63. P. Strittmatter, *J. Biol. Chem.*, **234**, 2665 (1959).
64. G. R. Drysdale, M. J. Spiegel, and P. Strittmatter, *J. Biol. Chem.*, **236**, 2323 (1961).

65. P. Strittmatter, *J. Biol. Chem.*, **237**, 3250 (1962).
66. P. Strittmatter, *J. Biol. Chem.*, **238**, 2213 (1963).
67. P. Strittmatter, *J. Biol. Chem.*, **239**, 3043 (1964).
68. P. Strittmatter, *J. Biol. Chem.*, **240**, 4481 (1965).
69. H. Beinert, *J. Biol. Chem.*, **225**, 465 (1957).
70. V. Massey and Q. H. Gibson, *Federation Proc.*, **23**, 18 (1964).
71. C. F. Strittmatter, in *Haematin Enzymes*, J. E. Falk, R. Lemberg, and R. K. Morton, Eds., Pergamon Press, London, 1961, p. 461.
72. B. Chance, in *Haematin Enzymes*, J. E. Falk, R. Lemberg, and R. K. Morton, Eds., Pergamon Press, London, 1961, p. 473.
73. S. Narasimhulu, D. Y. Cooper, and O. Rosenthal, *Life Sci.*, **4**, 210 (1965).
74. Y. Imai and R. Sato, *Seikagaku*, **37**, 9 (1965).
75. H. Remmer, J. B. Schenkman, R. W. Estabrook, H. Sasame, J. Gillette, D. Y. Cooper, S. Narasimhulu, and O. Rosenthal, *Mol. Pharmacol.*, in press.
76. W. Cammer, J. B. Schenkman, and R. W. Estabrook, *Biochem. Biophys. Res. Commun.*, **23**, 264 (1966).
77. B. W. Harding and D. H. Nelson, *J. Biol. Chem.*, **241**, 2212 (1966).

Oxidative Phosphorylation

P. D. BOYER, *Molecular Biology Institute and Department of Chemistry, University of California, Los Angeles, California*

I.	Introduction	193
II.	Basic Mechanistic Considerations	195
	A. Nature of Activation and Oxygen Loss	195
	1. P_i or ADP Activation; Concerted or Stepwise	195
	2. Direct or Indirect Loss of P_i Oxygen	196
	B. Possible Nature of Any Phosphorylated Intermediate(s)	200
	C. Possible Nonphosphorylated Covalent Precursors	201
	D. Other Possible Energized States	202
III.	Experimental Approaches	203
	A. Mitochondrial Composition Related to Phosphorylation	204
	B. Study of Net Phosphorylation Capacity	205
	1. P : O Ratios and Phosphorylation Sites	205
	2. Inhibition and Uncoupling of Oxidative Phosphorylation	207
	3. Phosphorylation by Mitochondrial Fractions	212
	4. Respiratory Control and Related Phenomena	214
	C. Exchange and Partial Reactions	215
	D. Energy-Linked Reductions	220
	E. Ion Transport, Swelling, and Contraction	221
	F. The Search for Intermediates	222
	1. Possible Existence and Amount of Intermediates	222
	2. Possible Intermediates from P_i	223
	3. Possible Intermediates from ADP	225
	4. Possible Nonphosphorylated Intermediates	226
	G. Relation to Other Energy-Linked Systems	227
	H. Studies with Model Systems	228
IV.	Summation	229
	References	231

I. Introduction

About 25 years ago, an important biochemical concept was becoming firmly established experimentally—namely, that in the formation of ATP dependent upon O_2 consumption, more than one ATP molecule could be formed for each atom of oxygen consumed. This process of oxidative phosphorylation, depicted by Reaction (1), where n may be up to 3 or

$$\text{Reduced cofactor} + \tfrac{1}{2}O_2 \qquad \text{Oxidized cofactor} + HOH$$
$$n P_i + n ADP \qquad\qquad n ATP + n HOH$$

(1)

more depending on the cofactor involved, is of paramount biological importance. For example, in the complete oxidation of glucose about 90% of all the energy made available for syntheses or energy-consuming processes in aerobic cells funnels through oxidative phosphorylation. Indeed, all formation of ATP in glucose oxidation is by oxidative processes, but that accompanying the oxidation of glyceraldehyde 3-phosphate and of α-ketoglutarate, in which cofactors are reduced, is usually termed "substrate-level" phosphorylation. Such formation of ATP differs from that of oxidative phosphorylation in a key aspect, namely that the oxygen atom which must be removed from P_i and ADP to form ATP appears in a substrate carboxyl group rather than in water (1,2). Indeed, another definition of oxidative phosphorylation is that it is the process by which water is eliminated from P_i and ADP to form ATP coupled to an enzymic oxidation-reduction reaction.

Although oxidative phosphorylation has been one of the most studied areas of biochemistry, the fundamental chemistry of the energy-coupling process has defied explanation. Much has been accomplished on the isolation and identification of the oxidation-reduction components and enzymes of mitochondria, and a multitude of diverse information about conditions and factors affecting oxidative phosphorylation has been assembled. Hopes for explanation arising from various experimental findings have usually been fleeting, and the desire for explanation has led to some overinterpretation and speculation clothed with importance and reality. Logical mechanisms would involve formation of intermediates with the oxidation-reduction components, with the P_i or ADP, or with other unidentified factors. As of this writing, no satisfactory evidence is available regarding the chemical nature of any intermediate compounds or stages that may exist.

A vast literature on oxidative phosphorylation has been accumulated, and it would be inappropriate in a volume such as this to attempt to review the various approaches and ramifications. Emphasis will be placed on the nature of the process under investigation and on those approaches which appear to give information pertinent to mechanism. Additional aspects are covered in other chapters in this volume, particularly by Klingenberg (p. 1) on the mitochondrial respiratory chain. Some facets of oxidative phosphorylation in microorganisms are discussed by Smith (p. 55), and the important and closely related process of photosynthetic phosphorylation by Arnon (p. 123). For additional information, the student is referred to recent reviews (3–9).

II. Basic Mechanistic Considerations

In this section attention will be directed toward various chemical possibilities and the nature of possible intermediate stages, with only limited consideration of the implications of present information. The latter will be considered in more detail in the next section on experimental approaches.

The overall process of oxidative phosphorylation may involve three distinct chemical reactions or groups of reactions. These are the *oxidation-reduction* reactions, the *water-formation* reaction or reactions, and the *phosphorylation* reaction or reactions. Common water-formation and phosphorylation mechanisms may be operative for the various oxidation-reduction reactions. Relationships among the three types of reactions are of extreme mechanistic importance. Particularly pertinent is whether all reactions can occur independently, or whether the water-formation reaction is a requisite part of the oxidation-reduction or of the phosphorylation reaction.

The following discussions are based on the probability that the chemistry involved in oxidative phosphorylation, and particularly in the phosphorylation reaction, will be found to be analogous to reactions already detected in biochemical systems or, to a lesser extent, other known chemical reactions. The possibility must always be kept in mind, however, that unknown or new types of mechanisms or coupling may be involved, and the experimenter should enter the area only if he feels that he can ask realistic and important questions.

A. Nature of Activation and Oxygen Loss

1. P_i or ADP Activation; Concerted or Stepwise

In the absence of any definitive proof for intermediate substances, the possibility must be considered that the first covalent compound formed, from either P_i or ADP, is ATP itself (10,11). No established analogies, biochemical or otherwise, exist for such a reaction. Concerted reactions have been suggested for ATP formation linked to amide bond cleavage (12), but experimental evidence suggests enzyme-bound acyl phosphates as intermediates (13). The established biochemical pattern for enzymic reactions of the type depicted by Reactions (2) and (3)

$$A\text{—}B + P_i + ADP \rightleftharpoons A + B + ATP \qquad (2)$$

$$A\text{—}B + PP_i + AMP \rightleftharpoons A + B + ATP \qquad (3)$$

involves formation of covalent intermediates of A or B and a group from ATP; for example, synthesis of fatty acyl CoA involving acyl adenylate (14). Eventual detection of some type of intermediate formed from P_i or ADP in oxidative phosphorylation thus still appears plausible.

In either a concerted or a stepwise process, P_i or ADP must lose an oxygen to water, and in this sense be activated. Formation of a covalent intermediate would be activation in the usual biochemical sense. As indicated in Section III, present evidence favors P_i as the component which undergoes "activation" and loses an oxygen that forms water. For simplicity of presentation and discussion, most of the balance of this chapter will be presented as if the oxygen loss were from P_i.

2. Direct or Indirect Loss of P_i Oxygen

A fundamental but as yet unanswered question is whether the loss of oxygen to form water accompanying ATP formation is direct or indirect. The latter means that at least one unknown substance or group gains an oxygen from P_i and subsequently loses it to form water in continued ATP synthesis. This would allow separation of the water-formation and phosphorylation reactions. A direct formation of water from P_i oxygen would mean phosphorylation and water formation must occur concomitantly.

Three different means for meeting the kinetic and thermodynamic requirements for a direct loss of P_i oxygen to form water can readily be visualized. These different means, that might act singly or in concert, are as follows:

(1) Development of a more effective nucleophile as a consequence of electron transport, and displacement of an —OH group from P_i as this nucleophile is phosphorylated.

(2) A change in the environment of one or more oxygens of the P_i such that electron withdrawal from the P atom makes it more susceptible to attack by a nucleophile already present.

(3) A change in the environment of an —OH group of P_i so as to make it a more favorable leaving group for water formation, thus promoting phosphorylation of a nucleophile already present.

As it is difficult to visualize a mechanism for increasing the nucleophilicity of a phosphoryl oxygen of ADP by electron transport, the first means would imply formation of a phosphorylated intermediate. The second means, a change in the properties of P_i oxygens as the source of the driving force, would allow ADP to serve as the nucleophilic phosphoryl acceptor, thus forming ATP without participation of a phosphorylated intermediate.

More specific examples of hypothetical nature may help illustrate various possibilities. One means of developing nucleophilicity is by proton loss from an appropriate group. Such a nucleophile might attack P_i, already bonded to a metal to favor departure of a hydroxyl group, to yield a phosphorylated intermediate as indicated by Reaction (4). Formation of O^{2-} as a leaving group from P_i is highly unlikely, and formation of

$$YH \xrightarrow[\text{electron transport}]{\text{proton loss} \atop \text{accompanying}} Y\colon^- \quad \longrightarrow \quad \overset{\text{H}}{\underset{\delta-}{\overset{\delta-}{O}}}\overset{\delta-}{\underset{\delta-}{\underset{O}{\overset{O}{P^{\delta+}}}}} M^{2+} \quad \longrightarrow \quad Y-\overset{O}{\underset{O^-}{\overset{\|}{P}}}-O^- + {}^+MOH \quad (4)$$

OH^-, possibly assisted by a metal, would appear more likely. A suggestion for involvement of imidazole groups with loss of a proton to give an imidazolide anion in a reaction akin to that depicted by Reaction (4) has been made elsewhere (15).

Promotion by favoring departure of oxygen with increased susceptibility of P_i to nucleophilic attack could involve non-heme iron as shown by Reaction (5). The reaction could involve transient metaphosphate formation.

$$HOPO_3{}^{2-}\cdot Fe^{2+} \xrightarrow{e-} \overset{\text{H}}{\underset{\delta-}{\overset{\delta-}{O}}}\overset{\delta-}{\underset{\delta-}{\underset{O}{\overset{O}{P^{\delta+}}}}} Fe^{3+} \xrightarrow{ADP} ATP + \quad {}^{2+}FeOH \quad (5)$$

$$\text{(P}_i\text{ coordinated with} \atop \text{ferrous iron)} \qquad\qquad\qquad\qquad \text{(hydroxide coordinated} \atop \text{with ferric iron)}$$

An attractive means of forming monomeric metaphosphate at the active phosphorylation site may be visualized as involving dianionic orthophosphate specifically bound so that its hydroxyl group is exposed to a localized proton or proton-donating group, with metaphosphate formation occurring as depicted by Equation (6). Alternatively, a localized proton could drive the phosphorylation without discrete metaphosphate formation. The phosphoryl group formed as P_i is dehydrated could be donated

$$\text{Enzyme} \quad \rightleftharpoons \quad \text{Enzyme} \qquad (6)$$

directly to ADP to give ATP. Proton production in oxidative phosphoryl-ation could accompany $FADH_2$ or $CoQH_2$ oxidation, or other oxidation-reduction reactions linked to proton addition and removal.

Information from studies of hydrolysis of phosphate esters is particularly pertinent to any mechanism involving a direct loss of oxygen to water. In hydrolysis with P—O cleavage, water oxygen adds directly to the P atom. Hydrolytic mechanisms may vary considerably with different esters (16,17) and have not been carefully studied with ATP. Some results with simple esters warrant mention. Hydrolysis of monomethyl phosphate and of the strained ethylene phosphate is accompanied by an incorporation of water oxygen into the unhydrolyzed ester (18), with ratios of hydrolysis to exchange of about 5. This implies that the hydrolysis may involve a transition state with five oxygens surrounding the P atom, as crudely depicted in Equation (7). Reversal of step (a) would give incorporation

of ^{18}O from water into unhydrolyzed ester.

Oxidative phosphorylation could involve an analogous transition state between P_i and ADP, with some means of shifting the equilibrium toward ATP by water removal.

Further discussion of means for direct or indirect formation of water from P_i oxygen can instructively be considered in relation to whether or not ADP is an essential reactant for water formation to occur. Schemes I and II depict mechanisms in which ADP is an essential reactant. Scheme III depicts mechanisms without ADP as a required reactant for water formation. (The $\sim$, in accordance with common biochemical usage, indicates a "high-energy" bond with apparent free-energy of hydrolysis roughly equivalent to pyrophosphate or acyl phosphate linkages.)

Indirect loss of oxygen to water would be expected if phosphorolysis, a type of reaction well known biochemically, were involved. Such a sequence

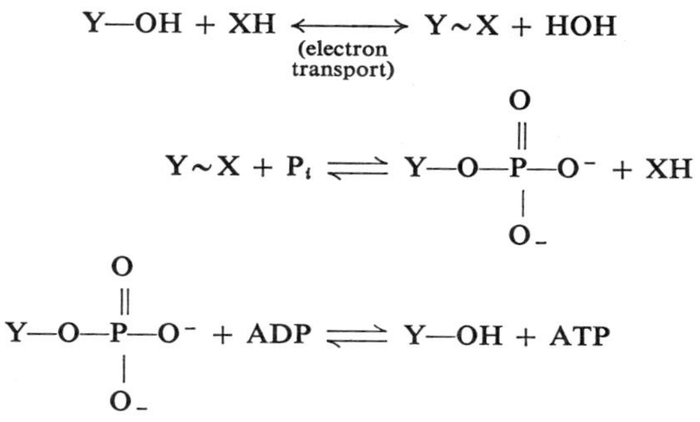

$$Y\text{—}OH + XH \underset{\substack{\text{(electron} \\ \text{transport)}}}{\longleftrightarrow} Y\sim X + HOH$$

$$Y\sim X + P_i \rightleftharpoons Y\text{—}O\overset{\displaystyle O}{\underset{\displaystyle O_-}{\overset{\displaystyle \|}{\text{—}P\text{—}}}}O^- + XH$$

$$Y\text{—}O\overset{\displaystyle O}{\underset{\displaystyle O_-}{\overset{\displaystyle \|}{\text{—}P\text{—}}}}O^- + ADP \rightleftharpoons Y\text{—}OH + ATP$$

Scheme I.   Indirect formation of water from P_i involving phosphorolysis.

would require at least three steps, such as depicted by Scheme I. The phosphorylation reactions would then be understandable, and major attention directed toward the nature of Y—OH and on how oxygen is lost from Y—OH to water. The possible nature of nonphosphorylated states that might be cleaved by P_i in the first step of Scheme I is considered in Section II-C.

Water formation from P_i by an indirect route as depicted in Schemes I and in II-B but not in Scheme III-B would require concomitant oxidation and reduction as well as ADP. Only when ADP is phosphorylated would the oxygen in Y—OH derived from P_i be available for water formation in the oxidation-reduction step. Direct routes of water formation is depicted in Schemes II-A and III-A would not require concomitant oxidation-reduction for water formation. Reactions giving direct formation of water but requiring concomitant oxidation-reduction do not have satisfactory analogy in known reactions and are thus not depicted here.

Hypothetical schemes of the type depicted by Scheme II-B have been suggested for oxidations of hydroquinone derivatives (reduced ubiquinone, vitamin E, reduced vitamin K), involving the formation of a carbonium ion or equivalent to which P_i adds. Subsequent phosphoryl transfer leaves an hydroxyl group which eventually forms water [see, for example, the schemes of Cohen, given by Racker (3), and of Boyer (19), and Todd (20)].

Experimental and other considerations make it likely that the mechanism of oxidative phosphorylation is encompassed by reactions such as depicted in Schemes I–III. Other possibilities obviously also exist. One suggestion in which phosphorylation precedes oxidation has been made for DPNH. This involves a postulated P_i addition to a double bond, phosphoryl transfer, then oxidation and water elimination by reduction (21).

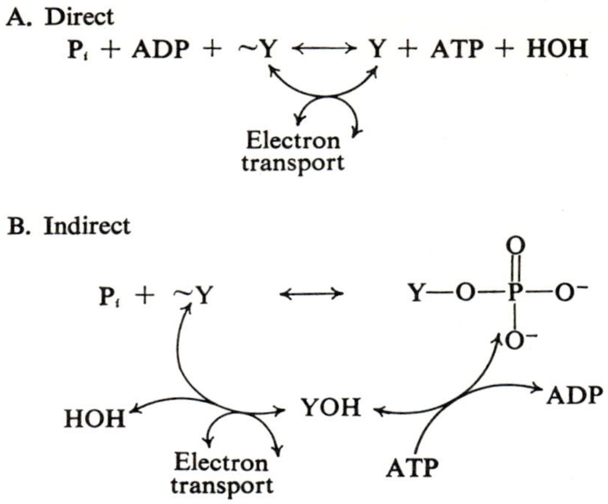

A. Direct

$$P_i + ADP + \sim Y \longleftrightarrow Y + ATP + HOH$$

Electron
transport

B. Indirect

$$P_i + \sim Y \longleftrightarrow Y-O-\overset{\overset{\displaystyle O}{\|}}{\underset{\underset{\displaystyle O^-}{|}}{P}}-O^-$$

HOH   YOH   ADP

Electron
transport
ATP

(or analogous reactions with $X \sim Y$ instead of $\sim Y$)

Scheme II.   Water formation from P_i with ADP as an essential reactant.

A. Direct

$$P_i + \sim Y \longleftrightarrow Y \sim P + HOH$$

ADP

Electron   Y
transport
ATP

B. Indirect

$$P_i + X \sim Y \longleftrightarrow XOH + Y-\overset{\overset{\displaystyle O}{\|}}{\underset{\underset{\displaystyle O^-}{|}}{P}}-O^-$$

Y   ADP

HOH
Electron   ATP
transport

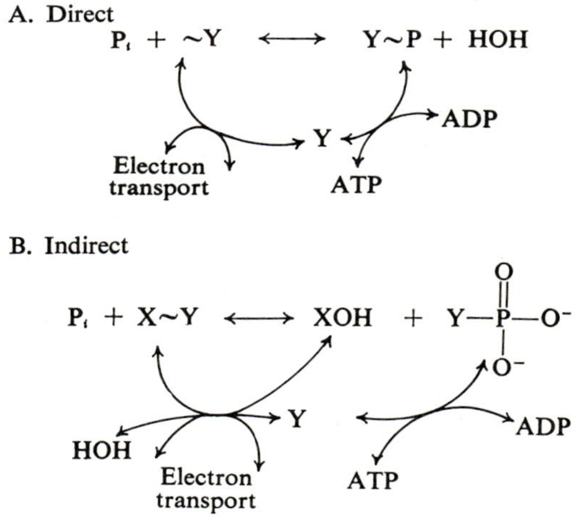

Scheme III.   Water formation from P_i without ADP as an essential reactant.

B. *Possible Nature of Any Phosphorylated Intermediate(s)*

Oxidative phosphorylation involving a phosphorylated intermediate or intermediates may be visualized as occurring by one of two distinct routes. For each of various oxidation steps a different phosphorylated inter-mediate could be formed. These could involve the oxidation-reduction components, unknown cofactors, or distinct "coupling" factors or enzymes. Alternatively, all steps could be coupled to a common "high-energy" compound or state, which is then phosphorylated to give a single phosphorylated intermediate.

Leading candidates for phosphorylated intermediates in oxidative phosphorylation are those already encountered in biological systems. These include acyl phosphates; N-phosphoryl derivatives, such as found in phosphohistidine, phosphoarginine, and phosphocreatine; phosphorylated enol, as in phospho(enol)pyruvate; pyro- or polyphosphate derivatives; and phosphorylated primary alcohol groups, as in phosphoserine. Thiophosphates, and phosphorylated phenol and hydroquinone derivatives, also merit consideration, although not as yet demonstrated in biological systems.

Because of the multiplicity of hypothetical intermediates and for other reasons discussed later, synthesis and testing of possible intermediates does not yet appear to be a promising approach.

C. Possible Nonphosphorylated Covalent Precursors

Oxidative phosphorylation lore is replete with other suggestions of nonphosphorylated "high-energy" intermediates (see reviews 3–9), mostly of vague chemical nature. The important evidence for the participation of a nonphosphorylated "high-energy" precursors in oxidative phosphorylation is given in Section III. Such a precursor could be a substance containing a labile covalent bond that is cleaved in a subsequent phosphorylation reaction. This would be in clear analogy with substrate-level phosphorylation in which oxidation precedes phosphorylation and gives rise to a labile covalent bonded intermediate.

The possible chemical nature of any nonphosphorylated precursors must be considered in relation to how oxygen is lost from P_i to HOH. Either such an intermediate must be able to induce a direct loss of oxygen from P_i to water, or one of the moieties of the intermediate must be able to accept an oxygen from P_i and lose this to water in a subsequent step, as for example, depicted by Scheme I. No biochemical or other analogous reactions are known to the author in which a covalent bond is cleaved in the presence of P_i to form a phosphoryl derivative and water. Indeed, even a hypothetical reaction of this type would appear to violate chemical principles. Thus if a covalent nonphosphorylated precursor is formed, probably one moiety of such a precursor retains oxygen from P_i when ATP is formed.

The nonphosphorylated precursors formed in substrate-level phosphorylations contain an acyl-S bond which is cleaved in the P_i uptake reaction. Such a linkage thus deserves consideration for oxidative phosphorylation, and carboxyl activation to form an acyl-S or acyl imidazole linkage has been suggested (15,22). Acyl imidazoles have not been demonstrated in biological systems, although acetyl imidazole is formed enzymically (23).

No results to date prove or strongly support the existence of more than one nonphosphorylated or phosphorylated intermediate in oxidative phosphorylation. Both Ernster (6) and Racker (24) have tended to favor one phosphorylated and one nonphosphorylated intermediate. Sanadi, in his review (9), favors multiple intermediates. Experimental support from Green's laboratory for prominent suggestions of separate site-specific precursors and phosphorylated intermediates (25) is now questionable (26). Many experimental findings can be rationalized by the concept of a single type of energized nonphosphorylated compound or state which, with or without formation of a phosphorylated intermediate, gives rise to ATP. Such a concept, however, imposes the need for a common chemistry coupled to reactions of widely spaced oxidation-reduction potentials.

One chemical feature common to various oxidation-reduction steps could involve proton withdrawal or addition to appropriate groups. All reactions, even though termed electron transfers, can be written as involving H^+ formation or release by suitable participation of coordinated or adjacent groups. Thus schemes based on proton withdrawal or addition may have special merit. Another means of introducing a common chemistry would be to invoke participation of a common oxidation-reduction intermediate at each step. This is difficult, particularly in the case of substances such as quinones whose oxidation-reduction potential has not been demonstrated to be markedly changed by environment. The increasing prominence assignable to non-heme iron in different parts of the respiratory chain, and the wide difference in oxidation-reduction potential that such iron may assume, presumably because of variable ligands present, suggests some consideration of the possible role of such iron in oxidative phosphorylation.

D. Other Possible Energized States

Precursor "high-energy" states other than those mentioned above deserve some consideration. An intriguing possibility, but one difficult to approach on an experimental basis, is that oxidation-reduction reactions lead to conformation change of proteins and associated structures and that this conformation change is in some manner coupled to a phosphorylation reaction. The latter could involve formation of various covalent derivatives. Conformation change accompanying cytochrome c oxidation and reduction has been suggested from enzyme digestion studies (27) and rotatory dispersion measurements (28,29). In the complex organized respiratory particles, various oxidation-reduction reactions could even be coupled to a common conformation site (22). Such would make oxidative phosphorylation, like active transport, dependent upon complex macromolecular organization and thus not attainable in solution at a molecular level.

Oxidation-reduction reactions, particularly in insoluble matrixes, give rise to the possibility of charge separation as an energized state. Boyer has directed attention to localized positive charge (possibly protons) as a means of driving ATP formation (10). One variant of such possibility has been suggested as a "chemiosmotic" mechanism, involving localized production of hydroxide ions and protons (30). The separation of H^+ and e^- or OH^- is then visualized as driving ATP formation. Charge separation has been invoked for ion transport as well as ATP formation (31).

All mechanisms with reasonable biochemical analogies are based on a probable stoichiometry between electron transport and phosphorylation; that is, in a tightly coupled system, the passage of two electron equivalents through a particular step is usually considered to result in formation of one ATP. Mechanisms involving formation of an ATP coupled to passage of one electron equivalent would appear to merit more consideration. Another unusual alternative deserves mention. Conceivably, nature may have evolved a mechanism in which there is no absolute stoichiometry between electron flux and phosphorylation, but in which phosphorylation yield is somehow maintained near to the thermodynamic possibility. This would imply increased yield when ADP and P_i are at relatively high and ATP at relatively low concentrations. Such seems unlikely on the basis of present knowledge, but it must be remembered that nature is ingenious and has had considerable time to perfect her systems.

III. Experimental Approaches

It is clear that any understanding of oxidative phosphorylation is interdependent with understanding of the participants in the oxidation-reduction reactions. Many parts of the respiratory chain and their interconnections remain obscure and will not be considered here. Attention will be directed toward means of studying the formation of phosphorylated or other closely associated "high-energy" states in mitochondria or particulates therefrom.

A. Mitochondrial Composition Related to Phosphorylation

The composition of the respiratory chain components of mitochondria is reviewed by Klingenberg (this volume), and evidences for integrated structural features are discussed therein. In addition to the catalytic proteins, chiefly those of the citric acid cycle and respiratory chain, mitochondria contain considerable protein of indefinite function, probably a structural protein (32), and phospholipid. Removal of most of the phospholipid may be achieved by acetone extraction with loss of oxidative and phosphorylative capacities. Addition of various phospholipids will restore oxidation, but not phosphorylation (33), but no specific role of lipids in phosphorylation has been uncovered.

Of particular interest is that mitochondria as isolated from tissues contain considerable amounts of adenine nucleotides and P_i (34). The ratios of intramitochondrial ADP and ATP are relatively constant even with wide fluctuation of extramitochondrial ratios (35). Not all intramitochondrial P_i is equally accessible for ATP formation. Compartmentation of P_i in mitochondria is a well-established phenomenon. For example, during net oxidative phosphorylation, $^{32}P_i$ may form extramitochondrial $AT^{32}P$ of higher specific activity than intramitochondrial P_i (36,22). Whether intramitochondrial nucleotides show similar localization is uncertain. Intramitochondrial ADP is an early phosphate acceptor of oxidative phosphorylation (34,37). Specificity for ADP in oxidative phosphorylation, however, appears to be a property of intact mitochondria. Löw has made the somewhat surprising demonstration that particles from mitochondria will directly phosphorylate IDP, GDP, and UDP, as well as ADP (38). Intact mitochondria are thus visualized as having either a transphosphorylation between endogenous ATP and exogenous ADP (37) or a specific interchange of ADP and ATP across a mitochondrial membrane (34).

Metals present in liver mitochondria isolated in sucrose have been determined in Vallee's laboratory (see ref. 7, p. 159) with results as given in Table I. The large amount of Mg^{2+} gives an environment for oxidative phosphorylation which is about $10^{-2}M$ in Mg^{2+}. Mg^{2+} may be the only divalent cation required for oxidative phosphorylation, but the small amount of Mn^{2+} present cannot be ignored.

Mitochondria which are aged in isotonic solutions or undergo swelling in the presence of various agents lose their capacity for phosphorylation

TABLE I
Metal Ions in Liver Mitochondria

Metal ion	$m\mu moles/mg$ protein
K^+	130
Na^+	6.3
Mg^{2+}	42
Ca^{2+}	5.6
Zn^{2+}	1.9
Fe^{2+}, Fe^{3+}	7.5
Mn^{2+}	0.4

but not for oxidation. Phosphorylation capacity can be partially restored by exposure to ATP. These and other studies point out the extreme structural lability of the phosphorylation reaction, but do not give specific information about phosphorylation mechanism.

The overall catalytic efficiency of the mitochondrial system is not strikingly high, perhaps reflecting the complexity of components that must be structurally associated to accomplish oxidation coupled to phosphorylation. Cytochrome oxidase may have turnover at $25°$ of about 20,000 (39) and is not rate limiting in oxidative phosphorylation reactions. In intact bovine heart mitochondria, oxygen uptake with excess substrate is about 35 gram atoms of oxygen per minute per 100,000 g of protein, equivalent to an ATP synthesis of about 100 moles/min per 100,000 g of mitochondrial protein.

B. Study of Net Phosphorylation Capacity

1. P:O Ratios and Phosphorylation Sites

Evidence for the occurrence of P:O ratios as high as 3 for the oxidation of pyruvate to CO_2 and water was presented in 1943 by Ochoa. As reviewed by Racker (3), subsequent studies have led to the concept that, for DPNH oxidation, a phosphorylation occurs in the reaction of DPNH with flavoprotein, a second in the oxidation of reduced flavoprotein by cytochrome c, and a third in the oxidation of reduced cytochrome c by oxygen. Stoichiometric experiments, principally by Lehninger and by Lardy and associates, first firmly established that oxidation of DPNH can give more than two molecules of ATP per molecule of DPNH oxidized, that oxidation of succinate, which reduces flavoprotein, can give nearly two, and that oxidation of reduced cytochrome c can give up to one molecule of ATP per oxygen atom consumed. The coupling of such oxidation-reduction steps to phosphorylations has been confirmed in intact mitochondria by the elegant methods of Chance based on observations of shifts in the steady-state levels of respiratory chain components (40,41). The sites of coupling indicated are clearly compatible with the observed oxidation-reduction potentials as given in Table II and the free-energy

TABLE II

Free-Energy Changes in Oxidations

Oxidation-reduction pair	Approx. potential, pH 7, 30°	Approx. $-\Delta F$ in kcal for oxidation by $\frac{1}{2}$ O_2
DPNH–DPN$^+$	-0.32	53
Succinate–fumarate	$+0.03$	36
Cytochrome c, Fe^{2+}–Fe^{3+}	$+0.26$	26
Theoretical O_2 electrode	$+0.82$	

change expected for oxidation by oxygen of 1 mole of DPNH or succinate or 2 moles of cytochrome c–Fe^{2+}, while maintaining an equal ratio of oxidized to reduced form.

The apparent free energy of hydrolysis of ATP from the equilibrium $K = [ADP][P_i]/[ATP]$, with water concentration incorporated into the equilibrium constant K, under physiological conditions is about 7.5 kcal/mole. From this it is sometimes assumed that in oxidation of DPNH about six ATP might be formed, since, from the relation $-\Delta = n\mathscr{F}\Delta E$, a transfer of two equivalents through 0.16 V would be equivalent to 7.5 kcal. However, the apparent $-\Delta F$ refers to standard $1M$ states of ADP, ATP, and P_i. The apparent free energy for formation of ATP under conditions that might exist in the cell would be much higher. Thus, if the ratio of ATP to ADP were one, and P_i were $10^{-3}M$, or $[ATP]/[ADP][P_i] = 10^3$, then about 11.5 kcal of energy would be necessary for each mole of ATP formed. Data of Klingenberg (42) and of Chance and Hollunger (43) show that oxygen uptake is depressed and carrier reduction appreciable when the $[ATP]/[ADP][P_i]$ ratio is such that about 12–15 kcal of energy would be required per mole of ATP synthesized.

Whether phosphorylation in the flavoprotein and cytochrome region is coupled to a one- or two-electron transfer process is an unsettled question.

Coupling of oxidation of reduced cytochrome c to a one-electron transfer mechanism appears energetically feasible, and would give a maximum P/O of 4 for DPNH oxidation. An interesting contribution in this regard was made by Hommes (44), who presented kinetic evidence for a dimeric organization of the cytochromes. The reaction of reduced cytochromes was second order in coupled mitochondria, but first order in the presence of 2,4-dinitrophenol.

2. Inhibition and Uncoupling of Oxidative Phosphorylation

Oxidative phosphorylation by mitochondria or particles therefrom is exquisitely sensitive to a wide variety of chemical agents and physical factors. This has made difficult meaningful correlations of structure to inhibitory action. Nonetheless, considerable information has been obtained from judicious use of inhibitors, and a number of these have been characterized as blocking or uncoupling certain steps in the process. The various distinct types of inhibition of ATP formation observed are as follows:

Respiratory inhibition
Primary uncoupling
Secondary uncoupling
Phosphorylation block

In addition, the overall process of net ATP accumulation or utilization may be inhibited by the following:

Increased ATP hydrolysis
Nucleotide transfer or interconversion block

Respiratory Inhibition. Agents such as cyanide, antimycin A, and rotenone that are potent respiration inhibitors obviously also inhibit oxidative phosphorylation. Use of oxidation inhibitors allows experiments on the phosphorylation process as coupled to isolated parts of the respiratory chain, and on whether oxidation-reduction is essential for the phosphorylation reactions.

An important question about energy-coupling mechanisms is raised by the interesting report that concentrations of antimycin (45) and of rotenone (46) less than stoichiometric with the cytochrome components of the respiratory chain can under some circumstances completely inhibit oxidations. Antimycin, but not rotenone, inhibits succinate as well as DPNH oxidation. This result could reflect interaction of antimycin at an energy-coupling site and raise the possibility that not only various electron-transport steps but even several chains might be coordinated with one site.

Antimycin sensitivity of respiration is not well understood. Results of Tappel (47) and Estabrook (45) are compatible with non-heme iron involvement. Antimycin effects can be reversed in an interesting manner. Exposure of mitochondria to hypotonic media at pH 6 abolishes the respiratory inhibition by antimycin and uncouples phosphorylation from oxidation. Sensitivity to antimycin and phosphorylation are regained by washing at pH 8 (50). This is suggestive of an ability of antimycin to act on a primary energy-conserving step rather than an electron-transport step. Oxygen and $P_i \rightleftharpoons$ ATP exchange reactions associated with phosphorylation are only partially inhibited by antimycin, indicating that "high-energy" forms can arise from ATP cleavage even though respiration is completely inhibited by antimycin (51,52).

Primary Uncoupling. This type of action implies dissociation of the oxidation reactions from generation of any type of "high-energy" compound or state. This may be the mode of action for 2,4-dinitrophenol, a possibility first clearly suggested by Lardy and Elvehjem (48). Experimental support for this possibility came from demonstration by Loomis and Lipmann (49) that 2,4-dinitrophenol allows respiration to occur without concomitant phosphorylation. The effect is readily reversed by removal of the 2,4-dinitrophenol (49). This uncoupling action of 2,4-dinitrophenol has since become the prototype for a variety of other uncoupling agents, but the mechanism of its uncoupling action is unexplained, and considerable speculation in this area has yielded little of value.

Several lines of experimental evidence suggest that 2,4-dinitrophenol and similar agents act prior to participation of P_i in the oxidative phosphorylation sequence. The ability of 2,4-dinitrophenol to almost abolish the $P_i \rightleftharpoons$ HOH exchange (10,53) rules out formation of a phosphorylated derivative followed by hydrolysis with attack of water oxygen on the P atom, but not formation of an $X—O—PO_3^{2-}$ with X—O bond cleavage. Lack of incorporation of water oxygen into the dinitrophenol (53) rules out other unlikely mechanisms. Respiration becomes independent of P_i and ADP concentration in the presence of 2,4-dinitrophenol (54), and this would not be expected if intermediate phosphorylated products were formed and hydrolyzed. Perhaps the most important evidence for primary uncoupling is that 2,4-dinitrophenol can uncouple energy-linked reductions of mitochondria or particles therefrom which can occur without participation of P_i (see Section III-D). The inability of P_i to compete with or overcome the action of 2,4-dinitrophenol (10) argues against an energized state that is dissipated or hydrolyzed by 2,4-dinitrophenol or, alternatively, forms a phosphoryl derivative with P_i. Present evidence favors at least one action of dinitrophenol being that of preventing formation of a "high-energy" substance or state from either electron transport or by energy derived from ATP hydrolysis. A possible means by which this could be accomplished would be if the primary coupling of oxidation depended upon proton withdrawal from a group, as depicted by Equation (4). In the tightly coupled system, the unprotonated group would be formed by electron transport or by a reaction sequence dependent upon ATP cleavage. In the 2,4-dinitrophenol uncoupled system, both electron transport and ATP cleavage would give the protonated group. Uncoupling by spatial dislocation as well as by 2,4-dinitrophenol could thus reflect simply the accessibility of a key group to protonation by water. The "high-energy" compound or state would thus not form. Such a mode of action is in clear contrast to schemes involving dissipation of a "high-energy" form by 2,4-dinitrophenol. The latter type of action, in terms of the presentation of this chapter, would place 2,4-dinitrophenol with secondary uncouplers.

Uncoupling by structural separation, mitochondrial aging or swelling, etc. can be most readily interpreted in terms of accessibility of water to a critical site, either to hydrolyze a key nonphosphorylated intermediate, as commonly suggested, or, as mentioned in the preceding paragraph, to merely allow water to act as a proton donor or in other manner prevent formation of a "high-energy" intermediate. Development of ATPase activity is readily understandable in terms of such primary uncoupling.

Secondary Uncoupling. This type of uncoupling implies that a primary "high-energy" state is formed but is dissipated by means other than by formation of ATP. This might be by wasteful or metabolically useful processes. Thus uncoupling by action of Ca^{2+} (see ref. 5) can represent use of oxidation energy for ion transport. Other mitochondrial "housekeeping" processes when stimulated might appear as uncoupling phenomena by utilizing energy independent of phosphorylation. Experimentally valid P/O ratios are thus likely to represent minimal values.

Another type of secondary uncoupling results from one action of arsenate. Arsenate likely forms an unstable arsenyl intermediate or arsenyl ADP which is spontaneously hydrolyzed by water. Its action is effectively blocked by oligomycin (55).

As noted previously, 2,4-dinitrophenol or spatial dislocation and related uncoupling might be recognized as secondary uncouplers when their mechanism becomes understood.

Phosphorylation Block. One of the most potent tools for the study of oxidative phosphorylation is the antibiotic, oligomycin. This was introduced to the phosphorylation field by Lardy (56) as a result of his intensive screening for inhibitors. Oligomycin can inhibit net phosphorylation,

$P_i \rightleftharpoons HOH$ exchange, $P_i \rightleftharpoons ATP$ exchange, $ADP \rightleftharpoons ATP$ exchange, and 2,4-dinitrophenol- or arsenate-activated ATPase, as well as energy-driven reactions linked to ATP cleavage (4–9). All these actions are consistent with the suggestion of Lardy and associates that oligomycin prevents or drastically slows the phosphorylation reaction of oxidative phosphorylation. This occurs without uncoupling, as demonstrated by the inhibition of oxygen uptake in "tightly coupled" mitochondria, and the release of this inhibition of oxygen uptake by 2,4-dinitrophenol. Such stimulation of oxygen uptake and the inability of oligomycin to inhibit respiration of uncoupled systems shows that this substance does not have a primary effect on oxidation-reduction reactions.

The means by which oligomycin exerts its effect is unknown. Prevention of the combination of either P_i or ADP or both with the active site, or blocking of interchange of these bound components with medium P_i and ADP would appear likely. The ability of oligomycin to stimulate energy-linked reductions by some preparations in absence of ATP (57,58) implies that it may also inhibit access of water to a senstive site.

Arsenate could also act to block phosphorylation by competing with a P_i site but not undergoing any reaction to form an intermediate. This would mean a phosphorylation block rather than secondary uncoupling forms the basis of arsenate action.

Increased ATP Hydrolysis. In damaged mitochondria or particulate systems from mitochondria, potent ATPase activity may be encountered. Such ATPase may hydrolyze newly formed ATP, and prevent any accumulation. The utility of hexokinase-glucose as a trapping system for demonstration of net oxidative phosphorylation probably reflects competition for the ATP formed. Particles capable of ATP formation will respire maximally without ADP or P_i added, and thus must show reduced phosphorylation because of primary or secondary uncoupling as well as by ATPase action.

Nucleotide Transfer or Interconversion Block. With intact mitochondria, either passage of adeninine nucleotides to and from the intramitochondrial site of phosphorylation, or interconversion of ATP and ADP at the mitochondrial surface occurs (34,35,37). The interesting inhibitor, atractyloside, has its principal locus of action on the reaction between intramitochondrial ATP and ADP of the reaction medium (37,59–63). Intramitochondrial phosphorylations and exchange reactions occur unabated in presence of atractyloside and the inhibitor is ineffective in particulate systems.

Other Aspects of Inhibitor Action. A number of inhibitors have one or more of the effects described above. Azide has multiple effects, acting as an inhibitor of electron transport, an uncoupling agent like dinitrophenol, and a phosphorylation blocking agent, like oligomycin (64). The latter action is well shown by an ability to inhibit the 2,4-dinitrophenol activated ATPase. Various alkylguanidine derivatives have effects akin to oligo-mycin, but in addition inhibit at other points between electron transfer and phosphorylation. Also, their effects depend upon the respiratory state in an unexplained fashion (65–67). A competition of guanidines and Mg^{2+} has been noted, with the important implication of a role for Mg^{2+} prior to a phosphorylation step (68). Heytler (69,70) introduced a very potent agent, m-Cl-carbonylcyanidephenylhydrazone,

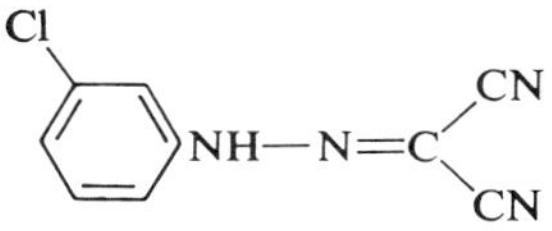

m-Cl-carbonylcyanidephenylhydrazone

that acts as an uncoupling agent in strikingly low concentrations. Other related derivatives are even more potent. The chemistry of their interaction is uncertain, but they will combine with 1,2- and 1,3-aminothiols as indicated by spectral shifts. Such thiols will prevent or reverse the action, and Heytler directs attention to the possibility of related structures at the phosphorylation site.

Various —SH group reagents are effective uncouplers, and studies of the potentiation of arsenite effect by 2,3-dimercaptopropanol (71), the effectiveness of γ-(p-arsenophenyl)-n-butrate (71), and of p-mercuri-benzoate (72) implicate an —SH group or possibly dithiol groups in a hydrophobic area. The —SH group reagents and Cd^{2+} appear to act between oxidation and phosphorylation steps. As with other tests with —SH reagents, it is difficult to know whether such effects reflect a primary role of the —SH group in the chemistry, or a secondary structural change accompanying reaction of the —SH group (73).

Uncoupling effects of phenols, particularly that of pentachlorophenol (77), as well as those of fatty acids, are counteracted by the binding of these agents by serum albumin. Binding of the phenols to hydrophobic areas of mitochondrial proteins is logical. Binding of inhibitory agents by serum albumin is probably the basis of its beneficial effect in damaged mitochondrial systems.

A wide variety of phenols have been tested, and the most interesting suggestion to emerge is that of Hemker (74,75) pointing to increased effectiveness correlated with increased lipid solubility and ability to lose a proton. A difference in optimum concentration of 2,4-dinitrophenol for inducing ATPase and stimulating oxygen uptake has led to the conclusion that 2,4-dinitrophenol acts by hydrolyzing a "high-energy" intermediate rather than by allowing respiration to proceed without the formation of any such intermediate (76). Reasonable alternative explanations for the data exist, however, including difference in rate-limiting steps in the two processes, effects of the different assay conditions, substrates concentrations, etc. on the susceptibility to 2,4-dinitrophenol, and possible multiple ATPases responsive to 2,4-dinitrophenol.

Complexities of action of inhibitors and uncouplers are shown by extensive studies of Lardy and associates (78). Both oligomycin and aurovertin are equally effective in inhibiting oxidative phosphorylation, $P_i \rightleftharpoons ATP$ and $P_i \rightleftharpoons HOH$ exchanges, and partially inhibitory amounts are additive. This gives strong evidence that both antibiotics are acting on a common phosphorylation site. In contrast, oligomycin inhibits all ATPases induced by a wide variety of uncoupling agents, while aurovertin enhances ATP hydrolysis induced by some agents and only partially inhibits that induced by 2,4-dinitrophenol. They conclude that ATP can irreversibly generate a category of "high-energy" intermediate(s) separate from that generated by oxidative phosphorylation. Some uncouplers discharge this type as well as an early intermediate of oxidative phosphorylation.

With respect to the action of various inhibitors of oxidation and phosphorylation, it is instructive to consider their effects in some purified enzyme systems. Thus an active complex of β-hydroxybutyrate dehydrogenase and lecithin is sensitive to uncouplers of oxidative phosphorylation and to oligomycin (79). Particularly striking are the results of Handler (see p. 301), showing effects of 2,4-dinitrophenol, amytal, antimycin A, and oligomycin with purified flavoprotein oxidases. Effects of the inhibitors can be localized to different steps in a sequential series of reactions of molybdenum, flavin, non-heme iron, and O_2. Such results offer encouragement that the apparently complex behavior of oxidative phosphorylation and electron transport may find its eventual explanation in a beautifully organized and simple pattern of association of a relatively few number of proteins.

3. Phosphorylation by Mitochondrial Fractions

One of the logical approaches to the possible understanding of oxidative phosphorylation is to fractionate the mitochondria into soluble components capable of net oxidative phosphorylation. This has not been achieved. All systems that have been obtained by fractionation of mito-chondria and that are capable of oxidative phosphorylation or other energy-linked functions have as a principal component particles containing the respiratory enzymes and associated components.

Three principal types of particle preparations are currently used for oxidation and phosphorylation studies. The first particulate preparation to be thoroughly studied was the so-called Keilin-Hartree preparation, made by grinding minced heart with sand in phosphate buffer [see King (80) for a recent description]. Recently, this preparation has been found to be capable of performing energy-linked reduction (81) and weak phosphorylations (82). The most widely used type of preparation, intro-duced by Kielley and Bronk (83), involves sonication of mitochondria. This includes the thoroughly studied "ETP_h" preparations of Green's laboratory (see ref. 84) for description of the composition and references for preparation). A third type of preparation involves use of detergents, such as the "digitonin particles" used extensively by Lehninger and associates (4,7).

When carefully prepared and assayed, particles are capable of phos-phorylation with good P:O ratios. They show a pattern of behavior to effects of uncoupling and phosphorylation blocking agents similar to those of intact mitochondria. Particles are useful in the search for inter-mediates because they have lost many of the accessory enzymes of mito-chondria. They also may allow a better control of environmental conditions at the phosphorylation site. Perhaps their principal use has been in the search for and assay of "coupling factor" proteins which are necessary for phosphorylation but not for oxidation upon addition to suitable particulate preparations. Linnane in 1958 gave a clear description of stimulation of phosphorylation by a protein fraction (85,86). Linnane's factor was not highly purified, but it appears to be closely related to the best-characterized coupling factor, namely the soluble ATPase isolated by Pullman et al. (87,24). This and other factors isolated in Racker's labora-tory appear to be required for phosphorylation at all sites, as established by well-documented experiments. The ATPase also acts as a coupling factor in restoring energy-linked reduction driven by ATP, but, significantly, is not required for energy-linked transhydrogenase driven by succinate oxidation (24). A closely related factor has been studied by Sanadi (9).

Other work has indicated that site-specific coupling factors may exist, although these may have quite indirect roles. Wadkins and Lehninger reported that their ADP–ATP exchange enzyme stimulated phosphoryla-tion associated with the terminal phosphorylation step (88). Subsequent kinetic studies indicated complex formation between the exchange enzyme and cytochrome c (89). Isolation from heart mitochondria of an apparent complex between the exchange enzyme and reduced cytochrome c has been reported (90). Relation of these results to oxidative phosphorylation is at present obscure.

Pinchot, in studies with bacterial systems, has shown oxidative formation of a particulate-bound DPN which will react slowly with P_i and ADP to give ATP (91). The activity is attributed to a "high-energy" bond between DPN and a site-specific coupling factor. The preparations are crude, and the possibility of reactions other than those of oxidative phosphorylation being involved looms large. The principal claims for site-specific coupling factors have come from Green's laboratory. Designation of these as different "ATP synthetases" (25), each involving formation of a specific phosphorylated protein, seemed premature, even prior to the questioning (26) of much of the experimental evidence supporting the concept.

The mode of action of various coupling factors is difficult to ascertain. They could help restore requisite structural integrity, or even remove inhibitors, instead of participating by furnishing substrate binding sites and catalytic groups. Of particular interest in this regard is the report of Lee and Ernster (92) that very low amounts of oligomycin will mimic coupling factors in restoring some phosphorylation ability to otherwise ineffective particles. The generality of the action of the ATPase coupling factor of Racker and associates and its capacity to act as an ATPase suggests, of course, that it plays a role in the terminal sequence of phosphorylation involving P_i, ADP, ATP. Pertinent to this is the requirement of the factor for the $P_i \rightleftharpoons$ ATP and $P_i \rightleftharpoons$ HOH exchanges, as well as net phosphorylation. Many questions remain unanswered, even if such a role proves correct. Formation of a phosphorylated enzyme is usually visualized by Racker (24) and others, but no evidence of such phosphoprotein formation has been reported. The lack of an ADP–ATP exchange in systems where the $P_i \rightleftharpoons$ ATP exchange is observed argues against phosphoprotein formation, but Racker has presented evidence that the ADP may remain bound to the enzyme most of the time under the assay conditions used (93). This would allow the $P_i \rightleftharpoons$ ATP but not the ADP $\rightleftharpoons$ ATP exchange to be observed. Consideration of how P_i loses oxygen is unfortunately omitted in most discussions.

4. Respiratory Control and Related Phenomena

In 1951 Rabinovitz, Stulberg, and Boyer (94) and Lardy and Wellman (95) independently observed an important phenomenon: that respiration by heart (94) or liver (95) mitochondria could be limited by absence of ADP as a P_i acceptor. This demonstrated an important metabolic control on oxygen uptake. Implicit in the observation was that oxygen uptake was depressed because of equilibrium considerations, and thus that the phosphorylations and oxidations of oxidative phosphorylation might be dynamically reversible. This has since been fully demonstrated. In addition, respiratory control gave an important tool, especially as used with the rapid polarographic determination of oxygen, for following phosphorylation associated with respiration. This approach has allowed probing of oxidative phosphorylation and uncoupling phenomena in a simple manner. The techniques have been elegantly exploited, particularly by Chance and colleagues (40,41,43), in association with simultaneous measurement of changes in light absorption, fluorescence, and pH. Designation of various metabolic states of mitochondria as used by Chance are discussed in the chapter by Klingenberg.

Logically, oxygen uptake should be controlled by the ratio $[ATP]/[ADP][P_i]$, and such control has been demonstrated by Klingenberg (42) and by Chance and Hollunger (43). Thus depletion of ADP or P_i or addition of ATP under appropriate conditions can decrease oxygen uptake, and the ratio of reduced to oxidized cytochrome c is a predictable function of the $[ATP]/[ADP][P_i]$ ratio. Estimation of an apparent $-\Delta F$ for the hydrolysis of the "high-energy" precursor to ATP from this observation (42) appears unwarranted, however, both because of the questionable assumption that a covalent compound susceptible to hydrolysis exists and, even if it does, because of the lack of any information about the ratio of the hydrolyzed to unhydrolyzed forms.

C. Exchange and Partial Reactions

Cohn, in 1953, reported the important observation that mitochondria catalyzed a rapid and dinitrophenol-sensitive exchange of water oxygens with oxygen of P_i (96). The possibility that such an exchange occurred by a reversal of the overall phosphorylation reaction of oxidative phosphorylation, $ADP + P_i \rightleftharpoons ATP + HOH$, led to the independent discovery of the $P_i \rightleftharpoons ATP$ exchange by Boyer et al. (97) and by Swanson (98). As mentioned earlier, the possibility of a dynamic reversal of oxidative phosphorylation was also suggested by the occurrence of respiratory control. A dynamic reversal of oxidative phosphorylation could give an $ADP \rightleftharpoons ATP$ exchange as well as a $P_i \rightleftharpoons ATP$ and $P_i \rightleftharpoons HOH$ exchange. An important implication of the $P_i \rightleftharpoons ATP$ exchange was that the energy derived from cleavage of ATP was somehow conserved for the resynthesis and this energy might be used to reverse electron flow (97). Such apparent reversal has since been documented by the occurrence of ATP-driven energy-linked reduction as discussed in Section III-D.

The phosphoryl group appearing in ATP or in glucose 6-phosphate during net oxidative phosphorylation has undergone more extensive oxygen exchange than the P_i of the reaction medium (99–101). This reflects an ATP $\rightleftharpoons$ HOH exchange that appears to be more diagnostic for oxidative phosphorylation than the $P_i \rightleftharpoons$ HOH exchange or the $P_i \rightleftharpoons$ ATP exchange. A number of enzymes will give a $P_i \rightleftharpoons$ ATP exchange, and these may appear in mitochondrial preparations (102). In addition, recent evidence suggests that in mitochondria and in particles, a second reaction, in addition to oxidative phosphorylation, may give a prominent $P_i \rightleftharpoons$ HOH exchange (103). As far as is known, only oxidative phosphorylation and photosynthetic phosphorylation give an ATP $\rightleftharpoons$ HOH exchange.

In all phosphorylation systems tested, the $P_i \rightleftharpoons$ HOH exchange and the ATP $\rightleftharpoons$ HOH exchange have been more rapid than the $P_i \rightleftharpoons$ ATP exchange (97,99–104,78). Various suggestions have been made to explain the inequalities of exchanges observed, such as dynamic reversal of a reaction of the type depicted by Equation (8)

$$P_i + {\sim}Y \rightleftharpoons Y{-}PO_3{}^{2-} + HOH \qquad (8)$$

at a rate greater than the dynamic reversal of the overall reaction, to explain the relatively high $P_i \rightleftharpoons$ HOH exchange, or the occurrence of a vague type of labile phosphorylated intermediate with exchangeable oxygens to explain the greater incorporation of water oxygen into ATP than into P_i. Different accessibility of substrates to compartments within mitochondria and even in particles could influence exchange rates.

It is perhaps instructive to point out that exchange inequalities do not demand separate reactions, special labile states, or limited mitochondrial compartments. Inequalities of exchanges have been observed and put on a sound theoretical basis with highly purified enzymes (105,106). Such inequalities may arise from ordered pathways of substrate addition or from substrate dissociation steps being slow with respect to the chemical bond forming and breaking steps. For example, in oxidative phosphorylation, possibilities of differential exchange rates can be evaluated by the simple diagram indicated in Scheme IV. This scheme shows interconversion of two forms of an enzyme–substrate complex, an $E \cdot P_i \cdot ADP$ and an $E \cdot HOH \cdot ATP$ form. Implicit in the scheme are mutually exclusive binding sites for ADP and ATP. Step 3 in this scheme indicates the interconversion of P_i and ADP to ATP and HOH coupled to a change in energy state from ${\sim}Y$ to $-Y$. The other steps represent association and dissociation

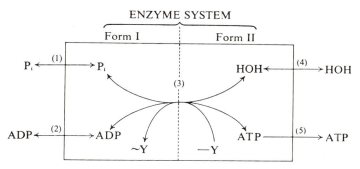

Scheme IV. Depiction of substrate interconversion and dissociation steps of oxidative phosphorylation.

of substrates from their binding sites on an enzyme surface. Step 4, water release and binding, may be regarded as sufficiently rapid to not limit oxygen exchanges. For these considerations, a water-binding site is not necessary and water could participate directly from solvent water. Either of these conditions would mean that the $P_i \rightleftharpoons HOH$ rate would always equal or exceed the $P_i \rightleftharpoons ATP$ rate. A rapid, dynamic reversal of step 3 compared to 1 or 5 would give oxygen exchanges exceeding the $P_i \rightleftharpoons ATP$ exchange rate, as has been observed. A slow step 3 in reconstituted systems of Racker (24) would explain occurrence of the $P_i \rightleftharpoons HOH$ and $P_i \rightleftharpoons ATP$ exchanges without appreciable $ADP \rightleftharpoons ATP$ exchange (see ref. 93).

Scheme IV involves a single mode of entry of water oxygen to explain both the $P_i \rightleftharpoons HOH$ and the $ATP \rightleftharpoons HOH$ exchanges. In such a simple scheme there is a theoretical limit to the extent that the $P_i \rightleftharpoons HOH$ and $ATP \rightleftharpoons HOH$ exchanges can exceed the $P_i \rightleftharpoons ATP$ exchange. In particles under some conditions this limit is exceeded, and the high $P_i \rightleftharpoons HOH$ exchange probably finds its explanation in an exchange separate from reversal of oxidative phosphorylation but dependent upon ATP cleavage (103).

Exchange patterns quite analogous to those of oxidative phosphorylation, but involving the —COOH of glutamate instead of water, together with P_i and ATP, have been observed with glutamine synthetase (105). This gives credence to the explanations offered for mitochondrial catalyzed exchanges. The availability of simple explanations for differences in exchange rates observed does not, of course, rule out more complicated schemes with various covalent intermediates. An important point is that postulation of more complicated events is unnecessary.

Catalysis of a rapid $P_i \rightleftharpoons HOH$ exchange independent of adenine nucleotides would give strong evidence for a direct loss of oxygen from P_i, giving a phosphoryl intermediate which then formed ATP, as depicted by Reactions (9) and (10).

$$\sim Y + HOPO_3{}^{2-} \rightleftharpoons Y—PO_3{}^{2-} + HOH \qquad (9)$$

$$Y—PO_3{}^{2-} + ADP + H^+ \rightleftharpoons YH + ATP \qquad (10)$$

Reversal of Reaction (9) would give a $P_i \rightleftharpoons HOH$ exchange independent of ADP or ATP participation.

If reversible ATP formation at a catalytic site is essential for oxygen exchange, a requirement of ADP for exchange should be demonstrable. Cooper (107) has shown a marked stimulation of oxygen exchange in digitonin particles by ADP addition, even though the particles tenaciously retain some bound AMP, ADP, and ATP. He suggests, on this basis, a concerted reaction for ATP formation. An earlier report that the $P_i \rightleftharpoons$ HOH exchange could be activated by respiration in absence of ATP (101) appears ascribable to the presence of considerable adenine nucleotide in the preparation used. Mitchell et al. have demonstrated a nearly complete dependency of the oxygen exchanges upon adenine nucleotide addition with particles prepared by sonication (103). They further showed that the rapid $P_i \rightleftharpoons ATP$ and $ATP \rightleftharpoons HOH$ exchanges required ADP, but that the nucleotide requirement for a rapid $P_i \rightleftharpoons HOH$ exchange could be met by ATP.

A requirement of ADP for the $ATP \rightleftharpoons HOH$ exchange is consistent with a concerted reaction for ATP formation, but may also have other explanations. These include conformation change dependent upon ADP addition, a transient metaphosphate formation, or some types of indirect mechanism for formation of water oxygen from P_i. One such mechanism is depicted in Scheme II-B.

The demonstration that ADP but not ATP will meet the requirement for a rapid $ATP \rightleftharpoons HOH$ exchange argues against formation of a pentavalent P atom in ATP (see Reaction (7)) as an explanation for the oxygen exchange associated with oxidative phosphorylation. In other mechanisms considered herein, with direct or indirect loss of P_i oxygen, the primary mode of entry of water oxygen in the exchange reaction would be at the P_i bound to the catalytic site.

Respiratory inhibition by cyanide and combinations of cyanide, amytal, rotenone, and antimycin A depress the $P_i \rightleftharpoons ATP$ (51,52) and the $P_i \rightleftharpoons HOH$ and $ATP \rightleftharpoons HOH$ exchanges (52) by intact mitochondria only slightly. A considerable depression in the $P_i \rightleftharpoons ATP$ exchange observed with digitonin particles led Lehninger and Wadkins to propose that oxidation-reduction changes in respiratory carriers were necessary for the $P_i \rightleftharpoons ATP$ exchange reaction (see ref. 4). This is not compatible with observations reported in the original descriptions of the $P_i \rightleftharpoons ATP$ exchange reaction. Sufficient cyanide to inhibit respiration as well as anaerobiosis in the presence of substrate failed to inhibit the $P_i \rightleftharpoons ATP$ exchange (97,98). The lack of any striking inhibition by combinations of respiratory inhibitors on the $P_i \rightleftharpoons ATP$ exchange and, particularly, on the $ATP \rightleftharpoons HOH$ exchange in intact mitochondria make it likely that those inhibitory effects observed are secondary, resulting from changes in rates of penetration or dissociation of reactants or damage to enzyme systems by treatments used, rather than reflecting a compulsory participation of oxidation-reduction processes in the reactions leading to P_i release and uptake and to water formation.

One of the important implications of the $ADP \rightleftharpoons ATP$ exchange reaction, studied extensively by Wadkins and Lehninger (88), is that it occurs by a reaction depicted by Equation (11), involving formation of a phosphorylated enzyme.

$$ATP + Enzyme \rightleftharpoons Enzyme—PO_3{}^{2-} + ADP \qquad (11)$$

The lack of a P_i requirement by a purified $ADP \rightleftharpoons ATP$ exchange enzyme favors this, but clear distinction of the purified enzyme from a nucleoside diphosphokinase activity or demonstration of a phosphoryl protein in the preparation has not been forthcoming. Of considerable interest is the recent report of formation of a phosphorylated enzyme by highly purified nucleoside diphosphokinase (108). This could correspond to the Wadkins-Lehninger enzyme. In digitonin particles, Cooper and Kulka noted equal rates of the oligomycin-sensitive $P_i \rightleftharpoons ATP$ and $ADP \rightleftharpoons ATP$ exchange reactions, and suggest that both reactions involve a common enzyme as a rate-limiting component (109).

An interesting enzyme has been obtained in low yield from mitochondria by Chiga and Plaut which catalyzes both the $P_i \rightleftharpoons ATP$ and $ADP \rightleftharpoons ATP$ exchanges (110). Catalysis of the $P_i \rightleftharpoons ATP$ exchange implies that energy from the ATP cleavage is somehow conserved by the enzyme. Such exchange is observed only with systems, such as succinate thiokinase or glutamine synthetase, in which the energy from ATP cleavage is used for covalent bond formation. The nature of the energy conservation is the Chiga-Plaut enzyme might have important implications for oxidative phosphorylation.

ATPase activities of mitochondria, mitochondrial fragments, and soluble preparations therefrom most likely arise from modification of the oxidative phosphorylation system. Uncoupled systems frequently show marked ATPase activity. Whether there are multiple ATPases separable from mitochondria, or whether the various ATPase activities are a reflection of the same catalytic protein assayed under different conditions, with different accessory substances present or with some subtle structural modifications remains a moot question. The topic has been ably reviewed by Racker (3), who favors the view of a single ATPase entity. Recent antibiotic studies by Lardy (78), however, suggest multiple ATPases. Other than the study of the ATPase coupling factor, discussed elsewhere, experiments with the ATPases, although of interest for other reasons, have contributed little to the understanding of oxidative phosphorylation.

The inhibitor sensitivity of the $P_i \rightleftharpoons HOH$ and the $P_i \rightleftharpoons ATP$ exchanges is pertinent to the suggestion of Brierley and Green (111) of a hypothetical "mesomerase" which supposedly catalyzes "the reversible transfer of the pyrophosphate bond from the pair of ligands on the one surface (of the mitochondrial membrane) to the pair of the other surface." This is taken to represent the locus of atractyloside inhibition. Among other things, this indefinite chemistry would necessitate reversible loss of oxygen from P_i as well as the $P_i \rightleftharpoons ATP$ exchange to occur at the membrane surface in an oligomycin-insensitive reaction. Such is contrary to experimental fact.

D. Energy-Linked Reductions

Utilization of energy derived from ATP to drive reductive synthesis is a well-established phenomenon in metabolic systems. An important implication of the discovery of the $P_i \rightleftharpoons ATP$ exchange by mitochondria was that the energy derived from the cleavage of ATP was somehow conserved for the rapid resynthesis, and that this energy might be used to reverse electron flow in mitochondria (97). Chance and Williams (40) in 1956 reported the discovery that ATP could drive the reduction of DPN by succinate in mitochondria. Such energy-linked reduction has since been thoroughly studied, principally by Chance, Ernster, Klingenberg, and Slater and their associates (see refs. 6, 9, and 41). The reduction of DPN, of α-ketoglutarate plus NH_3, of acetoacetate, or an energy-requiring reduction of DPN by TPNH with succinate as the reductant, have been most frequently studied. Comments herein will be limited to relationships to oxidative phosphorylation that have emerged.

The most important aspects of the energy-linked reduction studies are the probabilities that such reductions use at least part of the catalytic mechanism of oxidative phosphorylation and that under some circumstances the energy-linked reductions can occur without participation of ADP, ATP, or P_i (112,113,41). All systems capable of oxidative phosphorylation also perform energy-linked reductions. The similar catalytic mechanism is shown by the inhibition of the ATP-driven reduction by oligomycin, and the uncoupling by 2,4-dinitrophenol. The sensitivity of such processes to 2,4-dinitrophenol, as mentioned earlier, places the dinitrophenol action prior to reaction with P_i in oxidative phosphorylation.

Although oxidative phosphorylation and energy-linked reduction may have a common catalytic mechanism for coupling of oxidation-reduction to a high-energy compound or state, another route in addition to reversal of oxidative phosphorylation appears to be involved in ATP-driven reduction. Pronounced differences in the Mg^{2+} requirement and in inhibitor sensitivity of energy-linked reduction as driven by ATP and of exchange reactions characteristic of oxidative phosphorylation have been observed (103). Such results give strong support to the probability that the initial step in the utilization of ATP for energy-linked reduction may occur without involving the same catalyst as the phosphorylation step of oxidative phosphorylation. The pattern appears similar to other metabolic sequences where routes of synthesis were initially regarded as reversals of degradation routes (e.g., fatty acid synthesis, phosphoenolpyruvate formation), but where alternative routes for the ATP-driven syntheses have since been found. Those involve cleavage of more ATP molecules than are formed in the degradative route. Utilization of ATP for ion transport and volume changes may also involve routes distinct from the phosphorylation reaction of oxidative phosphorylation, as mentioned in the next section.

The occurrence of energy-linked reduction in which part of the energy from succinate oxidation is used to drive DPN reduction by succinate with oligomycin present clearly indicates a lack of P_i, ADP, or ATP requirement for this process. The ability to use reduced cytochrome c for DPN reduction driven by ATP suggests that energy transfer among all three sites may occur without phosphorylation (see ref. 41). More definitive experiments in this regard would be reassuring, however. The concept of a single or interconvertible "high-energy" compound or state formed or interchanged without P_i, ADP, or ATP involvement is sufficiently important to demand convincing experimental support.

E. Ion Transport, Swelling, and Contraction

The phenomenon of ion accumulation by mitochondria has received considerable attention in the past few years (4–9). The important relationship to oxidative phosphorylation is akin to the energy-linked reductions, namely that ion transport appears to be driven by a nonphosphorylated "high-energy" compound or state derivable either from electron transport or from ATP. Best estimates of the stoichiometry involved are that the equivalent of one "high-energy" phosphate will cause transport of at least 2 Ca^{++} (114) or several K^+ or H^+ (114a). Respiratory inhibitors, lack of substrate, or uncoupling agents but not oligomycin inhibit the uptake driven by oxidation. Oligomycin inhibits uptake driven by ATP. A particularly interesting aspect is that ATP-driven ion transport can occur in particles incapable of catalyzing net oxidative phosphorylation (115). Such behavior suggests that ATP can drive ion transport by a route distinct from the phosphorylation reaction of oxidative phosphorylation.

Mitochondrial water uptake and swelling, thoroughly studied by Hunter, by Lehninger, and by Packer and their associates (see refs. 4, 7, and 41) presents some interesting and puzzling phenomena. At least two distinct types of swelling and contraction phenomena must be distinguished. Packer has shown a relatively rapid cycle of swelling and shrinking, paralleling the change in the steady-state level of respiratory carriers. This swelling requires active electron transport and is not inhibited by oligomycin (116). A close relation to the energy-conserving reactions of oxidative phosphorylation is thus indicated. A slower type of swelling can be reversed by respiration in presence of ADP or by ATP addition, and the effect of ATP is blocked by oligomycin (117). This has led to suggestion that the contraction may depend upon "high-energy" intermediates or states formed by either ATP or respiration (see refs. 4 and 7). The inability of 2,4-dinitrophenol to uncouple the ATP-driven contraction (118) makes such an interpretation at present unacceptable. The studies of Connelly and Lardy (119) with the antibiotics oligomycin and aurovertin show that ATP can effect the slow swelling and contraction by mechanisms not involving reversal of oxidative phosphorylation. They suggest three alternative routes for formation of the "high-energy" intermediates of oxidative phosphorylation, one by oxidation and two routes from ATP. One route from ATP involves reversal of oxidative phosphorylation and the other an independent irreversible reaction.

F. The Search for Intermediates

1. Possible Existence and Amount of Intermediates

For many biochemical processes, the detection of intermediates has given insight into the nature of the process. The search for possible intermediates in oxidative phosphorylation has thus been intensive—much more so than might be indicated by the literature, because unfruitful experiments usually are not published. Substances of unknown nature can be obscure and elusive, and search will continue until intermediate compounds or stages are found or can reasonably be regarded as ncn-existent.

Logical experiments designed to quantitate the amount of intermediates that might exist have been conducted by Eisenhardt and colleagues (120,121). They have demonstrated a rapid burst of ATP formation following ADP addition to mitochondria lacking ADP. VanDam has shown that a considerable rapid ATP synthesis may result from oxidation of accumulated DPNH (122). His methods, however, would have not detected the small, dinitrophenol-insensitive burst of ATP described by Eisenhardt and Rosenthal (121). This burst could result from accumulation of substances unrelated to oxidative phosphorylation such as succinyl CoA or the nonphosphorylated "high-energy" form of succinyl CoA synthetase. Thus at this stage there is no convincing evidence for existence of any nonphosphorylated "high-energy" intermediates. These experiments indicated an appreciable accumulation of some intermediates. Later experiments have, however, shown that most if not all of the "ATP-jump" results from the rapid oxidation of reduced nicotinamide nucleotides (122). Thus it appears that even in tightly coupled mitochondria there is little or no accumulation of any intermediates. Intermediates present in amounts about stoichiometric with the respiratory chain would, however, be difficult to detect experimentally.

2. Possible Intermediates from P_i

A considerable effort has been expended in various laboratories in the search for phosphorylated derivatives of oxidation-reduction components, other low-molecular weight cofactors, or proteins which might serve as intermediates in oxidative phosphorylation. In other enzymic systems leading to ATP formation, the phosphorylated precursors of ATP, such as 1,3-diphosphoglycerate, phosphoenolpyruvate, aspartyl phosphate, acetyl phosphate, carbamyl phosphate, and protein-bound phosphohistidine, are sufficiently stable and prevalent to allow isolation by techniques available. It is thus of some significance that no phosphorylated derivative of oxidized or reduced forms of DPN, FMN, FAD, ubiquinone, vitamin K, vitamin E, or cytochromes has been isolated in mitochondria under circumstances which provide firm evidence that the substance is an intermediate in oxidative phosphorylation. Some suggestions of intermediates of DPN (123) and of vitamin K (124) have not been followed by convincing experimental evidence. These considerations and results, and other findings discussed elsewhere in this chapter, make it unlikely that phosphorylated oxidation-reduction cofactors participate in oxidative phosphorylation.

A frequently considered possibility is that of phosphorylation of an enzyme or "coupling factor" protein as a step in the phosphorylation process. A search for such intermediates led to the discovery of protein-bound phosphohistidine and demonstration of its participation in synthesis of ATP by mitochondria (15,125). The principal phosphorylated protein in a soluble system from mitochondria has, however, been found to be part of, or closely associated with, succinate thiokinase (126,127), and present evidence neither proves nor disproves a role for phosphohistidine in oxidative phosphorylation (128). Reports of a phosphorylated coupling factor in the cytochrome c region (25) appear to have alternative explanations (26). Indications of a phosphorylated coupling factor acting in the cyt b-c site (129) must await further demonstration of distinction from mitochondrial proteins reversibly phosphorylated by ATP (126,130) and more complete examination of the chemical nature and metabolic properties of the substance.

The only mitochondrial phosphoprotein which has been established as reaching approximate maximal labeling from $^{32}P_i$ at the time the rate of $AT^{32}P$ labeling becomes maximal is bound phosphohistidine. Studies of labeling rates of ATP and bound phosphohistidine, contrary to claims by Slater et al. (131,132), do not rule out participation of bound phosphohistidine in oxidative phosphorylation (133,134). Present data are largely reconcilable, however, with a role of mitochondrial phosphohistidine in the succinate thiokinase reaction (128). During the time period in which the rate of ATP labeling becomes maximal, only very small amounts of comparatively acid-stable (phosphoseryl-type) phosphoproteins are formed, and similarly small amounts of lipid phosphates are labeled (133,135). Further, both of these fractions continue to greatly increase in labeling for some time. Such labeling rates are not consistent with an intermediate function. Phosphoserine-containing proteins have been postulated as intermediates in oxidative phosphorylation (130,136), but the above-mentioned studies, and demonstrations that most if not all such labeling comes from $AT^{32}P$ (135), given little support to such suggestion.

An experiment of excellent design to attempt to find evidence for a phosphorylated intermediate and indirect loss of P_i oxygen to water was conducted by Itada and Cohn (104). In the arsenolysis of ATP by the 3-phosphoglycerate kinase and glyceraldehyde 3-phosphate dehydrogenase reactions, ^{18}O from arsenate is transferred, via the carboxyl of 3-phosphoglycerate, to P_i. This is in accord with the participation of the acyl phosphate as an intermediate in ATP formation. Thus if a similar type of reaction occurred in oxidative phosphorylation, arsenolysis of ATP by mitochondria might result in ^{18}O transfer from arsenate to phosphate. The results (104) showed only a slight amount of such transfer, but sufficient that participation of an Y—OH group as a phosphorylated intermediate must still be considered. The very low rate could reflect a rapid interchange of Y—OH and water oxygens, thus removing most of the ^{18}O derived from arsenate prior to phosphorylation by ATP.

3. Possible Intermediates from ADP

Although in overall oxidative phosphorylation by mitochondria (10) or by particles (Hill and Boyer, unpublished), oxygen is lost from water to P_i, such a loss, as discussed earlier, may be indirect (10). One type of indirect loss could involve participation of an activated ADP structure (38). Thus if ADP remained firmly bound as indicated in the reaction pattern of Scheme V, the primary loss of oxygen would be from the bound ADP,

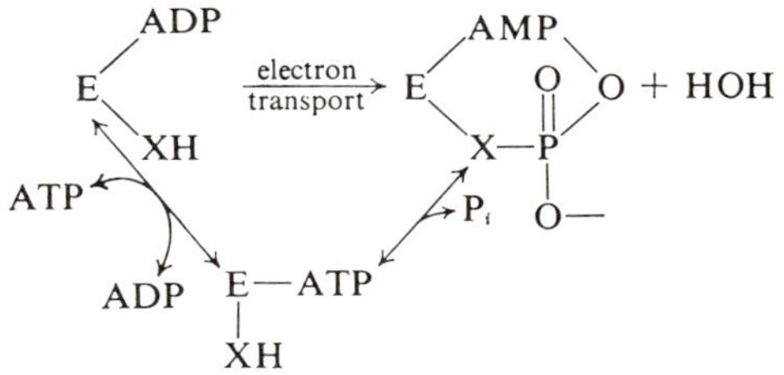

Scheme V. A hypothetical covalent bond formation by ADP which would give indirect loss of P_i oxygen to water.

but this would be regenerated from P_i, and in the repetition of the cycle P_i would lose oxygen to water. This appears experimentally unlikely. Mitchell and Boyer (unpublished) have observed no indication of preferential incorporation of ^{18}O from water into bound nucleotides of particles as compared to nucleotides in solution. An ADP or other oxygen-containing substance group firmly bound and present in amounts only about stoichiometric with the respiratory chain cannot, however, at this stage be eliminated as an intermediate.

Another intriguing but speculative mode of ADP participation has been suggested from the work of Skulachev (137,138). With aged mitochondrial preparations he has demonstrated oxygen uptake stimulation by ADP addition but not by 2,4-dinitrophenol. This does not appear to represent any type of respiratory control by ADP lack, and has been interpreted as indicating participation of the adenine ring in an oxidation-reduction sequence with concomitant phosphorylation. At this stage, satisfying chemical evidence for the postulated oxidation-reduction reaction of the adenine ring is lacking, and alternative explanations for the stimulation of uptake appear likely.

Evidence for an ADP activation in an unknown reaction comes from the demonstration of a CoA-dependent incorporation of water oxygens into ADP by a crude *E. coli* extract (139). Reactions which exchange water oxygen with nucleotide phosphoryl groups are unusual. Such observations thus serve to keep alive hypotheses for possible ADP activation in oxidative phosphorylation and other enzymic reactions.

4. Possible Nonphosphorylated Intermediates

The designation "nonphosphorylated intermediates" refers to hypothetical intermediates which are not formed from either the P_i or ADP destined to become ATP.

By careful spectral measurements, Chance has searched for spectral changes in respiratory carriers of mitochondria under different metabolic states. The only change detected has been in an unknown form of a cytochrome, designated cytochrome b_{555} from the position of an absorption band at 77°K. This absorption band appears in tightly coupled mitochondria in the absence of ADP and the presence of sulfide as an oxidation inhibitor. Uncoupling agents cause the compound to disappear. The tenuous suggestion is made that the compound might represent a "high-energy" intermediate of oxidative phosphorylation. This could be so, but at this stage the lack of any detected change in absorption spectra of all the other oxidation-reduction components (40,140) in coupled contrasted to uncoupled conditions seems to favor the view that "high-energy" complexes of various carriers are not formed as part of the oxidative phosphorylation sequence.

In earlier studies, Chance and colleagues have postulated intermediates of carriers with an inhibitor, I, and complexes with electron carriers, designated $C \sim I$, appear in schemes of various investigators. Frequency of use of this designation may give the suggestion an unwarranted appearance of validity. Although such complexes are plausible, it must be emphasized that any convincing experimental evidence for their existence as discernible chemical entities or for the existence of a special substance, I, is lacking. The occurrence of respiratory control and the dynamic reversal of oxidative phosphorylation make the accumulation of reduced carriers under some conditions a certainty from thermodynamic considerations, irrespective of the number or type of intermediate steps involved. None of the observations on the nature of accumulated reduced carriers demand any specific combination of the carriers, covalent or otherwise.

Means of probing the nature of the nonphosphorylated "high-energy" compound(s) or state(s) associated with energy-linked reductions and probably with oxidation phosphorylation are singularly lacking. If covalent linkages are involved, such as acyl–S or acyl–imidazole, appropriate inhibitors might reveal their existence, provided that methodology could be made sufficiently sensitive. Covalent linkages, such as the pyrophosphate linkage, offer little chance for cleavage by inhibitors at neutral pH and temperature. Nonetheless, additional studies of the inhibitor sensitivity of energy-linked reduction could prove fruitful.

If the "high-energy" forms represent nothing more complicated than loss of a proton in an environment not accessible to water, or separation of charges [e.g., H^+ and OH^- as suggested by Mitchell (30)], their detection by usual chemical means will be difficult.

G. Relation to Other Energy-Linked Systems

The possibility of analogies between contractile processes, myosin interaction with ATP, and oxidative phosphorylation has caught the attention of a number of investigators (see refs. 4–9, 141, 142). In one case, covalent bond cleavage is used to cause conformation change; in the other conformation change accompanying electron transport could cause covalent bond formation. The analogy with myosin is strengthened by the uncoupling of mitochondria and the activation of myosin ATPase by —SH group reagents, and by 2,4-dinitrophenol (141,143,144) as well as the ability of both mitochondrial (10,53,104) and myosin systems (145,146) to catalyze incorporation of more than one solvent oxygen into P_i cleaved from ATP and to catalyze an $P_i \rightleftharpoons HOH$ exchange (142). The analogy from oxygen exchanges is weakened from the observation that with myosin 2,4-dinitrophenol accelerates but p-mercuribenzoate inhibits the oxygen exchange, while with mitochondrial both agents inhibit the exchange (142a).

Other suggestions have been made that ATP-driven transport, such as in the kidney or in salt-excreting glands, may have mechanistic similarity to oxidative phosphorylation. Although such suggestions for relationships between contraction, transport, and phosphorylation coupled to energy-yielding process may have validity, they are of little value to the explanation of mechanism at the present time. With all systems, the amount of real information and understanding is meager, and the investigator finds himself attempting to explain one poorly understood system by analogy with another of a similar degree of understanding.

The process of photophosphorylation, discussed in the chapter by Arnon, likely involves basic mechanisms quite similar to oxidative phosphorylation. Light energy is used to drive oxidation-reduction reactions which are coupled to ATP synthesis. At this stage, photo-phosphorylation is no better understood than oxidative phosphorylation, and no defined intermediates have been clearly identified. The system has considerably different properties than oxidative phosphorylation, and may be more revealing for study in some aspects of coupling of oxidation-reduction with phosphorylation. One observation that promises to give considerable insight into the coupling mechanism is that temporary exposure of chloroplasts to low pH will allow a subsequent small burst of ATP formation in the dark near neutral pH (147). This has been interpreted as supporting the Mitchell hypothesis of membrane separation of H^+ and OH^- as a driving force for oxidative phosphorylation (147).

Other oxidative systems of interest are the nitrate oxidizing systems and other oxidative reactions of some bacteria. The soluble system of Brodie capable of phosphorylation coupled to malate oxidation (148) as well as the particulate DPNH system (149) could prove particularly valuable.

H. Studies with Model Systems

The most widely studied model systems involve formation and oxidation of quinol derivatives, such as the demonstration by Clark et al. of forma-tion of ADP accompanying oxidation of a quinol phosphate by Br_2 in the presence of AMP in anhydrous formamide (150). Of potential interest is the report from Wieland's laboratory (151) that digitonin particles would slowly oxidize reduced coenzyme-Q_6 monophosphate with formation of ATP without apparent intermediate formation of P_i. Localized produc-tion of intramitochondrial P_i and normal oxidative phosphorylation still needs consideration, however. Several other observations warrant mention. A phosphorylation reaction accompanies the non-enzymic oxidation of 2-methyl-1,4-naphthohydroquinone diphosphate (152). Washed bacterial particles give ATP upon oxidation of a synthetic 6-chromanyl phosphate of vitamin K_1 (153). Quinone methines have been considered as inter-mediates in oxidative phosphorylation (154). All these interesting studies must be balanced against experimental results with mitochondrial systems catalyzing oxidative phosphorylation. No components rapidly labeled from $^{32}P_i$ having properties expected for CoQ phosphates, vitamin K phosphates, or related compounds have been found. Also, as discussed earlier, the bulk of the evidence favors the phosphorylation reactions as separate from the oxidation reactions.

An unusual model system has been described by Brinigar and Wang (155). Oxidation of formyl prophyrin in presence of imidazole, AMP, P_i, and CO gave a small amount of ADP. Brinigar and Wang suggest a mechanism based on one-electron transfer. In this, as well as in other model systems, reaction with some component other than P_i might be plausible, and any reactions in which known respiratory chain components are oxidized with formation of any type of covalent bond structure are of potential value. The best model systems for the phosphorylation reactions of oxidative phosphorylation may prove to be other enzymic reactions in which P_i is taken up. For non-enzymic reactions, oxidative reactions which may form suitable nonphosphorylated "high-energy" compounds or states appear to warrant more attention.

IV. Summation

To the reader of the preceding pages, it must be clear that firm knowledge about oxidative phosphorylation, other than that the remarkable process occurs, is disappointingly meager. Well-recognized aspects include observations that minimal values for ATP synthesis are about 3, 2, and 1 for each DPNH, succinate, or reduced cytochrome c oxidized, respectively; that mitochondria may show respiratory control requiring concomitant phosphorylation for oxygen uptake; and that reduced electron-transport chain components accumulate when phosphorylation is blocked.

Attempts to fractionate mitochondrial systems have yielded complex particulates containing respiratory components, and one or more protein coupling factors. Soluble systems capable of oxidative phosphorylation have not been obtained. Whether the coupling factors have specific catalytic properties is uncertain, and suggestions of site-specific coupling factors are at this time untenable. The best-characterized coupling factor has ATPase properties, and appears essential for all energy-conserving sites. This factor may serve as the terminal phosphorylation enzyme for all oxidative phosphorylation, but could have a quite independent mode of action.

Mitochondria are capable of other energy-linked processes—namely, energy-requiring reduction, ion transport, or volume change—which can be driven either by ATP cleavage or by electron transport. The phosphorylation reaction of oxidative phosphorylation, the various electron-transfer steps, and the other energy-linked processes appear to share a common "high-energy" form or state. Whether this "high-energy" form or state represents a covalent compound, a protein conformation change, or other means for energy interconversion is unknown. Multiple compounds or states or something as simple as localized positive charge or protons could be involved. ATP-driven energy-linked processes can also occur by routes probably not involving reversal of the phosphorylation reaction of oxidative phosphorylation.

Mitochondria or particles therefrom readily lose their ability to phosphorylate although retaining capacity for oxidations. Net formation of ATP dependent upon oxygen uptake by mitochondria can be stopped by inhibition of oxidation, by uncoupling oxidation from formation of a high-energy compound or state (primary uncoupling) or by dissipation of the high-energy compound or state (secondary uncoupling), by blocking the phosphorylation reaction, by rapid hydrolysis of ATP formed, or by blocking interchange of extra- and intramitochondrial nucleotides. The widely used uncoupling agent 2,4-dinitrophenol acts as a primary or possibly a secondary uncoupler, and the antibiotic oligomycin blocks the phosphorylation reaction.

In oxidative phosphorylation, an oxygen from P_i forms water and ADP probably acts as the primary nucleotide acceptor of a phosphoryl group. The overall process involves an oxidation-reduction reaction, a water-formation reaction, and a phosphorylation reaction. Rapid $P_i \rightleftharpoons ATP$, $P_i \rightleftharpoons HOH$, and $ATP \rightleftharpoons HOH$ exchanges catalyzed by mitochondria or particles therefrom provide means for study of the phosphorylation and water-formation reactions. Differences in rates of these exchanges may reflect different rates of substrate dissociation from the phosphorylation site rather than participation of discreet covalent intermediates. Both rapid oxygen-exchange reactions require adenine nucleotides. ATP meets the requirement for a rapid $P_i \rightleftharpoons HOH$ exchange, and oxidative phosphorylation as well as a separate exchange coupled to ATP cleavage may contribute to the $P_i \rightleftharpoons HOH$ exchange. The rapid $P_i \rightleftharpoons ATP$ and $ATP \rightleftharpoons HOH$ exchanges require ADP, indicative that reversible ATP formation is necessary for the water-formation reaction. Whether formation of water from P_i oxygen occurs by a direct or an indirect mechanism remains an important unsolved problem.

The occurrence of rapid $P_i \rightleftharpoons ATP$ and $ATP \rightleftharpoons HOH$ exchanges without concomitant oxidation-reduction reactions indicates that the phosphorylation and water-formation reactions are independent from the oxidation-reduction reaction. This view also receives support from the occurrence of energy-linked reduction without requiring P_i or adenine nucleotides and from the failure of oxidations in absence of adenine nucleotides to activate a $P_i \rightleftharpoons HOH$ exchange.

Several observations are consistent with a single or similar water-formation and phosphorylation reactions for various electron-transfer steps. These include the evidence for a common or interconvertible non-phosphorylated "high-energy" precursor, the similar effect of inhibitors and uncouplers on phosphorylation coupled to different oxidations, and the ability of coupling factors to restore all phosphorylations.

Intensive search has not disclosed the presence of any phosphorylated intermediates in the process, or the accumulation of any appreciable amount of nonphosphorylated precursor. Consideration must be given to the possibility that oxidations give rise to a "high-energy" state present in amounts only about stoichiometric with respiratory chain components, and that ATP is the first phosphorylated product formed from P_i.

Acknowledgment

Preparation of this review and unpublished results reported herein were supported in part by grants from the U. S. Public Health Service and the U. S. Atomic Energy Commission.

References

1. W. H. Harrison, P. D. Boyer, and A. B. Falcone, *J. Biol. Chem.*, **215**, 303 (1955).
2. L. P. Hager, *J. Am. Chem. Soc.*, **79**, 4868 (1957).
3. E. Racker, in *Advances in Enzymology*, Vol. 23, F. F. Nord, Ed., Interscience, New York, 1961, p. 323.
4. A. L. Lehninger and C. L. Wadkins, *Ann. Rev. Biochem.*, **31**, 47 (1962).
5. V. Massey and C. Veeger, *Ann. Rev. Biochem.*, **32**, 597 (1963).
6. L. Ernster and C. Lee, *Ann. Rev. Biochem.*, **33**, 729 (1964).
7. A. L. Lehninger, *The Mitochondrion*, Benjamin, New York, 1964.
8. D. E. Griffiths, in *Essays in Biochemistry*, Vol. 1, P. N. Campbell and G. D. Greville, Eds., Academic Press, New York, 1965, p. 91.
9. D. R. Sanadi, *Ann. Rev. Biochem.*, **34**, 21 (1965).
10. P. D. Boyer, *Proc. Intern. Symp. Enzyme Chem. Tokyo, 1957*, Maruzen, Tokyo, 1958.
11. C. Cooper, *Biochemistry*, **4**, 335 (1965).
12. J. M. Buchanan and S. C. Hartman, in *Advances in Enzymology*, Vol. 21, F. F. Nord, Ed., Interscience, New York, 1959, p. 199.
13. E. Khedouri, V. P. Wellner, and A. Meister, *Biochemistry*, **3**, 824 (1964).
14. L. T. Webster, Jr. and F. Campagnari, *J. Biol. Chem.*, **237**, 1050 (1962).
15. P. D. Boyer, *Science*, **141**, 1147 (1963).
16. J. R. Cox and O. B. Ramsey, *Chem. Rev.*, **64**, 317 (1964).
17. *Chem. Soc. (London) Spec. Publ.*, **No. 8** (1957), "Phosphoric Esters and Related Compounds."

18. P. C. Haake and F. H. Westheimer, *J. Am. Chem. Soc.*, **83**, 1102 (1961).
19. P. D. Boyer, "Vitamin E," in *The Enzymes*, Vol. 3, P. D. Boyer, H. Lardy, and K. Myrbäck, Eds., Academic Press, New York, 1960, Chap. 18, p. 353.
20. A. Todd, *Nature*, **187**, 819 (1960).
21. R. M. Burton and N. O. Kaplan, *Arch. Biochem. Biophys.*, **101**, 139, 150 (1963).
22. P. D. Boyer, in *Oxidases and Related Redox Systems*, T. E. King, H. S. Mason, and M. Morrison, Eds., Wiley, New York, 1965.
23. E. R. Stadtman and F. H. White, *J. Am. Chem. Soc.*, **75**, 2022 (1953).
24. E. Racker and T. E. Conover, *Federation Proc.*, **22**, 1088 (1963).
25. G. Webster and D. Green, *Proc. Natl. Acad. Sci., U.S.*, **52**, 1170 (1964).
26. D. E. Green, statement read at meeting of the American Society of Biological Chemists, Atlantic City, April 1965.
27. M. Nozaki, T. Yamanaka, T. Horio, and K. Okunuki, *J. Biol. Chem.*, **44**, 453 (1957).
28. D. D. Ulmer, *Biochemistry*, **4**, 902 (1965).
29. D. W. Urry and P. Doty, *J. Am. Chem. Soc.*, **87**, 2756 (1965).
30. P. Mitchell, *Nature*, **191**, 144 (1961); *Biochem. Soc. Symp.*, **No. 22**, 142 (1963); P. Mitchell and J. Moyle, *Nature*, **208**, 147 (1965).
31. D. L. Millard, J. T. Wiskich, and R. N. Robertson, *Proc. Natl. Acad. Sci. U.S.*, **52**, 996 (1964).
32. S. H. Richardson, H. O. Hultin, and S. Fleischer, *Arch. Biochem. Biophys.*, **105**, 254 (1964).
33. S. Fleischer, G. Brierly, H. Klouwen, and D. B. Slautterback, *J. Biol. Chem.*, **237**, 3264 (1962).
34. P. Siekevitz and V. R. Potter, *J. Biol. Chem.*, **215**, 237 (1955).
35. B. C. Pressman, *J. Biol. Chem.*, **232**, 967 (1958).
36. R. K. Crane and F. Lipmann, *J. Biol. Chem.*, **201**, 245 (1953).
37. H. W. Heldt, H. Jacobs, and M. Klingenberg, *Biochem. Biophys. Res. Commun.*, **18**, 174 (1965).
38. H. Löw, I. Vallen, and B. Alm, in *Energy-Linked Functions of Mitochondria*, B. Chance, Ed., Academic Press, New York, 1963, p. 5.
39. L. Smith and P. W. Camerino, *Biochemistry*, **2**, 1432 (1963).
40. B. Chance and G. R. Williams, in *Advances in Enzymology*, Vol. 17, F. F. Nord, Ed., Interscience, New York, 1956, p. 65.
41. B. Chance, Ed., *Energy-Linked Functions of Mitochondria*, Academic Press, New York, 1963.
42. M. Klingenberg, *Biochem. Z.*, **335**, 263 (1961).
43. B. Chance and G. J. Hollunger, *J. Biol. Chem.*, **236**, 1577 (1961).
44. F. A. Hommes, *Arch Biochem. Biophys.*, **107**, 78 (1964).
45. R. N. Estabrook, *Biochim. Biophys. Acta*, **60**, 236 (1962).
46. L. Ernster, G. Dallner, and G. F. Azzone, *J. Biol. Chem.*, **238**, 1124 (1963).
47. A. L. Tappel, *Biochem. Pharmacol.*, **3**, 289 (1960).
48. H. A. Lardy and C. A. Elvehjem, *Ann. Rev. Biochem.*, **14**, 1 (1945).
49. W. F. Loomis and F. Lipmann, *J. Biol. Chem.*, **173**, 807 (1948).
50. E. E. Jacobs and D. R. Sanadi, *J. Biol. Chem.*, **235**, 531 (1963).
51. H. Löw, P. Siekevitz, L. Ernster, and O. Lindberg, *Biochim. Biophys. Acta*, **29**, 392 (1958).
52. P. D. Boyer, L. L. Bieber, R. A. Mitchell, and G. Szabolsci, *J. Biol Chem.*, **241**, 5384 (1966).

53. G. R. Drysdale and M. Cohn, *J. Biol. Chem.*, **233**, 1575 (1958).
54. P. Borst and E. C. Slater, *Biochim. Biophys. Acta*, **48**, 363 (1961).
55. F. Huijing and E. C. Slater, *J. Biochem.*, **49**, 493 (1961).
56. H. A. Lardy, D. Johnson, and W. C. McMurray, *Arch. Biochem. Biophys.*, **78**, 587 (1958).
57. H. Löw, H. Kreuger, and D. M. Ziegler, *Biochem. Biophys. Res. Commun.*, **5**, 231 (1961).
58. L. Danielson and L. Ernster, *Biochem. Z.*, **338**, 188 (1963).
59. A. Kemp, Jr. and E. C. Slater, *Biochim. Biophys. Acta.*, **92**, 178 (1964).
60. A. Bruni, S. Luciani, and A. Contessa, *Nature*, **201**, 1219 (1964).
61. A. Bruni, S. Luciani, A. Contessa, and G. F. Azzone, *Biochim. Biophys. Acta*, **82**, 630 (1964).
62. P. V. Vignais and P. M. Vignais, *Biochem. Biophys. Res. Commun.*, **14**, 559 (1964).
63. A. Bruni and G. F. Azzone, *Biochim. Biophys. Acta*, **93**, 462 (1952).
64. H. E. Robertson and P. D. Boyer, *J. Biol. Chem.*, **214**, 295 (1955).
65. B. C. Pressman, *J. Biol. Chem.*, **238**, 401 (1963).
66. J. B. Chappell, *J. Biol. Chem.*, **238**, 410 (1963).
67. G. Schafer, *Biochim. Biophys. Acta*, **93**, 279 (1964).
68. B. C. Pressman and J. K. Park, *Biochem. Biophys. Res. Commun.*, **11**, 182 (1963).
69. P. G. Heytler, *Biochemistry*, **2**, 357 (1963).
70. R. A. Goldsby and P. G. Heytler, *Biochemistry*, **2**, 1142 (1963).
71. A. L. Fluharty and D. R. Sanadi, *Biochemistry*, **2**, 519 (1963).
72. W. W. Kielley, *Proc. Intern. Congr. Biochem., 5th Congr., 1961*, **5**, 378 (1963).
73. P. D. Boyer, in *The Enzymes*, P. D. Boyer, H. Lardy, and K. Myrbäck, Eds., Academic Press, New York, 1959, Chap. 11.
74. H. C. Hemker, *Biochim. Biophys. Acta*, **63**, 46 (1962).
75. H. C. Hemker, *Biochim. Biophys. Acta*, **81**, 9 (1964).
76. H. C. Hemker, *Biochim. Biophys. Acta*, **81**, 1 (1964).
77. E. C. Weinbach and J. Garbus, *J. Biol. Chem.*, **240**, 1811 (1965).
78. H. A. Lardy, J. L. Connelly, and D. Johnson, *Biochemistry*, **3**, 1961 (1964).
79. P. Jurtshuk, Jr., I. Sekuzu, and D. E. Green, *J. Biol. Chem.*, **238**, 3595 (1963).
80. T. E. King, *J. Biol. Chem.*, **234**, 2342 (1961).
81. D. W. Haas, *Biochim. Biophys. Acta*, **89**, 543 (1964).
82. J. Kettman, *Biochem. Biophys. Res. Commun.*, **19**, 237 (1965).
83. W. W. Kielley and J. R. Bronk, *Biochim. Biophys. Acta*, **23**, 448 (1957).
84. D. E. Green and D. C. Wharton, *Biochem. Z.*, **338**, 335 (1963).
85. A. W. Linnane, *Biochim. Biophys. Acta*, **30**, 221 (1958).
86. A. W. Linnane and E. B. Titchner, *Biochim. Biophys. Acta*, **39**, 469 (1960).
87. M. E. Pullman, H. S. Penefsky, A. Datta, and E. Racker, *J. Biol. Chem.*, **235**, 3322 (1960).
88. C. L. Wadkins and A. L. Lehninger, *Federation Proc.*, **22**, 1092 (1963).
89. R. P. Glaze and C. L. Wadkins, *Biochem. Biophys. Res. Commun.*, **15**, 194 (1964).
90. L. Laturaze and P. V. Vignais, *Biochim. Biophys. Acta*, **92**, 184 (1964).
91. G. B. Pinchot and M. Hormanski, *Proc. Natl. Acad. Sci. U.S.*, **48**, 1970 (1964).
92. C. Lee and L. Ernster, *Biochem. Biophys. Res. Commun.*, **18**, 523 (1965).
93. H. Zalkin, M. E. Pullman, and E. Racker, *J. Biol. Chem.*, **240**, 4011 (1965).
94. M. Rabinovitz, M. P. Stulberg, and P. D. Boyer, *Science*, **114**, 641 (1951).

95. H. A. Lardy and H. Wellman, *J. Biol. Chem.*, **195**, 215 (1952).
96. M. Cohn, *J. Biol. Chem.*, **201**, 735 (1953).
97. P. D. Boyer, A. B. Falcone, and W. H. Harrison, *Nature*, **174**, 401 (1954).
98. M. A. Swanson, *Biochim. Biophys. Acta*, **20**, 85 (1956).
99. P. D. Boyer, W. W. Luchsinger, and A. B. Falcone, *J. Biol. Chem.*, **223**, 405 (1956).
100. M. Cohn and G. R. Drysdale, *J. Biol. Chem.*, **216**, 831 (1955).
101. P. C. Chan, A. L. Lehninger, and T. Enns, *J. Biol. Chem.*, **235**, 1790 (1960).
102. A. B. Falcone and P. Witonsky, *J. Biol. Chem.*, **239**, 1954 (1964).
103. R. A. Mitchell, R. W. Hill, and P. D. Boyer, *J. Biol. Chem.*, in press (1967).
104. N. Itada and M. Cohn, *J. Biol. Chem.*, **238**, 4026 (1963).
105. D. J. Graves and P. D. Boyer, *Biochemistry*, **1**, 739 (1962).
106. P. D. Boyer and E. Silverstein, *Acta Chem. Scand.*, **17**, 195 (1963).
107. C. Cooper, *Biochemistry*, **4**, 335 (1965).
108. N. Mourad and R. E. Parks, Jr., *Biochem. Biophys. Res. Commun.*, **19**, 312 (1965).
109. R. G. Kulka and C. J. Cooper, *J. Biol. Chem.*, **237**, 936 (1962).
110. M. Chiga and G. W. E. Plaut, *J. Biol. Chem.*, **234**, 3059 (1959).
111. G. Brierley and D. E. Green, *Proc. Natl. Acad. Sci. U.S.*, **53**, 73 (1965).
112. L. Ernster, *Proc. Intern. Congr. Biochem., 5th Moscow, 1961*, **5**, 115 (1963).
113. A. M. Snoswell, *Biochim. Biophys. Acta*, **60**, 143 (1962).
114. C. S. Rossi and A. L. Lehninger, *J. Biol. Chem.*, **239**, 3971 (1964).
114a. B. C. Pressman, *Proc. Natl. Acad. Sci., U.S.*, **53**, 1076 (1965).
115. F. D. Vasingron and J. W. Greenwalt, *Biochem. Biophys. Res. Commun.*, **15**, 133 (1964).
116. L. Packer, in *Energy-Linked Functions of Mitochondria*, B. Chance, Ed., Academic Press, New York, 1963.
117. D. Neubert and A. L. Lehninger, *Biochim. Biophys. Acta*, **62**, 556 (1962).
118. A. L. Lehninger, *J. Biol. Chem.*, **234**, 2186, 2465 (1959).
119. J. L. Connelly and H. A. Lardy, *Biochemistry*, **3**, 1969 (1964).
120. L. Schachinger, R. Eisenhardt, and B. Chance, *Biochem. Z.*, **333**, 182 (1960).
121. R. H. Eisenhardt and O. Rosenthal, *Science*, **143**, 476 (1964).
122. K. Van Dam, *Biochim. Biophys. Acta*, **128**, 377 (1966).
123. D. E. Griffiths and R. Chaplain, *Biochem. Biophys. Res. Commun.*, **8**, 497 (1963).
124. P. J. Russell, Jr., and J. Ballantine, *J. Biol. Chem.*, **235**, 226 (1960).
125. M. DeLuca, K. E. Ebner, D. E. Hultquist, G. Kreil, J. B. Peter, R. W. Moyer, and P. D. Boyer, *Biochem. Z.*, **338**, 512 (1963).
126. R. A. Mitchell, L. G. Butler, and P. D. Boyer, *Biochem. Biophys. Res. Commun.*, **16**, 545 (1964).
127. G. Kreil and P. D. Boyer, *Biochem. Biophys. Res. Commun.*, **16**, 551 (1964).
128. O. Lindberg, J. J. Duffy, A. W. Norman, and P. D. Boyer, *J. Biol. Chem.*, **240**, 2851 (1965).
129. R. Beyer, *Biochem. Biophys. Res. Commun.*, **17**, 184 (1964).
130. S. Sperti, L. A. Pinna, M. Lorini, V. Moret, and N. Siliprandi, *Biochim. Biophys. Acta*, **93**, 284 (1964).
131. E. C. Slater, A. Kemp, Jr., and J. M. Tager, *Nature*, **201**, 781 (1964).
132. E. C. Slater and A. Kemp, Jr., *Nature*, **204**, 1268 (1964).
133. L. L. Bieber, O. Lindberg, J. Duffy, and P. D. Boyer, *Nature*, **202**, 1316 (1964).
134. P. D. Boyer, *Nature*, **207**, 409 (1965).

135. L. L. Bieber and P. D. Boyer, *J. Biol. Chem.*, **241**, 5375 (1966).
136. K. Ahmed and J. D. Judah, *Biochim. Biophys. Acta*, **71**, 295 (1963).
137. V. P. Skulachev, *Nature*, **198**, 444 (1963).
138. V. P. Skulachev, *6th Intern. Congr. Biochem.*, *6th Congr.*, *New York*, *1964*, *Abstr. Symp. X*, p. 758.
139. K. G. Reid and R. A. Smith, *Arch. Biochem. Biophys.*, **109**, 358 (1965).
140. B. Chance, C. P. Lee, and B. Schoener, Info. Exch. Group 1, Memo 356, May 6, 1965.
141. J. J. Blum and D. R. Sanadi, *J. Biol. Chem.*, **239**, 452, 455 (1964).
142. M. E. Dempsey, P. D. Boyer, and E. S. Benson, *J. Biol. Chem.*, **238**, 2708 (1963).
142a. L. Sartorelli, H. Fromm, and P. D. Boyer, *Biochemistry*, **5**, 2877 (1966).
143. D. Gilmour and M. Griffiths, *Arch. Biochem. Biophys.*, **72**, 302 (1957).
144. S. V. Perry and J. B. Chappell, *Biochemistry*, **65**, 469 (1957).
145. H. M. Levy, E. M. Ryan, S. S. Springhorn, and D. E. Koshland, Jr., *J. Biol. Chem.*, **237**, PC1730 (1962).
146. R. G. Yount and D. E. Koshland, Jr., *J. Biol. Chem.*, **238**, 1713 (1963).
147. A. T. Jagendorf and E. Uribe, *Proc. Natl. Acad. Sci.*, *U.S.*, **55**, 170 (1966).
148. A. Asano and A. F. Brodie, *Biochem. Biophys. Res. Commun.*, **13**, 423 (1963).
149. P. Murphy and A. F. Brodie, *J. Biol. Chem.*, **239**, 4292 (1965).
150. V. M. Clark, D. W. Hutchinson, and A. Todd, *Nature*, **187**, 59 (1960).
151. W. Gruber, R. Hohl, and T. Wieland, *Biochem. Biophys. Res. Commun.*, **12**, 242 (1963).
152. G. E. Tomasi and R. D. Dallam, *J. Biol. Chem.*, **239**, 1604 (1964).
153. A. Asano, A. F. Brodie, A. F. Wagner, P. E. Wittreich, and K. Folkers, *J. Biol. Chem.*, **237**, PC2411 (1962).
154. R. Erickson, A. Wagner, and K. Folkers, *J. Am. Chem. Soc.*, **85**, 1535 (1963).
155. W. S. Brinigar and J. H. Wang, *Proc. Natl. Acad. Sci. U.S.*, **52**, 699 (1964).

PART II

THE OXIDIZING ENZYMES

Flavocoenzymes: Chemistry and Molecular Biology

ANDERS EHRENBERG, *Medicinska Nobelinstitutet,*
Biokemiska Institutionen, Stockholm, Sweden
AND
PETER HEMMERICH, *Institut für anorganische Chemie*
an der Universität Basel, Basel, Switzerland

I.	Introduction: Definitions and Nomenclature.	239
II.	Flavoquinone	244
III.	Flavohydroquinone	248
IV.	Flavosemiquinone.	250
V.	Molecular Complexes	253
VI.	Flavin–Metal Interaction	254
VII.	Redox Mechanisms and Kinetics	261
	References	262

I. Introduction: Definitions and Nomenclature

The term "flavin" (which has no chemical or biological relationship with "flavone") applies to the redox-active chromophore of a colored respiratory enzyme class, the flavoproteins. The group's name is derived from the latin *flavus* (yellow), because of the color in the isolated form of the first characterized flavoproteins, but the term is somewhat misleading, since neither flavins nor flavoproteins are invariably yellow, nor is a yellow color in respiratory systems invariably indicative of flavins. The yellow color pertains in dilute solution to the oxidized state of the flavin prosthetic group only and may be overshadowed by other colored constituents in flavoproteins, in particular heme or non-heme iron. In high concentrations, the reduced form shows a yellow to orange-red appearance.

Actually the flavin group is part of the nonprotein constituent of flavoproteins, the flavocoenzymes. These differ from the prosthetic groups of most other oxidizing enzymes in that their binding to the apoprotein may vary considerably from one enzyme to the other. Thus with some flavoproteins, the coenzyme may be removed reversibly, with others only irreversibly and in at least one important case (succinic dehydrogenase) it cannot be removed at all without complete destruction of the apoprotein.

Flavocoenzymes (Fig. 1) consist of the heteroaromatic isoalloxazine group with methyl substituents in positions 7 and 8, which is further linked to a five-carbon hydroxylic monophosphorylated side chain, which in turn may be coupled to an adenine nucleotide.

The generally adopted name of these units, "flavin nucleotides" ("FMN" for flavin mononucleotide, "FAD" for flavin adenine dinucleotide) is misleading, since the flavin side chain is not a sugar (ribose), but a polyol (ribitol). This implies that the flavin side chain is stable to hydrolysis. Hence, the correct names for the two flavocoenzymes are riboflavin-5′-monophosphate (for FMN) and riboflavin adenosine diphosphate (for FAD).

By mild acid hydrolysis the P—O—P bridge of FAD is cleaved to give FMN, from which in turn (upon prolonged treatment) the phosphate group is removed to give riboflavin. This was the first flavin derivative to be isolated in a pure state and was at that time known as vitamin B_2.

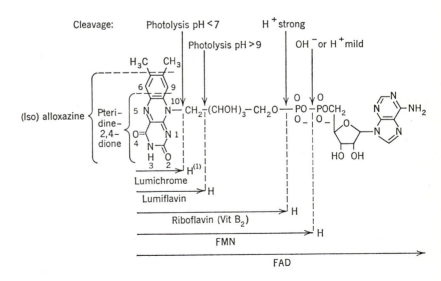

Fig. 1. Structure and cleavage loci of flavocoenzyme (2).

While the flavin side chain does not undergo hydrolysis, it may be easily degraded by photodecomposition. This reaction appears to be without biological importance, since the photolability of flavins is readily quenched in the presence of most proteins and even 20-fold diminished in FAD because of intramolecular transfer of excitation energy between the flavin and adenine moieties. Any work on systems containing free FMN or riboflavin, however, has to be conducted in reduced light, except under very acid conditions.

The photolysis of flavins is by now well known to be a photoinduced disproportionation, which implies that the heteroaromatic nucleus is able, in its excited (triplet) state, to dehydrogenate its own side chain. This can, of course, be prevented by external addition of any better electron-donating agent, tertiary amines such as EDTA, which then cause photoreduction of the nucleus without destruction of the side chain. This photoreduction, as any reduction of flavins, may be reversed by oxidants, in particular O_2. In fact, riboflavin in illuminated aerobic solutions behaves as an amino acid oxidase. Hence, electron donation from protein to protein-bound photoexcited flavin may occur at any time when flavoproteins are exposed to light, resulting in a flavin-catalyzed photodenaturation of the protein.

The first isolable intermediate product of flavin photolysis (earlier called "deuteroflavin") has recently been shown to have structure I (Fig. 2). The same product may be obtained from riboflavin by periodate oxidation. This aldehyde is further degraded to a methyl group under alkaline conditions to give the simplest representative of flavins (II), lumiflavin (so called because of its extremely strong green fluorescence, which is stronger than that of riboflavin, while the absorption spectra of the two compounds are the same).

On the other hand, under neutral to mildly acid conditions, the side chain is completely removed to give *lumichrome* (III), a nearly colorless compound of strong blue fluorescence which has no flavin properties, that is, its redox potential is much lower than that of flavin. Since lumi-flavin as well as lumichrome can be synthesized from *alloxane* (IV) and suitable *o*-phenylenediamines, the corresponding classes of compounds have been called *alloxazines* and *isoalloxazines*, respectively (Fig. 2). This nomenclature does not give any information of biological importance, since the biosynthesis of flavin proceeds via *purines* (adenosine) and *pteri-dines* (V). Furthermore, the parent compound "isoalloxazine" does not exist (dissociable H at N-3, N-10), even in tautomeric equilibrium with alloxazine (dissociable H at N-1, N-3) (Fig. 1), to any measurable extent.

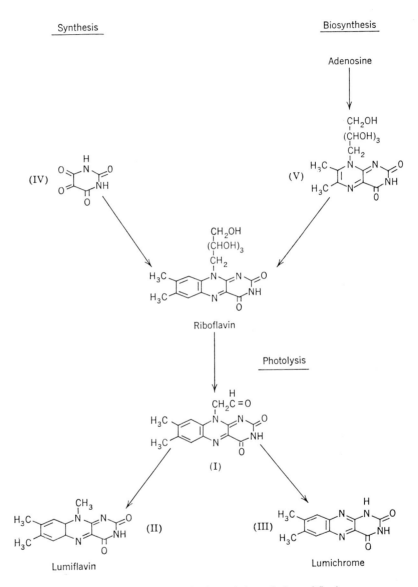

Fig. 2. Synthesis, biosynthesis, and degradation of flavins.

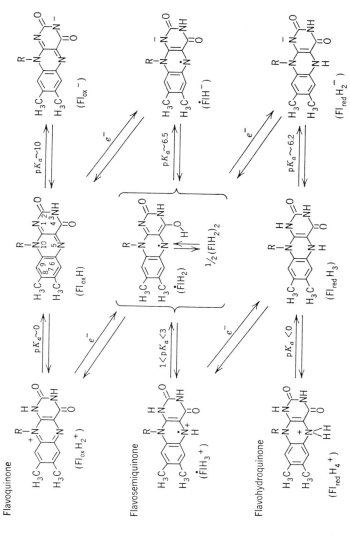

Fig. 3. Flavin species, as they occur at different pH and redox states. Approximate buffer regions of the singlet species are given.

Consequently, flavins cannot be obtained by introduction of a suitable side chain into preformed alloxazine, either chemically or enzymically. A reasonable systematic definition of a flavin would therefore be "10-substituted 7,8-dimethylbenzopteridine-2,4-dione."

The apparent biological function of flavin is that of a carrier of redox equivalents. Care should be taken to avoid specifications as "electron equivalents" and "hydride equivalents," since flavin acts upon each, and it even converts them into each other. Hence, active flavin oscillates between different oxidation states. As can be seen from polarography, this oscillation exhibits—even in the absence of protein—strict thermodynamic reversibility. This is indicative of a "quinoid" system.

It is obvious that in each valence state flavin has completely specific chemical and physical properties. It is therefore necessary to assign oxidation states to any phenomenon encountered in flavin chemistry, i.e., to find out which constituent of a complicated flavin mixture is the reactive one in acid–base and disproportionation equilibria. Such a differentiation is possible, as long as no secondary redox system (which might form stable complexes with flavin) is involved in the reaction, e.g., Fe(II,III), —SH, or a second flavin moiety.

In the following, therefore, a separate section is dedicated to each flavin oxidation state. In analogy to benzoquinone, we speak in terms of flavoquinone for the oxidized state, flavosemiquinone or flavin radical for the half-reduced state, and flavohydroquinone for the totally reduced state. Care is taken to avoid the term "semiquinone" or "radical" in all cases where paramagnetism has not been explicitly established. The generally used term "leucoflavin" is abandoned, because flavohydroquinone is not colorless, in contrast to what is implied by the term "leuco." The flavin species and their interrelationships are listed in Figure 3.

II. Flavoquinone

In the presence of oxygen and in the absence of reducing agents, flavoquinone is the only stable oxidation state of flavin. Consequently its importance is easily overemphasized when compared with that of the other states. We represent the neutral flavoquinone species as $Fl_{ox}H$, where H means the proton in position 3, which dissociates at pH > 9 to give Fl_{ox}^-. $Fl_{ox}H$ exhibits a very characteristic absorption spectrum with maxima at 445 and 373 mμ for riboflavin and FMN (450 and 375 mμ for FAD), which is generally known as "the flavin spectrum." Deprotonation at N-3 does not have a pronounced influence on the spectrum, which shows that there is negligible π conjugation of electrons for this part of the pyrimidinoid ring of the flavin nucleus.

Protonation of flavoquinone, to give $Fl_{ox}H_2^+$, occurs at pH < 1 on N-1 (not on N-5), and results in a drastic hypsochromic shift, which eliminates most of the "flavin color." The acidity of this cation makes it unstable at physiological pH. Its possible biological importance in the case of free FMN and FAD can only be that of a transient intermediate. However, in the case of a flavoquinone-iminol, e.g.,

the acidity of the cation $Fl_{ox}RH^+$ is about four orders of magnitude lower. Such esters have been prepared and shown to be very high in hydrolytic energy at the quinone level but not at the hydroquinone level. Thus there is a possible mechanism for conversion of redox energy into hydrolytic energy, which would not involve lipophilic benzoquinones.

At the same time, the properties of these esters show that the dissociable proton of free $Fl_{ox}H$ is located exlusively at N-3 and that tautomers are not present to any measurable extent.

Base-catalyzed alkylation of flavoquinone results in 100% substitution at N-3, provided certain precautions are taken: the reaction has to be conducted in a water-free medium, since otherwise the preformed $Fl_{ox}R$ is immediately split by action of OH$^-$ at C=O-(2). Strong bases and heat have to be avoided since they cause reactions at CH_3-8 (see below). Substitution in position 3 is useful for obtaining flavins highly soluble in polar ($R = CH_2COO^-$) or unpolar ($R = CH_2C_6H_5$) media. Up to now there has been no indication of biological activity for these compounds.

In the case of the nondissociable flavocoenzyme in succinic dehydrogenase (SD), however, the (apparently covalent) fixation cannot occur through any of positions 1, 2, 3, 4, 5, and 10 for the following reasons. The SD coenzyme liberates urea upon alkaline hydrolysis, which is only expected with 1,3-unsubstituted flavins. Since removal of the ribityl side chain does not break the protein bond, position 10 is also excluded, and all possible substitutions by alkyl, aryl, alkoxy, amino and mercapto

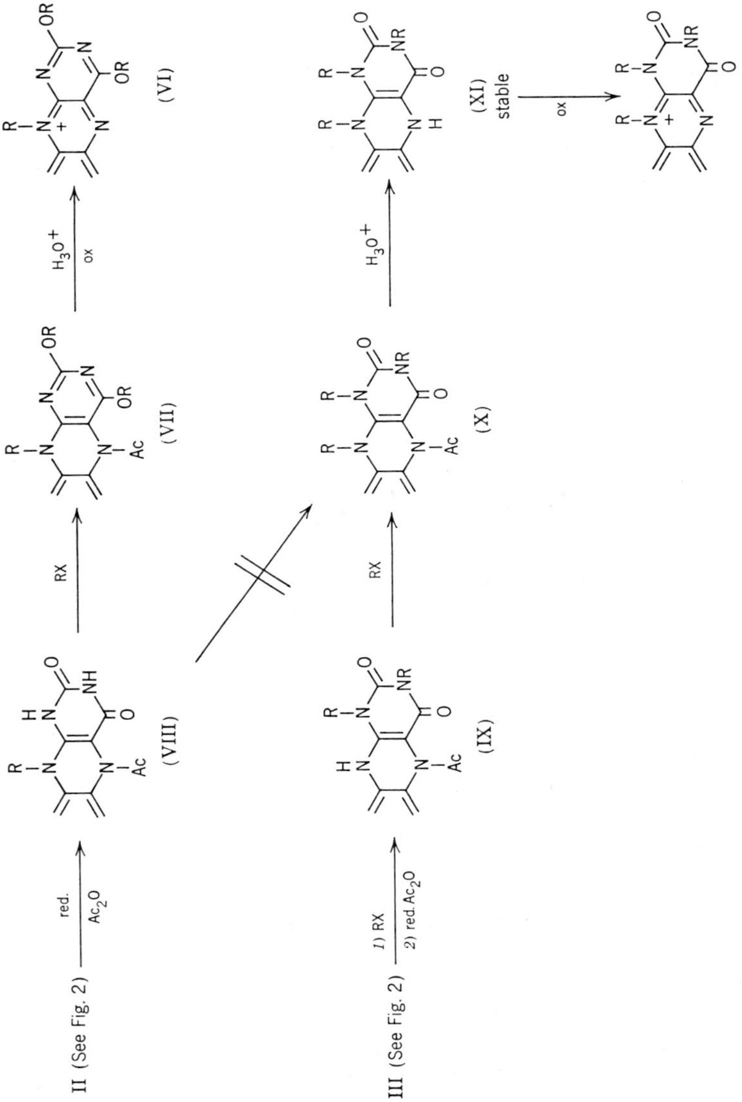

Fig. 4. *O*- and *N*-Alkylation of flavohydroquinones.

residues have been carried out in positions 2, 4, and 5 and shown to be reversible by either hydrolysis or mild oxidation. None of these flavin derivatives has so far been shown to possess any biological activity.

As mentioned above, however, there is one additional functional group in flavoquinones, CH_3-8. This methyl group appears to have a noticeable acidity, due to the strong electron deficiency of the nucleus, which might compete with that of nitrotoluenes or toluquinones. Hence, under strongly basic non-aqueous conditions—say BaO in CH_3OH—flavoquinone is dimerized irreversibly, whereby CH_3-8 is oxidized to give 8-CH_2—CH_2-8' and the nucleus is reduced, as is known from the conversion of nitro-toluenes into stilbenes. Similar CH_3 groups as possible active centers have been discussed as occurring in ubiquinone and the K vitamins. Indeed there is one type of flavin derivative (VI; Fig. 4) which undergoes (CH_3-8) self-condensation even at pH 7 in phosphate buffer. Furthermore, CH_3-8 can undergo Knoevenagel-like condensations with aldehydes.

Apart from pH there are other environmental influences, which alter the flavoquinone spectrum appreciably. First, flavins are subjected to strong solvatochromism. This is easily overlooked in free flavins, since these are soluble only in very polar media. In contrast to that, the environment of protein-bound flavocoenzyme may be rather lipophilic. Hence, 3-benzyllumiflavin in benzene might be a better spectral model for protein-bound flavin than FMN in aqueous solution. Characteristic of flavoquinones in nonpolar environments is the pronounced shoulder at 485 mμ, which is also found with many flavoenzymes.

Second, there may be hydrophobic bond interactions with neighboring electron-rich aromatic systems, e.g., adenine in FAD. Hence, there is a distinct red shift in the spectrum of FAD compared to FMN. The stability of such complexes is very low, and certainly the effect does not contribute appreciably to the energy contained in a coenzyme–apoenzyme bond, but its spectral influence is obvious and so is its importance for flavin kinetics. The same might occur between FMN and protein residues like tyrosine or tryptophane, e.g., in the Old Yellow enzyme.

Still more sensitive to all kinds of environmental influences is the fluorescence of flavoquinone which is partially quenched by the adenine in FAD. Fixation to protein usually, but not always, quenches it completely. The fluorescence yield will be altered not only by the well-known strong activators of radiationless transitions, such as ions of heavy atoms and energy-conducting π-electron systems, but by any factor chemically influencing the flavoquinone excited singlet.

The first excited state of flavoquinone $Fl_{ox}H^*$ differs from the semi-quinone anion $\dot{F}lH^-$ in the first approximation only by a hole in the second occupied electronic level. Since the chemical qualities are determined primarily by the distribution of the electron(s) in the orbitals of highest energy, $Fl_{ox}H^*$ and $\dot{F}lH^-$ should have much in common. Hence, it is easily understood that the excited state is a stronger base than the ground state of flavoquinone. This is reflected by the pH dependence of flavoquinone fluorescence.

III. Flavohydroquinone (Fig. 3)

From infrared spectra it appears that the acid protons in $Fl_{red}H_3$ are bound at N-1 and N-3, as might be expected. But, contrary to the case of flavoquinone, prototropy to the "iminol" tautomers apparently does not involve a large increase in energy.

This is reflected by the fact that flavohydroquinone is easily alkylated at the carbonyl oxygens and that the "iminol esters" (VII; Fig. 4) thus formed are not easily hydrolyzed, unlike the 5-acetyl group. N-3 competes with CO-2 and CO-4 in alkylation, but no substitution can be achieved at N-1. This might be due primarily to steric hindrance exerted by the neighboring methyl group at N-10. However, starting from an alloxazine prealkylated at N-1 (IX), alkylation of N-10 is easy, although there should be steric hindrance to at least the same extent. Hence, the failure of N-1 in $Fl_{red}H_2^-$ to accept an alkyl might also reflect a relatively high electron density at the carbonyl oxygens.

The proton which dissociates most readily from $Fl_{red}H_3$ must be situated at N-1, since alkylation at N-3 does not alter the acidity of flavohydroquinone ($pK_a \sim 6$). This shows again that N-3 is a position of the flavin nucleus which, to a first approximation, is not affected in its chemical behavior by change of oxidation state of the whole system.

On the other hand, the position undergoing the most drastic change upon reduction is N-5. Among the most important facts in flavohydroquinone chemistry is the nearly complete lack of both acidic and basic qualities in the NH-5 group. This problem is intimately connected with the conformation of flavohydroquinone and the absorption spectra of the flavohydroquinone species.

To begin with the cation, $Fl_{red}H_4^+$ ($pK_a \leqslant 0$) it is apparent that the system is folded about the central axis (butterfly wing conformation), N-5 being a quaternary ammonium tetrahedral center. In this state, flavohydroquinone does not absorb at wavelengths higher than about 400 mμ. Upon deprotonation to give the neutral $Fl_{red}H_3$, a drastic increase

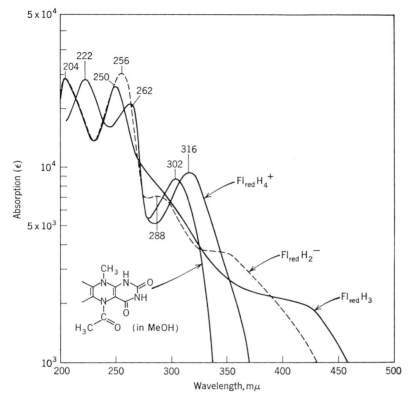

Fig. 5. Absorption spectra of flavohydroquinone species (in water of suitable pH, cf. Fig. 3 if not indicated otherwise) (10). Note that absorption less then $\epsilon = 10^3$ is not plotted.

in long-wavelength absorption occurs (Fig. 5), which extends up to 600 mμ and exceeds in this region even the absorption of flavoquinone. Consequently, while dilute flavin solutions are "decolorized" upon reduction, more concentrated ones change in appearance from yellow to light red. The color of flavohydroquinone therefore appears to be a measure of its *tricyclic* delocalization of electrons, that is, of its planarity. Upon further deprotonation, occurring at pH 6–7, to give the anion $Fl_{red}H_2^-$, the color is bleached again (Fig. 5) due to the fact that the full charge in the pyrimidine part of the system prevents delocalization of electrons from N-5 and thus enhances the folding of the molecule.

The planarity of flavohydroquinone thus may vary considerably even in enzyme systems. Sandwich formation with neighboring acceptor systems such as adenine might help to flatten the molecule, or steric requirements might enhance the bent form which will increase both acidity and basicity of $Fl_{red}H_3$.

The bent form is also favored by steric overcrowding in 1,10-dialkyl-flavohydroquinones (XI; Fig. 4) and, for electronic reasons, by acylation of N-5. Any influence of this kind also protects the molecule from autoxidation, which apparently occurs through the planar state. Hence 1,10-dialkylflavohydroquinones (XI) can be obtained in pure form as slightly yellow crystals, whereas 5-acylflavohydroquinones (VII–X) may be kept indefinitely as colorless crystals, but may be hydrolyzed in very weakly acid or basic medium even at room temperature. Under these conditions iminolester bonds in $Fl_{red}HR_2$ (cf. reaction VII → VI, Fig. 4) are not affected.

The apparent inertness of the hydrogen at N-5 may be the cause for the slow exchange of deuterium which is observed in some reduced flavoproteins.

IV. Flavosemiquinone

In partially reduced flavin solutions, flavosemiquinone molecules exist in equilibrium with the oxidized and reduced forms. For the neutral species (pH 3–5) this equilibrium may be written

$$Fl_{ox}H + Fl_{red}H_3 \overset{K_R}{\rightleftharpoons} 2\dot{Fl}H_2 \qquad \text{(cf. Fig. 3)}$$

The equilibrium constant K_R is called the radical formation constant as the equilibrium has been written. Its inverse value, $K_D = 1/K_R$, is the radical disproportionation constant.

In the region pH 8–9 there is a minimum yield of free radicals in a half-reduced solution, since the predominant oxidized and reduced forms are $Fl_{ox}H$ and $Fl_{red}H_2^-$, which enhance the disproportionation of $\dot{Fl}H^-$ and $\dot{Fl}H_2$, respectively. In case of FMN this minimum radical yield is less than 1% in a 5mM solution. At pH 12 the yield has increased to more than 2% (independent of concentration) corresponding to $K_R \approx 2 \times 10^{-3}$ for the anions. With decreasing pH the yield is increasing and reaches a constant level (2.8% radicals in 5mM half-reduced FMN) in the range pH 6 to pH 3. In even more acid media the radical yield increases rapidly. At pH < 0, yields in excess of 50% and $K_R > 10$ have been determined.

The semiquinone of flavin can normally be obtained only as $\dot{Fl}H_3^+$ in a reasonably pure form. The properties of $\dot{Fl}H_2$ and $\dot{Fl}H^-$ are very difficult to study because of the masking effect of the high concentrations of the accompanying oxidized and reduced species and various dimolecular complexes (see below). The formation of the latter complexes may be avoided by using flavin derivatives soluble in nonpolar media. The disproportionation has been avoided with suitably alkylated (e.g., in positions N-3 and CO-4) model compounds which cannot be oxidized to the quinone level.

These experimental difficulties are partly counterbalanced by the paramagnetism of the free radicals, which distinguishes them from other flavin species. This property has made it possible to obtain a wealth of information by means of electron spin resonance (ESR) spectroscopy. In this connection it is important to emphasize that evidence of this basic property, paramagnetism, has to be given in each single flavoprotein oxidoreduction, in order for it to be likely that flavin semiquinones participate.

Partly reduced acid flavin solutions exhibit a characteristic reddish color. This has been shown to be due to the light absorption spectrum of $\dot{F}lH_3^+$, which has maxima at about 360 and 490 mμ. Deprotonation at pH ~ 2 gives $\dot{F}lH_2$ for which the longest wavelength absorption has been demonstrated to be shifted to the region 560–620 mμ. Nonpolar media shift the absorption towards the longer wavelength. The intensity of the absorption has been estimated to be of the order of 3000–4000 M^{-1} cm^{-1}. Spectral properties of $\dot{F}lH_2$ at shorter wavelengths have not been possible to determine. The spectrum of $\dot{F}lH^-$ is even less well known. All results so far indicate, however, that its light absorption, if any, in the region 550–650 mμ must be much weaker than that of $\dot{F}lH_2$.

Also, the ESR spectrum (derivative of the absorption) varies in a characteristic way with pH as is shown with FMN in Figure 6. The anion $\dot{F}lH^-$ shows the simplest spectrum with 12 evenly spaced lines. In the case of $\dot{F}lH_2$ and $\dot{F}lH_3^+$ more hyperfine lines are seen and their spacing is not uniform.

The radicals of lumiflavin (II, Fig. 2) give ESR spectra that are better resolved than those of FMN and other compounds having more than one carbon atom in the N-10 substituent. By means of isotopic substitutions (D for H, ^{15}N for ^{14}N) in various positions of lumiflavin it has been possible to determine which nuclei participate in the observed hyperfine interaction with the unpaired electron. From these results, information about the spin density distribution has been deduced.

In this way it has been shown that there is a very small, but still detectable, spin density localized on the pyrimidinoid nitrogens in $\dot{F}lH^-$ and that it has decreased below detectability in $\dot{F}lH_2$ and $\dot{F}lH_3^+$. Alkylation in

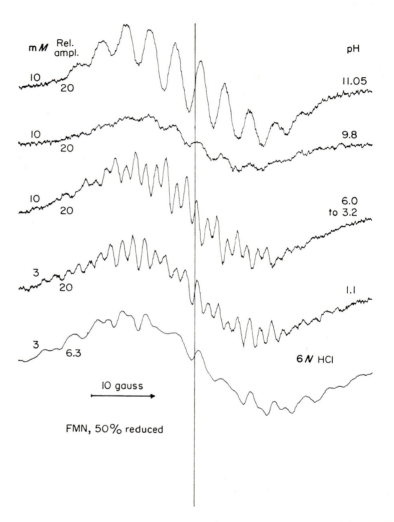

Fig. 6. ESR spectra of FMN free radicals at various pH values. 50% reduction with $Na_2S_2O_4$. Flavin concentrations and relative spectrometer amplifications are given. A Varian X-band spectrometer Model V-4500 equipped with a 100 kc/sec field modulation unit has been employed (6).

position 3 has only a very slight influence on the ESR spectrum. Thus the unpaired spin is virtually isolated from the pyrimidinoid nitrogens.

The ESR work shows that in $\dot{F}lH^-$ the spin density is preferentially localized on ring atoms in positions 5, 6, 8, and 10. In $\dot{F}lH_2$ and $\dot{F}lH_3^+$ these positions still carry large spin densities, but a smearing out to position 9, and perhaps also to position 7, has been demonstrated. These results indicate that, disregarding permanent charges and polarizabilities, positions 5, 8, and 10 would be preferable for radical reactions and one-electron transfer.

Comparison of the ESR spectra of lumiflavin and 10-ribityl compounds has shown that in the case of $\dot{F}lH^-$ and $\dot{F}lH_2$ the two methylene protons of the ribityl compounds give very different hyperfine couplings. This indicates that the ribityl chain is held in a fixed position by a hydrogen bond to the isoalloxazine nucleus.

In $\dot{F}lH_2$ there are two dissociable protons. Only one of these protons takes part in the hyperfine interaction, as is demonstrated by exchange experiments with D_2O. The ESR-"inactive" proton is likely to be attached on N-3. Of the remaining positions for the "active" proton, N-1 is unlikely because of its vanishing spin density. Iminol formation in position 4 or direct attachment on N-5 (or a tautomer of the two) would both be compatible with the observed hyperfine interaction.

Isotropic hyperfine structure can be observed by means of ESR on the coenzyme radicals which have a rapid enough rotational motion in solution to average out the anisotropic terms. Unfortunately, these anisotropic effects mask the isotropic structure when a coenzyme radical is tightly bound to a protein molecule.

V. Molecular Complexes

Dimolecular complexes of charge-transfer type have been demonstrated to be formed easily between flavin molecules of different redox states. These complexes have broad light absorptions with maxima in the region 800–1000 mμ. The tail of the absorption at 600 mμ and shorter wavelengths seriously hampers the study of the light absorption of the radical species in this region. The ESR spectra of the complexes in which one of the components is a flavin radical do not show any resolved hyperfine structure, since the unpaired electron is delocalized over both components. Complexes of this (sandwich) type have not been shown to be of any significance for flavoenzymes.

The light absorption of FAD at 450 mμ as compared to that of FMN is slightly decreased and the peak is slightly broadened and shifted towards longer wavelengths. Simultaneously the fluorescence is partly quenched. These changes are due to complex formation between the isoalloxazine and adenine moieties.

Similar changes are brought about by a number of aromatic compounds. They are also frequently seen when flavin coenzyme is bound to a protein (e.g., in Old Yellow enzyme and in D-amino acid oxidase) where (hetero)-aromatic amino acids could be the complexing partners.

This broadening has frequently been considered as caused by formation of charge-transfer complexes. Recent work indicates, however, that this characterization is untenable. The complex is absolutely dependent on water and might be described as a ternary complex of flavin, aromatic compound, and water. The 450-mμ band remains a single transition and the small changes in shape and position depend on a partial restoration of the vibrational structure.

VI. Flavin–Metal Interaction

It follows from what has been said about chemical differences in flavin species, that the metal affinity of flavins has to be discussed separately for each oxidation state.

To begin with flavohydroquinone, it has been discussed above that N-5 does not coordinate with cations since this would imply complete folding and loss of resonance in the tricyclic system. Displacement of the proton at N-5 by a metal ion, on the other hand, cannot be expected, since the acidity of this proton is unmeasurably low. Consequently, bidentate coordination of $Fl_{red}H_2{}^-$ should not be favored over monodentate coordination: in other words, no chelate effect can be expected for flavohydroquinone. This has been confirmed experimentally.

Also in flavoquinone, as has been discussed above, N-5 does not exhibit any basicity. Furthermore, prototropy to produce a 4-iminol compound requires considerable energy in this oxidation state. Hence, oxinate-like metal chelates of flavoquinone (XIII, Fig. 7) will not be stable, unless chelation is accompanied by back-donation of electrons from the metal to the ligand. The term "back-donation" is a key term in metal coordination chemistry; it means that, in addition to a normal σ-acceptor bond formed between a metal ion and a ligand lone pair, there is a delocalization of d electrons through a π bond from the metal toward unfilled ligand orbitals of suitably low energy. Clearly, such interactions occur more easily in lower metal valencies as exemplified in the metal carbonyls. A difficulty arises, however, when not only the metal, but also the free

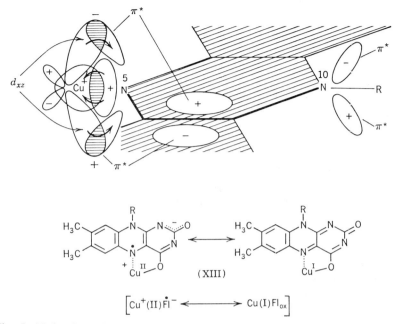

Fig. 7. Molecular orbital scheme of metal-to-ligand charge transfer in valence-mesomeric flavin-copper chelate.

ligand itself, shows redox activity, as does flavin. In such a case, metal and ligand valency in a complex cannot be well defined.

The lowest stable valencies of physiologically important redox-active metals are Cu(I), Co(II), Fe(II), Mo(V), viz., d^{10}, d^7, d^6, d^1 configurations. The donating power of these ions towards a given mono- or bidentate ligand depends, of course, on the total state of coordination, i.e., of the "auxiliary" ligands in an octahedral metal–flavin complex. Most of these ligands might be provided by the protein; but water, bound substrate or the product of enzymic catalysis, and PO groups from the nucleotide side chain also have to be considered.

The strength of the field exerted by these auxiliary ligands will determine whether the metal exists in a high or low spin state, which in turn drastically alters the chemical and physical behavior of the whole system. This demonstrates the complexities involved in flavin–metal interaction.

As said above, interaction of flavoquinone cannot be observed with the common solvated metal ions in a polar environment. The affinity of the metal ion towards the solvent exceeds all other interactions. But if we change to a less polar medium, e.g., acetonitrile, weak complex formation of flavoquinone with cations might be forced to take place in the presence of a large excess of metal ion, even with ions completely devoid of donating power, such as Mg^{2+}. On the other hand, even Zn^{2+} donates electrons to flavoquinone under those forced conditions (Fig. 7). The difficulty is that in the absence of complete structural analysis one never knows exactly how polar a protein medium at a certain active site of an enzyme might be. But it can generally be said that flavoquinone has very weak, if any, affinity towards redox-inactive metal ions in the physiological environment.

The same is true for iron in its high spin state, above a certain pH. Below pH of about 7, hydrated Fe^{3+} will oxidize reduced flavin. This fact is also of minor biological significance, since "free hydrated iron" is not a natural redox partner of flavin. Iron–protein complexation, however, will shift these equilibria to the same extent as the strength of the ligand field is increased.

The best model for flavin–metal interaction is therefore obtained with a very polarizable cation such as Cu^+, where different states of multiplicity are not possible. Cu^+ (like isoelectronic Ag^+) forms a very strong complex with flavoquinone in aqueous solution according to Reaction (1) in Figure 8. The complex CuFl is stable indefinitely as long as autoxidation is prevented. It seems that this autoxidation is dependent on the second equilibrium of dissociation (Reaction (2), Fig. 8), which in turn is followed by disproportionation. Equilibrium (2) favors the left side, but when $Fl_{red}H_2^-$ is continuously removed by oxygen, this can be overcome, since all the reactions involved are rather fast.

The back-donation exerted by Cu^+ towards Fl_{ox}^- is illustrated in terms of molecular orbitals in Figure 7.

Cu^+ is well suited as a model cation because of its low coordination number (maximum of 4), which allows only 1 to 2 additional solvent molecules in CuFl, if Fl can be shown to be a bidentate ligand. This has been done in two ways.

First, "isoflavins" (carrying CH_3 groups in positions 6, 7 instead of 7, 8) do not form the deep red CuFl, because of steric hindrance of coordination at N-5. Second, Fl_{ox}^- entering the coordination sphere of Cu^+ displaces two monodentate ligands which can be quantitatively estimated.

If one removes more H_2O from CuFl(aq.) by carrying out the complex formation in a less polar solvent, the formation of (hydrated) Cu^{2+} will be still more suppressed in Reaction (2), which renders CuFl more stable toward air.

This shows that in a protein medium containing a redox-active metal it is no longer *a priori* certain that active flavin will shuttle between flavo-

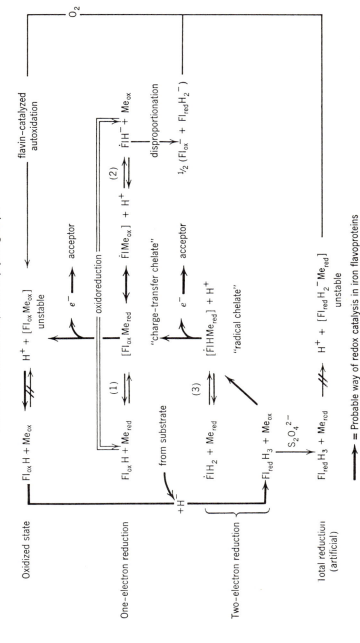

Fig. 8. Electron flow sheet in flavin-metal systems. The natural electron flow in metalloflavoproteins, as it appears most reasonable from the behavior of the models, is indicated by bold arrows. (After Hemmerich and Spence, in ref. 3.)

quinone and flavohydroquinone states. It is as likely that "total reduction" from the semiquinone to the hydroquinone state is blocked (because of the general lack of metal ion affinity in $Fl_{red}H_2^-$), or that "full oxidation" to the quinone state plus "oxidized" metal is blocked (because of the lack of metal-ion affinity of $Fl_{ox}H$ toward oxidized metals).

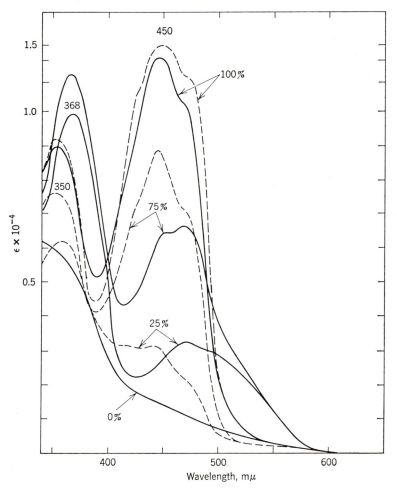

Fig. 9. Spectra of peracetylriboflavin, 1.25mM, in CHCl₃, reduced by shaking with aqueous $S_2O_4^{2-}$, at various degrees of reoxidation (indicated in per cent) by air (11). Excess of base (triethylamine, $2 \times 10^{-2}M$) is present as buffer. (——) with 5mM Zn^{2+} (as acetate) present; (– – –) without metal.

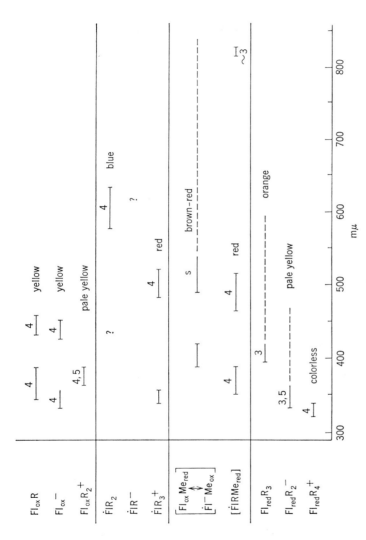

Fig. 10. Main areas of visible absorption of monomeric species. End absorptions stretching further out toward the infrared are indicated by dashed lines. Approximate maximal molar extinctions per centimeter are given by the numbers in log units. Indication of colors applies to about 1mM solutions. (After Hemmerich and Spence, in ref. 3.)

Generally speaking, metal affinity of flavins is inherent only in the semi-quinone, because only the semiquinone exhibits sufficient basicity at N-5 *and* low 3,4 prototropic energy simultaneously. Consequently the complexation of flavoquinone through back-donation from reducing soft cations has to be considered as chelation and reduction in one step. Hence we may differentiate two kinds of flavosemiquinone–metal chelates.

(*1*) "Charge-transfer chelates" involving ligand–metal charge transfer (Fig. 8) (in particular with Fe(III), Mo(VI), Cu(II), and Ag(II)—only the last two are stable in the presence of water). Spin-pairing and, hence, diamagnetism might be expected from all of them except MoFl. At least no well-defined ESR signals have been observed.

(*2*) "Radical chelates" with Mn(II), Fe(II), Co(II), Ni(II), Zn(II), and Cd(II). All of them might be expected to retain their radical character

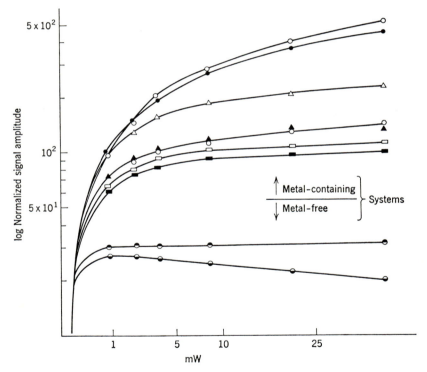

Fig. 11. Saturation with microwave power of ESR signals from flavin radicals at −170°. Signal amplitude in arbitrary units is plotted against the square root of microwave power (in milliwatts) incident on the cavity. (Redrawn from H. Beinert and P. Hemmerich, *Biochem. Biophys. Res. Commun.*, **18**, 212 (1965).)

in the complex. This has been demonstrated by ESR with diamagnetic Zn(II) and Cd(II).

Since in both categories of chelates the ligand is essentially flavosemiquinone, no drastic differences in electronic spectra are found. Furthermore, the difference of one proton does not have such spectral influence, since the proton is invariably fixed at N-3.

Because only half-reduced flavin has any appreciable metal affinity, disproportionation equilibria are reversed by addition of metal ions towards the comproportionated (semiquinoid) state (Fig. 9). Hence, metal chelation can stabilize half-reduced flavin thermodynamically as well as kinetically. Though other mechanisms for radical stabilization in metal-free flavoproteins may exist, only the metal chelation mechanism is understood in terms of molecular structure.

Once more, it must be emphasized that structural assignment of oxidation states and metal binding in complex flavoproteins cannot be based on visible spectra alone, but also upon magnetic criteria and comparison with model systems.

A survey of visible absorption of identified free (not protein bound) flavin species is given in Figure 10.

The fact that there is indeed direct interaction between flavin radical and metal involved in the reaction mechanism of metal flavoproteins has been established by paramagnetic resonance relaxation of flavosemiquinone through the metal. The relaxation of flavosemiquinone in metal flavoproteins is enhanced by an order of magnitude compared with metal-free flavoproteins. This effect can be simulated with model flavins in pH regions where chelation does not yet remove all the free $\dot{F}lH_2$ in favor of the ESR-inactive chelate (Fig. 11).

VII. Redox Mechanisms and Kinetics

The redox potentials of FMN and FAD are practically identical. This shows that the interaction of adenine with flavin it not sufficient to change the redox properties. Binding of FMN to the apoproteins of Old Yellow enzyme has, however, been demonstrated to effect an increase in redox potential of nearly 100 mV. This means that $FMNH_2$ is extremely tightly bound to the proteins, which might be due to the fact that the bent $FMNH_2$ fits closer to the protein "surface" than the planar FMN does.

The flavin is capable of being reduced and oxidized in one- or two-equivalent steps. The mechanism is largely dependent on the reducing or oxidizing agent in the particular case. Figure 3 shows the primary products when the redox equivalents are electrons or hydrogen atoms.

Two-equivalent reduction or oxidation is achieved by transfer of two electrons or, perhaps more likely, of one hydride equivalent.

In solution these reactions are rapid, and since free radicals are also formed by the likewise rapid disproportionation reaction, it is experimentally very difficult to demonstrate which type of reaction is taking place. Available experimental information in this field is therefore scarce. Rapid flow kinetics have indicated that during reduction of riboflavin by dithionate the semiquinone is formed from riboflavin in a one-equivalent reaction. Reoxidation of dithionite-reduced FMN or FAD by oxygen seems, however, to involve a complex formation with oxygen and simultaneous removal of two hydrogen atoms.

References

General Reviews

1. H. Beinert, in *The Enzymes*, Vol. 2, P. D. Boyer, H. Lardy, and K. Myrbäck, Eds., Academic Press, New York, 1960, p. 360.
2. P. Hemmerich, C. Veeger, and H. C. S. Wood, *Angew. Chem.*, **77**, 699 (1965); *Angew. Chem. Intern. Ed.*, **4**, 671 (1965).
3. E. C. Slater, Ed., *Flavins and Flavoproteins*, Elsevier, New York, 1966.

Biosynthesis

4. G. W. E. Plaut, *J. Biol. Chem.*, **211**, 111 (1954).

Photochemistry

5. B. Holmström, *Arkiv Kemi*, **22**, 329 (1964).

ESR Spectra

6. A. Ehrenberg, G. Eriksson, and F. Müller, in *Flavins and Flavoproteins*, E. C. Slater, Ed., Elsevier, New York, 1966.
7. A. Ehrenberg, G. Eriksson, and P. Hemmerich, in *Oxidases and Related Redox Systems*, Vol. I, T. E. King, H. S. Mason, and M. Morrison, Eds., Wiley, New York, 1965, p. 179.

Absorption Spectra

8. G. Weber, in *Flavins and Flavoproteins*, E. C. Slater, Ed., Elsevier, New York, 1966.
9. H. A. Harbury, K. F. La Noue, P. A. Loach, and R. M. Amick, *Proc. Natl. Acad. Sci. U.S.*, **45**, 1708 (1959); H. A. Harbury and K. A. Foley, *ibid.*, **44**, 662 (1958).
10. K. H. Dudley, A. Ehrenberg, P. Hemmerich, and F. Müller, *Helv. Chim. Acta*, **47**, 1354 (1964).

Metal Chelation

11. P. Hemmerich, F. Müller, and A. Ehrenberg, in *Oxidases and Related Redox Systems*, Vol. I, T. E. King, H. S. Mason, and M. Morrison, Eds., Wiley, New York, 1965, p. 157.

Mechanisms of Flavoprotein Catalysis[*]

GRAHAM PALMER and VINCENT MASSEY, *Biophysics Research Division, Institute of Science and Technology, and Department of Biological Chemistry, The University of Michigan, Ann Arbor, Michigan*

I.	Spectra of Oxidized Flavins and Flavoproteins . . .	264
II.	Spectra of Flavin and Flavoproteins at the Semiquinoid Oxidation State	267
III.	Substrate-Produced Long-Wavelength Absorption Bands . .	274
IV.	Charge-Transfer Absorptions of Flavins and Flavoproteins .	277
V.	Spectra of Reduced Flavins and Flavoproteins . . .	280
VI.	Fluorescence of Flavoproteins	281
VII.	Magnetic Resonance Studies on Metal-Free Flavoproteins .	284
VIII.	Reactivity of Reduced Flavins and Flavoproteins with O_2 . .	286
IX.	Steady-State Kinetics of Flavoprotein Systems . . .	287
X.	Flavoprotein Reaction Mechanisms	289
	A. General Considerations	289
	B. Enzymes Functioning Solely between the Fully Oxidized and Fully Reduced Forms . . .	290
	C. Enzymes Involving True Semiquinoid Intermediates .	291
	D. Enzymes Involving Fully Reduced Flavin and Intermediates	292
	E. Enzymes Involving Substrate-Complexed Flavin Semiquinone	293
	F. Enzymes Functioning at the Semiquinoid Level, but also Involving Another Redox Group in the Enzyme . .	295
XI.	General Conclusions Regarding Mechanisms	296
	Abbreviations	297
	References	297

Great advances in our understanding of flavoproteins and flavoprotein catalysis have been made in the last ten years, largely because of the extensive use of static and rapid-reaction spectrophotometry. As a result of this work it is evident that many different spectral forms of flavins and

* The preparation of this review and much of the unpublished work reported here was made possible through support by the United States Public Health Service (Grant Numbers GM 12176 and GM 11106).

flavoproteins exist at all stages (oxidized, semiquinone, and reduced) of reduction. Examination of kinetics of spectral changes of a flavoprotein induced by electron donors and acceptors, the stopped-flow technique, has in several instances permitted the delineation of spectral intermediates and their importance in catalysis. However, a complete understanding of these phenomena requires an unequivocal assignment in chemical terms of the observed spectral forms. Unfortunately, our knowledge in this area is very limited. In the first part of this article we wish to review the current state of knowledge concerning the spectral properties of flavins and flavoproteins (more in the hope of accentuating the deficiencies of our knowledge and thereby stimulating further studies, than of providing much elucidation), while the latter part of the article is devoted to general aspects of flavoprotein catalysis.*

I. Spectra of Oxidized Flavins and Flavoproteins

The two prominent visible absorption bands of oxidized flavins appear to be $\pi-\pi^*$ transitions. Thus, the oscillator strengths for the visible bands are of the order of 0.1–0.2. These are values typical of allowed $\pi-\pi^*$ transitions and one to two orders of magnitude greater than found with $n-\pi^*$ transitions. Furthermore, flavins are characteristically very fluorescent, yet Kasha (1) has established that lowest singlet $n-\pi^*$ transitions are devoid of fluorescence. In addition, it would appear that the two visible absorption bands are due to electronic transitions in different parts of the isoalloxazine ring structure. Thus, Weber et al. (2) have shown the fluorescence polarization spectrum of 3-methyl tetraacetyl riboflavin in the viscous inert solvent, mineral oil, to be independent of wavelength in the regions 300–380 mμ ($p = 0.18$) and 400–490 mμ ($p = 0.25$) with a change between 380 and 400 mμ. The plot of polarization as a function of wavelength reflects the relative orientation of the transition moment associated with absorption and emission. This should remain constant for each electronic transition. Thus, these data indicate that each of the two characteristic absorption bands are due to single electronic transitions.

A characteristic feature of the visible spectra of free flavins that is mirrored in many flavoproteins is the resolution often seen in the 450 mμ

---

* It is not our intention of providing extensive descriptions of the various flavoprotein systems and for this information the interested reader is referred to *The Enzymes*, Vol. 7, and the recent I.U.B. Symposium on Flavins and Flavoproteins (cf. refs. 36 and 83).

band. As first shown by Harbury et al. (3) with 3-methyl lumiflavin, the transition from polar to nonpolar solvents leads to the appearance of shoulders in this band around 430 and 480 mμ. Figure 1 illustrates this phenomenon in a very pronounced form, showing the difference in spectrum of 3-methyl tetraacetyl riboflavin dissolved in water and in a mixture of 1 part benzene and 24 parts mineral oil. As the fluorescence polarization is independent of wavelength in the range 400–500 mμ, it is clear that this detail in the spectrum is due to the resolution of vibrational structure that is blurred in polar solvents such as water and methanol. While not clearly shown, some vibrational detail appears in the "370 mμ band" in the inert solvent. As first pointed out by Harbury et al. (3), the wavelength maximum of this band is very dependent on the polarity of the solvent, shifting from about 370 mμ in water to about 330 mμ in solvents such as benzene and CCl₄. This shift is in marked contrast to the much smaller solvent shifts observed with the longer wavelength band. In nonpolar solvents this latter transition exhibits the characteristic "polarization red shift" (4) on increasing the refractive index of the solvent. However, when we change from the inert solvents to polar solvents which are able to partake in various interactions with the chromophoric molecule, e.g., H bonding, dipole–dipole, the rationalization of the spectral shifts is not obvious. These multiple effects of the solvents thus preclude any satisfactory analysis with the exception of the qualitative observation that

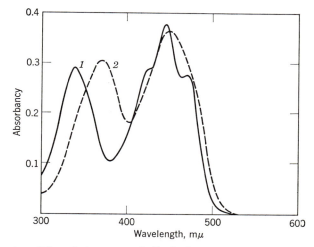

Fig. 1. Spectra of 3-methyl tetraacetyl riboflavin (3.2 × 10⁻⁵M), in a mixture of 1 part benzene and 24 parts mineral oil (curve 1) and in water (curve 2). Greull and Massey, unpublished.

non-polar solvents result in a marked resolution of spectrum of the 450 mμ band of simple flavins (shoulders around 430 and 480 mμ). This has led to the widespread interpretation that in flavoproteins exhibiting similar detail (cf. Table II) the flavin must be in a hydrophobic environment. While this interpretation may be correct, caution should be used in its general application, for the origin of these effects does not preclude the existence of at least one specific hydrogen bond from the protein to the flavin in the "hydrophobic-type" flavoprotein spectrum, nor does it specify whether the hydrophilic spectra are due to protein or solvent. In fact, it may be argued that the "hydrophilic type" of spectrum indicates that the flavin is accessible to the solvent. In the case of D-amino acid oxidase it was observed by Yagi and Ozawa (33) that the addition of benzoate transforms the spectrum from the unresolved to the resolved type. This has been interpreted as benzoate inducing a change in conformation of the protein so that the flavin, which in the native enzyme is in a hydrophilic environment, is now in a hydrophobic environment of the protein and presumably not accessible to solvent. Alternatively, it could be argued that the bound benzoate lies on the surface of the flavin, thus excluding the aqueous solvent and providing a lower dielectric to the flavin. The effects of over 50 compounds on the spectrum of D-amino acid oxidase have recently been examined by Massey and Ganther (34). Many compounds other than benzoate were found to induce benzoate-like spectral changes. The one important feature for binding was the presence in the molecule of a single carboxyl group; spectral shifts similar to those of benzoate were produced by simple compounds such as sorbic acid and glycollic acid. Furthermore, while a great number of halogen-, methyl-, methoxy-, and nitro-substituted benzoates *all* produced benzoate-type spectral effects, *ortho*-substituted hydroxy and amino benzoates did not produce spectral resolution and actually caused the 455 mμ band of the native enzyme to shift to lower wavelength. These observations would seem to preclude the latter explanation above and suggest that the phenomenon is the consequence of conformational changes attendant on breaking a hydrogen bond by the benzoate in the native enzyme. It is considered that a hydrogen bond from a charged amino group in the protein to a carbonyl residue of the isoalloxazine ring stabilizes the protein in a conformation whereby the flavin is primarily in a polar environment (conceivably by being exposed to the solvent, water). On binding with benzoate and other carboxylate anions, this hydrogen bond is abolished, with a consequent change in protein configuration which buries the flavin in a nonpolar milieu. In the case of the hydroxy- and aminobenzoates mentioned, a resolved spectrum is not obtained, presumably because the bound derivative can now substitute an additional hydrogen bond to the flavin and thus maintain the original configuration. This interpretation is supported by the ionization behavior of the protein-bound FAD in the native and benzoate-complexed enzyme (34).

II. Spectra of Flavin and Flavoproteins at the Semiquinoid Oxidation State

The previous chapter on free flavins documents the absorption characteristics of the cationic and neutral semiquinone species of free flavins (35). Unfortunately no unequivocal assignment of spectrum can yet be made to the anionic semiquinone, as this is one which, by virtue of a pK in the physiological pH range, might be expected to be encountered with flavoproteins. Evidence has been obtained recently (36) which suggests that the anion radical may have an absorption spectrum very different from that of the neutral form. Figure 2 shows the spectrum of the free radical form of glucose oxidase produced by anaerobic titration with dithionite, at pH 6.3 and pH 8.5. In both cases practically complete conversion to the semiquinone was obtained as judged by quantitative EPR spectroscopy (36). It is evident that the spectrum produced at pH 8.5 is very different from that at pH 6.3. The latter, designated as type A in Table I and Figure 2, is very similar to that assigned to the neutral semiquinone species of free flavins (37) and is characteristically blue in color. The spectrum produced at pH 8.5 (designated as type B in Table I) is characteristically reddish in color. In view of the pH dependence of the spectrum it is tempting to assign the type A spectrum to the neutral flavin radical and type B to the flavin anion radical. An inspection of the absorption spectra of other flavoproteins reported in the literature shows a fairly clear-cut distribution into semiquinone spectra resembling either type A or type B. Thus the spectrum of the radical species of D-amino acid oxidase is qualitatively and quantitatively almost identical to that of glucose oxidase type B (36). Spectra with elements characteristic of type B have also been reported on substrate reduction of the electron-transferring flavoprotein (38) and on dithionite titration of L-amino acid oxidase (39). Type A spectra are seen with the Shethna flavoprotein on reduction with dithionite (40) and on reoxidation of reduced cytochrome b_5 reductase with potassium ferricyanide (41). Similarly, type A spectra are seen on substrate reduction of TPNH-cytochrome c reductase (18) and ferredoxin-TPN reductase (26,27). Type A spectra are also seen on titration of the acyl CoA dehydrogenases by $Na_2S_2O_4$ (38). While much work remains to be done in this field, we regard the discovery of semiquinone species with a type B

TABLE

Characteristic Properties

Enzyme	Pros-thetic group	Mol. wt.	Moles flavin per mole protein	Spectral type and λ maxima, mμ	Flavin fluorescence	Substrate Electron donors
Old yellow enzyme (yeast)	FMN	102,000	2	Resolved; 280, 383, 464	None	TPNH
D-Amino acid oxidase (pig kidney)	FAD	92,000 and polymers	2	Unresolved; 274, 370–380, 455	Slight; ca. 17% that of FAD	D-Amino acids
L-Amino acid oxidase (*Crotalus adamanteus*)	FAD	140,000	2	Partially resolved, 275, 390, 462		L-Amino acids
Glucose oxidase (*Aspergillus niger*)	FAD	186,000	2	Unresolved; 278, 383, 452	None	β-D-Glucose; number of D-sugars
Glucose oxidase (*Penicillium amagasakiense*)	FAD	154,000	2	Unresolved; 278, 380, 460	None	β-D-Glucose; number of D-sugars
Glycollic acid oxidase (spinach)	FMN	140,000	2	Unresolved; 274, 340, 445	None	Glycollate
Long-chain L-α-hydroxy acid oxidase (pig kidney)	FMN			Resolved; 377, 452		Long-chain L-α-hydroxy acids
Pyruvate oxidase (*E. coli mutant*)	FAD	265,000	4	Highly resolved; 370, 438 ("shoulder" peaks at 460 mμ)		Pyruvate
DPNH peroxidase (*S. faecalis*)	FAD	120,000 (flavin content)		Unresolved; 450	None	DPNH and DPNH analogs
Butyryl CoA dehydrogenase (pig liver)	FAD	Uncertain 120,000–220,000	Probably 2	Unresolved; 270, 360, 445 (yellow form); 266, 357, 426, 710 (green form)	None	C_4 to C_8 acyl CoA
Palmityl CoA dehydrogenase (pig liver)	FAD	Uncertain; 120,000–220,000	Probably 2	Unresolved; 275, 370, 447	None	C_4 to C_{16} acyl CoA

I

of Flavoproteins

Specificity	Spectral characteristics of semiquinone (chemical reduction)	Long-wavelength band produced by substrate	Comments	Ref.
Electron acceptors				
Cyt c; O_2; methylene blue; $K_3Fe(CN)_6$	Type B	Transient; type B spectrum produced by TPNH plus TPN	Also shift of spectrum of oxidized enzyme 10mμ to longer wavelengths with TPN.	5, 84
O_2	Type B; peaks at 370, 400, 490 mμ EPR signal	Different spectra with different amino acids; max 500–560 mμ. Transient	Spectrum passes to resolved type on complexing with large number of carboxylic acids	6
O_2; phenazine methosulfate; methylene blue	Type B	Transient; different with different amino acids	Also contains some carbo-hydrate	7, 39, 85, 86
O_2	Type A at pH 6.3 (max at 570 mμ). Type B at pH > 7.6 EPR signal	None detectable	Not inhibited by mercurials. Contains 16% carbo-hydrate; also unknown chromophore.	8
O_2; dyes	None observed	None observed	Inhibited strongly by mercurials. Enzyme from *P. notatum* appears similar.	9
O_2; indophenol dyes; thionine; methylene blue			Also acts as polymer at high ionic strength.	10
O_2; DCIP				11
$K_3Fe(CN)_6$; cyt b_1			Not to be confused with the pyruvate dehydrogenase complex.	12
H_2O_2; O_2		Stable. Vary with electron donor.	The long-wavelength inter-mediates are generally stable even in presence of $Na_2S_2O_4$.	
Electron-transferring flavoprotein; phenazine metho-sulfate; $K_3Fe(CN)_6$	Type A spectrum. EPR signal	Stable. 570 mμ max: Type A	Also exists in green (oxidized) form with strong 710 mμ absorption max.	13
Electron-transferring flavoprotein, phena-zine methosulfate, $K_3Fe(CN)_6$	Elements of Type A spectrum	Strong long-wavelength absorption, distinct shoulder around 550 mμ		14

(*continued*)

TABLE

Characteristic Properties

Enzyme	Pros-thetic group	Mol. wt.	Moles flavin per mole protein	Spectral type and λ maxima, $m\mu$	Flavin fluorescence	Substrate Electron donors
General fatty acyl CoA dehydro-genase (pig liver)	FAD	Uncertain; 120,000–220,000	Probably 2	Unresolved; 275, 365, 442	None	C_6 to $C_{>16}$ acyl CoA
Electron-transferring flavoprotein (pig liver)	FAD	70,000	1	Highly resolved; 270, 375, 437.5 ("shoulder" peaks at 460 $m\mu$)	Slight; ca. 10% of FAD	Reduced acyl CoA dehydro-genases; re-duced sarco-sine dehydro-genase
DPNH cyt b_5 reductase (calf-liver microsomes)	FAD	40,000	1	Highly resolved; 275, 390, 461	None	DPNH and DPNH analogs
TPNH cyt c reductase (pig liver microsomes)	FAD	40,000 (flavin content)	Probably 2	Resolved; 379, 454		TPNH
Lipoyl dehydro-genase (pig heart) (beef liver) (yeast)	FAD	102,000	2	Highly resolved; 274, 355–370, 455	Very strong; (fluorescence consider-ably greater than that of free FAD)	Dihydrolipoyl compounds, DPNH, and DPNH analogs
Glutathione reductase (yeast)	FAD	116,000	2	Highly resolved; 274, 379, 463	None	TPNH
Vitamin K reductase (beef liver)	FAD	52,000	1	Resolved; 275, 380, 455		DPNH, TPNH
Thioredoxin reductase (E. coli)	FAD	68,000	1	Resolved	Quite fluorescent	TPNH, DPNH
Ferredoxin-TPN reductase (spinach)	FAD	44,000 (gel filtration)	1	Resolved; 275, 385, 456	None	Reduced ferre-doxin, TPNH

I (*continued*)

of Flavoproteins

specificity Electron acceptors	Spectral characteristics of semiquinone (chemical reduction)	Long-wavelength band produced by substrate	Comments	Ref.
Electron-transferring flavoprotein, phenazine methosulfate, $K_3Fe(CN)_6$		Stable; spectra different with different substrates		15
DCIP, cyt c_1, cyt c, $K_3Fe(CN)_6$ quinones		Probably Type B. Seen with acyl CoA and sarcosine and their respective dehydrogenases		16
Cyt b_5, $K_3Fe(CN)_6$, DCIP	Strong transient absorption on $K_3Fe(CN)_6$ oxidation of reduced enzyme; resembles Type A	Weak flat absorption in fully reduced enzyme (NADH)	Tyrosine hydroxyl involved in FAD binding. FMN also reactivates apoenzyme. Protein SH group required for DPNH binding	17
Cyt c, $K_3Fe(CN)_6$, neotetrazolium dyes, DCIP		Stable with limited TPNH. Type A shows some EPR signal.	Long-wavelength intermediate very resistant to oxidation by O_2 or cyt c.	18
Lipoyl compounds, DPN and DPN analogs, $K_3Fe(CN)_6$, menadione, DCIP	Same as that produced by substrate; no EPR signal.	Peaks at 340, 440 mμ and distinct shoulder 540 mμ. Very stable even with excess substrate.	Protein disulfide forms part of active center in conjunction with FAD. Long-wavelength absorption band produced by substrate is half-reduced form of enzyme. (Enzyme from *E. coli* differs in not having *stable* long-wavelength band.)	19–22
Oxidized glutathione		Peaks at 358, 448 mμ and distinct shoulder at 540 mμ. Very stable even with excess substrate.	Protein disulfide forms part of active center in conjunction with FAD. Long-wavelength band very similar in properties to that of lipoyl dehydrogenase.	23
Vitamin K_2. Various benzo- and naphthoquinones			FAD not tightly bound; can be removed by dialysis.	24
Thioredoxin				25
cyt f, ferredoxin, TPN, DPN, $K_3Fe(CN)_6$, menadione, DCIP		Stable with TPNH: Type A		26, 27

(*continued*)

TABLE

Characteristic Properties

Enzyme	Prosthetic group	Mol. wt.	Moles flavin per mole protein	Spectral type and λ maxima, mμ	Flavin flourescence	Substrate Electron Donors
Glyoxylate carboligase (E. coli)	FAD	Unit molecular wt, 100,000 (flavin content)		Well resolved; 275, 380, 445	Slight	Glyoxylate
Shethna flavoprotein (Azotobacter)	FAD			Well resolved	None	
Oxynitrilase (almonds)	FAD	ca. 75,000	1	Resolved; 275, 388, 457		Benzaldehyde and HCN
Lactic oxidative decarboxylase (M. phlei)	FMN	260,000	2			Lactate
L-6-Hydroxynicotine oxidase (Arthrobacter oxydans)	FAD	140,000	4	Resolved; 365, 440	Slight	L-6-hydroxynicotine, L-6-hydroxy-nor-nicotine
Flavodoxin (Cl. pasteurianum)	FMN	15,000	1	Resolved; 272, 372, 443		TPNH, "activated" hydrogen (hydrogenase) Na$_2$S$_2$O$_4$

spectrum as one which may serve to dispel much of the confusion in this field. The initial impetus for the search for involvement of semiquinones in flavoprotein catalysis came from the early work of Beinert (42) and others (42,44), in which it was clear that flavin semiquinone absorbed strongly at 570 mμ. By now it has become routine to monitor for possible semiquinone intermediates by following absorbance changes in the neighborhood of this wavelength. From Figure 2, it is evident that the type B semiquinone has negligible absorbance at 570 mμ, and could easily be missed if this were the only wavelength used for observation.

The assignment of the type A spectrum to the neutral semiquinone and type B to the anionic semiquinone is now much more certain. It has been shown with glucose oxidase (85) that production of radical over a wide range of pH results in complete production of type A spectrum at

I (*continued*)

of Flavoproteins

specificity ——————— Electron acceptors	Spectral characteristics of semiquinone (chemical reduction)	Long-wavelength band produced by substrate	Comments	Ref.
			No evidence of redox role of flavin in the reaction glyoxylate → hydroxy-malonic semialdehyde.	28
	Highly stable, max. 570 + 620 mμ, Type A		Physiological function not known.	29
	Type B		No evidence for redox role of flavin.	30, 62, 35
Oxygen		Stable: probably type A		32
Oxygen			Crystalline	87
TPN		Type A (reduction with TPNH)	Produced instead of ferre-doxin when growth medium low in iron salts	88

pH values below 6.0 and complete production of type B spectrum at pH values above 9.0. A plot of the extinction coefficient at 570 mμ as a function of pH falls from a value of 3000 mole^{-1} cm^{-1} at low pH to one of 600 at high pH. At intermediate pH values the curve fits that of a dissociation of pK 7.4–7.5. Furthermore, a spectrum similar to type B has been found by Ehrenberg, Hemmerich and Müller (private communication) for the anionic, radical in model flavin compounds.

Recently, the authors (85) have described a new method for the quantitative generation of flavoprotein semiquinones. This involves a photoreduction under anaerobic conditions using ethylenediamine tetra-acetate as electron donor. With this method, unambiguous semiquinone spectra of either type A or type B have been obtained with a large number of flavoproteins.

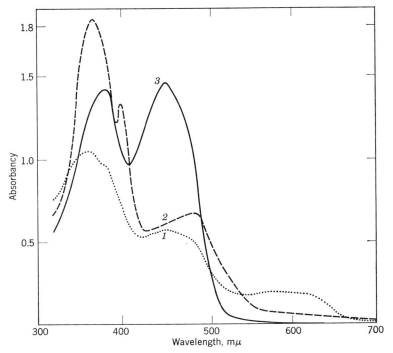

Fig. 2. Spectra of the two semiquinoid species of glucose oxidase produced by anaerobic titration with $Na_2S_2O_4$ at pH 6.3 and at pH 8.5. Type A spectrum is that of the semiquinone form at pH 6.3 (curve *1*); type B spectrum is that of the semiquinone produced at pH 8.5 (curve *2*). [From Massey et al. (36), with modification.] Curve *3* is oxidized enzyme.

III. Substrate-Produced Long-Wavelength Absorption Bands

With the exception of lipoyl dehydrogenase, all the long-wavelength absorption produced in flavoproteins on partial reduction with dithionite (or catalytic reduction with H_2 and Pt) is correlated with free radical as measured by EPR spectroscopy. Thus, these absorption bands represent true semiquinoid flavin forms of the enzymes. With lipoyl dehydrogenase [and presumably with glutathione reductase in view of its remarkable similarities to lipoyl dehydrogenase (23)], the long-wavelength absorption produced by partial titration with dithionite is clearly due to a half-reduced form of the enzyme, since full reduction is obtained by the addition of more dithionite. In this case, absolutely no free radicals can be detected by EPR spectroscopy. Identical spectra are obtained by the addition of reducing substrates to lipoyl dehydrogenase and glutathione reductase. With these two enzymes it is clear that a protein disulfide functions in conjunction with the flavin at the active center, so that a total of four electron-reducing equivalents can be accepted by the active center to produce the fully reduced enzyme. Thus, the red intermediate produced by the addition of exactly two reducing equivalents has to be explained by interaction of the two prosthetic groups. Three possibilities are evident: (*1*) interaction between oxidized flavin and dithiol, (*2*) interaction between reduced flavin and disulfide, and (*3*) interaction between flavin semiquinone and sulfur radical. Possibility (*1*) has been favored by Sanadi and co-workers (45) and also recently by Kosower (46), who considers that the species may represent a charge-transfer complex between oxidized flavin and sufhydryl anion. Possibility (*3*) has been favored by the authors and co-workers (19,47), largely on the basis of the similarity of the absorption spectrum to that of a flavin semiquinone and also because the addition of arsenite to enzyme reduced with a *stoichiometric* amount of DPNH causes this spectrum to change to one almost identical with that of oxidized enzyme, i.e., because possibility (*1*), achieved by addition of arsenite, has a spectrum similar to that of oxidized enzyme. Clearly, in the absence of further evidence no unequivocal assignment is possible to this intermediate. Either explanation would account for the lack of a free-radical signal in EPR spectroscopy. Our present concept is that the intermediate probably is one in which the flavin semiquinone and sulfur radical have interacted to give a weak covalent bond, as shown in Reaction (1). The role of this intermediate in catalysis will be considered in a later section.

$$ \tag{1} $$

With most other flavoproteins where detailed studies have been made, it is clear that intermediates seen on the addition of substrate, either transient or stable in nature, have different spectra than the true semiquinoid forms produced chemically. Furthermore, these substrate-produced forms appear to be singularly devoid of EPR-detectable radicals, with the exception of TPNH cytochrome *c* reductase, which does exhibit some free-radical character (18). Figure 3 illustrates this phenomenon with the enzyme D-amino acid oxidase. On partial reduction with dithionite this enzyme gives an intense free-radical signal in the EPR spectrometer and a visible absorption almost identical to that labeled type B in Figure 2. However, on reduction with D-amino acids the intermediate spectra shown in Figure 3 are obtained. As well as being totally different from the true semiquinone, it is evident that the spectra of the substrate-produced intermediates vary depending on the substrate used, suggesting that some contribution to the spectrum is made by the substrate or some product of the substrate. The suggestion has been made that these spectra are due to interaction of flavin radical and amino acid radical (48) possibly of the same type as proposed above for lipoyl dehydrogenase [see Reaction (2)].

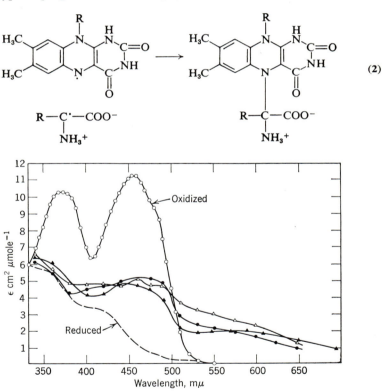

Fig. 3. Spectra of the transient intermediates observed on the anaerobic reduction of D-amino acid oxidase with substrates (▲) proline, (△) methionine, and (●) alanine. [After Massey and Gibson (48).]

Again, the unequivocal assignment of these spectra in chemical terms is not yet possible; as will be considered in the following section, considerable evidence exists for charge-transfer bands in this enzyme.

Historically, the first evidence that long-wavelength bands produced by substrates with flavoproteins are not necessarily due to the uncomplexed semiquinoid forms came from the work of Dolin on DPNH peroxidase (31,49). With this enzyme, there are no data available to permit comparison of dithionite intermediate spectra and substrate-produced spectra, and no EPR studies have been made. Nevertheless, it is clear from the fact that different reduced pyridine nucleotides yield different stable spectra that substrate interactions exist in the spectral forms of this enzyme. Similar conclusions can also be reached with the acyl CoA dehydrogenases, which on titration with dithionite yield EPR-detectable free radicals with spectra different from those produced by substrate, where no rapidly produced EPR-detectable free radical is obtained (38). Whether these instances are due to flavin radical–substrate radical interactions as proposed above for D-amino acid oxidase, or to charge-transfer absorptions cannot be decided at this stage. Although the number of cases is now considerable where no correlation may be observed between the production of long-wavelength absorption by substrate and the appearance of free radicals, we should be cautious in concluding that this is generally true. As already mentioned, TPNH cytochrome c reductase does give a free-radical signal on reaction with TPNH (18). The optical spectrum produced is similar to that of type A of Figure 2. An inspection of Table I reveals that two enzymes reduced by substrate exhibit either type A or type B spectra and so might be expected to be true semiquinones (ferredoxin TPN reductase, electron-transferring flavoprotein). Unfortunately, no EPR studies have yet been reported with these enzymes which would confirm or deny this hypothesis.

IV. Charge-Transfer Absorptions of Flavins and Flavoproteins

The interpretation of long-wavelength absorption bands of flavoproteins is complicated and at the same time clarified by the recognition that charge-transfer bands can be observed in the same spectral region where semiquinones absorb light. The earliest and so far best documented example of a charge-transfer band in a flavoprotein is with lipoyl dehydrogenase (50). When fully reduced with DPNH in the presence of arsenite (or in the absence of arsenite at low temperatures), this enzyme exhibits a broad absorption band centered around 710 mμ. That this was due to a charge-transfer interaction between enzyme-FADH$_2$ and DPN was shown in several ways. First, the band is abolished by *Neurospora* DPNase, an enzyme which is specific for hydrolyzing the nicotinamide-ribose linkage of oxidized pyridine nucleotides. Second, enzyme fully reduced with TPNH in the presence of arsenite does not exhibit any long-wavelength absorption, but this is produced by the addition of DPN. Third, other oxidized pyridine nucleotides do produce the band, with wavelength maxima consistent with the redox potential of the nucleotide. Thus, with desamino DPN (whose redox potential is the same as that of the DPN couple, viz. -320 mV) the band is also centered at 710 mμ, whereas with thionicotinamide DPN (with a redox potential of -285 mV) the band is centered at 740 mμ. Furthermore, similar absorption bands have been found in model systems. $FMNH_2$ and DPN give rise to an absorption band centered around 580 mμ. Finally, the position of the wavelength maxima in the complexes formed between $FMNH_2$ and N-methyl 3-nicotinamide and N-methyl 4-nicotinamide is in good agreement with the energy differences found with other charge-transfer bands involving these compounds (36). It should be emphasized that although this charge-transfer band is formed with lipoyl dehydrogenase [and a similar one with glutathione reductase (23), under certain conditions] it plays no role in the normal catalysis of the enzyme, since its formation is quite a slow process (50). A similar band, also with a maximum at 710 mμ, is found with the green form of butyryl CoA dehydrogenase (13). However, in this case the enzyme flavin is obviously in the oxidized form. We would like to speculate that the green color in this instance is due to a charge transfer from a group of the protein (conceivably a sulfhydryl group) to the flavin. It would be of interest to determine if this band is abolished on addition of mercurial to the enzyme. Another example of long-wavelength absorption in an oxidized flavoprotein is with the old yellow enzyme. Rutter and Rolander (51) found that when this enzyme is adjusted to pH 10 with alkaline ammonium sulfate, it turns green with maximum absorption at 680 mμ. Other examples of probable charge-transfer absorptions involving oxidized enzyme are provided by D-amino acid oxidase and compounds such as 2-amino and 3-amino benzoate (34,52). These and other examples are summarized in Table II. Finally, again with respect to D-amino acid oxidase, reasonable evidence exists for charge-transfer absorption involving fully reduced flavin and imino acids (36). Figure 4 shows a complex formed between oxidized enzyme and the cyclic imino acid, piperidine carboxylic acid, and the spectral changes occurring on reduction of this complex with the corresponding amino acid, pipecolic acid. From the series of isosbestic points it is evident that this system involves only two components. Therefore, if the not unreasonable assumption is made that the original spectrum

TABLE II[a]

Charge Transfer Complexes of Flavins and Flavoproteins

Complex	Absorption maximum, mμ	Ref.
FMNH$_2$–FMN	900 (pH 6.3)	50,44
FMNH$_2$–N-methyl 3-nicotinamide	510	50
FMNH$_2$–N-methyl 4-nicotinamide	610	Massey, unpublished
FMNH$_2$–DPN	580	50
Lipoyl dehydrogenase–FADH$_2$–DPN	710	50
D-Amino acid oxidase–FAD–Δ' pyrolline 2-carboxylate	615	34
–FAD–Δ' piperidine 2-carboxylate	640	34
–FAD–indole-2 carboxylate	530	34
–FAD–2-amino benzoate	565	34
–FAD–3-amino benzoate	530	34
–FAD–2,5-dihydroxybenzoate	< 530	34
–FADH$_2$–Δ' piperidine 2-carboxylate	535	36
Butyryl CoA dehydrogenase (green form) FAD–unknown group	720	13
Glutathione reductase FADH$_2$–TPN	720	23

[a] After Massey et al. (36).

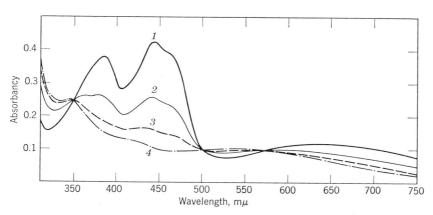

Fig. 4. The effect of pipecolic acid added anaerobically to a mixture of D-amino acid oxidase and piperidine 2-carboxylate (36). Conditions: 0.1M pyrophosphate, pH 8.5 at 17°. (*1*) Oxidized enzyme plus 2.17 moles of piperidine 2-carboxylate. Then D-pipecolic acid: (*2*) 2.17 moles, (*3*) 4.35 moles, (*4*) 52 moles.

was due to charge transfer between the oxidized enzyme and piperidine carboxylic acid, then it follows that the final spectrum is due to charge transfer between the reduced enzyme and piperidine carboxylic acid. This absorption is similar to that found on addition of pyruvate and NH_4^+ to fully reduced enzyme (53) which has previously been ascribed to the enzyme–substrate complex (54). In view of the results shown in Figure 4, it would seem more reasonable to ascribe such bands to charge transfers between fully reduced enzyme and product. The similarity between the catalytic intermediates shown in Figure 3 and the static spectrum of Figure 4 obviously suggest that the intermediates may represent flavin in the fully reduced form, rather than in the semiquinoid form as previously suggested (48). However, kinetic evidence exists for the formation of intermediates with spectral properties more akin to type B of Figure 1 before the development of those shown in Figure 3 (36). These rapidly formed intermediates probably represent the true catalytic intermediates of the enzyme reaction and may well show true semiquinoid character.

In summary, it is evident that charge-transfer absorptions occur in the wavelength region traditionally ascribed to flavin semiquinone; hence, considerable experimental evidence is required to distinguish between the various possibilities discussed. Only in the case that quantitative correlation is obtained between optical absorption and EPR signal is it safe to conclude that flavin semiquinones are involved. If no such correlation is found, the absorption may be due to charge transfer involving either oxidized or reduced flavin, or to radical–radical interactions that may well result in the formation of weak covalent bonds (see below).

V. Spectra of Reduced Flavins and Flavoproteins

Apart from the charge-transfer absorptions involving reduced flavin already considered, the optical absorptions of reduced flavins and flavoproteins do not appear to be particularly informative. The spectrum consists of a series of weak overlapping bands which are probably due to the several $n–\pi^*$ transitions to be anticipated in the isoalloxazine nucleus. Characteristically, the absorption of flavin in the reduced form at 450 mμ is decreased roughly 10 times from that of the oxidized flavin. A useful test in flavoproteins is to determine whether substrate is able to reduce the absorption at 450 mμ to the same extent as can dithionite; if this is not achieved, then complexities should be expected. In this way a new chromophoric group (as yet unidentified) has been discovered in glucose oxidase (55).

VI. Fluorescence of Flavoproteins

The intense green fluorescence of free flavins is well known; the emission maximum is at 530 mμ and the quantum yield varies from 0.25 (riboflavin, lumiflavin, FMN) to about 0.025 (FAD) in the neutral range. As the pH is decreased the fluorescence of FAD increases and at pH 3 approaches that of FMN. This property is very useful for the identification of FAD, viz. the increase in fluorescence on lowering the pH and the increase in fluorescence at pH 7 on incubation with nucleotide pyrophosphatase (converting FAD to FMN). At extremes of pH, the fluorescence of all flavins disappears; in the alkaline range this is due to the ionization of the 3-imino group (pK 10.4) but in the acid range it is due to quenching of fluorescence by H$^+$. Whereas the quantum yield of the riboflavin and FMN shows the usual negative dependence on temperature—a consequence of thermal deactivation processes—that of FAD is essentially independent of temperature (56). The reason for this anomalous behavior resides in the lack of fluorescence when FAD is in the folded or hairpin configuration (57). As this configuration is probably stabilized by hydrogen bonds the absence of any effect of temperature on the fluorescence yield is due to a fortuitous compensation by two opposite effects as the temperature is increased: the increase in fluorescence accompanying the conversion of the hairpin to the open configuration and the decrease in fluorescence attendant to the thermal quenching process. In support of this, Velick has shown (56) that the fluorescence behavior of FAD in the solvent methyl carbitol is identical with that of FMN in water. In this solvent FAD is expected to be in the open configuration.

Figure 5 shows the effect of solvent on this fluorescence, using 3-methyl lumiflavin. Two aspects should be noted: the increase in fluorescence yield as the polarity of the medium is decreased, and the development of a new emission band around 505 mμ in the solvents such as benzene. This development of fine structure thus mirrors the development of vibrational detail in the last absorption band of the optical spectrum, which has been discussed in a previous section. In this respect, it is interesting that lipoyl dehydrogenase, which shows vibrational structure in its last absorption band, also shows vibrational structure in its fluorescence emission spectrum (58). Unfortunately, very few flavoproteins exhibit the flavin fluorescence in reasonable yield. Those best known are lipoyl dehydrogenase (58), electron-transfer flavoprotein (59), D-amino acid oxidase (60,61), and thioredoxin reductase (25). [In passing, it should be noted that many other flavoproteins exhibit traces of green fluorescence; for instance, pig liver aldehyde oxidase emits maximally at 530 mμ with a

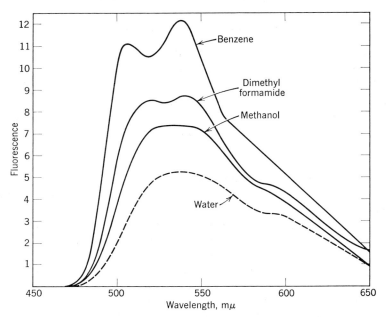

Fig. 5. The effect of solvent on the fluorescence emission spectra of 3-methyl lumiflavin. The spectra have been corrected for phototube response at different wavelengths, and are all for the same concentration of flavin. (Data of Greull and Massey, unpublished.)

vanishingly small quantum yield although polarization of fluorescence measurements indicate that the flavin is bound to a large molecule (62). Whether these minute fluorescences are due to the natural coenzyme or to traces of flavin contaminants is an open question.]

In the case of flavoproteins in which the fluorescence is quenched, it has been widely assumed that this implies linkage of the protein to the flavin through the 3-imino group of the isoalloxazine nucleus. The rationale for this assumption appears to be based on an erroneous translation of a paper by Kuhn and Rudy (63) which in fact correctly reports the fluorescence of 3-substituted flavins, and on the observation that ionization of the 3-imino group leads to loss of fluorescence. As substantiated by Harbury and Foley (64), 3-substituted flavins are fluorescent, and recent comparative studies in the authors' laboratories (65) have shown that the fluorescence yield is not affected by substitution at the 3-position (cf. Fig. 5 for the fluorescence of 3-methyl lumiflavin). To our knowledge no convincing evidence for interaction of the protein and the 3-imino group of the isoalloxazine nucleus has yet been presented for any flavoprotein. One may inquire into the reason for this lack of fluorescence in the enzymes and there would appear to be at least three obvious phenomena which would account for these effects. First, there is the formation of complexes between the isoalloxazine ring and some aromatic moiety— either the adenine ring or some group in the protein, e.g., tryptophan, for it is a well-known fact that such complexes are nonfluorescent. Second, we have the possibility in the metalloflavoproteins of fluorescence quenching by paramagnetic ions, the presumably close juxtaposition of metal and flavin facilitating this process. The third possibility of the quenching of fluorescence on combination with protein has been advanced by Weber (57) and involves interaction with tyrosine residues. This is based on the observation of fluorescence quenching of free flavins by phenols. With DPNH cytochrome b_5 reductase, Strittmatter has provided convincing evidence that a protein tyrosine residue is involved in linkage with the flavin (66).

In those enzymes which exhibit fluorescence these effects must be absent. In particular it follows that the flavin must be in the open, and not the closed, configuration. Of the fluorescent enzymes, lipoyl dehydrogenase is probably the most thoroughly studied (58). The excitation spectrum follows the absorption spectrum very closely in the 300–500 mμ range, but at lower wavelengths there is considerable emission accompanying the protein peak, indicating that intramolecular energy transfer between the aromatic amino acids and the isoalloxazine ring has occurred. Relative to its coenzyme, the fluorescence emission of the enzyme differs in that it exhibits both increased intensity (about five times as great at 5°) and the emission maximum is at 520 mμ. Furthermore, as mentioned above, there is a pronounced shoulder at 540 mμ. The fluorescence intensity exhibits a strong negative temperature dependency much more than even that of FMN. The decrease of this fluorescence with temperature gives a linear Arrhenius plot with activation energy 4.8 kcal/mole (58). This value, which is the classic one for the strength of a hydrogen bond, suggests that the open configuration is stabilized by the H-bonding of the adenine moiety to the protein rather than to the isoalloxazine and that the breaking of this bond by thermal energy allows the formation of the hairpin configuration with a consequent quenching of the fluorescence.

The available data with the other two enzymes are quite meager. The first reports on the fluorescence properties of D-amino oxidase indicated that it was equally as fluorescent as the free flavin, for on addition of the apoprotein to FAD no change was observed in the intensity of the fluorescence emission (60). However, more recent measurements (61) indicate that the holoenzyme has about one-fifth the fluorescence of the coenzyme. Interestingly, the fluorescence of the holoenzyme is decreased about five times by benzoate; if the benzoate were forming a complex with the flavin, complete quenching of the fluorescence would have been expected. Thus, unless the holoenzyme contains some free flavin, one must assume that the binding of the benzoate is to the protein. This is directly relevant to the influence of benzoate on the optical spectrum of the enzyme (see above).

No data are available on the quantitative aspects of the fluorescence of ETF. However, it has been observed that butyryl CoA dehydrogenase does not quench the fluorescence. Thus, any complex formed between these two proteins is unlikely to be at the flavin. Polarization of fluorescence measurements on this system would be extremely valuable in that they should throw light on whether such a complex does exist. As of this time, there has been little if any kinetic work utilizing the fluorescence of these enzymes. For instance, one would hope to study the quenching of fluorescence on addition of substrate and observe whether it would disappear before or with the appearance of a long-wavelength band. The former case would of course be evidence for the formation of an enzyme–substrate complex prior to the electron-transfer event. As fluorescence measurements are much more sensitive than absorbancy measurements, extremely rapid early events involving reaction with the substrate could be slowed down to a range accessible to the stopped-flow spectrophotometer simply by diluting the reaction mixture.

VII. Magnetic Resonance Studies on Metal-Free Flavoproteins

The great advances made in the understanding of the electron paramagnetic resonance of free flavosemiquinones is reviewed earlier in this volume. Unfortunately, the situation in the enzymes is much more difficult for two reasons. First, the relatively slow molecular movement of proteins in solution (or the absence of appreciable movement in the frozen states) severely restricts the time averaging of any local inhomogeneities in the magnetic fields due to the molecules themselves, with the consequence that the hyperfine structure, so valuable in determining the electronic geography of these radicals, is broadened to such an extent that the individual lines fuse and one is left with a plain featureless resonance. Thus, any investigation of the kind so elegantly performed by Ehrenberg and his colleagues (cf. ref. 35) is not feasible.

The second problem is that the presence of substrate-induced free radicals in metal-free flavoproteins is the exception rather than the rule, although many of these enzymes can be persuaded to exhibit quite substantial yields of radical with appropriate manipulation (67). As we have discussed above, these observations were extremely disappointing in view of the fact that the long-wavelength absorption traditionally ascribed to flavosemiquinones was usually present. However, the recognition that complexes of the enzyme flavin with various components of the reaction mixture might also exhibit optical absorption in the green and red regions of the spectrum has somewhat clarified the picture.

One other rationale which has been advanced to explain the lack of EPR signal is the interaction of two adjacent radicals (19,48). This can be of two forms: a *magnetic* dipole–dipole interaction which couples the two electrons to produce an $S = 1$ (triplet) state. The separation of the energy levels of this state can be a function of the orientation of the molecule to the applied magnetic field which the consequence that the resonance is smeared out over a wide range of applied field. This effect is compounded by the shortening of the relaxation time also produced by this magnetic interaction (68).

The alternative is an *electrostatic* interaction between the two unpaired spins which is called the exchange interaction (68). Usually this interaction leads to a pairing of the two electrons with a reduction in the bulk magnetism of the system. However, a thermally accessible triplet state exists and hence at conventional temperatures it should be possible to detect this.

The distinction between the two possible mechanisms can be made: although both might be expected to yield $\Delta_m = 2$ triplet resonances at half-field, the former case should exhibit full magnetism when examined by magnetic susceptibility methods while the latter should show a substantial reduction in magnetism (due to the diamagnetic state), and this may show a marked temperature dependence.

Studies on the effect of power saturation on the line shape of flavoprotein radical spectra have produced rather puzzling results (36,36a). It has been found that increasing the microwave power incident on the sample produces very unusual changes in line shape which are most easily interpreted by assuming that the observed line shape is composed of two species—a narrow line of peak-to-peak width 15–19 gauss which saturates easily and a broad line of width 70–80 gauss which saturates with difficulty, if at all. The width of the narrow line can be correlated with the optical spectrum of the species, being 15 gauss for the anionic (type B) radical and 19 gauss for the neutral (type A) species. (It will be interesting to see if these linewidth variations are also obtained with model flavins.)

The above phenomenon is independent of modulation frequency (400 cps and 100 kc/sec), but measurements at the K band (35 Gc/sec) suggest the presence of at least three species (36a).

The effect appears to be independent of temperature over the range 1–54°C, there being no perceptible changes in line shape. However, at low temperatures, the phenomenon is virtually abolished, although small line-shape anomalies persist in the wings of the spectrum even at very low powers (e.g., 10 μW) at, say, −70°C. Experiments in collaboration with H. Beinert and T. Vanngard have failed to yield any evidence for triplet resonances at half field. At this time it is not obvious what causes these effects. It may be that this is some obscure artifact arising when the spin-lattice and rotational relaxation time are comparable; more interesting is the possibility that we are actually observing different chemical species, although what these species could be is difficult to decide.

VIII. Reactivity of Reduced Flavins and Flavoproteins with O_2

Free reduced flavins are reoxidized very readily by molecular oxygen. One of the most challenging aspects of flavoprotein research is an understanding of why this reactivity is retained with some flavoproteins (the oxidases) but is very greatly diminished with the majority of flavoproteins. In a detailed study of the oxidation of $FMNH_2$ by oxygen, Gibson and Hastings observed that the reaction was autocatalytic (69). This, they pointed out, was most readily interpreted on the basis of the reaction being stimulated by the product FMN, presumably by formation of an FMN–$FMNH_2$ complex. This complex, by virtue of its greater planarity compared to that of free $FMNH_2$, may be expected to react more readily with oxygen than the uncomplexed leucoflavin. However, it was found that a satisfactory analysis of the results would only be made when both pathways were considered (69). The hypothesis of greater reactivity of fully reduced flavin than of semiquinoid flavin with O_2 would appear to be consistent with the observation that glucose oxidase functions exclusively between the fully oxidized and fully reduced levels in catalysis (70); this enzyme has a greater rate of reaction with O_2 than any other flavoprotein so far studied. However, considerable reactivity with O_2 is displayed by D- and L-amino acid oxidases; both of these appear to function through semiquinone intermediates, and in the latter case experimental evidence strongly suggests that the fully reduced enzyme is less reactive with O_2 than the semiquinoid enzyme (71). As already pointed out, the absolute identity of the intermediates of these enzymes has not yet been established; until this is accomplished, any simple explanation of oxygen reactivity must remain in abeyance.

IX. Steady-State Kinetics of Flavoprotein Systems

With only a few exceptions the kinetic behavior of flavoprotein enzymes can be typified by reference to lipoyl dehydrogenase (19). When reciprocal initial velocity is plotted against reciprocal acceptor (B) concentration at some fixed concentration of substrate (AH_2), a straight line is obtained. Repeating the experiment with several values of AH_2 yields a family of parallel lines. If the intercepts on the ordinate ($1/V$ axis) are plotted versus the reciprocal of their respective concentration of AH_2, the secondary plot is also linear, the intercept on the ordinate being $1/V_{max}$ at infinite concentration of both substrates and the intercept on the abscissa being the negative reciprocal of the Michaelis constant for AH_2. Utilizing the same data, one can make the primary plot a function of AH_2 and the secondary plot a function of B, thus obtaining the Michaelis constant for B. Clearly, one can obtain V_{max} for the back reaction and the Michaelis constant for A and BH_2 in a like manner.

The most simple mechanism that yields this behavior has been discussed theoretically by Alberty (72) and Dalziel (73). Formally, it is

$$E + AH_2 \underset{k_2}{\overset{k_1}{\rightleftarrows}} E \cdot AH_2 \underset{k_4}{\overset{k_3}{\rightleftarrows}} EH_2 + A$$

$$EH_2 + B \underset{k_6}{\overset{k_5}{\rightleftarrows}} EH_2 \cdot B \underset{k_8}{\overset{k_7}{\rightleftarrows}} E + BH_2$$

where E and EH_2 represent oxidized and reduced forms of the enzyme (EH_2 does not necessarily imply full reduction of the flavin—it is merely a convenient shorthand expression). The definition of the various kinetic constants in terms of rate constants is shown in Table III. As the mechanism is symmetrical, the corresponding parameters for the reverse reaction can be obtained by substituting the appropriate rate constants in the above equations by inspection. The kinetic equation for the forward reaction is:

$$v = \frac{V_{max}}{1 + K_{AH_2}/[AH_2] + K_B/[B]}$$

The absence of a term in both substrates is the reason for the parallel line plots. While, at first sight, this kinetic behavior of yielding parallel

TABLE III

Kinetic Parameters for Flavoprotein Mechanisms

Kinetic parameter	Mechanism		
	Simple flavoprotein	Theorell Chance	D-Amino acid oxidase
V_{max}	$\dfrac{k_3 k_7}{k_3 + k_7}$	k_5	$\dfrac{k_3 k_7}{k_3 + k_7}$
K_{AH_2}	$\dfrac{k_7(k_2 + k_3)}{k_1(k_3 + k_7)}$	$\dfrac{k_5}{k_1}$	$\dfrac{k_7(k_2 + k_3)}{k_1(k_3 + k_7)}$
K_B	$\dfrac{k_3(k_6 + k_7)}{k_5(k_3 + k_7)}$	$\dfrac{k_5}{k_3}$	$\dfrac{k_7(k_3 + k_4)}{k_5(k_3 + k_7)}$
K_0	—	$\dfrac{k_2 k_5}{k_1 k_3}$	$\dfrac{k_2 k_4 k_7}{k_1 k_5(k_3 + k_7)}$

Lineweaver-Burk plots will give an unambiguous differentiation from the mechanisms found with pyridine nucleotide-linked enzymes such as alcohol dehydrogenase, Dalziel (74) pointed out that under certain restricted conditions the latter systems might also exhibit parallel plots. Consider the classic mechanism for alcohol dehydrogenase (75):

$$E + AH_2 \underset{k_2}{\overset{k_1}{\rightleftharpoons}} E \cdot AH_2$$

$$E \cdot AH_2 + B \underset{k_4}{\overset{k_3}{\rightleftharpoons}} EA + BH_2$$

$$EA \underset{k_6}{\overset{k_5}{\rightleftharpoons}} E + A$$

This yields a kinetic equation of the form

$$v = \frac{V_{max}}{1 + K_{AH_2}/[AH_2] + K_B/[B] + K_0/[AH_2][B]}$$

with the definition of the kinetic constants shown in Table III. Inasmuch as this equation contains a term in both substrates, one will find that the Lineweaver-Burk plots delineated above will be convergent. However, by reference to Table III, it can be seen that if k_2 is negligible when compared to $k_1 k_3$, then the term in $[AH_2][B]$ vanishes and parallel-line kinetics will prevail.

Similar results are obtained with more complex systems. For example, with D-amino acid oxidase, the following reaction mechanism has been suggested (48):

$$E + AH_2 \underset{k_2}{\overset{k_1}{\rightleftharpoons}} EAH_2 \underset{k_4}{\overset{k_3}{\rightleftharpoons}} EH^{\cdot} \cdot AH^{\cdot}$$

$$EH^{\cdot} \cdot AH^{\cdot} + O_2 \underset{k_6}{\overset{k_5}{\rightleftharpoons}} EA + H_2O_2$$

$$EA \underset{k_8}{\overset{k_7}{\rightleftharpoons}} E + A$$

This gives rise to a rate equation of the same form as alcohol dehydrogenase with the somewhat different definitions of the kinetic constants (Table III). The condition that makes the last term in the denominator vanish is that $k_2 k_4 \ll k_1 k_3 k_5$. From the data of Massey and Gibson (48) and using the parameters of Table III, one can show in fact that $k_2 k_4 / k_1 k_3 k_5 \approx 10^{-7}$, that is, the coefficient of the last term is negligibly small and the kinetic equation thus reduces to one yielding parallel lines. This restriction thus serves to remove the recent objection of Dixon and Kleppe (76), who argued against the above mechanism on the grounds that parallel Lineweaver-Burk plots are obtained. Their second criticism, viz., that the ratio V_{max}/K_{O_2} $[= k_3 k_5 / (k_3 + k_4)]$ is independent of the amino acid substrate can be answered by assuming $k_4 \ll k_3$. The expression then reduces to k_5, which is the rate of reaction of the spectroscopically detectable intermediate with oxygen. As k_3 is 50,000 min^{-1} for alanine, this requires $k_4 < 5000$ min^{-1}, which is not unreasonable. The restriction that $k_4 < k_3$ is clearly compatible with the requirement that $K_{AH_2 \cdot O_2}$ be very small for k_2/k_1 is known to be 0.1. From the parameters of Table III and the data of Massey and Gibson, we can calculate for alanine $k_1 = 1.3 \times 10^5$ M^{-1} min^{-1}; $k_2 = 1.3 \times 10^4$ min^{-1}; $k_3 = 50,000$ min^{-1}; $k_4 = 5000$ min^{-1}; $k_5 = 3.3 \times 10^6$ M^{-1} min^{-1}; $k_7 = 500$ min^{-1}.

X. Flavoprotein Reaction Mechanisms

A. General Considerations

One of the outstanding unsolved problems in the field of biological oxidation is the mechanism by which the two-electron transfers occurring in the early stages of the respiratory chain are linked to the one-electron transfer occurring in the cytochrome region of the respiratory chain.

Since the recognition that flavins and particularly flavoproteins may exist as stable semiquinones, the possibility has been obvious that this crossover may occur at the flavoprotein level. Although the flavoproteins intimately concerned in the respiratory chain (succinic dehydrogenase and DPNH dehydrogenase) are forbiddingly complicated enzymes (probably involving metals and labile sulfide in their reaction mechanisms), a consideration of the reaction pathways followed by the more simple flavoproteins should add considerably to an understanding of what may occur in the more complex enzymes. A survey of present knowledge of simple flavoproteins reveals that no general pathway of catalysis is followed. Some enzymes appear to function without the detectable intervention of semiquinones; others appear to function between the oxidized and semiquinone levels; others between the fully reduced and semiquinoid levels; and others utilizing all redox levels of the bound flavin. Before considering individual examples, we wish to consider the problem of how many flavin molecules constitute an active center. An inspection of Table I reveals that most purified flavoproteins contain two molecules of flavin per molecule of protein. The only exceptions to this are the enzymes DPNH-cytochrome b_5 reductase; vitamin K reductase; the electron-transferring flavoprotein, ferredoxin-TPN reductase; and oxynitrilase, which all contain one molecule of flavin per molecule of protein. In these cases, the active center must therefore consist of whatever groups the protein contributes and one molecule of flavin. In the other enzymes serious consideration has to be given to whether the two flavin molecules interact with each other in the catalytic process. In the case of L-amino acid oxidase, this would appear plausible. With D-amino acid oxidase the evidence is overwhelmingly against such interaction, as it is also with lipoyl dehydrogenase and glutathione reductase. In general, we would consider it more likely that the possession of two flavin groups per molecule of protein has no more significance than the multiple binding sites for coenzyme found with several pyridine nucleotide-linked dehydrogenases, i.e., that the bulk of these enzymes have two independent active centers. However, the possibility of cooperative interaction is a real one and searching study, not yet performed with any enzyme, is required for the evaluation of this possibility.

B. Enzymes Functioning Solely between the Fully Oxidized and Fully Reduced Forms

Formally, a mechanism of this sort may be written as in Scheme I. The only well-authenticated example of a mechanism of this kind is

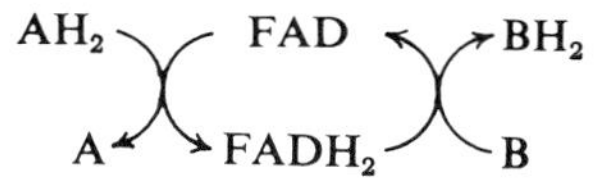

$$AH_2 \diagdown \diagup FAD \diagdown \diagup BH_2$$
$$A \diagup \diagdown FADH_2 \diagup \diagdown B$$

Scheme I

provided by glucose oxidase, which has been studied by Gibson et al. (70) and by Nakamura and Ogura (77). Using rapid reaction spectrophotometry, both groups of workers were unable to detect any spectral intermediates between fully oxidized and fully reduced flavin. This was the case whether the anaerobic reduction of the enzyme was followed by rapidly reacting substrates such as glucose and 2-deoxyglucose or by slowly reacting substrates such as xylose. Furthermore, in turnover experiments involving both sugar and oxygen, no intermediate forms of the enzyme were detectable. In addition, a true semiquinone of the enzyme was prepared by partial titration with $Na_2S_2O_4$ or by catalytic reduction with Pt and H_2. Enzyme prepared in this way was quite unreactive to reduction by substrate (70). This observation eliminates the possibility of an intermediate of the type $\left|\begin{smallmatrix}-FADH^{\cdot}\\-FADH^{\cdot}\end{smallmatrix}\right.$ being formed even transiently in catalysis.

While rapid reaction studies are obviously necessary, and have not yet been performed, it seems likely that a mechanism of this sort may be followed also by glycollic acid oxidase and L-α-hydroxy acid oxidase, since no spectral intermediates have yet been detected with these enzymes.

C. Enzymes Involving True Semiquinoid Intermediates

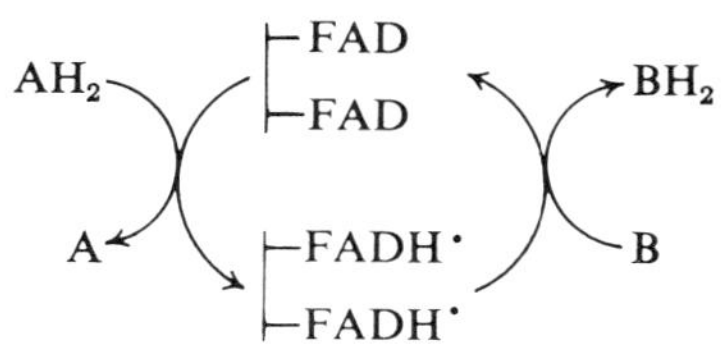

$$AH_2 \diagdown \diagup \begin{matrix}-FAD\\-FAD\end{matrix} \diagdown \diagup BH_2$$
$$A \diagup \diagdown \begin{matrix}-FADH^{\cdot}\\-FADH^{\cdot}\end{matrix} \diagup \diagdown B$$

Scheme II

A mechanism of the sort shown in Scheme II has been suggested for snake venom L-amino acid oxidase by Wellner and Meister (71). This was based on the early observation of Beinert (78) that a long-wavelength intermediate was formed transiently on the addition of L-leucine, which could be due to a semiquinone of the type shown above. However, no EPR studies have been conducted with this enzyme, so as with practically every flavoprotein, the exact nature of the intermediate remains in some doubt.

This enzyme is very susceptible to inhibition by excess substrate (79); Wellner and Meister have suggested that this is due to the conversion of the intermediate to the less reactive fully reduced enzyme at high substrate concentrations (Scheme III). This interpretation is elegantly supported

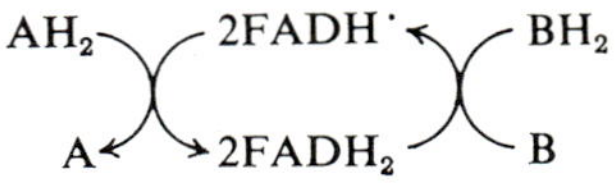

$$AH_2 \diagdown \diagup 2FADH^{\cdot} \diagdown \diagup BH_2$$
$$A \diagup \diagdown 2FADH_2 \diagdown \diagdown B$$

Scheme III

by the kinetic analysis of Wellner and Meister (71) and by rapid reaction studies of Gibson (80). While the kinetic representation of mechanism appears beyond doubt, we consider that more experiments are required to define the semiquinoid nature of the intermediate. Veeger (39) observed on titration with dithionite a spectrum of type B (Fig. 2) very similar to that found with D-amino acid oxidase. If the mechanism suggested in Scheme III is correct, this form of the enzyme should be reduced rapidly by substrate, and so clearly differentiate the mechanism from the type found with D-amino acid oxidase.

Recent work (85,86) has in fact demonstrated that the mechanism proposed by Wellner and Meister is incorrect and that this enzyme probably has a reaction mechanism basically similar to that of D-amino acid oxidase. The true semiquinoid form of the enzyme was produced in quantitative yield by illumination in the presence of EDTA; this species was completely resistant to further reduction by amino acid, thus disproving the Wellner–Meister mechanism. In further analogy to D-amino acid oxidase, the spectrum of the transient intermediate formed by substrate was shown by rapid reaction spectrophotometry to be completely different from that of the free-semiquinoid form, and furthermore to differ with different L-amino acid substrates (86).

D. Enzymes Involving Fully Reduced Flavin and Intermediates

Another example of an enzyme involving a true semiquinoid intermediate would appear to be given by TPNH cytochrome c reductase (18). In this case, however, it would appear that the flavin functions in catalysis between the fully reduced and semiquinoid levels (Scheme IV). This enzyme has been studied by rapid reaction techniques which show that the predominant reaction in catalysis is the reduction of cytochrome c (and

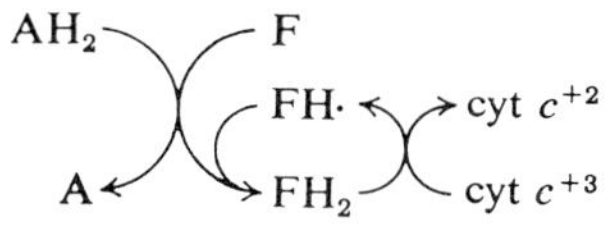

$$AH_2 \diagdown \diagup F$$
$$\diagdown \diagup FH\cdot \leftharpoonup \rightharpoonup cyt\ c^{+2}$$
$$A \diagup \diagdown \rightarrow FH_2 \diagdown \diagup cyt\ c^{+3}$$

Scheme IV

ferricyanide) at the expense of conversion of fully reduced enzyme to a form with long-wavelength absorption. This form is very resistant to further oxidation with O_2, cytochrome c, or ferricyanide but is oxidized by menadione (18). Such resistance to oxidation by a simple semiquinone would appear rather unlikely; it suggests that although the flavin may be at the semiquinoid oxidation level, it is covalently bound either to a product of the reaction or some group in the protein. In this connection, as carefully noted by Masters et al. (18), the amount of radical detectable by EPR spectroscopy is only about 7% of the total flavin content. Clearly, much further work remains to be done to delineate the real nature of the intermediate, particularly since cytochrome c was found to abolish the EPR-detectable radical although it does not lead to disappearance of the long-wavelength absorption.

The related microsomal enzyme, DPNH cytochrome b_5 reductase, would appear to function by a similar mechanism (41). This enzyme is being considered separately in detail elsewhere in this volume, so we shall mention only the salient points here. In a series of elegant papers, Strittmatter has demonstrated that reduction of the enzyme by DPNH yields a fully reduced form of the enzyme; the DPNH reaction involves binding with a protein sulfhydryl group. During reoxidation with ferricyanide a much stronger absorption band is found, with an absorption spectrum similar to that of type A (Fig. 2). No EPR studies have yet been reported and it is not yet known whether a similar long-wavelength band of the enzyme is produced on reaction with the natural acceptor cytochrome b_5. Flavin reduction proceeds at the same rate as the overall reaction and the rate-limiting step is tentatively identified as the cleavage of the carbon–hydrogen bond at position 4A in the nicotinamide bond preceding reduction of the flavin (41).

E. Enzymes Involving Substrate-Complexed Flavin Semiquinone

An example of this type of mechanism is that proposed for D-amino acid oxidase (48) (Scheme V). The reaction does not involve a simple semiquinoid form of the enzyme, since this form (produced by dithionite or

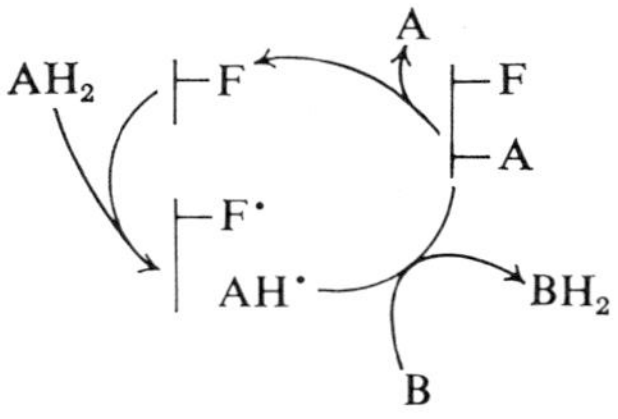

Scheme V

catalytic reduction with H_2 and Pt) reacts only extremely slowly with amino acids. Similarly, as discussed previously, the spectra of intermediates produced on reduction of the enzyme by substrate (Fig. 3) are very different from that of the simple semiquinoid enzyme (type B spectrum, Fig. 2). Although parallel Lineweaver-Burk kinetics are found, the experimental results cannot be explained satisfactorily in terms of two separate half-reactions involving the production of a substrate or product-free reduced enzyme and reaction of this form with O_2 in the second half-reaction. A detailed kinetic analysis of the enzyme reaction has been made with three different substrates: D-alanine, D-methionine, and D-proline. These studies suggest the mechanism shown in Scheme VI, where AA

$$\left|\begin{matrix}\text{—FAD}\\ \\ \end{matrix}\right. + \text{AA} \underset{k_2}{\overset{k_1}{\rightleftharpoons}} \left|\begin{matrix}\text{—FAD}\\ \text{—AA}\end{matrix}\right. \underset{k_4}{\overset{k_3}{\rightleftharpoons}} \left|\begin{matrix}\text{—FADH}^{\cdot}\\ \text{—AA}^{\cdot}\end{matrix}\right.$$

$$\left|\begin{matrix}\text{—FADH}^{\cdot}\\ \text{—AA}^{\cdot}\end{matrix}\right. + O_2 \underset{k_6}{\overset{k_5}{\rightleftharpoons}} \left|\begin{matrix}\text{—FAD}\\ \text{—IA}\end{matrix}\right. + H_2O_2$$

$$\left|\begin{matrix}\text{—FAD}\\ \text{—IA}\end{matrix}\right. \underset{k_8}{\overset{k_7}{\rightleftharpoons}} \left|\begin{matrix}\text{—FAD}\\ \\ \end{matrix}\right. + \text{IA}$$

Scheme VI

represents amino acid, $AA^{\cdot}$ a radical form of the amino acid, and IA, imino acid. As already discussed in the section on kinetics, k_4 is probably negligible (k_6 also would be expected to be very small). An analog computer solution of this mechanism and the introduction of known rate constants gives predicted time courses of production and utilization of the form $\left|\begin{matrix}\text{—FADH}^{\cdot}\\ \text{—AA}^{\cdot}\end{matrix}\right.$ under various conditions which are in remarkable agreement with experimental results. Thus we can be confident that a reaction scheme formally similar to that given above and involving a colored intermediate (which in turn must have some contribution from the substrate) must be operating with this enzyme. As already discussed in previous sections, there is still some doubt as to whether the intermediate involved is as shown in Scheme VI or whether it is a charge transfer complex between enzyme, $FADH_2$, and imino acid. However, the bulk of the evidence is in favor of the former. As discussed previously, the intermediate is probably a real chemical compound produced by covalent bond formation between the flavin semiquinone and amino acid radical.

Although no kinetic work has been done which permits a real conclusion, we would like to suggest that similar mechanisms apply in the case of the acyl CoA dehydrogenases. Like D-amino acid oxidase, partial reduction of these enzymes gives an EPR-detectable radical, whereas substrates do not. A point of further similarity is that the spectra given by different substrates differ significantly (81). Furthermore, very tightly bound enzyme substrate complexes have been found (13).

F. Enzymes Functioning at the Semiquinoid Level, but also Involving Another Redox Group in the Enzyme

The classical example of this type of mechanism is lipoyl dehydrogenase (19) (Scheme VII). Recent evidence has shown that glutathione reductase

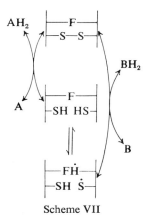

Scheme VII

function by an essentially identical mechanism (23). The spectral characteristics of the lipoyl dehydrogenase intermediate have been discussed in previous sections. Direct evidence for the participation in catalysis of a disulfide–dithiol redox couple has been obtained by amperometric titration of sulfhydryl groups; two extra sulfhydryl groups can be titrated after reduction with DPNH (82). Overwhelming indirect evidence from spectral studies and the effects of arsenite and mercurial (reviewed in ref. 83) also indicate that this redox active disulfide–dithiol couple is involved together with the flavin in the long-wavelength intermediate produced by substrate. This absorbing species is vitally concerned in catalysis. In the first detailed rapid spectrophotometric study of a flavoprotein (19), it was shown that the rate-limiting step is the reoxidation of the intermediate by oxidized lipoyl compounds. The fully reduced form of the enzyme is catalytically inert in reaction with lipoic acid and lipoyl derivatives (19). As discussed in detail in previous sections, the unequivocal identification of this intermediate in chemical terms yet remains to be made. However, the bulk of evidence points to it being a compound of the flavin and sulfur formed by covalent bonding between flavin radical and sulfur radical. The recent finding of an essentially similar reaction mechanism with glutathione reductase points to the possibility that the more general importance of redox active disulfide groups in flavoproteins. In view of the similarities in reactions catalyzed, we would like to suggest that thioredoxin reductase and ferredoxin TPN reductase may function by similar mechanisms.

XI. General Conclusions Regarding Mechanisms

In the examples listed above it is evident that different flavoproteins deal in several different ways with the overall two-electron transfer reactions which they catalyze. Some proteins such as glucose oxidase do this simply by oscillating between the fully oxidized and fully reduced levels of the flavin during catalysis. In other flavoproteins, such as L-amino acid oxidase, the enzyme appears to function most effectively with the flavin at the semiquinone level. In this case, the problem of catalyzing an overall 2-electron oxidoreduction appears to be solved by using a pair of flavin molecules in one active center. [*Note added in proof:* While recent work (86) has shown that L-amino acid oxidase does not function via this mechanism, it is feasible that enzymes may be found which use this route. However, to date no examples are available.] In enzymes such as DPNH cytochrome b_5 reductase and TPNH cytochrome c reductase, it would appear that the flavin is initially reduced to the $FADH_2$ level by substrate and then is reoxidized by the cytochromes in single electron oxidation steps. In the case of D-amino acid oxidase, the flavin appears again to function at the semiquinoid level, in this case at the expense of the production of substrate radical which then interacts with the flavin radical. Reoxidation in this instance by molecular O_2 is accomplished by the oxygen accepting electrons both from the flavin radical and from the amino acid radical. This is an interesting mechanistic example, because it is the reverse of the crossover phenomenon that must exist in the electron transfer chain. In

D-amino acid oxidase the initial reduction of the enzyme is a one-electron reduction followed by a two-electron reoxidation of both the enzyme and substrate radicals. Finally, with lipoyl dehydrogenase the flavin also appears to function at the semiquinoid level, the other electron from the substrate in this case being accepted by a second prosthetic group in the enzyme—a redox reactive disulfide. The example provided by lipoyl dehydrogenase probably provides the closest analogy with the respiratory chain flavoproteins succinic dehydrogenase and DPNH dehydrogenase. (These enzymes are being considered in detail elsewhere in this volume.) It would seem to us very likely that the substrate reduction of these enzymes will probably resemble that of lipoyl dehydrogenase, one electron being accepted by the flavin and the other by an iron sulfide linkage ("labile" sulfide). In the case of lipoyl dehydrogenase, the flavin radical and sulfur radical interact to give a species more suitable to two-electron transfer than to one-electron transfer. However, with the metalloflavoproteins, sufficient stabilization of the sulfur radical by interaction with the adjacent metal is probably obtained, so that no bond formation between flavin radical and sulfur radical occurs. This situation could then be ideal for single one-electron transfers to the next acceptor in the respiratory chain, and may well represent the way in which crossover from two-electron to one-electron transfer is achieved.

Abbreviations

FMN and FMNH$_2$	Oxidized and reduced forms of flavin mononucleotide
FAD and FADH$_2$	Oxidized and reduced forms of flavin adenine dinucleotide
DPN$^+$ and DPNH	Oxidized and reduced forms of diphosphopyridine nucleotide
TPN$^+$ and TPNH	Oxidized and reduced forms of triphosphopyridine nucleotide
A and AH$_2$	Oxidized and reduced forms of substrate
B and BH$_2$	Oxidized and reduced forms of acceptor
EPR	Electron paramagnetic resonance
ETF	Electron transfer flavoprotein
DCIP	2,6-Dichlorophenol-indophenol

References

1. M. Kasha, in *Light and Life*, W. D. McElroy and B. Glass, Eds., The Johns Hopkins Press, Baltimore, 1961, p. 31.
2. G. Weber, V. Massey, and G. Greull (unpublished).
3. H. A. Harbury, K. F. Lanoue, P. A. Loach, and R. M. Amick, *Proc. Natl. Acad. Sci. U.S.*, **45**, 1708 (1959).
4. N. S. Bayliss and E. G. McCrae, *J. Phys. Chem.*, **58**, 1002 (1954).
5. E. Haas (unpublished): data presented in ref. 38.

6. V. Massey, H. Ganther, P. E. Brumby, and B. Curti, in *Oxidases and Related Redox Systems*, T. E. King, H. S. Mason, and M. Morrison, Eds., Wiley, New York, 1965, p. 335.

7. D. Wellner and A. Meister, *J. Biol. Chem.*, **235**, pc 12 (1960); and *J. Biol. Chem.*, **236**, 2357 (1961).

8. B. E. P. Swoboda and V. Massey, *J. Biol. Chem.*, **240**, 2209 (1965).

9. K. Kusai, *Ann. Rept. Sci. Works; Fac. Sci. Osaka Univ.*, **8**, 341 (1960).

10. N. A. Frigerio and H. A. Harbury, *J. Biol. Chem.*, **231**, 135 (1958).

11. J. C. Robinson, L. Keay, R. Molinari, and I. W. Sizer, *J. Biol. Chem.*, **237**, 2001 (1962).

12. F. R. Williams and L. P. Hager, *J. Biol. Chem.*, **236**, pc 36 (1961).

13. E. P. Steyne-Parvé and H. Beinert, *J. Biol. Chem.*, **233**, 843, 853 (1958).

14. J. G. Hauge, F. L. Crane, and H. Beinert, *J. Biol. Chem.*, **219**, 727 (1956).

15. F. L. Crane, S. Mii, J. G. Hauge, D. E. Green, and H. Beinert, *J. Biol. Chem.*, **218**, 701 (1956).

16. F. L. Crane and H. Beinert, *J. Biol. Chem.* **218**, 717 (1956).

17. P. Strittmatter and S. F. Velick, *J. Boil. Chem.*, **228**, 785 (1957).

18. B. S. S. Masters, H. Kamin, Q. H. Gibson, and C. H. Williams, Jr., *J. Biol. Chem.*, **240**, 921 (1965).

19. V. Massey, Q. H. Gibson, and C. Veeger, *Biochem. J.*, **77**, 341 (1960).

20. C. J. Lusty, *J. Biol. Chem.*, **238**, 3443 (1963).

21. A. Wren and V. Massey, *Biochem. J.*, **89**, 471 (1963).

22. C. H. Williams, Jr., *J. Biol. Chem.*, **240**, 4793 (1965).

23. V. Massey and C. H. Williams, Jr., *J. Biol. Chem.*, **240**, 4470 (1965).

24. F. Märki and C. Martius, *Biochem. Z.*, **333**, 111 (1960).

25. E. C. Moore, P. Reichard, and L. Thelander, *J. Biol. Chem.*, **239**, 3445 (1964).

26. M. Shin and D. Arnon, *J. Biol. Chem.*, **240**, 1405 (1965).

27. G. Zanetti and G. Forti, *J. Biol. Chem.*, **241**, 279 (1966).

28. N. Gupta and B. Vennesland, *J. Biol. Chem.*, **239**, 3787 (1964).

29. Y. I. Shethna, P. W. Wilson, and H. Beinert, *Biochim. Biophys. Acta*, **113**, 225 (1966).

30. W. Becker, U. Benthin, E. Eschenhof, and E. Pfeil, *Biochem. Z.*, **337**, 156 (1963).

31. M. I. Dolin, *J. Biol. Chem.*, **235**, 544 (1960).

32. W. B. Sutton, *J. Biol. Chem.*, **226**, 395 (1957).

33. K. Yagi and T. Ozawa, *Biochim. Biophys. Acta*, **56**, 413 (1962).

34. V. Massey and H. Ganther, *Biochemistry*, **4**, 1161 (1965).

35. P. Hemmerich and A. Ehrenberg (this volume).

36. V. Massey, G. Palmer, B. E. P. Swoboda, C. H. Williams, Jr., and R. H. Sands, in *Flavins and Flavoproteins*, E. C. Slater, Ed., Elsevier, Amsterdam, 1966, p. 133.

36a. G. Palmer and V. Massey, unpublished results.

37. P. Hemmerich and J. T. Spence, in *Flavins and Flavoproteins*, E. C. Slater, Ed., Elsevier, Amsterdam, 1966, p. 82.

38. H. Beinert and R. H. Sands, in *Free Radicals in Biological Systems*, M. S. Blois, Jr., H. W. Brown, R. M. Lemmon, R. O. Lindblom, and M. Weissbluth, Eds., Academic Press, New York, 1961, p. 17.

39. C. Veeger, Discussion, in *Flavins and Flavoproteins*, E. C. Slater, Ed., Elsevier, Amsterdam, 1966.

40. Y. I. Shethna, H. Beinert, and P. Hemmerich, unpublished results; quoted in P. Hemmerich, C. Veeger, and H. C. S. Wood, *Angew Chem.*, **77**, 1 (1965).

41. P. Strittmatter, in *Flavins and Flavoproteins*, E. C. Slater, Ed., Elsevier, Amsterdam, 1966, p. 325.
42. H. Beinert, *J. Am. Chem. Soc.*, **78**, 5323 (1956).
43. A. Ehrenberg, *Arkiv Kemi*, **17**, 97 (1962).
44. Q. H. Gibson, V. Massey, and N. M. Atherton, *Biochem. J.*, **85**, 369 (1962).
45. R. L. Searls and D. R. Sanadi in *Light and Life*, W. D. McElroy and B. Glass, Eds., The Johns Hopkins Press, Baltimore, 1961, p. 157.
46. E. Kosower, in *Progress in Physical Organic Chemistry*, Vol. 3, S. G. Cohen, A. Streitweiser, Jr., and R. W. Taft, Eds., Interscience, New York, 1966.
47. V. Massey and C. Veeger, *Biochim. Biophys. Acta*, **48**, 33 (1961).
48. V. Massey and Q. H. Gibson, *Federation Proc.*, **23**, 18 (1964).
49. M. I. Dolin in *Flavins and Flavoproteins*, E. C. Slater, Ed., Elsevier, Amsterdam, 1966, p. 171.
50. V. Massey and G. Palmer, *J. Biol. Chem.*, **237**, 2347 (1962).
51. W. Rutter and B. Rolander, *Acta Chem. Scand.*, **11**, 1663 (1957).
52. C. Veeger, D. V. Dervatanian, J. F. Kalse, A. deKok, and J. F. Koster, in *Flavins and Flavoproteins*, E. C. Slater, Ed., Elsevier, Amsterdam, 1966, p. 242.
53. H. Kubo, H. Watari, and T. Shifa, *Bull. Soc. Chim. Biol.*, **41**, 981 (1959).
54. K. Yagi and T. Ozawa, *J. Biochem. (Tokyo)*, **54**, 204 (1963).
55. B. E. P. Swoboda and V. Massey, in *Flavins and Flavoproteins*, E. C. Slater, Ed., Elsevier, Amsterdam, 1966, p. 263.
56. S. Velick, in *Light and Life*, W. D. McElroy and B. Glass, Eds., The Johns Hopkins Press, Baltimore, Md., 1961, p. 108.
57. G. Weber, *Biochem. J.*, **47**, 114 (1950).
58. G. Palmer, G. Weber, and V. Massey, unpublished, quoted in ref. 83.
59. H. Beinert, in *The Enzymes*, Vol. 7, 2nd ed., P. Boyer, H. Lardy, and K. Myrbäck, Eds., Academic Press, New York, 1963, p. 467.
60. E. Walaas and O. Walaas, *Acta Chem. Scand.*, **10**, 122 (1956).
61. V. Massey (unpublished).
62. G. Palmer (unpublished).
63. R. Kuhn and H. Rudy, *Ber.*, **69**, 2557 (1936).
64. H. A. Harbury and K. A. Foley, *Proc. Natl. Acad. Sci. U.S.*, **44**, 662 (1958).
65. G. Greull and V. Massey (unpublished).
66. P. Strittmatter, *J. Biol. Chem.*, **236**, 2329 (1961).
67. H. Beinert and P. Hemmerich, *Biochem. Biophys. Res. Commun.*, **18**, 210 (1965).
68. H. Beinert and G. Palmer, in *Advances in Enzymology*, Vol. 27, F. F. Nord, Ed., Interscience, New York, 1965, p. 106.
69. Q. H. Gibson and J. W. Hastings, *Biochem. J.*, **83**, 368 (1962).
70. Q. H. Gibson, V. Massey, and B. E. P. Swoboda, *J. Biol. Chem.*, **239**, 3927 (1964).
71. D. Wellner and A. Meister, *J. Biol. Chem.*, **236**, 2357 (1961).
72. R. Alberty, in *Advances in Enzymology*, Vol. 17, F. F. Nord, Ed., Interscience, New York, 1956, p. 1.
73. K. Dalziel, *Acta Chem. Scand.*, **11**, 1706 (1957).
74. K. Dalziel, *Biochem. J.*, **84**, 244 (1962).
75. H. Theorell and B. Chance, *Acta Chem. Scand.*, **5**, 1127 (1951).
76. M. Dixon and K. Kleppe, *Biochim. Biophys. Acta*, **37**, 368 (1965).
77. T. Nakamura and Y. Ogura, *J. Biochem. (Japan)*, **52**, 216 (1962).
78. H. Beinert, *J. Biol. Chem.*, **225**, 465 (1957).
79. T. P. Singer and E. B. Kearney, *Arch. Biochem. Biophys.*, **29**, 190 (1950).

80. Q. H. Gibson, in *Flavins and Flavoproteins*, E. C. Slater, Ed., Elsevier, Amsterdam, 1966, p. 236.
81. F. L. Crane, S. Mii, J. G. Hauge, D. E. Green, and H. Beinert, *J. Biol. Chem.*, **218**, 707 (1956).
82. G. Palmer and V. Massey, *Biochim. Biophys. Acta*, **58**, 349 (1962).
83. V. Massey in *The Enzymes*, Vol. 7, 2nd ed., P. D. Boyer, H. Lardy, and K. Myrbäck, Eds., Academic Press, New York, 1963, p. 275.
84. T. Nakamura, J. Yoshimura, and Y. Ogura, *J. Biochem. (Tokyo)*, **57**, 554 (1965).
85. V. Massey and G. Palmer, *Biochemistry*, **5**, 3181 (1966).
86. V. Massey and B. Curti, *J. Biol. Chem.*, in press.
87. K. Decker and V. D. Dai, personal communication.
88. E. Knight, A. J. D'Eustacio, and R. W. F. Hardy, *Biochim. Biophys. Acta*, **113**, 626 (1966).

Metalloflavoproteins

K. V. Rajagopalan and P. Handler, *Department of Biochemistry,*
Duke University Medical Center, Durham, North Carolina

I.	Absorption Spectra of Iron Flavoproteins	302
II.	Physical Properties and Chemical Composition . . .	306
III.	Catalytic Activities	308
	A. Xanthine Oxidase Type Enzymes	308
	B. Aldehyde Oxidase	311
	C. Dihydroorotic Dehydrogenase	311
IV.	Xanthine and Aldehyde Oxidases	312
	A. Role of Electron Carriers in Catalysis	312
	B. EPR Studies of Metalloflavoproteins	320
	1. "Single Turnover"	321
	2. Reoxidation of Reduced Enzyme	322
	3. Anaerobic Titration	323
V.	Dihydroorotic Dehydrogenase	327
VI.	The Non-Heme Iron Complex	332
VII.	Concluding Remarks	334
	References	335

Metalloflavoproteins are enzymes which contain, in addition to a flavin, a metal which is essential to catalytic function. To date, the only such metals which have been identified are non-heme iron and molybdenum. Non-heme iron-containing flavoproteins include the succinate- and DPNH-linked dehydrogenases of mitochondria and the DPNH- and TPNH-specific dihydroorotic dehydrogenases of bacteria. Nitrate reductases from various sources have been characterized as molybdoflavoproteins. Xanthine oxidases of diverse origin and hepatic aldehyde oxidase are the only known flavoproteins which contain both molybdenum and non-heme iron. The possibility that the metals might be catalytically functional in these enzymes prompted much research which, however, proved inconclusive, owing to lack of penetrating techniques especially applicable to these studies and to inadequate understanding of the nature of the atypical absorption spectra of metalloflavoproteins.

In recent years the application of electron spin resonance techniques has provided a powerful tool for probing the behavior of non-heme iron and molybdenum, both paramagnetic metals, in enzymes containing these metals. Further, recognition of a broad class of non-heme iron proteins with characteristic absorption spectra has greatly facilitated the clarification of the absorption spectra of the more complex metalloflavoproteins. These recent developments have permitted much more definitive understanding of the role of metals in metalloflavoproteins than was possible a few years ago.

By far the most extensive studies on metalloflavoproteins have been directed at xanthine oxidase. A recent review (1) has adequately summarized most of the results of such studies. Accordingly the present article will be restricted to the most recent studies of this enzyme. Most of the significant studies of hepatic aldehyde oxidase and dihydroorotic dehydrogenase have been reported only recently and have not been summarized. A comparative résumé of studies on all three enzymes will therefore be attempted here. Since a review on nitrate reductase has appeared only recently (2) and the respiratory-linked dehydrogenases are dealt with elsewhere in this volume, the present discussion will be limited to hepatic aldehyde oxidase, xanthine oxidase, and dihydroorotic dehydrogenase. Application of EPR spectroscopy has provided considerable information concerning these enzymes, but no detailed presentation of these studies will be attempted here, since several recent reviews have dealt with them extensively (3,4).

It is believed that the purposes of this article would best be served not by summary documentation of the many isolated bits of information accumulated by diverse investigations of these enzymes but by selective consideration of information which contributes significantly to the understanding of critical aspects of the enzymology of these proteins, e.g., protein structure, nature of the active site, kinetic behavior, functional role of metals, and internal electron transport sequence.

The purpose of comparative studies of closely related enzymes is usually to identify significant common features from similarities in physical and chemical properties. Inevitably, however, the dissimilarities become equally prominent and frequently are as informative as the similarities. The present essay, therefore, will not only search for "unity in diversity," but will also focus attention on the diversities themselves.

I. Absorption Spectra of Iron Flavoproteins

The flavoprotein nature of milk xanthine oxidase was recognized independently by Ball (5) and by Corran et al. (6). Both laboratories suggested the presence of additional chromophoric groups in the enzyme.

The nutritional studies of Westerfeld and co-workers (7,8) led to the discovery of the presence of molybdenum and iron in the enzyme; but their attempts to correlate the unidentified chromophoric group with iron were inconclusive. Meanwhile, Mahler and Elowe (9) had detected the presence of iron in purified preparations of mitochondrial DPNH-cytochrome c reductase, and suggested that the anomalous absorption spectrum of the enzyme could be attributed to an iron complex similar to that in the iron-binding globulin of plasma (transferrin or siderophilin). Shortly thereafter, Singer, Kearney, and Bernath (10) obtained a highly purified preparation of succinic dehydrogenase from pig heart mitochondria and demonstrated it to be an iron flavoprotein. The atypical absorption spectrum of the enzyme was attributed to iron–protein bonds.

A number of other flavoproteins with anomalous absorption spectra have since been reported, including xanthine oxidase from calf liver (11); xanthine dehydrogenases from chicken liver (12), pigeon kidney (13), *Clostridium cylindrosporum* (14), and *Micrococcus lactilyticus* (15); aldehyde oxidase from rabbit liver (16); and dihydroorotic dehydrogenase from *Zymobacterium oroticum* (17). The common feature of all these enzymes is the presence of protein-bound non-heme iron. Differential spectrophotometric studies of the atypical nature of the absorption spectra (as compared to simple metal-free flavoproteins) of milk xanthine oxidase, hepatic aldehyde oxidase, and dihydroorotic dehydrogenase revealed that the degree of anomaly is quantitatively related to the non-heme iron content of these enzymes (18) and that all exhibit an essentially identical non-flavin absorption spectrum per atom of iron. Flavin-free iron-containing proteins have been prepared from each of these proteins by precipitation with 80% methanol, in the cold. Figure 1 shows the absorption spectra of the native oxidized forms of milk xanthine oxidase and dihydroorotic dehydrogenase; Figure 2 presents the absorption spectra of the iron proteins prepared from them. These absorption spectra are remarkably similar to that of the photosynthetic pyridine nucleotide reductase isolated from spinach by San Pietro (spinach ferredoxin) (19) and signal the occurrence of a specific form of protein-bound iron widely distributed in extremely diverse systems.

The studies of Rajagopalan and Handler (18) further indicated that the characteristic absorbance of iron flavoproteins in the region 500–700 mμ is entirely due to the chromophoric non-heme iron. This fact coupled with the proportionality of abnormal absorbance at 450 mμ to the non-heme iron content provides a direct means of arriving at iron-to-flavin ratios in purified iron-flavin proteins. The expected ratios of absorbancies at 450 and 550 mμ for different iron-to-flavin ratios are given in Table I.

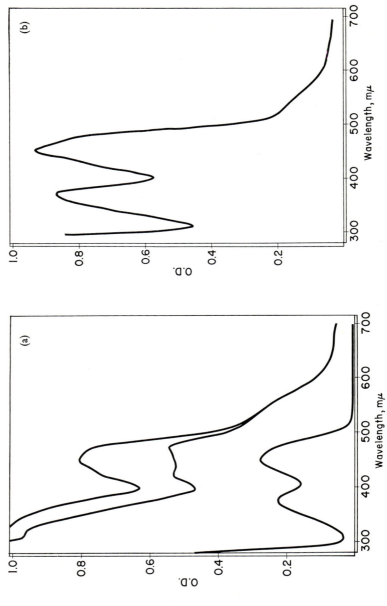

Fig. 1. (a) Absorption spectra of milk xanthine oxidase (top curve); equivalent concentration of FAD (bottom curve); and the difference (middle curve). (b) Absorption spectrum of dihyroorotic dehydrogenase.

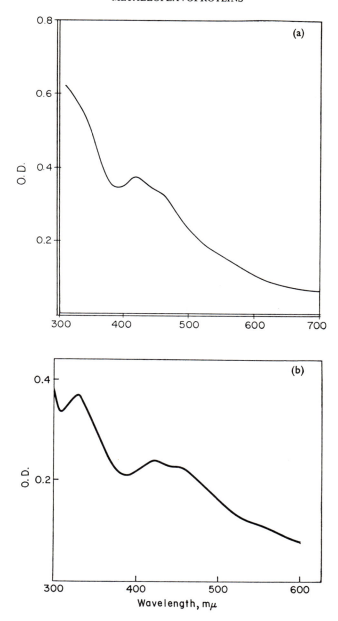

Fig. 2. Absorption spectra of flavin-free iron proteins from (a) milk xanthine oxidase and (b) dihydroorotic dehydrogenase.

TABLE I

Absorption Characteristics of Iron Flavoproteins

	Fe/flavin ratio	E_{450}/E_{550} ratio
1	(Dihydroorotic dehydrogenase)	5.9
4	(Xanthine oxidase)	3
4	(Aldehyde oxidase)	3
1		6[a]
2		4[a]
3		3.4[a]
4		3[a]

[a] Theoretical, based on E_{450} of 11,300 for flavin and E_{450} of $\sim$5900 and E_{550} of $\sim$2900 for iron.

The observed values for bovine milk xanthine oxidase, aldehyde oxidase, and dihydroorotic dehydrogenase are also given. It will be apparent that the iron-to-flavin ratio in an iron flavoprotein could be determined by merely quantitating the absorption spectrum of the enzyme if no other chromophores are present. It should be emphasized, however, that this technique is applicable only to those cases where the iron component has the same absorption characteristics as in the enzymes mentioned above.

II. Physical Properties and Chemical Composition

Crystalline milk xanthine oxidase has been found to have an E_{280}/E_{450} ratio of 5.0 (1), with a molar extinction coefficient of 70,000 at 450 mμ (1). While the molybdenum content of xanthine oxidase preparations from milk has been variable, it has been suggested that the fully active enzyme is most likely to have a Mo:FAD:Fe ratio of 2:2:8 per mole. Keilley, who purified a xanthine oxidase from calf liver (11), found that the absorption spectrum of the enzyme was similar to that of milk xanthine oxidase but on the basis of E_{280}/E_{450} of 10, concluded that the molar Mo:FAD:Fe content was 1:1:4. More recently, however, xanthine oxidase has been purified from pig liver by Brumby and Massey (20) with an E_{280}/E_{450} ratio of 5:1 (21). The enzyme from mammalian liver, therefore, seems to have a cofactor composition identical to that of milk xanthine oxidase. Xanthine dehydrogenases have also been isolated from chicken liver (12) and pigeon kidney (13). While the latter preparations were apparently similar, their absorption spectra were significantly different from those of milk or mammalian liver xanthine oxidase. The chicken liver enzyme was reported to be electrophoretically homogeneous and on the basis of an apparent Mo:FAD:Fe content of 1:1:8 a molecular weight of 480,000 was assigned to it (12). A similar Mo:FAD:Fe content was reported for the pigeon kidney enzyme (13). Recently, however, chicken liver xanthine dehydrogenase has been further purified and shown to have an absorption spectrum quite similar to that of milk xanthine oxidase with E_{450}/E_{550} = 3 (15). Since the chicken liver enzyme and the milk enzyme sedimented identically in sucrose gradient centrifugation, it was apparent that they had similar molecular weights, thus indicating a molar Fe:FAD ratio of 4 in the avian enzyme. Purified xanthine dehydrogenases from *Micrococcus lactilyticus* (15) and *Clostridium cylindrosporum* (14) also have absorption characteristics indicating an Fe:FAD ratio of 4. Thus it is now obvious that xanthine oxidase type enzymes from all sources have similar gross molecular properties, with molecular weights of about 280,000 and Fe:FAD ratios of 4.

In view of this newly recognized similarity, it is of great interest that the enzymes from different sources exhibit widely differing electron acceptor specificities. Molecular oxygen is an efficient electron acceptor for the mammalian enzymes and, perhaps, functions as such physiologically. Xanthine dehydrogenases from avian tissues and from the anaerobic bacteria *C. cylindrosporum* and *M. lactilyticus* do not transfer electrons to oxygen to any significant extent. The physiological acceptor for the avian enzymes seems to be DPN (12,22), while for the enzyme from *M. lactilyticus* and also, presumably, for the clostridial enzyme, ferredoxin serves as an efficient electron acceptor (23). The principal and, seemingly, physiological acceptor of each of the above groups of enzymes cannot serve as an efficient acceptor for the other two groups. In view of the evidence for the involvement of non-heme iron in the reduction of molecular oxygen by milk xanthine oxidase and hepatic aldehyde oxidase (24), the diversity of specificities of the different enzymes for electron acceptors is very intriguing and promises to be a fruitful line of investigation.

Highly purified aldehyde oxidase from rabbit liver is quite similar to xanthine oxidase in its gross molecular properties (16). It has a molecular weight of about 300,000 and a molar content of Mo:FAD:Fe of 2:2:8. Its absorption spectrum is quite similar to that of xanthine oxidase with E_{450}/E_{550} = 3. While considerable amounts of coenzyme Q_{10} have been detected in purified aldehyde oxidase, the significance of this quinone to the catalytic function of the enzyme is unknown at present. There is, however, considerable presumptive evidence for the involvement of lipid material in the structural integrity of the enzyme (16) which can be inhibited by a variety of inhibitors that are without effect on the xanthine oxidases.

Aldehyde oxidase was isolated from pig liver in 1940 and identified as a flavoprotein (25). Mahler et al. (26) subsequently reported the presence of molybdenum and the absence of non-heme iron in this enzyme but, on the basis of spectral studies, implicated a heme component in the catalytic activity of the enzyme. Later studies by Palmer revealed the identity of the heme with contaminating catalase activity (27). Palmer further noted that when the enzyme was reduced anaerobically, by substrate or dithionite, bleaching occurred in the region 500–600 mμ. Since this bleaching is characteristic of iron flavoproteins, it would appear that, contrary to the earlier report (26), non-heme iron is indeed present in pig liver aldehyde oxidase. In view of the molecular similarities of this group of enzymes as detailed earlier, it would not be surprising if the iron-to-flavin ratio in pig liver aldehyde oxidase were 4, as in the rabbit enzyme.

Dihydroorotic dehydrogenase, the least complex of the soluble metalloflavoproteins, was obtained from *Zymobacterium oroticum* in crystalline form by Friedman and Vennesland (17). The enzyme was shown to contain 1 mole of FAD, 1 mole of FMN, and 2 g atoms of non-heme iron per 62,000 g of protein. As mentioned earlier, the absorption spectrum of the enzyme displays a degree of anomaly commensurate with its non-heme iron content. Recent studies (15) have revealed a molecular weight of about 120,000, suggesting that the enzyme may be a dimer of the basic unit mentioned above. While the more thoroughly studied dihydroorotic enzyme is specific for DPN$^+$/DPNH and has considerable oxidase activity, a quite similar enzyme specific for TPN$^+$/TPNH and with only low oxidase activity has also been isolated from a bacterial source (28).

III. Catalytic Activities

A. Xanthine Oxidase Type Enzymes

The ability of the xanthine oxidase type enzymes to oxidize diverse purines and aldehydes as well as a variety of other compounds has long been recognized (1).

In view of the physical complexity of xanthine and aldehyde oxidases, their dual specificities (in each case nonspecific for any aldehyde, partial specificity for nitrogen-containing heterocycles) pose the problem of whether the two general classes of substrates are oxidized at the same or independent sites on these enzymes. It is abundantly clear that common sites are employed. Thus, kinetically, salicylaldehyde and xanthine are competitive substrates for milk xanthine oxidase (29). Urea and guanidine are formal competitive inhibitors against both classes of substrates and the K_i values are identical for the two classes (30). Atebrine, which is a noncompetitive inhibitor of xanthine oxidase, is competitive for aldehyde oxidase against all types of substrate (31). Again, cyanide inactivates xanthine and aldehyde oxidases equally against both major classes of substrate. No available evidence suggests independently active sites. It follows, therefore, that any mechanism proposed to account for the mechanism of action of these enzymes must adequately explain this dual specificity.

Although the rate of DPNH oxidation by purified milk xanthine oxidase is extremely slow, it has been reported that at low pH and with ferricyanide as electron acceptor, activity towards DPNH is considerably enhanced (32). The known lability of the enzyme at low pH suggests that the observed enhancement of DPNH oxidase activity might be the result of a significant physical alteration of the native enzyme. In any case, the insensitivity of the DPNH oxidase activity of xanthine oxidase strongly suggests that this reaction occurs at a binding site and by a mechanism which is independent of the major substrate binding and oxidizing sites of the enzyme.

The effect of pH on the activity of milk xanthine oxidase has been more extensively investigated by Greenlee and Handler (33). They observed that between pH 9.4 and pH 11 the V_{max} value for the oxidation of xanthine was invariant whereas K_m for xanthine increased markedly with increasing pH. Similar effects of pH on K_m were observed with other purine and aldehyde substrates. Greenlee and Handler further noted that at elevated pH the enzyme acquired the ability to oxidize a new class of substrates, aromatic heterocyclic compounds containing a quaternary nitrogen function. K_m values for any of this group of substrates, e.g. N'-methylnicotinamide, decreased with increasing pH, exactly opposite to the behavior of purine and aldehyde substrates. Since N'-methylnicotinamide competitively inhibited oxidation of xanthine at pH 10.8, a common binding site was inferred. It was further concluded that the effect of pH was the result of titration of an ionizable group on the enzyme with a pK_a of about 10.7. Fridovich (34) has extended these studies to the pH range 4.5–11.5 and has concluded that the singly ionized monovalent anionic form of xanthine is the actual substrate for xanthine oxidase and that neither the un-ionized nor the doubly ionized form is acted upon by the enzyme.

Studies on substrate specificity of xanthine oxidase from liver or of xanthine dehydrogenases from avian tissues or bacteria have not been as extensive as with the milk enzyme. Invariably, however, all these enzymes are capable of oxidizing a variety of purines and aldehydes. The calf liver enzyme has been reported to have a high DPNH diaphorase activity

(11), but in view of the inhomogeneity of the enzyme used it is not possible to assess the significance of the observation, particularly since, in the absence of the dye oxidant, there was no appreciable aerobic oxidation of DPNH. Xanthine dehydrogenase from *C. cylindrosporum* has been reported not to utilize pyridine nucleotides either as substrates or as acceptors (14). An interesting point of difference between the clostridial enzyme and the mammalian enzymes is the fact that unsubstituted purine is converted solely to 8-hydroxypurine by the former but to uric acid via hypoxanthine (6-hydroxypurine) and xanthine (2,6-dihydroxypurine) by the latter. Thus, the specific locus of oxidation of the purine ring structure is determined primarily by the nature of the catalytic site of the enzyme rather than by the distribution of electron density of the substrate molecule, as had been suggested by Pullman and co-workers (35).

Fridovich (36) has investigated the kinetics of aerobic oxidation of xanthine by milk xanthine oxidase at varying substrate and O_2 concentrations. His findings were most compatible with a "ping-pong" mechanism (37) which may be represented as follows:

Xanthine	Urate	O_2		H_2O_2
$\downarrow$	$\uparrow$	$\downarrow$		$\uparrow$
E	(EX $\rightleftharpoons$ FU)	F	(FO$_2$ $\rightleftharpoons$ EH$_2$O$_2$)	E

In this scheme E and F correspond to oxidized and reduced enzyme, respectively, but without specification of the nature of the reduced enzyme. Such a mechanism predicts that the K_m for either substrate will be dependent on the concentration of the other substrate and provides an explanation for the finding by Dixon and Thurlow in 1924 (38) that inhibition of xanthine oxidase by excess xanthine was dependent on the concentration of the acceptor. Fridovich has found that, at pH 10, apparent K_m(xanthine) is $1.5 \times 10^{-4}M$ at infinite oxygen concentration while apparent $K_m(O_2)$ is $8 \times 10^{-5}M$ at infinite xanthine concentration. The mechanism of excess substrate inhibition of xanthine oxidase is not well understood. It is of interest that Greenlee and Handler (39), who found inactivation of xanthine oxidase by binding of benzaldehyde to amino groups of the protein, observed that enzyme partially inactivated in this manner showed considerably less susceptibility to excess substrate inhibition. They further noted that in the course of dinitrophenylation of xanthine oxidase with dinitrofluorobenzene, 50% of the catalytic activity was relatively resistant to inactivation. On the basis of these studies they postulated that there are two binding sites for xanthine on the enzyme molecule and that binding of xanthine to both sites might explain excess substrate inhibition.

B. *Aldehyde Oxidase*

Hepatic aldehyde oxidase, which acts on both aldehyde and *N*-hetero-cyclic substrates, was early distinguishable from xanthine oxidase by its inability to oxidize xanthine (25). In general, aldehyde oxidases, particularly that from rabbit liver, exhibit the same dual specificity of substrates which is so remarkable for milk xanthine oxidase (40). Indeed, the first substrates recognized for aldehyde oxidase were aromatic heterocyclic compounds with ternary or quaternary nitrogen atoms, certain purines, and, later, several aldehydes (40). Although their general substrate behavior is similar, there are significant differences between the characteristic substrates of the two enzymes. Typically, compounds like *N′*-methylnicotinamide are oxidized by aldehyde oxidase at physiological pH but only at elevated pH ranges by xanthine oxidase. Purine is oxidized exclusively to 8-hydroxy-purine by aldehyde oxidase, whereas xanthine oxidase converts it sequentially to hypoxanthine, xanthine, and uric acid. Furthermore, quinones such as menadione, which are excellent electron acceptors for xanthine oxidase, are powerful inhibitors of aldehyde oxidase. With this exception, the electron acceptor specificities of the two enzymes are similar.

Palmer (41), who studied the kinetics of hog liver aldehyde oxidase with acetaldehyde as substrate and with 2,6-dichlorophenol indophenol and cytochrome *c* as acceptors, concluded that the data with either acceptor excluded the existence of free reduced enzyme as a catalytic intermediate and indicated the involvement of a ternary complex. The kinetic data obtained with cytochrome *c* may not be meaningful, however, if reduction of cytochrome *c* by the hog liver enzyme is dependent on the presence of oxygen as in the case of the related enzymes discussed below.

C. *Dihydroorotic Dehydrogenase*

Dihydroorotic dehydrogenase catalyzes reversible dehydrogenation between dihydroorotate and DPN^+ as first shown by Kornberg et al. (42). Friedmann and Vennesland (17), who crystallized the enzyme, observed that it also served as DPNH oxidase. Cysteine was found to "activate" the enzyme and to facilitate bleaching of its visible color by dihydroorotate.

Table II presents the results of a study on the specificity of the enzyme toward analogs of orotate (43). With DPNH as reducing substrate the enzyme catalyzes the reduction of 5-halogen derivatives of orotic acid, while 5-methyl and 5-amino derivatives are poor substitutes. In fact, at similar concentrations 5-fluoroorotate and 5-bromoorotate are even more

TABLE II

Substrate Specificity of Dihydroorotic Dehydrogenase at the Orotate Site[a]

Substrate	Relative rates	% Inhibition
Orotate	100	0.0
5-Fluoroorotate	196	—
5-Bromoorotate	155	—
5-Iodoorotate	78	2.0
5-Methylorotate	10.5	50.0
5-Aminoorotate[a]	1.7	54.0
5-Nitroorotate[b]	6.0	26.0
2-Thioorotate	2.0	0.5
5-Fluorouracil	0.0	5.0

[a] Activity was assayed under anaerobic conditions at 25° spectrophotometrically (340 mμ), in 133mM sodium phosphate, pH 6.5 containing 0.33 μmole of substrate, 0.12 μmole DPNH, 0.1 μmole EDTA, 13 μg of cysteine-treated enzyme in a total volume of 3.0 ml. For studying inhibitory effects, 0.33 μmole orotate derivative was added to standard incubation mixture containing 0.33 μmole orotate.

[b] Corrections were made for the absorbance change at 340 mμ on reduction of these substrates.

effective than orotate as electron acceptors. The reduction of orotate is strongly inhibited by the simultaneous presence of 5-methylorotate or 5-aminoorotate and somewhat less by 5-nitroorotate. The ineffectiveness of these derivatives as electron acceptors could not, therefore, be the result of poor affinity for the enzyme. In fact, the dissociation constant for the methylorotate–enzyme complex, determined by difference spectrophotometry (see below), was quite similar to that of the orotate–enzyme complex. It would appear that the methyl and amino substituents act as electron donors to the pyrimidine ring and thus reduce its ability to accept electrons from the enzyme. The greater effectiveness of 5-fluoro- and 5-bromo-orotate as electron acceptors is compatible with the tendency of the halogen substituents to act as electron-withdrawing groups.

IV. Xanthine and Aldehyde Oxidases

A. Role of Electron Carriers in Catalysis

The first compelling evidence for a divergence in behavior between simple flavoproteins and xanthine oxidase emerged from the demonstration by Fridovich and Handler (44) that addition of inorganic sulfite to a system containing xanthine and milk xanthine oxidase resulted in the oxidation of sulfite to sulfate by a chain propagation mechanism. A variety of other enzymes including flavoproteins such as L- and D-amino acid oxidases acting on their respective substrates were incapable of initiating sulfite oxidation. A free radical of molecular oxygen, the superoxide anion (O_2^-) has therefore been implicated in the xanthine oxidase mechanism.

Fridovich and Handler (45) also conducted a detailed study of the role of oxygen in the reduction of cytochrome c by xanthine oxidase, a phenomenon originally observed by Horecker (46). The absolute requirement for oxygen in this reaction was confirmed by Fridovich and Handler, who also demonstrated that the K_m for oxygen for cytochrome c reduction was considerably higher than for formation of H_2O_2 (Fig. 3). Further confirmation of the indirect nature of cytochrome c reduction came from the demonstration that tiron (4,5-dihydroxy-m-benzenesulfonic acid) and the globin of myoglobin inhibited specifically the reduction of cytochrome c without affecting the rate of oxidation of substrate by the enzyme. Significantly, the initiation of sulfite oxidation by xanthine oxidase was also inhibited by tiron and myoglobin.

Related to these phenomena is the ability of xanthine oxidase to induce the chemiluminescence of dimethyl biacridylium (DBA) or luminol in the presence of xanthine and oxygen, first observed by Totter et al. (47). The involvement of oxygen free radicals in this reaction has also been postulated (48). Tiron and myoglobin were found to be potent inhibitors

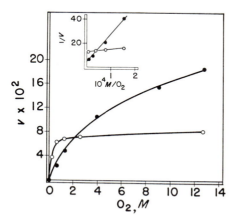

Fig. 3. Effect of O_2 tension on oxygen reduction (○) and cytochrome c reduction (●) by milk xanthine oxidase.

of the chemiluminescence reaction also (49). Thus the ability of xanthine oxidase to generate oxygen radicals from molecular oxygen manifests itself in the three phenomena described above; viz., initiation of sulfite oxidation, oxygen-mediated cytochrome c reduction, and inducement of chemiluminescence from DBA or luminol. Among a multitude of enzymes tested, only aldehyde oxidase and dihydroorotic dehydrogenase were found to mimic xanthine oxidase in eliciting the oxygen radical-dependent reactions (50). Inhibition of these reactions by tiron and myoglobin was a feature common to the three enzymes. It might be mentioned parenthetically that Miller and Massey have presented strong evidence indicating that inhibition by tiron is the result of the chemical interaction of this compound with oxygen free radicals (51). The inability of simple flavoproteins to generate oxygen radicals, coupled with the presence of iron in the three enzymes, therefore led to the conclusion that iron atoms in these enzymes are the site of reduction of molecular oxygen, as shown in Figure 4 (24).

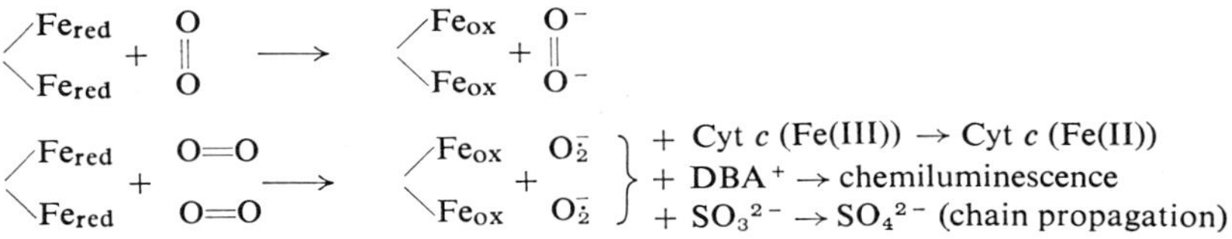

Fig. 4. Postulated mechanism of reduction of molecular oxygen by iron flavoproteins.

Rabbit liver aldehyde oxidase is susceptible to inhibition by a wide variety of compounds (16). A detailed study of the effects of a series of inhibitors on the reduction of a series of electron acceptors revealed the existence of a linear sequence of at least four electron carriers in the enzyme, as depicted in Figure 5 (52), and clearly indicated that several components in addition to flavin must be involved in the internal transport sequence of the enzyme. These inhibition studies revealed that reduction of molecular oxygen occurred terminally in the sequence. The inhibition pattern for cytochrome c reduction was identical with that for oxygen reduction, in conformity with the oxygen dependence of the former.

The sensitivity of aldehyde oxidase to steroids, the detergent Triton X-100, menadione, oligomycin, and antimycin A renders it unique among known flavoproteins and, in sum, constitutes the most cogent argument that the ubiquinone associated with this enzyme is indeed an integral aspect of its electron-transport system. Particularly noteworthy is the fact that milk xanthine oxidase is *not* sensitive to the inhibitors shown in Figure 5. Nevertheless, in view of the many similarities in the activities of the two

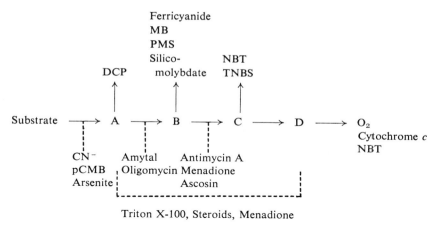

Fig. 5. Electron transport in hepatic aldehyde oxidase.

enzymes, it was suggested that there must be a similar sequence of carriers, albeit probably less complex, in xanthine oxidase also (52).

Xanthine oxidase and aldehyde oxidase are unique among oxidative enzymes in that, as shown in Figure 6, the reactions catalyzed by them are, formally, hydroxylations of substrates in which the hydroxyl function derives from the elements of water. Since, other than nitrate reductase, these are the only dehydrogenases presently known to contain molybdenum, it has been suggested that this metal may serve at the active site of these enzymes where it facilitates hydroxylation. Strong presumptive indication that such is in fact the case came from studies on the mechanism of progressive inhibition of these enzymes by methanol (31), Figure 7. Kinetic studies showed that the locus of methanol inhibition was the substrate binding site of these enzymes and, further, that progressive inactivation of the enzyme by methanol occurred only in the presence of substrate. Inactivated enzyme, reisolated by ammonium sulfate precipitation, could slowly be reactivated on prolonged dialysis. It was concluded that the inactivation

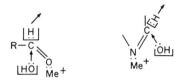

Fig. 6. Postulated mechanism for substrate activation and hydroxylation by aldehyde oxidase and xanthine oxidase.

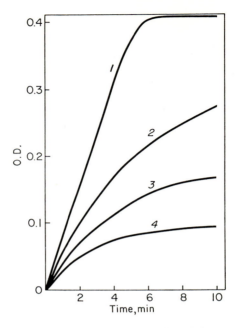

Fig. 7. Inactivation of milk xanthine oxidase by methanol. Curve *1*, control (xanthine, $5 \times 10^{-5}M$); curves *2*, *3*, and *4*, plus 0.5, 1.0, and $1.5M$ methanol, respectively.

was the consequence of binding of methanol to molybdenum in a lower valence state. Subsequent studies on reisolated, methanol-inactivated aldehyde and xanthine oxidases by EPR revealed complicated signals of Mo(V) in the samples, indicating that molybdenum was indeed the component susceptible to methanol inactivation (15) (see Fig. 8).

The inhibition of milk xanthine oxidase by cyanide was studied by Dixon and Keilin (53), who observed that the enzyme was protected against such inhibition by its substrates. These studies have been extended to aldehyde oxidase (31), leading to a significant correlation. In the absence of substrate, aldehyde oxidase is even more sensitive to cyanide than is xanthine oxidase and is comparably inhibited at cyanide concentrations one order of magnitude less than those used for xanthine oxidase. As shown in Figure 9, this enzyme is strikingly protected against cyanide inhibition by both quinoline (an excellent substrate) and atebrine, a structurally related analog, which serves as a competitive inhibitor of aldehyde oxidase with respect to all classes of substrates. In contrast, N'-methylnicotinamide, which is an excellent substrate—indeed the only

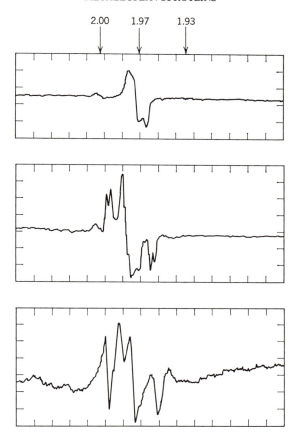

Fig. 8. EPR spectra of native aldehyde oxidase (top), methanol-inactivated aldehyde oxidase (middle), and methanol-inactivated xanthine oxidase (bottom).

known physiological substrate—but which bears a quaternary nitrogen, is relatively ineffective in protecting the enzyme against cyanide. These data, taken in concert, suggest the presence at the substrate binding site of a cyanide-sensitive group, presumably a metal, which, in some manner, binds to the tertiary nitrogen of appropriate heterocycles or to the carbonyl oxygen of aldehydes. While the latter are in place, reaction with cyanide is impossible. N'-Methylnicotinamide, then, affords no protection because its quaternary nitrogen prevents ligation to the metal. It is noteworthy, however, that the quaternary nitrogen results in partial withdrawal of electrons from the adjacent carbon (which is hydroxylated by

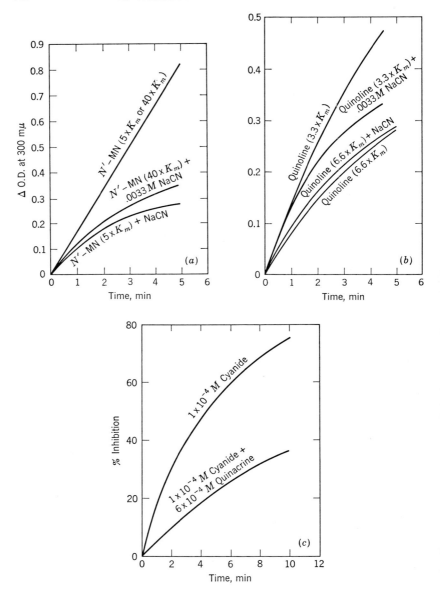

Fig. 9. Relative effects of (a) N'-methylnicotinamide and (b) quinoline on cyanide inactivation during catalysis by aldehyde oxidase. (c) Effect of atebrine on progress of inactivation of aldehyde oxidase during preincubation with cyanide.

the enzyme) in much the same fashion as would ligation of a metal ion to the tertiary nitrogen of heterocycles or to the carbonyl oxygen function.

Once inactivated by cyanide, incubation of xanthine or aldehyde oxidase with any class of substrate fails to bleach the enzyme at any wavelength. Thus the binding of cyanide to the enzyme prevents electron transfer from substrate to either the flavin or iron moieties, indicating that the cyanide binding site physically precedes flavin and iron in the electron transport sequences of these enzymes.

These concepts are strengthened by the properties of the xanthine dehydrogenase of *Micrococcus lactilyticus* (54). While resembling the milk enzyme in many respects, this enzyme is at once insensitive to cyanide and extremely sensitive to methanol, and even in the native state its molybdenum moieties appear to be largely, or exclusively, in the reduced [Mo(V)] condition (see below). Thus it appears highly likely that molybdenum is at the substrate binding site, that it is Mo(VI) which is attacked by cyanide and thus inactivated, and that it is Mo(V) which can be attacked by methanol.

Further studies on aldehyde oxidase indicated the presence of a transient sulfhydryl function at the "active site" (31). Thus, *p*-chloromercuribenzoate and arsenite were competitive inhibitors in the presence of substrate, yet did not inactivate the enzyme when preincubated in the absence of substrate, at the levels used. Low levels of dinitrofluorobenzene caused rapid inactivation of the enzyme but, again, only in the presence of substrate. From these observations it was concluded that the molybdenum atom at the active site may be linked, in some manner, to a sulfhydryl group in the native enzyme, with the latter becoming a free sulfhydryl group on reduction of enzyme by substrate. A similar interaction between molybdenum and a sulfhydryl group in xanthine oxidase has recently been suggested by Bray (55) on the basis of studies on inhibition by iodoacetamide. Bray, however, believes that the sulfhydryl group does not exist free during the catalytic activity of the enzyme.

Though the chemistry of the reactions catalyzed by aldehyde oxidase and xanthine oxidase is identical, there are several marked differences in behavior between the catalytic regions of the two enzymes. Thus pCMB is a competitive inhibitor of aldehyde oxidase, but inhibits xanthine oxidase noncompetitively (21). The effect of arsenite on xanthine oxidase also contrasts with that on aldehyde oxidase. Preincubation of xanthine oxidase with arsenite leads to rapid inactivation (56), whereas in the presence of substrate it is a weak uncompetitive inhibitor (31). Further, dinitrofluorobenzene has no effect on xanthine oxidase activity at pH 7.8, even in the presence of substrate (39). These differences in behavior between these enzymes indicate that it is presently futile to attempt to interrelate the catalytic mechanisms of these two enzymes. Further discussion of this aspect of the enzymology of these enzymes is presented below.

B. EPR Studies of Metalloflavoproteins

The presence of flavin and the paramagnetic metals iron and molybdenum in xanthine and aldehyde oxidases has rendered these enzymes prime candidates for studies by the technique of electron paramagnetic resonance spectroscopy (EPR). The considerable volume of information which has been accumulated on these enzymes through such studies will be summarized below. For more detailed discussion of EPR studies on these enzymes, the reader is referred to recent review articles (3,4).

Three types of EPR signals have been observed in xanthine oxidase (57) as well as in aldehyde oxidase (58) after anaerobic reduction by their respective substrates: a signal at $g = 2.00$ ascribed to an organic free radical, a signal at $g = 1.97$ ascribable to molybdenum(V), and an asymmetric signal at $g = 1.93$ identified with the reduced form of the non-heme iron component (3). Since the appearance of these signals on reduction of enzyme by its substrate would not, of itself, constitute evidence for participation of the signal-yielding components in the normal catalytic activity of the enzyme, Bray (59,60) devised techniques for studying the behavior of these signals within time intervals compatible with catalysis. One technique consists of rapid mixing of samples, e.g., enzyme and substrate, for specified times, followed by instantaneous quenching of the reaction by freezing the mixture at liquid nitrogen temperature (59). In conjunction with other experimental approaches such as refined techniques for attaining anaerobiosis, reflectance spectroscopy, and power saturation studies on the EPR signals, this technique has permitted EPR studies on enzymes by a variety of experimental approaches. Convincing evidence has emerged from these studies for the catalytically functional role of molybdenum and iron in xanthine oxidase (61) and in aldehyde oxidase (62). The catalytic turnover number (moles of substrate oxidized per mole of enzyme) for both xanthine oxidase and aldehyde oxidase is about 5 per second at 25° and pH 7.5–8.5, calculated on the basis of two independent sites on the enzyme protein, each site consisting of 1 Mo, 1 FAD, and 4 Fe. The turnover time, therefore, would be of the order of 200 msec. EPR studies have not only revealed that the rates of reduction and reoxidation of molybdenum, flavin, and iron in these enzymes are compatible with their turnover times but, in some instances, have also permitted kinetic separation in time among the redox

components, thereby indicating the sequence of the internal electron transport chains of these enzymes. A few of these experiments are described
below.

1. "Single Turnover" (61)

In this type of experiment, xanthine oxidase was mixed with an equivalent amount of xanthine for varying periods of time in the presence of
excess oxygen and the samples then frozen for EPR measurements, as
described above. Since the concentration of xanthine was sufficient to
saturate the available binding sites, the observed catalytic events would
reflect the synchronous activity of the total population of the enzyme
molecules. The data obtained by Bray, Palmer, and Beinert (61) with
xanthine oxidase are depicted in Figure 10. Four types of signals were
observed, two ascribed to Mo(V), one to a free radical, and the other to
iron. The kinetics of the appearance and disappearance of these signals,
reflecting the reduction and reoxidation of the components reponsible
for the signals, were quite different from one another. Of the two types
of molybdenum signals, the one identified as molybdenum δ was maximal
at 15 msec after mixing and was the fastest to arise and decay. The molybdenum β, free-radical, and iron signals were maximal at 40, 45, and 100
msec, respectively. By the time the iron signal was maximal the other

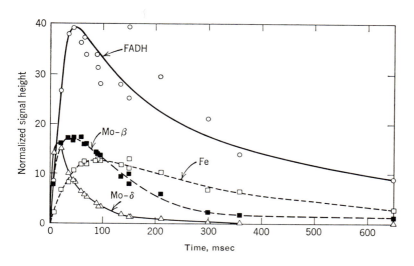

Fig. 10. Signal heights in "single turnover" experiment with xanthine oxidase. (○)
Free radical; (■) molybdenum β; (△) molybdenum δ; (□) iron.

signals had already started decreasing. These data strongly suggest that molybdenum is the component closest to the substrate binding site and that iron is the farthest and, hence, closest to the site of reaction with the electron acceptor, molecular oxygen. The free-radical signal, which was assumed to represent flavin semiquinone (FADH), showed kinetics of appearance and disappearance suggesting that the flavin was between molybdenum and iron in the sequence. Thus, on the basis of this and other experiments, Bray, Palmer, and Beinert concluded that the electron transport sequence in xanthine oxidase is Mo $\rightarrow$ FAD $\rightarrow$ Fe.

2. Reoxidation of Reduced Enzyme

Anaerobic incubation of aldehyde oxidase with excess substrate results in the formation of reduced enzyme in which the EPR signals of Mo(V)

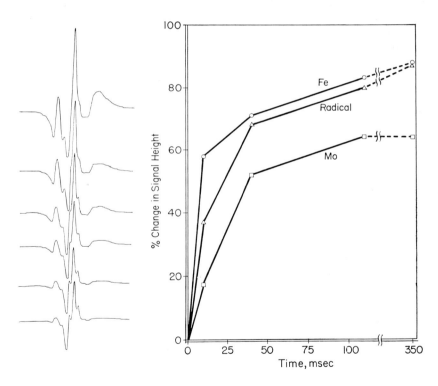

Fig. 11. Changes in signal height during reoxidation of prereduced aldehyde oxidase.
Abscissa: Time of reoxidation.

and iron are near maximal. A small free-radical signal is also present. Admission of oxygen results in reoxidation of the enzyme to the steady state in which all the EPR signals are much smaller. The kinetics of decay of these signals have been determined and the data are shown in Figure 11. At the earliest stage of reoxidation the iron signal shows the greatest degree of reoxidation, with Mo(V) being the least reoxidized. The small free-radical signal was at a stage between those of Mo(V) and iron. With time, all the signals approached dimensions observable at steady state. These data show that the iron component is closest to the site of reaction with oxygen, and the molybdenum, the farthest. Thus this experiment with aldehyde oxidase leads to the same general conclusion as that described for xanthine oxidase from the single-turnover experiment.

3. Anaerobic Titration

The substrates of xanthine and aldehyde oxidases are two-electron donors. If, upon reduction by substrate, the two atoms of molybdenum and eight atoms of iron were able to lodge one electron each and if the two FAD molecules were capable of accepting two electrons each, it would be possible to introduce a total of at least 14 electrons into one molecule of either enzyme, in the absence of extraneous electron acceptors. Thus three to four molecules of substrate would be required, each of which would donate a pair of electrons in order to effect full reduction of an enzyme *unit* comprising one molybdenum, one FAD, and four iron atoms. If thermodynamic equilibrium is not attained at any stage between substrate and enzyme reduced to partial levels, titration of enzyme with increasing, stoichiometric, levels of substrate should lead to increasing degrees of reduction of enzyme until saturation is attained, with three to four molecules of substrate per enzyme unit. Such experiments have been performed with aldehyde oxidase, using spectrophotometric as well as EPR techniques to evaluate the degree of reduction of the enzyme. Figure 12 shows the absorption spectra of hepatic aldehyde oxidase in the native state and after equilibrium with increasing reducing equivalents of N'-methylnicotinamide. It will be seen that the extent of bleaching occasioned by 4 moles of substrate per mole of enzyme flavin approximates that observed after addition of dithionite, indicating that thermodynamic equilibrium for this system is greatly in favor of fully reduced enzyme. The loss of absorbancy in the region 500–600 mμ shows that the non-heme iron components also serve as electron acceptors. Since, however, no measurement was made of the unreacted substrate, if any, the absolute stoichiometry is uncertain.

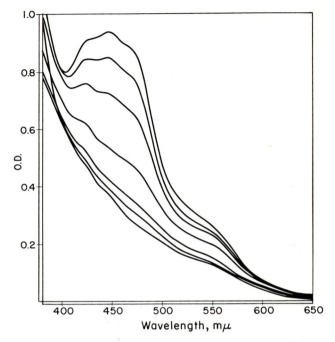

Fig. 12. Spectrophotometric changes on anaerobic titration of aldehyde oxidase with N'-methylnicotinamide at 25°. Top curve, oxidized enzyme; bottom curve, reduced with dithionite; intermediate curves, with increasing amounts of substrate, top to bottom. The next to bottom curve represents addition of approximately 4 moles of substrate per mole of flavin.

Figure 13 presents data obtained from a similar titration but employing measurement of EPR signals for evaluation of the extent of enzyme reduction, allowing the system to attain thermodynamic equilibrium after addition of each increment of substrate. Although full reduction of FAD is not observable by EPR, this technique provides the only method for monitoring changes in the molybdenum component, since the latter makes no detectable contribution to the optical absorption spectrum of aldehyde oxidase. The EPR data also indicate that maximal reduction of the enzyme, measured by the appearance of the iron and molybdenum signals, occurs on reaction with about 4 moles of N'-methylnicotinamide per mole of enzyme flavin as in the spectrophotometric experiment. The electrons derived from attack by the first 1–2 molecules of substrate are seen to lodge preferentially in the iron complex. (The free-radical signal

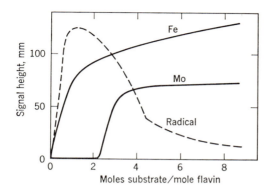

Fig. 13. Development of EPR signals at thermodynamic equilibrium during anaerobic titration of aldehyde oxidase with N'-methylnicotinamide.

represents only 5% of that expected for stoichiometric conversion of the flavin of the enzyme to a semiquinone.) The molybdenum component is stably reduced only after addition of yet further equivalents of substrate. (This thermodynamic situation does not preclude reduction of molybdenum as the initial kinetic event.) Thus, it is likely that all the flavin was fully reduced before permanent reduction of the molybdenum component.

Quantitation of the EPR signals has been somewhat difficult. The best available estimates indicate that the iron signal elicited by excess substrate is about 75% of that achievable with dithionite. Thus the EPR data also show that the equilibrium of the system,

$$\text{enzyme} + \text{substrate} \rightleftharpoons \text{reduced enzyme} + \text{oxidized substrate}$$

lies far in favor of reduced enzyme. Another observation arising out of these studies is that, in the partially reduced enzyme, the iron component tends to be the most reduced while the molybdenum component is little reduced, if at all. As a reflection of the potential difference along the internal electron transport chain of the enzyme, these data establish that molybdenum is nearest to the electron-withdrawing end (from the substrate) and iron closest to the electron-donating end (to O_2). These findings are in excellent agreement with the postulated electron transport sequence of Mo $\rightarrow$ FAD $\rightarrow$ Fe.

A similar experiment has also been performed with xanthine oxidase, with the enzyme exposed to different levels of xanthine for 550 msec prior

to EPR measurements. Under these conditions there was no thermodynamically distinguishable separation in the development of molybdenum
and iron signals, possibly because the 550 msec reaction time was insufficient for development of equilibrium. Another possible explanation is
that in the case of milk xanthine oxidase, equilibrium with substrate is
achieved when enzyme is only partially reduced. The fact that the degree
of bleaching of the absorption spectrum of milk xanthine oxidase by
xanthine amounts only to about 60% of that produced by dithionite (1)
might be another reflection of this phenomenon.

While application of EPR spectroscopy to the study of xanthine and
aldehyde oxidases has provided invaluable information, some unexpected
complexities have also been brought to light. Chief among them is the
valence state of molybdenum in the oxidized and reduced forms of these
enzymes. Milk xanthine oxidase exhibits no EPR signal in the oxidized
state, while the largest signal of Mo(V) so far detected in the reduced
enzyme has been estimated to represent about 25% of the theoretical
yield based on the molybdenum content of the enzyme (61). On the other
hand, native aldehyde oxidase always exhibits a signal at $g = 1.97$,
presumably ascribable to Mo(V), the magnitude of which is about 5%
of the total if all the molybdenum in the enzyme were Mo(V). On reduction
with substrate the size of the signal at $g = 1.97$ increases about fivefold,
indicating that more, but not all, of the molybdenum is in the Mo(V)
form on reduction. The situation has become more complicated with the
recent observation that xanthine dehydrogenase from *Micrococcus lactilyticus* exhibits a large signal at $g = 1.97$ in the oxidized state which is
altered very little on reduction with substrate (15). The comparative
studies on these enzymes has raised the question whether alternate
reduction of Mo(VI) to Mo(V) followed by reoxidation is an essential
characteristic of their functional behavior. Alternatively, one might consider that Mo(V) is the working form of this component throughout the
catalytic cycle in the steady state and that it serves primarily to accept
hydroxyl ions from the medium and direct these into the substrate, rather
than to conduct electrons. If the molybdenum is in close proximity to a
disulfide bond at the active site and is in equilibrium with it to varying
degrees, i.e.,

$$
\left[\mathrm{Mo(V)} \begin{array}{c} \diagup S \\ \mid \\ S \diagdown \end{array} \right]
\rightleftharpoons
\left[\mathrm{Mo(VI)} {-} S \begin{array}{c} \diagup \\ \\ S^- \diagdown \end{array} \right]
+ 2e^-
\left[\mathrm{Mo(V)} \begin{array}{c} \diagup S^- \\ \\ S^- \diagdown \end{array} \right]
$$

the molybdenum would exhibit varying extents of Mo(V) characteristics or none at all. The degree of enhancement of the Mo(V) EPR signal would then depend on the extent of the above-mentioned interaction in the oxidized form of the enzyme. The observed existence of transient sulfhydryl groups at the active sites of these enzymes would also be explainable on this basis. It is obvious that while comparative studies on related enzymes have led to unexpected differences in those aspects of their behavior where similarities might have been expected, such studies are of great importance to the ultimate construction of a unified hypothesis for their mechanism of action.

V. Dihydroorotic Dehydrogenase

Dihydroorotic dehydrogenase from *Zymobacterium oroticum* catalyzes the following reactions (17):

$$DPNH + H^+ + Orotate \rightleftharpoons Dihydroorotate + DPN^+ \tag{1}$$

$$DPNH + H^+ + O_2 \longrightarrow DPN^+ + H_2O_2 \tag{2}$$

$$Dihydroorotate + O_2 \longrightarrow Orotate + H_2O_2 \tag{3}$$

Activation of the enzyme by cysteine has been shown to be obligatory for optimal utilization of orotate or dihydroorotate as substrate in Reactions (1) and (3) (17). Using cysteine-preactivated enzyme, Miller and Massey (63) have conducted kinetic analyses of the above reactions with dihydroorotic dehydrogenase. Their results show that catalysis by the enzyme involves two binary complexes and two forms of the enzyme, oxidized and reduced. This mechanism is identical with that reported for milk xanthine oxidase (36).

The differential effect of cysteine on the reactions catalyzed by dihydroorotic dehydrogenase has prompted further investigation of the phenomenon. Aleman et al. (15) found that air oxidation of the enzyme or treatment with one mole of *p*-chloromercuribenzoate per mole of enzyme flavin resulted in marked loss of activities involving orotate or dihydroorotate as substrate without affecting the DPNH oxidase activity. Since the activities restored by cysteine are identical with those lost on air oxidation, the presence of an essential sulfhydryl group in the catalytic region of the enzyme was inferred.

Evidence for the role of the iron component in the transfer of electrons to molecular oxygen in this enzyme has already been detailed. In further exploration of the possible interactions between dihydroorotic dehydrogenase and its substrates, Aleman et al. (15) found that addition of orotate, a nonreducing substrate, caused subtle changes in the absorption spectrum of the enzyme. Figure 14 shows a series of difference spectra between enzyme–substrate mixture and enzyme alone. The difference spectrum is strongly indicative of modified environment of flavin and points to interaction of orotate with one of the flavin molecules in the enzyme. The enzyme–orotate complex has a dissociation constant of $2 \times 10^{-4}M$ which is in favorable relationship to the kinetic K_m of $1.8 \times 10^{-3}M$ for orotate reported by Miller and Massey (63).

On the basis of these studies, Aleman et al. (15) have postulated a basic functional unit in dihydroorotic dehydrogenase as shown in Figure 15. The dashed area in the lower sequence indicates the site of the unstable sulfhydryl group in the air-oxidized enzyme. The upper sequence shows that cysteine treatment renders the sulfhydryl group functional and

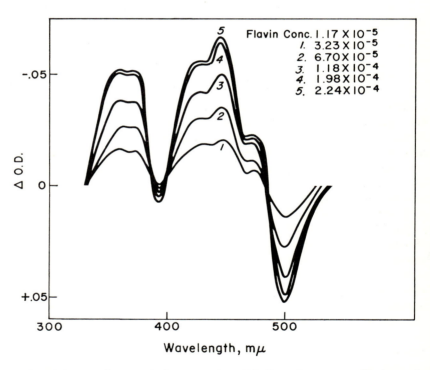

Fig. 14. Spectrophotometric demonstration of binding of orotate to dihydroorotic dehydrogenase. Difference spectra at 25° between enzyme plus increasing amounts of orotate in the blank compartment and enzyme plus corresponding volumes of buffer in the sample compartment in Cary Model 14 recording spectrophotometer.

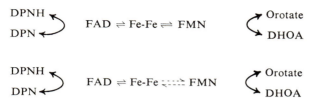

Fig. 15. Electron transport in dihydroorotic dehydrogenase.

establishes the link in the overall electron-transport sequence. The identification of FAD and FMN at the sites indicated is arbitrary.

Friedmann and Vennesland (17) noted that anaerobic addition of dihydroorotate to the enzyme resulted in partial bleaching of the absorbance at 454 mμ and a concomitant increase in absorbance at longer wavelengths. Rajagopalan et al. (58), employing absorption and reflectance spectrophotometry as well as EPR spectroscopy, observed that the emergence of long-wavelength absorbance on reduction of dihydroorotic dehydrogenase with DPNH or dihydroorotate paralleled the development of an EPR signal at $g = 2$ characteristic of organic free radicals. The species absorbing at long wavelengths was thus identified as a flavin semiquinone. In fact this was the first instance of a flavoprotein in which correlation has been observed between long-wavelength absorbance and EPR free-radical signal. At the same time, spectrophotometric observation of iron reduction, obvious in the case of aldehyde and xanthine oxidases as bleaching of absorbance at wavelengths longer than 520 mμ, has been precluded in dihydroorotic dehydrogenase by the compensating increase in absorbance on formation of the semiquinone.

Miller and Massey (63) and Aleman et al. (15) have studied the anaerobic reduction of the enzyme by titration with reducing substrates. Figure 16 shows the data obtained by Aleman et al. Addition of DPNH, up to two moles per mole of functional unit (1 FAD, 1 FMN, and 2 Fe), resulted in very nearly linear decrements in absorbancy at 454 mμ and increments in semiquinone formation, as measured at 620 mμ. Further, large increases in DPNH/enzyme ratios slowly evoked additional bleaching, tending ultimately to full reduction, just as obtained with dithionite. On the other hand, while one mole of dihydroorotate affected as much reduction as did the first mole of DPNH, further increases in dihydroorotate produced relatively little change. Since the potential of neither the enzyme nor the orotate/dihydroorotate couple is closely known, this inability to reduce the enzyme further with dihydroorotate might have been considered a

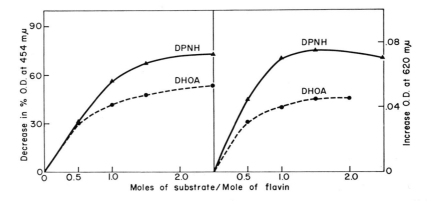

Fig. 16. Spectrophotometric changes during anaerobic titration of dihydroorotic dehydrogenase with DPNH and dihydroorotate (15).

reflection of the governing thermodynamic equilibria. However, as shown in Table III, reduction of the enzyme by dihydroorotate is actually enhanced and accelerated by the concomitant presence of DPN$^+$, suggesting

TABLE III

Effect of Substrates on the Absorbancy of Dihydroorotic Dehydrogenase at 454 mμ

	Relative O.D. at 454 mμ
Oxidized enzyme[a]	100
Enzyme + 0.5 mole of DPNH at 10 sec	68
Enzyme + 0.5 mole of DPNH + 5 moles of orotate at 10 sec	92
Enzyme + 1.0 mole of DPNH at 10 sec	38
Oxidized enzyme[b]	100
Enzyme + 0.5 mole of DHOA at 10 sec	87
Enzyme + 0.5 mole of DHOA at 20 min	71
Enzyme + 0.5 mole of DHOA + 2.5 moles of DPN at 10 sec	67
Enzyme + 0.5 mole of DHOA + 2.5 moles of DPN at 20 min	67
Enzyme + 7.5 moles of DHOA at 10 sec	68
Enzyme + 7.5 moles of DHOA at 20 min	38
Enzyme + 7.5 moles of DHOA + 0.1 mole of DPN at 10 sec	38

[a] Conditions were anaerobic, pH 6.5, enzyme containing 0.0256 μmole of flavin + addition of 2.0 μmoles of cysteine were placed in the side arm.

[b] Conditions as above, except pH of the system was 8.2 in this case.

an allosteric effect of DPN$^+$ binding which permits electron transfer from the first to the second flavin moiety of a functional unit.

Miller and Massey have employed the stopped flow technique for some rapid reaction studies on the kinetics of absorbance changes at 454 and 600 mμ on reaction of dihydroorotic dehydrogenase with DPNH and dihydroorotate (63). The extent of rapid changes in absorbance produced by DPNH indicated that besides reduction of flavin to the semiquinone level, bleaching of another chromophoric group had occurred. This chromophore had previously been identified as non-heme iron by Handler et al. (50). The experiments of Miller and Massey also showed that the rapid anaerobic bleaching produced by dihydroorotate was insufficient to represent reduction of flavin even to the semiquinone level. On this basis, they concluded that with dihydroorotate only the non-heme iron complex was reduced. This conclusion may not be valid, however, in view of the demonstration by Aleman et al. (15) that in the presence of DPN$^+$, bleaching of the enzyme by dihydroorotate was not only enhanced but occurred rapidly. Thus, a full understanding of the interactions of the enzyme with its substrates requires rapid reaction studies of the type conducted by Miller and Massey which include observation of the enzyme under catalytic conditions.

Aleman et al. (15) have also applied EPR spectroscopy to the study of the enzyme. In these studies, as shown in Figure 17, the appearance of the $g = 1.93$ and $g = 2.00$ signals proceeded in parallel during reduction by DPNH and by dihydroorotate and during reoxidation of the DPNH-reduced enzyme by orotate. In no instance was there evidence of an enzyme form in which the iron or flavin component was preferentially reduced. It seems clear that both flavin and iron participate in the catalytic cycle of this enzyme and that each pair of electrons which enters is distributed between one iron atom and one flavin molecule. But both organic substrates would be expected to accept or donate electron pairs and there is excellent evidence of orotate binding at a flavin. It is disappointing, therefore, that these studies have failed to reveal a kinetic difference between the flavin semiquinone signal and the iron signal. It is conceivable that in this enzyme, electron transfer between flavin and iron occur even at the liquid nitrogen temperature at which the samples for EPR spectroscopy are held for prolonged periods after substrate–enzyme interaction is halted by such freezing. While the EPR studies and most of the spectrophotometric studies have failed to provide evidence for a sequence of electron carriers in this enzyme, a rapid reaction experiment conducted by Miller and Massey on the reoxidation, by oxygen, or DPNH-reduced enzyme might be of significance (63). In this experiment the flavin semiquinone, measured

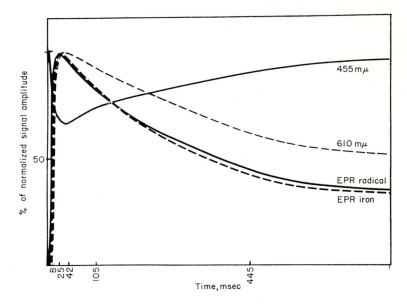

Fig. 17. EPR study of the kinetics of dihydroorotic dehydrogenase during a "single turnover." Enzyme, DPNH, and orotate in molar ratios of 1:2:5 were reacted at 22.5° for indicated times prior to rapid-freeze quenching.

at 600 mµ, remained constant in amount until most of the DPNH was oxidized, and then disappeared rapidly with full reoxidation of the flavin. In contrast, the absorbance at 454 mµ showed a small but rapid initial increase whereas the major change, corresponding to full reoxidation of the enzyme, occurred subsequently in conjunction with the disappearance of 600 mµ absorbance. In view of the previously documented evidence for the role of iron in reduction of oxygen, it is tempting to speculate that the rapid initial increase in absorbance at 454 mµ reflects the faster reoxidation of the iron. If this were so, this experiment would have provided the first separation in sequence between flavin and iron as electron carriers in dihydroorotic dehydrogenase.

VI. The Non-Heme Iron Complex

The ferredoxins of bacteria, the PPNR (photosynthetic pyridine nucleotide reductase) type of iron proteins from plants, bacteria, and mitochondria and the more complex iron flavoenzymes constitute a class of proteins in which related iron complexes of great stability function as redox carriers. Surprisingly, the oxidation-reduction potentials of these complexes vary over a wide range, from -450 mV in the bacterial and plant proteins to $+350$ mV in a ferredoxin-like protein isolated from *Chromatium* (64). The common characteristic of all these proteins is the liberation of H_2S, on denaturation by acid or heat, in an amount usually stoichiometrically equivalent to the iron content, although somewhat higher sulfide-to-iron ratios have been reported in DPNH dehydrogenase preparations from mitochondria (65). The optical rotatory dispersion spectrum of spinach PPNR was shown by Ulmer and Vallee to display several large superimposed Cotton effects (66). In this laboratory, the metalloflavoproteins discussed herein as well as a non-heme iron protein from mitochondria (67) have been shown to display optical rotatory dispersion spectra remarkably similar to that of PPNR with extremely large changes in dispersion observed, in the region of the absorption maximum due to the oxidized form of the iron complex, when these enzymes are reduced by their substrates or by dithionite. It is thus obvious that the aspect of molecular structure responsible for the characteristic absorption spectra of the non-heme iron components of these proteins is optically active, indicating that the ligand environment of the iron atoms possesses octahedral asymmetry or at least an asymmetry imposed by a rather distorted tetahedral structure. Interestingly, clostridial ferredoxin, which is markedly different from PPNR in its optical absorption spectrum, nevertheless displays an optical rotatory dispersion spectrum quite similar to that of PPNR (68). Any hypothetical structure for the non-heme iron complex must, therefore, also account for the observed asymmetry of the structure.

Mention has been made of the EPR signal at $g = 1.94$ observed on reduction of the non-heme iron component of the metalloflavoproteins. A detailed discussion of the probable valences of the iron in the oxidized and reduced states and of the possible chemical basis for the observed signal has recently been presented by Beinert and Palmer (4). The available evidence indicates that in the oxidized complex the iron is in the low-spin ferrous state and that the introduction of an electron into the complex results in its delocalization between the iron and one or more of its ligands. In this concept, it is the complex as a whole, and not the central iron atom or one of the ligands exclusively, which serves as the electron sink. The iron in the reduced complex would have the formal valence state of Fe(I). The coexistence of "labile sulfur" with iron in all the proteins of this group renders this sulfur a prime candidate for sharing the newly introduced and delocalized electron in the complex.

The importance of the iron–sulfur bond to the structural integrity of the complex was demonstrated with ferredoxin (69) and PPNR (70) by the ability of mercurials to eliminate the characteristic absorption spectra of these proteins. Brumby, Miller, and Massey (21) have extended these observations to xanthine oxidase and have shown that the contribution by non-heme iron to the absorption spectrum of the enzyme is abolished by incubation with mercurials as well as by high concentrations of urea. Under these conditions the degree of inactivation of the enzyme corresponded to the extent of loss of protein-bound flavin, iron, and labile sulfide. These workers also carried out, anaerobically, amperometric determination of sulfhydryl groups and disulfide bonds of the oxidized and xanthine-reduced forms of the enzyme. They observed that reduction of the enzyme caused the formation of about four sulfhydryl groups per mole of flavin, with a corresponding decrease in the apparent disulfide content. On the basis of these findings Brumby et al. have postulated the involvement of a unique type of disulfide group in the structure of the iron complex as shown in Scheme I.

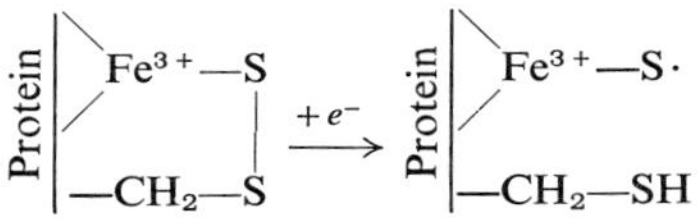

$$\text{Protein} \begin{array}{c} \diagdown \\ \diagup \end{array} \begin{array}{l} Fe^{3+}\!-\!S \\ | \\ -CH_2\!-\!S \end{array} \xrightarrow{+e^-} \text{Protein} \begin{array}{c} \diagdown \\ \diagup \end{array} \begin{array}{l} Fe^{3+}\!-\!S\cdot \\ \\ -CH_2\!-\!SH \end{array}$$

Scheme I

The disulfide is visualized to be linked to the protein through the iron atom in the Fe(III) state at one end and a cysteine residue at the other. Introduction of an electron into the system is presumed to reduce the disulfide group with the formation of a free sulfhydryl group and an iron–sulfur radical.

This provocative theory may need to be reconsidered in view of the evidence indicating that the iron in the complex exists as Fe(II) rather than Fe(III) and of possible further reduction of the iron–sulfur radical to iron sulfide. Such further reduction could make it incompatible with the observed maximal formation of the $g = 1.94$ signal in xanthine oxidase on reduction with dithionite.

VII. Concluding Remarks

It will be evident that the class of metalloflavoproteins includes some of the most complex of all known enzymes. Indeed, aldehyde oxidase may be the most complex of known enzymes. Much effort has been devoted to demonstration that the metal components of this group are catalytically functional rather than unusual forms of molecular cement. Each has proved to be a miniature electron-transport system although the absolute transport order has proved somewhat elusive. Much has been learned of their structure and function; yet most of the classical questions of the enzyme chemist remain unanswered. The flavin binding sites, metal ligands, substrate binding sites, proximity of metal to metal and of metal to flavin remain unknown. And despite the great efforts of the last five years, there is, as yet, no satisfactory description of the structure and function of a metalloflavoprotein.

References

1. R. C. Bray, in *The Enzymes*, Vol. 7, P. D. Boyer, H. A. Lardy, and K. Myrbäck, Eds., Academic Press, New York, 1963, p. 533.
2. A. Nason, in *The Enzymes*, Vol. 7, P. D. Boyer, H. A. Lardy, and K. Myrbäck, Eds., Academic Press, New York, 1963, p. 587.
3. H. Beinert, in *Non-Heme Iron Proteins: Role in Energy Conversion*, A. San Pietro, Ed., The Antioch Press, Yellow Springs, Ohio, 1965, p. 23.
4. H. Beinert and G. Palmer, in *Advances in Enzymology*, Vol. 27, F. F. Nord, Ed., Interscience, New York, 1965, p. 105.
5. E. G. Ball, *J. Biol. Chem.*, **128**, 51 (1939).
6. H. S. Corran, J. G. Dewan, A. H. Gordon, and D. E. Green, *Biochem. J.*, **33**, 1694 (1939).
7. D. A. Richert and W. W. Westerfeld, *J. Biol. Chem.*, **203**, 915 (1953).
8. D. A. Richert and W. W. Westerfeld, *J. Biol. Chem.*, **209**, 179 (1956).
9. H. R. Mahler and D. G. Elowe, *J. Biol. Chem.*, **210**, 165 (1954).
10. T. P. Singer, E. B. Kearney, and P. Bernath, *J. Biol. Chem.*, **223**, 599 (1956).
11. R. K. Kielley, *J. Biol. Chem.*, **216**, 405 (1955).
12. C. N. Remy, D. A. Richert, R. J. Doisy, I. C. Wells, and W. W. Westerfeld, *J. Biol. Chem.*, **217**, 293 (1955).
13. E. J. Landon and C. E. Carter, *J. Biol. Chem.*, **235**, 819 (1960).
14. W. H. Bradshaw and H. A. Barker, *J. Biol. Chem.*, **235**, 3620 (1960).
15. V. Aleman, S. T. Smith, K. V. Rajagopalan, and P. Handler, in *Flavins and Flavoproteins*, E. C. Slater, Ed., Elsevier, New York, 1966.
16. K. V. Rajagopalan, I. Fridovich, and P. Handler, *J. Biol. Chem.*, **237**, 922 (1962).
17. H. C. Friedmann and B. Vennesland, *J. Biol. Chem.*, **235**, 1526 (1960).
18. K. V. Rajagopalan and P. Handler, *J. Biol. Chem.*, **239**, 1509 (1960).
19. A. San Pietro, in *Light and Life*, W. D. McElroy and B. Glass, Eds., The Johns Hopkins Press, Baltimore, Md., 1961, p. 631.
20. P. E. Brumby and V. Massey, *Biochem. J.*, **89**, 46p (1963).
21. P. E. Brumby, R. W. Miller, and V. Massey, *J. Biol. Chem.*, **240**, 2222 (1965).
22. D. B. Morrell, *Biochim. Biophys. Acta*, **18**, 221 (1955).

23. R. C. Valentine, R. L. Jackson, and R. S. Wolfe, *Biochem. Biophys. Res. Commun.*, **7**, 453 (1962).
24. P. Handler, in "Enzyme Models and Enzyme Structure," *Brookhaven Symp. Biol.*, **15**, 263 (1962).
25. A. H. Gordon, D. E. Green, and V. Subrahmanyan, *Biochem. J.*, **34**, 764 (1940).
26. H. R. Mahler, R. Mackler, D. E. Green, and R. M. Bock, *J. Biol. Chem.*, **210**, 465 (1954).
27. G. Palmer, *Biochim. Biophys. Acta*, **56**, 444 (1962).
28. S. Udaka and B. Vennesland, *J. Biol. Chem.*, **237**, 2018 (1962).
29. V. H. Booth, *Biochem. J.*, **29**, 1732 (1935).
30. K. V. Rajagopalan, I. Fridovich, and P. Handler, *J. Biol. Chem.*, **236**, 1059 (1961).
31. K. V. Rajagopalan and P. Handler, *J. Biol. Chem.*, **239**, 2027 (1964).
32. F. Bergel and R. C. Bray, *Biochem. J.*, **73**, 182 (1959).
33. L. Greenlee and P. Handler, *J. Biol. Chem.*, **239**, 1090 (1964).
34. I. Fridovich, *Arch. Biochem. Biophys.*, **109**, 511 (1965).
35. A-M. Perault, C. Valdemoro, and B. Pullman, *J. Theoret. Biol.*, **2**, 180 (1961).
36. I. Fridovich, *J. Biol. Chem.*, **239**, 3519 (1964).
37. W. W. Cleland, *Biochim. Biophys. Acta*, **67**, 104 (1963).
38. M. Dixon and S. Thurlow, *Biochem. J.*, **18**, 976 (1924).
39. L. Greenlee and P. Handler, *J. Biol. Chem.*, **239**, 1096 (1964).
40. W. E. Knox, *J. Biol. Chem.*, **163**, 699 (1946).
41. G. Palmer, *Biochim. Biophys. Acta*, **64**, 135 (1962).
42. I. Lieberman and A. Kornberg, *Biochim. Biophys. Acta*, **12**, 223 (1953).
43. V. Aleman, Ph.D. dissertation, Duke University, 1965.
44. I. Fridovich and P. Handler, *J. Biol. Chem.*, **235**, 1836 (1961).
45. I. Fridovich and P. Handler, *J. Biol. Chem.*, **237**, 916 (1962).
46. B. L. Horecker and L. A. Heppel, *J. Biol. Chem.*, **178**, 683 (1949).
47. J. R. Totter, V. J. Medina, and J. L. Scoseria, *J. Biol. Chem.*, **235**, 238 (1960).
48. J. R. Totter, E. C. deDugros, and C. Riveiro, *J. Biol. Chem.*, **235**, 1839 (1960).
49. L. Greenlee, I. Fridovich, and P. Handler, *Biochemistry*, **1**, 779 (1962).
50. P. Handler, K. V. Rajagopalan, and V. Aleman, *Federation Proc.*, **23**, 30 (1964).
51. R. W. Miller and V. Massey, *J. Biol. Chem.*, **240**, 1466 (1965).
52. K. V. Rajagopalan and P. Handler, *J. Biol. Chem.*, **239**, 2022 (1964).
53. M. Dixon and D. Keilin, *Proc. Roy. Soc. (London)*, **B119**, 159 (1936).
54. S. T. Smith and P. Handler, unpublished data.
55. R. C. Bray, in *Flavins and Flavoproteins*, E. C. Slater, Ed., Elsevier, New York, 1966.
56. J. M. Peters and D. R. Sanadi, *Arch. Biochem. Biophys.*, **93**, 312 (1961).
57. G. Palmer, R. C. Bray, and H. Beinert, *J. Biol. Chem.*, **239**, 2657 (1964).
58. K. V. Rajagopalan, V. Aleman, P. Handler, W. Heinen, G. Palmer, and H. Beinert, *Biochem. Biophys. Res. Commun.*, **8**, 220 (1962).
59. R. C. Bray, *Biochem. J.*, **81**, 189 (1961).
60. R. C. Bray and R. Pettersson, *Biochem. J.*, **81**, 194 (1961).
61. R. C. Bray, G. Palmer, and H. Beinert, *J. Biol. Chem.*, **239**, 2667 (1964).
62. K. V. Rajagopalan, P. Handler, G. Palmer, and H. Beinert, to be published.
63. R. W. Miller and V. Massey, *J. Biol. Chem.*, **240**, 1453 (1965).
64. R. G. Bartsch, in *Bacterial Photosynthesis*, H. Gest, A. San Pietro, and L. P. Vernon, Eds., The Antioch Press, Yellow Springs, Ohio, 1963, p. 315.

65. C. J. Lusty, J. M. Machinist, and T. P. Singer, *J. Biol. Chem.*, **240**, 1804 (1965).
66. D. D. Ulmer and B. L. Vallee, *Biochemistry*, **2**, 1335 (1963).
67. J. S. Rieske, D. H. MacLennan, and R. Coleman, *Biochem. Biophys. Res. Commun.*, **15**, 338 (1964).
68. V. Aleman and P. Handler, unpublished observation.
69. W. Lovenberg, B. B. Buchanan, and J. C. Rabinowitz, *J. Biol. Chem.*, **238**, 3899 (1963).
70. D. L. Keister and A. San Pietro, *Arch. Biochem. Biophys.*, **103**, 45 (1963).

The Respiratory Chain-Linked Dehydrogenases*

Thomas P. Singer, *Biochemistry Department, University of California School of Medicine and Molecular Biology Division, Veterans Administration Hospital, San Francisco, California*

I.	Introduction .	339
II.	General Properties .	340
III.	Linkage to the Respiratory Chain	342
IV.	Artificial and Physiological Electron Acceptors in the Assay of the Respiratory Chain-Linked Dehydrogenases	344
V.	DPNH Dehydrogenase .	347
VI.	DPNH-Coenzyme Q Reductase	356
VII.	Succinate Dehydrogenase	361
	A. Background	361
	B. Assay and Chemical Determination .	362
	C. Nature and Composition of Preparations .	363
	D. Properties	363
	E. Evolutionary Development of Succinate Dehydrogenase	367
	F. Reconstitution	369
VIII.	Concluding Remarks	373
	References .	375

I. Introduction

The term "respiratory chain-linked dehydrogenases" has come into use in recent years to denote a group of flavoproteins which are structurally and functionally associated with the terminal electron transport system and which serve as ports of entry into the respiratory chain for reducing equivalents originating from oxidizable metabolites. In the earlier literature

* The original studies reported here were supported in part by grants from the National Heart Institute (HE 10027) and the National Science Foundation (GB 3762) and by Contract No. 5024(00) between the Office of Naval Research and the University of California.

these enzymes were referred to as "cytochrome-reducing dehydrogenases" but it is now recognized that this term is misleading, since some important members of this group (e.g., succinate and DPNH dehydrogenases) fail to react with cytochromes in the isolated state and their interaction with the cytochrome system in mitochondria almost certainly involves mediation by additional components. The enzymes usually classified in this group are succinate and DPNH dehydrogenases, α-glycerophosphate dehydrogenase, choline dehydrogenase, and the electron-transferring flavoprotein (ETF). In addition, recent evidence (1) suggests that the D($-$) lactate cytochrome reductase and the L($+$) lactate dehydrogenase of yeast mitochondria and the L-galactono-γ-lactone dehydrogenase of higher plants may be logically included in this classification. Of these enzymes only the first two are of ubiquitous occurrence in mitochondria and they will be used to illustrate some of the unique properties of this type of enzyme. For a more complete discussion of the individual enzymes the reader is referred to a recent review (1).

II. General Properties

It has been known for several decades that typical members of this group of enzymes are so intimately bound to the "insoluble matrix" of cells as to make their extraction uncommonly difficult. In fact, for many years it was assumed that enzymes like succinate dehydrogenase are catalytically active only when structurally bound to the cytochrome system and that any attempt to extract and study them as individual protein entities is bound to be futile. The extraction and purification of succinate dehydrogenase from mammalian cells (2) paved the way toward recognition of the fact that, although in most cases special and often elaborate procedures have to be used to obtain these dehydrogenases free from other components of the respiratory chain and in soluble form, they may be isolated, characterized, and studied as individual flavoproteins and that the essential characteristics of these enzymes do not necessarily change on separation from their mitochondrial environment. By now all known respiratory chain-linked dehydrogenases have been isolated in partially or extensively purified form and their characteristics as individual enzymes have been extensively studied.

The generalizations valid for the entire group are few indeed. They are all flavoproteins; hence the general properties of flavoproteins (see Palmer and Massey, this volume) hold true also for this group of enzymes. With two known exceptions [ETF and the D($-$) lactate dehydrogenase of yeast] they are metal flavoproteins; hence many of their catalytic properties are consequences of this fact (cf. Rajagopalan and Handler, this volume). Succinate, DPNH, α-glycerophosphate, and choline dehydrogenases contain non-heme iron, while $D(-)$ lactate cytochrome reductase is a zinc flavoprotein (3). As in several other metal flavoproteins, the characteristic reactivities of the flavin prosthetic group are masked in the native structure of the enzyme. None of these enzymes react at a significant rate with molecular O_2, a fact of considerable physiological importance in energy conservation. With the exception of $D(-)$ lactate cytochrome reductase and, possibly, of ETF, they also fail to reduce cytochrome c appreciably, although on suitable conformation changes DPNH dehydrogenase may be transformed into DPNH cytochrome c reductase (see below). The substrate specificity of these flavoproteins tends to be pronounced: most of them react only with one substrate pair of physiological occurrence.

The characteristics mentioned above clearly do not set apart this group of enzymes from other flavoproteins. What, then, are the unique features of these dehydrogenases which justify their classification in a separate group?

As already mentioned, the main distinguishing feature of this group of enzymes is their structural and functional linkage to the respiratory chain of mitochondrial membranes. One consequence of this fact is that one often has to resort to special techniques to extract these flavoproteins in relatively unmodified form. Thus α-glycerophosphate, choline, and DPNH dehydrogenases and the $D(-)$ lactate cytochrome reductase of yeast become extractable only on digestion with phospholipase A, which may hydrolyze some of the lipids responsible for binding the flavoproteins to the respiratory chain.* Further, significant solubilization of succinate dehydrogenase from mammalian or yeast cells requires desiccation with an organic solvent, followed by brief exposure to alkaline pH (1). Once removed from their insoluble lipid environment these flavoproteins often become remarkably labile, as if the maintenance of their native conformation depended on a stable union with other mitochondrial constituents. Thus it is well known that while the activity of succinate dehydrogenase from heart muscle may be preserved for several days in the frozen state in particulate preparations, once solubilized the enzyme tends to deteriorate in a matter of hours even when great care is taken to stabilize it. While in this case it might be argued that the gradual inactivation in the soluble

---

* It has been questioned (4) that phospholipase A is the agent responsible for the solubilization of respiratory chain-linked flavoproteins. While the final answer has to await the isolation of pure phospholipase A from *N. naja* venom, the evidence in favor of this hypothesis is strong and there is no clear-cut evidence against it.

state is the result of the conditions used in its extraction (desiccation with organic solvents and alkaline extraction) with other enzymes, such as DPNH or choline dehydrogenases, which are extracted by specific enzymes under very mild conditions (5,6) the lability of the soluble enzyme to all types of treatments is still very much higher than of particle-bound preparations.

An important property of most respiratory chain-linked flavoproteins is their high selectivity for electron acceptors. In their native habitat these enzymes pass electrons from the substrate to cytochromes b and c_1 and then via the respiratory chain to O_2. The identity of the immediate reaction partners of the individual flavoproteins is not known in most cases, although with DPNH and succinate dehydrogenases the weight of evidence suggests that it is coenzyme Q. At least in the case of mammalian enzymes,* once the flavoproteins are extracted in soluble form, they show no reaction with CoQ_{10} or cytochromes b or c_1. It is impossible to decide in most cases whether this lack of reactivity between presumed reaction partners is due to preparative modification of the flavoprotein or of the cytochrome in question, to the removal of an essential component (e.g., lipid) required for catalytic interaction, or to the circumstance that rapid interaction between the components requires an integrated structure which is not established on simple mixing of the components. In the case of the DPNH and succinate dehydrogenase segments of the respiratory chain, available evidence suggests that an absolute requirement for lipids exists in the interaction of the flavoproteins with CoQ (7–9). In this connection it may not be fortuitous that reconstitution of the respiratory chain even from *particulate* fragments has been successful only in cases where the system was fragmented at points where readily extractable and replaceable components function (CoQ_{10} or cytochrome c (10)).

The difficulty in reproducing the physiological events of electron transfer with soluble preparations was directly responsible for the fact that for many years our understanding of this important group of enzymes lagged behind knowledge of readily extractable flavoproteins and has been the main cause of the apparent contradictions in the literature about the respiratory chain-linked dehydrogenases. These facets are discussed in Section IV.

III. Linkage to the Respiratory Chain

Direct information on the nature of the chemical bonds responsible for the tight linkage of these dehydrogenases to the respiratory chain is still

---

* D(−) Lactate cytochrome reductase and L(+) lactate dehydrogenase may conceivably react directly with cytochrome c in the cell.

lacking. The circumstance that of a variety of methods tested only phospholipase A has been found to extract in soluble form a number of respiratory chain-linked flavoproteins (5,6,11) and is the method of choice for the solubilization of other enzymes in this group (12) suggests the involvement of phospholipid bridges. Although succinate dehydrogenase is neither solubilized nor inactivated by phospholipase A, a possible role of lipids in the attachment of this enzyme to the cytochrome system is suggested by the fact that prior desiccation with an organic solvent is a prerequisite for the extraction of this enzyme in significant yield on exposure to alkaline pH (13).

Alternative views on the attachment of respiratory chain-linked flavoproteins to the cytochrome system have been advanced by King (14) and Klingenberg (15). King has been studying the alkali inactivation of the succinate oxidase system and its reversal by soluble succinate dehydrogenase (cf. Section VII). He reported that on exposure of a heart muscle preparation to pH 10 and centrifugation at this pH an extract was obtained which contained succinate dehydrogenase active both in phenazine methosulfate assays and in reconstituting the succinate oxidase activity of alkali-treated heart muscle preparations. On the other hand, if the extract was neutralized prior to centrifugation, it contained no succinate dehydrogenase by these criteria. From these data it was concluded that the linkage between the dehydrogenase and the respiratory chain may not be covalent in nature but may be due to hydrogen bonding or simple electrostatic attraction.

Since these conclusions were based on catalytic assays, which measure an indeterminate fraction of the succinate dehydrogenase content (13), Kimura and Singer (16) repeated the experiments and determined the extraction of succinate dehydrogenase at various pH values by a chemical method (covalently bound flavin), which measures the concentration of both active and inactive enzymes. It was found that whether the centrifugation is performed before or after neutralization, following exposure to pH 10, only a trace of succinate dehydrogenase is extracted: the bulk remains attached to the insoluble particles but is inactivated by the alkali and thus escaped detection in King's experiments. This is in accord with the fact, observed by many workers, that the enzyme becomes extractable only after treatment with lipid solvents. Thus there appears to be no experimental basis for assuming that the forces uniting the enzyme with the cytochrome chain are hydrogen bonds or electrostatic attraction or similar, relatively weak interaction.

Klingenberg (15) postulates that the interaction of components in the respiratory chain depends on protein–protein interaction and is determined by the primary structure of the individual components: lipids and structural protein are regarded as a rather unspecific matrix for these interactions. The shortcoming of this interesting suggestion is that it is contrary to the general experience that components of the respiratory chain (with the exception of cytochrome c and coenzyme Q) are separable from each other only under rather special conditions and their extraction as individual proteins in true solution in some cases has not even been accomplished, while most attempts to reconstitute a functional unit from individual components or to reinsert a component in the original position have met with failure.

IV. Artificial and Physiological Electron Acceptors in the Assay of the Respiratory Chain-Linked Dehydrogenases

As we have seen, the physiological reaction partner of most of the respiratory chain-linked flavoproteins is not known for certain. Even in the case of succinate and DPNH dehydrogenases, where the weight of evidence suggests CoQ_{10} as the natural oxidant, one must remember that in mitochondria the reaction between flavoprotein and endogenous CoQ is lipid-dependent and probably requires a certain degree of spatial organization. Hence the addition of a dispersion of CoQ_{10} to a purified flavoprotein devoid of lipids probably does not simulate the conditions of the physiological reaction very closely.

As a result of these limitations, most workers in the field have been using artificial electron acceptors for the assay of the individual flavoproteins. When using a dye for the assay of a respiratory chain-linked flavoprotein in the course of attempts at isolation, one is faced with the problem that until the enzyme is separated from all other components of the respiratory chain and significantly purified, one cannot be certain that the particular dye is indeed capable of reacting *directly* with the flavoprotein and of measuring its full activity. On the other hand, the success of isolation hinges primarily on the availability of a reliable assay method.

While at first glance this might appear to be a hopeless dilemma, in practice one can usually eliminate all but one or two commonly available electron acceptors on the basis of the fact that even in the particulate source they measure less activity than is required for the known rate of oxidation of the substrate via the complete chain or that the reaction with the particular dye is sensitive to inhibitors whose loci of action in the cytochrome system are known to be at some distance from the flavoprotein. Then with the artificial electron acceptors which pass these tests one tries to elaborate conditions under which the process of extracting the enzyme from the particles and purifying it neither creates nor destroys catalytic activity unaccountably. Once the isolation of the flavoprotein is accomplished and its properties become known, it may become possible to devise a more nearly physiological assay procedure and thereby check the validity of the conclusions based on the use of artificial oxidants.

The sequence described is essentially that which has been used in the isolation of DPNH dehydrogenase. The initial isolation of the flavoprotein was accomplished by the use of ferricyanide under special conditions (5). Alternative assay procedures failed to meet the criteria listed (17). Following the isolation of the enzyme it was found that highly purified preparations gave the same EPR signal at $g = 1.94$ on reduction by substrate as mitochondria or even whole cells and that the catalytic activity in the ferricyanide assay paralleled that calculated from the rate of appearance of the $g = 1.94$ signal in a satisfactory manner (18).

One problem inherent in the use of artificial electron acceptors is their unspecificity. This has created no difficulty in the case of enzymes like succinate or choline dehydrogenases, since there appears to be only one flavoprotein in mitochondria for the oxidation of each of these substrates. On the other hand, several mitochondrial flavoproteins are actually or potentially capable of oxidizing DPNH to a greater or lesser extent; hence measurement of the DPNH–ferricyanide interaction in mitochondria would tend to register the activities of all flavoproteins capable of catalyzing this reaction. Fortunately, the activity of the respiratory chain-linked enzyme is much higher than that of other known mitochondrial enzymes capable of DPNH oxidation (at least in heart muscle) and in derivative particles, such as ETP or ETP_H, the enzyme in question is responsible for most of the DPNH–ferricyanide activity.

The use of artificial electron acceptors for the assay of this group of enzymes has also been criticized (19,20) on the grounds that they fail to distinguish between fully active enzyme preparations and those which have undergone some subtle preparative modification. The basic observation reported by King (19,20) is that on storage of soluble succinate dehydrogenase activity in the phenazine methosulfate assay decays more slowly than does the ability of the preparation to combine with alkali-inactivated heart muscle particles (which lack succinate oxidase activity) and to restore their ability to oxidize succinate (cf. Section VII). This author, in fact, coined the meaningless term "phenazine reductase" (20) to denote a stage in the decay of the enzyme where it is no longer capable of restoring the activity of alkali-treated heart muscle particles but is still capable of catalyzing succinate oxidation in the standard catalytic assay.

The present author's views have been detailed elsewhere (1,13) and may be summed up as follows. Succinate dehydrogenase from aerobic sources, once extracted, is a notoriously unstable enzyme. On storage it loses its various catalytic activities at different rates. Thus, some 10 years ago it was reported (21) that activity in the succinate–phenazine methosulfate assay decays far faster than that in the fumarate–FMNH$_2$ assay. The underlying changes do not appear to be major ones, such as dissociation into subunits would be, since they have not been detected by physical measurements but may be viewed as a type conformational alteration. The difference in the rates of inactivation noted by King (19,20) is only a special case of this phenomenon and has no bearing on the use of artificial electron acceptors, since instances are not uncommon in the literature where differential inactivation of an enzyme toward one natural substrate but not another has been recorded.

It has been proposed (19) that "only physiological reactions, like the reconstitution [i.e., reactivation of alkali-treated particles] can be used with certainty in the final judgement of the nativity of the enzyme isolated . . .." This appears to this author to be somewhat naive. First, it would seem unlikely that any preparation of succinate dehydrogenase hitherto described (all of which use rather drastic methods for extraction) would yield a preparation which is completely "native." In fact, analysis of the preparations (20) which are considered active in the reconstitution test has revealed (22) the existence of multiple forms of succinate dehydrogenase and only a very small fraction of the total succinate dehydrogenase content was capable of reactivating alkali-treated preparations in the succinoxidase assay. Second, the reconstitution test advocated is far from physiological: it is not even a reconstitution in the sense of replacing a missing component in a multienzyme system, but a more complex phenomenon (16). A physiological assay for purified succinate dehydrogenase would be of great value indeed, but so far it remains a distant goal which has not been quite reached.

The use of naturally occurring substances, even components of the respiratory chain, as electron acceptors in the assay of respiratory chain-linked flavoproteins is also not devoid of pitfalls. As described in Section V, DPNH dehydrogenase may be readily fragmented into smaller units capable of reducing cytochrome c rapidly, which the unmodified enzyme hardly reacts with. DPNH-cytochrome c reductases isolated from heart and liver mitochondria are not naturally occurring enzymes but modification products of the dehydrogenase, created by the extraction procedure, such procedures having been selected so as to yield maximal cytochrome c reductase activity (23,24). Similarly, the recently described soluble DPNH-ubiquinone reductase (DPNH-CoQ reductase) (25,26) appears to be a modification product created by the extraction process, in which a new reaction site for CoQ emerges, one which is probably not identical with the physiological site (9,27).

V. DPNH Dehydrogenase

The flavoprotein was first solubilized from a nonphosphorylating submitochondrial particle, ETP, from heart muscle by means of phospholipase A (5,28). Activity was followed by the ferricyanide method under rigidly defined conditions (17,29). Subsequently, the enzyme was also isolated from the phosphorylating particle, ETP_H, from mitochondria, and from liver. It was shown that (1) no activity is destroyed or created by the extraction procedure, (2) that the reaction site of ferricyanide, under the conditions specified, is with the flavoprotein itself, and (3) that DPNH–ferricyanide activity parallels the rate calculated from measurements of the rate of appearance of the EPR signal at $g = 1.94$ on addition of substrate.

The purified enzyme is a high molecular weight flavoprotein containing FMN, non-heme iron, and "labile sulfide" as constituents. The ratio of FMN:Fe:S is about 1:16–18:27 (28,30). Since homogenous preparations have not been obtained (the enzyme is present in minute amounts in mitochondria and appears to exist in several polymeric forms in purified preparations), estimates of the molecular weight are based on FMN content, corrected for ultracentrifugally detectable impurities. Assuming 1 mole of FMN per mole of enzyme, the molecular weight is estimated to be 550,000–800,000, in accord with sedimentation velocity data (28).

As is true of other iron flavoproteins (Rajagopalan and Handler, this volume), the enzyme shows a generalized absorption over the entire visible range, quite unlike that of simple flavoproteins. There are no distinct features except for a maximum near 410 mμ. These features are related to the unusually high non-heme iron:flavin ratio, since most of the absorption is due to iron–protein bonds, rather than flavin. The difference spectrum after reduction by DPNH is shown in Figure 1.

Estimates of the turnover number range from 0.55 to 1.4×10^6 per min; the uncertainty is due to the presence of polymeric forms in highly purified preparations and to the uncertainty of the molecular weight (28,30). Taking the lower value, this is one of the highest values recorded for a flavoprotein and the highest one for any constituent of the mammalian respiratory chain. Thus, although the molar concentration of DPNH dehydrogenase in mammalian mitochondria is at least an order of magnitude

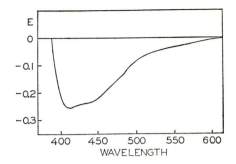

Fig. 1. Difference spectrum of DPNH dehydrogenase obtained immediately after the addition of $8 \times 10^{-4}M$ DPNH to the enzyme. Negative values denote decolorization. From Ringler et al. (5).

lower than that of any of the cytochromes [ref. 27, and Klingenberg, this volume), in view of the high turnover number the potential activity of the dehydrogenase is not likely to become rate-limiting during the events of electron transport via the respiratory chain.

Both purified and particle-bound forms of the enzyme catalyze the oxidation of a number of analogs of DPNH at appreciable rates as well as transhydrogenations between DPNH and DPN analogs (21). Even TPNH is oxidized but at only 0.03% of the rate of DPNH (31); hence the reaction may be of little biological significance. The specificity of the enzyme for electron acceptors is quite pronounced (Table I) and is

TABLE I

Acceptor Specificity of DPNH Dehydrogenase

	Relative rate at V_{max}[a]	
Electron acceptor	Beef heart enzyme	Beef liver enzyme
Ferricyanide	100	100
DCIP	0.5	~0.5
Cytochrome c	0.02	0.014
Cyrochrome b_5	—	0.009
Menadione	~0.2	0.1
Lipoamide	0.06	0.3

[a] Each assay was performed at 30° under optimal conditions for the particular electron acceptor. For conditions see reference 1.

essentially the same in soluble, purified preparations and in particle-bound complexes.

The properties of the highly purified, soluble enzyme parallel those of its particulate counterpart in more complex preparations in all respects studied except two. Table II compares the properties of the purified

TABLE II

Comparison of DPNH Dehydrogenase and DPNH-CoQ Reductase
(Complex I)

Property[a]	DPNH dehydrogenase	DPNH-CoQ reductase
Form	Soluble	Particulate
Non-heme iron:FMN ratio	16–18	17–18
"Labile" S/Fe ratio	1.5–1.6	1.0
Substrate-reducible EPR signal at $g = 1.94$	+	+
Reactivity with cyt. c	Trace	Very low
Fragmentation and conversion to cytochrome reductase by heat or acid–ethanol	+	+
Turnover number, $Fe(CN)_6^{3-}$	$1.3–1.5 \times 10^6$	6.5×10^5
Turnover number, CoQ_1	$0.5–0.7 \times 10^3$	$9.5 \pm 0.3 \times 10^3$ 11.7×10^3
Turnover number, CoQ_1, in presence of rotenone	$0.5–0.7 \times 10^3$	$0.7–1 \times 10^3$
CoQ_{10} content	0	4.5 mμmoles/mg
Lipid content	0	0.22 mg/mg

[a] All turnover numbers are at 30° per mole of FMN. From reference 32.

enzyme with those of the particulate DPNH-CoQ reductase ["complex 1" of Hatefi et al. (33)], the simplest particle-bound form of the enzyme known, which is active in reconstituting electron transport in the DPNH oxidase chain (10). In addition to the parameters noted in Table II, the absorption spectra and several other kinetic constants of the two preparations are nearly the same. The two differences are the absence of lipid, including CoQ_{10}, from the soluble enzyme and its presence in the complex, and the lack of rotenone- and amytal-sensitive CoQ_1 reductase activity in the soluble enzyme, in contrast with the particle. More recent studies in this laboratory (27) have shown that the differences between soluble dehydrogenase and complex I in regard to CoQ reductase activity are also evident when coenzyme Q_{10}, the naturally occurring homolog, is used.

The lack of CoQ reductase activity in the purified dehydrogenase is very likely related to the absence of lipids. It has been demonstrated that very brief digestion of particles with cobra venom leads to inactivation of the rotenone-sensitive DPNH–CoQ reaction and that this is reversed by the addition of phospholipids (8,9). On somewhat longer contact with the venom the inactivation becomes irreversible: presumably, under these conditions other lipids, including endogenous CoQ_{10}, are lost which may not be readily replaceable. If the reaction of external CoQ were with the built-in CoQ_{10} of the respiratory chain, the purified enzyme would not be expected to catalyze the reduction of external CoQ. Thus, except for the reduction of external CoQ, a complex reaction in which the role of lipids and endogenous CoQ have not been elucidated, the isolated flavoprotein shows no significant differences from its behavior in more complex systems.

The most remarkable property of the dehydrogenase and a clue to the origin of the many low molecular weight DPNH oxidizing preparations which have been extracted from mitochondria is the unusual lability of the enzyme. A variety of treatments, ranging from exposure to temperatures above 30° through contact with low concentrations of urea or thiourea to prolonged incubation with DPNH, lead to fragmentation of the high molecular weight enzyme into polypeptide chains of varying chain length accompanied by labilization and loss of most of the iron and sulfide, consequent changes in absorption spectrum and EPR behavior, alteration or loss of the catalytic properties of the native enzyme, and the emergence of activities not seen prior to treatment with the modifying agent (21,23,24,31).

One of the most convenient agents for the fragmentation of the enzyme to lower molecular weight derivatives is heat at neutral pH. The course of the reaction is illustrated in Figure 2. It may be seen that ferricyanide activity and substrate-inducible EPR signal at $g = 1.94$ decay at identical rates. At the same time (Table III) DPNH-cytochrome c reductase and rotenone-insensitive CoQ_1 reductase activities emerge. In general, under the influence of most protein-modifying agents, ferricyanide and EPR activities are the first to disappear; the rise in cytochrome c, CoQ_1, and DCIP reductase activities is a secondary process. The activities revealed by the conformation change almost certainly occur at the flavin level, since most of the iron and labile sulfide are lost in the process and what remains is inactive in EPR assays, while ferricyanide reduction in intact particles and in the unmodified dehydrogenase appears to involve and may proceed via iron (Fig. 2 and ref. 18). While in the course of thermal

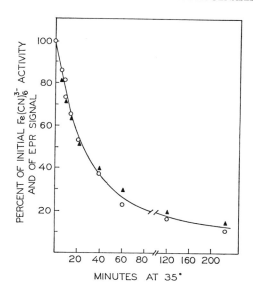

Fig. 2. Parallelism of loss of $Fe(CN)_6^{3-}$ activity (○) and of EPR signal (▲) on thermal inactivation of DPNH dehydrogenase at 35° (1). Purified DPNH dehydrogenase (27.3 mg/ml) was incubated in $0.03M$ phosphate, pH 7.8, in the dark and samples were removed and assayed for DPNH dehydrogenase activity and for substrate-induced EPR signal at the times indicated.

TABLE III

Emergence of CoQ_1 Reductase Activity on Thermal Degradation of DPNH Dehydrogenase

Time at 35°, min	μmoles DPNH oxidized/min per ml at 30° [a]		
	CoQ_1	Cyt. c	$Fe(CN)_6^3$
0	4.37	2.47	4500
20	8.55	14.7	2000
40	12.9	27.6	1000
60	20.6	39.3	537
120	24.6	74.8	342
160	25.0	69.8	270

[a] DPNH dehydrogenase (specific activity = 417) was incubated at 35° in $0.03M$ phosphate, pH 7.8, at 10.9 mg protein/ml in air.

degradation or fragmentation by acid–ethanol at elevated temperatures over 99% of the original ferricyanide activity is lost, even the small activity which remains may occur at a new site (presumably flavin) unmasked by the conformation change, since in the resulting fragments the DPNH–ferricyanide reaction is not affected by many reagents which are strongly inhibitory in more intact systems (Table IV, last two columns).

TABLE IV

Comparison of Characteristics of DPNH Oxidation in Different Preparations

Property	Preparation		
	ETP or "DPNH oxidase"	Soluble DPNH dehydrogenase	Mackler, DeBernard, Mahler et al., DPNH cytochrome reductase
Substrate-induced EPR signal at $g = 1.94$	+	+	0
Substrate-induced $g = 2.00$ signal	Weak	Weak	Strong
Stability of FMN linkage	Stable	Stable	Very labile
Inhibition by high DPNH concentration	+	+	0
Competition by DPN+, DPNH for ferricyanide	+	+	0
Cytochrome c reduction by DPNH (with amytal)	Trace	Trace	High
Inhibition of DPNH–$Fe(CN)_6^{3-}$ activity by —SH reagents at 0°	0	0	+
Temperature-dependent inactivation by —SH reagents	+	+	0
Relative rates of oxidation of:			
AcPyDPNH/DPNH	0.014	0.016	0.0011–0.0015
TNDPNH/DPNH	0.007	0.006	0.006–0.0024
DeamDPNH/DPNH	0.50	0.48	0.15–0.19
K_m for DPNH (mM)	0.105	0.108	0.011
Turnover number at 30° per mole of FMN	8×10^{5a}	$8{-}14 \times 10^5$	0.28×10^5

[a] For DPNH-CoQ reductase complex of Hatefi et al. (33). From reference 37.

One of the most interesting molecular transformations of DPNH dehydrogenase is that induced by reduced substrate. Apparently, prolonged contact with DPNH strains the tertiary and quaternary structure so that the relatively weak bonds which hold the enzyme in the original conformation break and fragmentation to a large array of products occurs (31). Among these products one finds peptides of relatively short chain length, fragments of molecular weight exceeding 100,000, and several endowed with DPNH–cytochrome reductase activity, one of which is apparently identical with the main product of heat or heat–acid–ethanol degradation. Figure 3 illustrates the kinetics of the loss of ferricyanide activity and of the creation of cytochrome reductase activity by DPNH in air at 0°. Anaerobically fragmentation still occurs, but, curiously, this is unaccompanied by changes in catalytic activity. It seems that anaerobically the substrate combines with and protects the active center, while major dissociation of the protein occurs.

One of the first signs of the substrate-induced conformation change is the appearance of sensitivity to —SH reagents, such as organic mercurials. The native enzyme, whether in the soluble, purified form or in particulate complexes, is relatively insensitive to inhibition by —SH combining

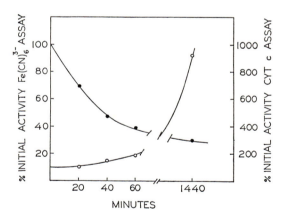

Fig. 3. DPNH-induced transformation of DPNH dehydrogenase to DPNH-cytochrome c reductase in air at 0° (31). Samples of the dehydrogenase (0.3 ml) in 0.03M phosphate, pH 7.8, were incubated with 0.03-ml aliquots of the same phosphate buffer (control) or of DPNH to give a DPNH concentration of 0.98mM in air at 0°. At the times indicated, assays were carried out for DPNH-ferricyanide and DPNH-cytochrome c activity, both at V_{max}. The ferricyanide and cytochrome c activities did not change during incubation for 24 hr in the absence of substrate.

substances, and there is no evidence that —SH groups would directly participate either in combination with the substrate or in intramolecular electron transport. There are nevertheless at least three types of —SH groups present in the enzyme which, after certain conformation changes, affect catalytic activity (34). One of these becomes reactive on brief incubation with DPNH and is regarded as one of the first signs of the DPNH-induced conformation change (34). (For an alternative view of this complex process, see ref. 35.)

Another type of —SH group combines with mercurials, such as PCMS or PCMB, even in the cold, but the resulting mercaptide retains full catalytic activity (36). On elevating the temperature in a range where the native enzyme is stable, however, activity in the DPNH–ferricyanide assay is gradually lost, at a rate and to an extent determined by the temperature (Fig. 4). The process is regarded as a conformation change affecting the non-heme iron site but not the flavin region, since the inactivated enzyme remains fully active in assays involving only the substrate and the FMN site, such as the transhydrogenase reaction, and since it is

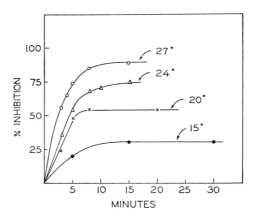

Fig. 4. Effect of p-chloromercuriphenylsulfonate (PCMS) on purified DPNH dehydrogenase (36). The enzyme (13.6 mg/ml, sp. act. = 320) was incubated at pH 7.8, 0°, with 0.385 mM PCMS for 5 min. Unreacted mercurial was removed at 0° on Sephadex G-25. The enzyme in the excluded fraction assayed immediately (0 time sample) showed no inhibition in the standard ferricyanide assay. Aliquots containing 0.06M triethanolamine buffer, pH 7.8 (at all temperatures), and 1 mg protein/ml were incubated at the temperatures stated and aliquots were cooled to 0° and assayed periodically in order to follow the development of inhibition. A sample maintained at 0° showed no inhibition even after 14 hr.

irreversible. This conformation change does not, however, entail fragmentation to lower molecular weight derivatives.

The unique feature of the fragmentations elicited by heat, proteolytic enzymes, urea, substrates, etc. is that the ability of the enzyme to oxidize DPNH is not completely lost but merely altered. Among the products of fragmentation one usually finds derived proteins of a molecular weight of approximately 80,000, which will oxidize DPNH and its analogs at appreciable rates, but the substrate specificity, kinetic constants, and electron acceptor specificity are greatly altered, as compared with the original enzyme. Several different modifying agents, such as heat, subtilisin, and exposure to ethanol at acid pH and elevated temperatures appear to yield the same derivative (23,24), the DPNH-cytochrome c reductase first isolated by Mahler et al. (38). By slight variations of the acid–ethanol procedure a variety of DPNH-oxidizing preparations from heart mitochondria have been described, bearing different names, and these were reported to have appreciably different properties (25,38–40). This has created a good deal of confusion, since no physiological role could be found for any of these enzymes.

On comparing these preparations by rigid criteria (23,24), however, the presumed differences could not be substantiated. Further, it could be shown that these proteins did not exist as such in the mitochondria but were created by the extraction procedure. On treating the purified soluble dehydrogenase by the same procedures as used in the extraction of cytochrome reductases, it was converted to an 80,000 molecular weight derivative, which on isolation proved to be indistinguishable from the Mahler-Mackler-De Bernard preparation of DPNH-cytochrome reductase. Table IV compares some of the catalytic properties of particle-bound DPNH dehydrogenase (as in ETP) with those of the soluble purified enzyme extracted by phospholipase A (DPNH dehydrogenase) and of the derived cytochrome reductases. It may be seen that the first two types of preparation have nearly identical properties, while the latter shows a very different set of properties. As already mentioned, the same picture emerges if one compares absorption spectra, non-heme iron, or labile sulfide content or other molecular properties.

Since several types of treatments result in the formation of an 80,000 molecular weight enzyme endowed with cytochrome c reductase activity, the suggestion has been repeatedly raised that this may be a subunit of the dehydrogenase, bound to other proteins in a multienzyme complex (the high molecular weight dehydrogenase) and that in this complex the cytochrome reductase and diaphorase activities are masked and other activities are modified. In this sense the dehydrogenase, treated as a discrete protein in this chapter, would be in reality a multienzyme complex. Recently (41), it has even been proposed that the flavoprotein itself contains only 2 atoms of iron and is in an integrated complex with an iron-rich protein, which is responsible for most of the non-heme iron, labile sulfide, and all of the EPR signal at $g = 1.94$.

There are several experimental observations which militate against this view.

1. The procedures which lead to the transformation of the dehydrogenase to the 80,000 molecular weight class product do not yield two proteins, a flavoprotein and a non-heme iron protein, but, depending on the method, dozens of polypeptides and proteins.

2. If the transformations described were merely a dissociation into constituent proteins, one would expect to find substantially the same products regardless of the method of dissociation. Actually, while some procedures yield the same DPNH-cytochrome reductase (among numerous other products), other agents (urea, thiourea, trypsin, DPNH) yield entirely different types of cytochrome-reducing fragments; substrate-induced fragmentation yields at least five chromatographically separable DPNH-cytochrome reductases.

3. The emergence of cytochrome c (or CoQ_1) reductase activity does not parallel the fragmentation; conditions have been found where fragmentation was extensive but no cytochrome c activity emerged and, conversely, cytochrome c activity arose without fragmentation.

In the author's opinion the interpretation which best fits all the available data is that the DPNH-cytochrome reductases of the 80,000 molecular weight class are stages in the fragmentation of the original protein, perhaps the only catalytically active fragments of sufficient stability to permit isolation.

VI. DPNH-Coenzyme Q Reductase

A recent development of considerable interest regarding DPNH dehydrogenase has been the report of Sanadi and co-workers (25,26,41) of the isolation of a low molecular weight soluble flavoprotein from heart mitochondria which catalyzes the oxidation of DPNH by coenzymes Q_1, Q_6, and Q_{10}. The material has been extracted from phosphorylating particles (ETP_H) by the heat–acid–ethanol process which had been used in the isolation of DPNH-cytochrome reductases. The DPNH–CoQ reaction is reported to be inhibited by amytal and rotenone, although, as expected from the method of extraction, the flavoprotein has lost the characteristic EPR signal at $g = 1.94$, as well as much of the ferricyanide activity, non-heme iron, and labile sulfide characteristic of the native enzyme. The inhibitor-sensitive CoQ reductase activity is regarded by these workers as the retention of the original reaction site of coenzyme Q and their preparation is viewed as a purified, functional form of the respiratory chain-linked DPNH dehydrogenase (41).

Since these reports are in conflict with much of what has been said in the preceding section and since the general question of flavoprotein–coenzyme Q interactions is of broad interest in the field of the respiratory chain-linked dehydrogenases, the identity and origin of the Sanadi DPNH-coenzyme Q reductase (DPNH-ubiquinone reductase) will be discussed in some detail.*

One of the first questions which arises is why the same extraction procedure applied to two closely related mitochondrial fragments (ETP and ETP_H) would yield a DPNH-cytochrome c reductase devoid of CoQ reductase activity in the first instance and a flavoprotein endowed with both activities in the second, as has been claimed (41). Since many readers of this book may be unfamiliar with technical details, it may be worth mentioning that both particles are prepared from beef heart mitochondria by differential centrifugation, ETP following exposure to pH 8.5 in the cold, ETP_H following sonication at pH 7.6. The only material difference between the two is that ETP_H retains some of the capacity for oxidative phosphorylation while ETP does not. In all other regards the catalytic properties of ETP and ETP_H are nearly identical (32). Sanadi et al. (41) emphasize that their preparation can only be extracted from phosphorylating particles. As may be seen in Table V, the yield of CoQ_6 reductase activity from phosphorylating and nonphosphorylating particles is quite comparable. Further, the demonstration (43) that a Keilin-Hartree preparation, traditionally regarded as nonphosphorylating, will catalyze oxidative phosphorylation after treatment with oligomycin, makes one wonder how meaningful are the differences in the phosphorylating capacities of submitochondrial particles.

In addition to comparable catalytic activities toward CoQ homologs, ferricyanide, and cytochrome c, the preparations from ETP and ETP_H have identical substrate specificities, chromatographic properties, and prosthetic group content. Further, both are inhibited 25–40% by 3.5 mM amytal (42), while $>90\%$ inhibition is obtained with much lower concentration of amytal when intact particles are assayed for CoQ_6 reductase

---

* The reader may note that the interpretations given in this chapter concerning the identity and functional state of the Pharo-Sanadi preparation are materially different from those presented by Klingenberg (first chapter in this volume).

TABLE V

Comparison of Soluble DPNH-CoQ Reductase Extracts from ETP, ETP_H, and SP_{HL}

| Expt. | Parent particle | Specific activity of extract[a] | | | Ratio, $Q_1 : Q_6 : Q_{10}$ | Yield[b] of soluble CoQ_6 activity/ g particle |
		CoQ_1	CoQ_6	CoQ_{10}		
1	ETP	24.9	10.9	6.2	100:44:25	138
	ETP_H	24.3	13.7	7.5	100:57:31	163
	SP_{HL}	21.7	14.1	9.5	100:65:44	101
2	ETP	26.6	14.3	17.6	100:54:66	189
	SP_{HL}	26.5	12.1	12.3	100:46:46	140

[a] μMoles DPNH oxidized/min per mg protein.

[b] μMoles DPNH oxidized per minute (total units). The soluble preparations were isolated, CoQ reductase activity was assayed, and extracts were prepared (15 min at 43°, pH 5.3 in 9% alcohol) exactly as per Pharo et al. (26). The slight variation in the CoQ_6 and CoQ_{10} reductase activities was not due to differences in particles but to the relative periods of storage at $-20°$. There is a tendency for particles to yield extracts of lower activity even on a few days' storage at $-20°$. SP_{HL} is a variant of ETP_H (26). Data from reference 42.

activity. The behavior of the extracts from the two sources toward rotenone in the CoQ_6 assay is also very similar (Fig. 5). Maximal inhibition (25–40%) occurs at a rotenone/*enzyme-bound* FMN ratio between 4.5 and 6 and disappears as the rotenone concentration is raised. In contrast, the physiological CoQ reduction in particles is inhibited over 95% by an amount of rotenone which approximates the DPNH dehydrogenase content and, more importantly, the inhibition is not relieved at higher rotenone concentrations.

These experiments and the fact that CoQ reduction in particles is lipid-dependent but in acid–alcohol extracts of both ETP and ETP_H does not require lipids, are not readily reconciled with the view that the reaction site of externally added CoQ homologs is the same (in terms of locus and conformation) in soluble samples of the reductase and in respiratory chain preparations. It is suggested that (*1*) there is presently no known difference between DPNH-cytochrome reductase and DPNH-CoQ reductase, and (*2*) the reaction site of long-chain CoQ homologs in acid–ethanol extracted preparations may not be the same as in intact preparations but is created or unmasked by the exposure to heat, ethanol, and acid pH, as is also true of cytochrome *c* reduction. The overriding interest of this interpretation

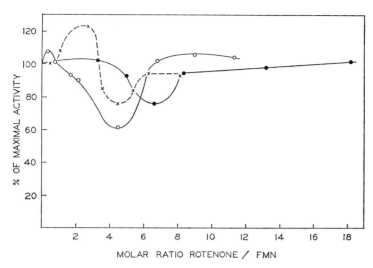

Fig. 5. Rotenone titration of soluble DPNH–CoQ reductase. Rotenone was added after DPNH and the reaction was started with enzyme. The rotenone:FMN ratios are for total FMN in the preparation; actually, part of the flavin is not enzyme bound in such extracts (see text); (○) acid–ethanol extract from ETP; (●) same from ETP$_H$; (×) lyophilized extract from ETP.

is that it would not only mean the creation of a new reaction site or structure not seen in the native enzyme by conformation changes, but one which mimics the physiological reaction in its partial sensitivity to rotenone and barbiturates. The possibility that the locus of CoQ reduction is the same in particles and in extracts but that the heat–acid–ethanol treatment modifies the conformation of this site so as to change the characteristics of CoQ reduction cannot be completely eliminated, however.

If the partially rotenone- and amytal-sensitive CoQ reduction observed in the Pharo-Sanadi preparation does not represent the physiological reaction as it occurs in mitochondria but is a new activity created or revealed by the modifying effects of heat, ethanol, and acid pH, then it should be possible to convert by the same treatment the purified, soluble dehydrogenase, which is devoid of significant activity toward long-chain CoQ homologs, to a low molecular weight fragment endowed with such activity.

Table VI shows that exposure of the purified flavoprotein to pH 5.3 and 9% alcohol at 43° causes a very large increase in CoQ_6 and CoQ_{10} reductase activities, along with the expected rise in CoQ_1 and cytochrome c reductase

TABLE VI

Conversion of DPNH Dehydrogenase to DPNH-CoQ Reductase[a]

	Fe(CN)$_6$$^{3-}$ activity		Cytochrome c reductase		CoQ$_1$ reductase		CoQ$_6$ reductase		CoQ$_{10}$ reductase		Total protein mg
	Units	Sp. act.	Units	Sp. act.	Units	Sp. act.	Units	Sp. act.	Units	Sp. act.	
Before conversion	471,000	471	92	0.092	279	0.28	22	0.022	16	0.016	1000
After conversion	8,185	138	3,176	53.6	1,777	29.9	900	15.2	720	12.2	5.9

[a] The flavoprotein after the Sephadex G-200 gel exclusion step (5) in 0.05M phosphate, pH 7.6 was treated at pH 5.3, 43°, in 9% ethanol for 15 min exactly as in the Pharo-Sanadi procedure. The suspension was chilled, neutralized to pH 6.8, and precipitated protein was removed by centrifugation. Units are μmoles DPNH oxidized/min per gram of original DPNHD. Sp. act. = units/mg protein. Protein was determined by biuret on the starting material and by the Lowry method on the extract after conversion.

activities and the almost complete disappearance of the characteristic activity of the native enzyme with ferricyanide. Thermal exposure to 43° alone or treatment with alcohol at pH 5.3 in the cold destroy ferricyanide activity without the emergence of CoQ_6 reductase activity. Thus simultaneous exposure to heat, acid, and alcohol is required for the transition of the flavoprotein to a state capable of reducing long-chain CoQ homologs. The CoQ reductase activity emerging after acid–ethanol treatment compares favorably with the best extracts of ETP_H derived by the same procedure (Table V).

The most interesting feature of the transformation of DPNHD to a DPNH-CoQ reductase under the influence of heat–acid–ethanol is that the CoQ_6 and CoQ_{10} reductase activities emerging are inhibited by amytal and rotenone to about the same extent (30–35% and 25–50%, respectively) as in extracts obtained from ETP_H. The titration curves with rotenone are also very similar to those given by the Pharo-Sanadi preparation. In contrast, the trace of CoQ reductase activity of untreated DPNH dehydrogenase is not inhibited by amytal or rotenone.

The complete transformation of DPNH dehydrogenase, an enzyme virtually devoid of CoQ reductase activity, to one which reduces long- and short-chain CoQ homologs and is inhibited by amytal and rotenone in the same manner as the soluble DPNH-CoQ reductase of Pharo and Sanadi strongly suggests that these CoQ reductase activities are probably not physiological but emerge as a consequence of conformation changes in the protein.

VII. Succinate Dehydrogenase

A. Background

Since the beginning of the century, succinate dehydrogenase has been one of the most widely studied enzymes. Despite numerous attempts at isolation, it was not obtained in soluble, purified form until 1954 (2). Studies prior to that time utilized particle or cell preparations of varying degrees of complexity and assay methods which we now recognize as unreliable. Despite these circumstances, many of the properties of the enzyme were correctly predicted and, in fact, these early studies on succinate dehydrogenase laid the foundation of our current knowledge of respiratory chain-linked flavoproteins and of their relation to the cytochrome system.

When it was discovered (2) that mammalian succinate dehydrogenase does not react directly with the electron acceptors which had been traditionally used for its assay (methylene blue, cytochrome *c*, etc.), while phenazine methosulfate does and, apparently, measures the full potential activity, the way was open to its purification (44) and isolation in nearly homogenous form from beef heart mitochondria (45). The enzyme was subsequently purified from pig heart (46), aerobic yeast (47) and bacteria (48,49).

Although the enzyme has not been purified extensively from other sources, its properties have been studied in many different forms of life, so that it has become possible to show that during evolution its properties have changed in accord with environmental conditions (cf. Section VII-E).

B. Assay and Chemical Determination

The method of choice (50) for the assay of the enzyme in higher organisms and aerobic yeast mitochondria, in the purified state as well as in complex systems, is the spectrophotometric determination of the reduction of 2,6-dichlorophenol indophenol (DCIP) with phenazine methosulfate (PMS) as catalyst (DCIP does not react directly with the enzyme):

$$\text{Succinate} + \text{PMS} \longrightarrow \text{Fumarate} + \text{PMSH}_2$$
$$\text{PMSH}_2 + \text{DCIP} \longrightarrow \text{PMS} + \text{DCIPH}_2$$

It is also possible to determine the content of succinate dehydrogenase chemically, owing to the fact that its flavin dinucleotide prosthetic group is covalently linked to the peptide chain (51). Thus it is possible to remove all noncovalently bound flavin from a sample by extraction, digest the residue with proteolytic enzymes, and determine the flavin peptide content fluorometrically at the pH of maximal fluorescence (Fig. 5) (52). The molar concentration of the flavin equals the succinate dehydrogenase content at least in preparations from mammalian heart, aerobic yeast, and probably also mammalian brain, since in these tissues all the covalently bound flavin appears to be linked to the dehydrogenase but in liver there appears to be considerable covalently linked flavin which is not associated with succinate dehydrogenase, while in other tissues the method has not been sufficiently explored. The great advantage of the chemical determination is that it measures active as well as nonactivated or inactivated forms of the enzyme (see below) and thus permits the determination of the stoichiometric relation of this flavoprotein to other components of the respiratory chain in whole cells, mitochondria, and derived particles and permits following the fate of the enzyme in the course of reconstitution (Section VII-F).

C. Nature and Composition of Preparations

On the basis of bound flavin content and physical measurements, the molecular weight of the heart enzyme is considered to be 200,000 (45). Homogeneous preparations contain 1 mole of covalently bound flavin per 200,000 g of protein (45) and 4 g-atoms of non-heme iron per mole of flavin (45,46). A value of 8 has also been reported (53) for soluble preparations isolated by a modification of the Wang procedure (46) but since the preparations analyzed were impure, it is not clear whether the extra iron belongs to the dehydrogenase or to impurities. The enzyme from yeast mitochondria also contains 48 atoms of iron per mole of covalently bound flavin (47).

The heart enzyme has also been purified as a particulate complex ("succinate-coenzyme Q reductase") and in this form it contains considerable lipid, 1 mole of cytochrome b (apparently in nonfunctional form), and 8 g-atoms of non-heme iron (54). In view of the complex nature of the preparation, it is again impossible to decide whether the extra iron is associated with the flavoprotein. The complex also contains 8–9 moles of labile sulfide per mole of bound flavin (30).

The linkage of flavin to the protein is acid stable and resists digestion by a number of proteolytic enzymes. Digestion by trypsin and chymotrypsin yields flavin peptides of various chain lengths (51,55); among these a hexapeptide has been obtained in pure form (51). Although no way has been found to convert the peptide to a free flavin, it appears virtually certain that the flavin constituent of the mitochondrial enzyme is FAD. The peptide is linked to the isoalloxazine via the C-terminal end but neither the nature of this terminal amino acid nor the point of attachment to the ring have been conclusively established.

Covalent attachment of the FAD to the protein results in a marked shift of the 375 mμ band in the absorption spectrum of the flavin to a shorter wavelength and there is complete quenching of the fluorescence at neutral pH, so that maximal fluorescence is at pH 3.2 (Fig. 6) (51).

D. Properties

The substrate specificity of the enzyme is quite restricted. Besides succinate itself, only monomethyl, monoethyl, and monochlorosubstituted succinates are oxidized. The action of the mammalian enzyme is readily

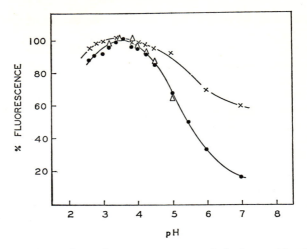

Fig. 6. Comparison of pH–fluorescence curves of flavin peptides from heart muscle and yeast mitochondria (68). Shaded circles, proteolytic digests of beef heart mitochondria, after hydrolysis to the mononucleotide level; ($\triangle$) same from aerobic yeast mitochondria (wild type); ($\times$) same from "petite" mutants. The ordinate denotes the relative degree of fluorescence with the value at the pH of maximal fluorescence taken as 100. The buffers were made by mixing $0.05M$ citric acid and $0.1M$ Na_2HPO_4.

reversible, but at 38° the maximal rate of succinate oxidation is some 60 times faster than that of the reduction of fumarate when each reaction is measured under optimal conditions (1). The turnover number, per mole of covalently bound flavin, is $17,000 \pm 1,000$ per min in the direction of succinate oxidation in intact preparations (1). In soluble preparations the turnover number is considerably lower (1) although there is a report (56) that phospholipids may increase the latter value to the level found in mitochondria. This finding requires further confirmation and extension, since phospholipase A does not influence the activity of succinate dehydrogenase in mitochondria, whereas it would be reasonable to expect an inhibition by the lipase if the primary dehydrogenase assay required the presence of phospholipids.

It should be mentioned that the activity of the dehydrogenase in mitochondria, as measured in the phenazine–DCIP assay, always exceeds the maximal rate of succinate oxidation via the respiratory chain, even if respiratory control is removed. It would seem, therefore, that the rate-limiting reaction in the mitochondrial oxidation of succinate is not succinate dehydrogenase activity (at least not the oxidation of succinate by the succinate dehydrogenase flavin) but some other event in electron transport.

Perhaps the most unique property of succinate dehydrogenase from mammalian and yeast cells is its activation by the substrate and by competitive inhibitors. The phenomenon, discovered by Kearney (57), consists of a reversible conformation change in the dehydrogenase from a form of low (perhaps no) activity to one of high activity on incubation with substances (succinate, malonate, inorganic phosphate, etc.) which combine at the active center. The activation has been observed in soluble, purified preparations and in particulate ones, including mitochondria, from a variety of animal tissues and from aerobic yeast.* Although different activators activate at different rates, the final extent of activation is the same, regardless of the nature of the activator. The energy of activation for the process is high (35,600 cal/mole). Thus at low temperatures (e.g., 15°), activation is too slow to be observable, while at 38° it may be complete in 1 or 2 min or less, depending on the nature and concentration of the activator. On removal of the activator, the enzyme is rapidly deactivated but can once again be activated on incubation with an activating agent (58).

The activation process appears to be a conformation change affecting a limited region of the flavoprotein, since its effects are manifest only in certain assays (those presumably involving catalytically functional non-heme iron moieties) such as the succinate–phenazine methosulfate reaction, or the succinate–O_2 and succinate–cytochrome c reactions in the respiratory chain, but not the fumarate–$FMNH_2$ reaction, which appears to be a direct interaction with the flavin (Table VII).

Thorn (59) has recently extended these observations and showed that a modest degree of activation occurs also at elevated temperatures in the absence of added substrates and that this activation is reversed on cooling the enzyme. Whether this "temperature activation" is related to the substrate activation is as yet uncertain, but there appear to be some qualitative differences between the two phenomena.

Since in mitochondria, ETP, and soluble preparations, succinate dehydrogenase is usually in the unactivated (or deactivated) state to a large extent, determination of succinate dehydrogenase activity without provisions for full activation may lead to serious error, particularly in rapid kinetic measurements. Although this has been repeatedly emphasized in

* There has been some confusion in the literature owing to the statement of Wang et al. (55) that while succinate activates the enzyme, phosphate merely removes heavy metal inhibition. It has been proven conclusively (see ref. 1 for a detailed summary) that phosphate and succinate activate by the same mechanism and that heavy metal inhibition is in no way involved in the process.

THOMAS P. SINGER

TABLE VII

Effect of Activation on the Succinate–PMS and the Fumarate–FMNH$_2$ Reactions[a]

Sample	Temperature of assay, °C	Specific activity (μmoles substrate/min per mg protein)	
		Succinate–PMS reaction	Fumarate–FMNH$_2$ reaction
ETP, 4 times washed	25	0.100	0.032
same, activated by 5 mM malonate	25	0.942	0.038
Keilin-Hartree, 4 times washed	25	0.124	0.023
same, activated by 2 mM malonate	25	0.404	0.018
Soluble enzyme, Wang et al. preparation	15	0.123	0.022
same, activated by 5 mM succinate	15	1.31	0.022

[a] The ETP preparation was washed 4 times by centrifugation with 0.05M imidazole buffer, pH 7.4, and was resuspended in the same buffer at a protein concentration of 19.8 mg/ml. For batch activation, a 5-ml. aliquot was incubated with 25 μl 1M malonate under N$_2$ for 30 min at 38°. After cooling to 0°, aliquots of both the malonate activated and unactivated samples were assayed without further activation. The Keilin-Hartree preparation was similarly treated, except that the protein concentration during malonate activation was 24.9 mg/ml. The Wang et al. (46) preparation was the gel eluate, precipitated with 0.55 saturated (NH$_4$)$_2$SO$_4$ and desalted on Sephadex G-25 (equilibrated with 0.02M imidazole buffer, pH 7.5 at 0°). A 1.56-ml aliquot (8.55 mg protein per ml) was incubated with 0.04 ml 0.2M succinate, pH 7.6, for 15 min at 25° for batch activation. Activities are corrected for inhibition by residual malonate, where present.

the literature (e.g., Thorn (59)), all too often workers in the field fail to control this important variable.

It is intriguing to speculate that the reversible activation by the substrate (or by fumarate or oxalacetate) may be of importance as one of the factors regulating the activity of succinate dehydrogenase in mitochondria.

There is a wealth of information concerning the action of inhibitors on this enzyme, which space does not permit to survey in this chapter (cf. ref. 1). Among the most widely studied are competitive inhibitors such as malonate and oxalacetate and —SH reacting substances. Recently, it has been reported that combination of the dehydrogenase with substrate competitors elicits characteristic spectral changes (60). The phenomenon was first described by Kearney (57), who observed characteristic spectral shifts on activating the dehydrogenase with malonate. It was subsequently found (58) that the spectral changes are not reversed on deactivating the enzyme by removing the activator. Further, they are not simple charge-transfer complexes, since although the deactivated enzyme binds (and is inhibited by) malonate, no spectral change occurs until after activation has taken place. The chemical basis of these spectral shifts remains for further studies to elucidate.

The iron moieties of the dehydrogenase are quite unreactive toward most iron chelators but after binding the —SH groups with suitable inhibitors, iron chelates are readily formed with attendant inactivation and conformation changes (61). 2-Theonyltrifluoroacetone, however, inhibits the succinate–CoQ reaction (but not dehydrogenase activity) in particles and in the succinate–coenzyme Q reductase complex (62). That at least some of the iron–labile sulfide moieties of the enzyme are in some way involved in the catalysis is suggested by the numerous reports of the presence of a substrate-inducible $g = 1.94$ signal in soluble and particle-bound preparations of the enzyme, starting with the observations of Beinert and colleagues (63,64). Since, however, there has been no correlation established between the rate of appearance of this signal and catalytic activity, as in the case of DPNH dehydrogenase and other metal flavoproteins (see Rajagopalan and Handler, this volume), it is not yet possible to conclude that the iron undergoes compulsory oxidoreduction in the course of the catalytic cycle.

E. Evolutionary Development of Succinate Dehydrogenase

A survey of the catalytic characteristics of succinate dehydrogenase from different forms of life (65) has revealed that the properties of the enzyme in each case appear to be excellently suited to the physiological needs of the cell. The dehydrogenase in typical aerobic cells such as heart, brain, or yeast grown in the presence of high O_2 tension and low glucose catalyzes the oxidation of succinate one to two orders of magnitude faster than the reduction of fumarate. Further, the K_m for succinate is relatively low and the inhibition by fumarate relatively slight, so that even low concentrations of succinate are efficiently utilized without interference by accumulated fumarate. The activation of the dehydrogenase by succinate and its rapid deactivation as the succinate concentration falls, a behavior which has been observed with succinic dehydrogenase from a variety of aerobic cells, further contributes to the efficiency of the oxidative process. Thus the enzyme is geared to its predominant metabolic role: the oxidation of succinate via the Krebs cycle, coupled to the energy conservation mechanism.

In many typically anaerobic forms of life [*M. lactilyticus* (48) or *Ascaris lumbricoides* (66)] the enzyme catalyzes the reduction of fumarate some two orders of magnitude faster than the oxidation of succinate. The K_m for fumarate is very low, the K_m and K_1 values for succinate are high, so that while traces of fumarate are efficiently reduced to succinate without product inhibition by accumulated succinate, conditions are not favorable for the oxidative process. All this is in accord with cellular needs, since in such organisms the cytochrome system is generally absent and fumarate serves as the terminal oxidant (65). As might be expected, the molecular properties of the enzyme from such sources are also markedly different from those of the mammalian enzyme: there is no activation by substrates; the flavin is not covalently bound, and the non-heme iron content is far greater than in the heart enzyme (48).

Facultative anaerobes which thrive in both aerobic and anaerobic conditions either have an "in between" type of succinate dehydrogenase, one which catalyzes succinate oxidation and fumarate reduction equally well, and has kinetic constants favorable to either process (49) or have two or more distinct and separate succinate dehydrogenases. Thus *E. coli* cells contain two succinate dehydrogenases, one resembling closely the heart enzyme in its properties, the other resembling the anaerobic type enzyme from *M. lactilyticus*. The two enzymes are under independent genetic control and it has been shown that the repression and derepression of their development is in accord with the physiological needs prevailing during growth (67).

Finally, yeast offers perhaps the most beautiful example of this evolutionary adaptation to variations in the O_2 tension of the environment (37,65,68). Cells grown in anaerobic conditions or in the presence of high glucose concentrations, so that fermentative conditions prevail, lack mitochondria and in such cells the synthesis of the typical aerobic type dehydrogenase is completely repressed. They contain in the cytoplasm, however, a fumarate reductase, i.e., an enzyme which rapidly reduces fumarate but is apparently incapable of oxidizing succinate (69). The enzyme is a copper–FAD protein of 62,000 to 64,000 molecular weight, which contains no iron and is not activated by succinate, and its kinetic constants are ideally suited for the reduction of fumarate. The enzyme occurs in several forms in the cytoplasm; there are apparently isoenzymes differing in charge, and one form, apparently a monomer, of 32,000 molecular weight has also been isolated. The genetic control of these isoenzymes appears to be the same but it is independent of the control of the biosynthesis of the mitochondrial enzyme. The latter is synthesized under conditions of full aerobiosis and is in every respect similar to the heart enzyme. In aerobiosis the synthesis of the cytoplasmic enzymes is partially repressed.

F. Reconstitution

It has been known for many years that on incubating a particle preparation from heart muscle (Keilin-Hartree preparation) at pH 9.0, succinoxidase activity and succinate–methylene blue activity disappear completely, while cytochrome oxidase activity is only partially destroyed (70). Keilin and Hartree suggested that the alkaline pH under these conditions inactivates succinate dehydrogenase. Later Keilin and King (71) succeeded in restoring the succinoxidase and succinate–methylene blue reductase activities of the alkali-treated particles by the addition of soluble succinate dehydrogenase. The fact that the reactivation entailed incorporation of the soluble enzyme was suggested by the observations that the resulting complex could be repeatedly washed by centrifugation without loss of activity. These findings have been confirmed in the author's laboratory (16).

Since these experiments used activity determinations exclusively, rather than chemical measurements of succinate dehydrogenase content, they did not distinguish conclusively among the following alternative mechanisms: (1) That the alkali treatment removes succinate dehydrogenase and the added soluble enzyme recombines with the deficient particle, thus restoring its catalytic activities; (2) that the alkali treatment does not remove the flavoprotein but merely inactivates it, and the added dehydrogenase combines with the particle and takes over the function of the inactivated flavoprotein; (3) that the alkali treatment does not extract the dehydrogenase but inactivates it (e.g., by oxidation of —SH groups) and the added soluble enzyme specifically reactivates the bound, inactivated flavoprotein molecule, without combining with the alkali-treated particle.

Since the covalent nature of the flavin–protein linkage (51) provides a specific chemical micromethod for determining succinate dehydrogenase in both active and inactive forms, Kimura et al. (16) were able to follow the flavoprotein during the cycle of alkali inactivation and reconstitution. As seen in Table VIII, the second alternative listed above appears to be correct, since the pH 9.3 treatment does not extract the dehydrogenase from the particles, but the added soluble enzyme is indeed reincorporated. When reactivation is complete, the amount of succinate

TABLE VIII

Catalytic Activities and Turnover Numbers in Alkali-Treated and Reactivated Preparations[a]

Soluble s.d. added (μmoles/mg Keilin-Hartree preparation)	Bound flavin in resulting complex (mμmoles/mg)	Specific activity	
		Succinic dehydrogenase	Succinoxidase
None (untreated)	0.129	1.47	0.897
None (AT)	0.115	0.198	0.041
0.132	0.145	0.556	0.232
0.263	0.175	0.926	0.406
0.395	0.197	1.22	0.582
0.652	0.205	1.39	0.655
0.980	0.211	1.39	0.607
1.31	0.207	1.31	0.625
2.61	0.209	1.45	0.614

[a] Each sample was assayed after isolation of the reaction product of the alkali-treated preparation plus soluble dehydrogenase by repeated washing in the ultracentrifuge. Succinic dehydrogenase activity refers to the spectrophotometric phenazine assay. Specific activity is μmoles/min per mg protein. AT, alkali-treated Keilin-Hartree preparation; s.d., succinic dehydrogenase. Data from Kimura and Singer (16).

dehydrogenase present is approximately double that present originally in the preparation.

The question whether the newly acquired flavoprotein or that originally present in the particle functions in this double-headed respiratory chain was resolved by Kimura and Hauber (22) in an ingenious manner. They added to an alkali-treated ETP preparation, in which the dehydrogenase had been deactivated by repeated washing, a fully activated sample of the soluble dehydrogenase at 17° (Fig. 7). Full succinoxidase activity appeared in less than a minute, the time necessary for combination of the components, while activation at this temperature requires hours. Hence it is the newly acquired molecule of flavoprotein, rather than that originally present, which donates electrons to the respiratory chain.

Another point of importance evident from Table VIII is that only a small fraction (usually 10–20%) of the succinate dehydrogenase molecules present in the soluble preparation are capable of combining with the alkali-treated Keilin-Hartree preparation. It has been shown (22) that strictly fresh preparations of the flavoprotein, isolated with rigorous care, still

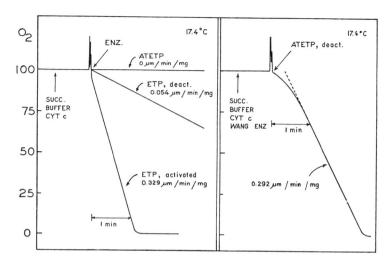

Fig. 7. Demonstration of the functioning of the newly acquired succinate dehydrogenase moiety of reconstituted samples in succinoxidase assay. Polarographic assays of succinoxidase activity at 17°, in the presence of 0.1M phosphate–0.06M succinate. The left side of the figure shows the activities of (1) beef heart ETP, deactivated by repeated centrifugation, (2) the same after 15 min activation with 80 mM succinate in N_2 (control sample for full activity), and (3) of an alkali-treated, deactivated ETP, prepared by exposing sample 1 to pH 9.4 for 60 min. Sample 3, denoted as ATETP, contains all of its original succinate dehydrogenase content in deactivated and inactivated form. The right side shows the rapid recombination of this ATETP with a sample of fully activated soluble dehydrogenase. Since nearly the same activity is reached within 1 min as in the untreated, activated control and since activation at 17° is very slow, the flavoprotein in the added soluble enzyme must be functioning in respiration.

contain several molecular forms of succinate dehydrogenase, which can be at least partially separated on molecular sieves. One form combines with the deficient particles but confers little or no catalytic activity, another produces high dehydrogenase activity but low oxidase activity in the resulting complex, and a third restores the original high dehydrogenase and oxidase activity. The prevalent species is a fourth form, which does not combine at all. It appears that the form which restores all the original activities, and hence may be the least modified, combines preferentially. Thus the "reconstitutively active" preparations of the enzyme show, in fact, considerable molecular heterogeneity, and their combination with the alkali-treated particle may result in the formation of different complexes with different properties, only one of which resembles the catalytic characteristics of the original particle (Fig. 8).

These views of the mechanism of the reconstitution process have been sharply disputed by King (20) and also differ from the interpretations presented by Klingenberg in this volume. The first area of disagreement concerns the question whether the conditions used in reconstitution tests actually dissociate the flavoprotein during the alkali treatment and thus produce a respiratory chain preparation free from succinate dehydrogenase or not. King has reported (20) that at pH 9.95 the flavoprotein is indeed dissociated from the respiratory chain, but treatment at this pH produces a particle which cannot be reconstituted, while at pH 9.3–9.4, where particles capable of reconstitution are isolated, the flavoprotein is not dissociated

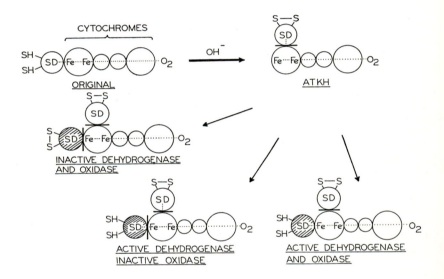

Fig. 8. Schematic representation of the events attending the alkali inactivation of succinoxidase and its reactivation with various forms of soluble succinate dehydrogenase. Alkali inactivation is visualized as resulting in a "displacement" of SD on the respiratory chain, as originally proposed by Keilin and King, with loss of dehydrogenase and oxidase activity, and a simultaneous oxidation of certain —SH groups, which increases the reaction of the enzyme with $FMNH_2$. The alkali-treated Keilin-Hartree preparation (ATKH) takes up various forms of succinate dehydrogenase, yielding a double-headed respiratory chain. The newly acquired dehydrogenase is shaded; interruption of electron transport is noted by a solid line. One form of soluble SD present in the preparation of Wang et al. combines stoichiometrically but confers no dehydrogenase or oxidase activity on the resulting complex, possibly because its —SH groups are oxidized. Another modified form combines and restores dehydrogenase activity but cannot transfer electrons to the cytochrome system. The third form restores all activities.

appreciably (cf. Table VIII and ref. 16).* Hence the interpretation that the reconstitution entails a reversible resolution of the respiratory chain with respect to the flavoprotein contradicts the experimental facts.

The second area of disagreement concerns the postulate (19,20) that the reconstituted preparation is in every respect identical with the original material. Apart from the fact that the reconstituted preparations contain twice the original content of succinate dehydrogenase, they also show significant differences from the original in having higher succinate dehydrogenase and fumarate reductase activities. These latter differences are due to the fact that the dehydrogenase molecule originally present is not entirely inactivated with regard to these assays and, in fact, its reactivity with FMNH₂ (in the fumarate reductase assay) is enhanced by oxidation of certain —SH groups during exposure to alkaline pH (16).

The third area of disagreement concerns the interpretation that the reconstitution test is a physiological assay for the dehydrogenase and is more reliable than catalytic activity determinations. This point has already been dealt with earlier in this chapter. Suffice it to add that the reconstitution test and catalytic activity determinations measure different properties of the enzyme. Catalytic assays indeed may not detect minor conformation changes which nevertheless prevent activity in the reconstitution test. Conversely, it has been shown (1,13) that the measurement of succinoxidase activity in the reconstitution test fails to detect modifications of the enzyme which are revealed by lowered turnover number in catalytic assays. The real problem is that a completely unmodified preparation of the dehydrogenase has not yet been obtained. However subtle the conformation changes which lead to differences in behavior in reconstitution tests (Fig. 8) may be, they are nevertheless real: even the most carefully isolated samples contain different forms of the dehydrogenase. Thus all available methods must be brought to bear on the problem if we are to define the form of the enzyme as it occurs in mitochondria.

VIII. Concluding Remarks

It is the purpose of this last section to call attention to some of the more interesting problems concerning the flavoproteins of the respiratory chain which remain to be explored in depth.

---

* King (14) reported that if the alkali-treated preparation is centrifuged while still at pH 9.3, the flavoprotein remains in the supernatant but recombines at neutral pH and is removed by centrifugation if the preparation is neutralized prior to centrifugation. This report could not be confirmed (16).

In recent years many laboratories have investigated the biosynthesis of these enzymes and attempted to correlate it with the formation of mitochondria. Yeast is the traditional experimental material for this type of study since on exposure of anaerobic cells to O_2 in a suitable medium the respiratory chain is formed rapidly, under readily controlled conditions, rising from undetectable amounts to the high level seen in aerobic cells. In a recent study (68) it was shown that while mitochondrial succinate and DPNH dehydrogenases and cytochrome oxidase are formed at essentially identical rates and their biosynthesis is complete within a few hours after the exposure of the cells to O_2, synthesis of the complete succinoxidase and DPNH oxidase chain takes very much longer. Thus the rate-limiting step in the synthesis of the respiratory chain may be the organization of the individual enzymes into a functional unit. While available information on the biosynthesis of the flavoproteins may be readily reconciled with current knowledge of protein synthesis and its control, the latter does not provide a mechanism for the assembly of the individual enzymes into a functional unit. The marked difference between the rates of synthesis of succinate dehydrogenase and of succinoxidase would suggest that the assembly of the proteins and lipids into a multienzyme system is not a spontaneous process, consequent upon the structure of the individual proteins, but a more complex one, the mechanism of which remains to be elucidated.

Another gap in our knowledge of this group of enzymes concerns the measurement of their concentration and of their redox state in cells, mitochondria, and other complex material. For succinate dehydrogenase a chemical determination is provided by bound flavin analysis, at least in certain types of cells, but no comparable method exists for the other respiratory chain-linked flavoproteins of mammalian cells. FMN content has been used by investigators in Green's laboratory and on the assumption that all the FMN in mitochondria or derived particles is associated with DPNH dehydrogenase, the stoichiometric relation of this enzyme to the various cytochromes and the "particle weight" of an "elementary particle" have been calculated (cf. Klingenberg this volume). Unfortunately, only a minor part of the FMN content of mitochondria, ETP, or ETP_H belongs to DPNH dehydrogenase (28) and thus the true concentration of the enzyme in such preparations is much lower than has been calculated.

The situation with respect to measurement of the redox state of the respiratory chain-linked flavoproteins is even less satisfactory. Several years ago Chance (72) proposed the use of double-beam spectrophotometry (at 465 mμ minus a reference wavelength) to follow oxidation-reduction of these flavoproteins in mitochondria. On isolation of the individual enzyme it became clear, however, that the method is of doubtful value, since the properties of these flavoproteins do not lend themselves to this type of determination. The limitations of the method have been discussed elsewhere (29). For the purposes of this chapter it is sufficient to point out that it is not known whether the respiratory chain-linked flavoproteins shuttle between oxidized and reduced states during catalysis but it is known that most of their color is due to non-heme iron and not flavin. Thus measurements of absorbancy changes at 465 mμ *may* include changes in the redox state of all flavins present, as well as changes in other colored components, and the probability is that most of the color change is due to oxidation-reduction of protein-bound iron. The fact that not all the non-heme iron in mitochondria is associated with flavoproteins further contributes to the ambiguity. Since no alternative method has been found, absorbancy changes in the 465 mμ region are still used by some laboratories to measure the "state of flavoproteins" in particles, despite the inherent danger of erroneous conclusions (29) as in early studies aimed at defining the site of action of amytal.

References

1. T. P. Singer, in *Comprehensive Biochemistry*, Vol. 14, M. Florkin and E. H. Stotz, Eds., Elsevier, Amsterdam, 1966, p. 127.
2. T. P. Singer and E. B. Kearney, *Biochim. Biophys. Acta*, **17**, 151 (1954).
3. T. Cremona and T. P. Singer, *J. Biol. Chem.*, **239**, 1466 (1964).
4. T. E. King, in *Oxygen in the Animal Organism*, F. Dickens and E. Neil, Eds., Pergamon Press, London, 1964, p. 215.
5. R. L. Ringler, S. Minakami, and T. P. Singer, *J. Biol. Chem.*, **238**, 801 (1963).
6. G. Rendina and T. P. Singer, *J. Biol. Chem.*, **234**, 1605 (1959).
7. F. Crane and B. Ehrlich, *Arch. Biochem. Biophys.*, **89**, 134 (1960).
8. S. Fleischer, A. Casu, and B. Fleischer, *Federation Proc.*, **23**, 486 (1964).
9. J. M. Machinist and T. P. Singer, *J. Biol. Chem.*, **240**, 3182 (1965).
10. Y. Hatefi, A. G. Haavik, L. R. Fowler, and D. E. Griffiths, *J. Biol. Chem.*, **237**, 2661 (1962).
11. R. L. Ringler, *J. Biol. Chem.*, **236**, 1192 (1961).
12. C. Gregolin and T. P. Singer, *Biochim. Biophys. Acta*, **67**, 201 (1963).
13. T. P. Singer and E. B. Kearney, in *The Enzymes*, Vol. VII, P. Boyer, H. A. Lardy, and K. Myrbäck, Eds., Academic Press, New York, 1963, p. 383.
14. T. E. King, *Biochim. Biophys. Acta*, **58**, 375 (1962).
15. M. Klingenberg, this volume, p. 1.
16. T. Kimura, J. Hauber, and T. P. Singer, *Nature*, **198**, 362 (1963).
17. S. Minakami, R. L. Ringler, and T. P. Singer, *J. Biol. Chem.*, **237**, 569 (1961).
18. H. Beinert, G. Palmer, T. Cremona, and T. P. Singer, *J. Biol. Chem.*, **240**, 475 (1965).
19. T. E. King, *Biochim. Biophys. Acta*, **47**, 430 (1961).

20. T. E. King, *J. Biol. Chem.*, **238**, 4037 (1963).
21. V. Massey and T. P. Singer, *J. Biol. Chem.*, **228**, 263 (1957).
22. T. Kimura and J. Hauber, *Biochem. Biophys. Res. Commun.*, **13**, 169 (1963).
23. H. Watari, T. P. Singer, and E. B. Kearney, *J. Biol. Chem.*, **238**, 4063 (1963).
24. T. Cremona, E. B. Kearney, M. Villavicencio, and T. P. Singer, *Biochem. Z.*, **338**, 407 (1963).
25. R. L. Pharo and D. R. Sanadi, *Biochim. Biophys. Acta*, **85**, 346 (1964).
26. R. L. Pharo, L. A. Andreoli, S. R. Vyas, and D. R. Sanadi, *J. Biol. Chem.*, **241**, 4771 (1966).
27. J. Salach, P. Bader, and T. P. Singer, *Biochem. Biophys. Res. Commun.*, in press.
28. S. Minakami, T. Cremona, R. L. Ringler, and T. P. Singer, *J. Biol. Chem.*, **238**, 1529 (1963).
29. T. Cremona and E. B. Kearney, *J. Biol. Chem.*, **239**, 2328 (1964).
30. C. J. Lusty, J. Machinist, and T. P. Singer, *J. Biol. Chem.*, **240**, 1804 (1965).
31. C. Rossi, T. Cremona, J. Machinist, and T. P. Singer, *J. Biol. Chem.*, **240**, 2634 (1965).
32. J. M. Machinist and T. P. Singer, *Proc. Natl. Acad. Sci. U.S.*, **53**, 467 (1965).
33. Y. Hatefi, A. G. Haavik, and D. E. Griffiths, *J. Biol. Chem.*, **237**, 1676 (1962).
34. H. Mersmann, J. Luthy, and T. P. Singer, *Biochem. Biophys. Res. Commun.*, **25**, 43 (1966).
35. D. D. Tyler, J. Gonze, R. W. Estabrook, and R. A. Butow, in *Non-heme Iron Proteins*, A. San Pietro, Ed., Antioch Press, Yellow Springs, Ohio, 1965, p. 447.
36. T. Cremona and E. B. Kearney, *J. Biol. Chem.*, **240**, 3645 (1965).
37. T. P. Singer, E. Rocca, and E. B. Kearney, in *Flavins and Flavoproteins*, E. C. Slater, Ed., Elsevier, New York, 1966, p. 391.
38. H. R. Mahler, N. K. Sarkar, L. P. Vernon, and R. A. Alberty, *J. Biol. Chem.*, **199**, 585 (1952).
39. B. Mackler, *Biochim. Biophys. Acta*, **50**, 141 (1961).
40. B. DeBernard, *Biochim. Biophys. Acta*, **23**, 510 (1957).
41. D. R. Sandi, L. R. Pharo, and L. A. Sordahl, in *Non-heme Iron Proteins*, A. San Pietro, Ed., Antioch Press, Yellow Springs, Ohio, 1965, p. 429.
42. T. P. Singer and J. Salach, *Federation Proc.*, **26**, in press.
43. C. P. Lee and L. Ernster, in *Regulation of Metabolic Processes in Mitochondria*, J. M. Tager, S. Papa, E. Quagliariello, and E. C. Slater, Eds., Elsevier, Amsterdam, 1966, p. 218.
44. T. P. Singer, E. B. Kearney, and N. Zastrow, *Biochim. Biophys. Acta*, **17**, 154 (1955).
45. T. P. Singer, E. B. Kearney, and P. Bernath, *J. Biol. Chem.*, **223**, 599 (1956).
46. T. Y. Wang, C. L. Tsou, and Y. L. Wang, *Sci. Sinica (Peking)*, **5**, 73 (1956).
47. T. P. Singer, V. Massey, and E. B. Kearney, *Arch. Biochem. Biophys.*, **69**, 405 (1957).
48. M. G. P. J. Warringa, O. H. Smith, A. Giuditta, and T. P. Singer, *J. Biol. Chem.*, **230**, 97 (1958).
49. T. P. Singer and F. J. S. Lara, in *Proceedings of the International Symposium on Enzyme Chemistry, Tokyo-Kyoto, 1957*, K. Ichihara, Ed., Academic Press, New York, 1958, p. 330.
50. O. Arrigoni and T. P. Singer, *Nature*, **193**, 1256 (1962).
51. E. B. Kearney, *J. Biol. Chem.*, **235**, 865 (1960).

52. T. P. Singer, J. Hauber, and E. B. Kearney, *Biochem. Biophys. Res. Commun.*, **9**, 146 (1962).
53. T. E. King, *Biochem. Biophys. Res. Commun.*, **16**, 511 (1964).
54. D. M. Ziegler and K. A. Doeg, *Arch. Biochem. Biophys.*, **97**, 41 (1962).
55. T. Y. Wang, C. L. Tsou, and Y. L. Wang, *Sci. Sinica (Peking)*, **7**, 55 (1958).
56. P. Cerletti, R. Strom, and M. G. Giordano, *Biochem. Biophys. Res. Commun.*, **18**, 259 (1965).
57. E. B. Kearney, *J. Biol. Chem.*, **229**, 363 (1957).
58. T. Kimura, J. Hauber, and T. P. Singer, *Biochem. Biophys. Res. Commun.*, **11**, 83 (1963).
59. M. B. Thorn, *Biochem. J.*, **85**, 116 (1962).
60. D. V. Dervatanian and C. Veeger, *Biochim. Biophys. Acta*, **92**, 233 (1964).
61. V. Massey, *Biochim. Biophys. Acta*, **30**, 500 (1958).
62. D. M. Ziegler, in *Biological Structure and Function*, Vol. 2, T. W. Goodwin and O. Lindberg, Eds., Academic Press, New York, 1961, p. 253.
63. H. Beinert and W. Lee, *Biochem. Biophys. Res. Commun.*, **5**, 40 (1961).
64. H. Beinert and R. H. Sands, *Biochem. Biophys. Res. Commun.*, **3**, 41 (1960).
65. T. P. Singer, in *Oxidases and Related Oxidation-Reduction Systems*, Vol. 1, T. E. King, H. S. Mason, and M. Morrison, Eds., Wiley, New York, 1965, p. 448.
66. E. Bueding, in *Control Mechanisms in Respiration and Fermentation*, B. Wright, Ed., Ronald Press, New York, 1963, p. 167.
67. D. Hirsch, M. Rasminsky, B. D. Davis, and E. C. Lin, *J. Biol. Chem.*, **238**, 3370 (1963).
68. T. P. Singer and J. Hauber, *Federation Proc.*, **24**, 297 (1965).
69. C. Rossi, J. Hauber, and T. P. Singer, *Nature*, **204**, 167 (1964).
70. D. Keilin and E. F. Hartree, *Proc. Roy. Soc. (London)*, **B129**, 277 (1940).
71. D. Keilin and T. E. King, *Proc. Roy. Soc. (London)*, **B152**, 163 (1960).
72. B. Chance, in *Enzymes: Units of Biological Structure and Function*, O. H. Gaebler, Ed., Academic Press, New York, 1956, p. 447.

Cytochromes[*]

QUENTIN H. GIBSON, *Cornell University, Division of Biological Sciences, Ithaca, New York*

I.	Cytochrome *c*	380
	A. Structure	380
	B. Comparative Studies of Cytochrome *c* from Different Sources	382
II.	Cytochrome *b*	383
III.	Cytochrome c_1	384
IV.	Cytochrome Oxidase	386
V.	Spectrophotometric Appearances in Soluble Preparations of Cytochrome Oxidase	388
VI.	The Ratio of Cytochrome a_3 to Cytochrome *a* . . .	392
VII.	Stoichiometry of Cytochrome Oxidase Reactions . . .	394
VIII.	Kinetics of Cytochrome Oxidase Reactions	395
IX.	Rates of Reaction of Cytochrome Oxidase with Oxygen and Ligands	397
X.	Mechanism of Cytochrome Oxidase Reactions . . .	402
XI.	Peroxidase	404
XII.	Mechanism of Action of Peroxidase	406
	A. Spectroscopic Observations on the Enzyme . . .	406
	B. Kinetics of the Interconversion of the Various Forms of Peroxidase	406
	References	410

The biological importance of cytochromes in the economy of animal tissues has been recognized in fact, though naturally not by the name, ever since the beginnings of the application of chemistry in enzymology; and the study of the cellular oxidases has been carried on by the means of observation available at any given time for a period which is now on the order of 80 years. The result of this activity is an extensive literature: Lemberg's survey of 1949 under the title *Haematin Compounds and Bile*

* The author was supported by the PHS grant GM 11231–03 during preparation of this manuscript.

Pigments (1) had over 3000 references, while a recent review by Margoliash, restricted to the subject of cytochrome *c* (2), has over 600 references, the majority of which come from recent literature. Thus, the preparation of an account of the cytochromes of reasonable length is difficult because of the abundance of material available.   To make the task possible at all, an arbitrary decision has been made to deal *in extenso* only with mammalian cytochrome oxidase as the type enzyme of its class and as the one about which the greatest amount of information is available. Within this area priority will be given to chemical enzymology at the expense of physiological considerations and of behavior of the cytochrome systems in relatively intact preparations, such as mitochondria.

I. Cytochrome *c*

An excellent and comprehensive review of all aspects of the chemistry and function of cytochrome *c* by Margoliash and Schejter has appeared recently (2).   Only certain aspects will therefore be mentioned here.

The unusual stability and ease of extraction and purification of this protein has resulted in the accumulation of more information on the structure and reactivity of cytochrome *c* than is available for almost any other protein. It has been obtained from higher plants, algae, fungi, some invertebrates, as well as from a number of fish, reptiles, birds, and mammals. Spectroscopically similar pigments have also been isolated from various bacteria, but these have in most cases been found to differ considerably from the mammalian type in reactivity, and will not be discussed here. The detailed studies of Margoliash and his co-workers (3–5) have made it plain, however, that lack of change in gross spectroscopic properties is not a sufficient criterion of the preservation of the native form of the molecule, since dimerization and further polymerization, as well as varying degrees of deamination, occur in the acid media usually used in isolation of the protein, and these changes result in the production of species with increased capacity to react with ligands such as carbon monoxide, and decreased reactivity with cytochrome oxidase.

A. Structure

The heme group of cytochrome *c* is linked to the protein part of the molecule by covalent linkages; this type of linkage has so far been found only in this protein, and in the cytochromoids from certain photosynthetic bacteria, among hemoproteins (6). Theorell (7,8) first isolated a sulfur-containing "porphyrin C," in which two cysteine molecules were linked to the protoheme, and the introduction of the silver salt method of thioether bond cleavage (9,10) allowed the unequivocal identification of the linkage of the protoheme through the α carbon of the vinyl side chains to the sulfur groups of the amino acid.

The information now available from the complete amino acid sequence of the protein (11–13) makes it possible to identify heme binding as taking place through the only two suitably placed cysteine residues, at positions 14 and 17 in the mammalian pigment. Some uncertainty still exists, however, about the nature of the fifth and sixth coordination bonds to the iron. Theorell and Åkesson first suggested that these were utilized in bonding to two histidine residues, from evidence obtained from titration curves (14,15). While it is generally agreed that one of the coordination positions is filled by an imidazole, other evidence suggests that a lysine group could provide the other. Margoliash et al. (3) have shown, from molecular models, that the histidine residue at position 18 and the lysine in position 13 could form the necessary bonds, providing that the protein chain in that region were in the fully extended form. As will be discussed later, there are considerable differences in the properties of the oxidized and reduced forms of the protein, and Harbury and Loach (16), on the basis of studies with small peptides from enzymic digests of cytochrome, have suggested that a change in coordinating groups may occur when the valence of the iron atom changes. Thus the ferric form of cytochrome c may be a diimidazole, while in the ferrous form the fifth and sixth coordination positions could be linked to one histidine and one lysine.

While the primary structure of cytochrome c has been completely elucidated by the work of Margoliash, Smith, Kreil, and Tuppy, there is as yet little information about the tertiary structure. The inaccessibility of the heme, especially in the ferrous form, to ligands clearly demonstrates that it must be located in a crevice in the native protein which is closed at physiological pH. The changes that can be observed when the crevice opens at alkaline or acid pH have been studied by a number of workers, since Keilin (17) first described the spectral changes that occur under such conditions [Ehrenberg and Theorell (18); George and Lyster (19)]. The latter workers concluded that the thermodynamic quantities measurable for the reaction of cyanide and azide at pH 4–5 are compatible with the suggestion that a heme–imidazole bond is being replaced, and is responsible for holding the crevice closed. The crevice is more tightly held in the ferrous form of the hemoprotein, and ligand reactions cannot be observed at all at neutral pH. The pK of crevice opening is nevertheless similar for oxidized and reduced compounds, and Schejter and George (20) have concluded that the large difference in ΔF for ligand binding between the two forms must be the result of large differences in the ΔF of crevice closure and suggest that considerable rearrangement of tertiary structure may be occurring.

There is, indeed, a mounting body of evidence for the occurrence of conformation changes in the cytochrome c molecule. The weak absorption band at 695 mμ is found only in the enzymically active, monomeric form (21), and its disappearance from the monomer can be correlated with such properties as the increase in CO binding capacity; these findings are interpreted as indicating changes in the protein chains at some distance removed from the linkage to the heme. Recently, Greenwood and Palmer (22) have shown that a biphasic reduction of monomeric cytochrome c by ascorbate occurs at pH values above 8; the experiments suggest that only one form, which is the only type present at pH 7, is reducible by ascorbate, and that the unreactive form which is relatively slowly formed at higher pH must first be transformed to the reactive form. This conversion is strongly influenced by temperature, with an activation energy 19 kcal/mole. The pH range in which this kinetic difference can be observed is similar to that in which a change in the rate of combination of cyanide with the ferric form [George and Tsou (23)], a change in the ferro-ferricytochrome c redox potential (24), and a spectrophotometrically detectable change in the 695 band occur. All these lines of evidence therefore indicate that a change in heme–protein linkage, involving considerable changes in overall protein structure, as indicated by the relatively large activation energies involved, occurs in the region pH 8.5–9.

B. Comparative Studies of Cytochrome c from Different Sources

A complete amino acid sequence has now been obtained for a number of mammalian type pigments isolated from different sources; for a full discussion of the findings and their bearing on evolutionary questions, the reader is referred to the reviews by Margoliash and Schejter (2) and Smith and Margoliash (25). All the pigments resemble each other closely in their reactivity with mammalian cytochrome oxidase, their redox potential, and in gross spectroscopic properties. However, minor differences do exist, and the yeast pigment, which has a greater reactivity with carbon monoxide and susceptibility to proteolysis than the mammalian, has no detectable band at 695 mμ. The characteristic spacing of two cysteine residues in the amino acid chain is present throughout, but in invertebrate pigments the acetyl group on the amino-terminal glycine residue is replaced by a chain of four or five amino acids, so that the heme linkage occurs at a point further along the amino acid chain. Differences exist in the basic amino acid in the position corresponding to 13 in the mammalian pigment, and greater variations are observed in the part of the chain further removed from the heme linkage, with the exception of an invariant region from residues 70 to 80.

II. Cytochrome *b*

By comparison with cytochrome *c*, cytochrome *b* from the mammalian respiratory chain has been poorly characterized, perhaps due mainly to the fact that the pigment is firmly bound to the mitochondria and special treatment is necessary to bring it into solution.

A value of about +75 mV is now generally accepted for the E_0' of cytochrome *b*, since the experiments of Chance (26) and Holton and Colpa-Boonstra (27) convincingly demonstrated that Ball's (28) original value of −40 mV was too low, due to his assumption that all the *b*-type pigment reacting with dithionite was functional in succinate oxidation. The presence of some pigment which shows kinetic anomalies in heart preparations has been explained as being due to disruption of electron-transport chains during preparation, resulting in its displacement from the electron pathway from succinate to cytochrome *c* (26) or to the presence of a second form of the pigment, termed *b'* by Slater and his group (29). In phosphorylating particles the kinetic anomalies are rarely seen, though some difference in apparent *b* function is seen in different mitochondrial preparations (e.g., Chance, ref. 30). It is possible that an explanation of some of these differences and anomalies may be obtained from the extensive studies of Green and his co-workers on the fragmentation of the electron-transport system (31). Cytochrome *b* appears as a constituent both of complex II, transferring electrons from succinate to coenzyme Q, and of complex III, mediating electron flow between CoQ and cytochrome *c*. Possibly the ratio of cytochrome *b* linked to these transfer systems is variable.

Extraction of mitochondrial cytochrome *b* in highly dispersed form [Sekuzu and Okunuki (32), Feldman and Wainio (33), Goldberger et al. (34)] has not led to the purification of the biologically active material. Goldberger et al. have obtained a "monomeric" *b* of apparent molecular weight 28,000 in the presence of cationic detergents, after separation from mitochondrial fragments by bile salt treatment, and have also reported (35) that the polymeric form can be depolymerized by alkaline incubation in the presence of a substance termed mitochondrial structural protein. The redox potential of the monomeric form, which was found to be

−340 mV and could therefore account for the lack of biological activity in enzyme systems, was raised by combination with structural protein to the point where reduction could be effected by mixing with members of the $CoQH_2$ series, but enzyme studies have not been published. The possibility that some pigment may be dissociated during preparation of mitochondria in an analogous way, with consequent change in redox potential, may also explain some of the kinetic anomalies. However, a study of the state of reduction of cytochrome pigments in whole muscle indicated a relatively small state of reduction of cytochrome c in the anoxic state, and suggests a low potential for at least some of the pigments in an intact organ.

Considerably more information is available about two non-mitochondrial b-type cytochromes, b_5 from microsomes (see Strittmatter, this volume), and cytochrome b_2 from baker's yeast. Since Bach, Dixon, and Zerfas (37) first noted that the absorption spectrum of this cytochrome followed lactic dehydrogenase during purification, the enzyme has been most thoroughly studied, by Morton and co-workers (38), by Boeri et al. (39), by Nygaard (40), and by Okunuki's group (41). Crystallization was achieved in 1954 (42), and the enzyme was conclusively shown to contain both flavin mononucleotide and protoheme in a 1:1 ratio, with a minimum molecular weight of 80,000. Whereas the heme prosthetic group of the native enzyme is readily removed by acid–acetone treatment, Morton and Shipley (43) have shown that treatment with PCMB, followed by acetone precipitation at −15° leads to the halving of the molecule, normally present as a dimer, and to a change in spectrum so that it closely resembles cytochrome c, and the heme becomes bound by thioether linkages. At the same time, the redox potential changes to about 280 mV and the pigment becomes active in assays for cytochrome c, Ferricyanide, dichlorophenolindophenol, and oxygen can act as electron acceptors for the crude enzyme; the reaction with oxygen seems to be greatly reduced in rate as the enzyme is purified, and may be attributed to a fraction of enzyme from which the flavin has been removed (38) or to the presence of non-heme iron in preparations made by methods other than Morton's.

III. Cytochrome c_1

Cytochrome c_1 was first described by Yakushiji and Okunuki (44). Independently, Keilin and Hartree (45) concluded that the altered position of the α-band absorption peak of cytochrome c *in vivo* could be accounted for by overlapping the bands of the c pigment already well known in solution with those of a second insoluble material with an absorption maximum at 553–554 mμ, which they termed cytochrome e. The absorption peaks could be resolved by spectroscopy at liquid air temperatures. Subsequently (46), they concluded that cytochromes c_1 and e were identical, and that the pigment was in all probability a functional member of the respiratory chain.

The similarity of the spectra of cytochromes c and c_1 made it difficult to assign a position in the chain to c_1 relative to the other components in intact mitochondria. In preparations which had been lysed and depleted of cytochrome c by water washing, Estabrook (47) was able to demonstrate electron transport from DPNH or succinate to cytochrome c_1, which, however, became reduced at a rate about five times as fast as cytochrome b. If antimycin A were present, cytochrome b became reduced, while c_1 remained oxidized, thus placing the site of action of the inhibitor at the point of interaction of cytochromes b and c_1. The anomalous behavior of cytochrome b in the uninhibited preparation was compared with the kinetic results on Keilin and Hartree preparations described by Chance (48). Assuming the same extinction coefficients for cytochromes c and c_1, a ratio of b/c_1 of about 5/3 was found. By investigating suspensions of intact mitochondria rapidly frozen while in different metabolic steady states, Chance (49) demonstrated that cytochrome c_1 became reduced under the same conditions as b and c, and to about the same extent.

The nature of the electron acceptor for cytochrome c_1 has been repeatedly investigated since Yakushiji and Okunuki (44) reported that c_1 was able to react slowly with cytochrome a. Only very slow reduction of cytochrome a was observed by Estabrook in water-washed mitochondria (47) in cyanide poisoned preparations, unless extraneous cytochrome c was added. The isolated cytochrome c_1 of Bomstein, Goldberger, and Tisdale (50) did not react with cytochrome oxidase preparations unless c was added, and the evidence that c_1 reacts only with c, probably by a chemical rather than enzymic reaction, is therefore strong for isolated preparations. The possibility of direct reaction of c_1 with a in undamaged preparations has nonetheless been revived recently by Jacobs, Andrews, and Crane (51), who have proposed an additional electron transport pathway, parallel to the accepted one, utilizing direct reaction between c_1 and a, on the basis of experiments with progressively extracted and disorganized systems.

A number of "soluble" preparations of cytochrome c_1 have been obtained from mitochondria by extraction with cholate (48,50,52–55). Of these, the best characterized is that of Bomstein et al. Their best material, which could be almost freed of cholate, had an iron content of 27 μA Fe per gram of protein, from which a minimum molecular weight of 37,000 could be calculated. Unlike isolated cytochrome b, the c_1 preparation was able to react with a number of enzyme systems, such as the microsomal cytochrome c reductase, although only at low rates.

IV. Cytochrome Oxidase

The history of the discovery of the cytochromes has been retold several times, perhaps most recently by Nicholls (56); no attempt will be made to repeat it here in full, though certain salient observations will be mentioned. As has already been said, the cytochrome oxidase has been studied at all times with interest and enthusiasm by the methods which have happened to be available, with the result that the advance of knowledge has been anything but an orderly process and has been associated with a number of major controversies, some of which continue at present. There are also examples of important discoveries which could not be appreciated at the time at which they were made because the general background of knowledge against which they could be appreciated did not exist. A prime example is the observation by MacMunn of the absorption bands associated with the cytochrome system (57,58). He observed that these bands were strengthened when reducing agents such as ammonium sulfide were added to tissue preparations and that they disappeared on oxidation with ferricyanide and hydrogen peroxide. These observations were the basis for his hypothesis that the pigments were associated with the uptake of oxygen by the tissues, as, in the end, proved to be the case. His experiments and the conclusions drawn from them were criticized by Levy (59) on the grounds that the spectroscopic appearances observed were due to mixtures of hemoglobin derivatives and degradation products of hemoglobin, and this point of view was supported by Hoppe-Seyler himself (60) in the controversy which followed. In retrospect, as pointed out by Nicholls, the dispute has fewer of the characters of a conflict between a senior authority in the wrong and a lesser known scientist in the right than is sometimes attributed to it. The real basis lay in the inability of the contestants to define the matters of dispute in chemical terms, or to obtain the substances in a sufficiently pure form to permit them to be characterized adequately, by either chemical or physical means. Thus, the debate was carried on at a time when hemoglobin was the only well-characterized hemoprotein known and when the differences between oxygenation and oxidation were ill understood. In such a context, and when it is borne in mind that MacMunn demonstrated the oxidation and reduction of his compounds in reactions with foreign chemical reagents rather than under physiological conditions, the position of Hoppe-Seyler appears much more reasonable; and an element of good fortune as well as of genius should be included in appraising MacMunn's proposals.

This celebrated controversy has been recapitulated at some length because it continues to have useful morals today. Although many of the compounds have been characterized with a precision impossible 80 years ago, they remain, nevertheless, imperfectly defined; and their properties and their behavior are imperfectly understood, while conflicting views are presented with an enthusiasm (which sometimes seems) inversely related to knowledge. After the Hoppe-Seyler–MacMunn controversy, no spectroscopic observations on the cytochrome system were published until the work of Keilin in 1925 (61). This, however, was conducted against a very different scientific background. Keilin, armed with the progress in chemistry during a period of almost 40 years, was able to make his observations under physiological conditions, and was able to show that the compounds he observed underwent spectrophotometric changes as the result of the variation of the external conditions of the tissues.

In the interval, considerable advances were made in the understanding of the significance and the function of cytochrome oxidase even though chemical work on the enzyme itself remained, relatively speaking, at a standstill. Thus, Ehrlich (62) observed in 1885 that animal tissues cause the formation of indophenol from α-naphthol and dimethyl-p-phenylenediamine in the presence of oxygen; and Battelli and Stern (63) were able to correlate the activity of tissues in indophenol formation with their succinoxidase activity and the uptake of the oxygen by the tissue. By hindsight, the importance of this correlation and its relation to the work of Keilin appear obvious, but they were less evident at that time, since the concept of the respiratory chain had not been developed and the relation between substrate-specific dehydrogenation and the uptake of oxygen was far from clear.

It was about the time of Keilin's work that the first chemical evidence of the nature of the indophenol oxidase became available, through detailed studies of the reversal of inhibition by carbon monoxide on illumination with light from various wavelengths. The experiments allowed the spectrum of the carbon monoxide compound of the respiratory enzyme to be determined and showed that the spectrum was that of a hemoprotein (64). In the very next year, Keilin (17) showed that the oxidation of a cytochrome c preparation by particles obtained from heart muscle was sensitive to carbon monoxide, and that the inhibition could be reversed by light. From these experiments he concluded that the respiratory enzyme and the indophenol oxidase of heart muscle might be one and the same. With the appearance of the papers of Keilin and Hartree (65,66), what may be regarded as the historical aspects of the study of cytochrome oxidase came to an end, and the current stage of its study was ushered in. In these papers Keilin and Hartree showed that the spectroscopic appearances were consistent with the presence of two distinct components containing cytochrome a, one of which reacts with carbon monoxide and with cyanide while the other does not react with either of them. The components were named a_3 and a for the reactive and nonreactive portions, respectively, and the proposal was made that the reaction with oxygen involves the sequential action of both components, cytochrome a_3 reacting first with oxygen and subsequently oxidizing cytochrome a. Their observations were carried out exclusively with a visual low-dispersion spectroscope, but in general outline they have been confirmed by most, but not all, workers in the field up to the present day. Their work also is of interest in that they made the first use of bile salts for partial clarification of extracts containing cytochrome oxidase, a technique which has been much employed more recently. Keilin and Hartree also noted that their preparations contained copper but were unable to obtain spectrophotometric evidence which would justify them in assigning a functional role to the metal.

The historical aspect of the subject may perhaps be rounded out by mentioning that a series of researches by Chance and his collaborators (see 67,68) have placed beyond doubt the identification of Keilin's cytochrome a_3 with the respiratory enzyme of Warburg and have shown the functional importance of the enzyme in intact mitochondria, as well as in heart muscle preparations. This work has depended very extensively on the development by Chance of appropriate dual wavelength spectrophotometric methods (see 69).

From the point of view of enzyme purification and the study of the properties of cytochrome oxidase preparations, the most notable event in the last decade or so has been the widespread use of optically clear or nearly clear preparations of cytochrome oxidase stabilized by the addition of either cholate or nonionic detergents, and occasionally by tissue lipids (70–73). In succeeding sections, an attempt will be made to deal systematically with certain of the properties of cytochrome oxidase, particularly as exhibited by preparations of the kind just mentioned.

V. Spectrophotometric Appearances in Soluble Preparations of Cytochrome Oxidase

The earlier attempts to obtain difference spectra for the oxidation of the cytochrome oxidase components a_3 and a were superseded by the development of Yonetani (71) of his cyanide method for separating the contributions of the two components. The method is based on the principle that the cyanide compound with oxidized cytochrome a_3 cannot readily be reduced by dithionite. If, then, a preparation of cytochrome oxidase to which cyanide has been added is subsequently reduced with dithionite, the difference spectrum obtained corresponds substantially to that of cytochrome a. By subtraction of this difference spectrum from the difference spectrum for the reduction of cytochrome oxidase in the absence of cyanide (which corresponds to the reduction of a_3 and a), the difference spectrum of a_3 alone may be obtained. The validity of this method depends, of course, upon the assumption that the cytochrome a_3–cyanide compound is quite uninfluenced by dithionite. Although this is true for experiments carried out reasonably rapidly, Lemberg and co-workers (74) have shown that slow reduction of the ferric–a_3–cyanide compound does in fact occur, and may be responsible for some of the variation shown by the difference spectra. The main outlines of the difference spectra, however, as described in the work of Yonetani, have been confirmed on a number of occasions, though, as will be seen, some of the details continue under discussion.

Not only have the findings of Yonetani been confirmed using analogous methods, but similar spectra have also been obtained by other chemical approaches. Thus, for example, Horie and Morrison (75) have made use of the protective effect of carbon monoxide on the oxidation of cytochrome a_3 to achieve a separation of difference spectra. In their experiments, cytochrome oxidase in the reduced form was equilibrated with carbon monoxide at a high partial pressure, thus forming the carbon monoxide–a_3 compound. On titration of the enzyme with ferricyanide, only cytochrome a became oxidized, and so the difference spectra for the oxidation of cytochrome a was obtained separately from that of cytochrome a_3. The results showed very good agreement with those obtained by Yonetani. A third method which yielded similar results was used by Gibson and Greenwood (76). These authors separated the spectra in time, making use of the slow dissociation of carbon monoxide from cytochrome a_3. They mixed the carbon monoxide compound of cytochrome oxidase with oxygen and followed the ensuing absorbance changes at a series of wavelengths with a rapid reaction apparatus. They analyzed their results in terms of two reactions, one of which was the oxidation of cytochrome a, while the other represented the dissociation of carbon monoxide from cytochrome a_3, at a rate determined independently from studies of the rate of replacement of carbon monoxide by nitric oxide in the preparation. In this case, too, the spectra agreed with those published by Yonetani. A very detailed study of these difference spectra was undertaken later by Lemberg and his co-workers (74), who used the cyanide method of Yonetani but who also separated the difference spectra in time using the differential rate of reduction of cyto-chromes a and a_3 on exposure to low concentrations of sodium dithionite. The reaction of cytochrome a with small concentrations of dithionite is substantially more rapid than the reaction of cytochrome a_3 and the reac-tion may be made slow enough so that the difference spectra can be recorded with an ordinary spectrophotometer. The study of a large number of pre-parations showed inconstant spectra for the oxidized and reduced forms of the enzyme, while the partition between a_3 and a was also variable; and quite appreciable ranges were observed for the isosbestic points of cyto-chromes a_3 and a. The general findings of all the work cited may be sum-marized by saying that at 444 mμ the contributions to the difference spectrum made by a_3 and a were approximately equal, while at 605 mμ the contribution of cytochrome a was of the order of 4 times that of cyto-chrome a_3. The total change in absorbance at these two wavelengths was approximately as 4.5:1, the change at 444 mμ being the greater.

The spectrum of the carbon monoxide complex of reduced cytochrome a_3 has also been deduced and has been shown to agree in form with the photo-chemical action spectrum obtained long ago by Warburg and more recently and more elegantly by Chance and his co-workers. In addition, some in-formation about the spectrum of the nitric oxide complex has been given by Okunuki (77). Finally, it is necessary to discuss the evidence for the "oxygenated" compound first described by Orii and Okunuki (78). It was found that if a preparation of cytochrome oxidase was reduced with dithion-ite and then shaken in air to destroy the excess dithionite, a stable spectrum appeared which had a maximum absorbance at 430 mμ in the Soret region. This band was attributed to an "oxygenated" complex. This interpretation has, however, been disputed. Thus, in 1963, Greenwood (73) showed that a very similar spectrophotometric appearance could be obtained by the addition of hydrogen peroxide to oxidized cytochrome oxidase and sug-gested that the 430 mμ band was due either to a peroxide complex or to the formation of cytochrome oxidase with iron in a higher oxidation state, though he noted that it was impossible to produce this spectrophotometric change by the addition of chloriridate to the preparations. He pointed out that the destruction of excess dithionite by shaking with air generates appreciable hydrogen peroxide in the medium and suggested that the observations of the Japanese school might satisfactorily be explained without postulating the formation of an "oxygenated" complex. The observations and many of the interpretations of Greenwood have very recently been confirmed *in extenso* by Lemberg and his co-workers (79). It is not necessary, however, to use dithionite in order to obtain spectra with some of the characters of the "oxygenated" compound, since the reoxidation of cytochrome oxidase preparations of the Yonetani type which have been reduced by means of ascorbate does not lead to a return of the initial spectrum (though the use of ascorbate is, of course, no guarantee that hydrogen peroxide is not formed in the solutions).

There is no convincing evidence that the "oxygenated" complex actually contains bound molecular oxygen. This has not been demonstrated gasometrically, nor has it been shown that the bands of reduced cytochrome oxidase appear upon exposure of the "oxygenated" complex to a vacuum, nor has it been demonstrated that oxygen may be displaced by carbon monoxide (admittedly, the affinity of carbon monoxide for the enzyme need not necessarily be greater than that of oxygen). It is perhaps significant that the first report of the "oxygenated" compounds stated that the quantitative relations of the complex were somewhat obscure (78).

One further method of obtaining difference spectra may be mentioned briefly at this point. In kinetic work Gibson and Greenwood (80) have made use of the photosensitivity of the carbon monoxide compound of cytochrome oxidase to study the reaction of reduced cytochrome oxidase with oxygen, removing the bound carbon monoxide photochemically and then following the rate of oxidation of the reduced cytochrome oxidase by molecular oxygen present in the solution. If the reaction is followed at different wavelengths, it is, of course, possible to obtain the difference spectra corresponding to the various phases of the kinetic reaction. It has been found that the reaction with oxygen may be analyzed into a rapid phase, which has been correlated with the oxidation of cytochrome a_3, and a slower phase which is independent of the concentration of oxygen and has been equated with the oxidation of cytochrome a. When the difference spectra obtained in this way are compared with the static difference spectra so far discussed, it is found that the relative contribution of cytochrome a_3 is apparently much greater than that observed in the static difference spectra. Thus, for example, at 444 mμ, where the ratio of the contributions to the static spectra is approximately 1:1, in the rapid kinetic difference spectra the ratio is more nearly 2.5:1. Unfortunately, obtaining spectra in this way is both difficult and laborious, and comparatively few complete spectra have been run. Individual determinations at selected points have, however, been made with a number of preparations and it seems that the reproducibility of the relative contributions of the spectra from one preparation to another is rather higher than that reported in the static experiments by Lemberg and his co-workers. The discrepancy appears to be a substantial one and an attempt to account for it will be made after discussion of the mechanism of action of the enzyme and the ratio of cytochromes a_3 to a in soluble preparations.

VI. The Ratio of Cytochrome a_3 to Cytochrome a

The ratio of cytochromes a_3 and a was for a considerable period a matter for speculation. It is important not only because it is basic to the mechanism of action of the enzyme but because it is a necessary piece of information which is required to assign spectrophotometric constants which may be derived from the difference spectra discussed in the previous section. So far, two methods have been used to determine this quantity. The first uses the definition of cytochrome a_3 as a material capable of combining with carbon monoxide when equilibrated with low partial pressures of gas, and a determination of the carbon monoxide binding capacity is regarded as equivalent to the determination of the content of cytochrome a_3. The second method depends upon the reductive titration of cytochrome oxidase using DPNH with phenazine methosulfate as mediator. A known number of electron equivalents may thus be introduced into an anaerobic solution of oxidized cytochrome oxidase and the extinction coefficients for the various components obtained. This titrimetric method depends upon a number of assumptions and requires detailed discussion.

The earliest gasometric measurements (80) showed that about one-third of the heme a in Yonetani-type preparations of cytochrome oxidase was able to combine with carbon monoxide, and, on the assumption that the preparations contained only active material, the conclusion was drawn that the ratio of cytochrome a_3 to a was $1:2$. Later, however, DPNH titrations suggested that the ratio might be more nearly $1:1$, and an investigation of the carbon monoxide binding capacity of soluble preparations made by the method of Griffiths and Wharton was undertaken (81). The new work gave average values similar to those of Gibson and Greenwood, and added the information that the carbon monoxide binding capacity of individual preparations has a wide scatter. The new and more sensitive methods used allowed particulate preparations to be studied as well, and in them the ratio was $1:1$ rather than the variable $1:2$ found for the soluble materials. Similar quantitative results both for Yonetani-type preparations and for particulate preparations have also been reported (82,83)

One explanation for the scatter of the results is that the ratio in the particles and in the native enzyme is $1:1$ but the soluble preparations contain substantial amounts of denatured enzyme which, while preserving some of the spectrophotometric properties of cytochrome oxidase, has lost the power of combining with carbon monoxide. A variation in the proportion of such denatured material would, of course, help to explain the scatter of the spectrophotometric results for different preparations made by the same method as noted especially by Lemberg and co-workers

(74), but also poses serious problems in the interpretation of difference spectra and in the assignment of extinction coefficients for the individual components. It seems reasonable to assign spectrophotometric coefficients for the a_3 difference spectra on the assumption that the amount of a_3 in a given preparation is measured by its carbon monoxide binding capacity. If such a criterion is applied, it is found that the extinction coefficients are greater than those previously derived from the difference spectra for the total heme a present in the cytochrome oxidase preparation. This finding seems to correlate with the difference spectra obtained by the flash photochemical experiments mentioned earlier, which also have tended to show that the extinction coefficient to be ascribed to cytochrome a_3 changes is about twice as large as previously believed. The reason for this correlation is that in the flash photochemical experiments attention is necessarily directed to the fractions of the enzyme which are capable of binding carbon monoxide, and it may be reasonably supposed that in the rapid reaction which follows exposure of the reduced preparation to oxygen, only the native enzyme, in the sense of having both carbon monoxide binding capacity and the associated capacity for rapid oxidation of cytochrome a, is taken into account. The problem remains open, however, and an attempt to specify extinction coefficients may well be premature.

The reductive titration experiments carried out by Van Gelder and Slater (84) and Van Gelder and Muijsers (85) require discussion, together with the results of oxidative titration experiments carried out in the stopped-flow apparatus by Gibson and Greenwood (80). In the titration experiments with DPNH and phenazine methosulfate which were carried out on an extended time scale and in which an approximation to equilibrium would be expected, the remarkable finding emerged that a straight-line relation was obtained between the observed change in absorbance and the amount of DPNH added at all wavelengths studied. This result may be explained in two different ways. One explanation would deny the separate existence of cytochromes a_3 and a and so simplify the interpretation of the titrimetric data. The second would involve the reduction at all times of equal amounts of cytochromes a_3 and a as each increment of DPNH was added to the solutions.

The second explanation, unlikely though it may appear *a priori* that equal amounts of a_3 and a should be reduced at all concentrations of added DPNH, is in agreement with the results of the stopped-flow titration experiments carried out by Gibson and Greenwood (80). In these experiments also there appeared to be (within the rather considerable errors of measurement) a linear relation between the amount of oxygen added and the change in absorbance at 445 mμ. This result was not anticipated by the authors and was commented upon by them in their publication. The implications of such an observation are both interesting and far reaching. It implies the existence of a functional redox unit in which all the components are present only in the oxidized or in the reduced state and suggests strong stabilizing interactions.

VII. Stoichiometry of Cytochrome Oxidase Reactions

Besides exhibiting stabilizing interactions, the titrations of cytochrome oxidase which have been carried out both reductively and oxidatively agree in showing that groups other than the hemes participate in electron exchange, and that there is one non-heme group for each heme group active in electron exchange. While some element of conjecture remains in the identification of the supplementary groups, evidence has accumulated which points to the participation of copper in cytochrome oxidase redox reactions. It has been mentioned that Keilin and Hartree (65,66) knew that their preparations contained copper but were unable to assign a role to it in electron transport. About 10 years or more ago, a group (86) of workers determined the copper content of clarified preparations stabilized with cholate and followed the copper-to-heme ratio as a function of the degree of enzyme purification. They concluded that copper was constantly present and that it was not removed from the enzyme during the purification. These findings were confirmed and extended by Griffiths and Wharton (72), who showed that in cytochrome oxidase preparations made with due care to exclude metals from the reagents, the ratio of iron to copper was exactly 1:1. They also showed that their preparations had a band in the infrared with a maximum at about 820 mμ which increased on the oxidation of the enzyme. Yonetani (87) also showed that his preparations contained copper; in some cases, however, the amount of copper was considerably more than 1 atom per heme molecule. A striking property of cytochrome oxidase is the firmness with which the metal is bound to the enzyme, and at least a considerable part of the enzyme copper appears to be inaccessible to chelating agents, though there has been much discussion on this point. The copper, when in the cupric state, is at least partly detectable by EPR methods (88). Some differences between the EPR spectrum of the Yonetani type of preparation and the spectrum of the Griffiths and Wharton type of preparation have been recorded and have been attributed to non-enzyme copper in some specimens of the Yonetani type of enzyme. This copper, unlike the enzyme-bound copper, shows fine structure on EPR examination and is accessible to binding by chelating agents. It is perhaps appropriate to emphasize that only one-half to one-third of the total enzyme copper has been visualized in EPR spectrometry as judged by the integration of the EPR traces.

A picture which shows interesting differences emerges when we turn to consider the relation of copper to the reductive titrations carried out by Van Gelder and his co-workers, as mentioned in connection with the stoichiometry of the cytochrome oxidase reaction. It was found that two electrons must be transferred for every heme *a* reduced, and it has been supposed that this additional electron is required to reduce cupric copper. It has been found possible to influence the slope of the titration curve, and hence presumably the number of effective electron-accepting groups other than the heme groups, by the addition of chelating agents to the cytochrome oxidase preparations. The binding agents particularly studied were EDTA and salicylaldoxime. These reagents were able to block access to one out of each pair of non-heme electron-accepting groups in the oxidized forms of cytochrome oxidase. Superficially, these experiments might seem to contradict the EPR experiments, but this is not necessarily so, since the EPR copper does not account for the total known to be present in the enzyme; and it may therefore be supposed that that part of the copper which is visible by EPR is inaccessible to chelating agents and vice versa. Unfortunately, it is not yet possible to give full consideration to these experiments which have appeared so far only in the form of preliminary communications.

VIII. Kinetics of Cytochrome Oxidase Reactions

Steady-state systems in which cytochrome oxidase is turning over have been studied by many investigators. The resulting literature is somewhat difficult to follow, and it is perhaps sufficient to restrict attention to some of the most recent work, particularly that of Smith and Conrad (89), Minnaert (90), and Yonetani (91). The problem has also been recently dealt with by Nicholls (92). Yonetani concludes that the results can be satisfactorily accounted for by supposing that reduced cytochrome *c* forms an intermediate complex with cytochrome oxidase whose breakdown is rate determining, and that oxidized cytochrome *c* can also combine with the enzyme to form an inhibitory enzyme–product complex. In the special case that the affinity of the oxidized and reduced *c* for the enzyme is equal, this scheme will explain the finding that the observed reaction is first order and that the specific rate constant declines with increasing concentration of cytochrome *c*. Such a scheme will also accommodate the observations of Estabrook (93) that cations, but not anions, compete with cytochrome *c* for combination with the oxidase. This interpretation also fits in well with rapid reaction studies by Gibson et al. (94), who found that when reduced cytochrome c is mixed with the oxidized cytochrome oxidase, there is a very rapid initial reaction which is apparently second order and which gives a rate constant as high as 3×10^7 M^{-1} sec^{-1} with partial reduction of cytochrome a and partial oxidation of cytochrome c. In a turnover system in which ascorbate is also present, the system does not proceed, however, towards the steady-state concentrations of the various components which would be expected from the initial rate of the reaction; instead, after the initial rapid reaction, the location of the rate-limiting step appears to shift from an initial position between ascorbate and oxidized cytochrome c to one between cytochrome a and cytochrome a_3, as defined spectrophotometrically. It is observed that in the absence of oxygen, the reduction of cytochrome a_3 is slow at the high concentrations which must be used for a rapid reaction experiment. A study of the kinetics of the cytochrome oxidase–cytochrome c initial reaction over a range of concentrations by a modification of the photochemical procedure used by Gibson and Greenwood (unpublished experiments) yields a limiting rate and equilibrium constant consistent with the formation of a complex of the type postulated by Yonetani (91) with similar numerical values for the coefficients. These findings would seem to be able to account for many of the discrepancies and anomalies which have been associated with the study of cytochrome oxidase action in systems with added exogenous cytochrome c.

Closely related to the kinetics of the oxidation of cytochrome c by cytochrome oxidase are the phenomena of the binding of basic proteins such as salmine and synthetic polyaminoacids such as polylysine which have been studied extensively by Person and his co-workers (95,96), by Smith and Conrad (89), and by Conrad-Davies and others (97). These workers have established that binding occurs readily between cytochrome oxidase and the polycations and that the resulting complexes are strongly inhibitory. The diversity of factors which influence the oxidation of cytochrome c by cytochrome oxidase, as well as the complication of its kinetics, has contributed to the difficulty of the assay of cytochrome oxidase and has led to anomalies such as apparently differing activities of cytochrome oxidase when assayed with cytochrome c as substrate and when assayed in a turnover system with low concentrations of cytochrome c, the reaction being followed by oxygen uptake.

Some features of the cytochrome oxidase reaction which can be examined in particulate preparations and in mitochondria have been studied by Chance and his collaborators. This extensive series of papers does not seem to have been reviewed recently. Comparatively little study has been devoted to the reaction of cytochrome c with cytochrome oxidase as such.

The work has been directed towards such basic questions as the determination of the constituents of, and the order of reaction in, the electron transport chain, and the effect and location of sites of control of this chain. It is clear, however, that the reactions of cytochrome oxidase and cytochrome c in the mitochondria and in ETP preparations are faster than those of cytochrome oxidase with exogenous cytochrome c; and turnover numbers much in excess of those which can be obtained with the soluble preparations have been reported (98). These differences between the behavior of the particle-bound respiratory chain and of the enzymes in solution may well be due to the presence of structured elements within the particles as suggested by Chance (99). In these systems the inhibitory effects described earlier do not appear to apply. It may well be a numerical coincidence, but the greatest rate obtained for particulate systems, reported by Smith (98), is of the same order as the limiting rate recently deduced by Yonetani (91) and the maximal rate obtained in solution systems for electron transfer within the cytochrome oxidase–cytochrome c complex (unpublished photochemical experiments of Gibson and Greenwood). There remains, however, an appreciable gap between understanding the behavior of solutions of cytochromes and the effects within the organized system.

IX. Rates of Reaction of Cytochrome Oxidase with Oxygen and Ligands

The rates of reaction of the enzyme with ligands and with oxygen have been comparatively little studied. This is due, in part, to the technical difficulty of following so rapid a reaction with adequate precision. Some data are, however, available both for the purified preparations and for the reactions of the enzyme as it occurs in mitochondria and in particulate preparations.

The rates of reaction of purified preparations of cytochrome oxidase with ligands have been measured; the reactions with carbon monoxide, cyanide, and nitric oxide appear to be quite straightforward, at least in the degree of detail with which they have so far been examined.

The reaction with carbon monoxide can conveniently be studied by the stopped-flow method, and has been reported to have a rate constant of 8×10^4 M^{-1} sec^{-1}, at 20° with an apparent activation energy of 7 kcal (80). Preparations stabilized with cholate, detergents, and tissue lipids have given closely similar results. The velocity of dissociation of carbon monoxide has been measured by replacing it with nitric oxide and again similar results have been obtained with various preparations; the rate constant is 0.025 sec^{-1} at 20°.

The reaction of reduced cytochrome oxidase with cyanide has a second-order combination rate constant of $1.2 \times 10^2 \, M^{-1} \sec^{-1}$. The dissociation velocity constant has been obtained by displacement of cyanide with carbon monoxide and is $0.08 \sec^{-1}$. The equilibrium between cyanide and reduced cytochrome oxidase has also been studied indirectly by determining the amount of free cytochrome oxidase in equilibrium with varying known concentrations of cyanide. On addition of carbon monoxide to such an equilibrium mixture, there is a rapid combination of carbon monoxide with free cytochrome oxidase which is followed by a slow displacement reaction. This allows both the equilibrium constant for the reaction with the cyanide and the rate of dissociation of cyanide from the complex to be obtained in a single series of experiments. The effect of pH on the reaction suggests that the effective cyanide species reacting with the oxidase is the un-ionized acid (80).

In the case of nitric oxide, only the velocity of combination of nitric oxide with reduced cytochrome oxidase has been measured. The rate is approximately $5 \times 10^7 \, M^{-1} \sec^{-1}$ at 20°.

So far as is known, the reaction of carbon monoxide with the oxidase is uncomplicated and is restricted to cytochrome a_3 only. The reaction of cyanide may involve groups other than the hemes, according to the experiments of Van Gelder and Slater (84) and Van Gelder and Muijsers (85), whose titrations indicate that cyanide may combine with copper as well as with the hemes. The reaction with nitric oxide has been insufficiently studied, but there is some kinetic information suggesting that nitric oxide may react elsewhere than with the hemes in that the replacement reaction of carbon monoxide by nitric oxide does not show an accurately first-order course when measured at all wavelengths, though the discrepancies are quite small (Gibson and Greenwood, unpublished results) when expressed as a proportion of the total absorbance change.

The rate and the mechanism of reaction with oxygen are, of course, the prime objects in the study of the enzyme, since they represent its physiological function. Here, more data are available than for the ligand-binding reactions which are of less immediate practical importance. In the case of preparations stabilized with wetting agents, the rate of reaction with oxygen has been followed by a number of techniques, and generally speaking, the agreement between methods and between preparations may be regarded as gratifying, particularly when the variability in the purity of the preparations is taken into account.

At low oxygen concentrations, the rate of oxidation of cytochrome oxidase is accurately proportional to the product of the concentrations of oxygen and of reduced enzyme. This is true of spectrophotometric studies over a range of measured rate constants extending from 10 to 500 sec^{-1}. The lower end of the range of measurement has been much extended by the elegant and ingenious work of F. Schindler (University of Pennsylvania, Ph.D. thesis), who has exposed cytochrome oxidase to low concentrations of oxygen and has measured the oxygen in solution by means of its influence in promoting light emission from luminous bacteria. In this way, effective measurement of oxygen concentrations of the order of $5 \times 10^{-8} M$ or less was possible, giving measured rates of oxidation of the order of 0.1 sec^{-1}. The second-order rate constant calculated by Schindler's method fits in well with the results of more conventional measurement, and suggests that there is no preliminary formation of a freely reversible complex between oxygen and cytochrome oxidase before oxidation of the iron. If there were formation of such a reversible complex, the observed rate of oxidation would reach a limit when all the enzyme was quickly converted into such a complex, in the case in which the rate of oxidation was supposed to be lower than the rate of formation of complex. If, on the other hand, the rate of oxidation were presumed to be much greater than the rate of formation of the complex, and the complex had a measurable rate of dissociation, then at low oxygen concentrations the observed rate of oxidation would drop below that predicted from the higher concentrations; and this is not observed, even over the very wide range of oxygen concentrations covered by Schindler.

When the concentration of oxygen is raised, a much less simple picture emerges. The reaction becomes kinetically complex, and the extent of the individual phases which can be observed is a function of wavelengths of observation. The account which follows is chiefly taken from Gibson and Greenwood (80). One detail of the experimental method employed in most of these investigations has turned out subsequently to be of importance in the interpretation of the results and must therefore be described. In the regular rapid kinetic procedure, especially in the earlier experiments reported in 1963, the rapid reaction was recorded photographically with an oscilloscope. Then, after an interval of a few seconds (usually between 5 and 20 sec after initiation of the reaction), an additional line was added to the oscilloscope trace to represent the final point of the reaction. Although this procedure is entirely satisfactory in dealing with pure substances, it is less so in circumstances where the purity of the reagents is unknown or where secondary reactions may occur. In later investigations, it was found that in the preparations of cytochrome oxidase used for the kinetic work there were some slow secondary changes after the main reaction had been completed. After this observation was made, the mode of recording the results was altered so that the absorbance changes were measured from the beginning rather than from the end of the reaction. This procedure only became available when the dead time of the photochemical method for following the reaction was reduced to a few microseconds (100) (approximately 5 μsec), since with the longer dead times (approximately 150 μsec) of the older apparatus there was serious uncertainty about the initial absorbance values shown in the recording. Although the difference in the results which arises from this change in procedure is not great, it is significant in the discussion of the details of the mechanisms of the cytochrome oxidase–oxygen reaction.

The broad outlines of the experimental results obtained with high concentrations of oxygen are quite clear and are illustrated in Figure 1. There is an initial rapid oxygen concentration-dependent reaction which is followed by a slower concentration-independent reaction. The distribution of the total absorbance change between the two phases differs with wavelength in a way which is consistent with the assignment of the first rapid step to the oxidation of a_3 by oxygen and with the assignment of the second to the oxidation of cytochrome a by cytochrome a_3.

The oxidation of copper is also known to take place comparatively rapidly when reduced cytochrome oxidase is mixed with oxygen. This has been shown both for Yonetani-type preparations (101) and for Griffiths and Wharton-type preparations in experiments using the rapid-quenching method of Bray (102). These experiments could not be carried out on a

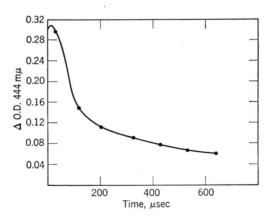

Fig. 1. The absorbance change observed at 444 mμ in an 8-cm path on firing a photolysis flash of 180 joules at a mixture containing $8 \times 10^{-7}M$ cytochrome oxidase, $2 \times 10^{-8}M$ cytochrome c, $1 \times 10^{-2}M$ sodium ascorbate, and $6.5 \times 10^{-5}M$ oxygen. The temperature was 28°, and $0.1M$ phosphate buffer, pH 7.4, was used.

time scale sufficiently rapid to allow the direct correlation of changes in enzyme copper as measured by the EPR signal with the changes in hemes a_3 and a, but they do show that the reduced enzyme copper can be oxidized in periods of a few milliseconds at most. Very recently it has been found that a substantial part of the absorption band at 820 mμ which is present in the oxidized enzyme but which is bleached on reduction can, under suitable circumstances of oxygen concentration, be shown to change at a rate intermediate between that of heme a_3 and heme a. If the identification of the 820 mμ absorption band of cytochrome oxidase as being primarily due to enzyme-bound copper is sustained (72), it would follow that the location of at least a part of the copper in terms of kinetics has been established.

Finally, one additional kinetic feature has been observed in the reaction of cytochrome oxidase with molecular oxygen. When the concentration of oxygen was raised to the highest values which can be obtained with conventional equipment, that is, using solutions equilibrated with one atmosphere of oxygen, the rate of oxidation of cytochrome a_3 did not keep pace with the increase in the concentration of oxygen (100). This result is apparently not due to the shortcomings of the photochemical method, since in experiments in which nitric oxide was used instead of oxygen no such kinetic limitations were observed. There is as yet no evidence which would permit the identification of the rate-limiting process, although it may be said with some assurance that it is not associated with a major spectrophotometric change.

One further reaction of a cytochrome oxidase derivative with oxygen, the reaction of the carbon monoxide derivative with oxygen, raises a point of theoretical interest in relation to the electron transport chain. The simplest prediction for this reaction is that the presence of the carbon monoxide molecule would protect the cytochrome oxidase against attack from molecular oxygen until it dissociated and, as the velocity of dissociation is known, the rate of oxidation would also be determined. When the experiment was performed, it was found, however, that the rate of oxidation greatly exceeded the calculated value (76). Detailed investigations showed that the overall changes could be accounted for in terms of two reactions: (*1*) the oxidation of cytochrome a and (*2*) the oxidation of the a_3–carbon monoxide complex. The only reasonable explanation of the results appears to be that once a molecule of cytochrome a_3 loses its carbon monoxide and becomes oxidized it is able to interact with molecules of cytochrome a other than those immediately associated with it; that is, that electron flow from a given molecule of cytochrome a is not restricted to a particular molecule of cytochrome a_3; or, in other words, the electron transport chain may branch in the a–a_3 region. Such branching might occur either between individual enzyme molecules, electron transport occurring at the time of collision, or within complexes of several enzyme molecules with a lifetime long compared with the rate of electron transport. This question was examined numerically by Gibson, Greenwood, Wharton, and Palmer (94), who concluded that both processes were active in preparations of the enzyme made by the method of Yonetani. The phenomenon is restricted to circumstances where the overall oxidation reaction is very slow and can only be observed in the case of the cytochrome oxidase after poisoning with carbon monoxide and possibly also with cyanide. In the normal un-inhibited enzyme, electron transfer occurs rapidly within individual units, as is shown by the independence of the rate for a_3 and a oxidation over about a 100-fold range of concentration of enzyme (103). The problem of the exchange of electrons between cytochrome oxidase molecules has not yet been effectively investigated in preparations of the oxidase which are not stabilized by wetting agents, and the results may well be complicated by the presence of substantial amounts of unreactive enzyme.

X. Mechanism of Cytochrome Oxidase Reactions

Comparatively little is known of the mechanism of cytochrome oxidase reactions, and the proposals which have been put forward have been largely conjectural. For this reason it seems appropriate to leave the discussion at a phenomenological level rather than to attempt an account of the various schemes which have been proposed. Some of the more elaborate of these, such as the mechanism proposed by Gibson and Green-wood (80), which attempted to take into account stoichiometry distorted by appreciable amounts of denatured enzyme in their preparations have been overtaken by improved experimentation, while others lacked experi-mental foundation from the beginning.

There is, nevertheless, one group of relatively recent observations which bears on the mechanism. In titrations of the enzyme no evidence was found, either in oxidative titrations using molecular oxygen or in reductive titra-tions using DPNH with phenazine methosulfate as mediator, of any differential oxidation or reduction of cytochrome a and a_3. In the circum-stances of these two sets of titrations it appeared that the only stable com-pounds present were fully oxidized enzyme and fully reduced enzyme. This finding need not apply to kinetic situations, but in view of the im-portance of the point, Gibson and Greenwood (unpublished) have sought specifically to examine the mechanism of interaction of a_3 and a during the reaction with oxygen, using the improved photochemical method and following the reaction at wavelengths where the time course of the oxidation of a_3 alone and of the oxidation of cytochrome a alone may be observed. The reaction was then followed over a wide range of oxygen concentrations at the two appropriate wavelengths. At high oxygen concentrations the a_3 concentrations change rapidly, as already deduced from experiments at 445 mμ, whereas the a absorbance change took place at the comparatively low rate of about 700 sec^{-1}. The concentration of oxygen was then cut down so that the rate of the second-order reaction between oxygen and reduced cytochrome a_3 was well below 700 sec^{-1}. The changes at the wavelengths monitoring a_3 and a were closely similar. There was no evidence of a lag in the changes at the wavelength monitoring a_3, and hence oxidation of a is not associated with reduction of a_3. This result differs from that in earlier stopped-flow experiments, where on reinvestigation it appears that slow secondary changes had modified the time course, even though internally consistent results had been obtained not only with several preparations made by Yonetani's method but also with preparations made by the method of Greenwood. If the new results are correct, it becomes logical to write mechanisms in which four electrons are transferred to oxygen in a "unit" process without the generation outside the enzyme molecule of radical forms (e.g., O_2^-). Radicals may occur, of course, within the enzyme–oxygen complex, as was indeed suggested by the experiments of Handler and Fridovitch (104), who showed that the oxidation of sulfite may be initiated during the turnover of cytochrome oxidase.

To summarize, the position with regard to cytochrome oxidase is satisfactory in that, for example, there is reasonable agreement on the form of the spectra of the enzyme and its derivatives. Some discussion on the interpretation of these spectra continues. One group of workers headed by Okunuki believes that the changes interpreted as due to two cytochromes a_3 and a are really due to changes within a single cytochrome, cytochrome a, and that the appearances observed by Yonetani in his cyanide difference method are to be attributed to reactions of cyanide with the formyl group of heme a. There are also differences about the existence and significance of an oxygenated complex of hemoglobin. In neither case is there disagreement about the experimental findings, but the interpretations are in dispute.

Turning to the stoichiometry of the enzyme, a substantial measure of agreement has appeared in very recent times on the amount of carbon monoxide which can be bound by cytochrome oxidase preparations of given spectrophotometric characteristics. There is also good agreement on the extinction coefficients of heme a in the various derivatives of the enzyme, and the metal contents of different preparations of the enzyme also agree well. In the case of cytochrome a_3 and a, on the other hand, extinction coefficients as obtained by static titrations and by kinetic titrations show differences of the order of 100%.

In the kinetic area, there is good general agreement on the turnover numbers which may be expected in the presence of various concentrations of cytochrome c, and the results now correlate satisfactorily with those in intact preparations of the enzyme. The broad outlines of the kinetic events using purified preparations of the enzyme appear to have been reasonably firmly established, and again correlation with at least some of the findings with mitochondria and electron transport preparations has been successfully achieved. There remain, however, points of difference and some internal inconsistency in the detailed interpretation of the kinetic results with the purified preparations.

Looking at the picture as a whole, it seems that there is agreement on first-order effects in every case; but there continues to be discussion where second-order effects are concerned. It is the opinion of the author that the contradictions which remain are due chiefly to the nature of the preparations themselves, and it would seem to follow that advances in the next few years may be expected to come from progress in enzyme preparations rather than from a refinement of physical methods for examining the enzyme. Alternatively, new and more powerful methods may offer hope for the future since progress with cytochrome oxidase has been as closely linked with instrumentation as that in any other area of enzymology.

XI. Peroxidase

Although the systematic enzymological study of peroxidase started later than that of cytochrome oxidase, the task of elucidating its action has proved easier; and little new work has been carried out on the enzyme in recent years. Thus, Kuhn, Hand, and Florkin (105) observed porportionality between the activity of their peroxidase preparations and absorbance in the Soret region. Later the enzyme was purified from horseradish and was crystallized by Theorell in 1942 (106). The enzyme also enjoys a special place in enzymological history because it was the object of the first study in which observations were made on the state of the enzyme during catalysis, since it was with peroxidase that Chance carried out the first stopped-flow study ever made of an enzyme during turnover; and it is interesting to recall that he also employed an analog computer to evaluate his results (107). These experiments antedated by some 15 years the general extension of this particular field of study. Like cytochrome c, peroxidase has been the subject of several excellent reviews (108) and has recently had a book devoted to it. It is therefore appropriate to deal with it in summary fashion and to concentrate on the enzyme made from horseradish. The rather different enzymes available from other sources and especially non-hematin peroxidases such as the enzyme of Dolin (109), which was extracted from *Streptococcus faecalis* and which proved to have flavin as its prosthetic group, will not be considered further, nor will the organic chemical aspects of the peroxidase reaction be included.

While the physiological function of peroxidases in plants is uncertain, the reaction commonly studied in the laboratory is coupled reduction of hydrogen peroxide and the oxidation of one of a large series of donors, which includes such substances as ascorbate, quinones, ferrocyanide, and cytochrome *c* as well as the leuco forms of many dyestuffs. Thus, in the first studies of Chance, the donor used was leukomalachite green (110). In general, the reactions of the enzyme are inhibited by cyanide but not by carbon monoxide. The reaction is with the uncharged species of the donor, and this constitutes a point of distinction from the flavin peroxidase of Dolin which has been shown to react with either the species AH^- or A_2^-.

Horseradish peroxidase, like all the plant heme peroxidases, has protoheme as its prosthetic group. It is loosely bound as compared with other hemoproteins and is readily picked up by hemoglobin globin. This easy separation of protein and prosthetic group has permitted studies of the recombination reaction between the apoprotein and heme. Again, in contrast with hemoglobin (111), the rate of recombination is very slow according to the work of Theorell and Maehly (112). Their experiments were carried out with the oxidized protoheme, and there is some possibility that the rate observed may have been limited by the velocity of breakdown of polymeric materials present in the solutions. It would be of interest to restudy this problem using the more modern methods which have become available in the interim.

In view of its potential importance in understanding the mechanism of HRP, the yeast cytochrome *c* peroxidase must also be mentioned. In recent work (113), the enzyme was found to have a minimum molecular weight of 49,000, and a maximal turnover number of 2.5×10^3 sec^{-1} at $23°$ and pH 6.0. Its absorption spectra were closely similar to those of HRP. HRP exists in several forms, which can be separated by electrophoresis (106). One of these, the major component, contains carbohydrate which behaves as an integral part of the molecule. The total amino acid composition has been given by the methods of Moore and Stein (114), in work by Morita and Kameda (115) and cited by Paul (108). The molecular weight is 4×10^4 and hemin content 1.35%.

XII. Mechanism of Action of Peroxidase

A. *Spectroscopic Observations on the Enzyme*

Although four spectrophotometrically distinguishable compounds of the enzyme have been described and have been named complexes I through IV, respectively, only two of these appear to participate in the ordinary catalytic cycle. Complex III, a bright red form analogous in some ways to reduced hemoglobin or myoglobin, is formed upon the addition of large excess of hydrogen peroxide. The spectra of complexes I, II, and III in the Soret region and complexes I and II in the visible region are shown in Figure 2. Complex IV, which has a band at 675 mμ, is formed after extended exposure to hydrogen peroxide (116). It has been suggested that this compound is related to the bile pigments and is a degradation product of the enzyme.

B. *Kinetics of the Interconversion of the Various Forms of Peroxidase*

The kinetics of the appearance of complex I, the first product known to be formed on mixing horseradish peroxidase with hydrogen peroxide, may be studied in the absence of an added electron donor. In these circumstances the complex I formed in the reaction is stabilized and has a life which permits spectrophotometry. The rate constant for its formation has been found to

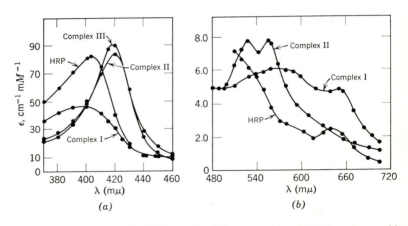

Fig. 2. (*a*) Soret and (*b*) visible bands of free peroxidase (HRP) and peroxidase combined with hydrogen peroxide to form the primary (I), secondary (II), and tertiary (III) complexes (51).

be approximately $1 \times 10^7 M^{-1} \sec^{-1}$. Clearly, if the conversion of complex I to complex II at the rate which is observed in the absence of donor formed in part of the peroxidatic mechanism, the turnover number of the enzyme would be much lower than that actually observed. In fact, however, as has been shown by Chance (116), the conversion of complex I to complex II is much accelerated by the addition of donor. This reaction is also second order, and, with individual donors, may be very rapid indeed, yielding a second-order rate constant as great as that for the primary reactions of hydrogen peroxide with the free enzyme. Chance (117) has shown that in the presence of small concentrations of hydrogen peroxide and of a suitably slowly reacting donor, complex I is an intermediate in the formation of complex II, the concentration of complex I rising and falling while the concentration of complex II first begins to rise after an initial lag period. Further, the rate of formation of complex II becomes independent of peroxide concentration above a certain level. Finally, the cycle of enzyme action is completed by the reduction of complex II to reform the oxidized enzyme.

The results so far may be summarized phenomenologically by the scheme

$$\text{HRP} + \text{peroxide} \underset{k_2}{\overset{k_1}{\rightleftarrows}} \text{complex I}$$

$$\text{complex I} + \text{donor} \longrightarrow \text{complex II}$$

$$\text{complex II} + \text{donor} \longrightarrow \text{HRP} + \text{oxidized donor}$$

based on the experiments of Chance (117,118). The scheme as written, however, contains a number of ambiguities and, of course, sheds no light on the chemical nature of the various intermediate compounds. While there is no dispute about the observations, there has been and continues to be uncertainty about what they mean. With regard to the nature of complex I, the spectrum, as pointed out by Brill and Williams (119), has similarities with that of a bile pigment; and it has been proposed that the spectrum of complex I is due to a mixture of a bile pigment-like material and another compound in which the molecule of peroxide or of hydroperoxide is attached to the sixth coordination position of the heme iron. Titration data suggest that complex I contains both oxidizing equivalents of the peroxide molecule. Titration experiments of Chance (120) show that complex I can be converted into complex II with the transfer of one electron, while George (121) has found that the conversion of complex II to the oxidized form of peroxidase again requires the transfer of one electron equivalent. To accommodate these findings, the reaction scheme was rewritten, as follows (118):

$$\text{complex I} + \text{AH}_2 \longrightarrow \text{complex II} + \text{AH}$$

$$\text{complex II} + \text{AH}_2 \longrightarrow \text{HRP} + \text{AH}$$

A different stoichiometry was proposed by George (121–123). His titrations suggested that complex II could be produced by titrating peroxidase with half a molecule of peroxide and that complex II, as Chance had found, was one oxidation equivalent above the free peroxidase. He therefore proposed a mechanism for the reaction

$$2HRP + H_2O_2 \longrightarrow 2 \text{ complex II} + H_2O$$

$$2 \text{ complex I} + AH_2 \longrightarrow 2HRP + A$$

The contradictions contained in these sets of results have remained unresolved up to the present time and can scarcely be clarified by inspection of the literature. Recently, however, Yonetani (113,124) purified cytochrome c peroxidase from yeast and examined a number of its properties, as well as the stoichiometry of the reactions in which it participates. This enzyme is spectrophotometrically very similar to the enzyme from horseradish but has the advantage of yielding a more stable complex II which is well suited for stoichiometric determinations. In the new work it has been found that complex II is formed quantitatively in titration of the enzyme with peroxide and that the full number of oxidizing equivalents of the peroxide are retained in the product of the titration, which is complex II, not complex I. Complex II is capable of oxidizing two molecules of cytochrome c, not one. This particular titration offers experimental advantages over those used with horseradish peroxidase since the spectrophotometric conditions for the determination of cytochrome c are more favorable than those, for example, for ferrocyanide where the reaction must be followed by the changes produced in the bands of the enzyme itself. In very recent work by rapid flow methods Yonetani was unable to detect the transient formation of complex I when the new preparation of the yeast enzyme was allowed to react with H_2O_2. Although it is mere speculation, it is tempting to correlate the results with the yeast enzyme with the classical work on the horseradish peroxidase. Such a correlation may be achieved if complex I is regarded as an artifact which appears only in circumstances where an electron donor is absent from the reaction mixture. Such a formulation may be fitted in with the suggestion of Brill and Williams (119) that complex I has, at least in part, a structure analogous to that of a bile pigment. Certainly the kinetic relations are such that only very small concentrations of complex I will normally appear when substantial concentrations of peroxide and donor are allowed to react with the enzyme, and it may be that the loss of oxidizing equivalents to "endogenous" donor does not occur under conditions where the enzyme is turning over rapidly.

A closely associated problem is that of the chemical nature of the compounds described so far under the terms complex I, II, etc. In this area too, although there is good agreement on observation, there remains considerable uncertainty as to interpretation. The earliest formulations of Chance would call for complex I to be an enzyme–substrate complex which would, by definition, be regarded as containing elements of peroxide. In the case that complex I, on conversion to complex II, was undergoing not a change in redox state but rather an isomerization, as is implicit in some of the schemes discussed by Chance, the complex II would also be expected to contain the elements of peroxide. This formulation, however, runs counter to the scheme proposed by George, where water is split out in the first step. A series of ingenious experiments has been proposed and carried out by Chance and Schonbaum (125). They have pointed out that in the case when a substrate such as methyl or ethyl hydrogen peroxide is employed, the alcohol will either appear free in solution or will remain bound to the enzyme, depending upon the mechanism adopted. This alcohol will be expected to enter into an equilibrium formulation, and the titration of the enzyme to equilibrium will be expected to give different results, according to whether or not the elements are retained in the complex. Chance and Schonbaum concluded that the elements of the peroxide actually continued to be bound to the enzyme in complex II.

A different series of considerations was introduced by George as a result of his titrations of peroxidase with various oxidizing agents. He showed that with some strong oxidizing agents it was possible to obtain spectrophotometric appearances corresponding to those of complex I with materials such as bromiridate, while other less vigorous oxidizing agents give rise to complex II and in several cases to mixtures of complexes I and II. These observations support the thesis that complex I and complex II should not be regarded as substrate complexes, but are peroxidase compounds with iron in higher oxidation states. The interpretations of George have, however, been questioned on the grounds of stoichiometry. Chance and Fergusson (126) have found that considerably more bromiridate was required for the oxidation of HRP than would have been predicted from the results of titrations with peroxide. They suggested, therefore, that either bromiridate gave rise to the formation of complex I by the intermediary formation of hydrogen peroxide in a reaction between bromiridate and water or alternatively that bromiridate reacted indirectly via groups within the protein molecule.

In this matter, as in connection with cytochrome oxidase, it seems possible that some of the difficulties may result from lack of adequately pure enzyme preparations. Thus, as has already been mentioned, peroxidase is known to contain carbohydrate, though fractions are available which are free from this; and it does not seem to be excluded that the different fractions may have an inherently different reactivity with peroxide and with strong oxidizing agents. Furthermore, the other groupings in the protein may be able to react with the peroxidase iron to a variable degree. There seems, indeed, to be some evidence that the amount of endogenous donor is a function of the mode of preparation of the enzyme.

The kinetic work on the reactions of peroxidase may be regarded as exemplifying some of the problems which present themselves repeatedly in enzyme chemistry. The pioneer work of Chance may readily be duplicated today and kinetic experiments may show that a compound is or is not formed and destroyed at rates consistent with its participation in catalysis. Such a demonstration cannot give any information about the chemical nature of any of the compounds involved, and discussion must be based on evidence of a chemical character; so far as peroxidase is concerned, it seems that final conclusions remain for the future. Nevertheless, the fact that it is possible to discuss the mode of action of an enzyme in terms of chemical reaction mechanisms and to attempt to assign definite chemical structures to the intermediate compounds involved is in itself a great advance over the situation in chemical enzymology 30 years or more ago, when very few relevant observations had been made and none of these had been substantiated by studies of changes in the properties of the enzyme during turnover.

References

1. R. Lemberg and J. Legge, *Haematin Compounds and Bile Pigments*, Interscience, New York, 1949.
2. E. Margoliash and A. Schejter, *Advan. Protein Chem.*, in press.
3. E. Margoliash, N. Frohwirt, and E. Wiener, *Biochem. J.*, **71**, 559 (1959).
4. E. Margoliash and J. Lustgarten, *J. Biol. Chem.*, **237**, 3397 (1962).
5. E. Margoliash, *Brookhaven Symp. Biol.*, **15**, 266 (1962).
6. M. D. Kamen and R. G. Bartsch, in *Haematin Enzymes*, J. E. Falk, R. Lemberg, and R. K. Morton, Eds., Pergamon, New York, 1961.
7. H. Theorell, *Enzymologia*, **4**, 192 (1937).
8. H. Theorell, *Biochem. Z.*, **298**, 242 (1938).
9. K.-G. Paul, *Acta Chem. Scand.*, **4**, 239 (1950).
10. K.-G. Paul, *Acta Chem. Scand.*, **5**, 389 (1951).
11. E. Margoliash and E. L. Smith, *Nature*, **192**, 1121 (1961).
12. G. Kreil and H. Tuppy, *Nature*, **192**, 1123 (1961).
13. E. Margoliash, E. L. Smith, G. Kreil, and H. Tuppy, *Nature*, **192**, 1125 (1961).
14. H. Theorell and A. Åkesson, *J. Am. Chem. Soc.*, **63**, 1804, 1812, 1818 (1941).
15. H. Theorell, *J. Am. Chem. Soc.*, **63**, 1820 (1941).

16. H. A. Harbury and P. A. Loach, *J. Biol. Chem.*, **235**, 3640 (1960).
17. D. Keilin, *Proc. Roy. Soc. (London)*, **B106**, 418 (1930).
18. A. Ehrenberg and H. Theorell, *Acta Chem. Scand.*, **9**, 1193 (1955).
19. P. George and R. L. J. Lyster, *Proc. Natl. Acad. Sci. U.S.*, **44**, 1013 (1958).
20. A. Schejter and P. George, *Biochemistry*, **3**, 1045 (1964).
21. A. Schejter, S. C. Glauser, P. George, and E. Margoliash, *Biochim. Biophys. Acta*, **73**, 641 (1963).
22. C. Greenwood and G. Palmer, *J. Biol. Chem.*, **240**, 3660 (1965).
23. P. George and C. L. Tsou, *Biochem. J.*, **50**, 440 (1952).
24. N. Frohwirt, "The Relation between Structure and Enzymatic Activity of Cytochrome *c*," Thesis, The Hebrew University, Jerusalem.
25. E. L. Smith and E. Margoliash, *Federation Proc.*, **23**, 1243 (1964).
26. B. Chance, *J. Biol. Chem.*, **233**, 1227 (1958).
27. F. A. Holton and J. P. Colpa-Boonstra, *Biochem. J.*, **76**, 179 (1960).
28. E. G. Ball, *Biochem. Z.*, **295**, 262 (1938).
29. E. C. Slater and J. P. Colpa-Boonstra, in *Haematin Enzymes*, J. E. Falk, R. Lemberg, and R. K. Morton, Eds., Pergamon Press, Oxford, 1961.
30. B. Chance, in *Oxidases and Related Redox Systems*, T. E. King, H. S. Mason, and M. Morrison, Eds., Wiley, New York, 1965, p. 931.
31. D. E. Green, D. C. Wharton, A. Tzagoloff, J. S. Rieske, and G. P. Brierley, in *Oxidases and Related Redox Systems*, T. E. King, H. S. Mason, and M. Morrison, Eds., Wiley, New York, 1965, p. 1032.
32. I. Sekuzu and K. Okunuki, *J. Biochem. (Tokyo)*, **43**, 107 (1956).
33. D. Feldman and W. W. Wainio, *J. Biol. Chem.*, **235**, 3635 (1960).
34. R. Goldberger, A. L. Smith, H. Tisdale, and R. Bomstein, *J. Biol. Chem.*, **236**, 2788 (1961).
35. R. Goldberger, A. Pumphrey, and A. L. Smith, *Biochim. Biophys. Acta*, **58**, 307 (1962).
36. F. Jöbsis, *J. Gen. Physiol.*, **46**, 905 (1963).
37. S. F. Bach, M. Dixon, and L. G. Zerfas, *Biochem. J.*, **40**, 229 (1946).
38. R. K. Morton, J. McD. Armstrong, and C. A. Appleby, in *Haematin Enzymes*, J. E. Falk, R. Lemberg, and R. K. Morton, Eds., Pergamon Press, Oxford, 1961, p. 501.
39. E. Boeri and M. Rippa, in *Haematin Enzymes*, J. E. Falk, R. Lemberg, and R. K. Morton, Eds., Pergamon Press, Oxford, 1961, p. 524.
40. A. P. Nygaard, in *Haematin Enzymes*, J. E. Falk, R. Lemberg, and R. K. Morton, Eds., Pergamon Press, Oxford, 1961, p. 544.
41. T. Horio, J. Yamashita, T. Yamanaka, M. Nozaki, and K. Okunuki, in *Haematin Enzymes*, J. E. Falk, R. Lemberg, and R. K. Morton, Eds., Pergamon Press, Oxford, 1961, p. 552.
42. C. A. Appleby and R. K. Morton, *Nature*, **173**, 749 (1954).
43. R. K. Morton and K. Shipley, *Biochim. Biophys. Acta*, **96**, 349 (1965).
44. E. Yakushiji and K. Okunuki, *Proc. Imp. Acad. Tokyo*, **16**, 299 (1940).
45. D. Keilin and E. F. Hartree, *Nature*, **164**, 254 (1949).
46. D. Keilin and E. F. Hartree, *Nature*, **176**, 200 (1955).
47. R. W. Estabrook, *J. Biol. Chem.*, **230**, 735 (1958).
48. B. Chance, *Nature*, **169**, 215 (1952).
49. B. Chance, in *Haematin Enzymes*, R. Lemberg, J. E. Falk, and R. K. Morton, Eds., Pergamon Press, Oxford, 1961, p. 606.

50. R. Bomstein, R. Goldberger, and H. Tisdale, *Biochim. Biophys. Acta*, **50**, 527 (1961).
51. E. E. Jacobs, E. C. Andrews, and F. L. Crane, in *Oxidases and Related Redox Systems*, T. E. King, H. S. Mason, and M. Morrison, Eds., Wiley, New York, p. 784.
52. E. H. Bernstein and W. W. Wainio, *Arch. Biochem. Biophys.*, **91**, 138 (1960).
53. I. Sekuzu, Y. Orii, and K. Okunuki, *J. Biochem.* (*Tokyo*), **48**, 214 (1960).
54. S. Takemori and T. E. King, *Biochim. Biophys. Acta*, **64**, 192 (1962).
55. T. E. King and S. Takemori, *Biochim. Biophys. Acta*, **64**, 194 (1962).
56. P. Nicholls, in *The Enzymes*, Vol. 8, P. Boyer, H. Lardy, and K. Myrbäck, Eds., Academic Press, New York, 1963, p. 3.
57. C. A. MacMunn, *Phil. Trans. Roy. Soc.*, **177**, 267 (1896).
58. C. A. MacMunn, *J. Physiol.*, **8**, 57 (1887).
59. L. Levy, *Hoppe-Seyler Z.*, **13**, 309 (1889).
60. F. Hoppe-Seyler, *Hoppe-Seyler Z.*, **14**, 106 (1890).
61. D. Keilin, *Proc. Roy. Soc.* (*London*), **B98**, 312 (1925).
62. P. Ehrlich, *Das Sauerstoff-Bedurfnis der Organismus*, Hirschwald, Berlin, 1885.
63. F. Battelli and L. Stern, *Biochem. Z.*, **46**, 343 (1912).
64. O. Warburg and E. Negelein, *Biochem. Z.*, **204**, 495; **214**, 64; **214**, 101 (1929).
65. D. Keilin and E. F. Hartree, *Proc. Roy. Soc.* (*London*), **B125**, 171 (1938).
66. D. Keilin and E. F. Hartree, *Proc. Roy. Soc.* (*London*), **B127**, 167 (1939).
67. B. Chance, *J. Biol. Chem.*, **202**, 383 (1953).
68. B. Chance, *J. Biol. Chem.*, **202**, 407 (1953).
69. B. Chance, *Rev. Sci. Instr.*, **22**, 634 (1951).
70. W. W. Wainio, P. Person, B. Eichel, and S. J. Cooperstein, *J. Biol. Chem.*, **192**, 349 (1951).
71. T. Yonetani, *J. Biol. Chem.*, **235**, 845 (1960).
72. D. E. Griffiths and D. C. Wharton, *J. Biol. Chem.*, **236**, 1850 (1961).
73. C. Greenwood, *Biochem. J.*, **86**, 535 (1963).
74. R. Lemberg, T. B. G. Pilger, N. Newton, and L. Clarke, *Proc. Roy. Soc.* (*London*), **B159**, 405 (1964).
75. S. Horie and M. Morrison, *J. Biol. Chem.*, **238**, 1855 (1963).
76. Q. H. Gibson and C. Greenwood, *J. Biol. Chem.*, **239**, 586 (1964).
77. I. Sekuzu, S. Takemori, T. Yonetani, and K. Okunuki, *J. Biochem.* (*Japan*), **46**, 43 (1959).
78. Y. Orii and K. Okunuki, *J. Biochem.* (*Japan*), **58**, 489 (1963).
79. R. Lemberg and G. Mansley, Information Exchange Group Memo No. 309, Group No. 1 (1965).
80. Q. H. Gibson and C. Greenwood, *Biochem. J.*, **86**, 541 (1963).
81. Q. H. Gibson, G. Palmer, and D. C. Wharton, *J. Biol. Chem.*, **240**, 915 (1965).
82. W. Vanneste, *Biochem. Biophys. Res. Commun.*, **18**, 563 (1965).
83. G. Mansley, J. Stanbury, and R. Lemberg, Information Exchange Group Memo No. 335, Group No. 1 (1965).
84. B. F. Van Gelder and E. L. Slater, *Biochim. Biophys. Acta*, **73**, 663 (1963).
85. B. F. Van Gelder and A. D. Muijsers, *Biochim. Biophys. Acta*, **81**, 405 (1964).
86. W. W. Wainio, B. Eichel, P. Person, and S. J. Cooperstein, *J. Biol. Chem.*, **183**, 89 (1950).
87. T. Yonetani, *J. Biol. Chem.*, **236**, 1680 (1961).

88. H. Beinert, D. E. Griffiths, D. C. Wharton, and R. H. Sands, *J. Biol. Chem.*, **237**, 2337 (1962).
89. L. Smith and H. Conrad, in *Haematin Enzymes*, J. Falk, R. Lemberg, and R. Morton, Eds., Pergamon Press, Oxford, 1961, p. 260.
90. K. Minnaert, *Biochim. Biophys. Acta*, **50**, 23 (1961).
91. T. Yonetani and G. S. Ray, *J. Biol. Chem.*, **240**, 3392 (1965).
92. P. Nicholls, *Arch. Biochem. Biophys.*, **106**, 25 (1964).
93. R. W. Estabrook, in *Haematin Enzymes*, J. E. Falk, R. Lemberg, and R. K. Morton, Eds., Pergamon Press, Oxford, 1961, p. 276.
94. Q. H. Gibson, C. Greenwood, D. C. Wharton, and G. Palmer, *J. Biol. Chem.*, **240**, 888 (1965).
95. P. Person, P. T. Mora, and A. S. Fine, *J. Biol. Chem.*, **238**, 4103 (1963).
96. P. Person, H. Zipper, A. S. Fine, and P. T. Mora, *J. Biol. Chem.*, **239**, 4159 (1964).
97. H. Conrad-Davies, L. Smith, and A. R. Wasserman, *Biochim. Biophys. Acta*, **95**, 238 (1965).
98. L. Smith and N. Newton, in *Oxidases and Related Redox Systems*, T. E. King, H. S. Mason, and M. Morrison, Eds., Wiley, New York, 1965, p. 745.
99. B. Chance, in *Haematin Enzymes*, J. E. Falk, R. Lemberg, and R. K. Morton, Eds., Pergamon Press, Oxford, 1961, p. 597.
100. Q. H. Gibson and C. Greenwood, *J. Biol. Chem.*, **240**, PC957 (1965).
101. Q. H. Gibson, C. Greenwood, and N. M. Atherton, *Biochem. J.*, **86**, 554 (1963).
102. R. C. Bray, *Biochem. J.*, **81**, 189 (1961).
103. Q. H. Gibson and C. Greenwood, *J. Biol. Chem.*, **240**, 2694 (1965).
104. P. Handler and I. Fridovitch, *J. Biol. Chem.*, **236**, 1836 (1961).
105. R. Kuhn, D. Hand, and M. Florkin, *Z. Phys. Chem.*, **201**, 255 (1931).
106. H. Theorell, *Arkiv. Kemi, Mineral. Geol.*, **16A**, No. 2 (1942).
107. B. Chance, in *Advances in Enzymology*, Vol. 12, F. F. Nord, Ed., Interscience, New York, 19, p. 153.
108. K. G. Paul, in *The Enzymes*, Vol. 8, P. D. Boyer, H. Lardy, and K. Myrbäck, Eds., 2nd ed., Academic Press, New York, 1963, p. 227.
109. M. I. Dolin, *J. Biol. Chem.*, **225**, 557 (1957).
110. B. Chance, *J. Franklin Inst.*, **228**, 459 (1940).
111. Q. H. Gibson and E. Antonini, *Biochem. J.*, **77**, 328 (1960).
112. H. Theorell and A. L. Maehly, *Acta Chem. Scand.*, **4**, 422 (1950).
113. T. Yonetani and G. S. Ray, *J. Biol. Chem.*, **240**, 4503 (1965).
114. S. Moore and W. H. Stein, *J. Biol. Chem.*, **192**, 663 (1951).
115. Y. Morita and K. Kameda, *Bull. Agr. Chem. Soc. (Japan)*, **23**, 28 (1959).
116. B. Chance, *Arch. Biochem. Biophys.*, **21**, 416 (1949).
117. B. Chance, *Arch. Biochem. Biophys.*, **22**, 224 (1949).
118. B. Chance, *Arch. Biochem. Biophys.*, **41**, 416 (1952).
119. A. S. Brill and R. Williams, *Biochem. J.*, **78**, 253 (1961).
120. B. Chance, *Arch. Biochem. Biophys.*, **37**, 235 (1952).
121. P. George, *Nature*, **169**, 612 (1952).
122. P. George, *Biochem. J.*, **54**, 267 (1953).
123. P. George, *Biochem. J.*, **55**, 220 (1953).
124. T. Yonetani, *J. Biol. Chem.*, **240**, 4509 (1965).
125. B. Chance and G. Schonbaum, in *Haematin Enzymes*, J. E. Falk, R. Lemberg, and R. K. Morton, Eds., Pergamon Press, Oxford, 1961, p. 254.
126. B. Chance and R. R. Fergusson, *Science*, **122**, 466 (1955).

The Copper-Containing Oxidases

Bo G. MALMSTRÖM and LARS RYDÉN, *Department of Biochemistry, University of Gothenburg, Sweden*

I.	Introduction	415
	A. Discovery of Copper Enzymes	415
	B. Classification and Nomenclature	418
	C. Biological Function and Scope of Chapter	419
II.	Tyrosinase	419
	A. General Aspects	419
	B. Induction Period and Reaction Inactivation	421
	C. The Double Reaction Specificity	421
	D. The State of Copper	423
	E. Reaction Mechanism	425
	1. Substrates and Products	425
	(a) The Catecholase Activity	425
	(b) The Cresolase Activity	425
	2. Detailed Mechanism	426
III.	Ascorbic Acid Oxidase	428
	A. General Aspects	428
	B. The Reaction Inactivation	428
	C. The State of Copper	429
	D. Mechanism	430
IV.	Laccase	431
	A. General Aspects	431
	B. The State of Copper	431
	C. Specificity	433
	D. Mechanism	434
V.	Conclusion	435
	References	436

I. Introduction

A. Discovery of Copper Enzymes*

The first clear indications of the involvement of copper in oxidases came from the pioneering studies of Kubowitz (1) and of Keilin and

* References to many of the early studies mentioned briefly in this section can be found in the detailed reviews quoted at the end of Section I-C.

Mann (2) on tyrosinase and laccase. These enzymes, both of which catalyze the oxidation of certain phenolic substances, were first recognized toward the end of the 19th century. Yoshida and Bertrand had demonstrated that the darkening of the latex of the Japanese lacquer tree is due to an enzyme, which was named laccase by Bertrand. Bertrand and Bourquelot were also led to the discovery of tyrosinase in mushrooms by a darkening process. Bertrand suggested that these enzymes contain a metal prosthetic group, and he was consequently the first investigator who envisaged the intimate roles that we now know metal ions to play in cellular oxidation processes and in enzyme action in general.

Laccase and tyrosinase have since been shown to be widely distributed in both the plant and animal kingdoms. Thus, Keilin and Hartree (3) found a laccase in animal kidney, and a protein, ceruloplasmin, with weak laccase activity was discovered in mammalian plasma by Holmberg and Laurell (4). In addition to its occurrence in plants and microorganisms, tyrosinase has been found in pigmented tissues of higher animals (5).

Ascorbic acid oxidase has been shown to be a copper-containing enzyme. The discovery that plant materials have the ability to catalyze the oxidation of ascorbic acid was made by Szent-Györgyi as early as 1928, and he later studied the enzyme from cabbage leaves in greater detail (6). However, as the oxidation of ascorbic acid can be brought about by Cu^{2+} as well as by products of tyrosinase action, many investigators doubted that a specific enzyme is responsible for this activity in plant tissues. This led to a controversy, which it took over ten years to settle. A number of investigators—for example, Lovett-Janison and Nelson in 1940 (7)—made decisive contributions by developing purification methods by which it could be shown that a protein with nondialyzable copper was much more efficient as a catalyst than Cu^{2+}. Incontrovertible evidence that ascorbic acid oxidase is a specific copper enzyme has since been provided, particularly by Dawson and his co-workers (8; see also ref. 9).

In addition to its involvement in the three groups of enzymes mentioned so far, copper has also been claimed to be a constituent of a number of other oxidases, e.g., uricase (10), monoamine oxidase (11), galactose oxidase (12), and dopamine β-hydroxylase (13). Except in the case of uricase, there is strong evidence for the essentiality of the metal, but its exact role in the catalytic reaction is not well understood (14,15). In addition, other mechanistic aspects have not been extensively studied, and therefore these enzymes will not be discussed in any detail in the present chapter.

In this connection it may be well to point out that studies on the mechanism of action of copper-containing oxidases have to a large extent been hampered by difficulties in getting appreciable amounts of highly purified preparations. These difficulties stem from, among other things, a frequently low content of the enzyme in the biological material, the catalytic formation of large amounts of pigments in the tissues or extracts, and the lability of the functional state of copper to denaturing influences. It should also be noticed that Cu^{2+} is a common contaminant in water and chemicals. As all proteins interact quite strongly with Cu^{2+}, a low content of copper in a protein preparation does not necessarily imply a specific association. However, with all enzymes treated here it has been shown that the copper-free apoenzyme is inactive but can be reactivated with copper ions.

TABLE I

Properties of Some Copper-Containing Oxidases

Enzyme	Source	Molecular weight	Copper content, %	Number of atoms per molecule	Other properties	Ref.
Tyrosinase	Common	128,000	0.20	4		16
	mushroom	119,000			Four subunits	17
					At least five comp.	18
	Neurospora crassa	32,000	0.195	1	Crystalline	19
Laccase	Japanese lacquer tree	120,000	0.22	4	Nitrogen content 9.0%	20
		141,000	0.23–0.25	5.3	Nitrogen content 10.1%	21
	Polyporus versicolor	60,000	0.44	4	14% carbohydrate	22
	Human serum	160,000	—	8	Crystalline	23
Ascorbic acid oxidase	Yellow summer squash	150,000	0.34	8	—	24

Table I lists a number of relatively well-defined preparations of copper-containing oxidases and some of their properties. The enzymes listed all belong to the classes tyrosinase, laccase, or ascorbic acid oxidase, and only these groups of enzymes will be discussed in any detail in the present chapter.

The importance of copper in biological oxidations has been emphasized by the recent finding of this metal in preparations of cytochrome oxidase. Several lines of evidence indicate that copper is a functional component in this enzyme. However, cytochrome oxidase will not be discussed here, since, as a heme enzyme, it is considered in detail by Gibson in his chapter in this volume.

B. Classification and Nomenclature

As indicated by the discussion in Section I-A, the terms tyrosinase and laccase do not refer to two single enzymes but rather to groups of enzymes with relatively varying properties but distinguished by their substrate specificities. Tyrosinase* catalyzes the oxidation of o-diphenols to o-quinones and, as the name implies, the oxidation of monophenols as well. Since the end product is an o-quinone in the latter case also, one atom of oxygen must be incorporated in the substrate. For this reason, tyrosinase acting on a monophenol is in modern terminology referred to as a "mixed-function oxidase" or "mixed-function oxygenase." The term hydroxylase is also used. Readers not familiar with these terms and their definitions are referred to the chapter by Hayaishi.

It has only recently been established that the catalytic activities of tyrosinase can be associated with a single enzyme species. Earlier, the term phenolase complex was therefore often used in discussions concerned with impure preparations possessing these activities.

Laccase is a name for a group of enzymes with very broad substrate specificity. The best substrates are p-phenols or related substances such as p-phenylenediamine, but o-diphenols are also oxidized. A weak activity with m-phenols has been found. It has often been stated that laccase can be distinguished from tyrosinase through its inability to catalyze the oxidation of monophenols such as p-cresol. However, it has recently been shown (see Section IV-C) that the oxidation of p-cresol and other monophenols is catalyzed by fungal as well as latex laccase. Still, there is no doubt that laccase is distinct from tyrosinase, as evidenced by the relative oxidation rates with the different substrates as well as by differences in spectral properties, the state of copper, and the mechanism of oxidation (see Sections II and IV).

Ascorbic acid oxidase is easily distinguished from the other copper enzymes by its rather narrow specificity, L-ascorbic acid being the best substrate. It may be pointed out that its substrate is related to many substrates of tyrosinase and laccase, because it contains a dienol structure.

---

* This enzyme has also been known by a variety of other names, such as phenol oxidase, polyphenol oxidase, and catechol oxidase. The name suggested by the Enzyme Commission, o-diphenol oxidase, is highly misleading as it implies a lack of monophenolase activity (see Section II-C). The Commission's name for laccase, p-diphenol oxidase, is equally unsuitable, as it indicates a more narrow specificity than is actually found. Thus, the admittedly poor but well-understood terms tyrosinase and laccase will be used exclusively in the present chapter.

C. Biological Function and Scope of Chapter

The biological function of copper-containing oxidases is to a large extent poorly understood. It has been proposed that ascorbic acid oxidase and tyrosinase may function as terminal oxidases in plants, as their substrates can be hydrogen carriers. The widespread distribution of these enzymes suggests an essential function, but there is no evidence that these oxidations are coupled to phosphorylation. In addition, the ordinary cytochrome respiratory chain has been found also in microorganisms and higher plants. The participation of laccase and tyrosinase in cellular oxidation processes, in which O_2 is not the electron acceptor, has also been proposed. A clearly demonstrated function of tyrosinase is its role in melanin formation in pigmented tissues. This aspect was discussed in detail at a recent symposium (25). It has been suggested that laccase is involved in the formation of various natural products, e.g., lignin.

The biological function of copper enzymes will not be considered further in this chapter, and interested readers are referred to a review by Bonner (26). Instead, this article will be limited to questions of mechanism and its relation to the nature of the proteins and their prosthetic group.

As will be seen, tyrosinase and laccase appear to operate by distinctly different mechanisms, so that it becomes necessary to consider these enzymes separately. There are some indications that the mode of action of ascorbic acid oxidase is closely related to that of laccase, but as the types of information available about the two enzymes are rather different, ascorbic acid oxidase will also be treated by itself. Some of the newly discovered copper enzymes, such as galactose oxidase may represent a third type of mechanism, as will be briefly discussed in Section II-D.

A number of general reviews dealing with copper-containing oxidases are available (26–28). However, these reviews are all about ten years old, while a very rapid development has occurred within the last five years. More modern reviews can be found dealing with certain aspects, such as the state of the metal (29), or with individual enzymes (see Sections II–IV).

II. Tyrosinase

A. General Aspects*

Tyrosinase catalyzes the two types of reactions shown in Reactions (1) and (2). Catalysis of Reaction (1) is called the catecholase activity of the

* Two reviews dealing specifically with tyrosinase may be mentioned (16,30). In our opinion, the one by Kertész and Zito (16) is highly subjective as, in arguing for a given view, it leaves out much of the evidence for the other side.

$$ 2 \text{ (catechol)} + O_2 \longrightarrow 2 \text{ (o-quinone)} + 2 H_2O \qquad (1) $$

$$ \text{(p-cresol)} + O_2 + 2 H \longrightarrow \text{(4-methylcatechol)} + H_2O \qquad (2) $$

enzyme and that of Reaction (2) the cresolase activity, in accordance with two substrates commonly used. Through a coupling to Reaction (1), the enzyme may bring about the oxidation of any hydrogen donor, AH_2, as shown in Equation (3). The ability of the enzyme to catalyze two reactions

$$ (3) $$

so different from each other is an interesting property that has stimulated much research.

A prerequisite for rigorous studies on the mechanism of catalysis is the availability of appreciable amounts of pure enzyme. Early preparations were made by Kubowitz from potatoes (see ref. 1) and by Keilin and Mann (31) from mushrooms. Most work has been done with the mushroom enzyme, the purification of which has been greatly improved by the introduction of chromatographic methods by Smith and Krueger (18). A further improvement in purification has been accomplished by Bouchilloux et al. (17). The purification (19) of a very well-defined tyrosinase from *Neurospora crassa* has formed the basis of some recent major advances (see Section II-D). Tyrosinase from animal tissues has not been prepared in homogeneous form, but studies on a partially purified enzyme from hamster melanoma (32) confirm its narrower substrate specificity (5) compared to plant or fungal tyrosinase.

B. Induction Period and Reaction Inactivation

The cresolase activity of purified tyrosinase shows a very long induction period, but this disappears on the addition of catalytic amounts of catechol. This induction period is probably the cause of early reports of polyphenolases without cresolase activity.

Several explanations of the induction period have been advanced. It has been suggested that catecholase activity is necessary for cresolase activity, or that cresolase oxidation is catalyzed by its own reaction products. However, it has been shown (33) that the induction period may be abolished by nonspecific hydrogen donors, such as ascorbic acid, or DPNH (this is not true of animal tyrosinase, which requires a more specific hydrogen donor). On the other hand, the view that cresolase activity requires a hydrogen donor is not universally accepted. For example, Karkhanis and Frieden (34) claim to have isolated a protein which causes the induction period.

Tyrosinase is inactivated during the reaction. A possible explanation of this has recently been provided (35). It was shown, with isotopically labeled phenols, that substrate molecules become irreversibly bound to the enzyme during the reaction, thereby causing inactivation.

C. The Double Reaction Specificity

Two main types of hypotheses have been advanced to provide a chemical basis for the double reaction specificity of tyrosinase:

(1) The indirect or non-enzymic oxidation hypothesis, which assumes that monophenols are oxidized indirectly by the products of catecholase action.

(2) The direct or enzymic oxidation hypothesis, which assumes that both reactions are enzyme catalyzed.

If the latter hypothesis is accepted, three different possibilities arise:

(a) Two different enzymes are involved.

(b) There is one enzyme but it has two active sites.

(c) Both reactions occur at the same active site of one enzyme.

The existence of the induction period for cresolase activity could be explained on the basis of the indirect oxidation hypothesis, but this must be abandoned on several other grounds. One reason is the existence of different forms of the enzyme which give different ratios between the catecholase and cresolase activity. Fifteen years ago, Mallette and Dawson (36) found that this ratio varied if the enzyme was prepared in different ways. Their finding can be explained by the existence of four isoenzymes of mushroom tyrosinase, isolated by Bouchilloux et al. (17). These essentially pure forms of the enzyme showed different activities toward monophenols and *o*-diphenols, a result which it appears difficult to explain in terms of indirect oxidation.

Experiments on the exchange of copper in mushroom tyrosinase, with radioactive copper in the medium (37), also provide evidence against the indirect hypothesis. No exchange was found with resting enzyme or with enzyme during cresolase activity, while there was exchange during catecholase activity. As, in terms of the indirect hypothesis, the enzyme should carry out the same reaction in both cases, these results appear inconsistent with an indirect oxidation.

In our opinion, the results of Mallette and Dawson (36) alone are enough to eliminate the indirect hypothesis, and the other experiments described all indicate that the protein itself is involved in cresolase activity. Furthermore, the fact that the hydroxylation of monophenols always leads to substitution in the *ortho*, and not in the *para*, position favors the direct hypothesis, as do ^{18}O experiments (38) which show that the oxygen atom in the hydroxylated product is derived from O_2 and not from H_2O. Despite the overwhelming experimental evidence against the indirect hypothesis, Kertész (16) still adheres to it. He justifies this on the basis of kinetic experiments which he finds to be consistent with rate equations derived for the indirect oxidation mechanism. It deserves to be pointed out that this is an incorrect, or at least insufficient, use of kinetics to settle the controversy. What Kertész has shown is that indirect oxidation is a possible mechanism for explaining certain kinetic results, but kinetics can never provide definite proof. However, kinetics can eliminate certain mechanisms. Thus, until it can be shown that direct oxidation is inconsistent with kinetic results, rate measurements can do nothing to counterbalance the great weight of evidence in favor of a direct oxidation of monophenols by tyrosinase.

As pointed out earlier, in terms of direct oxidation we must choose from among three alternatives. Even if criteria of enzyme homogeneity are never absolute, both *Neurospora* (19) and mushroom (17) tyrosinase appear to be sufficiently pure to exclude the two-enzyme hypothesis.

Mallette and Dawson (36) found the cresolase activity in their preparations to be more easily destroyed than catecholase activity, which tends to support the concept of two active sites. However, as other findings provide evidence for a single site, it is tempting to try to find alternative explanations of their results. Since they did not work with the pure isoenzymes (17), differential destruction of α-tyrosinase, which has "high-cresolase" activity, could be the cause. Another possibility is the loss or destruction of a group required for cresolase activity but not for catecholase activity. The exchange studies already discussed are, however, most easily explained in terms of two active sites.

Among the evidence for one active site may be mentioned that the same inhibitors affect both activities to the same degree (39), and that monophenols inhibit diphenol activity or vice versa (40). The fact that a monomer, containing one copper atom per molecule, is active (see Section II-D) also supports the concept of one active site. However, the question cannot be regarded as completely settled, in view of the partially contradictory evidence.

D. The State of Copper

In 1938, Kubowitz (1) prepared apotyrosinase by treatment with CN^-, and he showed that the enzyme can be reactivated by Cu^{2+}. Furthermore, he found that tyrosinase in the presence of catechol gave a complex with CO, which contained one molecule of CO per two copper atoms (CO preferentially forms complexes with Cu^+). On the basis of these results he suggested that the enzyme contains Cu^{2+} which becomes reduced by substrate and reoxidized by O_2, and this view of tyrosinase action has been rather generally accepted for more than 20 years. Doubts as to its validity were first raised by experiments of Kertész (41) with the Cu^+ reagent, biquinoline, suggesting that all copper in the resting form of tyrosinase is in the Cu^+ state. Unfortunately, the technique used is open to severe criticisms as a method to determine the valence state of the metal *in situ* (42). However, physical techniques, such as magnetic susceptibility measurements and electron spin resonance (ESR), can provide more clear answers.

The ESR technique (43) can be used to detect the presence of paramagnetic substances (i.e., compounds containing unpaired electrons)—e.g., free radicals and certain ions of transition elements—and also to give information about the physical environment of the paramagnetic center. In the case of copper, the ion Cu^{2+}, having the outer electron configuration $3d^9$, has one unpaired electron and is paramagnetic, while Cu^+, having the configuration $3d^{10}$, is diamagnetic. In view of this, ESR has in recent years played a prominent role in the study of the valence and bonding of copper in enzymes.

Studies on mushroom tyrosinase (17, see also ref. 42) showed a variable Cu^{2+} signal, which, however, corresponded to only 10–20% of the total copper content. This suggests that the Cu^{2+} signal was derived from denatured molecules, an interpretation that is strengthened by measurements on crystalline *Neurospora* tyrosinase (19), in which case less than 1% of the copper was detected by ESR. It would thus appear that Kertész's claim that tyrosinase contains only Cu^+ is correct. It must, however, be stressed that the absence of an ESR signal is not definitive proof of the absence of a paramagnetic ion, as interacting paramagnetic centers, e.g., a pair of ions of Cu^{2+}, may give such a broad signal that it is not detectable, because of the low amplitude. In the case of *Neurospora* tyrosinase, which contains only one copper atom per molecule, a Cu^{2+}–Cu^{2+} interaction is excluded, and it appears safe to conclude that all copper in the enzyme is in the form of Cu^+. Still, confirmation by susceptibility measurements, which do not suffer from the same limitations as ESR, would be desirable.

The ligands of Cu^+ are unknown. The absence of cysteine and cystine in the *Neurospora* as well as the mushroom enzyme (17,19) excludes SH groups, and imidazole groups would then seem most probable (44). Kubowitz's experiments (1) with CO suggest that pairs of Cu^+ are found at the active site. If this is true, the *Neurospora* enzyme must dimerize to become active. While polymers can be formed, this interpretation is contradicted by results of Bouchilloux et al. (17). These authors found it possible to make monomers of a molecular weight of 34,000, and with one copper atom, from the mushroom enzyme. Production of monomer did not lead to any loss of activity (though it has not been excluded that dimerization occurs during the conditions of the activity measurements).

As copper in tyrosinase is in the Cu^+ state, the original suggestion of Kubowitz (1), that the substrate is oxidized by Cu^{2+} in the enzyme, must be abandoned and an alternative function of the metal found. Two possibilities seem to suggest themselves. First, the metal may have a structural role only and not partake directly in the catalytic reaction. A more attractive alternative is that the metal participates in the binding of substrates or O_2. An interaction with Cu^+ might labilize the oxygen molecule, which is kinetically rather stable (45). Binding of O_2 to Cu^+ is found in hemocyanin (46), but this leads to a strong charge-transfer absorption in the visible region of the spectrum. The absence of strong color in tyrosinase under aerobic conditions would thus speak against this suggestion, but if the degree of electron transfer is smaller, the absorption band may occur at a lower wavelength.

The substrates of tyrosinase are aromatic substances containing unfilled π orbitals. Such substances form complexes with transition metals preferentially in their lower valences (47). A complex in which Cu^+ interacts with benzene has recently been studied (48). These chemical facts make substrate binding an attractive possibility for the function of the metal.

Monoamine oxidase (14) and galactose oxidase (15) have been found, with the ESR technique, to contain Cu^{2+} which is not reduced by substrate. Thus, a structural function or involvement in substrate binding must be considered in these enzymes also. However, oxygen binding to Cu^{2+} is less likely.

E. Reaction Mechanism

1. Substrates and Products

(a) *The Catecholase Activity.* The primary product of catecholase action is an *o*-quinone and $0.5O_2$ is consumed per molecule of diphenol, as shown in Reaction (1). Further consumption of oxygen is due to non-enzymic reactions of the products, but this can be avoided by coupling the reaction to a hydrogen donor according to Equation (3) and measuring the formation of the dehydrogenated form of the donor.

An important problem in enzymic oxidation is the possible occurrence of free-radical intermediates. The mere demonstration of the presence of free radicals in the reaction medium is not sufficient evidence in the absence of kinetic data, as the radical may be produced in a side reaction. In the case of catechol, the well-known equilibrium shown in Reaction (4) is a

$$(4)$$

possible radical-producing side reaction. Mason et al. (49) studied the formation of radicals at two different pH values by the ESR technique. At the higher pH a radical was formed, but the radical kinetics could not be related to an enzymic production but rather to the equilibrium of Reaction (4). At the lower pH, no radical was detected, although under the same conditions a radical was formed in the peroxidase-catalyzed oxidation of catechol. It was thus concluded that a free radical is not the primary product in tyrosinase-catalyzed oxidation of catechol.

(b) *The Cresolase Activity.* As already mentioned (Section II-C), the oxygen in the product of hydroxylation of monophenols is derived from O_2 and not from H_2O, as was demonstrated by the classical studies of Mason et al. in 1955 (38). Important information about the mechanism of cresolase action can also be derived from the exchange experiments of Dressler and Dawson (37) (see Section II-C), as these indicate that the diphenol formed by Reaction (2) is not in equilibrium with free diphenol in solution. This conclusion is strengthened by experiments with radio-actively labeled tyrosine (50), and it must be thus concluded that the diphenol formed in Reaction (2) is not released from the enzyme but is directly oxidized to an *o*-quinone according to Reaction (1).

In view of the findings summarized, Reaction (2) may be rewritten as in Reaction (5).

$$E + AH_2 + \text{[structure: phenol with OH]} + O_2 \longrightarrow E + AH_2 + \text{[structure: o-quinone]} + H_2O \qquad (5)$$

2. Detailed Mechanism

Wood and Ingraham (51) have made detailed proposals for the mechanism of cresolase action from experiments with isotopic rate effects. They depict the hydroxylation as an electrophilic attack on the benzene ring. Phenol with tritium in the *ortho* position in relation to the hydroxyl group gave a small rate effect, which means that the second step of Reaction (6) is at least partially rate determining. As the effect was very

$$\text{[structure: phenol with OH and T]} \rightleftharpoons \text{[structure with OH, O, } \delta \text{, T]} \rightleftharpoons \text{[structure with OH, O]} + T \qquad (6)$$

small, the authors suggested that a basic group in the protein accepts the proton, thereby speeding up the reaction.

In 1957, Mason (39) made a detailed proposal for the different actions of tyrosinase, summarized in Figure 1. It appears that with slight modifica-tions this scheme can still accommodate all known experimental facts. Activation of the enzyme by the reduction of Cu^{2+} is not possible, as the metal is present as Cu^+ without the addition of reducing agents. Instead, we would like to suggest that the hydrogen donor AH_2 in Reaction (5) reacts with the enzyme to form a hydrogenated enzyme, as shown in the modified scheme of Figure 2. This may occur by the binding of AH_2 or by direct hydrogenation of the enzyme, but the latter alternative suffers from the difficulty that a suitable hydrogen acceptor in the protein is not known. If the enzyme itself is not hydrogenated, the scheme of Figure 2 could be explained in terms of two binding sites (with only one Cu^+ for both sites). In catecholase activity, both bind diphenol, while in cresolase activity one binds AH_2 and the other the monophenol.

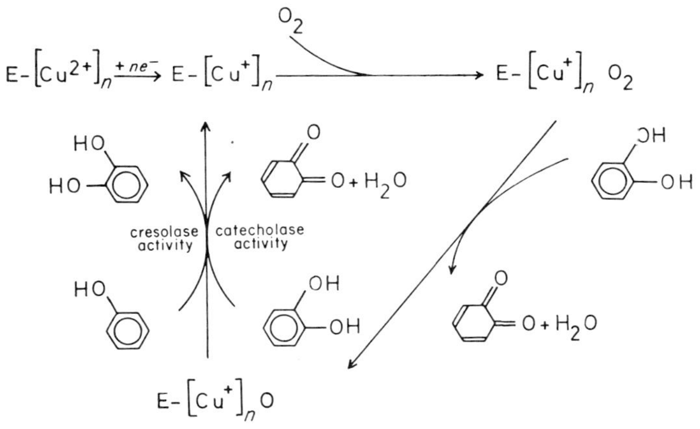

$$E-\left[Cu^{2+}\right]_n \xrightarrow{+ ne^-} E-\left[Cu^+\right]_n \xrightarrow{O_2} E-\left[Cu^+\right]_n O_2$$

cresolase activity | catecholase activity

$$E-\left[Cu^+\right]_n O$$

Fig. 1. Mechanism for tyrosinase action, according to Mason (39).

In the schemes of Figures 1 and 2, the electrophilic substituent could be the cupryl ion, CuO^+ (45). As a diphenol is not necessary for cresolase activity, we would like to depict the formation of this ion from $Cu^+ \cdot O_2$ to involve the hydrogenated enzyme instead, as shown in Figure 2. In cresolase action the hydrogenated enzyme is reformed from the diphenol, which does not dissociate from the enzyme (Section II-E-1b). In catecholase action, on the other hand, the hydrogens form H_2O with the oxygen of CuO^+, and nonhydrogenated enzyme is thus reformed. However, in this

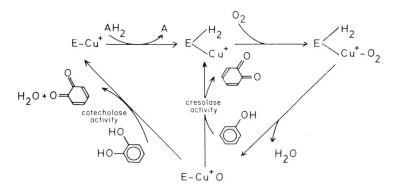

Fig. 2. Modified Mason scheme for tyrosinase action. ECu^+ represents the enzyme and AH_2 a diphenol or other hydrogen donor. For details see the text.

case the hydrogenated enzyme can be formed from the action of another substrate molecule.

Unfortunately, the picture of tyrosinase action summarized in Figure 2, while consistent with experimental facts, is largely hypothetical. For example, the formation of oxygenated and hydrogenated enzyme has not been directly demonstrated. Furthermore, there is no direct evidence for the involvement of Cu^+ in the binding of O_2 or substrate.

III. Ascorbic Acid Oxidase

*A. General Aspects**

The main reaction catalyzed by ascorbic acid oxidase is shown in Reaction (7). The most highly purified enzyme has been prepared from

$$
\begin{array}{c}
\underset{\text{HO}-\text{C}-\text{C}}{\overset{\text{O}}{\parallel}} \\
\text{HO}-\text{C}-\text{C} \\
\text{HOCH} \\
\text{CH}_2\text{OH}
\end{array}
\;+\; \tfrac{1}{2}\text{O}_2 \;\longrightarrow\;
\begin{array}{c}
\underset{\text{O}=\text{C}-\text{C}}{\overset{\text{O}}{\parallel}} \\
\text{O}=\text{C}-\text{C} \\
\text{HOCH} \\
\text{CH}_2\text{OH}
\end{array}
\;+\; \text{H}_2\text{O}
\tag{7}
$$

yellow squash (see Table I). In this case, the problem of getting sufficient amounts of purified enzyme is particularly pronounced, as less than a gram was obtained from one ton of squash.

Ascorbic acid oxidase shows a widespread occurrence in higher plants, and its presence in a bacterium has recently been reported (52). As stressed by Stark and Dawson (9), many systems, e.g., the cytochrome c–cytochrome oxidase system, catalyze the oxidation of ascorbic acid (cf. Section I-A), so that reports on the occurrence of this enzyme must be carefully evaluated.

B. The Reaction Inactivation

As in the case of other copper-containing oxidases, ascorbic acid oxidase is markedly inactivated during the catalytic reaction. With this

* A recent review of ascorbic acid oxidase has been written by Stark and Dawson (9).

enzyme the phenomenon has been extensively studied. Since it has been suggested that it may be caused by free-radical intermediates, a clear view of the cause of inactivation has significance for a discussion of the mechanism of the catalytic reaction. H_2O_2 has frequently been suggested as the inactivating agent, but no H_2O_2 has been detected in the reaction system. On the other hand, a protective effect by catalase and peroxidase is reversed by azide, which inhibits the H_2O_2-decomposing activity of these two enzymes (see ref. 9).

In 1962, Tokuyama and Dawson (53) reinvestigated the reaction inactivation in great detail. They found that it is caused by a substance which accumulates during the reaction. For example, a rapid reaction with excess enzyme does not lead to inactivation, while enzyme added in portions results in a greater inactivation than if all of it is added at once. Inert proteins protect the enzyme only if added at the beginning of the reaction, but catalase has an effect also when added during the reaction. All these findings indicate that H_2O_2 accumulates from a slow secondary process. It was also found that added H_2O_2 had an inactivating effect during the reaction, even if the amount was so small as to be barely detectable. It should be noted that added H_2O_2 has no effect on the resting enzyme (54).

The formation of H_2O_2 during the reaction has recently been related to the presence of Cu^+ in the enzyme (see Section III-C).

C. The State of Copper

Ascorbic acid oxidase from squash contains eight copper atoms (see Table I). The valence of the copper has not been determined by quantitative ESR or susceptibility measurements, but the presence of an ESR signal shows that at least part of the copper is in the Cu^{2+} state (55). The substrate reduces Cu^{2+} detected by ESR. The strong blue color also disappears on addition of substrate, but this does not necessarily indicate a reduction of Cu^{2+}, as charge-transfer complexes with Cu^+ can also have strong absorption in the same spectral region, oxyhemocyanin being an example (46).

Poillon and Dawson (24) have made a careful study of the valency problem with the aid of valency-specific reagents. Their results indicate that both Cu^{2+} and Cu^+ are present in the resting enzyme. Of the eight copper atoms, they found 6 Cu^{2+} and 2 Cu^+. It should, however, be noted that chemical methods for the determination of the valency of metals in proteins can be misleading, and a ratio of $6Cu^{2+}:2Cu^+$ has been found with ceruloplasmin despite the fact that magnetic measurements show the ratio to be $4Cu^{2+}:4Cu^+$ in the native protein (42).

In further experiments with a Cu^+-specific reagent (54), the reaction was carried out under conditions not leading to a loss of copper from the protein. Spectral measurements showed that Cu^+ in the protein had reacted, but this did not affect the activity or color of the enzyme, indicating that the presence of Cu^+ is not related to these properties. On the other hand, it was found that the reaction inactivation decreased when Cu^+ was blocked, and it was suggested that Cu^+ is responsible for the slow production of H_2O_2 during the reaction (see Section III-B).

Studies on the exchange of the copper in ascorbic acid oxidase with radioactive copper in the medium have been undertaken (56). In the resting enzyme there is no exchange, but there is rapid exchange in the presence of substrates. It was suggested that this is due to the reduction of Cu^{2+} to Cu^+, as Cu^+ generally forms weaker complexes than Cu^{2+}. While this is probably the correct interpretation, a conformation change on addition of substrate could also be responsible for the difference in exchange rate. That such a conformation change occurs is indicated by the fact that the two Cu^+ ions do not react with Cu^+ reagents unless substrate is added. It should be noted that even in the presence of substrate, only two Cu^+ react, so that if the six Cu^{2+} have all become reduced, the Cu^+ so formed must be unavailable for reaction.

As in the case of the other copper enzymes, the ligands involved in copper binding are unknown. On denaturation of the enzyme, copper is released and 12 SH groups become available, but these findings do not need to be related. A recent extensive study of the reactivation of the apoenzyme (57) showed that Cu^+, but not Cu^{2+}, can restore color and activity. The reconstitution was prevented by blocking the SH group. However, as this may lead to structural changes in the protein, it cannot be concluded that sulfur is a primary binding group.

D. Mechanism

As seen from Reaction (7), oxidation of ascorbic acid leads to the loss of two hydrogens. This, however, seems to occur in two one-electron steps with the formation of a free-radical intermediate, as shown by ESR measurements (55). In fact, it seems that the product of the enzymic reaction is a free radical, $AH\cdot$, and that the product shown in Reaction (7) is formed non-enzymically by a dismutation reaction:

$$2AH\cdot \longrightarrow A + AH_2 \tag{8}$$

Oxidation of ascorbic acid by the aqueous Cu^{2+} ion involves a valency change in the metal, but in this case H_2O_2, and not H_2O, is formed (see ref. 9). As discussed in Section III-C, there is some evidence that the enzymic reaction also involves the reduction of Cu^{2+} by the substrate. However, it should be noted that the mere demonstration that the substrate can reduce Cu^{2+} in the enzyme does not prove this to be part of the mechanism. It must at least be shown that the rate of reduction and reoxidation of the metal are rapid enough to account for the rate of the overall reaction.

IV. Laccase

A. General Aspects

Laccase has a broad specificity but the best substrates are *p*-diphenols and related substances, which are oxidized as shown in Reaction (9). The

$$\text{(9)}$$

enzyme was first prepared from the lacquer tree, but the best characterized preparation is now obtained from a fungus, *Polyporus versicolor* (see Table I). In both cases the enzyme contains appreciable amounts of carbohydrate, a content of 14% being found for the fungal enzyme.

Ceruloplasmin from human serum has been prepared in highly purified form (Table I). Its laccase activity is low, but this has had the advantage of allowing kinetic studies on the rate of reduction of Cu^{2+} on addition of substrate (58).

Laccase is probably better understood in regard to mechanism than any of the other copper-containing oxidases.

B. The State of Copper

The fungal enzyme, which has a total of four copper atoms per molecule, has been found, by ESR (59) and susceptibility (60) measurements, to contain two Cu^{2+} and two Cu^{+} in the resting state. Nakamura (61) claims that the latex enzyme, on the other hand, has only Cu^{2+}. However, he found the same extinction coefficient per gram-atom of copper as reported for the fungal enzyme. As the color is known to be related to specifically bonded Cu^{2+} (59), this appears hard to understand, since it would mean that, per gram-atom of Cu^{2+}, the color of the fungal laccase would be twice as intense as that of the latex enzyme. In fact, it has recently been reported (62) that the latex enzyme contains both Cu^{2+} and Cu^+ but not in a 1:1 ratio. However, a low nitrogen content (10%) and a variable number of copper atoms (4–6) per molecule have been reported for the type of preparations used (cf. Table I), which might indicate insufficient purity. In homogeneous preparations of ceruloplasmin the ratio of copper in the two valencies is close to 1:1 (60,63), as found also for fungal laccase. As well-defined preparations give this ratio, it is tempting to conclude that it is associated with laccase activity.

Cu^{2+} in laccase is bonded in an unusual manner, as is evidenced by the high extinction coefficient at the visible absorption maximum and by the unique ESR spectrum (64) shown for ceruloplasmin in Figure 3. The characteristic ESR spectrum is destroyed on denaturation (Fig. 3), and this finding has been used to show that the intense blue color is related to the unique bonding of Cu^{2+}. When this is destroyed, activity is also lost.

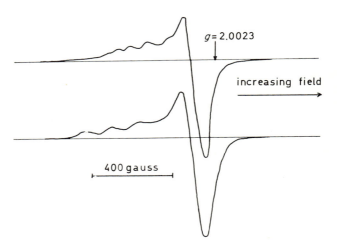

Fig. 3. The characteristic ESR spectrum of ceruloplasmin (top) compared with that of urea-denatured ceruloplasmin (bottom). The small lines (hyperfine structure) to the left of the main absorption line form a measure of the coupling between the unpaired electron and the nucleus and thus are influenced by the degree of electron delocalization. They are uniquely narrow in native ceruloplasmin and laccase but spread out on denaturation to give line widths and splittings of the same magnitude as given by most Cu^{2+} complexes (64).

The ESR spectrum has been interpreted (64) in terms of a high degree of delocalization of the unpaired hole on Cu^{2+}. As this should accept an electron when Cu^{2+} is reduced, this bonding should facilitate a mechanism involving a valency change.

The high extinction coefficient of laccase may be explicable on the basis of a charge-transfer state. As the charge-transfer absorption occurs at a relatively long wavelength (610 mμ), the energy which must be supplied to achieve a permanent oxidation-reduction reaction is small, which again should favor a mechanism involving a valency change.

It has been suggested (65) that the electron donor in the charge transfer is Cu^+. However, charge-transfer complexes of Cu^{2+} in proteins not containing Cu^+ are known (58). An alternative function of Cu^+ could be substrate binding through interactions with the π-electron system of the aromatic ring (58) (cf. Section II-D).

It has been found (61,66) that substrates reduce Cu^{2+} in laccase, and, as will be discussed later (Section IV-D), this valency change is involved in the mechanism of oxidation.

C. Specificity

It has long been known that laccase catalyzes the oxidation of p- as well as o-diphenols, but m-diphenols and monophenols have not been regarded as substrates for this enzyme. Recently, however, two separate groups (67,68) have shown that fungal as well as lacquer-tree laccase can catalyze the oxidation of monophenols, e.g., p-cresol. A weak activity towards m-diphenols has also been found (67).

The reason that the enzyme was earlier thought to lack monophenolase activity is probably the rapid inactivation of the enzyme during the oxidation of monophenols unless a protective agent is present (67). Gelatin and detergents can afford this protective effect, and in the presence of a diphenol, catechol, monophenols are also oxidized without inactivation of the enzyme. However, the catechol must be added before the monophenol substrate. Still, the oxidation of monophenol does not seem to be indirect (cf. Section II-D), but catechol has a protective influence. There is no inactivation during the oxidation of tyrosine, but this reaction is extremely slow.

The monophenol oxidation catalyzed by laccase appears to occur in quite a different manner from the tyrosinase-catalyzed oxidation. As will be discussed in the next section (Section IV-D), the primary product in laccase oxidation of diphenols is a free radical. This seems to be the case also in monophenol oxidation, which is not an oxygenation as in the case of tyrosinase (Section II-A). Instead, entirely different types of products are formed, e.g., the compound (1) in the case of p-cresol oxidation, which indicates that a free radical is the primary product.

(1)

In view of the double reaction specificity, the question of whether or not there are two different active sites arises, as in the case of tyrosinase. Inhibition studies (69) indicate that diphenol and monophenol oxidation involve the same active site of the laccase molecule.

D. Mechanism

Laccase-catalyzed oxidation definitely occurs by a mechanism involving reduction of Cu^{2+} by the substrate followed by reoxidation by O_2. The mechanism can be described by the two overall reactions given in Reactions (10) and (11), but these must be complex reactions involving several steps

$$ECu^{2+} + AH_2 \longrightarrow ECu^+ + (AH_2)^+ \tag{10}$$

$$ECu^+ + \tfrac{1}{4}O_2 + H^+ \longrightarrow ECu^{2+} + \tfrac{1}{2}H_2O \tag{11}$$

(ECu^{2+} represents the enzyme and AH_2 the substrate).

The first kinetic evidence for this mechanism was obtained by Nakamura (70) from susceptibility and spectrophotometric measurements. While he obtained the significant result that reoxidation does not involve a complex of the type $Cu^+\text{-}O_2$, as the rate of reoxidation is directly proportional to the concentration of O_2, the data did not allow a rigorous correlation of the rate of reduction and reoxidation with the rate of the overall reaction. However, in the case of the slower ceruloplasmin reaction, a detailed study of the kinetics of the valency change has been possible with the ESR technique (58). It could be shown that oxygen consumption occurred solely by a mechanism involving a valency change, and there can be little doubt that this mechanism is correct for all laccase-catalyzed oxidations.

A consequence of the mechanism just discussed is that all substances that can reduce Cu^{2+} in laccase must also be substrates, unless the product of oxidation inhibits the reoxidation of Cu^+. However, in cases where the reduction is slow, the reducing agent will be a poor substrate. This is the case with ascorbic acid, which must be the reason for the controversy (71,72) about whether or not this is a substrate for ceruloplasmin.

In agreement with the fact that ceruloplasmin is a poor laccase, Cu^{2+} reduction by all substrates is much slower than in fungal laccase (58). The basis for this is not understood, as the bonding of Cu^{2+} is identical in the two proteins, as evidenced by ESR (64). However, there are indications that Cu^{2+} in ceruloplasmin may be less available (65).

ESR measurements (58,73) have shown that the primary oxidation product is a free radical in laccase-catalyzed oxidations of diphenols. In Reaction (10) this is written as $(AH_2)^+$, and not $AH\cdot$, because with p-phenylenediamine, the best substrate of laccase, a characteristic ESR spectrum consisting of more than 75 lines has allowed the identification of the radical as $(NH_2C_6H_4NH_2)^+$. Kinetic measurements (58) show that the radical decomposes non-enzymically by a dismutation reaction [cf. Reaction (8)].

V. Conclusion

It appears natural to assume that copper found in many oxidases operates in the catalytic reactions by a mechanism involving a valency change, as suggested by Kubowitz (1) almost thirty years ago. However, the demonstration that tyrosinase contains only Cu^+ shows that this assumption can nevertheless be incorrect. Even in enzymes, such as laccase, where this mechanism appears valid, the situation is more complex, as they contain Cu^+ as well.

The discussions of mechanism presented in this chapter show that the copper-containing oxidases do not operate in a unified manner, but at least two types of mechanisms must be considered. Tyrosinase catalysis does not involve reduction of Cu^{2+} by the substrate. The product, quinone, is formed enzymically, most probably without the intermediate production of free radicals. Laccase, and probably also ascorbic acid oxidase, on the other hand, functions by a mechanism involving the transfer of one electron from the substrate to Cu^{2+}. In this case, the product of the enzymic oxidation is a semiquinone free radical, and the quinone is formed non-enzymically by a dismutation reaction.

Despite the wealth of information available about copper enzymes and their mechanism of action, we do not understand their intimate mode of action in terms of the detailed stereochemical and electronic events in the enzyme–substrate complexes. Among the many unsolved problems, it appears important to ascertain the exact role of Cu^+ found in all these enzymes. The question of the detailed mechanism of the reduction of O_2 is still one of most challenging problems for the copper enzymes as for all oxidases.

References

1. F. Kubowitz, *Biochem. Z.*, **299**, 32 (1938).
2. D. Keilin and T. Mann, *Nature*, **143**, 304 (1939).
3. D. Keilin and E. F. Hartree, *Proc. Roy. Soc. (London)*, **B119**, 114 (1935).
4. C. G. Holmberg and C. B. Laurell, *Acta Chem. Scand.*, **2**, 550 (1948).
5. A. B. Lerner and T. B. Fitzpatrick, *Physiol. Rev.*, **30**, 91 (1950).
6. A. Szent-Györgyi, *J. Biol. Chem.*, **90**, 385 (1931).
7. P. L. Lovett-Janison and J. M. Nelson, *J. Am. Chem. Soc.*, **62**, 1409 (1940).
8. F. J. Dunn and C. R. Dawson, *J. Biol. Chem.*, **189**, 485 (1951).
9. G. R. Stark and C. R. Dawson, in *The Enzymes*, Vol. 8, P. D. Boyer, H. Lardy, and K. Myrbäck, Eds., Academic Press, New York, 1963, p. 297.
10. H. R. Mahler, G. Hübscher, and H. Baum, *J. Biol. Chem.*, **216**, 625 (1955).
11. H. Yamada and K. T. Yasunobu, *J. Biol. Chem.*, **237**, 3077 (1962).
12. D. Amaral, L. Bernstein, D. Morse, and B. L. Horecker, *J. Biol. Chem.*, **238**, 2281 (1963).
13. E. Y. Levine, B. Levenberg, and S. Kaufman, *J. Biol. Chem.*, **235**, 2080 (1960).
14. H. Yamada, K. T. Yasunobu, T. Yamano, and H. S. Mason, *Nature*, **198**, 1092 (1963).
15. W. E. Blumberg, B. L. Horecker, F. Kelly-Falcoz, and J. Peisach, *Biochim. Biophys. Acta*, **96**, 336 (1965).
16. D. Kertész and R. Zito, in *Oxygenases*, O. Hayaishi, Ed., Academic Press, New York, 1962, p. 307.
17. S. Bouchilloux, P. McMahill, and H. S. Mason, *J. Biol. Chem.*, **238**, 1699 (1963).
18. J. L. Smith and R. C. Krueger, *J. Biol. Chem.*, **237**, 1121 (1962).
19. M. Fling, N. H. Horowitz, and S. F. Heinemann, *J. Biol. Chem.*, **238**, 2045 (1963).
20. T. Nakamura, *Biochim. Biophys. Acta*, **30**, 44 (1958).
21. T. Omura, *J. Biochem. (Tokyo)*, **50**, 264 (1961).
22. R. Mosbach, *Biochim. Biophys. Acta*, **73**, 204 (1963).
23. C. B. Kasper and H. F. Deutsch, *J. Biol. Chem.*, **238**, 2325 (1963).
24. W. N. Poillon and C. R. Dawson, *Biochim. Biophys. Acta*, **77**, 27 (1963).
25. V. Riley and J. G. Fortner, Eds., *The Pigment Cell. Molecular, Biological, and Clinical Aspects*, Ann. N.Y. Acad. Sci., **100** (1963).
26. W. D. Bonner, Jr., *Ann. Rev. Plant Physiol.*, **8**, 427 (1957).
27. C. R. Dawson and W. B. Tarpley, in *The Enzymes*, Vol. II, Part 1, J. B. Summer and K. Myrbäck, Eds., Academic Press, New York, 1951, p. 454.
28. T. P. Singer and E. B. Kearney, in *The Proteins*, Vol. II, Part A, H. Neurath and K. Bailey, Eds., Academic Press, New York, 1954, p. 124.
29. B. G. Malmström and J. B. Neilands, *Ann. Rev. Biochem.*, **33**, 331 (1964).
30. H. S. Mason, in *Advances in Enzymology*, Vol. 16, F. F. Nord, Ed., Interscience, New York, 1955, p. 105.
31. D. Keilin and T. Mann, *Proc. Roy. Soc. (London)*, **B125**, 187 (1938).
32. S. H. Pomerantz, *J. Biol. Chem.*, **238**, 2351 (1963).
33. R. C. Krueger, *Arch. Biochem. Biophys.*, **76**, 87 (1958).
34. Y. Karkhanis and E. Frieden, *J. Biol. Chem.*, **236**, PC 1 (1961).
35. B. J. Wood and L. L. Ingraham, *Nature*, **205**, 291 (1965).
36. M. F. Mallette and C. R. Dawson, *Arch. Biochem.*, **23**, 29 (1949).
37. H. Dressler and C. R. Dawson, *Biochim. Biophys. Acta*, **45**, 515 (1960).

38. H. S. Mason, W. L. Fowlks, and E. Peterson, *J. Am. Chem. Soc.*, **77**, 2914 (1955).
39. H. S. Mason, in *Advances in Enzymology*, Vol. 19, F. F. Nord, Ed., Interscience, New York, 1957, p. 131.
40. S. H. Pomerantz, *Biochem. Biophys. Res. Commun.*, **16**, 188 (1964).
41. D. Kertész, *Nature*, **180**, 506 (1957).
42. B. G. Malmström, in *Oxidases and Related Redox Systems*, T. E. King, H. S. Mason, and M. Morrison, Eds., Wiley, New York, 1965, p. 207.
43. G. M. Androes and M. Calvin, *Biophys. J.*, **2**, 217 (1962).
44. P. Hemmerich, Discussions comment to ref. 42, in *Oxidases and Related Redox Systems*, T. E. King, H. S. Mason, and M. Morrison, Eds., Wiley, New York, 1965, p. 216.
45. H. J. Bright, B. J. B. Wood, and L. L. Ingraham, *Ann. N.Y. Acad. Sci.*, **100**, 965 (1963).
46. F. Ghiretti, in *Oxygenases*, O. Hayaishi, Ed., Academic Press, New York, 1962, p. 517.
47. L. E. Orgel, *An Introduction to Transition Metal Chemistry—Ligand Field Theory*, Methuen, London, 1960.
48. R. W. Turner and E. L. Amma, *J. Am. Chem. Soc.*, **85**, 4046 (1963).
49. H. S. Mason, E. Spencer, and I. Yamazaki, *Biochem. Biophys. Res. Commun.*, **4**, 236 (1961).
50. K. Kim and T. T. Tchen, *Biochim. Biophys. Acta*, **59**, 569 (1962).
51. B. J. B. Wood and L. L. Ingraham, *Arch. Biochem. Biophys.*, **98**, 479 (1962).
52. W. A. Volk and J. L. Larsen, *Biochim. Biophys. Acta*, **67**, 576 (1963).
53. K. Tokuyama and C. R. Dawson, *Biochim. Biophys. Acta*, **56**, 427 (1962).
54. W. N. Poillon and C. R. Dawson, *Biochim. Biophys. Acta*, **77**, 37 (1963).
55. I. Yamazaki and L. H. Piette, *Biochim. Biophys. Acta*, **50**, 62 (1961).
56. R. J. Magee and C. R. Dawson, *Arch. Biochem. Biophys.*, **99**, 338 (1962).
57. Z. G. Penton and C. R. Dawson, in *Oxidases and Related Systems*, T. E. King, H. S. Mason, and M. Morrison, Eds., Wiley, New York, 1965.
58. L. Broman, B. G. Malmström, R. Aasa, and T. Vänngård, *Biochim. Biophys. Acta*, **75**, 365 (1963).
59. L. Broman, B. G. Malmström, R. Aasa, and T. Vänngård, *J. Mol. Biol.*, **5**, 301 (1962).
60. A. Ehrenberg, B. G. Malmström, L. Broman, and R. Mosbach, *J. Mol. Biol.*, **5**, 450 (1962).
61. T. Nakamura, *Biochim. Biophys. Acta*, **30**, 640 (1958).
62. W. E. Blumberg, W. G. Levine, S. Margolis, and J. Peisach, *Biochem. Biophys. Res. Commun.*, **15**, 277 (1964).
63. C. B. Kasper, H. F. Deutsch, and H. Beinert, *J. Biol. Chem.*, **238**, 2338 (1963).
64. B. G. Malmström and T. Vänngård, *J. Mol. Biol.*, **2**, 118 (1960).
65. W. E. Blumberg, J. Eisinger, P. Aisen, A. G. Morell, and I. H. Scheinberg, *J. Biol. Chem.*, **238**, 1675 (1963).
66. B. G. Malmström, R. Mosbach, and T. Vänngård, *Nature*, **183**, 321 (1959).
67. G. Fåhraeus and H. Ljunggren, *Biochim. Biophys. Acta*, **46**, 22 (1961).
68. G. Benfield, S. M. Bocks, K. Bromley, and B. R. Brown, *Phytochem.*, **3**, 79 (1964).
69. G. Fåhraeus, *Biochim. Biophys. Acta*, **54**, 192 (1961).
70. T. Nakamura, *Biochim. Biophys. Acta*, **42**, 499 (1960).

71. A. G. Morell, P. Aisen, and I. H. Scheinberg, *J. Biol. Chem.*, **237**, 3455 (1962).
72. E. Frieden, J. A. McDermott, and S. Osaki, in *Oxidases and Related Redox Systems*, T. E. King, H. S. Mason, and M. Morrison, Eds., Wiley, New York, 1965, p. 240.
73. T. Nakamura, in *Free Radicals in Biological Systems*, M. S. Blois, Jr., H. W. Brown, R. M. Lemmon, R. O. Lindblom, and M. Weissbluth, Eds., Academic Press, New York, 1961, p. 169.

Folate and B_{12} Coenzymes

F. M. HUENNEKENS, *Department of Biochemistry, Scripps Clinic and Research Foundation, La Jolla, California*

I.	Introduction	439
II.	Folate Coenzymes	440
	A. Structure, Nomenclature, and Chemistry	440
	B. Involvement in Oxidative Reactions	446
	1. Oxidoreduction of the Pyrazine Ring	446
	2. Tautomeric Forms of Dihydrofolate	449
	3. Dihydrofolate Reductase	452
	4. Thymidylate Synthetase	458
	5. Phenylalanine Hydroxylase	460
	6. Oxidoreduction of C_1 Units Attached to Tetrahydrofolate	460
	C. Involvement in Non-Oxidative Reactions	463
	1. Formate-Activating Enzyme	463
	2. Formyl- and Formimino-Transferring Enzymes	468
	3. Interaction of Aldehydes with Tetrahydrofolate	475
	4. Serine and Deoxycytidylate Hydroxymethylases	478
	5. Glycine Dehydrogenase	480
	6. Methionine Synthetase	481
III.	B_{12} Coenzymes	482
	A. Structure, Nomenclature, and Chemistry	482
	B. Chemical and Enzymic Synthesis of B_{12} Coenzymes	487
	C. Glutamate and Methylmalonyl CoA Mutase	492
	D. Diol Dehydrase	495
	E. Ribonucleotide Reductase	497
	F. Methionine Synthetase	498
	G. Methane Formation	502
	References	503

I. Introduction

Folic acid and vitamin B_{12} conform to the pattern established by other water-soluble vitamins in that they are converted to more complex coenzyme forms which serve as carriers for mobile metabolic groups (1) during the course of enzymic reactions. The folate coenzymes utilize covalent linkages at the N-5 and N-10 positions to carry one-carbon (C_1) fragments at the oxidation states of formate, formaldehyde, and methanol. The mobile metabolic groups carried by B_{12} coenzymes, on the other hand, are still largely uncertain, as is the site of attachment of these groups to the coenzyme. The present review will cover enzyme systems that utilize folate and B_{12} coenzymes, emphasizing, where possible, the role of the coenzyme in the mechanism of the reaction catalyzed.

II. Folate Coenzymes

A. Structure, Nomenclature, and Chemistry (2)

Folic acid (pteroylglutamic acid) contains three structural units: (1) the pterin (or pteridine) ring; (2) p-aminobenzoic acid; and (3) one or more glutamic acid residues, the latter being joined through γ-glutamyl linkages (Fig. 1a). Chemically synthesized folic acid, which contains only one glutamic acid, has been used for most biochemical studies. The numbering system for pertinent portions of the molecule is shown in the figure.

The field of one-carbon metabolism, mediated by folic acid coenzymes, began to take form in 1952 when Welch and Nichol (3) proposed that tetrahydrofolate (Fig. 1b) was the coenzymic form of folic acid and that its function was to "activate" C_1 units at the oxidation states of formaldehyde and formate. This hypothesis was strengthened by G. R. Greenberg's observation (4) that folinic acid (5-formyl tetrahydrofolic acid) was involved in the introduction of a formate unit into the C-2 position of

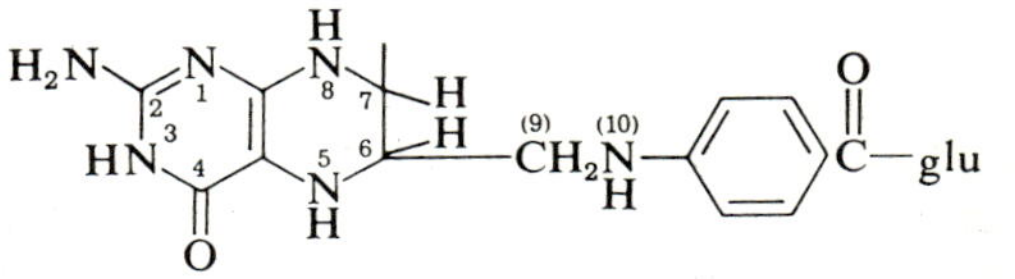

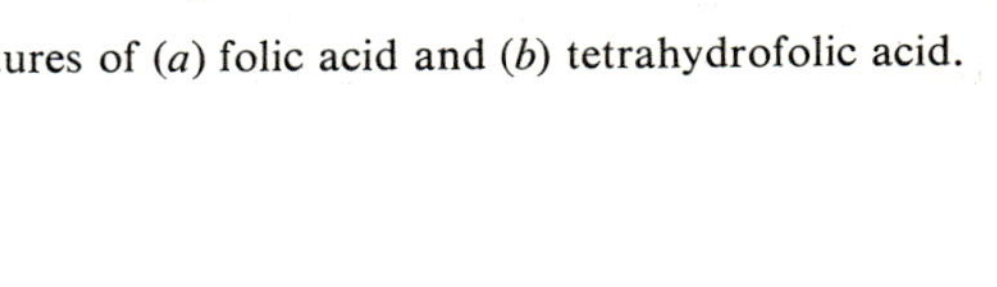

Fig. 1. Structures of (a) folic acid and (b) tetrahydrofolic acid.

purine, and by the discovery, made independently by Kisliuk and Sakami (5) and by Blakley (6), that tetrahydrofolate was a cofactor in the synthesis of serine from glycine and formaldehyde. The following decade witnessed the purification and characterization of a number of other tetrahydrofolate-dependent enzymes [reviewed by Huennekens and Osborn (7), Rabinowitz (8), Wright (9), Jaenicke (10), and Friedkin (11)].

Reduction of the pyrazine ring, as in the folate $\rightarrow$ tetrahydrofolate conversion, creates a new position in the molecule (N-5) to which C_1 groups can be attached covalently. More important, N-5 can also be utilized cooperatively with N-10 to permit a C_1 group to form a stable 5-membered ring with the coenzyme. Of the various possible combinations that could result from two sites of attachment on tetrahydrofolate and C_1 groups at three oxidation states, only six have been actually encountered in biological systems: 5-formyl, 10-formyl, 5-formimino, 5,10-methenyl, 5,10-methylene, and 5-methyl tetrahydrofolate (Table I).

TABLE I

C_1 Groups and Their Sites of Attachment to Tetrahydrofolate (24,46)

Site of attachment			C_1 unit	Oxidation state
N-5	N-10	N-5 and N-10		
N---N H HC=O	N---N H HC=O		Formyl	Formate
N---N H HC=NH			Formimino	Formate
		N------N ⊕\ / CH	Methenyl	Formate
		N------N \ / CH₂	Methylene	Formaldehyde
N---N H CH₃			Methyl	Methanol

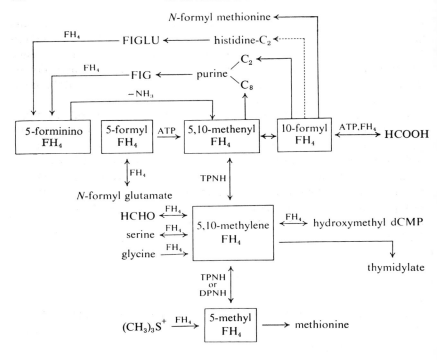

Fig. 2. Metabolic reactions involving folate coenzymes (228). FH_4 represents tetrahydrofolate.

Figure 2 summarizes the known reactions in which transfer of C_1 units between various donor and acceptor metabolites is mediated by folate coenzymes. The six C_1-tetrahydrofolate intermediates, listed above, are enclosed in boxes in order to emphasize their key roles. The formation or breakdown of these complexes, concomitant with the release or acceptance of C_1 units by various metabolites, is indicated by reaction arrows; double-headed arrows indicate reversible reactions. Metabolites containing potential C_1 units include: (1) inosinic acid, histidine, formiminoglutamate, formiminoglycine, N-formyl glutamate, and formate at the formyl level; (2) serine, deoxycytidylate, glycine, and formaldehyde at the next lower oxidation state; and (3) methionine and thymidylate at the methyl level. The occurrence of this integrated network for C_1 metabolism was suggested by early tracer experiments (2,7) which disclosed a facile transfer of C^{14} label between the β-carbon of serine, the C-2 and C-8 positions of purines, the C-2 position of histidine, the methyl groups of methionine and thymine, and free formate and formaldehyde.

Folate, 7,8-dihydrofolate, tetrahydrofolate, and the various C$_1$-tetra-hydrofolate adducts (Table I) are all accessible compounds via chemical or enzymic syntheses. Because of their amphoteric nature, folate coenzymes are moderately soluble only in water, dimethylformamide, and methanol, and they are difficult to crystallize. The amorphous materials, even when free from other folate compounds, are difficult to obtain completely pure on a dry weight basis. These factors are largely responsible for uncertainties in the reported values of physical constants, such as extinction coefficients, for folate coenzymes.

The lability of folate coenzymes necessitates special precautions in their storage and use. The reduced compounds are quite susceptible to air oxidation, and solutions of tetrahydrofolate are usually stabilized with 2-mercaptoethanol or ascorbate. The C^9—N^{10} linkage is also labile, a property that has been exploited for analytical purposes. Thus, tetrahydrofolate can be decomposed under controlled conditions (12) to yield p-aminobenzoylglutamate which is measured as a diazotizable amine by the Bratton–Marshall method. Alternatively, dihydrofolate is degraded by strong mineral acids, or even by phosphate at pH 7, to yield an unidentified product, or products, characterized by an absorbance maximum near 400 mμ (13).

The spectra of folate coenzymes can be analyzed in terms of contributions from two chromophoric groups, the pterin ring and the p-aminobenzoic acid moiety. Except when they are connected by a methenyl bridge (and, to a lesser extent, by a methylene bridge), the two chromophores act independently of each other. Thus, at neutral or basic pH values, the spectrum of N-methyl-p-aminobenzoylglutamate may be added to that of 6-methylpteridine, 6-methyl-7,8-dihydropteridine or 6-methyl-5,6,7,8-tetrahydropteridine to produce the spectra of folate, dihydrofolate, and tetrahydrofolate, respectively. This additivity is not observed in acid solution (14), probably due to the very low pK value of N-10 in the intact molecule (15) compared to the pK for this nitrogen atom in the model compound, N-methyl-p-aminobenzoylglutamate. Spectral characteristics of folate, dihydrofolate, tetrahydrofolate, and the known C$_1$ adducts with tetrahydrofolate are given in Table II. These spectra are useful for the identification of the compounds and for following enzyme-catalyzed reactions involving the conversion of one compound to another. Fluorescence spectra, also given in Table II, have not been employed as extensively for these purposes because folate coenzymes are often contaminated with highly fluorescent impurities (e.g., pteridine-6-carboxylic acid).

Certain of the structural and chemical features of folate coenzymes should be emphasized prior to discussing mechanisms. Thus, although the pyrazine ring itself is symmetrical, the C-6 and C-7 positions are not

TABLE II

Absorption and Fluorescence Spectra of Folate Coenzymes at pH 7[a]

Compound	Absorption maxima			Fluorescence maxima[b]		
	λ	ϵ, mM	Ref.	Excitation	Emission	Ref.
Folate	280	31.8	2,13	363	450–460[c]	24
	348	8.5				
7,8-Dihydrofolate	282	28.5	16	315	425	13,24
5,6,7,8-Tetrahydrofolate	297	28	17	305–310	360[d]	24
5-Formyl tetrahydrofolate	285	34	18	314	365	24
5-Formimino tetrahydrofolate	285	35.4	19	308	360	24
10-Formyl tetrahydrofolate	258	19	20,21	313	360	24
5,10-Methenyl tetrahydrofolate	345	26	22	370	470	24
5,10-Methylene tetrahydrofolate	290–295	32	23	—[e]		
5-Methyl tetrahydrofolate	290	31.5	13	—[e]		

[a] Values for 5,10-methenyl tetrahydrofolate are at pH 0.
[b] Aminco-Bowman spectrophotofluorometer.
[c] Weak fluorescence at pH 7.
[d] Maximum fluorescence at pH 3.
[e] Not tested.

equivalent because of the side chain at the former position and the unsymmetrical 2-amino-4-hydroxy pyrimidine ring fused onto the pyrazine structure; the hydroxyl group very probably exists largely in the keto form (25). The asymmetry of the pyrazine ring is clearly shown by the differences in reactivity between the N-5 and N-8 nitrogens; i.e., reduction of folate occurs preferentially at the N-8 end, whereas in tetrahydrofolate alkyl or acyl substitution occurs invariably at the N-5 position.

Another key structural feature is the potentially asymmetric position at C-6. In addition to the L-configuration at the α-carbon atom of the glutamate residue, an additional optical center at C-6 is created by reduction of the pyrazine ring. Tetrahydrofolate, prepared by the catalytic hydrogenation of folic acid, is a *dl*-mixture with respect to the C-6 position; only 50% of the material reacts in tetrahydrofolate-dependent enzyme systems (reviewed in 26). Enzymic reduction of 7,8-dihydrofolate, on the other hand, results in the synthesis of the *l*-L-diastereoisomer of tetrahydrofolate (27). The optical rotation of the latter product (ca. $-17°$) is approximately the same as that of naturally occurring 5-formyl tetrahydrofolate. For reasons not fully understood, the diastereoisomers of 5,10-methylene tetrahydrofolate, which can be separated by chromatography on TEAE-cellulose (28,29), have greatly enhanced rotations.

In tetrahydrofolate the relative basicities of N-5 and N-10 help to determine the reactivity of these positions toward various agents. Because tetrahydrofolate contains other prototropic groups (Table III) whose pK_a

TABLE III

pK_a Values for Prototropic Groups in Tetrahydrofolate

Group	pK_a	Ref.
N-1	1.2	15
N-3	—[a]	
N-5	4.8	15
	5.4[b]	14
	5.3[b]	30
N-8	—[a]	
N-10	-1.3	15
2-Amino	3.2[a]	31
4-OH	10.5	15
	10.5	14
Glutamate α-COOH	3.5	15
Glutamate γ-COOH	4.8	15

[a] No values available.
[b] Values obtained from model compounds.
[c] Value for 2-amino group in folic acid.

values are in the same range, it is somewhat difficult to obtain precise values for these two nitrogens. Earlier titrimetric and spectrophotometric work with model compounds gave estimated pK_a values of about 3 and 5, respectively, for N-10 and N-5 (32). Recently, Kallen and Jencks (15) have obtained evidence that the pK_a of N-10 is perhaps as low as -1.25 due to interaction of the protonated N-5 and N-1 groups with N-10. In any event, N-5 is considerably more basic than N-10, a finding that is in agreement with the theoretical calculations of Perault and Pullman (25). The basicity of these nitrogens, however, is only one of several factors that control the following experimental observations: (*a*) Formylation occurs more readily at N-10 while alkylation [e.g., with chloromethyluracil (33)] occurs more readily at N-5. (*b*) The interaction of formaldehyde with tetrahydrofolate to yield the 5,10-methylene derivative probably involves 5-hydroxymethyl tetrahydrofolate (34) as a transient intermediate. (*c*) 5,10-Methenyl tetrahydrofolate is cleaved hydrolytically on the C—N⁵ side. (*d*) 5,10-Methylene tetrahydrofolate is cleaved reductively at the C—N¹⁰ linkage.

B. Involvement in Oxidative Reactions

1. Oxidoreduction of the Pyrazine Ring

Because of its involvement in the biosynthesis of tetrahydrofolate, the oxidoreduction system (folate ⇌ dihydrofolate ⇌ tetrahydrofolate) has been studied intensively, with respect to both chemical and enzymic catalysis.

When folate is treated with hydrosulfite at room temperature, 7,8-dihydrofolate is produced in very good yield (16,35,36). This material, which has been used for most enzymic studies, has been termed "standard dihydrofolate" (37) to distinguish it from other possible tautomeric forms (Section II-B-2). At elevated temperatures (ca. 75°), hydrosulfite reduces folate completely to tetrahydrofolate (38). Catalytic hydrogenation of folate over platinum oxide in acidic solution, however, is the customary method for preparation of tetrahydrofolic acid (39,40). Alternatively, if the hydrogenation is carried out in neutral or alkaline solution, 7,8-dihydrofolate is the principal product (39). Borohydride, which readily reduces dihydrofolate, converts folic acid largely to the tetrahydro form although, under certain conditions (41), dihydrofolate can be made to accumulate in yields up to 30%.

Reoxidation of tetrahydrofolate usually gives rise to a mixture of dihydrofolate, folate, and degradation products (42), although treatment of tetrahydrofolate with iodine is reported to produce folate in good yield (43). 2,6-Dichlorophenolindophenol (44), or hydrogen peroxide in the

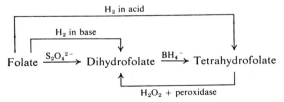

Fig. 3. Chemical routes for interconversion of folate, dihydrofolate, and tetrahydro-folate.

A. Reduction of folate and dihydrofolate by borohydride:

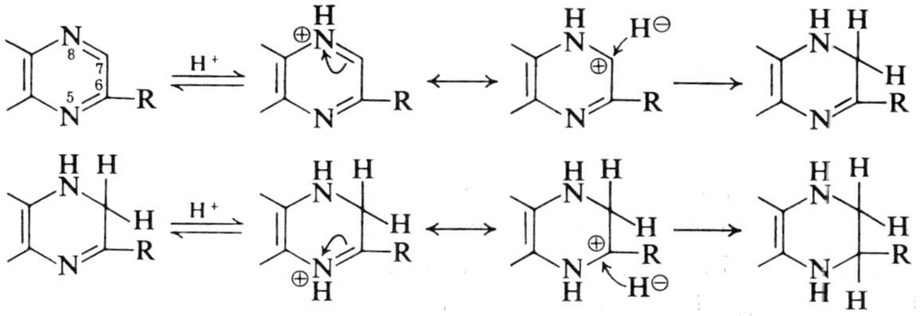

B. Reduction of folate by hydrosulfite:

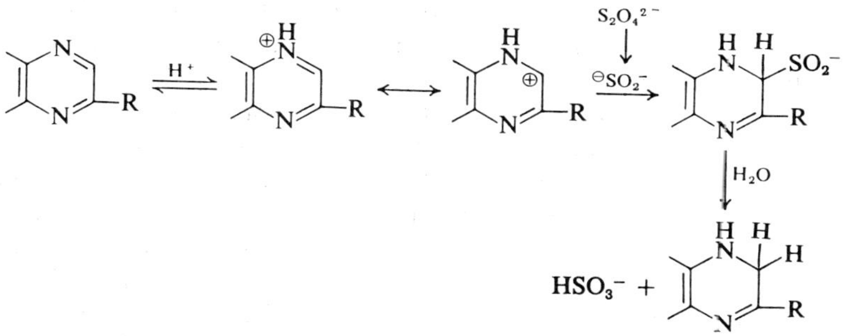

Fig. 4. Mechanisms for the reduction of folate and dihydrofolate (46).

presence of peroxidase (45), oxidizes tetrahydrofolate to dihydrofolate. The chemical routes for oxidation–reduction of folate, dihydrofolate (ignoring tautomeric forms), and tetrahydrofolate are summarized in Figure 3.

The stepwise reduction of folate to tetrahydrofolate by borohydride may be visualized in terms of the mechanism (46) outlined in Figure 4A. N-5 and N-8 in folate are pyridine-like nitrogens that ordinarily would appear to have little tendency to be protonated, especially at pH 7. However, because of resonance interaction with the pyrimidine ring, N-8 is more basic than N-5. Accordingly, in step 1 it is assumed that there is a displacement of electrons toward N-8 with a resulting positive charge at C-7. Introduction of the hydride ion into the latter position gives rise to 7,8-dihydrofolate. In the somewhat analogous case of DPN, the nitrogen atom initially bears a formal, rather than an induced, positive charge which can be rewritten at the C-4 position (Reaction (1)). Although DPN would be expected to be reduced more readily by hydride donors than folate,

$$\tag{1}$$

borohydride actually reduces DPN rather sluggishly and produces a mixture of the three possible isomers of DPNH (see Part II, chapter by Sund on the pyridine nucleotide coenzymes).

The facile reduction of folate by hydrosulfite occurs most probably by the mechanism outlined in Figure 4B, which is analogous to that postulated for the reduction of DPN. Heterolytic scission of the reductant gives rise to the powerful nucleophile $^-[SO_2]^-$, which reduces the C^7—N^8 double bond by a sequence that results in the eventual introduction at C-7 of a proton from water.

The rapid reduction of 7,8-dihydrofolate by borohydride (37,41), or by TPNH in the presence of dihydrofolate reductase (26), is not unexpected. The 5,6 double bond is a semiisolated azomethine linkage similar to that found in Schiff bases (e.g., amino acid–pyridoxal complexes); the latter are readily reduced with borohydride. Among the factors involved in reduction of 7,8-dihydrofolate, the basicity of N-5 more than offsets the steric hindrance by the side chain at C-6. An alternative mechanism has been proposed by Zakrzewski (47) in which reduction of dihydrofolate to tetrahydrofolate is preceded by an intramolecular rearrangement which permits the incoming hydrogen to add at C-7.

Nucleophiles other than H^- (e.g., CN^-, OH^-, HSO_3^-, hydroxylamine, acetone, and mercaptans) can also add to the *para* position of DPN (see Sund chapter, p. 603). These nucleophilic adducts (or possibly charge transfer complexes in some instances) tend to dissociate upon dilution or lowering the pH, although some are sufficiently stable to be isolated. The spectra of these adducts (λ_{max} ranging from 320 to 360 mμ) are remarkably like that of DPNH itself (λ_{max} at 340 mμ). It is of interest to note that the pyrazine ring of dihydrofolate (and to a lesser degree, of folate) is likewise receptive to nucleophilic agents: Dihydrofolate (λ_{max} at 280 mμ) undergoes a rapid change in absorption spectrum, indicative of adduct formation (λ_{max} at ca. 300 mμ), when admixed with CN^-, HSO_3^-, acetone, or hydroxylamine (48). Complex formation is a pH-dependent equilibrium reaction and, as yet, no adducts have been isolated.

Oxidation of tetrahydrofolate occurs preferentially at the N-5 end of the pyrazine ring, yielding 7,8-dihydrofolate. Appropriate substituents at the 5-position (as in 5-formyl, 5,10-methenyl, or 5,10-methylene tetrahydrofolate) protect against oxidation. Substitution of a methyl group at N-5, however, affects the course of oxidation. Thus, 5-methyl tetrahydrofolate is air-oxidized to a 5-methyl dihydrofolate (λ_{max} at 250 and 290 mμ), believed to be the 5,6-dihydro isomer on the basis of tracer data (41) and because it can be reduced to 5-methyl tetrahydrofolate without loss of optical activity at C-6 (49). Assuming that this structure is correct, 5-methyl-5,6-dihydrofolate, unlike 7,8-dihydrofolate, can be reduced non-enzymically by weak reductants such as DPNH or thiols (50).

2. Tautomeric Forms of Dihydrofolate

As shown in the top row of Figure 5, dihydrofolate can assume three tautomeric forms that differ only in the site of reduction in the pyrazine ring. Two additional "quinonoid" structures, proposed by Kaufman (51), are shown in the lower row. The latter have been invoked as possible structures for a transient intermediate that appears in the reaction catalyzed by phenylalanine hydroxylase (Section II-B-5) or when tetrahydrofolate is oxidized by dyes (44).

The early assignment of the 7,8-dihydro structure to "standard" dihydrofolate was based largely upon indirect chemical evidence (reviewed in ref. 26) and reinforced by the finding that greater than 90% of "standard" dihydrofolate was reduced in the reaction catalyzed by dihydrofolate reductase (Section II-B-3). The latter observation excluded the 5,6 isomer, which would have reacted only to the extent of 50% because the chemically synthesized material would have been a mixture of *dl* isomers with respect to the C-6 position. This argument would not

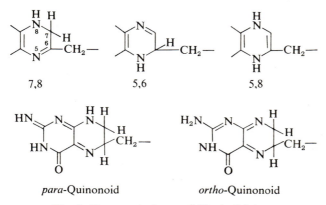

Fig. 5. Tautomeric forms of dihydrofolate.

apply if "standard" dihydrofolate were the 5,8 isomer. Authentic 5,8-dihydropterins, which have been synthesized by Pfleiderer and Taylor (52), exhibit anomalously high absorption maxima at 405–410 mμ; this is due perhaps not so much to the 5,8 structure as to various chromophoric substituent groups on the ring. On the basis of quantum-mechanical calculations, Perault and Pullman (25) have suggested that a 5,8 isomer would be particularly resistant to further reduction.

In order to search for alternate forms of dihydrofolate, Mathews and Huennekens (37) prepared the material by four routes: (1) reduction of folic acid with hydrosulfite; (2) reduction of folic acid with the pterin reductase from *Clostridium sticklandii*; (3) oxidation of tetrahydrofolic acid during the reductive synthesis of thymidylate; and (4) oxidation of tetrahydrofolic acid with indophenol. Each preparation of dihydrofolate was isolated from its reaction mixture by chromatography on DEAE-cellulose. All four dihydrofolate samples were identical with respect to absorption and fluorescence spectra, optical rotation, R_f values, and reduction with borohydride or with TPNH-dependent dihydrofolate reductases. These data suggest that "standard" dihydrofolate is the tautomer whose stability greatly exceeds that of the other possible forms (Fig. 5).

The possibility that "standard" dihydrofolate might be the 5,8 isomer was suggested by our work (53), later corrected (41), on the reduction of folate by borohydride and by Zakrzewski's observation (54) that reduction of folate with hydrosulfite in tritium-labeled water led to unlabeled dihydrofolate. The latter result seemed to exclude carbon-bound hydrogens at C-6 or C-7. Later, Pastore et al. (55) and Zakrzewski (47) showed that there is a large isotopic discrimination against tritium since, when the reaction was carried out in pure D_2O, one atom of non-exchangeable deuterium was found per mole of dihydrofolate. Moreover, Pastore et al. (55) found evidence for two methylene hydrogens in the proton magnetic resonance spectrum of "standard" dihydrofolate, a result clearly inconsistent with the 5,8 structure. Hillcoat and Blakley (56) have confirmed the PMR spectral data for dihydrofolate and Scrimgeour and Vitols (41) showed that the dihydrofolate, obtained by reducing folate with H^3-borohydride, contained one atom of non-exchangeable tritium.

Dihydrofolate is not too suitable for structural determinations because the glutamate carboxyls tend to obscure the properties of functional groups in the pyrazine ring. To overcome this difficulty, we have carried out studies on an authentic 7,8-dihydropteridine, containing a methyl group at C-6 instead of the *p*-aminobenzoylglutamate residue (14). 2-Amino-4-hydroxy-6-methylpteridine was prepared by (1) reductive cleavage of folic acid at the C^9—N^{10} linkage; (2) alkaline hydrolysis of 2,4-diamino-6-methylpteridine; and (3) condensation of 2,4,5-triamino-6-hydroxypyrimidine with N-(2-formyl-2-bromoethyl)pyridinium bromide in the presence of KI to yield N-(2-amino-4-hydroxy-6-pteridyl)methyl-pyridinium iodide, followed by conversion of the latter to the desired pteridine. Reduction of the pteridine with hydrosulfite or with zinc in NaOH yielded a crystalline dihydropteridine. The same dihydropteridine was also prepared by direct condensation of 2-amino-4-chloro-5-phenylazo-6-hydroxypyrimidine with aminoacetone semicarbazone hydrochloride followed by reductive ring closure.

Unlike dihydrofolate, in which the proton magnetic measurements are obscured by the glutamate methylene carbons, the model pteridine and dihydropteridine gave very clean and simple spectra (Fig. 6). The characteristic resonance band at $\tau = 1.08$ associated with a proton attached to a trigonal carbon in the pteridine was replaced in the dihydropteridine by a resonance band at $\tau = 4.84$ characteristic of methylene protons. Integration of the spectra confirmed the contribution of only one proton in the first case and of two protons in the second. The dihydropteridine must be the 7,8 tautomer since, in the 5,6-dihydro form, the methyl group at $\tau = 7.28$ would be a doublet. Finally, if reduction had produced the 5,8 isomer, there would still have been a reasonance signal, occurring at a low value, corresponding to a single proton attached to a trigonal carbon atom at C-7.

The model dihydropteridine was related to "standard" dihydrofolate in the following way. N-Methyl-p-aminobenzoylglutamic acid was admixed at pH 7 with an equimolar quantity of the 2-amino-4-hydroxy-6-methyl-6,7-dihydropteridine and the resultant spectrum was shown to be

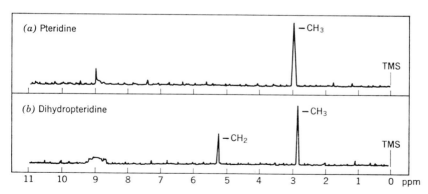

Fig. 6. Proton magnetic resonance spectra of a model pteridine and a 7,8-dihydro-pteridine (14). TMS = trimethylsilane.

nearly identical to that of dihydrofolate at the same concentration. The close similarity of the spectra was also observed when the experiment was repeated at pH 13, but not at pH 1. In the case of the mixed spectrum at the low pH, protonation of the secondary amino group ($pK_a = 2.8$) attached to the aromatic ring of N-methyl-p-aminobenzoylglutamate suppresses its contribution to the spectrum. In the case of dihydrofolic acid, this protonation does not seem to occur, presumably because the pK_a of N-10 is much lower (15) than in the model compound.

Thus, all lines of evidence now suggest that "standard" dihydrofolate is the 7,8 tautomer and that this form is considerably more stable than the others shown in Figure 5. The latter, however, may appear as transient intermediates in chemical or enzymic reactions involving folate or tetrahydrofolate (Fig. 3).

3. Dihydrofolate Reductase

Partially purified dihydrofolate (or folate) reductases, obtained from avian and mammalian liver, were able to catalyze the pyridine nucleotide-dependent reduction of both folate and dihydrofolate to tetrahydrofolate (Reactions (2) and (3)).*

$$\text{Folate} + 2\text{TPNH} + 2\text{H}^+ \longrightarrow \text{Tetrahydrofolate} + 2\text{TPNH}^+ \qquad (2)$$

$$\text{Dihydrofolate} + \text{TPNH} + \text{H}^+ \rightleftharpoons \text{Tetrahydrofolate} + \text{TPN}^+ \qquad (3)$$

As reviewed elsewhere (26,57), these enzymes shared the following properties: (*1*) a single enzyme catalyzed both reductive steps; (*2*) dihydrofolate was a better substrate than folate, hence the preference for "dihydrofolate reductase"; (*3*) TPNH† was the preferred reductant, although some activity was observed with DPNH; (*4*) the enzyme exhibited two pH optima (4.5 and 7.5); and (*5*) the enzyme was inhibited by extremely low concentrations of the folic acid antagonists, aminopterin and amethopterin.

---

*Using a highly purified dihydrofolate reductase, the equilibrium constant for Reaction (3), $K = [\text{tetrahydrofolate}][\text{TPN}^+]/[\text{dihydrofolate}][\text{TPNH}][\text{H}^+]$, has been shown (59) to be $K = 5.6 \times 10^{11}$ (or 5.6×10^4 at pH 7). Assuming a standard reduction potential for the TPN–TPNH couple ($E_0' = -0.32$ V), a value of $E_0' = -0.19$ V can be calculated for the dihydrofolate–tetrahydrofolate couple. The equilibrium constant for Reaction (2) (or for the folate → dihydrofolate step) has not yet been determined, but it is probably large since the reduction of folate to tetrahydrofolate goes to completion. A tetrahydrofolate dehydrogenase, partially purified from baker's yeast (71), is stated to catalyze the reverse of Reaction (2), a finding that is difficult to reconcile with the above equilibrium considerations.

†Dihydrofolate reductases from chicken liver (11), L1210 cells (11) and *S. faecalis* R (57a) transfer a hydride ion from the A-side of TPNH to dihydrofolate.

TABLE IV

Properties of Dihydrofolate Reductases

Source	K_m values,[a] M		Activators			pH optimum	Mol. wt.	Specific activity μmoles/min/mg protein
	Dihydrofolate	TPNH	Mercurials	Urea	K^+			
Chicken liver (59,60)	5×10^{-7}		Yes	Yes	—	4.5, 7.5	22,000	8
(61)	$(1.2–2.2) \times 10^{-7}$	1.8×10^{-6}	Yes	Yes	—[b]	4.0, 7.4	21,000–24,000	14–15
Sheep liver (62)	1.2×10^{-6}	1.7×10^{-6}	—	—	—	5.5	—	1.9
Calf thymus (63)	2.3×10^{-6}	3.3×10^{-5}	No	No	Yes	5.5	33,500	10
S. faecalis (64)	4.0×10^{-6}	4.6×10^{-5}	—	No[f]	No[f]	6.7	17,000–22,000	4.7
SF/O[c] (65)	1.3×10^{-5}	2.2×10^{-5}	—	—	—	6.2	—	2.4
SF/A$_k$°[c] (65)	1.3×10^{-5}	2.2×10^{-5}	—	—	—	6.7	—	4.9
D. pneumoniae (66,66a)	2.7×10^{-6}	1.7×10^{-5}	—	—	—	7.3	20,000	—
ame^r-3 (66a)	2.6×10^{-6}	1.6×10^{-5}	—	—	—	7.3	20,000	—
L. leichmanii (68)	6×10^{-6}	3.8×10^{-5}	No	No	—	6.3	$\sim15,000$	1.6
E. coli (67)	—	1.5×10^{-5e}	No	No	—	4–9	22,000	0.3
T-6 Bacteriophage induced enzyme (67)	—	5.0×10^{-5e}	No	No	—	7.0	31,000	1.6
Ehrlich ascites cells (69)	1.5×10^{-6}	5.6×10^{-6}	Yes	Yes	Yes	5.9, 7.6	20,200	2.5
L1210 cells, sensitive[d] (70)	—	2.0×10^{-5}	Yes	—	Yes	—	—	0.05[g]
resistant[d]	—	2.0×10^{-5}	Yes	—	(Na^+)	—	—	0.5[g]

[a] At pH 7.0–7.5, unless otherwise indicated. [b] Not reported. [c] SF/O and SF/A$_k$ are amethopterin-sensitive and amethopterin-resistant strains of S. faecalis. [d] Amethopterin-sensitive and resistant. [e] C. K. Mathews, unpublished data. [f] R. L. Blakley, unpublished data. [g] M. Friedkin, unpublished data. [h] Amethopterin-resistant strain.

Subsequently, dihydrofolate reductase has been obtained in highly purified form from several sources, including chicken liver (58–61), sheep liver (62), calf thymus (63), amethopterin-sensitive and amethopterin-resistant strains of *Streptococcus faecalis* (64,65), *Diplococcus pneumoniae* (66), T6 bacteriophage-infected and normal cultures of *Escherichia coli* B (67), *Lactobacillus leichmanii* (68), Ehrlich ascites cells (69), and L1210 cells sensitive or resistant to amethopterin (70). The properties of these enzymes are listed in Table IV. Purified dihydrofolate reductases are stable to lyophilization, to freezing and thawing, and to prolonged dialysis. The enzymes are colorless and have a single absorption maximum at 280 mμ with a pronounced shoulder at 295 mμ; no flavin or heme component is present. The lack of inhibition by a number of metal-binding agents (e.g., EDTA, cyanide, diethyldithiocarbamate, hydroxylamine, 8-hydroxyquinoline, and 1,10-phenanthroline) suggests that the enzymes do not contain bound metal ions.

As determined by filtration through Sephadex, or by sedimentation in a sucrose gradient, most dihydrofolate reductases have molecular weights of 20,000–22,000, values which are quite low for pyridinoprotein enzymes. The calf thymus (63) and the T6 viral (67) enzymes have molecular weights nearer 30,000. A much higher molecular weight protein with dihydrofolate reductase activity has been demonstrated in an amethopterin-resistant strain (ame^r-3) of *D. pneumoniae* (72). Preliminary evidence from this laboratory (60) indicates that the enzyme consists of a single polypeptide chain having one N-terminal and one C-terminal end. Calculations based on titration of the chicken liver enzyme with aminopterin (see below) have shown that the pure enzyme should have a turnover number of about 300 or a specific activity of about 15 μmoles of dihydrofolate reduced per minute per milligram of protein.

An interesting property of some, but not all, dihydrofolate reductases, discovered independently by Kaufman and Bertino, is the enhancement of catalytic activity when the protein is treated with urea (73), mercurials (74,75), or cations (76). The protein apparently occurs in a conformation that is not optimal for catalytic activity; alkylation of cysteine residue(s) allows the protein to assume a more favorable conformation. Urea probably activates the protein by breaking hydrogen bonds, while the cations may displace bound water. Thus, activation appears to involve an unfolding of the normal molecule (E_n) into one of several activated forms (E_a, E_a', E_a'', etc.), depending on the reagent used. The latter forms do not appear to have additional binding sites for substrates, nor do they bind the substrates more tightly (77); perhaps in the activated state the substrate sites are situated more advantageously for hydride ion transfer from

TPNH to dihydrofolate. As might be anticipated, the activated forms of the enzyme are more susceptible to digestion by trypsin than is the normal form (60). Conversely, addition of either substrate (TPNH or dihydrofolate) or of aminopterin causes the enzyme to become more resistant than usual to disruption. Thus, the protein probably assumes a more compact form (E_c) when substrates are bound. Measurements based upon quenching of the fluorescence of the tryptophan residues (excitation at 280 mμ and emission at 335 mμ) in the protein (78,79) have shown that addition of either substrate alters the protein conformation, as judged by the change in the binding constant for the other substrate (78). A hypothetical relationship between conformational forms of the enzyme is shown in Equation (4).

$$\left.\begin{matrix} E_a \\ E_a' \\ E_a'' \\ E_a'' \end{matrix}\right\} \underset{\xrightarrow{\hspace{2cm}}}{\overset{\text{urea, pCMB,}}{\underset{\text{or K}^+}{\xleftarrow{\hspace{2cm}}}}} E_n \underset{\xleftarrow{\hspace{2cm}}}{\overset{\text{substrates or}}{\underset{\text{inhibitors}}{\xrightarrow{\hspace{2cm}}}}} E_c \tag{4}$$

The ability of the enzyme to assume a variety of conformational forms may also provide an explanation for observations made during its purification. Although several thousandfold purification procedures have been reported, no dihydrofolate reductase has yet been obtained in crystalline form. This is somewhat surprising since the enzyme is quite stable and readily precipitated by ammonium sulfate at about 60% saturation. The specific activity also rises rather slowly during the latter stages of purification and there is a lack of symmetry between the protein profiles and those of enzymic activity when purified preparations are chromatographed on DEAE-cellulose or hydroxyapatite. On the other hand, examination of even partially purified preparations by starch gel electrophoresis, ultracentrifugation, or filtration through Sephadex does not reveal any gross inhomogeneity. The reductase (i.e., the form having the highest catalytic activity) thus appears to be contaminated by a protein, or proteins, having similar molecular weight and net charge. This could be merely an adventitous contaminant or it could be some form (or forms) of the enzyme having a decreased catalytic activity or no activity at all. If the latter explanation should prove to be true, it will be of interest to learn whether these forms can be interconverted reversibly. In view of the fact that the apparent level of the reductase rises as a consequence of amethopterin resistance (see below), the physiological control mechanism may utilize the principle of multiple forms possessing different catalytic activity or different affinity for the folate antagonists.

A somewhat different enzyme, present in C. sticklandii, reduces folate but not dihydrofolate (80). In the crude system this enzyme utilizes as the

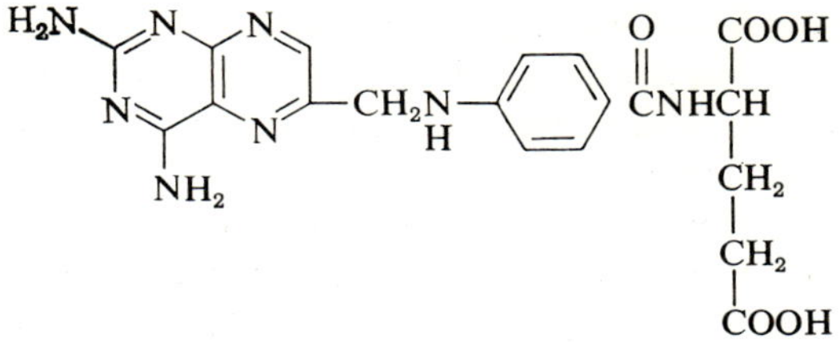

Aminopterin

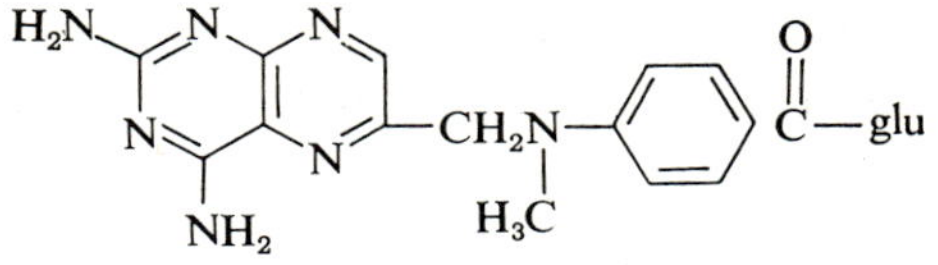

Amethopterin

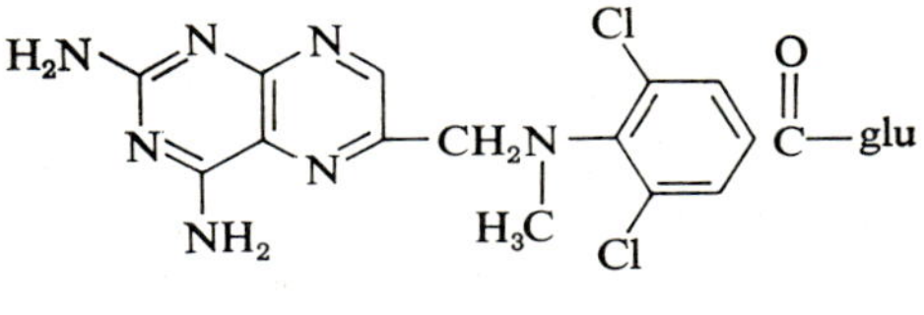

Dichloroamethopterin

Fig. 7. Structures of aminopterin, amethopterin, and dichloroamethopterin.

reductant the coenzyme A-dependent oxidation of pyruvate (Equation (5)), rather than TPNH or DPNH.

$$\text{Folate} + \text{pyruvate} + \text{CoA} \longrightarrow \text{Dihydrofolate} + \text{acetyl CoA} + CO_2 \qquad (5)$$

The enzyme is also insensitive to folic acid antagonists in contrast to a conventional TPNH-dependent dihydrofolate reductase which is also present in the organism. It has been shown recently that in the above reaction pyruvate dehydrogenase is linked to ferredoxin and reduced ferredoxin reduces folate, probably non-enzymatically (81).

In early nutritional studies (3) it was found that the utilization of folate, but not folinate, was markedly inhibited by 4-amino analogs of the vitamin such as aminopterin and amethopterin (Fig. 7). Further interest was aroused in these compounds when it was learned that they produce a dramatic, although transient, remission of symptoms in certain acute leukemic patients. As the details of the folate → folinate sequence became better understood, it was shown by Futterman and Silverman (82) and by Osborn et al. (83) that the antagonists were actually potent inhibitors ($K_I = 10^{-9}\ M$) of dihydrofolate reductase. Reduced forms of aminopterin

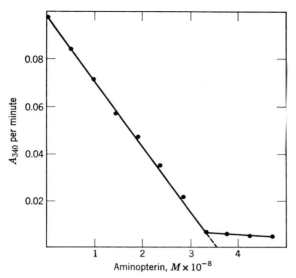

Fig. 8. Titration of dihydrofolate reductase with aminopterin. Standard assay system (59) with indicated concentrations of inhibitor.

(26,84,85) also inhibit the enzyme. With the chicken liver dihydrofolate reductase, dihydroaminopterin is about equal to aminopterin in inhibitory power while the tetrahydro compound has a much weaker affinity for the enzyme (26).

Werkheiser introduced the technique of "titrating" dihydrofolate reductase with a folate antagonist (86). Thus, in crude extracts or with purified enzymes, the activity decreases linearly upon the addition of amethopterin until nearly all of the enzyme has been titrated (cf. Fig. 8). Extrapolation of the linear portion of the plot to the x axis gives an amount of antagonist equivalent on a molal basis* to the enzyme being titrated. This type of stoichiometric inhibition requires that enzyme and inhibitor combine irreversibly. Work from Bertino's laboratory (87) has shown that the degree and nature of E–I interaction depends upon the pH and the ionic environment; at pH 4.5 the binding tends to be stoichiometric while at pH 7 it becomes slightly reversible. While the chicken liver reductase–inhibitor complex is not dissociated by prolonged dialysis against various buffers, it can be dissociated by chromatography on hydroxyapatite or DEAE-cellulose columns (59). This lability of the complex precludes the possibility of tagging the "active center" of the

*Recent studies (60,61,78) with purified dihydrofolate reductases have shown that enzyme and inhibitor do combine in equimolar amounts.

enzyme with H^3-aminopterin and degrading the complex to obtain an aminopterin-labeled peptide.

All of the dihydrofolate reductases (Table IV) are inhibited by very low concentrations of the folic acid antagonists. The intensity of inhibition is unusual inasmuch as covalent bonds are not involved in the enzyme–inhibitor interaction. Assuming a K_m value of $5.0 \times 10^{-7} M$ for the interaction of dihydrofolate with the chicken liver reductase and a K_I value of 1.0×10^{-9} for aminopterin, it can be calculated (88) that $\Delta F = -8.6$ kcal/mole and -12.3 kcal/mole for the binding of dihydrofolate and aminopterin, respectively, to the enzyme. The difference between ΔF values, i.e., 3.7 kcal/mole, is approximately that required for the formation of a single hydrogen bond. Thus, the 4-amino group of aminopterin must form such a hydrogen bond with a basic group on the protein whereas this possibility is denied to folate or dihydrofolate in which the 4-hydroxy group exists largely in the keto form (25).

The fact that dihydrofolate reductases are strongly inhibited by the folic acid antagonists is of considerable importance to cancer chemotherapy (reviewed in 89–91). In human leukemic leukocytes, dihydrofolate reductase is extremely sensitive to these drugs (92) and it is quite likely that this enzyme is the only protein in the cell that has any appreciable avidity for antifolate compounds (93). Since dihydrofolate reductase is involved, moreover, as a component of the "thymidylate synthesis cycle" (Section II-B-4), it is not surprising that folate antagonists repress DNA synthesis and leukocyte replication. In human leukemic patients, the acquired resistance to folate antagonists can be correlated to some degree with an increase in the level of dihydrofolate reductase. This unfortunate effect appears to be due to a drug-stimulated "induction" of new enzyme (94) rather than to a process of cell selection; the latter explanation is more likely to be applicable to the development of resistance in bacteria or cells in culture (reviewed in ref. 11).

4. Thymidylate Synthetase

Except for the synthesis of thymidylate, all the reactions in Figure 2 occur without any change in the oxidation state of the C_1 group being transferred. As shown by the laboratories of Greenberg, Friedkin, and Blakley, thymidylate synthesis (Reaction (6)) involves the transfer and reduction of the C_1 unit from 5,10-methylene tetrahydrofolate to the pyrimidine (reviewed in ref. 11).

5,10-Methylene tetrahydrofolate + deoxyuridylate $\longrightarrow$

dihydrofolate + thymidylate   (6)

5-Hydroxymethyl deoxyuridylate does not appear to be an intermediate in this reaction and there is no evidence for the involvement of a B_{12} coenzyme. Reduction of the C_1 unit to the methyl level is achieved at the expense of tetrahydrofolate being oxidized to dihydrofolate. Thymidylate synthetases have been partially purified from calf thymus (95), *E. coli* (96), and *S. faecalis* R (97).

In order for thymidylate to be synthesized continuously, Reaction (6), catalyzed by thymidylate synthetase, must be followed by the regeneration of 5,10-methylene tetrahydrofolate. The latter is accomplished by reconverting dihydrofolate to tetrahydrofolate via dihydrofolate reductase (Reaction (3)), followed by addition of the C_1 unit from a metabolite such as serine. This cyclic process for the biosynthesis of thymidylate contains the target enzymes for two important chemotherapeutic agents: thymidylate synthetase is extremely sensitive to fluoropyrimidines (98,99), while dihydrofolate reductase is inhibited by folic acid antagonists. Thymidylate synthetase is also inhibited by tetrahydrohomofolate, an analog of tetrahydrofolate containing an additional CH_2 group between the pyrazine ring and *p*-aminobenzoylglutamate (100).

Wahba and Friedkin (101) and Blakley et al. (57a) have shown that, during Reaction (6), tritium from the pyrazine ring of tetrahydrofolate is transferred without appreciable dilution to the methyl group of thymidylate. To account for this observation, Friedkin (102) has suggested a mechanism (Fig. 9) involving an intermediate in which deoxyuridylate and tetrahydrofolate are connected by a methylene bridge. Cleavage of this hypothetical intermediate by an internal shift of a hydride ion from the reduced pyrazine ring would yield thymidylate and 7,8-dihydrofolate. In order to explore this interesting hypothesis, we have recently synthesized 5-thyminyl

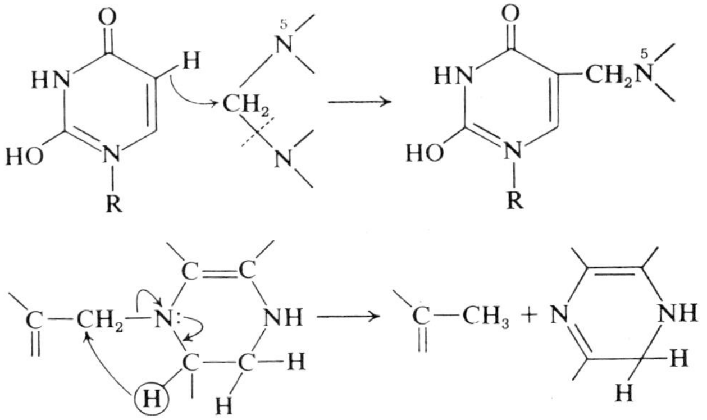

Fig. 9. Mechanism for reaction catalyzed by thymidylate synthetase (26,46).

tetrahydrofolate by the condensation of chloromethyluracil with tetra-hydrofolate (33). This model compound contains the key structural feature of the proposed intermediate in Figure 9, i.e., a methylene bridge between C-5 of the pyrimidine and N-5 of the reduced pyrazine ring, but it is not an exact model since it contains thymine rather than thymidylate. In the above synthesis, the preferential susceptibility of N-5 was not unexpected since treatment of tetrahydrofolate with biological alkylating agents, e.g., Chlorambucil,

$$HOOC(CH_2)_3 \underset{}{\bigcirc} N \overset{CH_2CH_2Cl}{\underset{CH_2CH_2Cl}{<}}$$

is known to yield predominantly the 5-substituted derivative (103).

5. Phenylalanine Hydroxylase

The hydroxylation of phenylalanine to yield tyrosine, according to Equation (7),

$$Phenylalanine + O_2 + TPNH + H^+ \longrightarrow Tyrosine + TPN^+ + H_2O \qquad (7)$$

is catalyzed by a system of enzymes that have been partially purified from mammalian liver (104). A similar system in adrenal tissue is respon-sible for the conversion of tyrosine to 3,4-dihydroxyphenylalanine (105). Kaufman has shown that reduced pteridines can replace a naturally occurring cofactor in these systems. 2-Amino-4-hydroxy-6-methyl tetra-hydropteridine is one of the more effective compounds among the group that includes reduced biopterin and tetrahydrofolate (104). The primary hydroxylation reaction is accompanied by the oxidation of the tetra-hydropteridine to the dihydro level. The latter (λ_{max} at 285 and 310 mμ) is an unstable compound [possibly one of the quinonoid structures (51) in Fig. 5] that rearranges rapidly, especially in the presence of phosphate, to the 7,8-dihydropteridine or is reduced to the tetrahydropteridine by a TPNH-dependent enzyme not identical to the customary 7,8-dihydrofolate reductase (44). Still unknown is the mechanism by which oxidation of the tetrahydropteridine brings about the conversion of oxygen to H_2O and the attacking species (probably OH^+) that becomes the hydroxyl group of tyrosine.

6. Oxidoreduction of C_1 Units Attached to Tetrahydrofolate

(a) *Interconversion of "Active Formate" and "Active Formaldehyde."* Folate coenzymes participate in still another type of oxidoreduction reaction, namely those involving the attached C_1 unit. The interconversion of "active formate" and "active formaldehyde," via Equation (8),

(5,10-Methylene tetrahydrofolate) + TPN$^+$ $\rightleftharpoons$
(5,10-methenyl tetrahydrofolate)$^+$ + TPNH (8)

was first encountered by Greenberg and Jaenicke (106,107) in their study of the conversion of the β-carbon of serine to C-2 of inosinic acid. Later, the enzyme responsible for the oxidative step, 5,10-methylene tetrahydrofolic dehydrogenase (originally called "hydroxymethyl tetrahydrofolic dehydrogenase" (108)), was partially purified from beef liver (108), chicken liver (109), baker's yeast (110), *E. coli* (111), and calf thymus (112). Each of these dehydrogenases is TPN-specific but a DPN-specific enzyme has been described in Ehrlich ascites tumor cells (113). The equilibrium constant for Reaction (8) is approximately 1. There is no net gain or loss of H$^+$ in this pyridine nucleotide-linked reaction since both methenyl tetrahydrofolate and TPN contain a pentavalent nitrogen.

Borohydride is an effective agent for the quantitative conversion of 5,10-methenyl tetrahydrofolate to the methylene derivative (32). A hydride ion mechanism (Step 1 in Fig. 10) is consistent with the fact that a pyridine

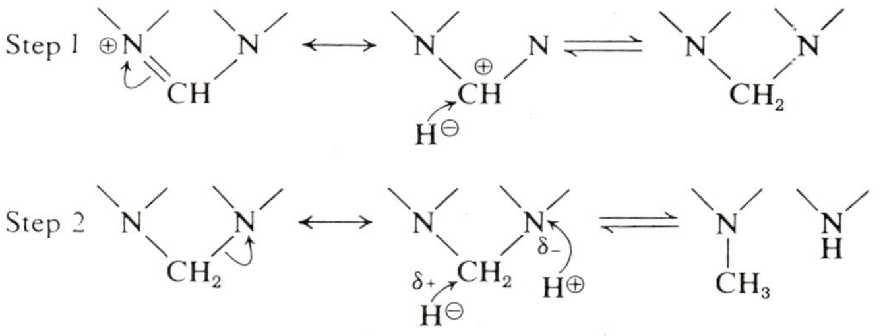

Fig. 10. Mechanism for the reduction of C_1 units attached to tetrahydrofolate (46,146).

nucleotide supplies the reducing power in the enzyme-catalyzed reaction. In principle, other nucleophiles might be expected to add to the bridge carbon of 5,10-methenyl tetrahydrofolate. It is difficult to test this possibility, however, since the alkaline pH needed to maintain most nucleophiles would cause methenyl tetrahydrofolate to open rapidly to the 10-formyl derivative (see Section II-C-2).

(b) *Interconversion of "Active Formaldehyde" and "Active Methyl."* The reduction of "active formaldehyde" to "active methyl" can be accomplished with borohydride, although at a slower rate than the reduction of methenyl tetrahydrofolate to the methylene derivative (114–116). A hydride ion mechanism can also be written for this reaction (Step 2 of Fig. 10).

Both enzymically and non-enzymically, the C—N linkage is cleaved exclusively on the N-10 side. Thus, enzymic synthesis of 5-methyl tetrahydrofolate (originally called "prefolic A" when it was first isolated from liver (117)) is catalyzed by a pyridine nucleotide-dependent reductase (Equation (9)) that is DPNH-specific in *E. coli* (118,119) and TPNH-specific in pig liver (116,119a).

$$5,10\text{-Methylene tetrahydrofolate} + \text{DPNH} + \text{H}^+ \longrightarrow$$
$$5\text{-methyl tetrahydrofolate} + \text{DPN}^+ \quad (9)$$

Although one might expect Reaction (9) to be reversible, the oxidation of 5-methyl tetrahydrofolate cannot be accomplished at the expense of DPN or TPN. The equilibrium constant may be unfavorable (119) or there may be a high activation barrier for formation of the five-membered ring. Donaldson and Keresztesy (117) showed, however, that the 5,10-methylene tetrahydrofolate reductase can catalyze the oxidation of 5-methyl tetrahydrofolate to tetrahydrofolate and HCHO by a different pathway not involving a pyridine nucleotide. Menadione and certain dyes serve as electron acceptors in this system. Katzen and Buchanan demonstrated that a flavoprotein with DPNH dehydrogenase activity can be detached from the methylene tetrahydrofolate dehydrogenase from *E. coli* (119). The electron transfer sequence was formulated as follows

(DPNH dehydrogenase is enclosed in the dashed line). The TPNH-dependent methylene tetrahydrofolate reductase from pig liver (119a) catalyzes Reaction (9), as well as the menadione-linked oxidation of 5-methyl tetrahydrofolate, TPNH, and, to a lesser extent, DPNH. All four activities are FAD-dependent and appear to be associated with the same protein.

The oxidation of 5-methyl tetrahydrofolate is somewhat similar to the oxidation of sarcosine to glycine and formaldehyde (Equation (11)). In the presence of serine hydroxymethylase and tetrahydrofolate, the products of Reaction (11) are converted to serine. Sarcosine dehydrogenase, a flavoprotein (120), is linked to the mitochondrial electron transport chain via the electron-transferring flavoprotein (121,121a).

The oxidation of methyl tetrahydrofolate or sarcosine might involve the N-oxide compound as an intermediate. This type of mechanism has been shown to occur in the oxidation of several N-methyl compounds catalyzed by iron chelates (122).

C. Involvement in Non-Oxidative Reactions

1. Formate-Activating Enzyme

Early tracer studies, reviewed in reference 7, revealed that formate was readily incorporated into the "active formate" pool (see Fig. 2) and thus into the various metabolites in equilibrium with this pool. Greenberg (4,107) first provided evidence for the ATP-dependent activation of formate to 10-formyl tetrahydrofolate by pigeon liver extracts:

$$HCOOH + \text{tetrahydrofolate} + ATP \xrightleftharpoons{Mg^{++}, K^{+}}$$
$$\text{10-formyl tetrahydrofolate} + ADP + P \quad (12)$$

The formate-activating enzyme (also called formyl tetrahydrofolate synthetase and tetrahydrofolic formylase) from *Clostridium cylindrosporum* was the first pteroprotein obtained in crystalline form (123,124), and it has been highly purified from *Micrococcus aerogenes* (125) and *Clostridium acidi-urici* (123). The enzyme has also been characterized from human erythrocytes (126) and *S. faecalis* (127). Although the equilibrium of Reaction (12) lies to the right ($\Delta F = -2$ kcal/mole (124)), the formation of ATP by the reversal of the reaction is probably important in the maintenance of the cellular economy of these purine-fermenting anaerobes (purine → formiminoglycine → 10-formyl tetrahydrofolate). A comparison of the properties of various formate-activating enzymes is made in Table V. Although the K_m values for ATP and tetrahydrofolate are quite similar for the bacterial enzymes, the *C. cylindrosporum* enzyme has a much more favorable K_m for formate and can be used for the quantitative estimation of this compound.

Reaction (12) is similar to a number of other reactions in which ATP is used as the energy source for bond formation between a carboxyl group and a base. Although the various "activating" enzymes have received

TABLE V

Michaelis Constants for Formate-Activating Enzymes

Source	K_m value, mM			Specific activity
	Formate	ATP	Tetrahydro-folate[a]	μmoles/min/mg protein
C. cylindrosporum (124)	6.7	0.29	0.26	420
M. aerogenes (125)	25.0	0.11	0.55	58
S. faecalis[b] (127) SF/A	6.3	0.097	0.27	—
SF/A$_k$	6.6	0.085	0.28	1.4
Human erythrocytes (126)	2.1	0.015	0.11	0.07

[a] Calculated for the *l*,L-diastereoisomer.

[b] SF/A and SF/A$_k$ are amethopterin-resistant mutant strains of *S. faecalis*.

much attention, the mechanisms by which they catalyze reactions similar to (12) have remained obscure. Two mechanisms have been suggested for formate activation via Reaction (12): (*a*) stepwise (128,129); and (*b*) concerted (130,130a). Using the highly purified enzyme from *M. aerogenes*, evidence has been obtained in support of a multistep mechanism:

$$\text{ATP} + \text{enzyme} \rightleftharpoons \text{enzyme-ATP} \qquad (13)$$

$$\text{Tetrahydrofolate} + \text{enzyme-ATP} \rightleftharpoons$$
$$\text{enzyme-phosphoryl tetrahydrofolate} + \text{ADP} \qquad (14)$$

$$\text{Formate} + \text{enzyme-phosphoryl tetrahydrofolate} \rightleftharpoons$$
$$\text{enzyme-formyl tetrahydrofolate} + \text{P} \qquad (15)$$

$$\text{Enzyme-formyl tetrahydrofolate} \rightleftharpoons \text{enzyme} + \text{formyl tetrahydrofolate} \qquad (16)$$

This mechanism assumes an obligatory order of the four reversible reactions: ATP must react first with the enzyme (Equation (13)), tetrahydrofolate must then react with the enzyme–ATP complex (Equation (14)), and formate can react only with enzyme-bound phosphoryl tetrahydrofolate (Equation (15)). The final step involves the dissociation of enzyme-bound 10-formyl tetrahydrofolate (Equation (16)). Thus, prior combination of ATP with the enzyme would be required before an adjacent site on the enzyme would be capable of binding tetrahydrofolate; in turn, the formation of phosphoryl tetrahydrofolate on the enzyme would be required before the third site could bind formate. The reverse reaction is likewise governed by the same restriction in terms of a compulsory order of events.

The above formulation rests upon the following experimental observations (129,131):

1. Formate-activating enzymes have been partially purified from a variety of microorganisms, vertebrate and invertebrate animals, and plants (132) and, in all cases, the enzyme catalyzes not only the overall reaction (12) as measured either by 10-formyl tetrahydrofolate or ADP formation, but also the initial reaction (14), as measured by ADP formation from ATP and tetrahydrofolate. In the latter instance, 10-formyl tetrahydrofolate is not formed, thus proving that none of the reactants is contaminated with HCOOH. It should be noted that the amount of enzyme required to demonstrate Reaction (14) is much larger than that required for Reaction (12), indicating that phosphoryl tetrahydrofolate may be a slowly dissociating, enzyme-bound intermediate.

2. During the enzyme-catalyzed interaction of ATP and tetrahydrofolate (in the absence of formate), the spectrum of the coenzyme (λ_{max} at 298 mμ) is replaced by that of a new species characterized by a decreased absorbancy at this wavelength (see Fig. 6, ref. 129). This partial reaction requires Mg^{++} and K^{+} and the extent of the spectral change depends upon the amount of enzyme used, again suggesting that the intermediate is enzyme-bound. Addition of formate causes the spectrum of the intermediate to be replaced instantaneously by that of 10-formyl tetrahydrofolate. Formation of the intermediate from tetrahydrofolate does not depress the amount of 10-formyl tetrahydrofolate produced when formate is eventually added, showing that the enzymic interaction between ATP and tetrahydrofolate is an obligatory part of the overall reaction and not a side reaction.

3. Reaction (12) is specific for all six components except P, which can be replaced to a large extent by arsenate.

4. Only ATP and 10-formyl tetrahydrofolate protect the enzyme against inactivation.

5. Reaction (12) in the forward direction is inhibited by high concentrations of enzyme, whereas the reverse reaction is favored under these conditions.

6. Each of the end products of Reaction (12) in the forward direction inhibits the reaction to a different extent, the order being 10-formyl tetrahydrofolate > ADP > P. As substrates for the reaction in reverse, these compounds may be arranged in the same order in terms of affinity for the enzyme.

7. Reaction (12) in the reverse direction is inhibited by each of the products in the order: ATP > tetrahydrofolate > formate. As substrates in the forward direction, these compounds may also be arranged in the same order in terms of affinity for the enzyme.

8. Maximal rates of exchange of P into ATP, ADP into ATP, and formate into 10-formyl tetrahydrofolate are achieved only when all components of Reaction (12) are present.

9. The enzyme has an inherent ATPase activity.

Reversible binding of ATP by the enzyme has also been postulated by Rabinowitz and Himes (130a) as a step in their mechanism for the formate-activating enzyme. According to the present formulation, only ATP and 10-formyl tetrahydrofolate of the six components in Equation (12) could be bound without prior binding of other components or intermediates, and, hence, these two compounds could be expected to protect the enzyme against inactivation.

Large amounts of enzyme, although favoring the equilibrium in the direction of enzyme–ATP complex formation, hinder the overall forward reaction at the final step by preventing the dissociation of the complex between 10-formyl tetrahydrofolate and the enzyme. Large amounts of enzyme, therefore, would favor the reaction in the reverse direction for the same reasons. This argument assumes that dissociation of enzyme-formyl tetrahydrofolate is greater than that of enzyme-ATP. In addition to Reactions (13)–(16) the enzyme may also catalyze Reactions (17) and (18).

$$\text{Enzyme-ATP} \rightleftharpoons \text{enzyme P} + \text{ADP} \qquad (17)$$

$$\text{Enzyme-phosphoryl tetrahydrofolate} \rightleftharpoons$$
$$\text{enzyme} + \text{phosphoryl tetrahydrofolate} \quad (18)$$

The observed ATPase activity can be attributed to Reaction (17), followed by breakdown of the E–P complex. Evidence for Reaction (18) is provided by the fact that the amount of intermediate produced from ATP and tetrahydrofolate in the absence of formate greatly exceeds the amount of enzyme. From the extremely rapid rate of conversion of the free intermediate to 10-formyl tetrahydrofolate upon the addition of formate, it may be concluded that Reaction (18) is reversible and that Reactions (15), (16), and (18) are among the most rapid in the entire sequence. Conversely, the inability to produce the phosphorylated intermediate by the phosphorolysis of 10-formyl tetrahydrofolate may be attributed to a low affinity of the enzyme–formyl tetrahydrofolate complex for P or to an unfavorable equilibrium for Reaction (15) in reverse.

Although the present mechanism accounts for some of the experimental observations (summarized above) and suggests a role for the enzyme as a reactant, it still does not explain all the exchange data (129,130a). For example, the exchange of formate into 10-formyl tetrahydrofolate and of ADP into ATP should proceed in the absence of ADP and formate, respectively. In both instances some exchange is actually observed under these conditions (cf. Table IV and Fig. 8 of ref. 130a), but *maximal* exchange is obtained only when all components of the overall reaction are present. The formate–10-formyl tetrahydrofolate exchange proceeds via Reactions (16) and (15) and the reasoning used previously in discussing the phosphorolysis of 10-formyl tetrahydrofolate may be applied, i.e., the coupled reactions may not proceed maximally unless the unfavorable equilibrium of Reaction (15) in reverse is shifted by removal of enzyme–phosphoryl tetrahydrofolate via Reactions (13) and (14). Similar arguments might be invoked for the ADP–ATP exchange. On the other hand, the exchange of P into ATP should require all components according to the mechanism shown in Equations (13)–(16); this has been observed by Rabinowitz and Himes (130,130a) and in our investigation (129).

If Reaction (12) is carried out in a stepwise manner as outlined in Reactions (13) through (16), a reasonable mechanism would be that shown in Figure 11A. In step 1, the nucleophilic attack by N-10 of tetrahydrofolate on the terminal P—O bond of ATP is consistent with mechanisms postulated for many other ATP-dependent reactions. The mechanism for the second step assumes that the enzyme facilitates a shift of electrons leading to the expulsion of the phosphoryl group by the incoming polarized formyl group. This mechanism, moreover, would predict the migration of O^{18} from the carboxyl group to the phosphate group during the overall reaction; such an effect has been observed in Reaction (12) catalyzed by

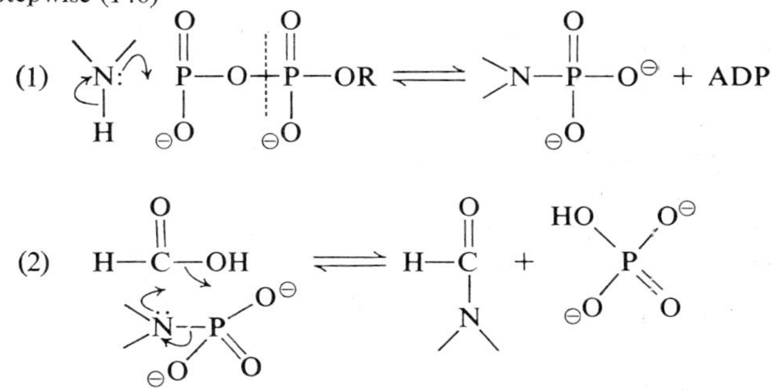

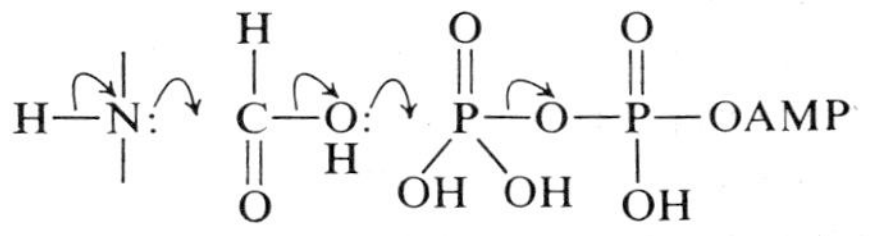

Fig. 11. Mechanisms for reaction catalyzed by formate-activating enzyme.

the *C. cylindrosporum* enzyme (130) and in the analogous reaction for succinate activation (133).

$$\text{Succinate} + \text{CoA} + \text{ATP} \rightleftharpoons \text{succinyl CoA} + \text{ADP} + \text{P} \tag{19}$$

The occurrence of enzyme-bound phosphohistidine (134) as an intermediate in the latter system lends support to the postulated phosphoryl tetrahydrofolate in the mechanism of Figure 11.

Rabinowitz and his colleagues (130,130a,135) have made a comprehensive study of the mechanism of Reaction (12) catalyzed by the formate-activating enzyme from *C. cylindrosporum*. In general, their experimental findings are quite similar to those summarized above for the *M. aerogenes* enzyme except for differences in the numerical constants. They find no evidence for intermediates, however, and believe that the formation of ADP and the "spectroscopic intermediate" from ATP and tetrahydrofolate is caused by traces of formate in the reactants. On the basis of the exchange experiments and the lack of intermediates, it has been suggested (130,130a,135) that Reaction (12) proceeds by a concerted, rather than a stepwise, reaction. It should be noted, however, that even an apparently concerted mechanism might consist of two reactions which could not be detected because the rate of disappearance of an intermediate might exceed its rate of formation. In attempting to apply a concerted type of mechanism to the formate-activating enzyme, two sequences of events may be suggested. The first proposal involves the formation of formyl phosphate in the transition state. The formation of formyl phosphate via a formokinase reaction has been demonstrated in *E. coli* (136), but this compound would appear to be eliminated as an intermediate in the *M. aerogenes* and *C. tetanomorphum* systems because of the failure of hydroxylamine at high concentrations to inhibit the overall reaction (125). Following Buchanan's concept (137), a second type of concerted mechanism may be envisioned (Fig. 11B) involving a series of nucleophilic displacements in which ATP acts as an acceptor for OH^- from formate.

2. Formyl- and Formimino-Transferring Enzymes

Reactions involving the transfer of formyl or formimino groups from (or to) tetrahydrofolate can be subdivided into two categories depending on the nature of the acceptor or donor.

(a) *Transfer of the C_1 Group Intramolecularly (from N-5 to N-10 or vice versa) on Tetrahydrofolate.* Although folinic acid (5-formyl tetrahydrofolate) was originally believed to be "active formate" in purine biosynthesis, subsequent work (described below) showed that the methenyl and 10-formyl derivatives are the true formyl donors. The ATP-dependent conversion of 5-formyl to 10-formyl tetrahydrofolate,

$$5\text{-Formyl tetrahydrofolate} + \text{ATP} \xrightarrow{\text{Mg}^{++}}$$
$$10\text{-formyl tetrahydrofolate} + \text{ADP} + \text{P} \quad (20)$$

a reaction discovered by G. R. Greenberg (4,138), accounts for the entrance of folinic acid into the principal area of formylation reactions.

Using a partially purified 5-formyl tetrahydrofolate cyclodehydrase (folinic isomerase) from *M. aerogenes* that has a pH optimum at 4, Kay et al. (139) obtained evidence for the occurrence of 5,10-methenyl tetrahydrofolate as an intermediate in Reaction (20); the same intermediate has been reported (140) for the sheep liver isomerase. On the other hand, no intermediate could be detected (139) with a partially purified chicken liver isomerase (pH optimum of 7) and the rate of conversion of the added methenyl compound to 10-formyl tetrahydrofolate was much slower than the rate of the overall isomerization reaction.

Assuming, however, that the methenyl compound is the intermediate, Reaction (20) can be separated into two steps:

$$5\text{-Formyl tetrahydrofolate} + \text{ATP} \longrightarrow$$
$$[5,10\text{-methenyl tetrahydrofolate}]^+ + \text{ADP} + \text{P} \quad (21)$$

$$[5,10\text{-Methenyl tetrahydrofolate}]^+ + \text{H}_2\text{O} \rightleftharpoons$$
$$10\text{-formyl tetrahydrofolate} + \text{H}^+ \quad (22)$$

As discussed below, the latter reaction proceeds at an appreciable rate non-enzymically and is also catalyzed by a specific enzyme, 5,10-methenyl tetrahydrofolate cyclohydrolase. The ATP-dependent cyclization reaction

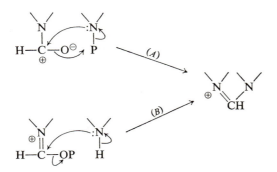

Fig. 12. Mechanisms for reaction catalyzed by folinic isomerase.

bears a superficial resemblance to Reaction (12). By analogy to the mechanism in Figure 11A, Reaction (21) might involve phosphorylation at N-10 and subsequent ejection of the phosphate by the 5-formyl group (Fig. 12A). An alternate mechanism (Fig. 12B) for ring closure could be envisioned in which ATP would react with the 5-formyl group, followed by an attack on the intermediate by the N-10 atom.

The conversion of 5-formimino tetrahydrofolate to the methenyl derivative (Equation (23)) is brought about by the enzyme, 5-formimino tetrahydrofolate cyclodeaminase (141–143).

5-Formimino tetrahydrofolate + 2H$^+$ $\longrightarrow$

[5,10-methenyl tetrahydrofolate]$^+$ + NH$_4$$^+$ (23)

In this instance, ATP is not required and ring closure is facilitated by the expulsion of ammonia (cf. mechanism in Fig. 12B).

Calculations by Kay et al. (139) have shown that the negative free energies of hydrolysis (to yield formate and tetrahydrofolate) of the formyl derivatives of tetrahydrofolate are in the relative order: 5,10-methenyl > 10-formyl ≫ 5-formyl. The ΔF value for 5-formyl tetrahydrofolate has been estimated to be -2 kcal/mole as compared to -6 kcal/mole for the 10-formyl derivative. The latter value is consistent with the observed reversibility of Reaction (12). The necessity for ATP to drive Reaction (20) to completion is also apparent from these data. Alternatively, Reaction (22) has an equilibrium constant that is not too far from unity and, as seen from the equation, the equilibrium position can be altered by varying the pH (20,142,143). Non-enzymic catalysis of Reaction (22) by a variety of anions (phosphate > Tris > maleate) has been studied by

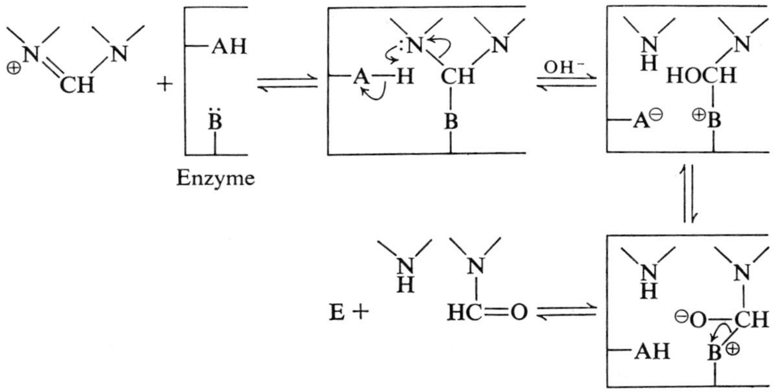

Fig. 13. Mechanism for reaction catalyzed by cyclohydrolase (146).

Hartman and Buchanan (144) and by Tabor and Wyngarden (143). Cyclohydrolase, which catalyzes Reaction (22) has been partially purified from several sources (141–145). Mechanisms for enzymic catalysis (146) of Reaction (22) are proposed in Figure 13.

Hydrolytic scission of 5,10-methenyl tetrahydrofolate occurs only on the N-5 side of the bridge carbon, whether the reaction is enzyme-catalyzed or base-catalyzed. The methenyl group causes N-5 or N-10 to assume a positive charge, and cleavage will occur, therefore, on the side in which the bond is more polarized as $N^{\delta-}$—$C^{\delta+}$. Calculations by Perault and Pullman (25), studies with model compounds (32), and direct measurements (15) have shown that N-5 is more basic than N-10, i.e., the former position is the more electron attracting, and bond polarization occurs preferentially on that side of the bridge carbon.

(*b*) *Transfer of the* C_1 *Group from N-5 or N-10 on Tetrahydrofolate to Water or to a Nitrogen Atom on the Acceptor.* Formyl groups can be delivered from tetrahydrofolate to appropriate acceptors, including water. Only one enzyme has been encountered for the latter activity, a beef liver deacylase (147) that is specific for 10-formyl tetrahydrofolate (but not the 5-formyl derivative):

$$\text{10-Formyl tetrahydrofolate} + H_2O \longrightarrow \text{formate} + \text{tetrahydrofolate} \qquad (24)$$

The enzyme requires the presence of catalytic amounts of TPN or TPNH, perhaps as an allosteric effector; DPN or DPNH are inactive. Similar deacylases for acetyl and succinyl CoA are known. The 10-formyl tetrahydrofolate deacylase may provide a means of regenerating free tetrahydrofolate in the absence of an acceptor system for the active formyl group.

The elucidation of the stepwise mechanism of purine biosynthesis by the laboratories of Buchanan (148,149) and G. R. Greenberg (4,107) provided some of the earliest evidence on the mechanism of transformylation reactions with nitrogenous bases as acceptors. Thus, 4-amino-5-imidazole carboxamide ribotide (AICAR) is converted to inosinic acid (IMP) via the intermediate, 4-formamido-5-imidazole carboxamide

$$(25)$$

$$(26)$$

ribotide (FAICAR) (Equations (25) and (26)). The enzymes catalyzing these reactions, AICAR transformylase and inosinicase, have been purified extensively from chicken liver (149).

The incorporation of formate into the C-8 position of the purine ring involves a similar mechanism. In this case, "active formate" has been identified as 5,10-methenyl tetrahydrofolate (144) and the acceptor is glycinamide ribotide (GAR) (Reaction (27)). In contrast to the readily reversible Reaction (25), Reaction (27), carried out by GAR transformylase, is apparently irreversible.

$$(27)$$

Although tracer experiments (reviewed in ref. 7) established that the C-2 position of histidine, like C-2 and C-8 of the purine ring, is derived from the one-carbon pool, ring closure in histidine biosynthesis does not follow the purine pattern, as described above. Instead, the N-1 and C-2 atoms of histidine originate from the corresponding segment of adenylic acid, N-3 from glutamine, and the remaining two carbons from C-1 and C-2 of ribose. Relevant portions of the pathway (150), established with enzymes from *Salmonella typhimurium*, are shown in Equation (28). ATP and 5-phosphoribosyl-1-pyrophosphate condense to form phosphoribosyl ATP (151). Several reactions (ring opening, loss of pyrophosphate, and conversion of the ribose chain to ribulose) transform the latter into a phosphoribulosylformimino derivative of AICAR which, in turn, reacts

Phosphoribosyl ATP   (28)

Imidazole glycerol
phosphate

with glutamine to yield (via an intermediate of unknown structure) imidazole glycerol phosphate and AICAR. The conversion of imidazole glycerol phosphate to histidine follows a well-established pathway (152) and AICAR is used to regenerate adenylic acid via inosinic acid (Reactions (25) and (26)).

Recently, a new formylation reaction has been discovered (153) that appears to be of considerable importance in protein synthesis in *E. coli*. 10-Formyl tetrahydrofolate (or possibly the 5,10-methenyl derivative) reacts with methionyl-sRNA to yield the *N*-formyl derivative of the latter (154,155):

10-Formyl tetrahydrofolate + methionyl-sRNA $\longrightarrow$
   tetrahydrofolate + *N*-formyl methionyl-sRNA   (29)

Blocking of the amino group thereby makes methionine ideally suited for the initiation of polypeptide chains; this is in agreement with the observation that methionine is the *N*-terminal amino acid in a number of bacterial proteins (156).

In addition to Equation (20), only one other reaction is known in which 5-formyl tetrahydrofolate participates directly. Silverman et al. (157)

have reported the occurrence of an enzyme catalyzing the reversible formylation of L-glutamate:

5-Formyl tetrahydrofolate + L-glutamate $\rightleftharpoons$

N-formyl-L-glutamate + tetrahydrofolate   (30)

The equilibrium of Reaction (30) strongly favors the formation of 5-formyl tetrahydrofolate. It is of interest that the purified enzyme also catalyzes the formation of 5-formimino tetrahydrofolate from FIGLU and tetrahydrofolate (see below).

The fermentation of purines by *C. acidi-urici* or *C. cylindrosporum* yields formiminoglycine (FIG), according to the sequence in Reaction (31) (158,159). The further degradation of FIG to glycine, NH_3, and formate

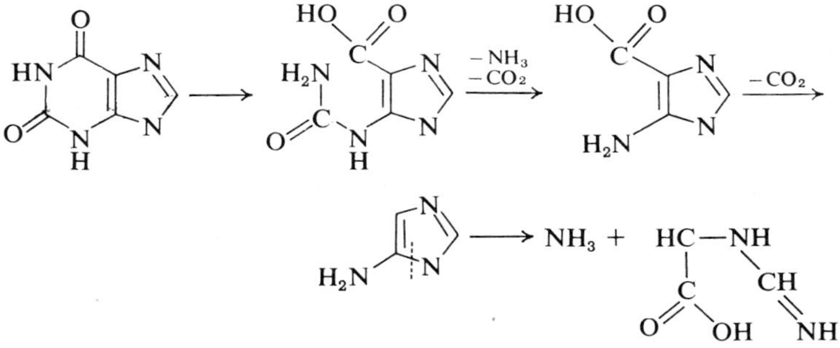

(31)

(160,161) proceeds via transfer of the formimino group to tetrahydrofolate (catalyzed by FIG formimino transferase) (Equation (32)), followed by Reactions (23) and (22). Reaction (32) is readily reversible and the equilibrium constant is near unity.

Formiminoglycine + tetrahydrofolate $\rightleftharpoons$

5-formimino tetrahydrofolate + glycine   (32)

The degradation of histidine, which is understood better than its biosynthesis, follows a pathway similar to that for purines. Through a

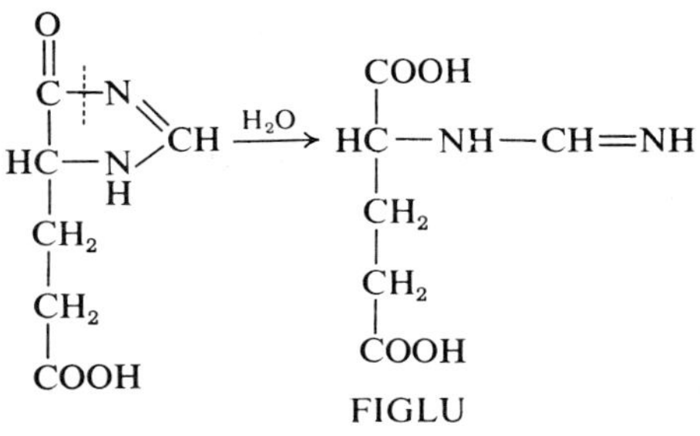

(33)

FIGLU

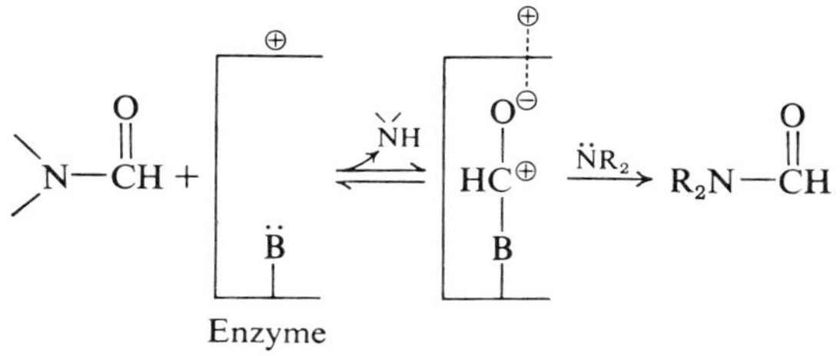

Fig. 14. Mechanism for transamidation involving formyl derivatives of tetrahydrofolate.

complex series of reactions involving urocanic acid as an intermediate, histidine is metabolized to imidazolonepropionic acid and the latter compound is then hydrolyzed to yield formimino-L-glutamic acid (FIGLU) (162,163) (Reaction (33)). FIGLU is converted to 10-formyl tetrahydrofolate, glutamate, and NH_3 by a sequence of reactions (164,165) similar to those encountered in the breakdown of FIG, i.e., Equation (34) followed

FIGLU + tetrahydrofolate $\longrightarrow$ 5-formimino tetrahydrofolate + glutamate (34)

by (23) and (22). While the decomposition of FIGLU follows a pathway parallel to that of FIG breakdown, the pig liver enzyme responsible for Reaction (34), FIGLU formimino transferase, cannot use FIG as a substrate, nor will the bacterial FIG transferase utilize FIGLU. The subsequent breakdown of 5-formimino tetrahydrofolate, however, is carried out with equal facility by the bacterial and mammalian systems.

The 5- and 10-formyl derivatives of tetrahydrofolate can be considered as amides and the transfer of the formyl group to various nitrogen acceptors (or to the base, water) is an example of transamidation. A mechanism (146) for this process, given in Figure 14, is similar to that advanced for the enzymic hydrolysis of peptides (166.) Reversal of the process, i.e., transfer of a formyl or formimino group to tetrahydrofolate, probably proceeds by a similar mechanism.

3. Interaction of Aldehydes with Tetrahydrofolate

The reversible synthesis of "active formaldehyde" from tetrahydrofolate and formaldehyde

HCHO + tetrahydrofolate $\rightleftharpoons$ 5,10-methylene tetrahydrofolate + H_2O (35)

can be carried out at a reasonable rate non-enzymically (32,34,43,167–174) and evidence has also been presented for the occurrence in avian liver of an enzyme that catalyzes the reaction (175). The structure of "active formaldehyde" was shown to be 5,10-methylene tetrahydrofolate on the basis of the following lines of evidence (32,171):

1. When the methylene tetrahydrofolic dehydrogenase (Reaction (8)) is freed from cyclohydrolase, 5,10-methenyl tetrahydrofolate (but not 5- or 10-formyl tetrahydrofolate) can be reduced to "active formaldehyde" and, conversely, only 5,10-methenyl tetrahydrofolate is produced by the enzymic oxidation of "active formaldehyde" (109). Similarly, only 5,10-methenyl tetrahydrofolate can be reduced chemically by borohydride to "active formaldehyde" (32).

2. A bell-shaped pH optimum curve for the chemical synthesis of "active formaldehyde" from HCHO and tetrahydrofolate is obtained having pK_a values of 3.0 and 5.4 (32). This curve was originally interpreted (146) as corresponding to the pK_a values of the N-10 and N-5 atoms, respectively, of tetrahydrofolate. The latter assignment is probably correct but recent studies by Kallen and Jencks have shown that pK_a for N-10 is much lower (15).

3. The equilibrium constant for Reaction (35) is about 10^4 at pH 7 (32,171). This is much greater than would be anticipated for a single linkage between HCHO and one nitrogen atom. Adduct formation of the type $CH_2(OH)_2 + R_2NH \rightleftharpoons H_2O + HOCH_2—NR_2$ usually has a K value of only $10^1–10^2$.

Kallen and Jencks (34) have recently reinvestigated Reaction (35) and have proposed a mechanism (Fig. 15) that is based on the assumption that the bell-shaped pH–rate profile is an expression of the kinetics of the reaction rather than the ionization constants of the reactants. The rate-determining step changes with pH: In alkaline solution the rate-determining step is the acid-catalyzed dehydration of a transient 5-hydroxy-methyl tetrahydrofolate and its conjugate base; in acid solution it is the attack on HCHO by N-5 of tetrahydrofolate. Secondary amines, such as

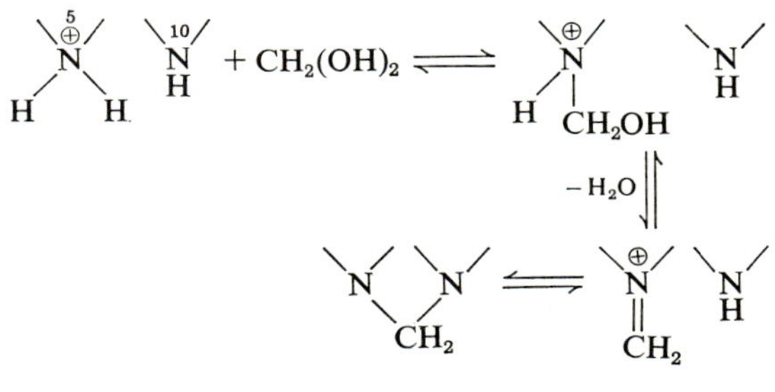

Fig. 15. Mechanism for non-enzymic interaction of formaldehyde with tetrahydrofolate.

morpholine and imidazole, catalyze Reaction (35) by a pathway that appears to involve the intermediate formation of cationic imines, $>N^{\oplus}=CH_2$.

5,10-Methylene tetrahydrofolate has an interesting pH stability toward aldehyde-trapping agents such as hydroxylamine (32). In an acidic medium, methylene tetrahydrofolate is readily decomposed as the equilibrium is shifted through the removal of free HCHO. The converse is true in basic solutions, in which there is little tendency for polarization of either C—N linkage.

The transfer of the C_1 group from "active formaldehyde" to an acceptor (see next section) very probably involves the opening of the bridge structure to form 5-hydroxymethyl tetrahydrofolate as a transient intermediate:

$$\begin{array}{ccc} \diagdown & & \diagdown \\ N-CH_2OH & \text{or} & N^{\oplus}=CH_2 \\ \diagup & & \diagup \end{array}$$

In either instance, it would seem likely that the polarized $C^{\delta+}$—$N^{\delta-}$ bond would be attacked by the carbanion form of the acceptor molecule.

An interesting corollary to the conversion of "active formaldehyde" to "active formate" is provided by a novel set of reactions in which glyoxylate replaces the C_1 unit (176,177). As shown in Equation (36),

$$\begin{array}{c} \underset{|}{CHO} + \underset{H\ \ H}{N---N} \rightleftharpoons \underset{\diagdown \diagup}{N-----N} \\ COOH \qquad\qquad\quad \underset{|}{CH} \\ \qquad\qquad\qquad\quad \underset{\big\downarrow [O]}{COOH} \end{array} \qquad (36)$$

$$\begin{array}{ccccccc} \underset{H\ \ |}{N---N} & \xleftarrow{-CO_2} & \underset{\diagdown\diagup}{^{\oplus}N-----N} & \xrightarrow{H_2O} & \underset{H\ \ |}{N---N} \\ \quad HC{=}O & & \underset{|}{C} & & \quad C{=}O \\ & & COOH & & \quad COOH \end{array}$$

glyoxylate adds to tetrahydrofolate to yield 5,10-carboxymethylene tetra-hydrofolate, an analog of "active formaldehyde." Unlike "active formaldehyde," in which the C_1 unit is stable to air oxidation, the glyoxyl-ate analog undergoes spontaneous oxidation to 5,10-carboxymethenyl tetrahydrofolate. The presence of the electron-donating carboxyl group on the bridge carbon apparently enhances the ability of the C—N linkage to be dehydrogenated. Other electron-rich groups should have the same effect, while inert alkyl groups or electron-withdrawing groups should stabilize the aldehyde adduct against air oxidation. Depending upon pH, 5,10-carboxymethenyl tetrahydrofolate can open selectively to yield 10-oxalyl tetrahydrofolate or it can decarboxylate to yield 5,10-methenyl tetra-hydrofolate. It is possible that the non-enzymic reactions shown in Equation (36) may be involved in the catabolism of glyoxylate, since a glyoxylate-activating enzyme, which catalyzes the first step in the sequence, has been partially purified from chicken liver (178).

4. Serine and Deoxycytidylate Hydroxymethylases

The enzymic interconversion of serine and glycine (Equation (37)) has been the subject of numerous investigations (reviewed in ref. 7) and, in fact,

$$\underset{\substack{| \\ OH \quad NH_2}}{CH_2-CH-COOH} + \text{tetrahydrofolate} \rightleftharpoons$$

$$\underset{\substack{| \\ NH_2}}{CH_2COOH} + \text{5,10-methylene tetrahydrofolate} \quad (37)$$

was the system in which tetrahydrofolate was first discovered as the coenzyme for C_1 metabolism. Alexander and Greenberg (179) and Blakley (180) obtained values of $K = 2.8 \times 10^3\ M^{-1}$ and $0.17 \times 10^3\ M^{-1}$, respectively, for the equilibrium constant of the overall reaction:

$$\text{Glycine} + \text{HCHO} \rightleftharpoons \text{serine} \quad (38)$$

the equilibrium constant for Reaction (35) is known (7) ($K \cong 10^4\ M^{-1}$), and taking an average value of $K = 10^3$ for Reaction (38), the constant for Reaction (37) can be calculated to be $K \cong 10$.

Serine hydroxymethylase (also called serine aldolase), a pyridoxal phosphate-dependent enzyme catalyzing Reaction (37), was partially purified from mammalian and avian tissues (179–182), bacteria (183), and plants (184). More recently, Schirch and Mason (185) have obtained the rabbit liver enzyme in nearly homogeneous form. The involvement of pyridoxal phosphate in the serine hydroxymethylase reaction was antici-pated by the classical studies of Snell and co-workers (186) on the mechan-ism of the non-enzymic cleavage of serine at elevated temperatures to glycine and HCHO in the presence of pyridoxal and alum. Although a metal ion is required in the non-enzymic model system, purified serine hydroxymethylases do not contain any metal ion and there is no stimulation when various metal ions are added.

Blakley (187) and Alexander and Greenberg (179) have reported K_m values of $2 \times 10^{-6}\ M$ and $3 \times 10^{-6} M$, respectively, for pyridoxal phosphate with rabbit liver and sheep liver serine hydroxymethylases. The K_m value for tetrahydrofolate (calculated for the l,L-diastereoisomer) has been given as $7 \times 10^{-5} M$ (179) and $2 \times 10^{-4} M$ (180) for serine

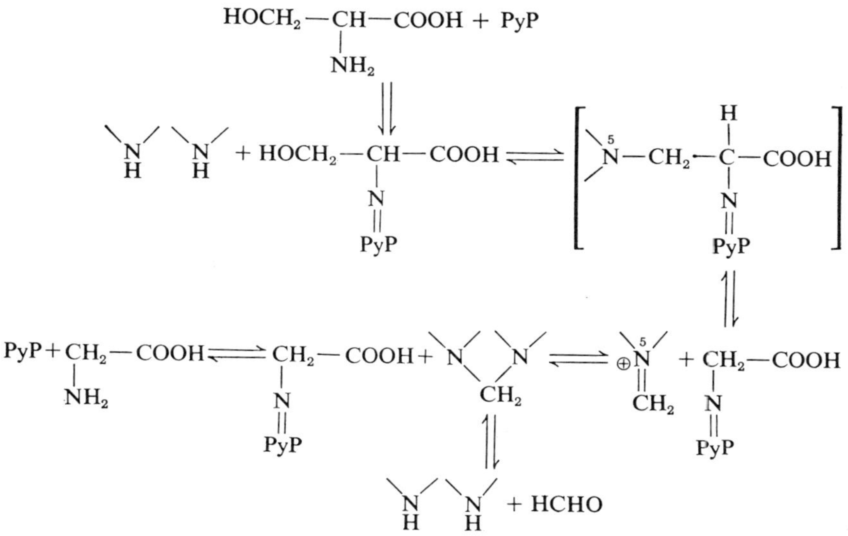

$$HOCH_2-CH-COOH + PyP$$
$$\quad\quad\quad\quad | $$
$$\quad\quad\quad\quad NH_2$$

Fig. 16. Mechanism for reaction catalyzed by serine hydroxymethylase. PyP = pyridoxal phosphate.

biosynthesis (Reaction (37) in reverse), while a value of $3.8 \times 10^{-4}M$ (181) has been obtained when the reaction was studied in the direction of serine breakdown.

Figure 16 outlines a mechanism for Reaction (37), catalyzed by serine hydroxymethylase, that embodies recent data reported by Schirch and Jenkins (188) and by Snell and his colleagues on the same reaction except that α-methyl serine is used as the substrate (189,190). Serine forms a Schiff base with pyridoxal phosphate and the resulting resonating system labilizes the bond between the α- and β-carbon atoms. Beta-elimination follows, and subsequent hydrolysis of the Schiff base releases the product, glycine. The mechanism assumes, moreover, that tetrahydrofolate reacts with the serine–pyridoxal phosphate complex prior to expulsion of the β-carbon.

Deoxycytidylate hydroxymethylase (191,192), an enzyme found in *Escherichia coli* after infection with the T-even bacteriophages, catalyzes the formation of 5-hydroxymethyl deoxycytidylate:

Deoxycytidylate + 5,10-methylene tetrahydrofolate $\rightleftharpoons$
$\quad\quad\quad\quad$ 5-hydroxymethyl deoxycytidylate + tetrahydrofolate (39)

Unlike the serine hydroxymethylase-catalyzed reaction, pyridoxal phosphate is not necessary in this reaction presumably because the hydroxymethyl acceptor, deoxycytidylate, already has considerable tendency to lose a proton at the C-5 position (cf. also the mechanism in Fig. 9).

5. Glycine Dehydrogenase

The oxidative decarboxylation of glycine according to Equation (40)

Glycine + tetrahydrofolate + DPN$^+$ $\longrightarrow$

$$5,10\text{-methylene tetrahydrofolate} + CO_2 + NH_4{}^+ + DPNH \quad (40)$$

was described by Sagers and Gunsalus (193) in cell-free extracts of *Peptococcus glycinophilus* and by Richert et al. (194) in avian liver preparations. In both systems there is a partial requirement for added pyridoxal phosphate, and glyoxylate cannot replace glycine as the substrate. Sagers and Klein (195) and Baginsky and Huennekens (196) fractionated the *P. glycinophilus* system into four protein fractions (P_1, P_2, P_3, and P_4), all of which were required to catalyze the overall reaction. Three of the fractions were shown to have special characteristics: P_1 contains bound pyridoxal phosphate; P_2 is a heat- and acid-stable protein; and P_3 is an FAD-containing flavoprotein (probably lipoyl dehydrogenase (196a)). The combination of P_1 and P_2 catalyzes the exchange of $^{14}CO_2$ with the carboxyl group of glycine (197). Baginsky and Huennekens (196) showed that the heat-stable protein and the flavoprotein are involved in the transfer of reducing power from glycine to DPN. These proteins, moreover, bear a close resemblance to thioredoxin and thioredoxin reductase (see Section III-E). According to a tentative mechanism suggested for this system (Fig. 17), oxidation and decarboxylation of glycine would occur before the rem-

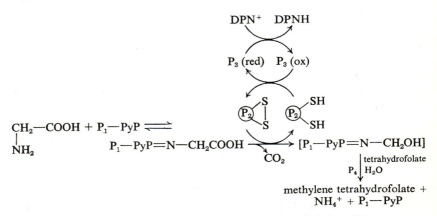

Fig. 17. Pathway for oxidative decarboxylation of glycine (196).

nant of the substrate interacts with tetrahydrofolate. Thus, DPNH formation should be observed in the absence of tetrahydrofolate and P_4, provided that a sufficiently large amount of P_1 is present. The mechanism also accounts for the observation by Klein and Sagers (197) that only P_1 and P_2 are required for the exchange of $^{14}CO_2$ with the carboxyl group of glycine. The proposal by Richert et al. (194) that glycine forms a Schiff base with pyridoxal phosphate during Reaction (40) is borne out by the isolation and detailed characterization of P_1 as a pyridoxal-containing protein [Klein and Sagers (198)]. Oxidation of such a pyridoxal glycine complex might involve the direct reduction of P_2

$$\text{\textcircled{P}}_2\overset{S}{\underset{S}{\big|}} \longrightarrow \text{\textcircled{P}}_2\overset{SH}{\underset{SH}{\big\langle}}$$

or it could proceed via expulsion of a proton from the α-carbon of glycine followed by attack of the carbanion upon the S—S bridge of P_2 to yield

$$\text{\textcircled{P}}_2\overset{\displaystyle SH}{\underset{\displaystyle S-CHCOOH}{\big\langle}}$$
$$N = PyP - P_1$$

Decarboxylation of the latter complex, followed by hydrolysis, would yield

$$CO_2,\ \text{\textcircled{P}}_2\overset{SH}{\underset{SH}{\big\langle}}\ ,\ \text{and}\ P_1 - PyP = N - CH_2OH$$

A similar complex involving a linkage between a protein-bound thiol group and the cobalt of vitamin B_{12a} has been postulated as a step in the reduction of B_{12a} to B_{12s} (see Section III-B).

6. Methionine Synthetase

At the lowest oxidation state of the C_1 unit, only the 5-methyl derivative of tetrahydrofolate occurs. This compound is the direct precursor of the methyl group of methionine via Reaction (41) catalyzed by methionine

5-Methyl tetrahydrofolate + homocysteine $\longrightarrow$ methionine + tetrahydrofolate

$$(41)$$

synthetase. As stated in Section II-6-B, 5-methyl tetrahydrofolate (originally called "prefolic A") was first isolated from natural sources by Keresztesy and Donaldson (117), and later synthesized chemically and enzymically in the laboratories of Buchanan (115), Sakami (114), and Keresztesy (116). 5-Methyl tetrahydrofolate can also be synthesized by a transmethylase present in a soil microorganism (199):

$$(CH_3)_3S^+ + \text{tetrahydrofolate} \longrightarrow (CH_3)_2S + \text{5-methyl tetrahydrofolate} \qquad (42)$$

Although Reaction (41) appears to be a straightforward transfer of the methyl group from methyl tetrahydrofolate to homocysteine, some indication of its complexity is shown by the fact that a coenzyme form of B_{12} is required (see Section III-F for a discussion of the mechanism).

III. B_{12} Coenzymes

A. Structure, Nomenclature, and Chemistry

Interest in vitamin B_{12} was greatly stimulated by the observation that it was effective in the treatment of pernicious anemia. The isolation of the vitamin as the cyanide complex in 1948 (200,201) was followed by intensive studies on its physical and chemical properties (reviewed in 202–204). Degradation procedures (205) and x-ray analysis (206), elucidated the structure shown in Figure 18. The "active center" of the molecule is the cobalt atom which is strongly chelated to four pyrrole nitrogens. These pyrroles, in turn, comprise a nearly planar corrin ring which is similar to the biologically important porphyrins except that one of the methine bridges is missing. The cobalt atom also forms two other linkages with groups at right angles to the plane of the corrin ring: The cyanide group is usually considered to be "above" the plane while the parent base (most commonly 5,6-dimethylbenzimidazole)* is "below" the plane. The structure of B_{12} may be written symbolically in order to emphasize its salient features:

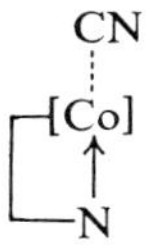

(a) the cobalt atom is trivalent and enclosed within a corrin ring ([ ]); (b) the cyanide group is attached ionically (---); (c) a nitrogen atom of

---

*Many analogs of vitamin B_{12} (e.g., pseudovitamin B_{12} which contains adenine as the parent base) have been prepared by supplying B_{12}-synthesizing microorganisms with the appropriate bases (207,208). Recently, Toohey (209) has obtained a cobalt-free B_{12} compound by a biosynthetic route.

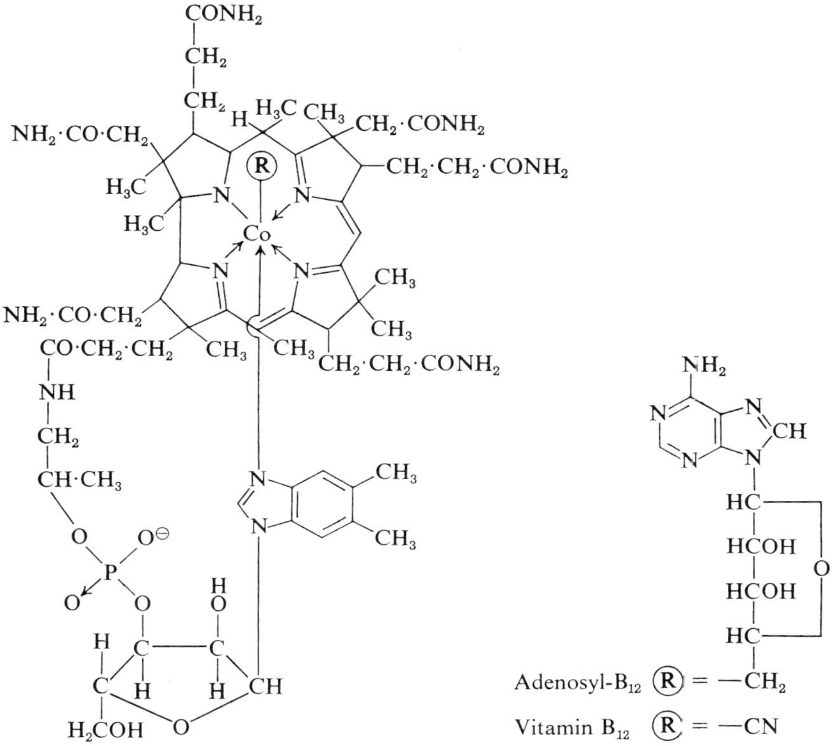

Fig. 18. Structures of vitamin B_{12} and adenosyl-B_{12}.

the parent base (N) is linked to the cobalt by a coordinate bond ($\rightarrow$); and (d) the parent base is joined by an extended structure ($\llcorner$ in the above diagram) to one of the pyrroles. In coenzyme forms of B_{12}, the cyanide group is replaced by an alkyl residue (adenosyl, methyl, etc.) linked covalently to the cobalt; otherwise the structure remains the same.

The first coenzyme form of B_{12} (adenosyl-B_{12}) was isolated from *Clostridium tetanomorphum* and *Propionibacterium shermanii* by Barker's group in 1958 (210–212); subsequently it was shown that more than 80% of the B_{12} in mammalian and avian liver also occurs in the coenzyme form (213). Another B_{12} coenzyme, methyl-B_{12}, has also been isolated from a patented microorganism (214) and from liver (215). Chemical studies (211,212,216–218) revealed that the Barker coenzyme differed from the vitamin only by replacement of the cyanide group by an adenine nucleoside. As in vitamin B_{12}, the oxidation state of cobalt in adenosyl-B_{12} is $+3$ (218a). The precise structure of the coenzyme, including the linkage of the

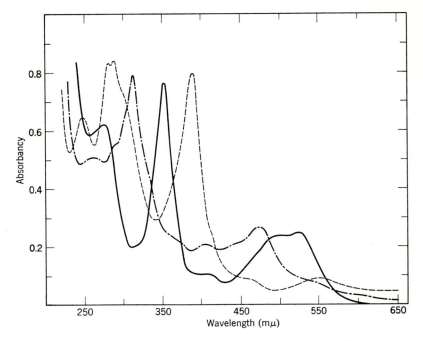

Fig. 19a. Spectra of (———) B_{12a}, (—·—) B_{12r}, and (– – –) B_{12s}.

adenosyl portion through its C-5′ position to the cobalt atom (Fig. 18), was again settled by x-ray analysis (219). Although coenzyme forms have been isolated with adenine, benzimidazole, and other bases affixed to the underside, the most commonly used form is that in which 5,6-dimethyl-benzimidazole is the parent base.

A number of comprehensive reviews on the chemistry and enzymology of vitamin B_{12} and its coenzymes are available (220–228). The nomenclature of vitamin B_{12} and its derivatives is rather complex [see discussions by Bonnett (223) and Beck (226)], and only a few points need to be emphasized. The most common form of vitamin B_{12}, in which dimethylbenzimidazole is the parent base, is "5,6-dimethylbenzimidazolylcyanocobamide"; this compound is often referred to by the trivial name "cyanocobalamin." The corresponding coenzyme (shown below) in which the cyanide group is

Ad
|
⌐[Co]
| ↑
└—N

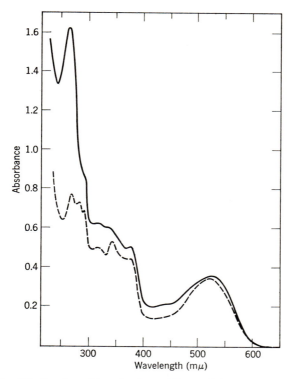

Fig. 19*b*. Spectra of (——) adenosyl-B_{12} and (– – –) methyl-B_{12}.

replaced by adenosine* is "dimethylbenzimidazole cobamide coenzyme" (abbreviated "DBC" or "adenosyl-B_{12}"). Methyl-B_{12} (or methyl-cobalamin) is analogous to adenosyl-B_{12} except that it has a methyl group on the top side of the corrin ring. Vitamin B_{12a} (also called "B_{12b}"), which is identical to B_{12} except that CN^- is replaced by OH^-, is more reactive in both chemical and enzymic systems and for this reason has been used in most investigations. B_{12r} and B_{12s} are reduced forms of B_{12a} in which the oxidation states of the cobalt are $+2$ and $+1$, respectively.

Vitamin B_{12} and its coenzymes have a number of interesting chemical properties. Reduction of B_{12a} with borohydride leads first to B_{12r} and then to B_{12s} (229); at pH 6 the reaction stops at the B_{12r} stage (230). These reactions can be followed qualitatively by noting the color of the solution

---

*Since deoxyadenosine usually refers to a nucleoside in which the methylene group is at C-2, it is perhaps less confusing to use "adenosyl-B_{12}" as a trivial name for this coenzyme rather than the more correct "5′-deoxyadenosyl-B_{12}."

(B_{12a} is red, B_{12r} is brown, and B_{12s} is gray-green) or quantitatively via the distinctive absorption spectra of these compounds (Fig. 19a). Catalytic hydrogenation of B_{12a} under carefully controlled conditions (231) or photolysis of alkyl derivatives of B_{12} under anaerobic conditions (see below) are also useful methods for generating B_{12r}. B_{12s} can also be prepared by reduction of B_{12a} with chromous acetate or zinc dust in acid (229).

In B_{12s} the cobalt may be considered formally to be in the unusual oxidation state of $+1$. As would be expected, this highly reduced compound is quite susceptible to air oxidation (as, to a lesser extent, is B_{12r}) and this point must not be overlooked when B_{12s} is invoked as an intermediate in enzymic mechanisms (see Sections III-B and III-F). B_{12s} slowly decomposes water to yield H_2 and B_{12r}; its half-life is about 1.5 hr at pH 8 (232). B_{12s} has been termed "hydridocobalamin" because of the possible equilibrium:

$$Co^{\ominus}H^+ \leftrightarrow Co^+H^{\ominus} \qquad (43)$$
$$(+1) \qquad\quad (+3)$$

Cobalt hydrides of this type have been prepared by hydrogenation of cobaltous pentacyanide (233) (Equation (44)). However, B_{12s} appears

$$Co(CN)_5 + \overset{\overset{\textstyle H_2}{\downarrow}}{H\cdot} \rightleftharpoons Co(CN)_5H \qquad (44)$$
$$(+2) \qquad\qquad (+3)$$

to be more like the species on the left side of Equation (43), since it does not contain deuterium when prepared in D_2O (234). This is consistent, too, with its behavior as a powerful nucleophile when reacting with alkyl halides (see Section III-B).

Alkyl derivatives of B_{12} are light-sensitive, provided that at least a trace of oxygen is present. Under these conditions, scission of the carbon–cobalt bond occurs homolytically to yield B_{12r} (235) and an alkyl free radical. In the case of adenosyl-B_{12}, the latter is eventually transformed to the stable compounds, adenosine-5'-aldehyde (236) and 8,5'-*cyclic*-adenosine (237). Methyl-B_{12} yields an even larger number of photolytic products arising from various recombinations of the methyl radical (238,239). The photolysis of hydroxyalkyl derivatives of B_{12} (240) and of 2',3'-isopropylidene-5'-deoxyuridinyl-B_{12} (241) has also been investigated with regard to reaction products.

The carbon–cobalt bond in adenosyl-B_{12} can also be cleaved by acid [forming B_{12}, adenine, and 2,3-dihydroxy-Δ-4-pentenal (242)] and by cyanide [forming dicyanocobalamin, adenine, and the cyanohydrin of the above pentenal (243)].

Intrinsic factor, a mucoprotein that binds vitamin B_{12} and facilitates its absorption through the intestine, has been used as an inhibitor for B_{12}-dependent enzyme systems (244), much in the same way that avidin has been employed for biotin-containing enzymes.

B. Chemical and Enzymic Synthesis of B_{12} Coenzymes

Alkyl derivatives of B_{12} can be synthesized chemically in good yield by the interaction of B_{12s} with compounds (RX) containing the alkyl group as a potential carbonium ion:

$$RX + \left[\begin{matrix} [\ddot{C}o]^{\ominus} \\ \uparrow \\ N \end{matrix}\right] \longrightarrow \left[\begin{matrix} R \\ | \\ [Co] \\ \uparrow \\ N \end{matrix}\right] + X^- \tag{45}$$

Thus, methyl-B_{12} has been prepared using methyl iodide, dimethyl sulfate, diazomethane, or adenosyl methionine (245–247). Except when steric hindrance is a factor (e.g., phenyl bromide and *tert*-butyl bromide), various alkylating agents (halides, sulfates, etc.) as well as bifunctional alkylating agents employed in cancer chemotherapy (248) have been used for this purpose (249). β-Cyanoethylcobalamin, prepared by the interaction of acrylonitrile with B_{12s}, is an interesting alkyl derivative of B_{12} that decomposes in alkaline solution to regenerate B_{12s} (Reaction (46)) (250).

$$\left[\begin{matrix} CH_2CH_2CN \\ | \\ [Co] \\ \uparrow \\ N \end{matrix}\right] + OH^- \rightleftharpoons \left[\begin{matrix} [\ddot{C}o]^{\ominus} \\ \uparrow \\ N \end{matrix}\right] + CH_2{=}CHCN + H_2O \tag{46}$$

Adenosyl-B_{12} (and other nucleoside analogs) are more difficult to synthesize since, after the requisite carbon–cobalt bond has been formed by reaction of B_{12s} with 2′,3′-O-isopropylidine-5′-tosyladenosine (245,246,249,251), the protective isopropylidine group must be removed by hydrolysis (Reaction (47)). The yield of adenosyl-B_{12} may be improved

2′,3′-Isopropylidine-5′-tosyladenosine + B_{12s} $\longrightarrow$

2′,3′-isopropylidine-5′-deoxyadenosyl-B_{12} $\xrightarrow{H_2O}$ 5′-deoxyadenosyl-B_{12} + acetone
$$\tag{47}$$

by a modification of this synthesis developed by Schmidt (252). *p*-Dimethylaminobenzaldehyde is used to block the 2′,3′ positions and, after the tosyl group has been placed on C-5, the blocking group is removed by acid hydrolysis to provide adenosine-5'-tosylate. The latter reacts with B_{12s} to produce adenosyl-B_{12} in yields of about 80%.

The spectra of adenosyl-B_{12} and methyl-B_{12} are shown in Figure 19b. Both spectra are nearly identical in the visible region, but the former has a higher absorbancy in the region of 260 mμ due to the adenosyl residue.

Schrauzer and his colleagues (253,254,254a) have recently synthesized a very interesting model of B_{12} in which dimethylglyoxime replaces the corrin ring as the chelator of the cobalt and pyridine (or another nitrogenous compound) is used as the parent base. The top side may then carry ionically linked cyanide (see below), covalently linked alkyl residues, or a pair of electrons in the state analogous to B_{12s}.

Although the enzymic conversion of vitamin B_{12} to the coenzyme, adenosyl-B_{12}, appears superficially to be a simple replacement of cyanide by an adenosyl group, the cofactor requirements alone suggest that it must be a complex process. Thus, it was found that extracts of *C. tetanomorphum* (255) or *P. shermanii* (256,257) could convert the vitamin to its coenzyme form when supplemented with ATP, a divalent metal ion (Mg^{++} or Mn^{++}), DPNH, and FAD. Tracer experiments with the *C. tetanomorphum* system established that ATP supplies the adenosyl group (255,258) and that tripolyphosphate (259) is the other reaction product:

$$\begin{array}{c} B_{12} \\ (\text{or } B_{12a}) \end{array} + \text{ATP} \xrightarrow[\text{Me}^{++}]{\text{DPNH, FAD}} \text{Ad-}B_{12} + \text{PPP} \qquad (48)$$

Peterkofsky and Weissbach (258) suggested that Reaction (48) might occur via a concerted mechanism since formation of adenosyl-B_{12} was paralleled by release of cyanide from the vitamin. The requirement for oxidoreduction cofactors seemed more consistent, however, with a stepwise mechanism, such as reduction of B_{12} followed by adenosylation of a reduced form of the vitamin. This latter hypothesis would also be consistent with the stepwise procedure used for the chemical synthesis of the coenzyme, as described above.

Although Brady et al. (260) were able to purify the adenosyl-B_{12}

synthesizing system (Equation (48)) several hundredfold from *P. shermanii* without observing any separation of activities, subsequent work from this laboratory (261) demonstrated that the *C. tetanomorphum* system could be resolved into two components, one responsible for the reduction of B_{12a} to B_{12s} (B_{12a} reductase) and the other for the conversion of B_{12s} to adenosyl-B_{12} (adenosylating enzyme). Aging of crude extracts inactivates the B_{12a} reductase but leaves the adenosylating enzyme intact. B_{12s}, generated *in situ* by reduction of B_{12a} with borohydride, can be used as the substrate for the adenosylating enzyme which does not appear to be appreciably inhibited by BH_4^- or its decomposition products.

Adenosylating enzyme has been purified from *C. tetanomorphum* (230) and used to follow Reaction (49) spectrophotometrically. The enzyme requires Mn^{++} (or, to a lesser extent, Mg^{++} or Co^{++}) for activity. B_{12s} cannot be replaced by B_{12a} or B_{12r} but ATP can be completely replaced

$$B_{12s} + ATP \xrightarrow{Mn^{++}} \text{adenosyl-}B_{12} + PPP \qquad (49)$$

by CTP and partially by UTP, GTP, and ITP. No evidence could be obtained for reversal of Reaction (49), even when devices were used to trap ATP or B_{12s}. It is possible that the product is actually an enzyme–PPP complex, which subsequently breaks down to E + PPP; failure to reverse the reaction would then be referable to an inability to regenerate E-PPP by simply admixing these two components.

The B_{12a} reductase, because of its lability and multiplicity of proteins, has been more difficult to purify. Following earlier studies (211,259,260) in which DPNH (sometimes supplemented with FAD) was employed as the reducing power for Reaction (48), we observed that dihydrolipoic acid could substitute for the reduced pyridine nucleotide (261). Because the thioredoxin system (Section III-E) can also be reduced by a TPNH-dependent flavoprotein, or non-enzymically with dihydrolipoic acid, the possibility was considered that the actual reductant for B_{12a} might also be a dithiol-containing protein. This assumption was greatly strengthened by our studies on the non-enzymic interaction of mono- and dithiols with B_{12a}, which should be described at this point.

The facile interaction of thiols with B_{12a} had been noted by several investigators (262–264). The glutathione–B_{12} complex, presumably having the structure shown below, is brown (λ_{max} at ~ 535 mμ), although, as explained subsequently, this is an exceptionally stable complex that does

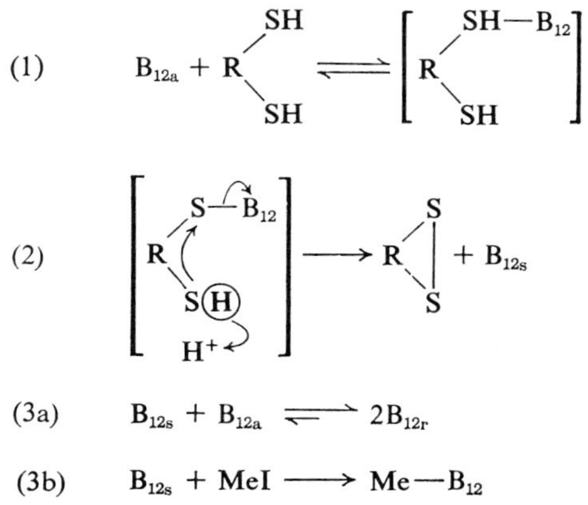

Fig. 20. Mechanism for the non-enzymic reduction of B_{12a} by dithiols. MeI = methyl iodide.

not react further. Other workers (222,264,265) have reported that thiols reduce B_{12a} to B_{12r}. Schmidt and Walker in this laboratory (266) have carried out a spectroscopic examination of the interaction of mono- and dithiols with B_{12a} and, on the basis of detailed kinetic measurements, the mechanism shown in Figure 20 has been proposed. For dithiols the first step is a rapid, reversible combination of one of the thiol anions with B_{12a}. The S–cobalt bond is probably more covalent than ionic. Steric considerations probably preclude formation of a 1:2 complex $[R—(S—B_{12})_2]$ between the thiol and B_{12}. Complex formation is followed by a rapid intramolecular reaction in which the adjacent free thiol group attacks the S–cobalt linkage forming the cyclic disulfide and expelling B_{12s}. The latter can react with excess B_{12a} to form B_{12r}* [as found also by Dolphin and Johnson (264)] or with a suitable acceptor, such as methyl iodide, to form the methyl-B_{12}. Monothiols react in much the same fashion except that the second step is slower unless high concentrations of the thiol are present. For reasons not yet understood, glutathione is a special case among the monothiols since its complex with B_{12a} is completely stable.

The enzymic system appears to follow the same pattern. Consistent with the earlier work of Hufham et al. (267), we have observed that when B_{12a} and DPNH are incubated with crude preparations of B_{12a} reductase (in the absence of adenosylating enzyme or ATP), B_{12r} is the sole product (268). When the adenosylating system is present, however, only adenosyl-B_{12} is produced. We interpret these observations according to the mechan-

*The equilibrium of this step has not yet been measured but it certainly lies well to the right.

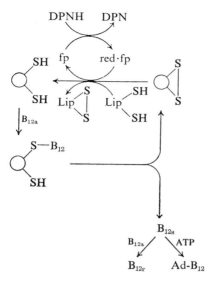

Fig. 21. Mechanism for the reaction catalyzed by B_{12a} reductase. fp = flavoprotein.

ism shown in Figure 21. A dithiol-containing protein, similar to thioredoxin, is assumed to be the vehicle for reduction of B_{12}. Moreover, B_{12s} may never exist in the free state, being produced transiently as needed by the acceptor system. B_{12s} would thus be shielded from oxidation by air. DPNH and the flavoprotein (or dihydrolipoic acid as a non-enzymic agent) may be necessary only to recycle the disulfide form of the protein to its reactive dithiol state.

Beck and his colleagues (269) have demonstrated that B_{12} is strongly

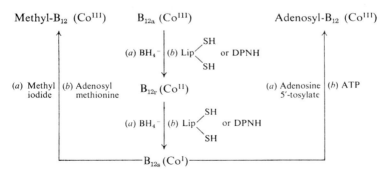

Fig. 22. Routes for the chemical and enzymic synthesis of B_{12} coenzymes. (a) Non-enzymic. (b) Enzymic.

bound to ribosomes in *L. leichmanii* and *E. coli*. It is therefore possible that the B_{12a} reductase system is present in the ribosomal fraction. Weissbach et al. (270) have recently suggested that ferredoxin is a reductant for B_{12}, although its effect might be referable to an involvement in the reduction of DPN.

Routes for the chemical and enzymic synthesis of adenosyl- and methyl-B_{12} are summarized in Figure 22. The chemical procedures are quite satisfactory, but the enzymes responsible for the enzymic transformations are still relatively unpurified and their mechanisms of action not yet elucidated.

C. Glutamate and Methylmalonyl CoA Mutase

Glutamate mutase is of historical interest because it was the first enzyme system in which a B_{12} coenzyme was implicated. Barker and his colleagues (210,271) established that the first step in the fermentation of L-glutamate by extracts of *C. tetanomorphum* involved an intramolecular rearrangement (272) leading to L-*threo*-β-methyl aspartate (Reaction (50)). The subsequent

$$
\begin{array}{ccc}
\underset{1}{\text{COOH}} & \underset{1}{\text{COOH}} & \text{COOH} \\
\underset{2}{\text{HCNH}_2} \xrightarrow{\text{Ad}-B_{12}} & \underset{2}{\text{HCNH}_2} \xrightarrow{-\text{NH}_3} & \text{C} \\
\underset{3}{\text{CH}_2} & \underset{4}{\text{HC}}\!-\!\underset{3}{\text{CH}_3} & \text{C}\!-\!\text{CH}_3 \\
\underset{4}{\text{CH}_2} & \underset{5}{\text{COOH}} & \text{COOH} \\
\underset{5}{\text{COOH}} & &
\end{array}
\qquad (50)
$$

L-glutamate   β-methyl   mesaconate
                       aspartate

step (β-methyl aspartate $\rightarrow$ mesaconate) is catalyzed by β-methyl aspartase, an enzyme that is present in large excess in the crude bacterial extracts. Since the product of this latter reaction, mesaconate, has an absorbancy at 230 mμ, its formation is frequently used to follow the glutamate mutase reaction (273).

The glutamate mutase activity can be abolished by irradiating, or by treating with charcoal, crude extracts of *C. tetanomorphum*. Activity can be restored by adding adenosyl-B_{12} ($K_m \cong 1.3 \times 10^{-5}M$)* and the use of such a reconstituted system greatly facilitated purification of the B_{12} coenzyme.

*When benzimidazole or adenine (rather than dimethylbenzimidazole) are the parent bases of the coenzyme, the K_m values are 2.3×10^{-7} and $1.2 \times 10^{-6}M$, respectively (274).

Glutamate mutase has still only been partially purified. After treatment with calcium phosphate gel, two separate proteins, both of which are required for activity, are obtained: a gel supernatant fraction ("S" fraction) and a gel eluate fraction ("E" fraction). The E fraction has been purified further (275). Neither protein contains any appreciable amount of bound pyridoxal phosphate and there is no stimulation when this cofactor is added. Tracer studies (275) have eliminated α-ketoglutarate and NH_4^+ as intermediates in Reaction (50).

By analogy with the glutamate mutase system, it was reasonable to assume that the similar transformation* of methylmalonyl CoA to succinyl CoA (Equation (51)) might also depend upon a B_{12} coenzyme.

$$
\begin{array}{ccc}
\text{COSCoA} & & \text{COSCoA} \\
| & & | \\
\text{H}_3\text{C—CH} & \underset{\longleftarrow}{\overset{\text{Ad}-\text{B}_{12}}{\rightleftharpoons}} & \text{CH}_2 \\
| & & | \\
\text{COOH} & & \text{CH}_2 \\
& & | \\
& & \text{COOH} \\
\text{methylmalonyl} & & \text{succinyl} \\
\text{CoA} & & \text{CoA}
\end{array}
\tag{51}
$$

Methylmalonyl CoA mutase, which catalyzes Reaction (51), has been highly purified from *P. shermanii* (278–280), sheep liver (281,282), and ox liver (283) and shown in all cases to be dependent on adenosyl-B_{12}. Unlike glutamate mutase, no separation of the enzyme into two fractions has been reported. The bacterial enzyme loses its bound coenzyme readily during purification but the mammalian counterpart must be repeatedly precipitated with acid ammonium sulfate before the requirement for adenosyl-B_{12} can be shown. Alternatively, it is possible to isolate an inactive holoenzyme that contains bound B_{12} (279,280,282), perhaps B_{12a}. This phenomenon has been encountered with several of the B_{12} enzymes. Such preparations, although visibly colored because of the protein-bound B_{12} that is present, still require *added* adenosyl-B_{12} for activity. It seems most likely that the added coenzyme activates the colorless apoenzyme present, although another possible explanation would be displacement of the inactive, bound form of B_{12} by adenosyl-B_{12}.

Despite many searching investigations, the mechanism of the mutase reactions is still obscure. Tracer experiments have shown that during isomerization of glutamate or succinyl CoA there is no equilibration of

---

*As demonstrated in the laboratories of Ochoa (276) and Wood (277), carboxylation of propionyl CoA leads to an isomer of methylmalonyl CoA (designated "a") which is converted by a racemase to methylmalonyl CoA ("b"), the substrate for Reaction (51).

hydrogens on the substrate with protons in water (284–286). Since it is likely that carbon–hydrogen bonds are in fact broken during rearrangement of the carbon skeleton, the hydrogen(s) must be sequestered in such a way that exchange with the solvent is prevented. Ingraham (287) has proposed a carbanion mechanism for Reactions (50) and (51) involving covalent linkage of the carbon skeleton to the top side of the cobalt, replacing the adenosyl residue; the latter absorbs a proton released from the substrate. It seems unlikely, however, that such an important linkage in the coenzyme (requiring ATP for its formation) would be jeopardized during operation of the coenzyme. A different mechanism, proposed by Whitlock (288) for Reaction (51), is based on the metal carbonyl-catalyzed hydrocarbonylation of olefins. In this mechanism, however, a carbonyl fragment is envisioned as occupying a position between the cobalt atom and one of the bonding pyrrole nitrogens with the remnant of the substrate on the bottom side of the corrin ring forming a linkage via its acyl group to the cobalt atom. For steric reasons the first part of this hypothesis does not appear feasible.

For the methylmalonyl CoA mutase reaction, Lynen and his colleagues (289) demonstrated by means of tracer experiments that the thioester group, rather than the free carboxyl group, migrates during the rearrangement (see below). Labeling experiments in Wood's laboratory (290) and by Phares et al. (291) have shown further that this group is transferred intramolecularly rather than intermolecularly.

$$
\begin{array}{c}
\text{H} \quad \overset{\displaystyle C \!=\! O}{\underset{\text{SCoA}}{\big|}} \\
\text{H}\!-\!\text{C}\!-\!\text{CH} \\
\text{(H)} \qquad \text{COOH}
\end{array}
$$

Sprecher and Sprinson (292) examined the stereospecificity of the glutamate mutase reaction. Replacement of the glycine group by a hydrogen (see below) results in an inversion of configuration at the β-carbon position; this would be consistent with, but does not prove, a carbonium ion mechanism.

$$
\begin{array}{c}
\text{COOH} \\
| \\
\text{H}\;\overset{\displaystyle H\text{C}NH_2}{\underset{}{\big|}} \\
\text{H}\!-\!\text{C}\!-\!\text{C} \\
\text{(H)} \qquad \text{COOH}
\end{array}
$$

D. Diol Dehydrase

Another type of B_{12}-dependent isomerase is diol dehydrase which catalyzes an internal oxidoreduction reaction (Reaction (52)). Ethylene

$$
\begin{array}{ccc}
\text{CH}_2\text{OH} & & \text{CHO} \\
| & \xrightarrow{\text{Ad-B}_{12}} & | \\
\text{HCOH} & & \text{CH}_2 \\
| & & | \\
\text{CH}_3 & & \text{CH}_3 \\
\text{1,2-propanediol} & & \text{propionaldehyde}
\end{array} \tag{52}
$$

glycol also serves as a substrate, being converted to acetaldehyde. The enzyme has been obtained in highly purified form from *Aerobacter aerogenes* and its properties studied by Abeles and his group (293–296). The favorable affinity for the coenzyme, and the ease with which Reaction (52) may be followed, have made this enzyme well suited for the assay of adenosyl-B_{12}.

The role of adenosyl-B_{12} in Reactions (50)–(52) has been puzzling. As with the other coenzymes derived from B vitamins, it seemed likely *a priori* that adenosyl-B_{12} carries some mobile metabolic group (1) during the course of the reaction. If this is true, however, the coenzyme–mobile group complex must have an absorption spectrum that is not appreciably different from that of the free coenzyme. During catalysis of Reactions (51) or (52) by highly purified enzymes, no changes in spectrum of the coenzyme [except for degradation (296,297)] have been reported. One would expect a reasonable steady-state concentration of such a complex to exist when the reversible reaction is run in one of the two directions.

Other internal oxidoreduction reactions (glucose-6-P $\rightleftharpoons$ fructose-6-P, methylglyoxal $\rightleftharpoons$ lactic acid), for which mechanisms can be written involving an internal shift of a hydride ion, do not require a B_{12} coenzyme. Why is this special coenzyme needed for Reaction (52)? The answer may lie in the fact that the original mechanism suggested by Brownstein and Abeles (293), shown below, does *not* assign to the coenzyme the function of carrying the mobile hydride ion. We have speculated recently that the

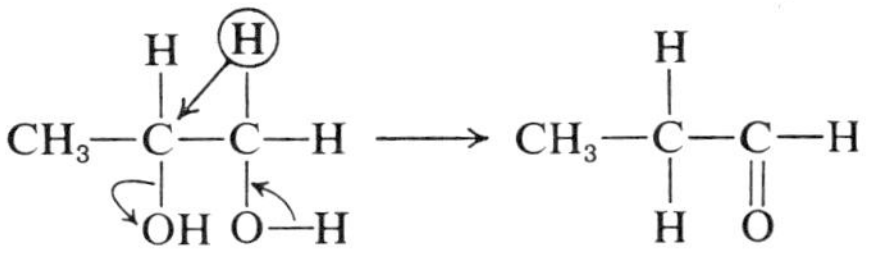

$$
\begin{array}{ccccc}
& \text{H} \;\; \text{(H)} & & \text{H} & \\
& | \,\diagup\, | & & | & \\
\text{CH}_3-\text{C}-\text{C}-\text{H} & \longrightarrow & \text{CH}_3-\text{C}-\text{C}-\text{H} \\
& | \quad\;\; | & & | \;\;\; \| & \\
& \text{OH}\;\; \text{O}-\text{H} & & \text{H}\;\;\; \text{O} &
\end{array}
$$

coenzyme carries the hydride ion, perhaps on the lower side of the corrin ring (228). To do this, the parent base on the underside would have to be protonated, enabling it to swing away temporarily. A hydride ion from the substrate would occupy this site until rearrangement in the skeleton of the substrate had created a new susceptible carbon atom to which the hydride ion could be returned.

Recent experiments from Abeles' laboratory have provided important evidence which, while not yet completely elucidating the mechanism, at least minimizes the number of alternative possibilities. Thus, when tritium-labeled propanediol and unlabeled ethylene glycol are used with diol dehydrase, both products, propionaldehyde and acetaldehyde, contain label (298); this clearly demonstrates a transfer of label from one substrate to the other and shows that the reaction is not necessarily intramolecular. In an extension of these elegant experiments (299), it was shown that when 1,2-propanediol-1-^{3}H is used as substrate, the adenosyl-B_{12} becomes labeled; moreover, the tritiated coenzyme can relabel the product at C-2. Finally, coenzyme labeled with tritium in C-5′ was prepared by an unambiguous chemical synthesis and shown to transfer its label to the product. Carbon-5 of the adenosyl-B_{12} is thus identified as the "active center" of the coenzyme and shown to be a transient carrier of hydrogen (probably as H^-) during isomerization of the substrate. The constant breaking of a carbon–hydrogen bond as part of the mechanism may well weaken the carbon–cobalt linkage and this could account for inactivation of the coenzyme as seen especially with diol dehydrase (296) (cf. also isolation of inactive holoenzyme forms of other B_{12} systems). A tentative mechanism for the diol dehydrase reaction is offered in Figure 23; similar mechanisms could be written for the two mutase reactions.

An enzyme somewhat similar to diol dehydrase is the ethanolamine deaminase (Equation (53)) studied in Stadtman's laboratory (300). The

$$\begin{array}{c} CH_2OH \\ | \\ CH_2NH_3 \\ \oplus \end{array} \longrightarrow \begin{array}{c} CHO \\ | \\ CH_3 \end{array} + NH_4^+ \qquad (53)$$

enzyme, isolated in homogeneous form, has a minimum molecular weight of 30,000. Depending on the purification procedure, the enzyme can be obtained in a pink or orange form. That either or both of these may be inactive holoenzymes is shown by the fact that, following treatment with

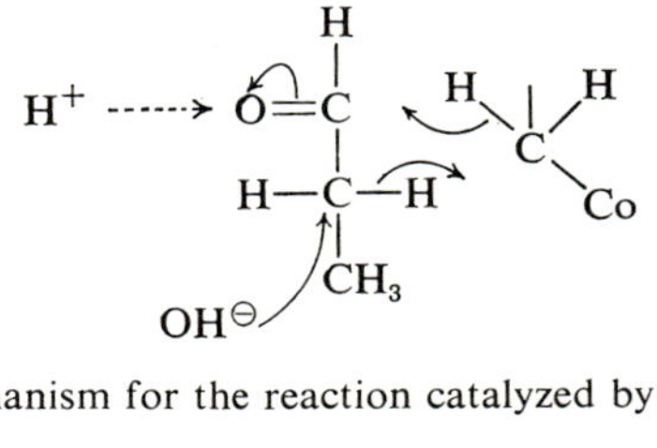

Fig. 23. Mechanism for the reaction catalyzed by diol dehydrase.

acid ammonium sulfate, activity of the preparation in the presence of added adenosyl-B_{12} ($K_m = 1.6 \times 10^{-6}M$) is increased severalfold.

E. Ribonucleotide Reductase

The direct conversion of ribonucleotides to deoxyribonucleotides, a process long suspected to occur, was finally demonstrated by Reichard (301) using a system from *E. coli*. Subsequent work by Reichard's laboratory, and by other groups (as discussed below) revealed that the *E. coli* system was complex and the sequence of electron flow for a typical substrate, CDP, is that shown in Equation (54).

$$\text{TPNH} \longrightarrow \text{thioredoxin reductase} \longrightarrow \text{thioredoxin} \longrightarrow \text{CDP} \qquad (54)$$

TPNH, the ultimate reductant, is used to reduce thioredoxin via thioredoxin reductase, a flavoprotein having a molecular weight of 60,000 and containing 1 mole of FAD per mole of protein (302). Thioredoxin is a smaller molecular weight (12,000) protein that is heat stable and acidic (303); the latter property, which is also characteristic of the ferredoxins, makes it easy to isolate thioredoxin by its retention on columns of DEAE-cellulose. Thioredoxin contains two vicinal thiol groups per mole; upon oxidation these form a cyclic disulfide (see Sections II-C-5 and III-B for discussions of similar systems). Reduced lipoic acid can be used to reduce thioredoxin non-enzymically, thereby obviating the need for the TPNH-dependent thioredoxin reductase. Reduced thioredoxin probably supplies its reducing power directly to C-2 of the nucleotide, but this is still not certain since two other protein fractions, B_1 and B_2, are also required for the overall reaction (304). It is possible that these proteins are concerned with the complex allosteric effects of various nucleotides observed on this system (304).

At one time, it was thought that the B_2 fraction contained B_{12} coenzyme but, as the system has been purified, no B_{12} could be detected microbiologically or by radioactivity in either B_1 or B_2. Added adenosyl-B_{12} is also without effect in the *E. coli* system. On the other hand, a comparable ribonucleotide reductase (which utilizes nucleoside triphosphates rather than diphosphates as the substrate), purified from derepressed cultures of *L. leichmanii** by the laboratories of Barker (306), Blakley (307–309), and Beck (310,310a), shows a definite requirement for adenosyl-B_{12} and utilizes the thioredoxin system. The existence of separate systems, one requiring and the other not requiring a B_{12} coenzyme, is reminiscent of separate types of

*Ribonucleotide reductase has also been found in Novikoff hepatoma (305).

methionine synthetase (Section III-F) which can operate with and without methyl-B_{12}.

The mechanism of the ribonucleotide reductase reaction is not yet understood. If reduced thioredoxin delivers a hydride ion, some other component of the system (A in the diagram below) may have to assist in the difficult task of expelling the $OH^{\ominus}$ (cf. mechanism in Fig. 11b).

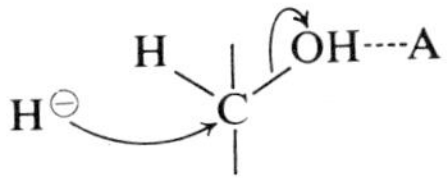

F. Methionine Synthetase

The terminal reaction in the biosynthesis of methionine (Reaction (41)) involves the methylation of homocysteine by 5-methyl tetrahydrofolate (114,115,311). Enzyme systems catalyzing this reaction have been partially purified from *E. coli* (49,312–314) and from mammalian (114,314–316) and avian liver (314).

A number of cofactors have been implicated in methionine synthesis. These include a protein-bound form of B_{12} (317–320), most probably methyl-B_{12} (312,314,321–324), DPNH plus FAD (both of which can be replaced by $FADH_2$), and ATP plus Mg^{++} (114,115,314,315,318,322,325,326). The function of ATP and Mg^{++} was established by our observation that in the pig liver system both of these factors could be replaced by *catalytic* amounts of adenosyl methionine (46,315). This finding was later confirmed for other mammalian (314) and bacterial (49,314,322,327) systems.

Tracer experiments have revealed that the methyl group of adenosyl methionine does not equilibrate with the actual methyl group being transferred from tetrahydrofolate to homocysteine (315). More likely, adenosyl methionine serves as the precursor for methyl-B_{12} (316,321,322):

$$\text{Adenosyl methionine} + B_{12s} \longrightarrow \text{adenosyl homocysteine} + \text{methyl-}B_{12} \quad (55)$$

Our hypothesis (316) to account for the unusual cofactor requirements for Reaction (41) is that methionine synthetase can exist in several forms. The active holoenzyme is assumed to contain bound methyl-B_{12} while the corresponding apoenzyme would require added methyl-B_{12}; no other cofactors are believed to be necessary under these conditions. Depending upon the source of the enzyme and the purification procedure used, two *inactive* forms of the holoenzyme are possible: one containing B_{12s} and the other B_{12a}. The B_{12a} form probably arises from the interaction of methyl-B_{12} with homocysteine:

$$\text{Methyl-}B_{12} + \text{homocysteine} \longrightarrow B_{12s} + \text{methionine} \quad (56)$$

This reaction, while certainly demonstrable with preparations of methionine synthetase, occurs at too slow a rate, in our opinion (316), to be a step in the overall reaction (Reaction (41)). The B_{12a} form would result from oxidation of B_{12s}, unless the latter were protected.

Reaction of these inactive holoenzymes, by the routes indicated in Figure 24, is believed to require a thioredoxin-like system. The B_{12s} form, existing as the thioprotein complex, requires only adenosyl methionine and the appropriate enzyme catalyzing Reaction (55).

Corollary evidence supporting the existence of a form of the enzyme that contains a stabilized B_{12s} comes from experiments from Weissbach's laboratory (325,328) and from our group (316). Methyl iodide or dimethyl sulfate can replace adenosyl methionine as a cofactor, presumably by resynthesizing methyl-B_{12} non-enzymically. Conversely, if the enzyme is incubated with propyl iodide, an inactive, but protected, form of the enzyme is obtained. Photolysis of the inactive enzyme, followed by treatment with adenosyl methionine, then yields a reactivated enzyme. One difficulty in understanding this experiment is that photolysis of propyl-B_{12} should yield B_{12r} which would be unable to react with adenosyl methionine (cf. the inability of ATP and adenosylating enzyme to react with B_{12r}). Although Johnson et al. (329) have shown that photolysis of methyl-B_{12} in the presence of cysteine produces S-methylcysteine in good yield

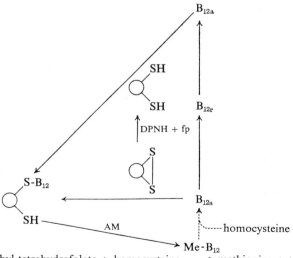

Fig. 24. Multiple forms of methionine synthetase (316). AM = adenosyl methionine. fp = flavoprotein.

($\sim 50\%$), the borohydride present may have reduced the nascent B_{12r} to B_{12s}. Similarly, in the photoreactivation of the propylated form of methionine synthetase, some component of the complex assay system (e.g., homocysteine) may be responsible for reduction of B_{12r} to B_{12s}.

If the inactive holoenzyme contains B_{12a}, regeneration of methyl-B_{12} would require oxidoreduction cofactors (DPNH + FAD) in addition to adenosyl methionine. The thiol protein would be the actual reductant, as shown in Figure 24, but it would be backed up by a pyridine nucleotide-

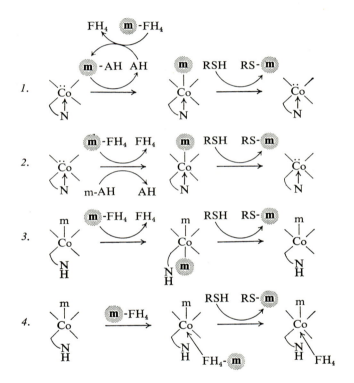

Fig. 25. Various mechanisms proposed for the reaction catalyzed by methionine synthetase (316). The following symbols are used: ⓜ, the methyl group originating from 5-methyl tetrahydrofolate; $\overset{..}{\underset{N}{Co}}$, B_{12s}; $\overset{\overset{m}{|}}{\underset{N}{Co}}$, methyl-$B_{12}$; m-AH, adenosyl methionine; AH, adenosyl homocysteine; m-FH_4, 5-methyl tetrahydrofolate; FH_4, tetrahydrofolate; RS-m, methionine; and RSH, homocysteine.

dependent flavoprotein system. It should be emphasized that this hypothesis, although attractive and in accord with many of the experimental observations, will require substantiation by isolation of the individual protein components and their employment in reconstructing the methionine synthetase system.

The above discussion has centered about the identity of the B_{12} coenzyme involved in methionine synthesis. If methyl-B_{12} is indeed the coenzyme, its function in the methyl transfer reaction (Equation (41)) is not yet clear. Woods and his colleagues (322) originally suggested an interesting mechanism (line 1 in Fig. 25) involving a series of methyl group transfers. That is, methyl-B_{12} was envisioned as the immediate precursor of the methyl group of methionine and the resultant B_{12s} was remethylated by adenosyl methionine. This sequence does not seem likely, however, because it requires that the energy-rich adenosyl methionine be reformed, in turn, by methyl tetrahydrofolate, a weak methylating agent. A variation of this mechanism (line 2) has been suggested by Weissbach's group (328), who also assume that methyl-B_{12} is the immediate precursor of methionine. B_{12s} is assumed to be remethylated by methyl tetrahydrofolate with adenosyl methionine acting as an activator or allosteric effector (323) in the reaction. The first step in this sequence, formation of methyl-B_{12} from B_{12s} and methyl tetrahydrofolate, has not yet been demonstrated, and no explanation is provided for how methyl tetrahydrofolate is activated for release of its methyl group.

In a third mechanism (line 3 of Fig. 25), methyl-B_{12} is considered to act as an intermediate carrier of the methyl group which is attached transiently to the *bottom* side of the coenzyme. Although not specifically suggested by them, this mechanism is supported by the report of Elford et al. (330) that Factor B (a derivative of B_{12} in which the parent base is missing) can accept a methyl group from methyl tetrahydrofolate. The fourth proposal (line 4), suggested by our laboratory (321,331), assumes that the parent base on the bottom side of methyl-B_{12} can be protonated, perhaps by a group on the enzyme, and thereby relinquish its coordination with the cobalt atom. Methyl tetrahydrofolate could then be complexed at this site in such a way that the unshared electron pair at N-5 would be directed toward the central cobalt atom. Such an interaction would labilize the N-methyl bond and facilitate the departure of CH_3^+ in order to methylate homocysteine.

Certain bacteria and plants contain a methionine synthetase that catalyzes Reaction (41) except that the triglutamate of 5-methyl tetrahydrofolate, rather than the monoglutamate, serves as substrate (318,332). These systems do not utilize a B_{12} coenzyme and it is possible that a metal ion (Me^{++}) might function in the same manner as envisioned for methyl-B_{12} in mechanism *4*:

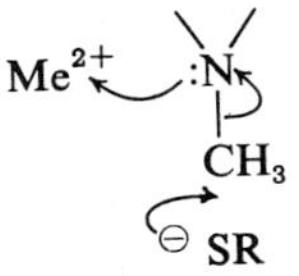

$$Me^{2+} \quad :N$$
$$CH_3$$
$$\ominus SR$$

G. Methane Formation

Certain bacteria, notably *Methanosarcina barkerii* and *Methanobacillus omelianskii*, produce methane. Recent studies from the laboratories of Stadtman (333,334) and Wolfe (335–337) have shown that ^{14}C-methyl tetrahydrofolate or ^{14}C-methyl-B_{12} can serve as a precursor of methane. It seems probable that methyl-B_{12} is cleaved reductively ([2H] in Reaction (57)) to yield CH_4 and B_{12s}.

$$\text{Methyl-}B_{12} + [2H] \longrightarrow CH_4 + B_{12s} \tag{57}$$

Regeneration of methyl-B_{12} could occur by the sequence:

$$\text{Methyl tetrahydrofolate} \xrightarrow{\quad} \text{methionine} \xrightarrow{ATP}$$
$$\text{adenosyl methionine} \xrightarrow{B_{12s}} \text{methyl-}B_{12} \tag{58}$$

Other alkyl B_{12} derivatives may undergo a similar type of reductive cleavage. Thus, E. Stadtman (338) and Wood (339) have isolated from *Clostridium thermoaceticum* carboxymethyl derivatives of B_{12} and have suggested that these may be the precursor of acetate in this bacterium (Reaction (59)). If this hypothesis is true, it is interesting to speculate

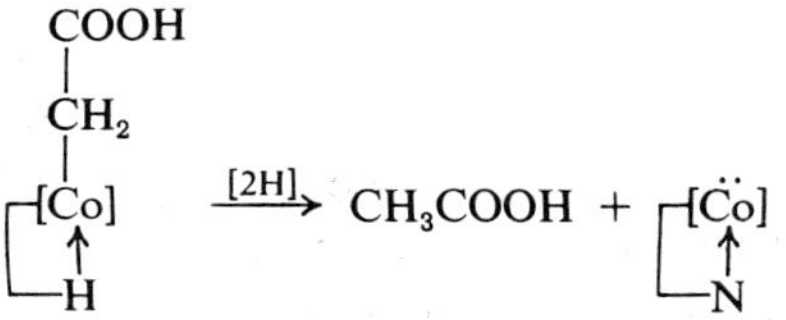

$$\begin{array}{c} COOH \\ | \\ CH_2 \\ | \\ [Co] \\ \uparrow \\ H \end{array} \xrightarrow{[2H]} CH_3COOH + \begin{array}{c} [Co] \\ \uparrow \\ N \end{array} \tag{59}$$

on the route for the *synthesis* of carboxymethyl-B_{12}, since a good leaving group would be necessary on the α-carbon of acetate in order to facilitate condensation with B_{12}; an alternate possibility would involve carboxylation of methyl-B_{12}.

Acknowledgments

Experimental work described herein was supported by grants from the National Cancer Institute (CA 6522) and the National Institute of Arthritis and

Metabolic Diseases (AM 08398), National Institutes of Health, Life Insurance Medical Research Fund (G-64-30), and National Science Foundation (GB-1934). The author is indebted to M. Baginsky, J. Brodie, H. Gigliotti, V. Gupta, S. Kerwar, J. Kirchner, G. Mell, P. Reyes, R. Schmidt, K. G. Scrimgeour, E. Vitols, D. Vonderschmitt, G. Walker, and J. Whiteley for helpful discussions of the mechanisms of folate- and B$_{12}$-dependent enzyme systems.

References

1. F. M. Huennekens, in *Currents in Biochemistry*, Vol. II, D. E. Green, Ed., Interscience, New York, 1956, p. 493.
2. E. L. R. Stokstad, in *The Vitamins*, W. H. Sebrell, Jr. and R. S. Harris, Eds., Academic Press, New York, 1954, p. 124.
3. A. D. Welch and C. A. Nichol, *Ann. Rev. Biochem.*, **21**, 633 (1952).
4. G. R. Greenberg, *Federation Proc.*, **13**, 745 (1954).
5. R. L. Kisliuk and W. Sakami, *J. Am. Chem. Soc.*, **76**, 1456 (1954); *J. Biol. Chem.*, **214**, 47 (1955).
6. R. L. Blakley, *Nature*, **173**, 729 (1954); *Biochem. J.*, **58**, 448 (1954).
7. F. M. Huennekens and M. J. Osborn, in *Advances in Enzymology*, Vol. 21, F. F. Nord, Ed., Interscience, New York, 1959, p. 369.
8. J. C. Rabinowitz, in *The Enzymes*, Vol. 2, P. D. Boyer, H. Lardy, and K. Myrbäck, Eds., Academic Press, New York, 1959, p. 185.
9. B. E. Wright, *Symposium on Vitamins in Metabolism* (4th Intern. Congr. Biochem., Vienna, 1958), p. 266.
10. L. Jaenicke, *Mechanismen Enzymatischer Reaktionen*, Springer-Verlag, Berlin, 1964, p. 212; *Experientia*, **17**, 481 (1961).
11. M. Friedkin, *Ann. Rev. Biochem.*, **32**, 185 (1963).
12. S. F. Zakrzewski, *J. Biol. Chem.*, **241**, 2957 (1966).
13. Unpublished observations from this laboratory.
14. J. M. Whiteley and F. M. Huennekens, *Biochemistry*, in press.
15. R. G. Kallen and W. P. Jencks, *J. Biol. Chem.*, **241**, 5845 (1966).
16. R. L. Blakley, *Nature*, **188**, 231 (1960).
17. F. M. Huennekens, C. K. Mathews, and K. G. Scrimgeour, in *Methods in Enzymology*, Vol. VI, S. P. Colowick and N. O. Kaplan, Eds., Academic Press, New York, 1963, p. 802.
18. H. E. Sauberlich, *J. Biol. Chem.*, **195**, 337 (1952).
19. J. C. Rabinowitz, in *Methods in Enzymology*, Vol. VI, S. P. Colowick and N. O. Kaplan, Eds., Academic Press, New York, 1963, p. 812.
20. M. May, T. J. Bardos, F. L. Barger, M. Lansford, J. M. Ravel, G. L. Sutherland, and W. Shive, *J. Am. Chem. Soc.*, **73**, 3067 (1951).
21. Y. Hatefi, Ph.D. Thesis, University of Washington, Seattle, 1956.
22. J. C. Rabinowitz, in *Methods in Enzymology*, Vol. VI, S. P. Colowick and N. O. Kaplan, Eds., Academic Press, New York, 1963, p. 814.
23. F. M. Huennekens, P. P. K. Ho, and K. G. Scrimgeour, in *Methods in Enzymology*, Vol. VI, S. P. Colowick and N. O. Kaplan, Eds., Academic Press, New York 1963, p. 806.
24. K. Uyeda and J. C. Rabinowitz, *Anal. Biochem.*, **6**, 100 (1963).

25. A. Perault and B. Pullman, *Biochim. Biophys. Acta*, **44**, 251 (1960).
26. F. M. Huennekens, *Biochemistry*, **2**, 151 (1963).
27. C. K. Mathews and F. M. Huennekens, *J. Biol. Chem.*, **235**, 3304 (1960).
28. B. T. Kaufman, K. O. Donaldson, and J. C. Keresztesy, *J. Biol. Chem.*, **238**, 1498 (1963).
29. B. V. Ramasastri and R. L. Blakley, *Biochem. Biophys. Res. Commun.*, **12**, 478 (1963).
30. A. Pohland, E. H. Flynn, R. G. Jones, and W. Shive, *J. Am. Chem. Soc.*, **73**, 3247 (1951).
31. S. F. Mason, in *Chemistry and Biology of Pteridines*, G. E. W. Wolstenholme and M. P. Cameron, Eds., Little, Brown, Boston, 1954, p. 81.
32. M. J. Osborn, P. T. Talbert, and F. M. Huennekens, *J. Am. Chem. Soc.*, **82**, 4921 (1960).
33. V. S. Gupta and F. M. Huennekens, *Biochemistry*, to be published.
34. R. G. Kallen and W. P. Jencks, *J. Biol. Chem.*, **241**, 5851 (1966).
35. S. Futterman, *J. Biol. Chem.*, **228**, 1031 (1957).
36. S. Futterman, in *Methods in Enzymology*, Vol. VI, S. P. Colowick and N. O. Kaplan, Eds., Academic Press, New York, 1963, p. 801.
37. C. K. Mathews and F. M. Huennekens, *J. Biol. Chem.*, **238**, 4005 (1963).
38. M. Silverman and J. M. Noronha, *Biochem. Biophys. Res. Commun.*, **4**, 180 (1961).
39. B. L. O'Dell, J. M. Vandenbelt, E. S. Bloom, and J. J. Pfiffner, *J. Am. Chem. Soc.*, **69**, 250 (1947).
40. Y. Hatefi, P. T. Talbert, M. J. Osborn, and F. M. Huennekens, *Biochem. Prep.*, **7**, 89 (1960).
41. K. G. Scrimgeour and K. S. Vitols, *Biochemistry*, **5**, 1438 (1966).
42. R. L. Blakley, *Biochem. J.*, **65**, 331 (1956).
43. R. L. Kisliuk, *J. Biol. Chem.*, **227**, 805 (1957).
44. S. Kaufman, *J. Biol. Chem.*, **236**, 804 (1961).
45. P. P. K. Ho, Ph.D. thesis, University of Washington, 1963.
46. K. G. Scrimgeour and F. M. Huennekens, in *Pteridine Chemistry*, W. Pfleiderer and E. C. Taylor, Eds., Pergamon, New York, 1964, p. 355.
47. S. Zakrzewski, *J. Biol. Chem.*, **241**, 2962 (1966).
48. K. Smith and K. G. Scrimgeour, *Federation Proc.*, **22**, 232 (1963).
49. A. R. Larrabee, S. Rosenthal, R. E. Cathou, and J. M. Buchanan, *J. Biol. Chem.*, **238**, 1025 (1963).
50. K. O. Donaldson and J. C. Keresztesy, *J. Biol. Chem.*, **237**, 3815 (1962).
51. S. Kaufman, *J. Biol. Chem.*, **239**, 332 (1964).
52. W. Pfleiderer and E. C. Taylor, *J. Am. Chem. Soc.*, **82**, 3765 (1960).
53. K. Smith, K. G. Scrimgeour, and F. M. Huennekens, *Biochem. Biophys. Res. Commun.*, **11**, 388 (1963).
54. S. Zakrzewski, *Federation Proc.*, **22**, 231 (1963).
55. E. J. Pastore, M. Friedkin, and O. Jardetsky, *J. Am. Chem. Soc.*, **85**, 3058 (1963).
56. B. L. Hillcoat and R. L. Blakley, *Biochem. Biophys. Res. Commun.*, **15**, 303 (1964).
57. G. H. Hitchings and J. J. Burchall, in *Advances in Enzymology*, Vol. 27, F. F. Nord, Ed., Interscience, New York, 1965, p. 417.
57a. R. L. Blakley, B. V. Ramasastri, and B. M. McDougall, *J. Biol. Chem.*, **238**, 3075 (1963).

58. S. F. Zakrzewski and C. A. Nichol, *J. Biol. Chem.*, **235**, 2984 (1960).
59. C. K. Mathews and F. M. Huennekens, *J. Biol. Chem.*, **238**, 3436 (1963).
60. G. P. Mell, K. Dus, and F. M. Huennekens, in preparation.
61. B. T. Kaufman and R. C. Gardiner, *J. Biol. Chem.*, **241**, 1319 (1966).
62. D. R. Morales and D. M. Greenberg, *Biochim. Biophys. Acta*, **85**, 360 (1964).
63. D. M. Greenberg, B. Tam, E. Jenny, and B. Payes, *Biochim. Biophys. Acta*, **122**, 423 (1966).
64. R. L. Blakley and B. M. McDougall, *J. Biol. Chem.*, **236**, 1163 (1961); **241**, 2995 (1966).
65. A. M. Albrecht, J. R. Palmer, and D. J. Hutchison, *J. Biol. Chem.*, **241**, 1043 (1966).
66. F. M. Sirotnak and D. J. Hutchison, *J. Biol. Chem.*, **241**, 2900 (1966).
66a. F. M. Sirotnak, G. J. Donati, and D. L. Hutchison, *J. Biol. Chem.*, **239**, 4298 (1964).
67. C. K. Mathews and K. E. Sutherland, *J. Biol. Chem.*, **240**, 2142 (1965).
68. D. Kessel and D. Roberts, *Biochemistry*, **4**, 2631 (1965).
69. J. R. Bertino, J. P. Perkins, and D. G. Johns, *Biochemistry*, **4**, 839 (1965).
70. E. R. Kashket, E. J. Crawford, M. Friedkin, S. R. Humphreys, and A. Goldin, *Biochemistry*, **3**, 1928 (1964).
71. M. E. Bush and K. O. Donaldson, *Biochem. Biophys. Res. Commun.*, **20**, 635 (1965).
72. F. M. Sirotnak and D. Hutchison, *Biochem. Biophys. Res. Curumun.*, **19**, 734 (1965).
73. B. Kaufman, *Biochem. Biophys. Res. Commun.*, **10**, 449 (1963).
74. B. Kaufman, *J. Biol. Chem.*, **239**, PC669 (1964).
75. J. Perkins and J. R. Bertino, *Biochem. Biophys. Res. Commun.*, **15**, 121 (1964).
76. J. R. Bertino, *Biochim. Biophys. Acta*, **58**, 377 (1962).
77. G. P. Mell, G. M. Schroeder, and F. M. Huennekens, *Federation Proc.*, **25**, 277 (1966).
78. J. P. Perkins and J. R. Bertino, *Biochemistry*, **5**, 1005 (1966).
79. G. P. Mell and F. M. Huennekens, in preparation.
80. B. E. Wright and M. L. Anderson, *J. Am. Chem. Soc.*, **79**, 2027 (1957).
81. K. G. Scrimgeour, K. S. Vitols, M. L. Norris, and H. J. Pashkar, *Arch. Biochem. Biophys.*, **119**, 159 (1967).
82. S. Futterman and M. Silverman, *J. Biol. Chem.*, **224**, 31 (1957).
83. M. J. Osborn, M. Freeman, and F. M. Huennekens, *Proc. Soc. Exptl. Biol. Med.*, **97**, 429 (1958).
84. R. L. Kisliuk, *Nature*, **188**, 584 (1960).
85. S. F. Zakrzewski, M. Hakala, and C. A. Nichol, *Biochemistry*, **1**, 842 (1962).
86. W. C. Werkheiser, *J. Biol. Chem.*, **236**, 888 (1961).
87. J. R. Bertino, B. A. Booth, A. L. Bieber, A. Cashmore, and A. C. Sartorelli, *J. Biol. Chem.*, **239**, 479 (1964).
88. C. K. Mathews, Ph.D. Thesis, University of Washington, Seattle, 1962.
89. F. M. Huennekens, J. R. Bertino, R. Silber, and B. W. Gabrio, *Exptl. Cell Res. Suppl.*, **9**, 441 (1963).
90. J. R. Bertino, *Cancer Res.*, **23**, 1286 (1963).
91. W. C. Werkheiser, *Cancer Res.*, **23**, 1277 (1963).
92. J. R. Bertino, R. Silber, M. Freeman, A. Alenty, M. Albrecht, B. W. Gabrio, and F. M. Huennekens, *J. Clin. Invest.*, **42**, 1899 (1963).

93. A. W. Schrecker and F. M. Huennekens, *Biochem. Pharm.*, **13**, 731 (1964).
94. J. R. Bertino, D. M. Donohue, B. Simmons, B. W. Gabrio, R. Silber, and F. M. Huennekens, *J. Clin. Invest.*, **42**, 466 (1963).
95. E. Jenny and D. M. Greenberg, *J. Biol. Chem.*, **238**, 3378 (1963).
96. M. Friedkin, E. J. Crawford, E. Donovan, and E. J. Pastore, *J. Biol. Chem.*, **237**, 3811 (1962).
97. R. L. Blakley, *J. Biol. Chem.*, **238**, 2113 (1963).
98. J. G. Flaks and S. S. Cohen, *J. Biol. Chem.*, **234**, 2981 (1959).
99. K-U. Hartman and C. Heidelberger, *J. Biol. Chem.*, **236**, 3006 (1961).
100. L. Goodman, J. DeGraw, R. L. Kisliuk, M. Friedkin, E. J. Pastore, E. J. Crawford, L. T. Plante, A. Al-Nahas, J. F. Morningstar, Jr., G. Kwok, L. Wilson, E. F. Donovan, and J. Ratzan, *J. Am. Chem. Soc.*, **86**, 308 (1964).
101. E. J. Pastore and M. Friedkin, *J. Biol. Chem.*, **237**, 3802 (1962).
102. M. Friedkin, in *The Kinetics of Cellular Proliferation*, F. Stohlman, Ed., Grune and Stratton, New York, 1959, p. 99.
103. V. S. Gupta, J. G. Ozols, and F. M. Huennekens, *Biochemistry*, in press.
104. S. Kaufman, in *Pteridine Chemistry*, W. Pfleiderer and E. C. Taylor, Eds., Pergamon, New York, 1964, p. 307; *J. Biol. Chem.*, **237**, PC2712 (1962).
105. A. R. Brenneman and S. Kaufman, *Biochem. Biophys. Res. Commun.*, **17**, 177 (1964).
106. L. Jaenicke, *Biochim. Biophys. Acta*, **17**, 588 (1955).
107. G. R. Greenberg, L. Jaenicke, and M. Silverman, *Biochim. Biophys. Acta*, **17**, 589 (1955).
108. Y. Hatefi, M. J. Osborn, L. D. Kay, and F. M. Huennekens, *J. Biol. Chem.*, **227**, 637 (1957).
109. M. J. Osborn and F. M. Huennekens, *Biochim. Biophys. Acta*, **26**, 646 (1957).
110. B. V. Ramasastri and R. L. Blakley, *J. Biol. Chem.*, **237**, 1982 (1962).
111. K. O. Donaldson, V. F. Scott, and W. Scott, *J. Biol. Chem.*, **240**, 4444 (1965).
112. Y. Yeh and D. M. Greenberg, *Biochim. Biophys. Acta*, **105**, 279 (1965).
113. K. G. Scrimgeour and F. M. Huennekens, *Biochem. Biophys. Res. Commun.* **2**, 230 (1960).
114. W. Sakami and I. Ukstins, *J. Biol. Chem.*, **236**, PC50 (1961).
115. A. R. Larrabee, S. Rosenthal, R. E. Cathou, and J. M. Buchanan, *J. Am. Chem. Soc.*, **83**, 4094 (1961).
116. J. C. Keresztesy and K. O. Donaldson, *Biochem. Biophys. Res. Commun.*, **5**, 286, 289 (1961).
117. K. O. Donaldson and J. C. Keresztesy, *J. Biol. Chem.*, **234**, 3235 (1959); **237**, 1298 (1962).
118. R. E. Cathou and J. M. Buchanan, *J. Biol. Chem.*, **238**, 1746 (1963).
119. H. M. Katzen and J. M. Buchanan, *J. Biol. Chem.*, **240**, 825 (1965).
119a. H. Gigliotti and F. M. Huennekens, unpublished observations.
120. W. R. Frisell, J. R. Cronin, and C. G. MacKenzie, *J. Biol. Chem.*, **237**, 2975 (1962).
121. G. P. Mell and F. M. Huennekens, *Federation Proc.*, **19**, 411 (1960).
121a. H. Beinert and W. R. Frisell, *J. Biol. Chem.*, **237**, 2988 (1962).
122. J. C. Craig, F. P. Dwyer, A. N. Glazer, and E. C. Horning, *J. Am. Chem. Soc.*, **83**, 1871 (1961).
123. J. C. Rabinowitz and W. E. Pricer, *J. Biol. Chem.*, **237**, 2898 (1962).

124. R. H. Himes and J. C. Rabinowitz, *J. Biol. Chem.*, **237**, 2903 (1962).

125. H. R. Whiteley, M. J. Osborn, and F. M. Huennekens, *J. Biol. Chem.*, **234**, 1538 (1959).

126. J. R. Bertino, B. Simmons, and D. M. Donohue, *J. Biol. Chem.*, **237**, 1314 (1962).

127. A. M. Albrecht, F. K. Pearce, and D. M. Hutchison, *J. Biol. Chem.*, **241**, 1036 (1966).

128. G. R. Greenberg and L. Jaenicke, in *Ciba Foundation Symposium on the Chemistry and Biology of Purines*, G. E. W. Wolstenholme and C. M. O'Connor, Eds., Little, Brown, Boston, 1957, p. 204.

129. H. R. Whiteley and F. M. Huennekens, *J. Biol. Chem.*, **237**, 1290 (1962).

130. R. C. Himes and J. C. Rabinowitz, *J. Biol. Chem.*, **237**, 2915 (1962).

130a. J. C. Rabinowitz and R. C. Himes, *Federation Proc.*, **19**, 963 (1960).

131. H. R. Whiteley, M. J. Osborn, and F. M. Huennekens, *J. Am. Chem. Soc.*, **80**, 757 (1958).

132. H. R. Whiteley, *Comp. Biochem. Physiol.*, **1**, 227 (1960).

133. L. P. Hager, *J. Am. Chem. Soc.*, **79**, 4864 (1957).

134. O. Lindberg, J. J. Duffy, A. W. Norman, and P. D. Boyer, *J. Biol. Chem.*, **240**, 2851 (1965).

135. K. Uyeda and J. C. Rabinowitz, *Arch. Biochem. Biophys.*, **107**, 419 (1964).

136. W. S. Sly and E. R. Stadtman, *J. Biol. Chem.*, **238**, 2639 (1963).

137. J. M. Buchanan, S. C. Hartman, R. L. Herrmann, and R. A. Day, *J. Cell. Comp. Physiol.*, **54**, Suppl. 1, 139 (1959).

138. G. R. Greenberg, *J. Am. Chem. Soc.*, **76**, 1458 (1954).

139. L. D. Kay, M. J. Osborn, Y. Hatefi, and F. M. Huennekens, *J. Biol. Chem.*, **235**, 195 (1960).

140. D. M. Greenberg, L. K. Wynston, and A. Nagabhushanam, *Biochemistry*, **4**, 1872 (1965).

141. J. C. Rabinowitz and W. E. Pricer, *J. Am. Chem. Soc.*, **78**, 5702 (1956).

142. J. C. Rabinowitz and W. E. Pricer, *J. Am. Chem. Soc.*, **78**, 4176 (1956).

143. H. Tabor and L. Wyngarden, *J. Biol. Chem.*, **234**, 1830 (1959).

144. S. C. Hartman and J. M. Buchanan, *J. Biol. Chem.*, **234**, 1812 (1959).

145. D. M. Greenberg, in *Methods in Enzymology*, Vol. 6, S. P. Colowick and N. O. Kaplan, Eds., Academic Press, New York, 1963, p. 386.

146. F. M. Huennekens, H. R. Whiteley, and M. J. Osborn, *J. Cell. Comp. Physiol.*, **54**, Suppl. 1, 109 (1959).

147. M. J. Osborn, Y. Hatefi, L. D. Kay, and F. M. Huennekens, *Biochim. Biophys. Acta*, **26**, 208 (1957).

148. L. Warren, J. G. Flaks, and J. M. Buchanan, *J. Biol. Chem.*, **229**, 627 (1957).

149. J. G. Flaks, M. J. Erwin, and J. M. Buchanan, *J. Biol. Chem.*, **229**, 603 (1957).

150. D. W. E. Smith and B. N. Ames, *J. Biol. Chem.*, **239**, 1848 (1964).

151. R. G. Martin, *J. Biol. Chem.*, **238**, 257 (1963).

152. B. N. Ames, in *Amino Acid Metabolism*, W. D. McElroy and H. B. Glass, Eds., Johns Hopkins Press, Baltimore, 1955, p. 357.

153. K. A. Marcker and F. Sanger, *J. Mol. Biol.*, **8**, 835 (1964).

154. J. M. Adams and M. R. Capecchi, *Proc. Natl. Acad. Sci., U.S.*, **55**, 147 (1966).

155. B. F. C. Clark and K. A. Marcker, *J. Mol. Biol.*, **17**, 394 (1966).

156. J. P. Waller, *J. Mol. Biol.*, **7**, 483 (1963).

157. M. Silverman, J. C. Keresztesy, and R. C. Gardiner, *J. Biol. Chem.*, **226**, 83 (1957).

158. J. C. Rabinowitz and W. E. Pricer, *J. Biol. Chem.*, **222**, 537 (1956).
159. R. D. Sagers and J. V. Beck, *J. Bacteriol.*, **72**, 199 (1956).
160. K. Uyeda and J. C. Rabinowitz, *J. Biol. Chem.*, **240**, 1701 (1965).
161. R. D. Sagers, J. V. Beck, W. Gruber, and I. C. Gunsalus, *J. Am. Chem. Soc.*, **78**, 694 (1956).
162. H. Tabor, in *Amino Acid Metabolism*, W. D. McElroy and H. B. Glass, Eds., Johns Hopkins Press, Baltimore, 1955, p. 373.
163. H. Waelsch and A. Miller, in *Amino Acid Metabolism*, W. D. McElroy and H. B. Glass, Eds., Johns Hopkins Press, Baltimore, 1955, p. 407.
164. A. Miller and H. Waelsch, *J. Biol. Chem.*, **228**, 397 (1957).
165. H. Tabor and J. C. Rabinowitz, *J. Am. Chem. Soc.*, **78**, 5705 (1956).
166. I. B. Wilson, F. Bergmann, and D. Hachmansohn, *J. Biol. Chem.*, **186**, 781 (1950).
167. L. Jaenicke, *Federation Proc.*, **15**, 281 (1956).
168. R. L. Kisliuk, *Federation Proc.*, **15**, 289 (1956).
169. R. L. Blakley, *Biochem. J.*, **72**, 707 (1959).
170. R. L. Blakley, *Biochem. J.*, **74**, 71 (1960).
171. R. L. Blakley, *Nature*, **182**, 1719 (1958).
172. F. M. Huennekens and M. J. Osborn, *Intern. Congr. Biochem., 4th, Abstr.*, Section 4, 1958, p. 54.
173. F. M. Huennekens, M. J. Osborn, and H. R. Whiteley, *Science*, **128**, 120 (1958).
174. G. P. Mell, K. G. Scrimgeour, and F. M. Huennekens, *Federation Proc.*, **18**, 287 (1959).
175. M. J. Osborn, E. N. Vercamer, P. T. Talbert, and F. M. Huennekens, *J. Am. Chem. Soc.*, **79**, 6565 (1957).
176. P. P. K. Ho, K. G. Scrimgeour, and F. M. Huennekens, *J. Am. Chem. Soc.*, **82**, 5597 (1960).
177. P. P. K. Ho, K. G. Scrimgeour, and F. M. Huennekens, in preparation.
178. P. P. K. Ho and F. M. Huennekens, in preparation.
179. N. Alexander and G. R. Greenberg, *J. Biol. Chem.*, **214**, 821 (1955); **220**, 775 (1956).
180. R. L. Blakley, *Biochem. J.*, **65**, 342 (1957).
181. K. G. Scrimgeour and F. M. Huennekens, in *Handbuch der physiologisch- und pathologisch-chemischen Analyse*, Vol. VI, K. Lang and E. Lehnartz, Eds., Springer-Verlag, Heidelberg, 1966, p. 181.
182. R. L. Kisliuk and W. Sakami, *J. Biol. Chem.*, **214**, 47 (1955).
183. B. E. Wright, *Biochim. Biophys. Acta*, **16**, 165 (1955).
184. A. P. Wilkinson and D. D. Davies, *Nature*, **181**, 1070 (1958).
185. L. G. Schirch and M. Mason, *J. Biol. Chem.*, **238**, 1032 (1963).
186. D. E. Metzler, M. Ikawa, and E. E. Snell, *J. Am. Chem. Soc.*, **76**, 648 (1954).
187. R. L. Blakley, *Biochem. J.*, **61**, 315 (1955).
188. L. G. Schirch and W. T. Jenkins, *J. Biol. Chem.*, **239**, 3801 (1964).
189. E. E. Snell, in *Chemical and Biological Aspects of Pyridoxal Catalysis*, E. E. Snell, P. Fasella, A. Braunstein, and A. Rossi-Fanelli, Eds., Pergamon Press, Oxford, 1963, pp. 1–12.
190. E. M. Wilson and E. E. Snell, *J. Biol. Chem.*, **237**, 3171 (1962).
191. C. K. Mathews, F. Brown, and S. S. Cohen, *J. Biol. Chem.*, **239**, 2957 (1964).
192. R. Somerville, K. Ebisuzaki, and G. R. Greenberg, *Proc. Natl. Acad. Sci. U.S.*, **45**, 1240 (1959).
193. R. D. Sagers and I. C. Gunsalus, *J. Bacteriol.*, **81**, 541 (1961).

194. D. A. Richert, R. Amberg, and M. Wilson, *J. Biol. Chem.*, **237**, 99 (1962).
195. R. D. Sagers and S. M. Klein, *Federation Proc.*, **24**, 219 (1965).
196. M. L. Baginsky and F. M. Huennekens, *Biochem. Biophys. Res. Commun.*, **23**, 600 (1966).
196a. M. L. Baginsky and F. M. Huennekens, *Arch. Biochem. Biophys.*, in press.
197. S. M. Klein and R. D. Sagers, *J. Biol. Chem.*, **241**, 197 (1966).
198. S. M. Klein and R. D. Sagers, *J. Biol. Chem.*, **241**, 206 (1966).
199. C. Wagner, S. M. Lusty, H. Kung, and N. L. Rogers, *J. Biol. Chem.*, **241**, 1923 (1966); **242**, 1287 (1967).
200. E. L. Smith, *Nature*, **161**, 638 (1948); **162**, 144 (1948).
201. E. L. Rickes, N. G. Brink, F. R. Koniusky, T. R. Wood, and K. Folkers, *Science*, **107**, 396 (1948).
202. K. H. Fantes, J. E. Page, L. F. J. Parker, and E. L. Smith, *Proc. Roy. Soc. (London)*, **B136**, 592 (1950).
203. E. L. Smith, *Vitamin B$_{12}$*, 2nd ed., Methuen, London, 1963.
204. K. Folkers and D. E. Wolf, *Vitamins Hormones*, **12**, 1 (1954).
205. A. R. Todd and A. W. Johnson, *Vitamins Hormones*, **15**, 1 (1957).
206. D. C. Hodgkin, J. Kamper, M. MacKay, J. Pickworth, K. N. Trueblood, and J. G. White, *Nature*, **178**, 64 (1956).
207. S. K. Kon and J. Pawelkiewicz, *Symp. No. XI, Intern. Congr. Biochem. 4th, Vienna, 1958*, p. 115.
208. D. Perlman, *Advan. Appl. Microbiol.*, **1**, 87 (1959).
209. J. I. Toohey, *Federation Proc.*, **25**, 1628 (1966).
210. H. A. Barker, H. Weissbach, and R. D. Smyth, *Proc. Natl. Acad. Sci. U.S.*, **34**, 1093 (1958).
211. H. A. Barker, R. D. Smyth, H. Weissbach, A. Munch-Petersen, J. I. Toohey, J. N. Ladd, B. E. Volcani, and R. M. Wilson, *J. Biol. Chem.*, **235**, 181 (1960).
212. H. A. Barker, R. D. Smyth, H. Weissbach, J. I. Toohey, J. N. Ladd, and B. E. Volcani, *J. Biol. Chem.*, **235**, 480 (1960).
213. J. I. Toohey and H. A. Barker, *J. Biol. Chem.*, **236**, 560 (1961).
214. K. Lindstrand, *Acta Chem. Scand.*, **19**, 1762 (1965).
215. K. Lindstrand, *Acta Chem. Scand.*, **19**, 1785 (1965).
216. H. Weissbach, J. N. Ladd, B. E. Volcani, R. D. Smyth, and H. A. Barker, *J. Biol. Chem.*, **235**, 1462 (1960).
217. J. N. Ladd, H. P. C. Hogenkamp, and H. A. Barker, *Biochem. Biophys. Res. Commun.*, **2**, 143 (1960).
218. A. W. Johnson and N. Shaw, *J. Chem. Soc.*, **1962**, 4608.
218a. H. P. C. Hogenkamp, H. A. Barker, and H. S. Mason, *Arch. Biochem. Biophys.*, **100**, 353 (1963).
219. P. G. Lenhert and D. C. Hodgkin, *Nature*, **192**, 937 (1961).
220. Symposium on B$_{12}$ Coenzymes, *Federation Proc.*, **25**, 1623 (1966).
221. D. Perlman, Ed., *Vitamin B$_{12}$ Coenzymes (Ann. N.Y. Acad. Sci., 112)*, 547–921 (1964).
222. J. A. Hill, J. M. Pratt, and R. J. P. Williams, *J. Theoret. Biol.*, **3**, 423 (1962).
223. R. Bonnett, *Chem. Rev.*, **63**, 573 (1963).
224. D. Dolphin, A. W. Johnson, R. Rodrigo, and N. Shaw, *Pure Appl. Chem.*, **7**, 539 (1963).
225. K. Bernhauer, P. Müller, and F. Wagner, *Angew. Chem. (Intern. Ed.)*, **3**, 200 (1964).

226. W. S. Beck, *New Engl. J. Med.*, **266**, 708, 765, 814 (1962).
227. H. Weissbach and H. Dickerman, *Physiol. Rev.*, **45**, 80 (1965).
228. F. M. Huennekens, *Progress in Hematology*, Vol. 5, E. B. Brown and C. V. Moore, Eds., Grune and Stratton, New York, 1966, pp. 83–104.
229. G. H. Beaven and E. A. Johnson, *Nature*, **176**, 1264 (1955).
230. E. Vitols, G. A. Walker, and F. M. Huennekens, *J. Biol. Chem.*, **241**, 1455 (1966).
231. H. Diehl and R. Murie, *Iowa State Coll. J. Sci.*, **24**, 555 (1952).
232. S. L. Tackett, J. W. Collat, and J. C. Abbott, *Biochemistry*, **2**, 919 (1963).
233. N. K. King and M. E. Winfield, *J. Am. Chem. Soc.*, **83**, 3366 (1961).
234. J. W. Collat and J. C. Abbott, *J. Am. Chem. Soc.*, **86**, 2308 (1964).
235. R. O. Brady and H. A. Barker, *Biochem. Biophys. Res. Commun.*, **4**, 373 (1961).
236. H. P. C. Hogenkamp, J. N. Ladd, and H. A. Barker, *J. Biol. Chem.*, **237**, 1950 (1962).
237. H. P. C. Hogenkamp, *J. Biol. Chem.*, **238**, 477 (1963).
238. H. P. C. Hogenkamp and T. G. Oikawa, *J. Biol. Chem.*, **239**, 1911 (1964).
239. H. P. C. Hogenkamp, *Biochemistry*, **5**, 417 (1966).
240. R. H. Yamada, T. Kato, S. Shimizu, and S. Fukui, *Biochim. Biophys. Acta*, **97**, 353 (1965).
241. A. W. Johnson, D. Oldfield, R. Rodrigo, and N. Shaw, *J. Chem. Soc.*, **1964**, 4080.
242. H. P. C. Hogenkamp and H. A. Barker, *J. Biol. Chem.*, **236**, 3097 (1961).
243. H. P. C. Hogenkamp, *Ann. N.Y. Acad. Sci.*, **112**, 552 (1964).
244. L. Ellenbogen, D. R. Highley, H. A. Barker, and R. D. Smyth, *Biochem. Biophys. Res. Commun.*, **3**, 178 (1960).
245. E. L. Smith, L. Mervyn, A. W. Johnson, and N. Shaw, *Nature*, **194**, 1175 (1962).
246. O. Müller and G. Müller, *Biochem. Z.*, **336**, 299 (1962).
247. W. Friedrich and E. Königk, *Biochem. Z.*, **336**, 444 (1962).
248. V. S. Gupta and F. M. Huennekens, *Arch. Biochem. Biophys.*, **106**, 527 (1964).
249. E. L. Smith, L. Mervyn, P. W. Muggleton, A. W. Johnson, and N. Shaw, *Ann. N.Y. Acad. Sci.*, **112**, 565 (1964).
250. R. Barnett, H. P. C. Hogenkamp, and R. H. Abeles, *J. Biol. Chem.*, **241**, 1483 (1966).
251. A. W. Johnson, L. Mervyn, N. Shaw, and E. L. Smith, *J. Chem. Soc.*, **1963**, 4146.
252. R. Schmidt and F. M. Huennekens, *Arch. Biochem. Biophys.*, **118**, 253 (1967).
253. G. N. Schrauzer and J. Kohnle, *Chem. Ber.*, **97**, 3056 (1964).
254. G. N. Schrauzer, R. N. Windgassen, and J. Kohnle, *Chem. Ber.*, **98**, 3324 (1965).
254a. G. N. Schrauzer and R. J. Windgassen, *J. Am. Chem. Soc.*, **88**, 3738 (1966).
255. H. Weissbach, B. Redfield, and A. Peterkofsky, *J. Biol. Chem.*, **236**, PC41 (1961).
256. R. O. Brady and H. A. Barker, *Biochem. Biophys. Res. Commun.*, **4**, 464 (1961).
257. J. Pawelkiewicz, B. Bartosinski, and W. Walerych, *Ann. N.Y. Acad. Sci.*, **112**, 638 (1964).
258. A. Peterkofsky and H. Weissbach, *Ann. N.Y. Acad. Sci.*, **112**, 622 (1964).
259. A. Peterkofsky and H. Weissbach, *J. Biol. Chem.*, **238**, 1491 (1963).
260. R. O. Brady, E. G. Castenara, and H. A. Barker, *J. Biol. Chem.*, **237**, 2325 (1962).
261. E. Vitols, G. Walker, and F. M. Huennekens, *Biochem. Biophys. Res. Commun.*, **15**, 372 (1964).

262. J. W. Dubnoff, *Biochem. Biophys. Res. Commun.*, **16**, 484 (1964).
263. F. Wagner and K. Bernhauer, *Ann. N.Y. Acad. Sci.*, **112**, 580 (1964).
264. D. H. Dolphin and A. W. Johnson, *J. Chem. Soc.*, **1965**, 2174.
265. J. L. Peel, *Biochem. J.*, **88**, 296 (1963).
266. R. Schmidt, G. Walker, and F. M. Huennekens, in preparation.
267. J. B. Hufham, R. C. Bergus, W. M. Scott, and J. J. Pfiffner, *J. Bacteriol.*, **88**, 538 (1964).
268. G. Walker, R. Schmidt, S. Murphy, and F. M. Huennekens, in preparation.
269. S. Kashket, J. L. Tave, and W. S. Beck, *Biochem. Biophys. Res. Commun.*, **3**, 435 (1960).
270. H. Weissbach, N. Brot, and W. Levenberg, *J. Biol. Chem.*, **241**, 317 (1966).
271. H. A. Barker, V. Rooze, F. Suzuki, and A. A. Iodice, *J. Biol. Chem.*, **239**, 3260 (1964).
272. A. Munch-Petersen and H. A. Barker, *J. Biol. Chem.*, **230**, 649 (1958).
273. H. A. Barker, R. D. Smyth, R. M. Wilson, and H. Weissbach, *J. Biol. Chem.*, **234**, 320 (1959).
274. H. A. Barker, F. Suzuki, A. Iodice, and V. Rooze, *Ann. N.Y. Acad. Sci.*, **112**, 644 (1964).
275. F. Suzuki and H. A. Barker, *J. Biol. Chem.*, **241**, 878 (1966).
276. R. Mazumder, T. Sasakawa, Y. Kaziro, and S. Ochoa, *J. Biol. Chem.*, **236**, PC53 (1961).
277. S. H. G. Allen, R. Kellermeyer, R. Stjernholm, B. Jacobson, and H. G. Wood, *J. Biol. Chem.*, **238**, 1637 (1963).
278. E. R. Stadtman, P. Overath, H. Eggerer, and F. Lynen, *Biochem. Biophys. Res. Commun.*, **2**, 1 (1960).
279. P. Overath, E. R. Stadtman, G. M. Kellerman, and F. Lynen, *Biochem. Z.*, **336**, 77 (1962).
280. R. W. Kellermeyer, S. H. G. Allen, R. Stjernholm, and H. G. Wood, *J. Biol. Chem.*, **239**, 2562 (1964).
281. P. Lengyel, R. Mazumder, and S. Ochoa, *Proc. Natl. Acad. Sci. U.S.*, **46**, 1312 (1960).
282. R. Mazumder, T. Sasakawa, and S. Ochoa, *J. Biol. Chem.*, **238**, 50 (1963).
283. J. R. Stern and D. L. Friedman, *Biochem. Biophys. Res. Commun.*, **2**, 82 (1690).
284. A. A. Iodice and H. A. Barker, *J. Biol. Chem.*, **238**, 2094 (1963).
285. J. D. Erfle, J. M. Clark, and B. C. Johnson, *Ann. N.Y. Acad. Sci.*, **112**, 684 (1964).
286. P. Overath, G. M. Kellerman, F. Lynen, H. P. Fritz, and H. J. Keller, *Biochem. Z.*, **335**, 500 (1962).
287. L. L. Ingraham, *Ann. N.Y. Acad. Sci.*, **112**, 713 (1964).
288. H. W. Whitlock, *J. Am. Chem. Soc.*, **85**, 2343 (1963).
289. H. Eggerer, E. R. Stadtman, P. Overath, and F. Lynen, *Biochem. Z.*, **333**, 1 (1960).
290. R. W. Kellermeyer and H. G. Wood, *Biochemistry*, **1**, 1124 (1962).
291. E. F. Phares, M. V. Long, and S. F. Carson, *Biochem. Biophys. Res. Commun.*, **8**, 142 (1962).
292. M. Sprecher and D. B. Sprinson, *Ann. N.Y. Acad. Sci.*, **112**, 655 (1964).
293. A. M. Brownstein and R. H. Abeles, *J. Biol. Chem.*, **236**, 1199 (1961).
294. R. H. Abeles and H. A. Lee, *J. Biol. Chem.*, **236**, 2347 (1961).
295. H. A. Lee and R. H. Abeles, *J. Biol. Chem.*, **238**, 2367 (1963).

296. R. H. Abeles and H. A. Lee, *Ann. N.Y. Acad. Sci.*, **112**, 695 (1964).
297. O. W. Wagner, H. A. Lee, Jr., P. A. Frey, and R. H. Abeles, *J. Biol. Chem.*, **241**, 1751 (1966).
298. R. H. Abeles and B. Zagalak, *J. Biol. Chem.*, **241**, 1245 (1966).
299. P. A. Frey and R. H. Abeles, *J. Biol. Chem.*, **241**, 2732 (1966).
300. B. H. Kaplan, *Federation Proc.*, **25**, 277 (1966).
301. P. Reichard, *J. Biol. Chem.*, **237**, 3513 (1962).
302. E. C. Moore, P. Reichard, and L. Thelander, *J. Biol. Chem.*, **239**, 3445 (1964).
303. T. C. Laurent, E. C. Moore, and P. Reichard, *J. Biol. Chem.*, **239**, 3436 (1964).
304. A. Holmgren, P. Reichard, and L. Thelander, *Proc. Natl. Acad. Sci. U.S.*, **54**, 830 (1965).
305. E. C. Moore and P. Reichard, *J. Biol. Chem.*, **239**, 3453 (1964).
306. R. L. Blakley and H. A. Barker, *Biochem. Biophys. Res. Commun.*, **16**, 391 (1964).
307. R. L. Blakley, *J. Biol. Chem.*, **240**, 2173 (1965).
308. E. Vitols and R. L. Blakley, *Biochem. Biophys. Res. Commun.*, **21**, 466 (1965).
309. R. L. Blakley, R. K. Ghambeer, P. F. Nixon, and E. Vitols, *Biochem. Biophys. Res. Commun.*, **20**, 439 (1965).
310. W. S. Beck, M. Goulian, A. Larsson, and P. Reichard, *J. Biol. Chem.*, **241**, 2177 (1966).
310a. M. Goulian and W. S. Beck, *J. Biol. Chem.*, **241**, 4233 (1966).
311. J. R. Guest, S. Friedman, and M. A. Foster, *Biochem. J.*, **84**, 93P (1962).
312. J. R. Guest, S. Friedman, and D. D. Woods, *Nature*, **195**, 340 (1962).
313. R. L. Kisliuk, *J. Biol. Chem.*, **238**, 1025 (1963).
314. H. Weissbach, A. Peterkofsky, B. G. Redfield, and H. Dickerman, *J. Biol. Chem.*, **238**, 3318 (1963).
315. J. H. Mangum and K. G. Scrimgeour, *Federation Proc.*, **21**, 242 (1962).
316. S. S. Kerwar, J. H. Mangum, K. G. Scrimgeour, J. D. Brodie, and F. M. Huennekens, *Arch. Biochem. Biophys.*, **116**, 305 (1966).
317. R. L. Kisliuk and D. D. Woods, *Biochem. J.*, **75**, 467 (1960).
318. J. R. Guest, C. W. Helleiner, M. J. Cross, and D. D. Woods, *Biochem. J.*, **76**, 396 (1960).
319. S. Takeyama, F. T. Hatch, and J. M. Buchanan, *J. Biol. Chem.*, **236**, 1102 (1961).
320. S. Takeyama and J. M. Buchanan, *Biochem. J.* (*Tokyo*), **49**, 478 (1961).
321. S. S. Kerwar, J. H. Mangum, K. G. Scrimgeour, and F. M. Huennekens, *Biochem. Biophys. Res. Commun.*, **15**, 377 (1964).
322. M. A. Foster, M. J. Dilworth, and D. D. Woods, *Nature*, **201**, 39 (1964).
323. N. Brot and H. Weissbach, *J. Biol. Chem.*, **240**, 3064 (1965).
324. S. Rosenthal, L. C. Smith, and J. M. Buchanan, *J. Biol. Chem.*, **240**, 836 (1965).
325. H. Weissbach, B. Redfield, and H. Dickerman, *Biochem. Biophys. Res. Commun.*, **17**, 17 (1964).
326. F. T. Hutch, A. R. Larrabee, R. E. Cathou, and J. M. Buchanan, *J. Biol. Chem.*, **236**, 1095 (1961).
327. S. Rosenthal and J. M. Buchanan, *Acta Chem. Scand.*, **17** (*Suppl. 1*), 288 (1963).
328. H. Weissbach, in *Transmethylation and Methionine Biosynthesis*, S. K. Shapiro and F. Schlenk, Eds., Univ. of Chicago Press, Chicago, 1965, p. 179.
329. A. W. Johnson, N. Shaw, and F. Wagner, *Biochim. Biophys. Acta*, **72**, 107 (1963).

330. H. L. Elford, R. E. Loughlin, and J. M. Buchanan, *Federation Proc.*, **23**, 480 (1964).
331. S. Kerwar, in *Transmethylation and Methionine Biosynthesis*, S. K. Shapiro and F. Schlenk, Eds., Univ. of Chicago Press, Chicago, 1965, p. 194.
332. D. D. Woods, M. A. Foster, and J. R. Guest, in *Transmethylation and Methionine Biosynthesis*, S. K. Shapiro and F. Schlenk, Eds., Univ. of Chicago Press, Chicago, 1965, p. 138.
333. B. A. Blaylock and T. C. Stadtman, *Biochem. Biophys. Res. Commun.*, **11**, 34 (1963).
334. B. A. Blaylock and T. C. Stadtman, *Biochem. Biophys. Res. Commun.*, **17**, 475 (1964).
335. M. J. Wolin, E. A. Wolin, and R. S. Wolfe, *Biochem. Biophys. Res. Commun.*, **12**, 464 (1963).
336. M. J. Wolin, E. A. Wolin, and R. S. Wolfe, *Biochem. Biophys. Res. Commun.*, **15**, 420 (1964).
337. J. M. Wood and R. S. Wolfe, *Biochem. Biophys. Res. Commun.*, **19**, 306 (1965).
338. J. M. Poston, K. Kuratomi, and E. R. Stadtman, *Ann. N.Y. Acad. Sci.*, **112**, 804 (1964).
339. L. Ljungdahl, E. Irion, and H. G. Wood, *Biochemistry*, **4**, 2771 (1965).

Ferredoxin and Photosynthetic Pyridine Nucleotide Reductase[*]

ANTHONY SAN PIETRO, *Charles F. Kettering Research Laboratory,*
Yellow Springs, Ohio

I.	Scope .	. 515
II.	Classification	. 516
III.	Properties .	. 518
IV.	Catalytic Activity .	. 521
	A. PPNR (Spinach Ferredoxin)	. 521
	B. *C. pasteurianum* Ferredoxin	. 525
V.	Proposed Structure of Ferredoxin .	. 526
VI.	Some Concluding Observations	. 528
VII.	Addendum .	. 528
	References .	. 529

I. Scope

Interest in non-heme iron proteins in which the iron atom is believed to undergo reversible oxidation-reduction during the course of their catalytic activity has intensified greatly in the past three years (1–7). The stimulus for the increased interest in these proteins has derived primarily from the detailed investigations of several processes, including mitochondrial electron transport, bacterial nitrogen assimilation, and photosynthetic electron transport. Information concerned with both the respiratory chain-linked dehydrogenases and metalloflavoproteins from mammalian or microbial sources is presented earlier in this volume (8,9) and will only be referred to here cursorily.

This chapter is concerned primarily with a special group of non-heme iron proteins of low molecular weight, called ferredoxins, which have considerable importance in electron transport in microorganisms and in chloroplasts. Two examples will be considered in detail: *C. pasteurianum*

[*] Contribution No. 204 of the Charles F. Kettering Research Laboratory. Some of the research described in this chapter was supported by the National Institutes of Health, Grant GM-10129. Literature survey completed December 1965.

515

ferredoxin and photosynthetic pyridine nucleotide reductase (PPNR), which is synonymous with spinach ferredoxin. Inasmuch as the studies of these ferredoxins have led to findings of more general interest, reference to related non-heme iron proteins will be included where applicable. A recent and more general review of iron proteins is available (10).

II. Classification

The term "ferredoxin" is used herein to denote those non-heme iron proteins of low molecular weight and of established biological function which contain non-heme iron and "labile sulfur" in essentially equivalent amounts (Table I, Type 1). It has been shown (11) very recently that the ferredoxins exhibit an electron paramagnetic resonance (EPR) signal at $g = 1.94$ in the reduced state at a temperature of 40°K. As stated by Palmer and Sands (11), "The signal can only be observed at temperatures significantly below those routinely employed which explains the failure to observe these resonances earlier."

There are a number of additional non-heme iron proteins (Table I, Type 2) which may be classified as "ferredoxin-like." They differ from the ferredoxins in molecular weight and the temperature sensitivity of the EPR signal at $g = 1.94$; that is, the "ferredoxin-like" proteins exhibit an EPR signal in the reduced state at a temperature of 98°K (12). Some of them are complex proteins which contain one or more cofactors, such as flavin, coenzyme Q, and molybdenum, in addition to the non-heme iron and "labile sulfur" (Table I, Type 2, examples 1–5); others are simpler proteins which do not contain additional cofactors (Table I, Type 2, examples 6–9).

It should be noted that the classification of non-heme iron proteins presented in Table I is essentially that suggested previously by Beinert (12) and extended by Racker (13). Although not indicated in Table I, the "labile sulfur" is liberated as hydrogen sulfide from all of them with cold dilute acid or upon denaturation by other methods.

Although non-heme iron proteins other than those listed in Table I have been described (for literature citations, see 1–4), they have not been characterized sufficiently to permit classification of them at this time.

A different method of classification of non-heme iron proteins, based on their absorption spectra, was considered by San Pietro and Black (2). On this basis, the numerous proteins purified from green plants, algae, photosynthetic bacteria, and anaerobic heterotrophic bacteria are again divisible into two types: the plant (or algal) type and the bacterial type

TABLE I

Classification of Non-Heme Iron Proteins

Type 1—Ferredoxins

Properties:

(*a*) Molecular weight below 20,000.

(*b*) Contain non-heme iron and "labile sulfur" in essentially equivalent amounts.

(*c*) EPR signal at $g = 1.94$ in the reduced state at a temperature of 40°K.

Examples:

(1) *C. pasteurianum* ferredoxin (12).

(2) Photosynthetic pyridine nucleotide reductase (11,12).

(3) Non-heme iron protein from *Chromatium* (12,14).

Type 2—"Ferredoxin-Like"

Properties:

(*a*) Molecular weight above 20,000.

(*b*) Pronounced EPR signal at $g = 1.94$ in the reduced state at a temperature of 98°K.

(*c*) Contain non-heme iron and "labile sulfur" in essentially equivalent amounts. In addition, the complex proteins of this type contain other cofactors such as flavin, coenzyme Q, and molybdenum.

Examples:

(1) DPNH dehydrogenase (8).

(2) Succinic acid dehydrogenase (8).

(3) Hepatic aldehyde oxidase (9).

(4) Dihydroorotic acid dehydrogenase from *Zymobacterium oroticum* (9).

(5) Milk xanthine oxidase (9).

(6) Non-heme iron protein of the cytochrome b and c_1 region of the respiratory chain (15).

(7) Non-heme iron protein from *Azotobacter* (12,16).

(8) Paramagnetic protein from *C. pasteurianum* (17).

(9) Non-heme iron protein of adrenal cortex (18–20).

(Fig. 1). Unfortunately, there is no correspondence between these latter two types of non-heme iron proteins (plant or bacterial) and the two types characterized in Table I. For example, a number of the ferredoxins which exhibit the EPR signal at $g = 1.94$ in the reduced state at a temperature of 98°K (Table I, Type 2) have absorption spectra similar to that of PPNR (see Section III). Yet PPNR itself does not show this EPR signal unless the temperature is decreased to 40°K (11).

Arnon (21) has recently suggested a definition of "ferredoxins" based on their oxidation-reduction potential. This kind of definition of a biological catalyst is rather restrictive and, to my knowledge, without precedence. If we consider the cytochromes, the classification of them as

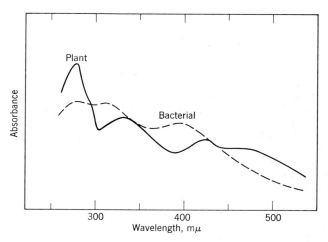

Fig. 1. Absorption spectra of spinach ferredoxin (24) and *C. pasteurianum* ferredoxin (25).

b-type, *c*-type, etc. is based on the nature of both the heme moiety and the linkages between the heme and protein but not on their oxidation-reduction potentials, which are quite variable. For example, the redox potentials for two *c*-type cytochromes, i.e., cytochrome *f* from green plant chloroplasts (22) and cytochrome-549 from *Anacystis nidulans* (23), are $+0.37$ and -0.26 V, respectively.

In the final analysis, the classification of non-heme iron proteins should be based on their chemical structure, i.e., the nature of the prosthetic group, rather than on a single chemical or physical property.

III. Properties

Both *C. pasteurianum* ferredoxin and spinach PPNR are proteins of low molecular weight and each exhibits a distinctive absorption spectrum in the oxidized form. The plant protein is red in color and its absorption spectrum is characterized by absorption maxima at 277, 330, and 420 mμ and a shoulder at 465 mμ (Fig. 1). In contrast, the bacterial protein is brown in color and exhibits absorption maxima at 280, 300, and 390 mμ. The striking dissimilarities between these two rather distinctive spectra could reflect a difference in the nature of the binding of the iron. The naturally occurring non-heme iron proteins may well be of a very heterogeneous nature and we should expect that numerous variants of iron binding exist among them.

Handler et al. (6) have noted the close resemblance in the region of 460 mμ between the absorption spectra of spinach PPNR and conalbumin. The binding of iron to conalbumin is thought to involve the phenolic hydroxyl groups of the tyrosyl residues of the protein. Since the spinach protein contains four tyrosyl residues and two atoms of iron per mole of protein (26,27), the attachment of iron may occur in part via the hydroxyl groups of the tyrosyl residues. A comparable situation is impossible for *C. pasteurianum* ferredoxin since it contains seven atoms of iron, but only one tyrosyl residue, per mole (25).

A number of the non-heme iron proteins listed in Table I have absorption spectra similar to that of spinach PPNR; for example, the flavin-free non-heme iron-containing proteins prepared from xanthine oxidase (9), dihydroorotic dehydrogenase (9), and aldehyde oxidase (9), succinate dehydrogenase (8), and the non-heme iron protein from adrenal cortex (18–20).

For the purposes of comparison, some additional properties of *C. pasteurianum* ferredoxin and spinach PPRN are presented in Table II.

TABLE II

Comparison of the Properties of *C. pasteurianum* Ferredoxin and Spinach PPNR

Property	*C. pasteurianum* ferredoxin	Spinach PPNR
Molecular weight	6000 (25)	13,000 (21)
Iron content (atoms per mole protein)	7 (25)	2 (21,27)
"Labile sulfur" content (moles per mole protein)	7 (25)	2 (27)
Oxidation-reduction potential at pH 7.55	-0.417 V (28)	-0.432 V (28)
Absorbance per atom of iron	~ 3000–4000 at 390 mμ (26)	5000 at 420 mμ (26)

They both contain non-heme iron and "labile sulfur" in a ratio 1:1. However, the absolute amount of non-heme iron and "labile sulfur" in each protein is quite different. Despite the vast difference in iron content of the two proteins, the absorbance per atom of iron in the visible region is comparable to that reported for other non-heme iron proteins (6). Upon loss of iron or "labile sulfur" (usually the removal of one is accompanied by loss of the other), both proteins lose their characteristic spectrum and catalytic activity (25–27).

Reduction of spinach PPNR with hydrosulfite leads to an immediate partial loss of the absorption of the protein in the visible region of the spectrum (29). The immediate decrease in absorption following hydrosulfite treatment corresponds to a loss of about 49 and 55% of the original absorbance at 420 and 465 mμ, respectively. The reduced protein can be reoxidized with oxygen, thereby regenerating the original absorption spectrum of the protein.

Treatment of ferredoxins with mercurials, such as pCMB or pCMS, results in a rapid loss in the visible spectra of the proteins. In the case of PPNR (26), there is a parallel loss in absorbance at each of the three absorption maxima in the visible region. After complete reaction, the percentage of the original absorbance remaining at 330, 420, and 465 mμ, was 40, 10, and 5%, respectively. Similar titrations have been performed by Katoh and Takamiya (30), who used mercuric chloride, copper sulfate, and silver nitrate, as well as pCMB, as the titrants.

Lovenberg et al. (25) suggested that sulfhydryl groups were involved in the binding of the iron in *C. pasteurianum* ferredoxin. Using isotopic iron, they observed exchange with the iron of the protein after treatment of the ferredoxin with sodium mersalyl; isotopic equilibrium was attained in five minutes. The reconstituted ferredoxin containing isotopic iron was essentially equivalent to native ferredoxin with respect to absorption spectrum, enzymic activity, iron and "labile sulfur" content, and electrophoretic mobility.

Both proteins are acidic in nature and are adsorbed very strongly on DEAE cellulose. This property has greatly simplified the purification of them. They both have a high content of acidic amino acids and a dearth of basic amino acids (25,26). The complete amino acid sequence of *C. pasteurianum* ferredoxin has been reported by Tanaka et al. (31,32) and will be considered below in relation to the model of the active site of *C. pasteurianum* ferredoxin proposed by Blomstrom et al. (33,34).

The oxidation-reduction potentials of these proteins are rather low and close to that of the hydrogen electrode (28). Arnon (21) has recently reported that the oxidation-reduction potentials of *Chromatium* and *Nostoc* ferredoxins are about −0.490 V (pH 7) and −0.405 V (pH 7.5), respectively.

The optical rotatory dispersion of PPNR has been measured by Ulmer and Vallee (35). The principal Cotton effect, which is positive and has an amplitude greater than 500°, appears to be associated with the absorption band of the protein at 420 mμ. Although the absorption band at 330 mμ is larger than that at 420 mμ, it gives rise to a smaller Cotton effect. In the presence of 1,10-phenanthroline, the iron is removed from the protein and the Cotton effects are lost. Fry and San Pietro (27) demonstrated that under conditions of pH and temperature at which PPNR is relatively stable, treatment with 1,10-phenanthroline results in loss of iron from the protein as well as a corresponding loss of "labile sulfur," absorption in the visible region of the spectrum, and biological activity.

IV. Catalytic Activity

A. PPNR (*Spinach Ferredoxin*)

The role of spinach ferredoxin as an electron carrier in several photo-reduction reactions catalyzed by chloroplasts is shown diagrammatically in Figure 2. The excitation energy, absorbed initially as two light quanta by the photochemical systems, appears in part as an excited electron which is used for the reduction of ferredoxin by some hitherto unknown mechanism. The resultant reduced ferredoxin is the first stable, chemically isolable, reductant formed by illuminated chloroplasts. It may be re-oxidized in the dark enzymically by TPN and TPN reductase or by nitrite and nitrite reductase. A dark non-enzymic oxidation of the reduced ferredoxin by oxygen, by a number of heme proteins, or by some component of the electron-transport chain which couples the two photochemical systems (Fig. 2, dashed arrow) is also possible.

There is a great deal of information available which is concerned with the two light reactions of photosynthesis as well as the thermochemical

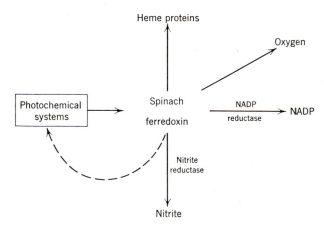

Fig. 2. Summary of photoreduction reactions (2).

process through which they are coupled and wherein phosphorylation is believed to occur. This evidence has been excellently reviewed by Hill (7), Arnon (21,36), Davenport (37), and Vernon and Avron (38) and will not be considered here. We will be concerned with the evidence which relates solely to the mechanism of catalysis by ferredoxin of TPN reduction by illuminated chloroplasts.

Chance and San Pietro (39) measured the kinetics of reduction of both TPN and spinach ferredoxin by illuminated chloroplasts and concluded that reduced ferredoxin was a permissible intermediate in TPN reduction. The apparent initial rate of reduction of ferredoxin was about three times that of TPN reduction on a molar basis. They further noted the rapidity of the dark oxidation of reduced ferredoxin (presumably by molecular oxygen); the initial rates of reduction in the light and subsequent oxidation in the dark were essentially equal. Thus, the true initial rate of reduction was probably greater than the measured rate.

In a recent study, Whatley et al. (40) measured the stoichiometry between ferredoxin and TPN. The ferredoxin was reduced anaerobically by illuminated chloroplasts and then TPN was added, in the dark, to oxidize the reduced ferredoxin. Whatley et al. showed that one mole of TPN oxidized two moles of reduced ferredoxin and thereby established that ferredoxin is a one-electron carrier. Similar results were obtained by Horio and San Pietro (41).

Fry et al. (29) demonstrated a reversible change in the valence state of one-half of the iron of spinach ferredoxin. As isolated, the protein contains two atoms of non-heme iron which are in the ferric valence state. Photoreduction of the protein by chloroplasts was accompanied by a concomitant change in the valence state of the iron. One-half of the iron was in the ferrous valence state when the protein was completely reduced. These results are in accord with the observations of Whatley et al. (40) that spinach ferredoxin functions as a single-electron carrier.

As noted earlier, the enzymic oxidation of reduced ferredoxin by TPN is mediated by a flavoprotein known successively as TPN-diaphorase (42,43), spinach pyridine nucleotide transhydrogenase (44,45), and TPN reductase (46,47). The evidence in support of the function of the flavoprotein may be summarized as follows: (1) the photoreduction of TPN with chloroplasts is inhibited by the antibody to TPN reductase even though excess spinach ferredoxin is provided (45); (2) both ferredoxin and TPN reductase are required for the dark reduction of TPN by hydrogen and hydrogenase (28); (3) chloroplasts from which the TPN reductase is removed, are unable to photoreduce TPN in the presence of only spinach ferredoxin but the activity is restored by subsequent addition of TPN reductase (46,48); and

(4) TPN reductase is reduced by hydrogen and hydrogenase via reduced ferredoxin and reoxidized by TPN (47).

To recapitulate, the mechanism of reduction of TPN by chloroplasts involves the initial photoreduction of ferredoxin, oxidation of reduced ferredoxin by TPN reductase to yield reduced TPN reductase, and, finally, reoxidation of the reduced TPN reductase by TPN. The reversible oxidation-reduction of the ferredoxin is accompanied by a change in the valence state of one-half of the iron (29). The flavin component [FAD (44)] of TPN reductase undergoes reversible oxidation-reduction in the transfer of electrons from reduced ferredoxin to TPN (47). Thus, the transfer of electrons (or hydrogen) from the photochemical system to the coenzyme involves the intermediate reversible oxidation-reduction of two proteins:

$$\text{Photochemical system} \longrightarrow \text{Ferredoxin} \longrightarrow \text{TPN reductase} \longrightarrow \text{TPNH}$$

$$\downarrow \qquad\qquad\qquad\qquad\qquad\qquad (1)$$

$$\text{Cytochrome } c$$
$$\text{(heme-proteins)}$$

On the other hand, the photochemical reduction of cytochrome c requires only ferredoxin; the electron transfer from reduced ferredoxin to cytochrome c is non-enzymic.

This type of mechanism was proposed independently in 1962 by Arnon's group (28) and by Lazzarini and San Pietro (49). Tagawa and Arnon (28) demonstrated that hydrogen in the presence of C. pasteurianum hydrogenase would replace light in TPN reduction. This dark enzymic reduction required ferredoxin and either chloroplasts or a soluble flavoprotein derived therefrom. The results obtained with the soluble system, in the complete absence of chloroplasts, provided definitive evidence that the physiological role of TPN reductase is the transfer of electrons from reduced ferredoxin to TPN.

Lazzarini and San Pietro (49) demonstrated an interaction between ferredoxin and TPN reductase in the absence of chloroplasts. Whereas neither protein alone showed significant TPNH–cytochrome c reductase activity, together they mediated a rapid TPNH-dependent reduction of cytochrome c. On the basis of these observations we suggested that the TPN reductase (transhydrogenase in ref. 49) interacts with the reduced pyridine nucleotide while the ferredoxin serves as an intermediary carrier between the TPN reductase and cytochrome c. We attempted to demonstrate the reduction of ferredoxin by TPNH and TPN reductase, under anaerobic conditions, but were unsuccessful. We ascribed the failure to observe the reduction of ferredoxin by TPNH to an unfavorable equilibrium. This was borne out by the report of Tagawa and Arnon (28) that the oxidation-reduction potential of spinach ferredoxin was about 0.1 V more negative than that of the pyridine nucleotides.

We should note that other examples of the joint catalysis by a ferredoxin and a flavoprotein of heme-protein reduction by reduced pyridine nucleotides have been reported. Shin et al. (46) observed the reduction of horseradish peroxidase by TPNH in the presence of spinach ferredoxin and TPN reductase. Yamanaka and Kamen (50,51) isolated both a TPN reductase and a ferredoxin from a facultative photoheterotroph, *Rhodopseudomonas palustris*, and a diatom, *Navicula pelliculosa*. The TPN reductase of *Navicula pelliculosa* catalyzed the reduction of cytochrome *c* by TPNH in the presence of added ferredoxin. With cytochrome-554 as the acceptor, although there was some activity in the absence of ferredoxin, the activity was greatly enhanced by addition of ferredoxin. In contrast, the TPN reductase of *R. palustris* functioned as a TPNH–cytochrome c_2 reductase, and addition of ferredoxin was not necessary.

The purification of a non-heme iron protein and a TPNH dehydrogenase flavoprotein from mitochondria of beef adrenal cortex has recently been described (18–20). The reduction of either cytochrome *c* or cytochrome P_{450} by TPNH required both the flavoprotein and the non-heme iron protein; neither protein was active in the absence of the other (18). The reduction of cytochrome P_{450} by the reconstituted system is of great interest, since this pigment is believed to be the terminal oxidase for adrenal cortex hydroxylations. The reaction sequence proposed (18) is the following:

$$\text{TPNH} \longrightarrow \text{Flavoprotein} \longrightarrow \text{Non-heme iron protein} \longrightarrow \atop \text{Cytochrome } P_{450} \longrightarrow O_2 \qquad (2)$$

As can be noted, there is great similarity between this sequence of reactions and that shown earlier (Reaction (1)) for the photosynthetic electron transport sequence to TPN.

It is interesting that the adrenal cortex non-heme iron protein is reducible aerobically by TPNH and its flavoprotein (18). This finding is consistent with the observed redox potential of the non-heme iron protein of $+0.15$ V at pH 7.4 (19). It is somewhat surprising, therefore, that Kimura and Suzuki (20) only reported reduction of the non-heme iron protein under anaerobic conditions by TPNH and its flavoprotein.

The photoreduction of nitrite by chloroplasts requires both ferredoxin and nitrite reductase, as does the dark reduction by hydrogen and hydrogenase (52). Paneque et al. (52) reported that spinach ferredoxin is a more effective electron carrier than *C. pasteurianum* ferredoxin in the photochemical system; the reverse is true for the dark reduction of nitrite by hydrogen and hydrogenase. They further demonstrated that the photochemical reduction of nitrite by chloroplasts is accompanied by the evolution of oxygen and the formation of ATP.

Huzisige and Satoh (53) reported that the photochemical reduction of nitrite by grana required a soluble enzyme which they designated as "photosynthetic nitrite reductase." Huzisige et al. (54) separated the "photosynthetic nitrite reductase" into two fractions: one is identical with spinach ferredoxin, the other is a flavoprotein. Both fractions are required to demonstrate nitrite reduction by illuminated grana. The flavoprotein contains FMN as its prosthetic group.

In concluding this section, it should be reiterated that the photoreduction of ferredoxin is the terminal step in the electron-transport chain of chloroplasts. According to present concepts (7,21,36,37), the reduction of ferredoxin by illuminated chloroplasts is accompanied by the formation of ATP and the evolution of oxygen in a molar ratio of $4:2:1$.

B. *C. Pasteurianum Ferredoxin*

This protein is required for nitrogen fixation in cell-free extracts of *C. pasteurianum* when either pyruvate or hydrogen is the reducing agent. It was isolated originally by Mortenson et al. (55), who showed that it was necessary for the phosphoroclastic cleavage of pyruvate:

$$CH_3COCOO^- + H^+ + HPO_4^{2-} \longrightarrow CH_3COPO_4^{2-} + CO_2 + H_2 \qquad (3)$$

Ferredoxin functions here as an electron carrier that couples pyruvate oxidation and hydrogenase (hydrogen evolution). In their original experiments, Mortenson et al. (55) observed that ferredoxin could replace methyl viologen, which was used originally by Shug and Wilson (56) as the electron carrier.

In the clostridial system, reduced ferredoxin, reduced either by pyruvate or hydrogen, is presumed to be the natural electron donor for nitrogen fixation via "nitrogenase." It is not known as yet whether reduced ferredoxin donates electrons directly to nitrogenase (17) or through the mediation of an intermediate carrier(s). It is important to note that Bulen et al. (57,58) have observed nitrogen fixation in cell-free preparations of *Azotobacter* in which a ferredoxin (or ferredoxin-like compound) function in nitrogen fixation has not been detected to date.

In view of the paucity of information available, it is impossible to discuss in any greater detail the mechanism of catalysis by ferredoxin of nitrogen fixation at this time. There is, however, a great deal of information indicating that ferredoxin is involved in the reduction of a variety of compounds, such as DPN, TPN, hypoxanthine, and nitrite, by hydrogen and hydrogenase. This information has recently been reviewed by Valentine (4). He proposed that DPN and TPN are reduced in a manner analogous to that described for the photoreduction of TPN (Reaction (1)). In this case, the reductant is hydrogen (and hydrogenase) rather than photo-excited electrons provided by the photochemical system of chloroplasts.

V. Proposed Structure of Ferredoxin

The model proposed for the active site of *C. pasteurianum* ferredoxin is shown in Figure 3. The chelate structure was proposed by Blomstrom et al. (33), and the amino acid sequence was determined by Tanaka et al. (31,32). In brief, the model assumes a linear arrangement of iron atoms which are attached to the cysteine residues of the protein. The "labile sulfur" and the sulfur atoms of cysteine both serve as sulfur bridges to bond the iron atoms to each other. The presumed proximity of the iron atoms would allow for a strong covalent interaction between them and thereby result in an efficient electron transport within (or through) the linear arrangement.

This model is consistent with much of the chemical and physical data on *C. pasteurianum* ferredoxin. It accounts for the numerical equivalence of "labile sulfur" atoms, iron atoms, and cysteine residues (25). The environment surrounding the two terminal iron atoms is different from that about the five interior iron atoms of the linear array. The loss of "labile sulfur" as hydrogen sulfide would be expected.

The cysteine residues are distributed equally between, and appear in, only two regions of the protein. Folding of the protein allows for a structure in which the cysteine residues of one region are complementary to those of the other region. The iron atoms would be held jointly by a cysteine from each region of the protein.

Sieker and Jensen (59) recently published the results of an x-ray investigation of the structure of ferredoxin from *Micrococcus aerogenes*. Their data indicate that the iron atoms of this ferredoxin are not arranged in one single long linear array but, rather, in several short linear arrays.

Blomstrom et al. (33,34) had earlier cautioned that, in the final analysis, only x-ray results are acceptable as definitive evidence with regard to the geometrical arrangements of the iron atoms in *C. pasteurianum* ferredoxin. The model shown in Figure 3 represents only a working hypothesis until such time as x-ray analyses are available. (Preliminary x-ray analyses are quoted in ref. 32.)

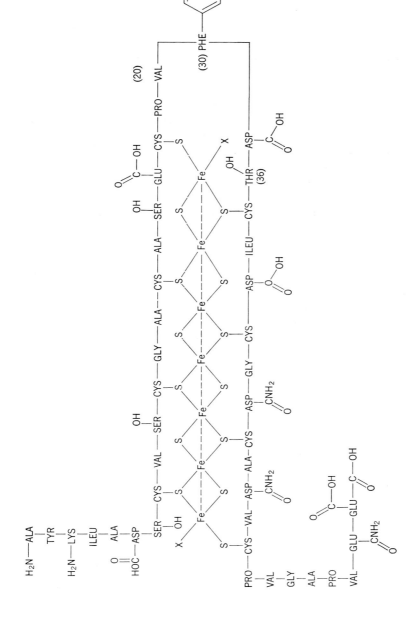

Fig. 3. Proposed structure of *C. pasteurianum* ferredoxin according to Tanaka et al. (32).

VI. Some Concluding Observations

1. Butow and Racker (60) have proposed that non-heme iron participates in mitochondrial oxidative phosphorylation. It is of interest to ask, therefore, whether or not non-heme iron may also be involved in photosynthetic phosphorylation. The two light reactions in photosynthesis are coupled by a thermochemical reaction postulated to operate between 0 and +0.4 V. This thermochemical reaction may be viewed as analogous to a portion of the electron-transfer sequence in the respiratory chain and is believed to be coupled with ATP formation.

2. The isolation of a non-heme iron protein from beef adrenal cortex mitochondria (18–20) is suggestive of the possible existence of other ferredoxin-like proteins in mammalian tissues. The trivial name adrenodoxin has been suggested (20).

3. Lovenberg and Sobel (61) have isolated a red protein, rubredoxin, from *C. pasteurianum* which substitutes for *C. pasteurianum* ferredoxin in the transfer of electrons from hydrogen to TPN. Oxidized rubredoxin exhibits absorption maxima at 280, 380, and 490 mμ. It contains one atom of iron per mole but no "labile sulfur." Since rubredoxin contains neither "labile sulfur" nor absorption maxima in the region of 300–350 mμ, it is tempting to speculate that the absorption of the non-heme iron proteins in this region is due to an iron–sulfur complex.

4. The available evidence suggests an interaction between non-heme iron proteins and flavoprotein(s). In the photosynthetic electron-transfer pathway, the non-heme iron and flavin are present in different proteins which interact with each other. In the case of the respiratory chain-linked dehydrogenases and metalloflavoproteins, these two constituents are part of the same protein (or complex of proteins).

VII. Addendum

Sobel and Lovenberg (62) investigated the characteristics of *C. pasteurianum* ferredoxin in a variety of oxidation-reduction reactions. They reported that "reduced ferredoxin when treated with mercurials liberates two more atoms of (ferrous) iron per mole than does oxidized ferredoxin; that there are no detectable free sulfhydryl groups in either species; that two electrons are transferred per mole of ferredoxin undergoing oxidation or reduction in several different reactions (reduction by illuminated chloroplasts; reduction by TPNH and spinach ferredoxin–TPN reductase; reduction by hydrogen and *C. pasteurianum* hydrogenase); and that the oxidation-reduction potential of ferredoxin varies systematically with pH (consistent with a value of $n = 2$)."

Palmer et al. (63) have shown that *C. pasteurianum* ferredoxin contains six atoms of iron; three are in the ferrous valence state and three in the ferric valence state. They point out that "this data, taken with the recent report of Sobel and Lovenberg [62] that clostridial ferredoxin is a two-electron acceptor, provides encouraging support for the model we have recently proposed [64] that the so-called $g = 1.94$ signal is due to low spin ferric iron in essentially tetrahedral symmetry, for on reduction one would expect to have only one ferric iron present. This would account for our observed quantitation (1.06 atoms of iron by EPR per mole of ferredoxin)."

The origin of the hydrogen sulfide (labile sulfur) has been shown by Malkin and Rabinowitz (65) not to be the sulfur atoms of the cysteine residues in the peptide chain of *C. pasteurianum* ferredoxin. Previously, Bayer et al. (66) proposed that the hydrogen sulfide arose by a β-elimination reaction from cysteine. If such a reaction occurred, the cysteine residues would have been converted to dehydroalanine and, finally, pyruvic acid, the hydrolysis product of dehydroalanine. Malkin and Rabinowitz (65) tested this hypothesis and could not obtain any evidence to support it. When ferredoxin was hydrolyzed under acidic conditions, no pyruvic acid was detected in the hydrolysate.

The most significant advance in this area is the reconstitution of *C. pasteurianum* ferredoxin reported recently by Malkin and Rabinowitz (67). The results of their studies are: "Apoferredoxin derivatives have been prepared by treatment of *Clostridium pasteurianum* ferredoxin with the mercurial, sodium mersalyl. An apoprotein has been isolated by passage of the mixture over a column of Chelex-100. Ferredoxin can be reconstituted from this preparation by the addition of ferrous ions and 2-mercapto-ethanol. No added source of 'inorganic' sulfide is required for reconstitution with this derivative. When this apoprotein is passed over a Sephadex G-25 column, reconstitution of ferredoxin from the isolated apoprotein requires the addition of both iron and a source of sulfide, as well as 2-mercaptoethanol. The reconstituted materials are identical to native ferredoxin in their activity in an enzymic assay, their iron and acid-labile sulfide content, and in their spectral characteristics."

Additional research reports and reviews have appeared (68–73).

References

1. A. San Pietro, Ed., *Non-Heme Iron Proteins: Role in Energy Conversion*, Antioch Press, Yellow Springs, Ohio, 1965.
2. A. San Pietro and C. C. Black, *Ann. Rev. Plant Physiol.*, **16**, 155 (1965).

3. L. E. Mortenson, *Ann. Rev. Microbiol.*, **17**, 115 (1963).

4. R. C. Valentine, *Bacteriol. Rev.*, **28**, 497 (1964).

5. L. E. Mortenson, H. F. Mower, and J. E. Carnahan, *Bacteriol. Rev.*, **26**, 42 (1962).

6. P. Handler, K. V. Rajagopalan, and V. Aleman, *Federation Proc.*, **23**, 30 (1964).

7. R. Hill, in *Essays in Biochemistry*, Vol. 1, P. N. Campbell and G. D. Greville, Eds., Academic Press, New York, 1965, p. 121.

8. T. P. Singer, this volume, p. 339.

9. K. V. Rajagopalan and P. Handler, this volume, p. 301.

10. B. G. Malmstrom and J. B. Neilands, *Ann. Rev. Biochem.*, **33**, 331 (1964).

11. G. Palmer and R. H. Sands, *J. Biol. Chem.*, **241**, 253 (1966).

12. H. Beinert, in *Non-Heme Iron Proteins: Role in Energy Conversion*, A. San Pietro, Ed., Antioch Press, Yellow Springs, Ohio, 1965, p. 23.

13. E. Racker, in *Non-Heme Iron Proteins: Role in Energy Conversion*, A. San Pietro, Ed., Antioch Press, Yellow Springs, Ohio, 1965, p. 325.

14. R. G. Bartsch, in *Bacterial Photosynthesis*, H. Gest, A. San Pietro, and L. P. Vernon, Eds., Antioch Press, Yellow Springs, Ohio, 1963, p. 315.

15. J. S. Rieske, D. H. MacLennan, and R. Coleman, *Biochem. Biophys. Res. Commun.*, **15**, 338 (1964).

16. Y. I. Shetna, P. W. Wilson, R. E. Hansen, and H. Beinert, *Proc. Natl. Acad. Sci. U.S.*, **52**, 1263 (1964).

17. R. W. F. Hardy, E. Knight, C. C. McDonald, and A. J. D'Eustachio, in *Non-Heme Iron Proteins: Role in Energy Conversion*, A. San Pietro, Ed., Antioch Press, Yellow Springs, Ohio, 1965, p. 275.

18. T. Omura, E. Sanders, D. Y. Cooper, O. Rosenthal, and R. W. Estabrook, in *Non-Heme Iron Proteins: Role in Energy Conversion*, A. San Pietro, Ed., Antioch Press, Yellow Springs, Ohio, 1965, p. 401.

19. K. Suzuki and T. Kimura, *Biochem. Biophys. Res. Commun.*, **19**, 340 (1965).

20. T. Kimura and K. Suzuki, *Biochem. Biophys. Res. Commun.*, **20**, 373 (1965).

21. D. I. Arnon, in *Non-Heme Iron Proteins: Role in Energy Conversion*, A. San Pietro, Ed., Antioch Press, Yellow Springs, Ohio, 1965, p. 137.

22. H. E. Davenport and R. Hill, *Proc. Roy. Soc. (London)*, **B139**, 327 (1952).

23. R. W. Holton and J. Myers, *Science*, **142**, 234 (1963).

24. R. Hill and A. San Pietro, *Z. Naturforsch.*, **18b**, 677 (1963).

25. W. Lovenberg, B. B. Buchanan, and J. C. Rabinowitz, *J. Biol. Chem.*, **238**, 3899 (1963).

26. K. T. Fry and A. San Pietro, in "Photosynthetic Mechanisms of Green Plants," Publ. 1145, National Academy of Sciences, National Research Council, Washington, D.C., 1963, p. 252.

27. K. T. Fry and A. San Pietro, *Biochem. Biophys. Res. Commun.*, **9**, 218 (1962).

28. K. Tagawa and D. I. Arnon, *Nature*, **195**, 537 (1962).

29. K. T. Fry, R. A. Lazzarini, and A. San Pietro, *Proc. Natl. Acad. Sci. U.S.*, **50**, 652 (1963).

30. S. Katoh and A. Takamiya, *Arch. Biochem. Biophys.*, **102**, 189 (1963).

31. M. Tanaka, T. Nakashima, A. Benson, H. F. Mower, and K. T. Yasunobu, *Biochem. Biophys. Res. Commun.*, **16**, 422 (1964).

32. M. Tanaka, A. Benson, H. F. Mower, and K. T. Yasunobu, in *Non-Heme Iron Proteins: Role in Energy Conversion*, A. San Pietro, Ed., Antioch Press, Yellow Springs, Ohio, 1965, p. 221.

33. D. C. Blomstrom, E. Knight, W. D. Phillips, and J. F. Weiher, *Proc. Natl. Acad. Sci. U.S.*, **51**, 1085 (1964).

34. W. D. Phillips, E. Knight, and D. C. Blomstrom, in *Non-Heme Iron Proteins: Role in Energy Conversion*, A. San Pietro, Ed., Antioch Press, Yellow Springs, Ohio, 1965, p. 69.

35. D. D. Ulmer and B. L. Vallee, *Biochemistry*, **2**, 1335 (1963).

36. D. I. Arnon, this volume, p. 123.

37. H. E. Davenport, in *Non-Heme Iron Proteins: Role in Energy Conversion*, A. San Pietro, Ed., Antioch Press, Yellow Springs, Ohio, 1965, p. 115.

38. L. P. Vernon and M. Avron, *Ann. Rev. Biochem.*, **34**, 269 (1965).

39. B. Chance and A. San Pietro, *Proc. Natl. Acad. Sci. U.S.*, **49**, 633 (1963).

40. F. R. Whatley, K. Tagawa, and D. I. Arnon, *Proc. Natl. Acad. Sci. U.S.*, **49**, 266 (1963).

41. T. Horio and A. San Pietro, *Proc. Natl. Acad. Sci. U.S.*, **51**, 1226 (1964).

42. M. Avron and A. T. Jagendorf, *Arch. Biochem. Biophys.*, **65**, 475 (1956).

43. M. Avron and A. T. Jagendorf, *Arch. Biochem. Biophys.*, **72**, 17 (1957).

44. D. L. Keister, A. San Pietro, and F. E. Stolzenbach, *J. Biol. Chem.*, **235**, 2989 (1960).

45. D. L. Keister, A. San Pietro, and F. E. Stolzenbach, *Arch. Biochem. Biophys.*, **98**, 235 (1962).

46. M. Shin, K. Tagawa, and D. I. Arnon, *Biochem. Z.*, **338**, 84 (1963).

47. M. Shin and D. I. Arnon, *J. Biol. Chem.*, **240**, 1405 (1965).

48. H. E. Davenport, *Nature*, **199**, 151 (1963).

49. R. A. Lazzarini and A. San Pietro, *Biochim. Biophys. Acta*, **62**, 417 (1962).

50. T. Yamanaka and M. D. Kamen, *Biochem. Biophys. Res. Commun.*, **19**, 751 (1965).

51. T. Yamanaka and M. D. Kamen, *Biochem. Biophys. Res. Commun.*, **18**, 611 (1965).

52. A. Paneque, J. M. Ramirez, F. F. Del Campo, and M. Losada, *J. Biol. Chem.*, **239**, 1737 (1964).

53. H. Huzisige and K. Satoh, *Botan. Mag. (Tokyo)*, **74**, 178 (1961).

54. H. Huzisige, K. Satoh, K. Tanaka, and T. Hayasida, *Plant Cell Physiol. (Tokyo)*, **4**, 307 (1963).

55. L. E. Mortenson, R. C. Valentine, and J. E. Carnahan, *Biochem. Biophys. Res. Commun.*, **7**, 448 (1962).

56. A. L. Shug and P. W. Wilson, *Federation Proc.*, **15**, 335 (1956).

57. W. A. Bulen, in *Non-Heme Iron Proteins: Role in Energy Conversion*, A. San Pietro, Ed., Antioch Press, Yellow Springs, Ohio, 1965, p. 261.

58. W. A. Bulen, R. C. Burns, and J. R. LeComte, *Proc. Natl. Acad. Sci. U.S.*, **53**, 532 (1965).

59. L. C. Sieker and L. H. Jensen, *Biochem. Biophys. Res. Commun.*, **20**, 33 (1965).

60. R. Butlow and E. Racker, in *Non-Heme Iron Proteins: Role in Energy Conversion*, A. San Pietro, Ed., Antioch Press, Yellow Springs, Ohio, 1965, p. 383.

61. W. Lovenberg and B. Sobel, *Federation Proc.*, **24**, 233 (1965); *Proc. Natl. Acad. Sci. U.S.*, **54**, 193 (1965).

62. B. E. Sobel and W. Lovenberg, *Biochemistry*, **5**, 6 (1966).

63. G. Palmer, R. H. Sands, and L. E. Mortenson, *Biochem. Biophys. Res. Commun.*, **23**, 357 (1966).

64. H. Brintzinger, G. Palmer, and R. H. Sands, *Proc. Natl. Acad. Sci. U.S.*, **55**, 397 (1966).

65. R. Malkin and J. C. Rabinowitz, *Biochemistry*, **5**, 1262 (1966).

66. E. Bayer, W. Parr, and B. Kazamaier, *Arch. Pharm.*, **298**, 196 (1965).

67. R. Malkin and J. C. Rabinowitz, *Biochem. Biophys. Res. Commun.*, **23**, 822 (1966).

68. B. B. Buchanan, *Structure Bonding*, **1**, 109 (1966).

69. S. Keresztes-Nagy and E. Margoliash, *J. Biol. Chem.*, **241**, 5955 (1966).

70. H. Bachmayer, K. T. Yasunobu, and H. R. Whiteley, *Biochem. Biophys. Res. Commun.*, **26**, 435 (1967).

71. J. F. Gibson, D. O. Hall, J. M. Thornley, and F. R. Whatley, *Proc. Natl. Acad. Sci., U.S.*, **56**, 987 (1966).

72. H. Bachmayer, L. H. Piette, K. T. Yasunobu, and H. R. Whiteley, *Proc. Natl. Acad. Sci., U.S.*, **57**, 122 (1967).

73. H. Matsubara, R. M. Sasaki, and R. K. Chain, *Proc. Natl. Acad. Sci., U.S.*, **57**, 439 (1967).

Quinones in Electron Transport

Frederick L. Crane, *Department of Biological Sciences, Purdue University, Lafayette, Indiana*

I.	Introduction	533
II.	Quinone Types	534
III.	Distribution of Quinones	538
IV.	Quinone Changes in Response to Growth Conditions	540
V.	The Function of Coenzyme Q in Mitochondria	543
	A. The Split-Pathway Interpretation of Solvent Extraction	548
	B. Oxidation-Reduction of Coenzyme Q	549
	C In Phosphorylating Systems	551
	D. In Intact Tissues	552
	E. Coenzyme Q Reductases	553
	F. Reduced Coenzyme Q_{10}-Cytochrome c Reductase	555
VI.	The Quinone Reductases	556
VII.	Vitamin K in Electron Transport	557
	A. Vitamin K in Bacteria	558
VIII.	Quinones in Chloroplasts	561
IX.	Quinones in Other Photosynthetic Systems	565
X.	Other Electron-Transport Systems	567
XI.	Electron Spin Resonance Studies	568
XII.	The Effect of Other Lipids on Quinone Function	568
XIII.	Quinones and Phosphorylation Mechanisms	569
XIV.	Genetic Studies	575
XV.	Summary	576
	References	577

I. Introduction

Several types of lipid soluble quinones have been found to be associated with electron-transport systems in plants, animals, and bacteria. Both naphthoquinones and benzoquinones are included and characteristically they all have a long terpenoid hydrocarbon chain attached to the quinone nucleus which makes them extremely lipophilic. The main types now recognized include members of the coenzyme Q (ubiquinone), vitamin K, desmethylvitamin K, plastoquinone, and tocopherylquinone groups. Much evidence is now available which indicates that these quinones function as redox carriers in the electron transport systems in which they are found, e.g., the terminal oxidase system of mitochondria and bacteria or the photosynthetic systems of chloroplasts and chromatophores. Besides playing a role as simple electron carriers, there is indication that large pools of reduced quinone in some way facilitate energy coupling processes which are supported by electron flow. The diversity of type and the large amount of quinone present in the various electron-transport systems when expressed in terms of the other catalytic components are rather surprising. The constant association of certain types of quinones with each type of system is also remarkable. Previous to the discovery of these terpenoid quinones, the fact that no type of quinone was found to have general distribution and the lack of quinones in animal tissue discouraged the view that these excellent redox agents would have a general biological significance (1,2).

As in all studies which encompass the diverse aspects of the biological realm there are only a few species in which intensive study of quinone chemistry and function has been made but many species in which the principles deduced from these studies are presumed to apply. The extension of principles in this way will probably turn out to be acceptable; but it should be remembered that there is evidence for a large variety of quinones and that the detailed operations of electron-transport systems are known in relatively few tissues. The lack of precise knowledge is especially noticeable in many lower organisms. It is in these lower organisms that quinone patterns deviate most strongly from the patterns found in higher plants and animals. It is conceivable that the study of quinone functions and electron-transport systems in lower organisms may not contribute greatly to an understanding of pathologic problems but it may provide interesting insight into the development of quinone types and the development of more efficient or specially adapted electron-transport systems.

II. Quinone Types

The quinones now known to be associated with electron transport function fall primarily into four major groups. The unknown quinones, which have shown up occasionally, may fit in as variations of these types when their structure is known. First in extent of distribution come the coenzyme Q group or ubiquinones which are based on a 2,3-dimethoxy-5 methylbenzoquinone nucleus with an unsaturated terpenoid side chain.

The members of the coenzyme Q group are designated by a subscript for the number of methylbutenyl (5-carbon) units in the terpenoid side chain as in coenzyme Q_{10}. The alternative nomenclature of ubiquinones indicates the number of carbon atoms in the side chain in parentheses after the name, e.g., ubiquinone(50) is the same as coenzyme Q_{10} (3). The most common members of this series have a terpenoid side chain which contains 30 to 50 carbons atoms with all isoprenoid units unsaturated as shown in structure I.

The only variations on this structure are in the quinones from fungi known as coenzyme Q_{10}-H10 and coenzyme Q_9-H9 shown in structure II (4). In close competition for universal significance is the vitamin K group.

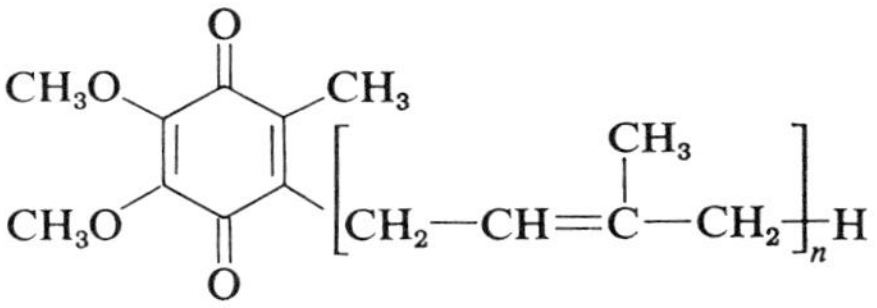

$$CH_3O \quad \overset{O}{\underset{O}{\bigcirc}} \quad CH_3 \qquad \left[CH_2-CH=\overset{CH_3}{\underset{|}{C}}-CH_2 \right]_n H$$

(I) $\quad n = 6\text{--}10$

$$CH_3O \quad \overset{O}{\underset{O}{\bigcirc}} \quad CH_3 \qquad \left[CH_2-CH=\overset{CH_3}{\underset{|}{C}}-CH_2 \right]_n CH_2-CH_2-\overset{CH_3}{\underset{|}{CH}}-CH_3$$

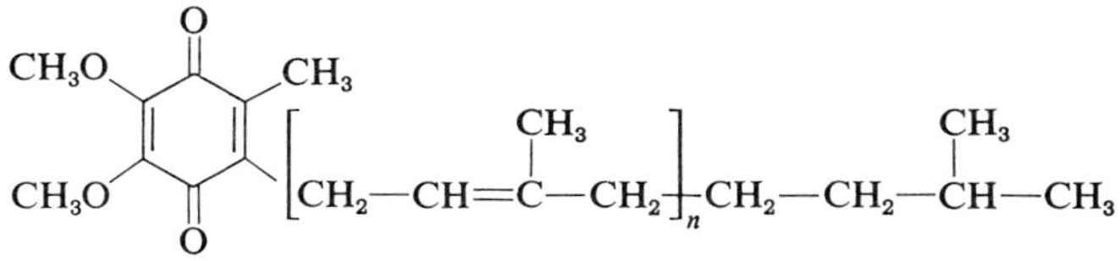

(II) $\quad n = 8\text{--}9$

With improved methods for lipid fractionation, these quinones are appearing in a great variety of organisms and it is quite likely that they are of wider distribution than the coenzyme Q group. These quinones have a simple naphthoquinone nucleus with a terpenoid side chain and usually a methyl group in the 2 position on the quinone ring. The best-known members of this group are vitamin K_1 (III) with a mono-unsaturated phytyl side chain and the members of the vitamin K_2 series which have terpenoid side chains similar to the coenzyme Q group with unsaturated isopentenyl units throughout (IV).

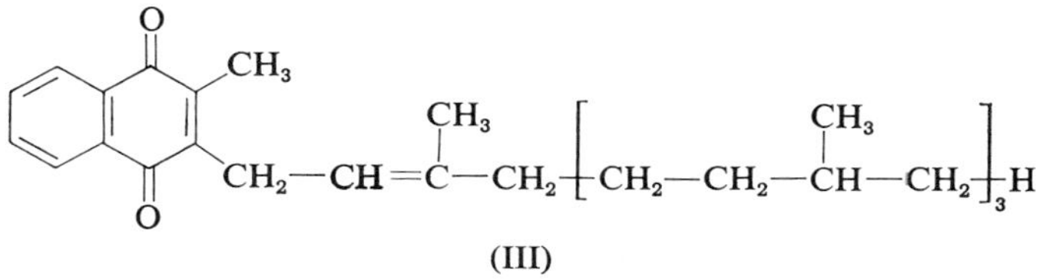

$$\overset{O}{\underset{O}{\bigcirc\!\bigcirc}} \quad CH_3 \qquad CH_2-CH=\overset{CH_3}{\underset{|}{C}}-CH_2 \left[CH_2-CH_2-\overset{CH_3}{\underset{|}{CH}}-CH_2 \right]_3 H$$

(III)

In addition to the vitamins K_1 and K_2 which have been known for many years, a series of variations in structure have been found. These include the desmethyl vitamin K series reported by Baum and Dolin (5) in *Streptococcus* and by Lester et al. (6) in *Hemophilus* (V) and possibly a type of naphthoquinone reported in spinach chloroplasts (7). There is also a series analogous to the coenzyme Q types with partially saturated side chains as for example: vitamin $K_{2(45)}H9$ (VIII) (8). In addition a rather unusual vinyl-substituted naphthoquinone, chlorobiumquinone, which was discovered in *Chlorobium* by Fuller (9) and elucidated by Fryden and Rappaport (10), may be considered to be a derivative of the vitamin K series (VI). A special group of 2,3-dimethylbenzoquinones is found in all oxygen-producing photosynthetic tissue. The principal example of this type is the plastoquinone A series of which only the members with 45-carbon and 20-carbon side chains have been reported from leaf tissue (VII) (11). There are, however, a whole series of quinones which seem to

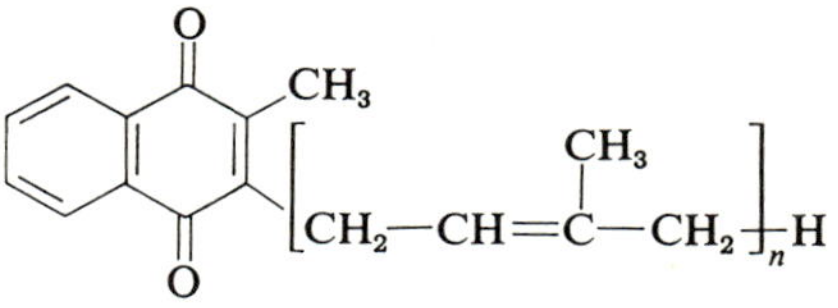

(IV)

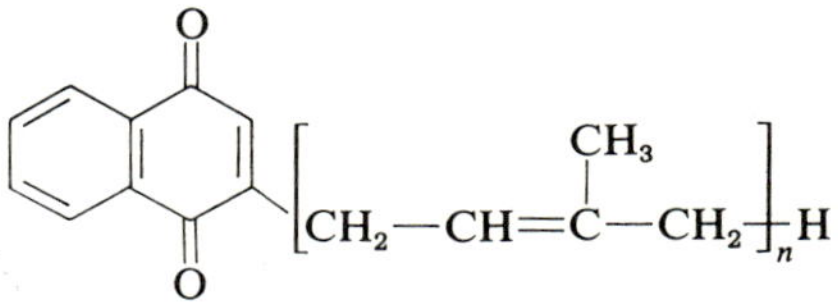

(V) $n = 7$

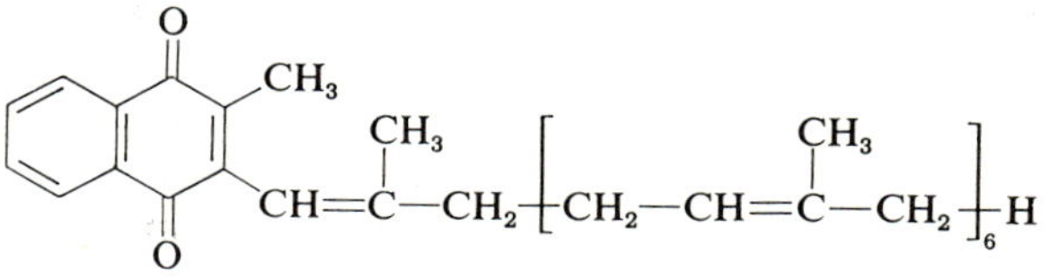

(VI)

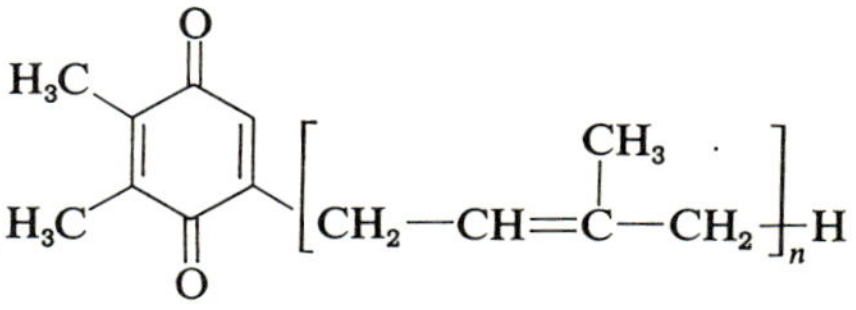

(VII) $n = 4, 9$

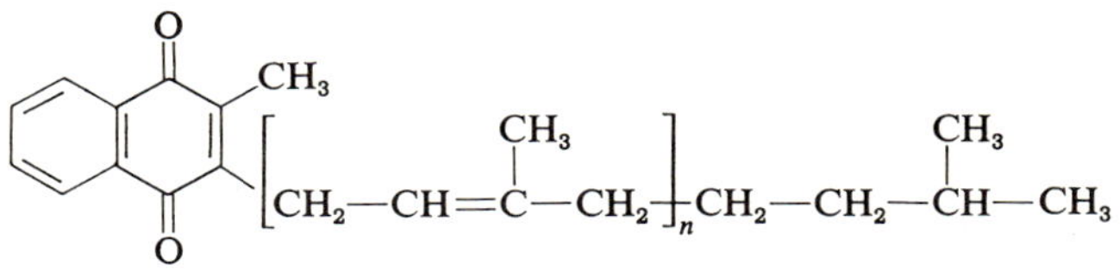

(VIII) $n = 8$

be related to the plastoquinone A series. The structure of these quinones is as yet unknown, so they have been designated plastoquinone B, C, and D (12,13). In the plastoquinone group, the ultraviolet spectra of A, B, C, and D are identical with an absorption maximum at 255 mμ (13). Structures of these quinones have been proposed (103). The tocopherylquinones are closely related to the plastoquinone series in that they are a series of p-benzoquinones with one, two, or three methyl groups on the quinone ring and a hydroxylated phytyl side chain. The fully substituted 2,3,5-trimethyl-6-(3-hydroxy)phytyl benzoquinone is α-tocopherylquinone (IX)

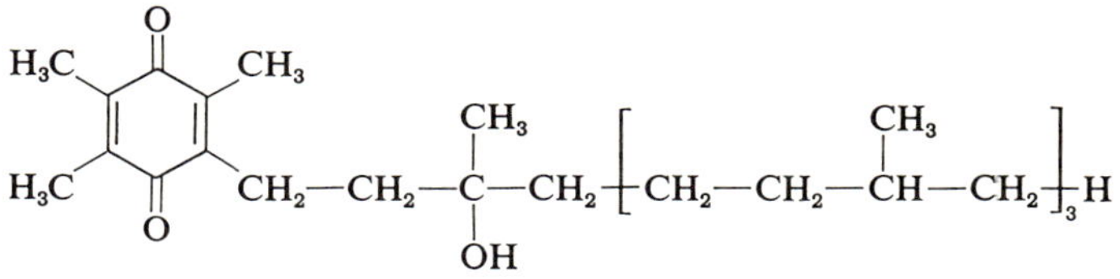

(IX)

whereas the 3,5-dimethyl is β-tocopherylquinone, 2,3-dimethyl is γ-tocopherylquinone, and 3-methyl is δ-tocopherylquinone. The existence of the tocopherylquinones *in vivo* has often been questioned on the basis that they occur in extremely small amounts compared to the corresponding chromenols, α-, β-, and γ-tocopherol, from which the quinones could arise by oxidation. There is mounting evidence, however, that these quinones may be of functional significance in some electron transport systems, and with new methods for study it will be possible to determine whether they derive from tocopherols during an electron flow process or whether their appearance is based on an oxidation artifact.

In addition to the quinones which fit into the four major groups, there are several quinones of lipophilic nature which are found in electron transport systems or which appear where none of the known types appear. It is possible that these quinones will fit into one of the major groups when their structure is known. Of these the most significant may be a quinone found in beef-heart mitochondria which may be related to coenzyme Q. This quinone shows a benzoquinone-type spectrum with a maximum about 273 mμ and is much more polar than any of the coenzyme Q group as revealed by silica gel chromatography. There is also other quinoid material in beef-heart mitochondria which shows a spectrum similar to the vitamin K series but is much more unstable than the standard vitamin K group. The presence of an unstable naphthoquinone or naphthoquinone derivative would explain the apparent presence in mitochondria of a vitamin K which is very difficult to isolate (14). Martius (3) has also reported material in mitochondria with properties similar to vitamin $K_{2(20)}$ which proved difficult to isolate. Some unusual forms of naphthoquinones, including a hydroxylated naphthoquinone, have also been found in the blue-green alga, *Anacystis nidulans* (15). A small amount of a hydroxylated or amino benzoquinone (X), related to coenzyme Q and named rhodoquinone, has been found in *Rhodospirillum* (16,100).

III. **Distribution of Quinones**

As mentioned in the Introduction, a member of one of the major groups of quinones is characteristic of each type of electron transport system. Since electron transport systems related to specific functions are often localized in certain organelles within a cell, the location of a quinone in an organelle suggests its function. Thus the terminal oxidation of substrates derived from pyruvate or fatty acids is carried out in the mitochondria of higher plants and animals. Photosynthesis in green plants is carried out in chloroplasts or in photosynthetic bacteria in small chromatophores. Oxidative detoxification processes occur in microsomes or ergastoplasm of liver cells, and terminal oxidation in bacteria is localized in the plasma membrane. As a preliminary to the study of quinone function it is well to determine the characteristic location of quinones within the cell to determine the process with which they may be associated. As a rule these localization studies can be relied on only for the major locus of quinone accumulation. The presence of a trace of quinone in other sites in the cell can easily be missed, especially if the amount of material available for cell fractionation is limited.

The characteristic location for the major quinones is shown in Table I. Coenzyme Q has been found in all mitochondria examined. The amount is

TABLE I

Distribution of the Major Quinones in Isolated Electron Transport
Systems

Quinone	Cell structure	Function observed
Coenzyme Q	Heart mitochondria	Succinoxidase, redox change
	Liver mitochondria	Succinoxidase
	Plant mitochondria	Redox changes
	Fungi mitochondria	DPNH–cyt c reductase
	Bacterial particles	Nitrate reductase
	Bacterial particles	Succinoxidase
	Microsomes	None
	Nucleii	None
	Aldehyde oxidase	Redox changes
	Bacterial chromatophores	Photoreduction
Vitamin K_1	Chloroplasts	None
Vitamin $K_{2(45)}$ 9H	*Microbacteria* ETP	DPNH oxidase
Vitamin $K_{2(35)}$	*Escherichia* ETP	DPNH oxidase
Desmethylvitamin K	*Hemophilus* ETP	DPNH tetrazolium reductase
	Chloroplasts?	None
Plastoquinone $A_{(45)}$	Chloroplasts	Hill reaction, redox changes
Plastoquinone B	Chloroplasts	Hill reaction
Plastoquinone C	Chloroplasts	TPN photoreduction
Plastoquinone D	Chloroplasts	Hill reaction, redox changes
α-Tocopherylquinone	Chloroplasts	Cytochrome c photoreduction, redox changes
β-Tocopherylquinone	Chloroplasts	Hill reaction
γ-Tocopherylquinone	Chloroplasts	None

usually severalfold greater than the other components of the electron-
transport system such as flavoproteins and cytochromes. Succinoxidase
particles from *E. coli* contain coenzyme Q together with vitamin K.
Coenzyme Q is also present in large amounts in the chromatophores of
certain photosynthetic bacteria as well as in the soluble molybdoflavo-
protein, aldehyde oxidase (17). The plastoquinones are found in chloroplasts. Vitamin K_2 is best represented in the particles which carry out
electron transport which are derived by breaking up bacterial cells.
α-Tocopherylquinone is present in significant amounts in chloroplasts
while it is difficult to find in mitochondria even though α-tocopherol is
found in both structures.

There are three principal systems in which quinone function has been
studied extensively. First there has been a detailed study of coenzyme Q
function in beef-heart mitochondria. The study of vitamin K function has

been concentrated in respiratory particles derived by fragmentation of *Mycobacteria*, and there have been corollary studies using *Escherichia coli* and *Hemophilus*. Finally, the study of plastoquinone function has been most extensive in spinach chloroplasts (*Spinacia*), although there have been studies by spectrophotometric techniques of *Chlorella* (chlorophyta) and *Anacystis* (cyanophyta). There has also been study of coenzyme Q function in the photosynthetic bacteria in connection with photosynthetic activity in these organisms.

In all these systems the study of quinone function has been carried out by means of a similar series of techniques. Techniques which have been applied in one or more of these systems are as follows:

(*1*) Spectrophotometric studies of changes in quinone in the intact cell or organelles such as mitochondria.

(*2*) Removal of the quinone to stop activity, followed by readdition of quinone to restore activity. Removal has been accomplished by solvent extraction, by irradiation with ultraviolet light to destroy quinone, or by growing cells under conditions which prevent quinone synthesis.

(*3*) Changes in quinone concentration in response to growth conditions such as the shift from anaerobic to aerobic growth or the shift from heterotrophic to autotrophic culture.

(*4*) Extraction of quinone from a system under conditions which would cause oxidation or reduction of the components to determine the oxidation-reduction state of the quinone.

(*5*) Oxidation-reduction studies of added quinone or hydroquinone.

(*6*) Isolation of enzymes or enzyme systems which catalyze oxidation or reduction of the quinone or hydroquinone.

(*7*) Electron spin resonance studies in an attempt to detect semiquinones as intermediates during the oxidation-reduction of quinones in enzyme systems.

(*8*) Cell-fractionation studies to determine what part of the cell contains quinone and if it is concentrated in the organelle associated with the proposed function.

(*9*) Comparative biochemical studies to determine if similar quinones are found in organisms which possess similar functional systems.

(*10*) Changes in electron transport function in quinone-deficient mutants.

IV. Quinone Changes in Response to Growth Conditions

Changes in metabolic patterns in microorganisms induced by variation in culture conditions are reflected by changes in quinone concentration (17).

The first evidence of this is shown by the disappearance of coenzyme Q in yeast cells cultured under anaerobic conditions. This effect is consistent with the disappearance of mitochondria as well as loss of the cytochromes of the terminal oxidase system in anaerobic yeast. The fact that coenzyme Q is undetectable in anaerobic yeast also raises a question about the role of the coenzyme Q which is proposed to be present in nuclei and microsomes of liver cells. One must consider that if coenzyme Q is a component of yeast nuclei it must disappear from anaerobic yeast nuclei as well as from mitochondria. Does this mean that the coenzyme Q in nuclei is involved in an oxidase system such as the system described by Penniall for liver nuclei? If so, does this system disappear under anaerobic conditions? A similar decrease in coenzyme Q content under anaerobic conditions has been reported by Lester and Crane (18) and by Kashket and Brodi (19) in *E. coli*. Bishop et al. (20) were not able to find this decrease in *E. coli* but did find a decrease of vitamin K content in certain other bacteria under anaerobic conditions. The discrepancy in results with *E. coli* may be related to variations in the strain of *E. coli* used or perhaps to different culture conditions. Itagaki's evidence that coenzyme Q_8 in *E. coli* is required for formate–nitrate reductase (21) would suggest that high levels of coenzyme Q might be maintained even under anaerobic conditions if nitrate reductase is active and nitrate is supplied.

An increased respiration is characteristic of thyrotoxic animal tissue. This increase is also reflected by an increase in coenzyme Q and other mitochondrial components in liver of rats treated with thyroxine which is presumably related to an increased synthesis of mitochondria (22). A similar increase of coenzyme Q is found in chickens in a cold environment where respiratory activity also increases (23). On the other hand, in a warm environment, increase in thyroxine no longer stimulates an increase of quinone.

A shift from autotrophic to heterotrophic growth in certain algae is also related to a dramatic change in quinone content. These changes support the role of coenzyme Q in oxidative electron transport and the role of plastoquinones in photosynthesis.

Fuller et al. (9) have shown that dark-grown *Euglena* contain a high level of coenzyme Q and a very low level of plastoquinone. A similar effect has been shown with *Chlorella* by M. D. Henninger (24). If *Chlorella* is grown in high concentration of glucose by the method of Stiller and Lee, the algae have good oxidative metabolism and are also able to utilize hydrogen directly to photoreduce CO_2, but complete photosynthesis is greatly inhibited. As shown in Table II, the algae grown on inorganic media which show normal photosynthesis contain the full complement of

TABLE II

Comparison of Quinone Content in Autotrophic and Glucose-Grown
Chlorella Cells[a]

Component	Autotrophic cells	Glucose-grown cells	% Change of quinone content, A − G
Chlorophyll	2%	1%	− 50
	μmoles/g dry wt.	μmoles/g dry wt.	
Plastoquinone A	0.261	0.045	− 80
Plastoquinone B	T	N.D.	− > 50
Plastoquinones C and D	0.044	T	− > 75
Vitamin K_1	0.019	T	− > 50
α-Tocopherylquinone	T	N.D.	− > 50
Coenzyme Q_{10}	0.044	0.089	+ 100
Hydroxynaphthoquinone	N.D.	0.104	+ > 100
Rate of photosynthesis			− 60

[a] T indicates 0.0005–0.0002 μmoles quinone per mg chlorophyll, which is below the level for accurate determination. N.D. indicates less than 0.0001 μmole per mg chlorophyll and no spot observed on thin-layer chromatograms.

plastoquinones and α-tocopherylquinone which are characteristic of chloroplasts of higher plants. When grown on glucose there is a loss of plastoquinones and α-tocopherylquinone which parallels the loss of chlorophyll and photosynthetic potential. On the other hand, coenzyme Q increases dramatically. A new quinone which appears to be identical with the hydroxylated naphthoquinone found in the blue-green algae *Anacystis nidulans* also appears in the glucose-grown cells. It is tempting to speculate that this hydroxylated naphthoquinone may be related to the hydrogen-dependent photoreductase system.

Changes in coenzyme Q content in the photosynthetic bacterium *Rhodospirillum* favors a role in photosynthesis. Geller (25) has shown a fourteenfold increase of coenzyme Q content when cells are grown in light as compared to cells grown in the dark, whereas there is little change in the cytochromes c_2 and *b*.

When *Rhodospirillum* cells are grown in the presence of diphenylamine, which inhibits syntheses of terpenoid chains, there is a decrease of coenzyme Q which is correlated with a decrease of photosynthesis. Activity in the chromatophores can be restored by short-chain homologs of coenzyme Q_9 (26).

There have been other reports of changes in coenzyme content of bacteria as a result of variation in growth conditions, but these have not been measured in relation to the electron transport systems (17).

V. The Function of Coenzyme Q in Mitochondria

The study of coenzyme Q function in mitochondria has encompassed the approaches described in the preceding section. Various aspects of these studies have been reviewed (3,27–30). In all these studies the evidence clearly shows a function for coenzyme Q in the electron transport system of mitochondria. The sequence of electron carriers which accomplish the oxidation of succinate or DPNH by molecular oxygen is shown diagrammatically in Figure 1. In this system, electrons are transferred successively from the substrates through oxidation-reduction of substrate-specific flavoprotein dehydrogenases which are associated with protein containing non-heme iron to coenzyme Q. The reduced coenzyme Q is then oxidized by the remaining members of the cytochrome chain including cytochromes c_1, c, a, and a_3. The redox potential of coenzyme Q, $E_0' = +98$ mV at pH 7.4 and 25°C, is consistent with a function between cytochromes b and c_1 (31) but the enzymic studies indicate a function before cytochrome b. As indicated in Figures 1 and 2 in the chapter by Klingenberg and as will be discussed below, the function of the cytochrome b and a large stoichiometric excess of coenzyme Q is ambiguous and may be partially on a sideline or bypass electron flow system. Evidence for the site of function for coenzyme Q comes mostly from studies of succinate oxidation. The function of coenzyme Q in the DPNH oxidase chain is not as well documented.

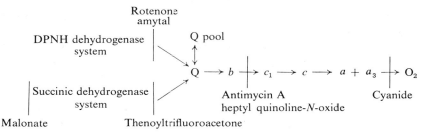

Fig. 1. Relation between sites of inhibitor action and the proposed site for coenzyme Q function in mitochondrial electron transport. See chapter by Klingenberg.

When mitochondria or particles derived from mitochondria which contain the electron transport system are extracted by shaking with solvents, coenzyme Q is extracted from the particles. As the coenzyme Q is extracted there is a loss of succinoxidase activity. If coenzyme Q is added back to the particles, the succinoxidase activity is restored.

Two types of solvent have been used for extraction and the pattern of activity changes is different for each, but the net result of inducing a coenzyme Q requirement is the same. Since acetone treatment gives more clear-cut results, it will be discussed first, whereas isooctane extraction will be discussed later in relation to changes in DPNH oxidation.

Extensive acetone extraction of wet or dry mitochondria leads to an almost complete extraction of coenzyme Q from the particles and a complete loss of DPNH and succinoxidase activity. If coenzyme Q is added in ethanolic solution to an aqueous suspension of the extracted particles, succinoxidase activity can be fully restored. If much of the phospholipid of the mitochondrion is removed during this extraction, then full activity may only be restored if phospholipid is added along with the coenzyme Q. Restoration of activity in this system is primarily restricted to members of the coenzyme Q group. There is some evidence, however, that certain hydroxylated benzoquinones may substitute for coenzyme Q in this type of restoration (32). Similar acetone extraction studies can be carried out with the succinic dehydrogenase complex to further localize the site of coenzyme Q function. This complex is derived from mitochondria by fragmentation with detergents such as amyl alcohol or deoxycholate. One fragment contains the succinic dehydrogenase, a large amount of coenzyme Q, and cytochromes b and c, and catalyzes oxidation of succinate by ferricytochrome c. Another fragment, which will be discussed later, contains the DPNH dehydrogenase, some cytochrome b and c_1, and a small amount of coenzyme Q together with a large amount of cytochromes a and a_3.

When the succinic dehydrogenase complex is extracted with acetone, coenzyme Q is removed and the enzymic activity is stopped. When coenzyme Q is added back, the succinate cytochrome c reductase activity is fully restored. An example of this effect is shown in Table III. It should be noted that residual succinoxidase in the preparation does not respond to specific quinones in exactly the same way as the reductase.

The only compounds which are able to restore both succinic cytochrome c reductase and succinoxidase activities extensively in acetone-extracted particles are quinones with the dimethoxymethylbenzoquinone nucleus as in coenzyme Q_{10} along with a hydrocarbon side chain in the 6 position of at at least 10 carbon atoms. In the untreated mitochondria the succinoxi-

TABLE III

Comparison of Succinic Cytochrome c Reductase and Succinic
Oxidase Restoration by Quinones (92)[a]

Quinone added	Succinic-cytochrome c reductase		Succinoxidase	
	μmoles min^{-1} mg^{-1}	% Inhibition by antimycin	μmoles min^{-1} mg^{-1}	% Inhibition by antimycin
None	0.00	—	0.16	100
2,3-Dimethoxy-5-methylbenzoquinone (Q$_0$)	0.15	0	0.00	—
2,3-Dimethoxy-5,6-dimethylbenzo-quinone (methyl Q)	0.22	0	0.32	30
Coenzyme Q$_1$	0.31	20	0.62	90
Coenzyme Q$_2$	0.37	75	0.72	100
Coenzyme Q$_3$	0.42	100	0.78	100
Coenzyme Q$_{10}$	0.61	100	0.50	100
Heptyl Q	0.31	0	0.29	100
Heptadecyl Q	0.09	100	0.37	100
Eicosahydro coenzyme Q$_{10}$	0.23	100		
Phytyl Q	0.35	87	0.44	100
Plastoquinone A$_{(45)}$	0.00	—	0.24	100
Menadione	0.18	0	0.36	62
2,3-Dimethyl-1,4-naphthoquinone	0.00	—	0.00	—
Vitamin K$_1$	0.00	—	0.00	—
Coenzyme Q$_{10}$ chromenol	0.00		0.00	
Unextracted enzyme	0.60		0.80	

[a] Dry SDC extracted 90 min at 20°C with dry acetone. All activity with cytochrome c added but no neutral lipid or phospholipid, so Q$_{10}$ is less effective than Q$_3$ in succinoxidase.

dase and succinic-cytochrome c reductase systems are inhibited by small amounts of antimycin A or nonylhydroxyl quinoline-N-oxide (NQNO). It has been shown in other studies that these inhibitors act at a site between cytochromes b and c_1 as indicated in Figure 1. The activity which is restored to the extracted particles by coenzyme Q$_{10}$ or its homologs is similarly inhibited by these agents. On the other hand, some quinones with short side chains are able to restore activity in the extracted system which is not sensitive to antimycin A or NQNO and thus institute electron flow which differs from the normal pathway through cytochrome c_1.

This ability for mediation of electron transfer between flavoproteins and cytochrome c by relatively water-soluble quinones is well known (3). A remarkable thing about the particulate electron transport systems is the fact that in the presence of coenzyme Q there is no leak directly to cytochrome c, but all activity is channeled through cytochrome c_1. Actually there are very few quinones among the many tested which permit a bypass around the antimycin site. It is also remarkable that the added coenzyme Q is restricted in its function to a pathway of electron transfer which resembles exactly that of the untreated particles in its inhibitor sensitivity.

Extraction studies have not been very successful in providing evidence for coenzyme Q function in the DPNH oxidase pathway. Acetone extraction of mitochondria or particles leads to complete loss of DPNH oxidase activity which is not restored by addition of coenzyme Q. Since DPNH ferricyanide reductase activity is destroyed by acetone treatment (101), it is expected that activity would not be restored. On the other hand, since a few minutes' treatment with ethyl ether will also inhibit DPNH oxidase activity without affecting ferricyanide reductase activity or cytochrome c oxidase activity, the break in the chain must come between the dehydrogenase and cytochrome c_1 (102). Since coenzyme Q does not restore activity, it is likely that another component necessary for DPNH oxidase is destroyed or disorganized somewhere in this middle region.

A small bit of evidence that some quinone may function in this system comes from the partial restoration of DPNH oxidase activity after ether extraction of mitochondria by addition of vitamin $K_{2(20)}$ (33). This report, however, has not shown any specificity or selective inhibitor studies to indicate restoration of the original system. There is also a report of restoration of DPNH oxidase in mitochondria from the ergot-producing fungus, *Claviceps*, by coenzyme Q after acetone extraction (34). The requirement for phospholipid to restore amytal-sensitive DPNH-coenzyme Q reductase activity in phospholipase-treated mitochondria (94,95) may imply that the failure to restore activity after extraction is related to the removal of phospholipids.

Isooctane extraction can also be used to remove coenzyme Q from mitochondria and derivative particles. The results are similar to those found with acetone extraction. Again, coenzyme Q restores activity with similar specificity with regard to quinone type and inhibitor sensitivity. DPNH oxidase activity that is lost after isooctane extraction cannot be restored by coenzyme Q but can be restored by cytochrome c. Cytochrome c will also restore succinoxidase in the isooctane-treated particles without addition of coenzyme Q, and coenzyme Q will partially restore succinoxidase activity without addition of cytochrome c. Isooctane extraction is much less efficient than acetone in removing coenzyme Q from the particles and it is probably in the extent of coenzyme Q extraction that much of the complication of isooctane extraction lies. Actually, much confusion has arisen in the literature as a result of isooctane extraction studies. The first complication comes from inhibition of electron transport by isooctane when it is added to particles without any extraction taking place. This type of inhibition can be reversed in a nonspecific fashion by any treatment which alters the binding of isooctane to the particles. Among the treatments which reverse isooctane inhibition is the addition of many fatty materials such as phytol, vitamin E, vitamin K, beef serum albumin, and coenzyme Q. Thus the reversal of inhibition is nonspecific and is easily detected by testing for specificity. The other effect accompanies the removal of one-half or more of the coenzyme Q in the particle. In this instance both succinoxidase and DPNH oxidase activity are inhibited and succinoxidase activity is restored only by members of the coenzyme Q group (or plastoquinone) or cytochrome c, but not by nonspecific lipid materials or serum albumin.

The simplest explanation for the dual effects of cytochrome c and coenzyme Q for restoration of succinoxidase after isooctane extraction of 50–80% of the coenzyme Q while the DPNH oxidase system is restored by cytochrome c alone is that the residual coenzyme Q may be part of the DPNH chain whereas any functional cytochrome c is available for succinate oxidation only after coenzyme Q is added. On the other hand, it may be that coenzyme Q does not function in the DPNH oxidase chain.

The results of isooctane extraction are much simpler to interpret after deoxycholate treatment of mitochondria under conditions which remove cytochrome c (35). Deoxycholate treatment induces a requirement for cytochrome c for succinate oxidation. If the deoxycholate treatment is followed by isooctane extraction, both coenzyme Q and cytochrome c are required for succinate oxidation. This procedure provided the first demonstration of an absolute requirement for coenzyme Q in succinate oxidation. The activity patterns in particles treated with the combination of deoxycholate and isooctane are similar to activities found after ether and acetone treatment, as shown in Table IV. It is perhaps unfortunate that the results of direct isooctane extraction are more complicated than the results of other treatments, because from electron microscope studies

TABLE IV

Restoration of Succinoxidase Activity after Various Types of Solvent
Extraction

Type of extraction	% of CoQ removed	Initial activity	Extracted activity	Restored activity		Cyt. c^a and CoQ$_{10}$	Ref.
				Cyt. c	CoQ$_{10}^a$		
Isooctane 0°	55	0.50	0.03	0.56	0.46	0.56	35
Isooctane 20°	88	1.20	0.05	0.96	0.80	1.00	35
Deoxycholate and isooctane	70	1.46	0.00	0.34	0.00	1.58	35
Acetone wet	99	0.11^b	0.00	0.02	0.05	0.50	44
Acetone dry	90	0.51	0.00	0.00	0.15	0.48	92

[a] Additional lipid fractions are required for full restoration of activity shown here.
[b] Mitochondria; all other preparations are ETP.

it appears that isooctane treatment does a minimum of damage to the inner structure of the cristae membranes as revealed by phosphotungstate staining (36). Acetone extraction, on the other hand, causes a breakup of the beaded structure of the membranes in mitochondrial cristae which is evident when studied by phosphotungstate staining. In contrast to this evidence of damage, mitochondria do not reveal any damage from acetone when studied by osmium fixation and sectioning technique (37).

A. The Split-Pathway Interpretation of Solvent Extraction

A proposal by Chance (38), that half the electron flow could go through coenzyme Q and half through cytochrome b, presented to account for the slow reduction of each component in relation to the overall rate of electron transport which is discussed in the next section, can also be used to rationalize the different effects of isooctane and acetone extraction. This interpretation is predicated on the removal of part of the cytochrome c by isooctane treatment or complete inactivation of the endogenous cytochrome c by acetone treatment so that added cytochrome c is required to restore, either fully or partially, electron flow between cytochrome c_1 and cytochrome oxidase.

Isooctane extraction has been used to remove up to 80% of the coenzyme Q. Cytochrome c restores activity completely and coenzyme Q restores activity partially, and there is no added effect of coenzyme Q over the cytochrome c level. This could be interpreted on the basis that the low level of coenzyme Q remaining in the particle supports full flow through the coenzyme Q site when sufficient cytochrome c is present to allow maximum activity through the cytochrome b site. When coenzyme Q is added alone, a maximum flow through the coenzyme Q pathway is seen.

It may further indicate that isooctane extraction favors establishment of a system for electron flow through the cytochrome b site at the expense of the coenzyme Q pathway. If treatment with deoxycholate precedes isooctane extraction, activity is only restored by coenzyme Q plus cytochrome c. This would mean that deoxycholate treatment destroys or prevents induction of a pathway through cytochrome b by isooctane treatment. Acetone extraction has a similar effect, which implies that the cytochrome b pathway is inactive, and activity can only be restored by coenzyme Q plus cytochrome c.

$$
F_s \longrightarrow Fe \begin{array}{c} \nearrow \, b \, \searrow \\ \searrow \, Q \, \nearrow \end{array} \overset{\text{Deoxycholate} \atop \text{acetone}}{\diagup} c_1 \longrightarrow c \longrightarrow a \; a_3 \longrightarrow O_2
$$

Since the cytochrome b–coenzyme Q region is a site at which phosphorylation occurs, it is interesting to consider how such a split pathway would contribute to energy coupling. It would also imply that the several components at other energy-coupling sites such as the cytochromes a and a_3 and copper in the oxidase region or the several DPNH dehydrogenase preparations and non-heme iron should be considered in terms of dual pathways rather than as linear systems. A discussion of the cytochrome oxidase region in terms of dual pathways has been presented by Jacobs et al. (39).

B. Oxidation-Reduction of Coenzyme Q

Further evidence for coenzyme Q function as an electron carrier in the electron-transport chain of mitochondria comes from studies of the state of oxidation or reduction of the quinone under conditions in which the various cytochromes of the chain are in the reduced or oxidized state (27,38,40).

Studies of this type have been carried out by stopping the electron transport and extracting the particles with solvent to determine the relative

amounts of quinone or hydroquinone present or by direct spectrophotometric observation of the coenzyme Q absorption band at 275 mµ in the
mitochondria. The latter procedure is made difficult by the large absorbance in this region from other mitochondrial components (e.g., protein)
and by the light scattering of the particles at this short wavelength. As a
result, studies of the latter type have been successful only with the development of specialized equipment and technique.

In the first studies of redox changes by the extraction technique it was
found that when mitochondria are incubated in air with no added substrate
the coenzyme Q is predominantly in the oxidized form, whereas if succinate
or DPNH is added as substrate and cyanide used to inhibit cytochrome
oxidase the quinone is predominantly in the reduced form. In early studies
it seemed that antimycin inhibited the reduction of coenzyme Q, but
subsequent studies have consistently indicated that antimycin A inhibits
oxidation of internal hydroquinone and that reduced coenzyme Q
accumulates in the presence of antimycin with either succinate or DPNH
as substrate.

Szarkowska and Klingenberg (40) find that reduced coenzyme Q
oxidation is not inhibited by antimycin A after mitochondria are uncoupled by addition of calcium, while cyanide still inhibits oxidation. If
antimycin is added without calcium, oxidation of reduced coenzyme Q is
inhibited.

Using similar extraction techniques, Redfearn and Pumphrey (41)
attempted to determine the rate of coenzyme Q reduction in mitochondria.
They found the rate of reduction to be lower than the overall rate of
succinate oxidation and on this basis proposed that coenzyme Q is not on
the main pathway of electron transport.

By direct spectrophotometric study of nonphosphorylating mitochondria, Chance (38) has also shown a maximum rate of Q reduction
equivalent to one-half the maximum rate of overall electron flow from
succinate to oxygen. These experiments are somewhat complicated by the
difficulties of the direct rapid spectrophotometry technique and interpretation of rapid reduction of quinone as compared to rapid formation of
fumarate. It is also noteworthy that part of the quinone is not completely
reduced until the oxygen is depleted. More than one functional site of
quinone could be indicated. The use of sulfide to inhibit overall electron
transport in order to observe quinone reduction introduces a complication
in assessing the results, since the conditions of the two assays are not
identical and the effect of the inhibitor on coenzyme Q reduction is unknown. Chance proposes that the changes observed may allow one-half
of the electrons to flow through coenzyme Q and the rest through cyto-

chrome *b* in a branched pathway. Such a possibility will have to be considered until refinements in measurement techniques improve the sensitivity of measurements in the ultraviolet region. As discussed previously, extraction studies can also be interpreted to support the branched chain concept.

On the other hand, Green et al. have shown that the rate of Q reduction by purified enzyme preparations is sufficient to account for the overall electron-transport rate (27).

Since there is a large amount of coenzyme Q in mitochondria, it may not be surprising that much of it is not on the direct pathway of electron flow. The sensitivity of the extraction studies is not adequate to measure the rapid turnover of a small portion of the coenzyme Q while a major part acts as an electron sink. Further evidence that only a small part of the coenzyme Q is necessary for maximum electron flow comes from the fact that mitochondrial fractions which catalyze a rapid electron flow have low levels of coenzyme Q. Thus the green particle prepared by deoxycholate treatment of mitochondria contains only 1/8 the level of coenzyme Q of the intact mitochondria or ETP and yet it has been shown to contain the full DPNH oxidase activity of the original ETP (42). Similarly, Takemori and King (43) have isolated a unit which catalyzes rapid succinic-cytochrome *c* reductase activity which contains only 1/10 the coenzyme Q level usually found in this type of particle. Furthermore, with this particle they have found that added coenzyme Q is capable of competitively reversing antimycin inhibition.

The amount of quinone which must be added back to acetone-extracted mitochondria in order to restore activity is also amazing small. For example, Lester and Fleischer (44) have shown full restoration by adding an amount of quinone equivalent to the amount originally in the particles used for assay. In view of the great chance for quinone to stick on the walls of the assay container as well as in other parts of the enzyme system, it is logical to assume that only a small fraction of the original quinone content is required to fully restore activity at the functional site.

C. In Phosphorylating Systems

The above studies indicate that a part of the coenzyme Q in mitochondria functions in a pathway of electron flow for DPNH and succinate oxidation in nonphosphorylating electron transport. What can we say about the function of the remainder of the coenzyme Q? Studies of the redox state of coenzyme under phosphorylating conditions indicate that the steady state of the remainder of the coenzyme Q is influenced by the action of the phosphorylating system. Hatefi (27) showed by the extraction technique that the presence of ADP favored the oxidation of the coenzyme Q pool, whereas the absence of ADP and presence of phosphate favored reduction. Similar studies with improved extraction technique by Redfearn and Pumphrey (41) gave similar results. Szarkowska and Klingenberg (40) and Chance (38) have made more sophisticated studies in which they have used both extraction technique and direct spectrophotometric measurements to determine the state of coenzyme Q. These studies again indicate that the presence of ADP under phosphorylating conditions causes oxidation of coenzyme Q, whereas in the absence of ADP reduced coenzyme Q accumulates. In these studies there is a difference depending on the substrate used in the change of the steady state of the quinone on shifting from the resting to the active state (no ADP *vs.* ADP added). In extraction studies with succinate as substrate, about 80% of the coenzyme Q is reduced in the resting state and there is at best 10% oxidation on addition of ADP. With DPN-linked substrates the coenzyme Q is only 60% reduced in the resting state; a shift to 40% reduced occurs with addition of ADP when the extraction technique is used. By direct spectrophotometry there is evidence of oxidation of 75% of the reduced coenzyme Q on shifting from the resting to the active state with succinate as substrate and a 30% oxidation with DPN-linked substrates. By improvement in extraction technique, both Chance (38) and Kroger and Klingenberg (45) have been able to show the same effect as that obtained with the spectrophotometric technique. The results presented would not be inconsistent with an improved rate of electron flow into the coenzyme Q pool, in response to ADP from DPN-linked substrates, together with an improved flow out giving a small net change. Since there is probably a phosphorylation site before coenzyme Q in the DPNH chain, such an effect would be expected. With succinate, electron flow to the quinone would not be controlled in the resting state, so full reduction would be favored. ADP would only stimulate outward flow of electrons from coenzyme Q, leading to strong oxidation. Furthermore, the addition of uncoupling agents causes conversion of the reduced coenzyme Q to the oxidized form. In both these studies there is no evidence for formation of ubichromenol or ubichromanol or any other form of coenzyme Q which could be considered as a phosphorylated or otherwise divergent form.

D. *In Intact Tissues*

Hoffmann et al. (46,47) have made an excellent study of the steady state of coenzyme Q in guinea pig heart *in vitro* and in isolated atria. They stopped the reactions by rapid freezing and used a modification of the low-temperature extraction procedure of Redfearn and Pumphrey for micro assays. A determination of total quinone on a thin-layer system indicated 90% recovery in the component which chromatographed as coenzyme Q_{10} after oxidation of the extract. Thus, all reduced quinone must be in forms which are easily oxidized to coenzyme Q_{10}. If we use oxidation of tocopherols as a model it is unlikely that coenzyme Q_{10} would be obtained by simple oxidation of the chromanol. This would indicate that no large amount of chromanol is present in the heart tissue.

In hearts *in vivo* it was found that 80–85% of coenzyme Q is in the reduced form. There was no increased oxidation in guinea pigs treated with thyroxine and adrenaline. There was no reproducible increase in reduction under anoxia, even though ATP levels were significantly decreased. From this and experiments with isolated heart tissue, Hoffmann et al. (46) conclude that the stress which can be produced *in vivo* was insufficient to cause change in redox level.

More dramatic changes in the redox state of coenzyme Q could be obtained in isolated guinea pig atria. Coenzyme Q in isolated atria was about 25% in the oxidized form. Under nitrogen the amount of coenzyme Q in the oxidized form decreased to 5%. Malonate, which inhibits succinic dehydrogenase, caused a shift to 45% oxidized. Amytal, which inhibits transfer of electrons from DPNH dehydrogenase to coenzyme Q, caused a shift to 40% oxidized. When a combination of malonate and amytal was added, 70% or more of the coenzyme Q was found oxidized. Adrenaline plus malonate caused oxidation of 80% of the coenzyme Q. The response to metabolic inhibitors was thus consistent with a role of coenzyme Q in electron-transport reactions. It was found also that ATP and creatinine phosphate decreased as various treatments shifted coenzyme Q to the oxidized state, which would be consistent with the view that a high level of reduced coenzyme Q is associated with energy-coupling processes. A comparison of the results in whole heart with the results obtained by Szarkowska and Klingenberg in mitochondria is shown in Table V.

E. Coenzyme Q Reductases

From the position of coenzyme Q function indicated in the electron chain, it can be predicted that it would be possible to isolate a succinic-coenzyme Q reductase and a DPNH-coenzyme Q reductase, both of which should be flavoproteins which would also contain non-heme iron and possibly some cytochrome *b*. The search for these enzymes has met various degrees of success.

TABLE V

Effects of Added Agents on the Redox State of Coenzyme Q in
Intact Heart and in Heart Mitochondria[a]

Treatment	Coenzyme Q, % reduced
Guinea Pig Atria	
None; aerobic	73
Under nitrogen	95
100mM malonate	52
2.5mM amytal	59
Malonate and amytal	30
Adrenaline	61
Malonate and adrenaline	20
Rat Heart Mitochondria	
Malate	3–24
Malate and KCN	45–70
Malate and antimycin	54–66
Succinate	89

[a] Data for guinea pig from Hoffmann et al. (47). Data for mitochondria from Szarkowska and Klingenberg (40).

A succinic coenzyme Q reductase has been isolated by Ziegler and Doeg (48). This enzyme catalyzes the rapid reduction of coenzyme Q_{10} by succinate. It contains 4.6 μmoles of flavin and 34 μmoles of non-heme iron per gram of protein. The major portion of the non-heme iron is reduced by succinate. It also contains 20% lipid and a small amount of b-type cytochrome. The cytochrome is not reduced by succinate. The activity of the enzyme in reduction of coenzyme Q is inhibited by thenoyltrifluoroacetone which presumably blocks the non-heme iron site. The succinic phenazine methyl sulfate reductase activity of the enzyme is not inhibited by thenoyltrifluoroacetone, which is consistent with the activity of the purified succinic dehydrogenase. The properties of this enzyme thus fit the predicted site of coenzyme Q function exactly.

Two enzyme preparations have been reported to fill the role of a DPNH-coenzyme Q reductase. Hatefi (30) reported a preparation from mitochondria which contains a large amount of coenzyme Q and 20% lipid in addition to non-heme iron and flavin. The enzyme causes reduction of dyes, ferricyanide, or coenzyme Q, and only the reduction of coenzyme Q is amytal sensitive. A drawback to identifying this preparation with the natural pathway from DPNH to coenzyme Q_{10} has been the fact that the enzyme only reduced short-chain homologs of coenzyme Q such as coenzyme Q_1 or Q_2 but not coenzyme Q_{10}. This failure has been attributed to the insolubility of coenzyme Q_{10}. The enzyme also retains a small amount of DPNH-cytochrome c reductase which is antimycin sensitive, which would indicate that the preparation is a fragment of the original chain.

A DPNH-coenzyme Q reductase preparation which can react with coenzyme Q_{10} has been reported by Pharo and Sanadi (49). This enzyme is a soluble flavoprotein and has been extensively purified. It contains only flavin and non-heme iron. It is possible that the latter enzyme is derived from the preparation of Hatefi by further fragmentation and that the ability to react with coenzyme Q_{10} depends upon removal of certain material from the Hatefi preparation. In any event, there is evidence for a DPNH-coenzyme Q_{10} reductase which would fit the proposed location for coenzyme Q_{10} in the electron-transport chain.

Machinist and Singer (94) have shown that the DPNH dehydrogenase obtained by phospholipase treatment of mitochondria can react with short-chain homologs of coenzyme Q. They also find that thermal denaturation or exposure to acid ethanol at $42°$ will increase the rate of reaction of this dehydrogenase with coenzyme Q_1. This implies that some coenzyme Q reductase activity in various preparations may be an artifact of the preparative procedure. These investigators also find that the amytal or rotenone sensitivity of coenzyme Q_1 reductase activity is removed by phospholipase treatment and restored by addition of phospholipids. Thus the amount and nature of phospholipids in a preparation may determine the characteristics of the reaction with coenzyme Q. The reader is referred to the chapter by Singer for further discussion of the types of DPNH dehydrogenase preparations.

F. Reduced Coenzyme Q_{10}-Cytochrome c Reductase

On the other side of the proposed site for coenzyme Q function it can be predicted that a unit containing cytochromes b and c_1 would be required to catalyze reduction of cytochrome c by the hydroquinone. Such a preparation has been obtained by Hatefi (30) by using detergent to fragment mitochondria. The fraction is very active in catalyzing the desired reaction and is sensitive to inhibition by antimycin A, 2-nonyl-4-hydroxy-quinoline-N-oxide and SN 5949. This enzyme contains about 4 mμmoles of cytochrome c_1, 7 mμmoles of cytochrome b, and 12 mμatoms of non-heme iron per milligram of protein as well as lipid. When coenzyme Q_2H_2

is added to the enzyme the cytochrome c_1 is fully reduced whereas the cytochrome b is only 50% reduced. This again points up the anomalous role of cytochrome b in nonphosphorylating electron-transport systems and the question of how much of it is involved in Q oxidation. Unfortunately, most of the work on this preparation has been done using coenzyme Q_2 rather than Q_{10}. In view of restoration studies which show coenzyme Q_2 producing a 20% antimycin-insensitive succinic cytochrome c reductase activity, this substrate could potentially set up an artificial system. Since the inhibitor sensitivity is excellent, however, it may be assumed that the system described is related to the pathway of coenzyme Q_{10} oxidation.

The isolation of the enzyme units described above is taken as evidence for a structural arrangement within the electron-transport system. By combining these units, Hatefi has been able to reconstruct a particle with DPNH and succinic cytochrome c reductase activity.

VI. The Quinone Reductases

A series of flavoproteins have been purified from animal, plant, and microbial cells which catalyze the oxidation of DPNH or TPNH or both in the presence of water-soluble quinones (50). Wosilait and Nason (96) found that certain of these dehydrogenase preparations showed specificity for certain types of quinones as acceptors, but Mahler et al. (97) found that the specificity may depend upon the presence or absence of metal in the enzyme. Since many other flavoprotein dehydrogenases react with quinones as well as with other oxidizing agents, the term quinone reductase is of doubtful value. There is also no good evidence for free water-soluble quinones within the cytoplasmic structures of cells, so the significance of quinones as electron acceptors is doubtful.

The most interesting of these enzymes is the so-called phylloquinone reductase or DT diaphorase which has been extensively studied by Ernster and co-workers (98). This enzyme mediates the oxidation of DPNH and TPNH by menadione and other quinones as well as dyes. The enzyme is unusual in that it is inhibited by very low $(10^{-9}M)$ concentrations of the nutritional vitamin K antagonist, dicoumerol. It is also characterized by an ability to oxidize either DPNH or TPNH. The enzymes from animal sources are activated by bovine serum albumin or Tweens (sorbitan monolaurate). The enzyme seems to be scattered about in large amounts in the cytoplasm of many types of cells, which implies that it is important for some unknown function. Of all species studied it seems to be absent only in pigeon liver. The fact that it is inhibited by dicoumerol has led to the suspicion that the natural acceptor *in vivo* would be vitamin K_1, but the isolated enzyme does not show significant activity when vitamin K_1 is substituted for water-soluble quinones.

The fatty nature of vitamin K could very well prevent proper interaction of enzyme and acceptor in the usual aqueous assay systems. Wosilait (51) has been able to show some vitamin K reductase activity with a DT diaphorase from dog liver by suspending the quinone in a detergent. A further example of detergent effect comes from the demonstration by Asano and Brodie (52) that a malate-vitamin K reductase from Myco-bacteria specifically requires phospholipid for activity. If specific phos-pholipids are required for interaction between fat-soluble quinones and their reductases, a rather systematic study of reaction conditions is needed before the natural acceptors can be determined.

VII. Vitamin K in Electron Transport

A series of proposals have been made for vitamin K function in phos-phorylation in animal mitochondria. The evidence reported has primarily been concerned with the efficiency of phosphorylation in vitamin K deficiency. A summary of this evidence has been made by Isler and Wiss (3,53). Attempts have also been made to destroy vitamin K by exposure of mitochondria to ultraviolet irradiation. Beyer has reported an effect of ultraviolet irradiation on the DPNH oxidase pathway which can be reversed by addition of vitamin K_1.

One of the major difficulties in determination of the role of vitamin K in mitochondria is caused by the very low levels of vitamin found. By biological assay, Green et al. (54) have found a small amount of vitamin K activity in liver mitochondria. Martius (3) has also found material in mitochondria deriving from labeled vitamin K fed to animals, which shows separation characteristics of vitamin $K_{2(20)}$ by countercurrent distribution. No material has been isolated which shows the spectral characteristics of vitamin K. On the other hand, Hall (33), working in this laboratory, has reported a material which shows spectral characteristics of a naphtho-quinone in heart mitochondrial lipids and has the chromatographic and redox properties of vitamin $K_{2(30)}$, although it proved to be much more labile than vitamin $K_{2(30)}$ in all isolation procedures. A similar material has been reported in beef-heart mitochondria by Sottacasa and Crane (14). In view of the many studies of Martius, Dallam, and Anderson and Beyer concerning effects of vitamin K (3,53), as well as the effects obtained by Hall in which vitamin K_2 acted to restore DPNH oxidase in ether-extracted mitochondria, further study in this area would be worthwhile. From the studies of Brodie (55) in which a clear role for vitamin K in the DPNH oxidase system of bacteria can be seen as well as the studies of White (56) in which desmethylvitamin K is shown to function in *Hemophilus* DPNH tetrazolium reductase activity, it is tempting to consider that a naphthoquinone, with its relatively low redox potential as compared to coenzyme Q, may function at an early point in the DPNH oxidase system. This site would presumably be before cytochrome *b* or coenzyme Q. The nature of the unstable naphthoquinone-like material in heart mitochondria suggests that it is better not to specify a vitamin K at this site, but to leave it open for variation in naphthoquinone chemistry. It would seem that the amount of quinone present will be small and that a pool of the magnitude found with coenzyme Q will not be present.

A. Vitamin K in Bacteria

A clear demonstration of vitamin K function has been achieved in several electron-transport systems from bacteria. Brodie (55) and his co-workers have demonstrated a requirement for vitamin K for oxidation of DPNH and associated phosphorylation in electron transport particles from *Mycobacterium phlei*.

Vitamin $K_{2(45)}9H$ is the natural quinone which has been isolated from the bacteria. Asano and Brodie (52) have made an extensive study of the electron-transport systems in the bacterial particles. Malate, DPNH, and succinate are oxidized by the particles and the oxidation of all three is stopped by exposure to 360 mμ ultraviolet light. The oxidation of malate and DPNH can be restored by adding back vitamin $K_{2(45)}9H$, but the succinate oxidase cannot be restored by adding back the vitamin; so an additional factor sensitive to ultraviolet light has been postulated for that system. A study of the cytochromes indicates a system similar to mitochondria with slight variations in the exact nature and relative amounts of the cytochromes. From these studies Asano and Brodie postulate electron transport from a malate dehydrogenase and from DPNH through $K_{2(45)}9H$ to the cytochromes as shown in Figure 2. Isolation of a phospholipid-requiring malate vitamin K reductase is consistent with their scheme. It is proposed that the additional ultraviolet-sensitive factor is a quinone of unknown nature exclusive to that system. From their studies which show a DPN requirement for succinate oxidation, it can also be suggested that succinate might be oxidized through the DPNH pathway in these particles. The proposed role for $K_{2(45)}9H$ would be analogous to the role proposed

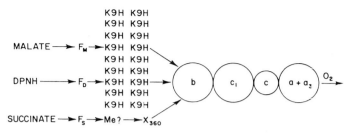

Fig. 2. Possible stoichiometric relation of vitamin $K_{2(45)}9H$ with other components of the electron transport system of *Mycobacterium phlei*. F, flavoprotein dehydrogenases; b, c, c_1, a, and a_3, cytochromes; K9H, $K_{2(45)}9H$; Me, possible metal site; X_{360}, component destroyed by 360 mμ light. Based on Asano and Brodie (52).

for coenzyme Q in mitochondria. There has also been difficulty getting agreement on the exact form of vitamin K involved in the system. Brodie et al. have shown that a 2-methylnaphthoquinone with at least one unsaturated isopentenyl group in the 3 position is necessary for restoration of the phosphorylating DPNH oxidase system after the natural vitamin K in the system is destroyed by irradiation with 360 mμ ultraviolet light (cf. Table VI).

TABLE VI

Effectiveness of Quinones with Modified Side Chains on Restoration of Electron Transport and Phosphorylation

System	Quinone	Restoration of function, %	
		Electron transport	Phosphorylation
M. phlei ETP	Vitamin K_1	80	120
	Dehydrovitamin K_1	50	0
Chloroplasts			
	Plastoquinone $A_{(45)}$[a]	51	51
	Plastoquinone $A_{(15)}$[a]	60	5

[a] Increase over control value.

A role for vitamin K in electron transport in *B. subtilis* has been described by Downey (57).

Escherichia coli is a rather unusual microorganism in that it contains large amounts of both coenzyme Q and vitamin K_2. Kashket and Brodie

(19) have shown that two types of electron-transport particles can be isolated from this system. A heavy fraction contains coenzyme Q and catalyzes both succinate and DPNH oxidation, whereas a light fraction contains only vitamin K and has only DPNH oxidase activity. The suggestion is obvious that the heavy particles resemble the mammalian mitochondrial system whereas the light particles are more like the *Mycobacterial* system with regard to quinone function in electron transport.

Itagaki (21) has proposed a role for coenzyme Q in the formate–nitrate reductase system of *E. coli*. Extraction of coenzyme Q_8 and vitamin K_2 from the particles by means of acetone leads to loss of activity. Addition of coenzyme Q_8 together with other lipid materials in the extract completely restores activity. On the other hand, vitamin K_2 is rather inefficient in restoration of activity. Thus coenzyme Q in *E. coli* may be involved in both the succinate and formate systems (Fig. 3).

A new version of the vitamin K story has arisen with the discovery of desmethylvitamin K in *Streptococcus* and *Hemophilus*. White (56) has shown that extraction of the desmethylvitamin K from electron transport particles of *Hemophilus* stops DPNH oxidase and DPNH dehydrogenase as measured by reduction of tetrazolium (cf. Fig. 4). Although DPNH oxidase could not be restored, White was able to restore DPNH tetrazolium reductase with the desmethylvitamin K. This indicates that the desmethylvitamin K is necessary to transfer electrons from the flavoprotein to the

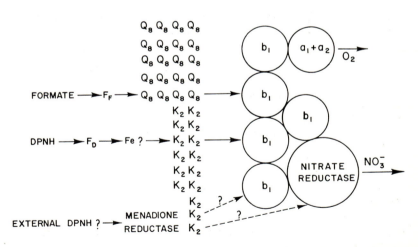

Fig. 3. Possible stoichiometric relation between quinones and other components in the heavy electron transport particles of *E. coli*. Symbols as in Figure 1. K_2, vitamin $K_2 40$. Based on Itagaki (21).

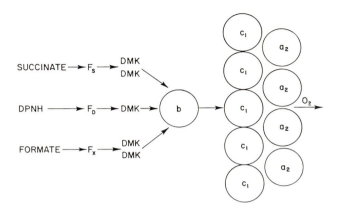

Fig. 4. Schematic diagram of the electron transport system in *Hemophilus* (56). DMK is desmethylvitamin $K_{2(35)}$. Other symbols as in Figure 1. Based on White (56).

tetrazolium. He has also shown reduction of the quinone in particles in the presence of DPNH. It would appear from these studies that the functional naphthoquinones in electron-transport systems may not all belong to the vitamin K series of compounds. Evidence for other forms of naphthoquinone has been reported in chloroplasts and blue-green algae. It may be that the traditional nutritional assay for vitamin K in controlling prothrombin time in K-deficient chicks is not always valid in assessing functional naphthoquinone content of electron transport systems.

VIII. Quinones in Chloroplasts

A role of vitamin K_1 in photosynthesis has been postulated ever since the discovery by Dam et al. (58) that vitamin K_1 is concentrated in the green parts of plants and in chloroplasts. It is ironic that although many quinones have subsequently been found in chloroplasts, for which evidence of function can be presented, no evidence for the function of vitamin K_1 has yet been achieved. Another quinone, plastoquinone, was found in leaves by Kofler in 1946 (59), during a search for vitamin K. Work with it was apparently dropped when it was found to have no vitamin K activity in biological assays. It was rediscovered during the course of our own survey work on coenzyme Q distribution. In contrast to coenzyme Q, which is in the mitochondrial fraction from leaves, this benzoquinone is localized in the chloroplasts. A more detailed examination of lipids from spinach chloroplasts has revealed no less than nine quinones (7,12,13). Several of these are present in very small amounts in comparison with other components of the photosynthetic unit (considered to be equivalent to 300 chlorophyll molecules). It is also not certain that all the quinones found in spinach are present in chloroplasts from other plants. The quinones which have been found and an estimate of the amount usually present in spinach chloroplast preparations are given in Table VII.

TABLE VII

Quinone Content of Spinach Chloroplasts

Quinone	Absorbance maximum in ethanol	R_f on TLC in chloroform[a]	μmoles quinone per mg chlorophyll[b]	Ref.
Plastoquinone A	255	0.90	0.120	12
Plastoquinone B	255	0.86	0.005	13
Plastoquinone C	255	0.67	0.013	12
Plastoquinone D	255	0.60	0.007	12
α-Tocopherylquinone	269,262	0.42	0.009	71
β-Tocopherylquinone	261	0.31	0.0005	70
γ-Tocopherylquinone	258	0.22	0.0002	70
Vitamin K_1	269,261,249,242	0.85	0.015	71
Desmethylvitamin K	261,248,242	0.69	0.0002	7
Coenzyme Q_{10}	275	0.76	0.0000	17

[a] Chromatography on 250 μ thin-layer silica gel with chloroform as developing solvent. Coenzyme Q_{10} is shown as a standard and is not present in chloroplasts.
[b] Values given are typical but vary somewhat with the season of year and the condition of the spinach.

Plastoquinone $A_{(45)}$ would be the original plastoquinone first described by Kofler. Plastoquinones B, C, and D are of unknown structure but do not seem to be simple isoprenologs of plastoquinone $A_{(45)}$. A homolog of plastoquinone $A_{(45)}$ with a 20-carbon side-chain plastoquinone $A_{(20)}$ has been isolated from horse chestnut leaves by Trebst (11) and has been found in chloroplasts from horse chestnut by Barr in this laboratory, but has not been found in spinach. Plastoquinones $A_{(45)}$, B, C, and D seem to be present in most leaf tissue which has been examined. Chromatographic techniques have been the method of detection. α-Tocopherylquinone is also widely distributed in leaves as is vitamin K-like material. The β, γ, and δ tocopherylquinones are present in such extremely small amounts that their further distribution has not been examined. The most likely quinones for further consideration for a role in chloroplasts are plastoquinones A, C, and D, α-tocopherylquinone, and vitamin K_1.

The first evidence of plastoquinone $A_{(45)}$ function in photosynthesis came from studies by Bishop on restoration of a Hill reaction activity in heptane-extracted chloroplasts. Other evidence came from the demonstration of oxidation-reduction changes in plastoquinone on exposure of chloroplasts to light. The Hill reaction in chloroplasts is represented by production of oxygen in light during reduction of an artificial acceptor such as a dye or ferricyanide. It can be measured either by determining oxygen release or dye reduction. The process is envisioned as an oxidation of water to produce a reductant which is transferred as electron flow through an electron-transport chain in the chloroplast to cause final reduction of the acceptor dye. Thus, in its general properties the Hill reaction is similar to a terminal oxidase system running backwards driven by the light energy captured in chlorophyll (60). It is perhaps not surprising that the system contains flavoproteins, non-heme iron and copper proteins, cytochromes, and quinones (61).

When freeze-dried chloroplasts are extracted with petroleum ether, plastoquinone $A_{(45)}$ can be extracted without removing much chlorophyll. When the quinone is added back to these extracted chloroplasts, Hill reaction activity is restored (62). These studies were extended by Krogmann (63), who showed that photosynthetic phosphorylation could also be restored by adding plastoquinone $A_{(45)}$ to the extracted system. Krogmann and Olivero (64) further showed that restoration of phosphorylation required members of the plastoquinone series possessing at least a 10-carbon unsaturated terpenoid chain, whereas Hill reaction activity could be restored with plastoquinones possessing shorter or saturated side chains (cf. Table VII). Some indication of the site of function for the plastoquinone $A_{(45)}$ is possible by use of a partial reaction devised by Vernon and Zaugg. The second light reaction is inhibited by the reagent chlorophenyldimethyl urea (CMU), thus stopping the overall Hill reaction. If ascorbate is used as a reducing agent in presence of a small amount of indophenol it is possible to show photoreduction of TPN without evolution of oxygen. When plastoquinone $A_{(45)}$ is extracted, this TPN reduction is not affected. Plastoquinone $A_{(45)}$ therefore must function at some site in the second light reaction or transfer electrons from the site of that reaction to the site of the first reaction (65), as shown in Figure 5. Levine and Smillie (66) have come to similar conclusions using plastoquinone-deficient mutants. Trebst (67) has shown that partial extraction of plastoquinone $A_{(45)}$ almost completely inhibits the photoreduction of ferricyanide whereas it does not affect reduction of TPN. Further extraction inhibits

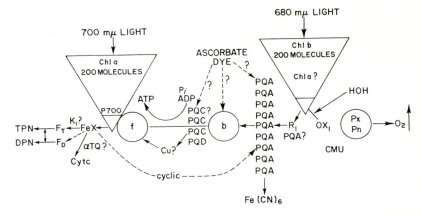

Fig. 5. A summary of possible sites for quinone function in chloroplasts. Various authors do not agree on the details of the electron transport system in chloroplasts (60,68,93). FeX, ferredoxin; R_1, primary acceptor for the first light reaction; OX_1, oxidant produced by second light reaction which then oxidizes water through Px and Pn to produce oxygen, peroxidase, and manganese have suggested function at this site which is also the region of CMU inhibition.

both reactions, and readdition of quinone restores both reactions. The restoration of TPN reduction is relatively specific to plastoquinones of the A type whereas ferricyanide reduction activity is rather nonspecifically restored by several quinones.

From these observations Trebst concludes there are two sites for plastoquinone A function. The first is a relatively specific site on the main pathway to TPN reduction and the other is a nonspecific side path detectable with ferricyanide. Witt and co-workers (68) have made a direct spectrophotometric study of oxidation-reduction changes in quinones and other components in the chloroplasts. They come to similar conclusions that a small amount of plastoquinone functions on the main pathway to TPN reduction whereas a large amount functions on a side path. This suggestion of a small active part and a large slowly reacting pool of quinone is strikingly reminiscent of the studies on coenzyme Q function in mitochondria.

Another observed oxidation-reduction change in plastoquinone A is also similar to effects observed in the oxidation state of coenzyme Q when mitochondria are uncoupled. Friend and Redfearn (69) have shown that much of the plastoquinone in phosphorylating chloroplasts is in the reduced form. If ammonium chloride, which acts as an uncoupling agent in chloroplasts, is added, the bulk of the plastoquinone A appears in the oxidized form after illumination. Again it appears that a pool of reduced quinone is most favorable to a tightly coupled phosphorylating system.

There has not been enough study of the other chloroplast quinones to decide how they may function in photosynthesis. Studies by Dilley and Crane (70) have shown that α-tocopherylquinone and plastoquinone D appear more in the oxidized form in light and shift to the reduced form in dark. These changes are inhibited by CMU. There is also evidence for a buildup of more total α-tocopherylquinone when TPN is present, which might indicate that the large pool of α-tocopheryl in chloroplasts can be mobilized in the presence of TPN in the light to produce more quinone.

Some evidence for the function of plastoquinones C and D and α-tocopherylquinone has come from restoration of activity in chloroplasts after they have been extracted by means of acetone (12). When dry chloroplasts are extracted with acetone, a large portion of all quinones is removed; this is in contrast to the effect of petroleum ether, which primarily extracts plastoquinone A. After acetone extraction the overall Hill reaction is not restored by plastoquinone A alone using either TPN or ferricyanide as acceptor. Restoration of ferricyanide reduction is best accomplished by addition of plastoquinone A plus β-tocopherylquinone and restoration of TPN reduction requires plastoquinones A, C, and D. If the first light reaction is studied by applying CMU inhibition and using ascorbate plus indophenol as reducing agent, it is found that TPN reduction activity has been lost by acetone extraction and it is only restored by addition of plastoquinone C. It should be noted that plastoquinone B has a stimulating effect on both of these reactions. The only effect of α-tocopherylquinone so far observed in the acetone-treated preparation is to restore reduction of cytochrome c (cf. Fig. 5) after acetone extraction. The sum of these preliminary studies would suggest that plastoquinone C and possibly α-tocopherylquinone function at a site closer to the first light reaction than plastoquinone A (71).

IX. Quinones in Other Photosynthetic Systems

The blue-green algae have an oxygen-evolving photosynthetic apparatus and some pigments similar to those of higher plants. Only three quinones have been found in the one species investigated, *Anacystis nidulans*. These are a plastoquinone which is very similar to plastoquinone $A_{(45)}$ but probably not identical with it, and two rather polar naphthoquinones (15). Amesz (72) has used direct spectrophotometric technique to examine changes in the quinones in light and dark. He finds clear evidence for reduction of the plastoquinone in light and some spectral changes which could indicate reduction of a small part of the total available naphthoquinone. Unfortunately, it is not known how much of each of the quinones is localized in the chlorophyll-bearing lamellae in the algae and how much is in other parts of the cell.

The purple sulfur bacteria such as *Rhodospirillum* and *Chromatium* carry out a photosynthesis which is dependent on electrons from thiosulfate or organic substrates for the photoreduction of DPN. Succinate can also be used as a substrate. These organisms are rich in coenzyme Q, which is concentrated in the chromatophores or photosynthetic units. Clayton (73) has made an extensive study of oxidation and reduction of coenzyme Q in chromatophores and found evidence that the quinone can function in photosynthesis.

The rapid reduction of coenzyme Q in light coincident with oxidation of cytochrome would indicate that the quinone may act as a primary electron acceptor from the activated chlorophyll (74). This is indicated by the alternative route in Figure 6. This primary site would be analogous to a proposed site for plastoquinone A as primary acceptor for the second light reaction in green plants (65). Evidence for action of quinone as a primary acceptor of electrons from chlorophyll also comes from observations that isolated chlorophylls can cause reduction of plastoquinone (75) or coenzyme Q (74). On the other hand, Duysens (60) has indicated that

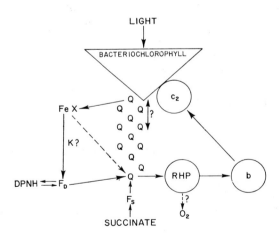

Fig. 6. Possible association of coenzyme Q_9 with other components of the chromatophores of *Rhodospirillum*. FeX, ferredoxin; RHP, c_2, b, cytochromes; F_D and F_S, flavoproteins. Based on Vernon (74).

another fluorescence-quenching substance may act as the primary acceptor with subsequent transfer to plastoquinone A. By analogy with the proposed role of ferredoxin in the first light reaction and the evidence for non-heme iron in the succinic-coenzyme Q reductase, one must certainly consider the possibility of a non-heme iron compound at the primary site in chloroplasts which would in effect be a quinone–photoreductase system carrying electrons from the photoreduced chlorophyll to the quinone. A scheme for electron transport during cyclic electron flow accompanied by phosphorylation in chromatophores of *Rhodospirillum*, as discussed by Vernon (74), is shown in Figure 6. The suggested site for quinone function between the flavoprotein or the non-heme iron protein, ferredoxin, and the heme protein, RHP, is very similar to the proposed site of coenzyme Q function in animal mitochondria. All attempts in which coenzyme Q has been extracted from the particles and then added back to restore activity have failed.

Rudney (26) has produced quinone-deficient particles by growing cells in the presence of diphenylamine, which inhibits the synthesis of terpenes. The chromatophores from these deficient cells could be reactivated by adding coenzyme Q_2 or Q_5 but not with the natural long-chain quinones.

A small amount of another quinone, rhodoquinone, has been isolated from *Rhodospirillum rubrum* (16). It has a hydroxyl or amino group replacing one methoxyl group of coenzyme Q. No evidence for function has been reported.

X. Other Electron-Transport Systems

Coenzyme Q is not exclusively localized in mitochondria. There is excellent evidence that it is also present in the light membrane fraction obtained from certain types of cells (76,78). This microsomal fraction probably contains several types of membranes and it carries certain specialized electron-transport systems such as those involved in oxidative detoxification in liver cells or steroid dehydrogenation reactions in adrenal cells. At present there is no evidence for coenzyme Q function in any of these systems.

The purified enzyme aldehyde oxidase has been shown by Handler and co-workers (77) to be a miniature electron-transport system in that it contains flavin, non-heme iron, molybdenum, and coenzyme Q. They have presented evidence that coenzyme Q functions in the electron-transport system of this enzyme.

Jayaraman and Ramasarma (78) have found coenzyme Q not only in microsomes but also in the nuclear fraction of liver cells. Since oxidative enzyme systems have been found in nuclei (79), it is possible that the co-enzyme Q is associated with those systems.

The presence of vitamin K in certain obligate anaerobic bacteria poses a puzzle which may open a new approach to quinone function (80). One can ask whether these anaerobes have an electron-transport system which uses the quinone, or whether the vitamin K in these organisms functions in some other way.

The function of coenzyme Q_8 in the formate–nitrate reductase systems of *E. coli* suggests a possible site for function of these quinones in pathways other than to oxygen. There are other anaerobes such as *Clostridium* in which no quinones have been found. A comparative study of oxidation-reduction processes in different types of anaerobic bacteria could very well reveal some important differences in electron-transport processes.

XI. Electron Spin Resonance Studies

The association of quinones and cytochromes in electron-transport systems would seem to favor oxidation or reduction by single electron steps which would momentarily form a semiquinone. Definite electron spin resonance signals would be expected from the semiquinone. Several electron spin resonance signals have been detected in both mitochondria and chloroplasts. It has not been possible to assign any such signals to coenzyme Q. There have been suggestions that one of the signals from chloroplasts may be caused by a semiquinone of plastoquinone, but it is more likely to be associated with the P700 pigment system (8). Improvements in sensitivity and resolution of electron spin resonance equipment may be necessary before definite evidence for semiquinones can be obtained.

XII. The Effects of Other Lipids on Quinone Function

Although the primary binding of coenzyme Q and other fat-soluble quinones might be at nonpolar sites of the structural proteins in the respective membrane systems, there is evidence that other lipids in the membrane structure influence the function of the quinone. After extensive extraction of mitochondria with either isooctane or acetone, other neutral lipids and phospholipids are removed along with coenzyme Q. Certain of these lipids must be added back along with the coenzyme Q in order to restore full electron-transport activity.

After extensive isooctane extraction it was found that both a phospholipid and a neutral lipid fraction were necessary in order to get maximum restoration of activity with added coenzyme Q_{10}. The most effective phospholipid was the cephalin fraction from mitochondria, whereas the lecithin fraction was relatively ineffective. In addition to the phospholipid a fraction which is eluted from a silicic acid column shortly after coenzyme Q_{10} but before the phospholipids appear was also found to be necessary for coenzyme Q_{10} function. This fraction was referred to as NL II (28).

After acetone extraction a requirement for other lipids also appears. Crane and Ehrlich (82) have shown that a lipid from the acetone extract is required in order to get maximum activity with coenzyme Q_{10} but is not necessary when short side-chain analogs such as coenzyme Q_3 are used to restore activity. The nature of the factor was undefined, but it was eluted from a Decalso column only with ethanol and was therefore more polar than the NL II fraction which was necessary in isooctane extraction studies. Lester and Fleischer (37,44) have also shown that phospholipid is required to restore activity with coenzyme Q_{10} after acetone extraction. In their studies they found no specificity for phospholipid type.

A different lipid factor called NL I was found to stimulate succinic ferricyanide reductase activity when it was added along with coenzyme Q_{10} after acetone extraction. NL I did not act like NL II in restoration of succinoxidase activity (82).

Certain other lipids are thus necessary for coenzyme Q activity and appear to modify the pathway of electron flow through the system. The type of quinone which can act at the coenzyme Q site is also affected by the other lipids present in the system.

Requirements for phospholipid to activate the malate vitamin K reductase of *M. phlei* as well as a requirement for reactivation of mitochondrial DPNH oxidase (94,95) after phospholipase treatment have been discussed previously.

XIII. Quinones and Phosphorylation Mechanisms

There have been several suggestions for quinone intermediates as the primary site for coupling electron flow to formation of ATP. Many of these proposed phosphorylated intermediates are quite effective in forming ATP in model systems. Evidence for such intermediates in the respiratory or photosynthetic particles has been difficult to find.

Quinol phosphates have been shown by Todd (83) to donate phosphate to AMP with formation of ADP. The reaction requires a phosphorylated hydroquinone which is not formed directly without an activated phosphate.

Clark and Todd (3) have proposed that a quinol phosphate could be formed from the naturally occurring quinones by the formation of a tertiary carbonium ion and nucleophilic addition of a phosphate ion which upon reduction would lead to formation of the quinol phosphate (reaction II, Fig. 7).

As an alternative the chromanol can be subjected to addition of phosphate with charge separation (84). Reduction to the chromanol will produce the chromanol phosphate (reaction I, Fig. 7).

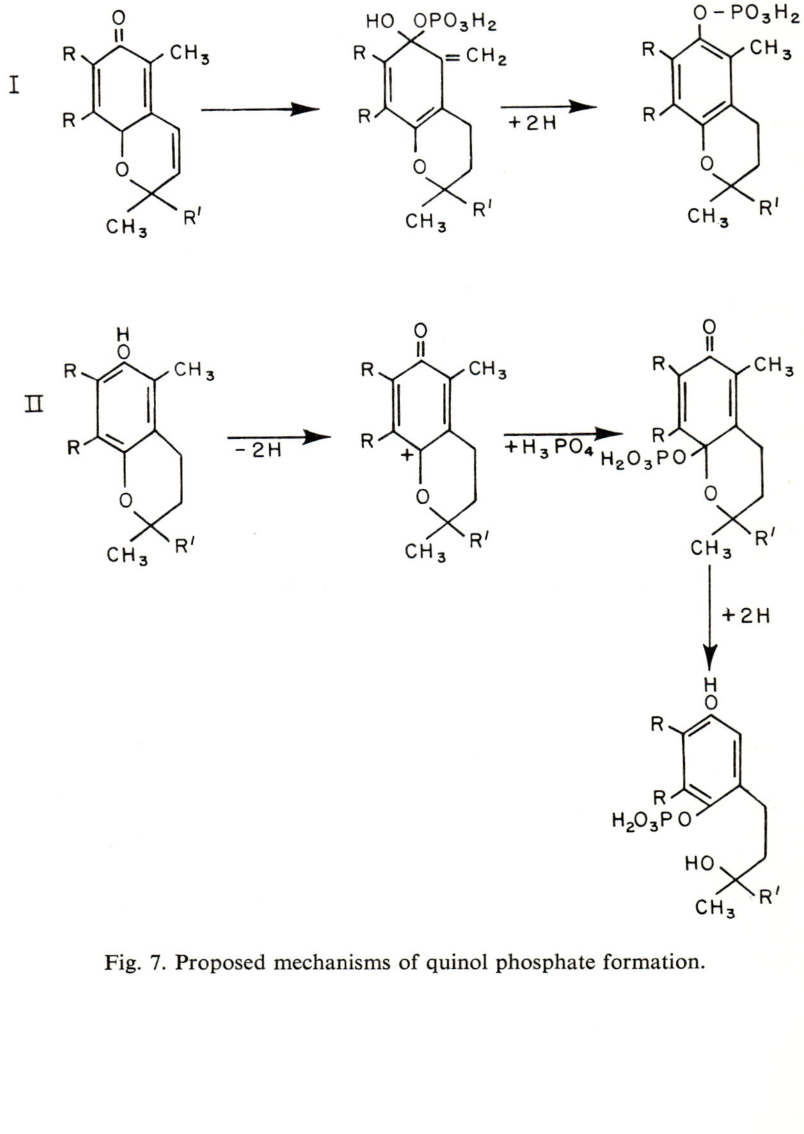

Fig. 7. Proposed mechanisms of quinol phosphate formation.

A third variation on this approach has been to propose a methine phosphate formation on the 5-methyl group of the chromanol (85). The phosphate can then be transferred to ADP during an oxidation.

In an extension of this approach, Moore and Folkers (86) have incorporated the non-heme iron of mitochondria into a complex with 2 moles of coenzyme Q. They have shown that the presence of iron favors formation of the coenzyme Q_{10} chromanol in the test tube. They have proposed initial addition of phosphate to the iron in the chelate with subsequent transfer to the methyl group on the chromanol as an approach to phosphorylation. A system of the type proposed by Moore and Folkers is shown in Figure 8. This mechanism would utilize the ring methyl group between a carbonyl group and the isoprenoid side chain which is common to vitamin K, coenzyme Q, and α-tocopherylquinone. Plastoquinones and desmethylvitamin K's do not have this group. It is possible that the 5-methyl group in plastoquinone would fulfill the requirement for such a

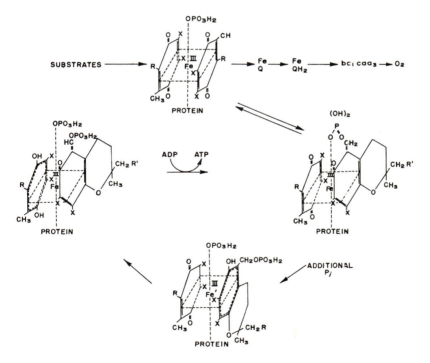

Fig. 8. Possible association of non-heme iron and quinone during chromanol formation or phosphorylation. Based on a proposal by Moore and Folkers (86).

methyl group, but there is nothing available for this function in the desmethyl naphthoquinones. Since phosphorylation processes have not been thoroughly studied in the *Hemophilus* and *Streptococcus* strains which contain the nonmethylated quinones, it is possible that electron transport through the quinone is not associated with phosphorylation in these organisms. The actual addition of phosphate to the quinone may not be the only significant reaction of a complex of this type. One can also feel than an iron–quinone complex would have a variety of reduction states available with graded redox potential, which could be of great advantage in attempting to postulate a mechanism for reversed electron flow from succinate to DPN (88). For example, the presence of ATP, when electron flow to oxygen is blocked, may encourage a greater degree of reduction at certain sites in the complex, which would permit reduction of DPN as if the high-potential non-heme iron protein had been converted to a low-potential type analogous to the ferredoxin of chloroplasts. An association of quinone with some other component of chloroplasts has also been invoked by Rumberg et al. (68) as a possible explanation of the absorbance band at 518 mμ in chloroplasts. This band disappears when quinone is removed and is restored when plastoquinone $A_{(45)}$ is added back. One might consider whether this band represents a hydroquinone non-heme iron chelate.

At present there is no evidence for the proposed structures in phosphorylating systems. Possible encouragement for formation of a chromanol or of at least a hydroquinone–iron complex comes from the pool of reduced quinone which appears to accumulate in electron-transport systems under phosphorylating conditions and the appearance of large amounts of quinone when phosphorylation is uncoupled. Further encouragement comes from the fact that all the major natural quinones in electron transport have the β-unsaturated position in the terpenoid side chain which favors chromanol formation. Brodie (54) has found that the unsaturated side chain is necessary to restore phosphorylation, whereas analogs with saturated side chains only restore electron transport (cf. Table VI). Similar studies have not been possible in mitochondria because no system for adding quinones to restore a phosphorylating system have been found. The differences which Ramasarma et al. (87) find in the metabolism of saturated and unsaturated coenzyme Q analogs fed to rats indicates that the unsaturated quinone is more actively involved in metabolic processes.

It should also be noted that among quinones found in lower concentration, tocopherylquinones and chlorobium quinone do not have the β-unsaturated side chain. They could be converted easily to such a form, however, or as minor components they may be indicative of alternative electron pathways.

Since the actual amount of total quinone recovered as quinone or hydroquinone has not been determined in experiments with coupled mitochondria and chloroplasts, it is possible that a part of the large pool of reduced quinone would be in a stabilized form as represented by the chromanol. When uncoupling occurs, the formation of chromanol or iron chelate would be reduced and the oxidation-reduction would then involve only the quinone and hydroquinone.

There is only limited evidence for chromanols in electron-transport systems. As mentioned above, α-tocopherol is found in both mitochondria and chloroplasts. There is no evidence that it undergoes any enzymic changes in mitochondria and only suggestions of possible changes, based on the appearance of excess α-tocopherylquinone in light, in chloroplasts. Brodie has presented evidence for the chromanol of vitamin $K_{2(45)}9H$ (naphthotocopherol) in *Mycobacteria* particles during phosphorylation. The conditions used for extracting this compound (reducing conditions and strong acid) indicate that it is probably an artifact. No evidence for a chromanol (XI) of coenzyme Q has been found. On the other hand, the corresponding oxidized form, ubichromenol (XII) has been found in many

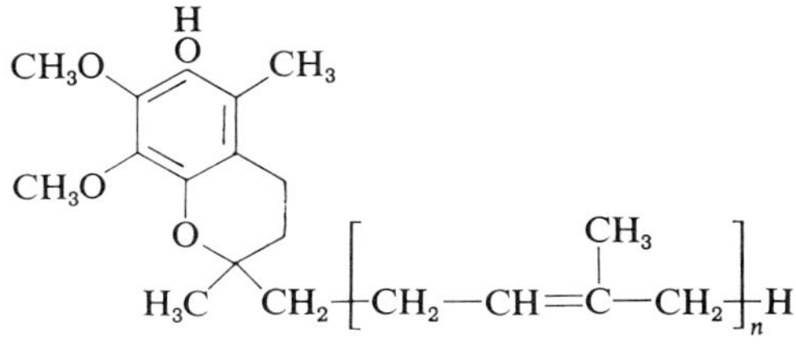

(XI) $n = 5$–9

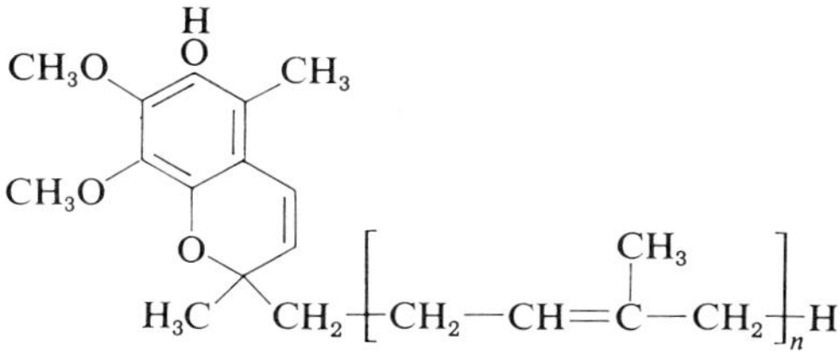

(XII) $n = 5$–9

tissues in both animals and fungi (17). There is no evidence that this compound is involved in electron transport or phosphorylation. However, the ubichromenol content of certain tissues varies independently of coenzyme Q levels and responds to changes in metabolic conditions. The analogous derivative of plastoquinone, solanochromene, has not been found in green leaves. It has been isolated only from cured tobacco, which suggests that it is produced during the drying process.

If any chromanols are involved in phosphorylation processes they are obviously hard to detect. We have previously shown that the chromanol of hexahydrocoenzyme Q_4 is unstable when incubated aerobically with mitochondria (28). The studies with phosphorylating systems suggest a large pool of some reduced form of quinone, but it may be that this is not held in a single chromanol form. One can postulate an energized derivative of the chromanol and hope that advances in chromanol chemistry will suggest better approaches to the mystery of the reduced quinone pool. On the other hand, it may be that the activation of phosphate does not involve the quinones and that the large pool of hydroquinone is useful in some other way in electron-transport systems. The study of other energy-coupling processes, such as ion transfer, which do not involve phosphate transfer may open up new possibilities for quinone function (88). Mitchell (99) has suggested quinone function as a hydrogen ion carrier across the mitochondrial membrane associated with phosphorylation.

One other objection to quinone function in phosphate activation lies in the concentration of coenzyme Q at one site in mitochondria. The existence of a site for formation of ATP between cytochrome c and oxygen is well established. The inhibition of coenzyme Q oxidation by antimycin A would preclude coenzyme Q function in the oxidase region of the chain. Thus the bulk of coenzyme Q may not be involved in the last phosphorylation site. There are some alternatives which may lead to a solution of this dilemma that would be favorable to the quinones in phosphorylation hypothesis. First, we cannot be absolutely sure that a small amount of quinone does not function in the cytochrome oxidase region. In some of the early studies of coenzyme Q function some evidence was submitted for partial antimycin inhibition of coenzyme Q reduction. This was overwhelmed by later evidence to the contrary. Whittaker and Redfearn (89) have also presented evidence for antimycin-sensitive coenzyme Q_2 reduction which is probably related to the fact that coenzyme Q_2 stimulates antimycin-insensitive succinic cytochrome c reductase activity after acetone extraction of the succinic dehydrogenase complex. The effects of isooctane extraction could also be interpreted along the line of a coenzyme Q function in the oxidase region, since succinoxidase is stopped without much effect on succinic cytochrome c reductase when only one-half the coenzyme Q is removed. This system is also characteristically fully restored by exogenously added cytochrome c. On the other hand, if the isooctane-soluble phospholipid

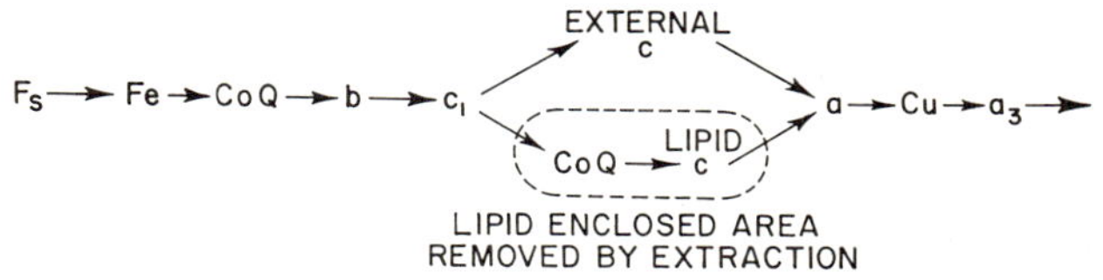

Fig. 9. Possible lipid enclosed pathway for coenzyme Q function in the cytochrome oxidase region of mitochondria. Symbols as in Figure 1.

complex with cytochrome c is added, coenzyme Q is necessary for full activity. This would suggest that the lipid-enclosed electron-transport pathway to cytochrome oxidase requires both lipid-bound cytochrome c and the easily removed coenzyme Q, whereas the pathway from succinate dehydrogenase to cytochrome c may involve coenzyme Q, which is not easily removed by isooctane. A system of this type is diagrammed in Figure 9.

The second alternative is that another quinone may function at this site. Sottocasa and Crane (14) have recently presented evidence for two other quinones besides coenzyme Q in heart mitochondria. Unfortunately there is no evidence yet that one of these might function in the cytochrome oxidase region. If evidence for quinone function in the cytochrome c oxidase region could be found it would make proposals for quinone function in phosphorylation more feasible. On the other hand, several sites for quinone function would not be incompatible with a direct role of quinones in energy coupling before phosphate transfer.

XIV. Genetic Studies

Studies of mutants which do not form a certain quinone provide another approach to the role of quinones. Mutants of algae have been obtained which promise to be valuable for the study of plastoquinone function in photosynthesis.

Levine et al. (66) have obtained a mutant of *Chlamydomonas* which is deficient in plastoquinone A. This mutant has greatly impaired photosynthesis and is grown heterotrophically. The overall Hill reaction in the algae is not functional, but photoreduction of TPN can still be achieved when indophenol and ascorbate are added as an artificial reductant system to replace the water-splitting light reaction. This effect is consistent with the other experiments described which indicate that plastoquinone A acts either as the initial electron acceptor for the second light reaction or in the flow of electrons from the second to the first reaction.

A mutant of *Scenedesmus* has been obtained by Bishop and Gaffron (90) which retains the ability to photoreduce CO_2 in the presence of hydrogen (first light reaction) but has lost the overall photosynthetic reaction starting from water. This again indicates a site for plastoquinone A somewhere between reaction one and reaction two.

No reports have been made of mutants which lack coenzyme Q, although it would appear that *E. coli* or *Proteus vulgaris*, which contain both coenzyme Q and vitamin K, would be likely candidates for such study. The fact that certain bacteria are stimulated in growth by vitamin K may in effect indicate a vitamin K-deficient mutant, but electron transport in these organisms has not been studied (80).

A comparison of coenzyme Q_6 content in wild type and a respiratory-deficient yeast showed no significant difference in quinone content (91).

XV. Summary

Coenzyme Q, vitamin K_1, vitamin K_2, desmethylvitamin K_2, plasto-quinones, and tocopherylquinones have been found in electron-transport systems from animals, plants, and microorganisms. Evidence for the functioning of these quinones in the electron-transport system in each of these systems has been presented. The pattern of evidence for function in each system follows similar lines and in the overall view constitutes an overwhelming body of evidence for quinone function in the main pathways of electron transport. In each of the major systems—mitochondria, bacterial particles, and chloroplasts—a large amount of coenzyme Q, naphthoquinone, or plastoquinone, respectively, seems to be concentrated at one site in the chain far in excess of the stoichiometric requirements of the other components of the chain. In phosphorylating systems a large part of this excess quinone seems to accumulate in the reduced form, which indicates that only a small amount of quinone participates at one time in the main pathway of electron flow.

The coenzyme Q in mitochondria appears to transfer electrons from the non-heme iron flavoprotein dehydrogenase units to the cytochrome part of the chain. A similar role for the naphthoquinones in bacterial particles is indicated. In photosynthetic organisms the major site of quinone function appears to be in the transfer of electrons from the region of activated reduced chlorophyll to the cytochromes.

Besides the principal quinones, smaller amounts of other quinones have been found in mitochondria, bacterial chromatophores, and chloroplasts.

The study of these quinones is just beginning, and evidence for function has only been presented in studies of chloroplasts.

Theories have also been proposed for the function of quinones, especially in the chromanol form, in phosphorylation processes. At present there is no clear evidence for the formation or function of these chromanol forms, although the existence of a pool of reduced quinone in phosphorylating systems would favor further studies in this area. Suggestions have also been made that quinones may function in association with non-heme iron which may favor chromanol formation or changes in redox potential.

References

1. R. H. Thomson, *Naturally Occurring Quinones*, Butterworths, London, 1957.
2. R. H. Thomson, in *Comparative Biochemistry*, Vol. 3A, M. Florkin and H. S. Mason, Eds., Academic Press, New York, 1962, p. 302.
3. G. E. W. Wolstenholme and C. M. O'Connor, Eds., *Ciba Symposium on Quinones in Electron Transport*, Churchill, London, 1961.
4. W. V. Lavate, J. R. Dyer, C. M. Springer, and R. Bentley, *J. Biol. Chem.*, **240**, 524 (1965).
5. R. H. Baum and M. I. Dolin, *J. Biol. Chem.*, **238**, PC4109 (1963).
6. R. L. Lester, D. C. White, and S. L. Smith, *Biochemistry*, **3**, 949 (1964).
7. M. McKenna, M. D. Henninger, and F. L. Crane, *Nature*, **203**, 524 (1964).
8. P. H. Gale, B. H. Arison, N. H. Trenner, A. C. Page, Jr., K. Folkers, and A. F. Brodie, *Biochemistry*, **2**, 200 (1963).
9. R. C. Fuller, R. M. Smillie, N. Rigopoulos, and V. Yount, *Arch. Biochem. Biophys.*, **95**, 197 (1961).
10. B. Frydman and H. Rapoport, *J. Am. Chem. Soc.*, **85**, 823 (1963).
11. A. Trebst and H. Eck, *Z. Naturforsch.*, **16b**, 44 (1961).
12. M. D. Henninger and F. L. Crane, *Biochemistry*, **2**, 1168.
13. M. D. Henninger, R. Barr, and F. L. Crane, *Plant Physiol.*, **41**, 696 (1966).
14. G. L. Sottocasa and F. L. Crane, *Biochemistry*, **4**, 305 (1965).
15. M. D. Henninger, H. N. Bhagavan, and F. L. Crane, *Arch. Biochem. Biophys.*, **110**, 69 (1965).
16. J. Glover and D. R. Threlfall, *Biochem. J.*, **85**, 14P (1962).
17. F. L. Crane, "Distribution of Ubiquinones," in *Biochemistry of Quinones*, R. A. Morton, Ed., Academic Press, New York, 1965, p. 183.
18. R. L. Lester and F. L. Crane, *J. Biol. Chem.*, **234**, 2169 (1959).
19. E. Kashket and A. F. Brodie, *Biochim. Biophys. Acta*, **40**, 550 (1960).
20. D. H. L. Bishop and H. K. King, *Biochem. J.*, **85**, 550 (1962).
21. E. Itagaki, *J. Biochem. (Tokyo)*, **55**, 432 (1964).
22. D. N. Moury and F. L. Crane, *Biochemistry*, **3**, 1068 (1964).
23. D. N. Moury, Ph.D. thesis, Purdue University, 1964.
24. M. D. Henninger, *Biochem. Biophys. Res. Commun.*, **19**, 233 (1965).
25. D. M. Geller, *J. Biol. Chem.*, **237**, 2947 (1962).
26. H. Rudney, *J. Biol. Chem.*, **236**, PC39 (1961).

27. Y. Hatefi, in *Advances in Enzymology*, Vol. 25, F. F. Nord, Ed., Interscience, New York, 1963, p. 275.
28. F. L. Crane, *Biochemistry*, **1**, 510 (1962).
29. F. L. Crane, in *Progress in the Chemistry of Fats and Other Lipids*, Vol. VII, R. T. Holman, Ed., Pergamon Press, London, 1964, Part 2, p. 267.
30. Y. Hatefi, in *The Enzymes*, Vol. 2, P. D. Boyer, H. Lardy, and K. Myrbäck, Eds., Academic Press, New York, 1960.
31. V. Moret, S. Pinamouti, and E. Fornasari, *Biochim. Biophys. Acta*, **54**, 381 (1961).
32. H. Ozawa, K. Mourose, S. Natari, H. Ogawa, and K. Yamaguchi, *Biochim. Biophys. Acta*, **86**, 395 (1964).
33. W. P. Cunningham, C. L. Hall, and F. L. Crane, *Proc. Intern. Congr. Biochem.*, *6th, New York, 1964*, **8**, 647 (1964).
34. T. A. Anderson, *Biochim. Biophys. Acta*, **89**, 540 (1964).
35. F. L. Crane, C. Widmer, R. L. Lester, and Y. Hatefi, *Biochim. Biophys. Acta*, **31**, 476 (1959).
36. W. P. Cunningham, C. L. Hall, F. L. Crane, and M. L. Das, *Federation Proc.*, **24**, 296 (1965).
37. D. E. Green and S. Fleischer, in *Horizons in Biochemistry*, M. Kasha and B. Pullman, Eds., Academic Press, New York, 1962, p. 381.
38. B. Chance, in *Biochemistry of Quinones*, R. A. Morton, Ed., Academic Press, New York, 1965, p. 460.
39. E. E. Jacobs, E. C. Andrews, and F. L. Crane, in *Oxidases and Related Redox Systems*, T. E. King, H. S. Mason, and M. Morrison, Eds., Wiley, New York, 1965.
40. L. Szarkowska and M. Klingenberg, *Biochem. Z.*, **338**, 674 (1963).
41. E. R. Redfearn and A. M. Pumphrey, *Biochem. J.*, **76**, 64 (1960).
42. F. L. Crane, Y. Hatefi, R. L. Lester, and C. Widmer, *Biochim. Biophys. Acta*, **25**, 220 (1957).
43. S. Takemori and T. E. King, *Science*, **144**, 852 (1964).
44. R. L. Lester and S. Fleischer, *Biochim. Biophys. Acta*, **47**, 358 (1961).
45. A. Kroger and M. Klingenberg, *Proc. Intern. Congr. Biochem., 6th, New York, 1964*, **8**, 657 (1964).
46. P. C. Hoffmann, H. W. Kunz, W. Schmid, and M. Siess, *Biochem. Z.*, **339**, 548 (1964).
47. P. C. Hoffmann, H. W. Kunz, and M. Siess, *Biochem. Z.*, **339**, 559 (1964).
48. D. M. Ziegler, in *Symposium on Biological Structure and Function*, T. W. Goodwin and O. Lindberg, Eds., Academic Press, New York, Vol. 2, 1961, p. 253.
49. R. L. Pharo and D. R. Sanadi, *Biochim. Biophys. Acta*, **85**, 346 (1964).
50. C. Martius, "Quinone Reductases," in *The Enzymes*, Vol. 7, P. D. Boyer, H. Lardy, and K. Myrbäck, Eds., Academic Press, New York, 1963, p. 517.
51. W. D. Wosilait, *Federation Proc.*, **20**, 1005 (1961).
52. A. Asano and A. F. Brodie, *J. Biol. Chem.*, **239**, 4280 (1964).
53. O. Isler and O. Wiss, *Vitamins Hormones*, **17**, 53 (1959),
54. J. P. Green, E. Sondergaard, and H. Dam, *Biochim. Biophys. Acta*, **19**, 182 (1956).
55. A. F. Brodie, *Federation Proc.*, **20**, 995 (1961).
56. D. White, *J. Biol. Chem.*, **240**, 1387 (1965).

57. R. J. Downey, *J. Bacteriol.*, **88**, 904 (1964).
58. H. Dam, in *Advances in Enzymology*, Vol. 2, F. F. Nord, Ed., Interscience, New York, 1942, p. 317.
59. M. Kofler, in *Festschrift E. C. Barrell*, Hoffman–LaRoche, Basle, 1946.
60. L. M. N. Duysens, *Progr. Biophys.*, **14**, 81 (1964).
61. R. B. Park and J. Biggens, *Science*, **144**, 1009 (1964).
62. N. I. Bishop, *Proc. Natl. Acad. Sci. U.S.*, **45**, 1696 (1959).
63. D. W. Krogmann, *Biochem. Biophys. Res. Commun.*, **4**, 275 (1961).
64. D. W. Krogmann and E. Olivero, *J. Biol. Chem.*, **237**, 3292 (1962).
65. D. I. Arnon and F. L. Crane, in *Biochemistry of Quinones*, R. A. Morton, Ed., Academic Press, New York, 1965, p. 433.
66. R. P. Levine and R. L. Smillie, *J. Biol. Chem.*, **238**, 4052 (1963).
67. A. Trebst, in *Photosynthesis Mechanisms in Green Plants*, Natl. Acad. Sci.—Natl. Res. Council Publ. 1145, Washington, D.C., 1963.
68. B. Rumberg, P. Schmidt-Mende, J. Weisard, and H. T. Witt, in *Photosynthesis Mechanisms in Green Plants*, Natl. Acad. Sci.—Natl. Res. Council Publ. 1145, Washington, D.C., 1963, p. 18.
69. J. Friend and E. R. Redfearn, *Phytochem.*, **2**, 397 (1963).
70. R. A. Dilley and F. L. Crane, *Plant Physiol.*, **39**, 33 (1964).
71. R. A. Dilley, M. D. Henninger, and F. L. Crane, in *Photosynthesis Mechanisms in Green Plants*, Natl. Acad. Sci.—Natl. Res. Council Publ. 1145, Washington, D.C., 1963, p. 273.
72. J. Amesz, *Biochim. Biophys. Acta*, **79**, 257 (1964).
73. R. K. Clayton, *Biochem. Biophys. Res. Commun.*, **9**, 49 (1962).
74. L. P. Vernon, *Ann. Rev. Plant Physiol.*, **15**, 73 (1964).
75. S. Okayama and Y. Chiba, *Nature*, **205**, 172 (1965).
76. S. Leonhauser, K. Leybold, K. Krisch, H. Standinger, P. H. Gale, A. C. Page, Jr., and K. Folkers, *Arch. Biochem. Biophys.*, **96**, 580 (1962).
77. P. Handler, K. V. Rajagopalan, and V. Aleman, *Federation Proc.*, **23**, 30 (1964).
78. J. Jayaraman and T. Ramasarma, *Arch. Biochem. Biophys.*, **103**, 258 (1963).
79. R. Penniall, W. D. Currie, N. R. McConnell, and W. R. Bibb, *Biochem. Biophys. Res. Commun.*, **17**, 752 (1964).
80. R. J. Gibbons and L. P. Engle, *Science*, **146**, 1307 (1964).
81. H. Beinert, B. Kok, and G. Hoch, *Biochem. Biophys. Res. Commun.*, **7**, 209 (1962).
82. F. L. Crane and B. Ehrlich, *Arch. Biochem. Biophys.*, **89**, 134 (1960).
83. A. Todd, *Nature*, **187**, 819 (1960).
84. I. Chmielewska, *Biochim. Biophys. Acta*, **39**, 170 (1960).
85. K. Folkers, J. L. Smith, and H. W. Moore, *Federation Proc.*, **24**, 79 (1965).
86. H. W. Moore and K. Folkers, *J. Am. Chem. Soc.*, **86**, 3393 (1964).
87. V. C. Joshi, J. Jayaraman, C. K. Ramakrishna Kurup, and T. Ramasarma, *Indian J. Biochem.*, **1**, 7 (1964).
88. B. Chance, Ed., *Energy-Linked Functions of Mitochondria*, Academic Press, New York, 1963.
89. P. A. Whittaker and E. R. Redfearn, *Biochem. J.*, **92**, 36P (1964).
90. N. I. Bishop and H. Gaffron, *Biochem. Biophys. Res. Commun.*, **8**, 471 (1962).
91. H. R. Mahler, G. Neiss, P. O. Slonimski, and B. Mackler, *Biochemistry*, **3**, 893 (1964).

92. K. S. Ambe and F. L. Crane, *Biochim. Biophys. Acta*, **43**, 30 (1960).
93. J. A. Bassham, in *Advances in Enzymology*, Vol. 25, F. F. Nord, Ed., Interscience, New York, 1963, p. 39.
94. J. M. Machinist and T. P. Singer, *J. Biol. Chem.*, **240**, 3182 (1965).
95. S. Fleischer, A. Casu, and B. Fleischer, *Federation Proc.*, **23**, 486 (1964).
96. W. D. Wosilait and A. Nason, *J. Biol. Chem.*, **206**, 255 (1954); **206**, 271 (1954).
97. H. R. Mahler, A. S. Fairhurst, and B. Mackler, *J. Am. Chem. Soc.*, **77**, 1514 (1955).
98. L. Ernster, in *Biological Structure and Function* (IUB/IUBS Symposium), Vol. 2, T. W. Goodwin and O. Lindberg, Eds., Academic Press, New York, 1961, p. 139.
99. P. Mitchell, *Biochem. Soc. Symp.*, **22**, 142 (1962).
100. H. W. Moore and K. Folkers, *J. Am. Chem. Soc.*, **87**, 1409 (1965).
101. T. P. Singer, personal communication.
102. W. P. Cunningham, F. L. Crane, and G. L. Sottocasa, *Biochim. Biophys. Acta*, **110**, 265 (1965).
103. B. C. Das, M. Lounasmaa, C. Tendille, and E. Lederer, *Biochem. Biophys. Res. Commun.*, **21**, 318 (1965); W. T. Griffiths, J. C. Wallwork, and J. F. Pennock, *Nature*, **211**, 1037 (1966).

Oxygenases (Oxygen-Transferring Enzymes)

OSAMU HAYAISHI, *Department of Medical Chemistry, Kyoto University Faculty of Medicine, Kyoto, Japan*

I.	Introduction—Oxidation and Oxygenation	581
II.	Historical Survey.	582
III.	Method	584
IV.	The Classification of Oxygenases	585
	A. Dioxygenases and Monooxygenases	585
	B. Further Classification	585
V.	Distribution and Relative Activities of Oxygenases and Oxidases	588
VI.	Physiological Significance of Oxygenases	589
VII.	General Properties and Mechanism of Action.	591
	A. General	591
	B. Dioxygenases Containing Inorganic Iron.	591
	C. Tryptophan Pyrrolase (Heme-Containing Dioxygenases)	593
	D. Monooxygenases.	594
	1. General Mechanism of Hydroxylation.	594
	2. The Mechanism of Monooxygenase Reactions	596
	3. The Mechanism of Double Hydroxylation	599
VIII.	Concluding Remarks	600
	References .	601

I. Introduction—Oxidation and Oxygenation

Biological oxidation processes may occur in three seemingly different ways: (*1*) By the removal of an electron, as when a ferrous ion (Fe^{2+}) is converted to a ferric ion (Fe^{3+}) [Reaction (1)], (*2*) by the removal of hydrogen, as when alcohol is oxidized to aldehyde [Reaction (2)], or (*3*) by the addition of oxygen to a molecule, as when carbon is oxidized to carbon dioxide [Reaction (3)].

$$Fe^{2+} \longrightarrow Fe^{3+} + e \tag{1}$$

$$CH_3CH_2OH \longrightarrow CH_3CHO + H_2 \tag{2}$$

$$C + O_2 \longrightarrow CO_2 \tag{3}$$

Although the fundamental principle governing these three types of reactions is basically the same, they are catalyzed by different types of enzymes in the cell. The properties and mechanism of action of the enzymes catalyzing the transfer of electrons and hydrogen atoms have been described in the preceding chapters. In this chapter a group of enzymes which catalyze the third type of reaction—the addition of oxygen—will be discussed.

It should be emphasized, however, that the transfer of electrons is involved in all oxidation and reduction reactions. Oxidation is defined as removal of electrons with an increase in valency, whereas reduction is defined as a gain in electrons with a decrease in valency. When carbon and oxygen interact to form covalent $C\!=\!O$ bonds [Reaction (3)], electrons are shared between both atoms. Since these electrons are somewhat closer to the oxygen nucleus than to the carbon nucleus, however, the oxygen has partially gained electrons and is reduced while the carbon has partially lost electrons and is oxidized.

II. Historical Survey

Studies on the biological oxidation processes were first initiated by Lavoisier about 200 years ago. The early investigators generally agreed that the combination of a substrate with oxygen was the essential characteristic of a biological oxidation process. The formation of an oxide was designated as oxidation, and reduction was defined as the removal of oxygen from an oxide. Biological tissues were found to contain a number of catalysts which facilitate this process by activating oxygen. In 1897 Bach and Engler independently suggested that organic peroxides might be the active form of oxygen because of their ability to oxidize other compounds. Shortly thereafter, Bach and Chodat extended this theory by proposing that oxygen initially reacts with an acceptor, A, to produce an organic peroxide [Reaction (4)], which then reacts with a substrate, S, to form an oxide [Reaction (5)].

$$A + O_2 \xrightarrow{\text{oxygenase}} A\begin{array}{c} O \\ | \\ O \end{array} \tag{4}$$

$$A\begin{array}{c} O \\ | \\ O \end{array} + S \xrightarrow{\text{peroxidase}} AO + SO \tag{5}$$

The enzymes which catalyze Reactions (4) and (5) were named "oxygenase" and "peroxidase," respectively, by these authors.

It soon became apparent, however, as a result of an extensive series of studies by Wieland and others, that biological oxidation can proceed in the complete absence of oxygen. Its essential feature, therefore, is not the direct addition of oxygen to the substrate but the removal of hydrogen or electrons from the substrate. The enzymes which catalyze the latter reactions were designated "dehydrogenases." When molecular oxygen serves as the immediate electron acceptor, the enzymes were called "oxidases." During the last several decades, many dehydrogenases have been isolated and purified, and some have been crystallized from animal and plant tissues as well as from microorganisms. In 1932 Wieland made the following statement in his famous book, *On the Mechanism of Oxidation.*

> Limiting ourselves to the chief energy-supplying foods, we have in this class carbohydrates, amino-acids, the higher fatty acids, and glycerol. There is no known example among them of an unsaturated compound in the case of which it is necessary to assume direct addition of oxygen, that is, additive oxidation.

According to his view, biological oxidation proceeds exclusively by the removal of electrons or hydrogen atoms from substrates and the direct addition of molecular oxygen need not be assumed. Oxygen might still be incorporated into substrates by hydration reactions involving water, but prior and subsequent oxidative processes would involve the removal of hydrogen and electrons. This theory of biological oxidation was generally accepted until 1955, when the enzymic incorporation of molecular oxygen into substrates was established through the use of a heavy isotope of oxygen, ^{18}O.

In 1955 Mason and his co-workers observed that an oxygen atom derived from molecular oxygen was incorporated into 4,5-dimethyl-catechol when 3,4-dimethylphenol was oxidized by a phenolase complex. Oxygen of the water molecule was not used [Reaction (6)].

$$3,4\text{-Dimethylphenol} + \tfrac{1}{2}O_2 \longrightarrow 4,5\text{-Dimethylcatechol} \tag{6}$$

Concurrently, Hayaishi and associates, using $^{18}O_2$ and $H_2{}^{18}O$, established that the two atoms of oxygen inserted into catechol by the action of pyrocatechase were derived exclusively from atmospheric oxygen [Reaction (7)]. These findings obviously upset Wieland's concept that molecular

$$\text{Catechol} + O_2 \longrightarrow cis,cis\text{-Muconic acid} \tag{7}$$

oxygen could act only as an ultimate electron acceptor and that all the oxygen atoms incorporated into substrates were derived from the oxygen atoms of water.

Both these reactions involve "oxygen fixation" into a substrate molecule, and are similar to oxygenation reactions known to occur by chemical or photochemical processes. Since the enzymes involved in these reactions are different in properties and cofactors from previously known dehydrogenases and oxidases, it was inferred that molecular oxygen must be activated in some way prior to its incorporation into organic substrates. In view of these considerations, Bach and Chodat's term "oxygenase" was selected to designate a group of enzymes which catalyze the incorporation of molecular oxygen into various substrates (1).

III. Method

In order to establish that a given enzyme is an oxygenase, the enzyme-catalyzed incorporation of molecular oxygen into the reaction product must be demonstrated. Usually, two parallel experiments are run, which differ only in that oxygen-18 is present as a component of water in one case and as atmospheric oxygen in the other. The reaction products are isolated, purified, and crystallized, if possible, and then are pyrolyzed in order to convert oxygen atoms stoichiometrically to CO_2. The CO_2 thus obtained is analyzed by a mass spectrometer, and the percentage of ^{18}O incorporated into the product is calculated. This calculation involves a comparison between the observed enrichment of ^{18}O in the product and the theoretical enrichment, which would be attained if all the oxygen atoms inserted into the molecule were derived from the isotopic source of the medium. Corrections should be made for any nonisotopic oxygen atoms initially present in the substrate molecule and also for non-enzymic exchange of oxygen between the substrate or reaction product and water during incubation and isolation. Oxygen atoms in carbonyl or aldehyde groups are especially labile and readily exchange with oxygen atoms of water even at neutral pH. It is, therefore, necessary to carry out isolation procedures rapidly and under the mildest possible conditions in order to minimize this complication. Oxygen analysis is usually carried out by the procedure of Rittenberg and Ponticorvo. The technique and various modifications of it are described in detail in an excellent review by Samuel (2).

IV. The Classification of Oxygenases

A. Dioxygenases and Monooxygenases

As mentioned above, enzymes which catalyze the incorporation of molecular oxygen into various substrates are called "oxygenases." Two subclasses may be defined. Enzymes which catalyze the addition of *both* atoms of oxygen to a molecule of substrate, S, as shown in Reaction (8), are designated "dioxygenases" (synonyms: true oxygenases, oxygen transferases) (3). Pyrocatechase, originally studied by Hayaishi, is of this type.

On the other hand, enzymes which incorporate only *one* atom of oxygen into a substrate are called "monooxygenases" [Reaction (9)]. Phenolase, first examined by Mason, belongs to this group.

$$S + O_2 \longrightarrow SO_2 \qquad (8)$$
$$S + \tfrac{1}{2}O_2 \longrightarrow SO \qquad (9)$$

It was soon realized that monooxygenases function only in the presence of an appropriate electron donor, which serves to reduce the other oxygen atom to water. The stoichiometry of such a reaction is shown by Reaction (10), in which AH_2 is an electron donor and S is a substrate.

$$S + O_2 + AH_2 \longrightarrow SO + H_2O + A \qquad (10)$$

Monooxygenases were previously called mixed function oxidases by Mason, because they functioned both as oxidases and as oxygenases (4). The term "hydroxylase" has also been used to designate this group of enzymes, particularly those involved in steroid hydroxylation. Since a diverse group of reactions such as dealkylation, decarboxylation, deamination, and *N*- or *S*-oxide formation are also catalyzed by enzymes of this class, "monooxygenase" seems to be the most appropriate generic term.

B. Further Classification

Since information concerning the mechanism of action of oxygenases is sparse, any classification scheme is necessarily arbitrary. For convenience,

a scheme based on the properties of the enzymes and their required cofactor
is proposed (Table I).

TABLE I

Classification of Oxygenases

Classification	Cofactors	Examples
Dioxygenases	Inorganic iron	Pyrocatechase, metapyrocatechase, 3-hydroxyanthranilate oxygenase, homogentisate oxygenase
	Heme	L-Tryptophan pyrrolase, D-tryptophan pyrrolase
Monooxygenases	Pyridine nucleotides	Kynurenine hydroxylase, steroid hydroxylases, squalene oxidocyclase
	Flavin nucleotides	Salicylate hydroxylase, diketocamphane lactonizing enzyme
	Heme (P_{450})	Aromatic hydroxylase, steroid 11-β-hydroxylase, steroid 21-hydroxylase
	Ascorbic acid and copper	Phenolase, dopamine β-hydroxylase
	Pteridine derivatives	Phenylalanine hydroxylase, tyrosine hydroxylase, glycerylether hydroxylase, tryptophan 5-monooxygenase
	Substrate itself	Lactic oxidative decarboxylase, inositol oxygenase, lysine oxygenase, arginine oxygenase

Dioxygenases may be divided into two subgroups depending on the
nature of iron in the active center. A large number of enzymes, which
contain inorganic iron as the only cofactor, catalyze the insertion of two
atoms of oxygen into various aromatic compounds. Since the benzene ring
of various mono- or diphenolic compounds is oxidatively cleaved by these
enzymes, they are often referred to as phenolytic dioxygenases. Pyrocatechase (catechol 1,2-oxygenase) and metapyrocatechase (catechol
2,3-oxygenase) are typical phenolytic dioxygenases.

$$\text{Catechol} \xrightarrow{\text{Pyrocatechase}} \textit{cis,cis}\text{-Muconic acid} \tag{11}$$

$$ \text{(12)} $$

Catechol   cis,cis-α-Hydroxymuconic
                                             semialdehyde

The other group of dioxygenases contains heme instead of inorganic iron. L-Tryptophan 2,3-oxygenase (pyrrolase) has been isolated from liver and *Pseudomonas*, and D-tryptophan 2,3-oxygenase has been found in the intestinal mucosa. L- and D-Tryptophan are converted to L- and D-formyl-kynurenine, respectively.

$$ \text{(13)} $$

Tryptophan   Formylkynurenine

Monooxygenases may be further classified on the basis of the electron donor involved. Since the primary reductant is not known in most cases, reclassification will soon be required. In most monooxygenase reactions, TPNH functions as a specific electron donor, although in certain instances DPNH serves as the specific electron donor. Recent evidence indicates that the oxidation of reduced pyridine nucleotides is often mediated by another electron carrier, such as FAD, FMN, or one of several pteridine derivatives. For example, FAD or FMN is reduced by reduced pyridine nucleotides, and the resultant $FADH_2$ or $FMNH_2$ serves as a direct reductant for salicylate hydroxylase and diketocamphane lactonizing enzyme, respectively. Similarly, dihydropteridines are reduced by reduced pyridine nucleotides, and the tetrahydropteridines thus produced serve as reducing agents for phenylalanine hydroxylase, tyrosine hydroxylase, and glycerol ether monooxygenase. Although ascorbic acid serves as a reductant for phenolase and dopamine β-hydroxylase, cuprous ion may be the ultimate reductant of oxygen in these reactions. P_{450}, a heme-like pigment, is assumed to function in microsomal hydroxylation reactions including steroid 21-hydroxylase and aromatic hydroxylase, and in steroid 11-β-hydroxylase of adrenal mitochondria. A non-heme iron protein has recently been shown to be involved in the transfer of electrons from TPNH via a flavoprotein to P_{450}.

Although the above-mentioned reactions require the presence of added electron donors, the substrate itself may serve as an electron donor for certain monooxygenases. For example, in the oxidative decarboxylation of lactic acid, an atom of oxygen is introduced into the product, acetic acid. Simultaneously, a pair of electrons is withdrawn from the same substrate molecule and transferred via enzyme-bound FMN to the second oxygen atom, which is reduced to water. In this case, FMN only acts catalytically and is not needed in stoichiometric amounts.

$$CH_3CHOHCOOH + {}^{18}O_2 \longrightarrow CH_3CO^{18}OH + CO_2 + H_2^{18}O \qquad (14)$$

Similar reactions have recently been described with lysine or arginine as substrate. Lysine oxygenase, which was purified from *Pseudomonas* and crystallized, was shown to catalyze the conversion of L-lysine to δ-aminonorvaleramide. Arginine oxygenase, described by Thoai and co-workers, converts L-arginine to γ-guanidinobutyramide.

V. Distribution and Relative Activities of Oxygenases and Oxidases

The oxygenases are widely distributed in nature; they have been isolated from animals, plants, and microorganisms. Aerobic microorganisms such as *Pseudomonas*, *Mycobacteria*, and *Nocardia* are rich sources of oxygenase while anaerobic microorganisms are devoid of this activity.

In cells, oxygenases are not associated with any one specific constituent of the cell, but are distributed in various cellular fractions. For example, homogentisate dioxygenase and 3-hydroxyanthranilate dioxygenase are both found in the soluble, supernatant fraction of liver cells, whereas kynurenine 3-hydroxylase and the cholesterol side-chain cleavage enzyme are located in the mitochondrial fraction. A nonspecific aromatic hydroxylase and several hydroxylases associated with steroid synthesis are found in the microsomes.

Since the rate at which oxygen is reduced through the conventional electron-transport system in the mitochondria is much faster than the summated rates of oxygenase reactions under physiological conditions, oxygenase activities may be masked unless the cytochrome system is removed or inhibited by some means. For example, the apparent activity of kynurenine hydroxylase is enhanced three- to tenfold by the addition of $10^{-3}M$ cyanide or azide. This effect may be partly caused by the unique requirement of this enzyme for certain monovalent anions for maximum activity, but may also result from inhibition of the cytochrome system, which makes more reduced nicotinamide nucleotide available for hydroxylase action.

The cytochrome system, terminating in cytochrome oxidase, exhibits an affinity for oxygen 10 to 100 times higher than that of most oxygenases and other oxidases, whereas the affinity of all these enzymes towards reduced nicotinamide nucleotides is about the same. At low steady-state concentrations of oxygen in the cell, therefore, oxygen would be preferentially used by the cytochrome system (5).

VI. Physiological Significance of Oxygenases

The biological role of oxygenases appears to be different from that of oxidases and dehydrogenases, which have been described in the preceding chapters. Dehydrogenases and oxidases, which catalyze the transfer of hydrogen atoms and electrons, are chiefly involved in energy metabolism, namely, the generation of adenosine triphosphate. On the other hand, oxygenases are mainly involved in the synthesis or metabolism of essential metabolites and in the oxidation of foreign compounds. They play an important role in the metabolism not only of various aromatic compounds and other cyclic compounds but also of simple aliphatic compounds. Indeed, oxygenases act on a wide variety of compounds, which include amino acids, sugars, fats, hydrocarbons, hemoglobin, steroids, vitamins, hormones, and toxic and carcinogenic substances.

The role of oxygenases in tryptophan metabolism may serve as a useful example. Tryptophan is hydroxylated by a monooxygenase to give 5-hydroxytryptophan, which is then decarboxylated to form the hormone, serotonin. The major route of tryptophan degradation is initiated by tryptophan 2,3-dioxygenase (tryptophan pyrrolase), which cleaves the indole ring with the insertion of two atoms of oxygen and the ultimate formation of kynurenine by way of formylkynurenine. Kynurenine may be further transformed in three alternative ways. First, it may be hydroxylated at the 3 position, a reaction catalyzed by a specific hydroxylase of mitochondria. 3-Hydroxykynurenine is then transformed into 3-hydroxy-anthranilic acid, which is further cleaved by a specific dioxygenase. The resulting compound can either be converted to acetyl-CoA through various intermediates including glutaryl-CoA, or can cyclize to form picolinic and quinolinic acids. Quinolinic acid may react with phosphoribosylpyrophosphate to form nicotinic acid ribotide, the precursor of DPN. Second, kynurenine may be transformed to kynurenic acid or, third, to anthranilic acid. The degradation of both of these compounds also involve a number of oxygenase reactions. It is noteworthy that most, if not all, of the oxidative reactions of tryptophan metabolism involve oxygen fixation.

Similarly, oxygenases catalyze many reactions in the metabolism of phenylalanine and tyrosine. The enzymic formation of tyrosine from phenylalanine and the formation of adrenaline, nonadrenaline, melanin, and thyroxine involve consecutive hydroxylation and oxygenation reactions. A number of genetic defects of these physiologically important monooxygenases and dioxygenases in the pathway of aromatic amino acid metabolism have been described. For example, phenylketonuria is known to be caused by a genetic defect in phenylalanine hydroxylase, whereas alkaptonuria is caused by the genetic defect of homogentisate dioxygenase.

The biosynthesis of cholesterol, the formation of steroid hormones and bile acids, and the interconversion of various steroid hormones all require numerous oxygenation reactions. More specifically, the cyclization of squalene into lanosterol is the first reaction in sterol synthesis which requires TPNH and molecular oxygen. Thereafter, four oxygenase reactions presumably occur in the conversion of lanosterol to cholesterol. Furthermore, the cleavage of the side chain of steroids, aromatization, and epoxidation reactions are all known to be catalyzed by similar enzymes.

Certain aliphatic compounds can also serve as substrates for oxygenases. In the formation of unsaturated fatty acids from saturated fatty acids, hydroxylation initially occurs and is rapidly followed by dehydration to yield the isolated double bond. The hydroxylated intermediate seems to be tightly bound to the enzyme. Recently several oxygenases were shown to be involved in the biosynthesis of prostaglandins from unsaturated fatty acids. Omega oxidation of certain fatty acids is also catalyzed by monooxygenases which require TPNH and oxygen. The oxidative decarboxylation of lactate, lysine, and arginine has also been shown to be accompanied by the incorporation of molecular oxygen.

The direct incorporation of atmospheric oxygen takes place not only in the oxidation of organic compounds, but also in the biological oxidation of an inorganic compound, sulfur. Suzuki recently demonstrated that molecular ^{18}O is incorporated into thiosulfate by a cell-free extract of *Thiobacillus thiooxidans*, as shown in Equation (15), and that a catalytic amount of reduced glutathione is required for this process.

$$2S + O_2 + H_2O \xrightarrow{\text{GSH}} H_2S_2O_3 \qquad (15)$$

Other physiologically important reactions are also catalyzed by oxygenases. For example, the metabolism of various drugs, toxic compounds, and carcinogens, the degradation of heme, and the biosynthesis of alkaloids in plants are also catalyzed by a variety of oxygenases. These reactions have been discussed in detail in other review articles (1).

VII. General Properties and Mechanism of Action

A. General

Although molecular oxygen has two unpaired electrons (triplet state) and is paramagnetic, it does not behave as a typical biradical and is a rather inert oxidizing agent. It has generally been assumed that molecular oxygen must, therefore, be activated in order to react with various substrates. About 40 years ago, Warburg proposed a theory of oxygen activation catalyzed by iron-containing enzymes (Atmungsferment or Oxygen Übertragendes Ferment). His theory in many ways is reminiscent of that of Bach and his associates, inasmuch as Warburg assumed that the primary reaction in cell respiration is the reaction between molecular oxygen and enzyme bound iron, as follows:

$$\text{Enz—Fe} + O_2 \longrightarrow \text{Enz—Fe—}O_2 \tag{16}$$

The oxidation of various substrates (S) might then occur according to Equation (17).

$$\text{Enz—Fe—}O_2 + 2S \longrightarrow \text{Enz—Fe} + 2SO \tag{17}$$

Recent experimental evidence with highly purified dioxygenases is consistent with this view that metallic ions are involved in the activation of molecular oxygen. On the other hand, the role of metals in monooxygenase reactions has not been clearly established. Iron and copper may play an essential role in at least some monooxygenase reactions, whereas these metals may not be involved in others.

B. Dioxygenases Containing Inorganic Iron

Several dioxygenases have recently been obtained in either crystalline or highly purified forms and have been shown to be homogeneous upon ultracentrifugation and electrophoresis. These include metapyrocatechase, pyrocatechase, and 3,4-dihydroxyphenylacetate-4,5-dioxygenase. In these highly purified enzymes, inorganic iron seems to be the only cofactor involved in such reactions and presumably binds not only oxygen but also the substrate.

By analogy with the oxygenation of hemoglobin, it was initially thought that oxygen is first bound to ferrous ion (Fe^{2+}) and is subsequently transferred to a substrate. Although oxyhemoglobin had generally been assumed to be a loose and reversible binding between oxygen and the chelated ferrous ion, recent molecular orbital calculations suggest that an electron is partially transferred from iron to oxygen in the complex, thereby activating the oxygen molecule. $Fe^{2+} \cdot O_2$ would then be in equilibrium with $Fe^{3+} \cdot O_2^-$ (3). Although the detailed chemistry and the exact nature of the enzymically activated oxygen remain uncertain, circumstantial evidence points to the fact that either $O_2^- \cdot$ or $\cdot O_2H$ is the active species in oxygenation reactions as well.

Orthophenanthroline or α,α'-dipyridyl, a well-known chelating agent for ferrous ion, inhibits metapyrocatechase competitively with respect to the substrate catechol. During catalysis, changes in the valence state of iron in crystalline metapyrocatechase have been observed by the use of the electron paramagnetic resonance technique. Such experiments as well as the observed protection of the enzyme by catechol indicate that the site of attachment of the substrate catechol is closely associated with ferrous iron.

A mechanism for metapyrocatechase action which is consistent with these results is shown in Figure 1. Molecular oxygen first combines with divalent iron to form $Fe^{2+} \cdot O_2$ which is in equilibrium with $Fe^{3+} \cdot O_2^-$. Since catechol is known to chelate with trivalent iron, the substrate will preferentially combine with $Fe^{3+} \cdot O_2^-$ to form a ternary complex on the enzyme surface. Within this complex a proton may be dislodged, and an electron will be transferred from catechol to ferric iron, reducing the latter

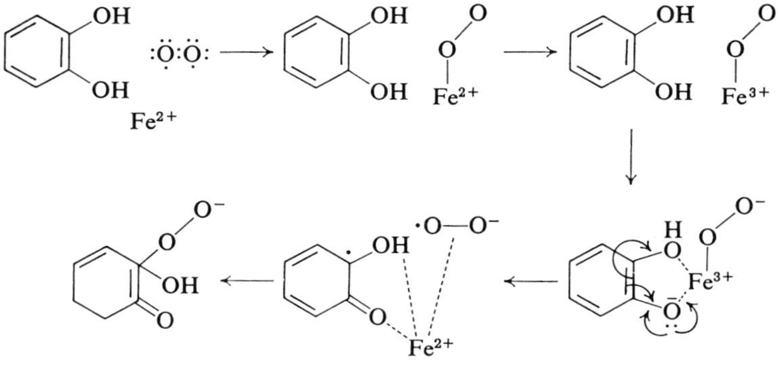

Figure 1

to a divalent state. The anionic oxygen radical and the substrate radical will then react with each other to form an oxygenated product. These transformations are summarized in Figure 1.

Although the electron transfers have been schematically represented as a sequence of reactions, the entire event is presumed to take place by a concerted mechanism within a ternary complex of ferrous ion, oxygen, and substrate on the enzyme surface. Once the catechol is oxygenated, it would be easy to visualize the electron transfers which might be involved in the cleavage of the ring system [Reaction (18)].

$$\text{(18)}$$

Pyrocatechase, which yields cis,cis-muconic acid rather than α-hydroxy-muconic semialdehyde, is unique and differs from all the other oxygenases containing non-heme iron in two respects. First, pyrocatechase has a distinct red color, whereas all other phenolytic dioxygenases are colorless. Second, the iron in native pyrocatechase appears to be in the ferric state. In electron spin resonance studies a signal at a g value of about 4.2, characteristic of iron in the trivalent state, is observed with the native enzyme. This signal disappears upon the addition of catechol and appears again when the substrate is used up. It was therefore inferred that the iron in pyrocatechase is in the trivalent state and binds with catechol first. A partially reduced iron then binds with oxygen to form a ternary complex, and the subsequent series of reactions will take place in a manner similar to that described for metapyrocatechase.

C. Tryptophan Pyrrolase (Heme-Containing Dioxygenases)

Both L- and D-tryptophan pyrrolases contain heme instead of inorganic iron and catalyze oxygenative cleavage of the indole ring of tryptophan to form formylkynurenine as shown in Reaction (13). Knox and his co-workers have claimed that only the Fe^{2+} form of tryptophan pyrrolase is catalytically active. On the other hand, Feigelson and his collaborators have shown that the highly purified apoenzyme can become fully active upon addition of hematin (trivalent iron) to the apoenzyme without exogenous reductant. This discrepancy might be reconciled by assuming that the preparation used by Feigelson is similar to pyrocatechase in that the substrate tryptophan can easily reduce hematin, while the more

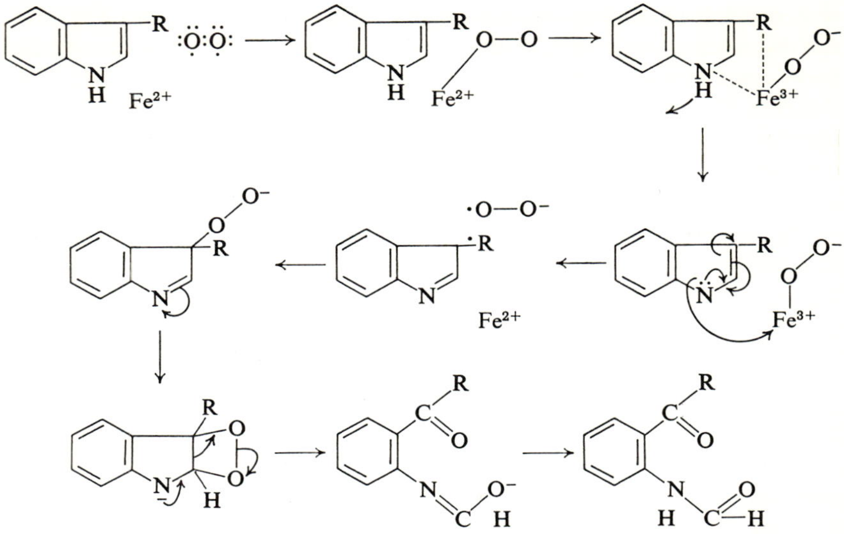

Figure 2

purified preparation which was used by Knox has lost this property of being easily reduced by the substrate. Further experiments, however, are necessary to elucidate this interesting problem.

Spectroscopic analysis and inhibition experiments with KCN and CO indicate that the hematin coenzyme is first reduced to a divalent form, which combines with oxygen and then with tryptophan to form a ternary complex. Within this ternary complex, one electron from the lone pair on the nitrogen of the indole ring is used to reduce the ferric ion, and subsequent electron shifts enable the anionic oxygen radical to attack the β-carbon of the indole ring to form an oxygenated intermediate. The subsequent rearrangement will take place in a manner similar to the metapyrocatechase reaction. The mechanism is briefly outlined in Figure 2.

D. Monooxygenases

1. General Mechanism of Hydroxylation

The hydroxylation of various aromatic compounds, steroids, and foreign organic compounds occurs commonly in biological systems. Many of these hydroxylation reactions are catalyzed by monooxygenases, but others rather involve the hydration of a double bond of the heterocyclic or unsaturated compounds followed by a dehydrogenation. The source of the hydroxylic oxygen in these cases is not molecular oxygen but water, as shown in Equation (19). The formation of 6-hydroxynicotinic acid from nicotinic acid, of barbituric acid from uracil, and of β-keto acids from α,β unsaturated fatty acids are reactions of this type.

$$S + H_2O \longrightarrow SH_2O$$
$$SH_2O + A \longrightarrow SO + AH_2$$
$$\overline{S + H_2O + A \longrightarrow SO + AH_2} \tag{19}$$

In this type of hydroxylation, many compounds other than oxygen can function as electron acceptors, and the entire reaction can proceed under anaerobic conditions. Usually, oxidized DPN or TPN serves as the physiological electron acceptor in these dehydrogenaticn reactions, although some dyes may be substituted for them in *in vitro* systems. The great majority of hydroxylation reactions, however, are catalyzed by monooxygenases, which require the presence of oxygen and an electron donor. In certain instances, oxygenase and hydrase types of hydroxylation may occur successively. For example, when kynurenic acid (**1**) is converted to 7,8-dihydroxykynurenic acid (**4**), the primary reaction is catalyzed by an oxygenase, which requires DPNH and oxygen. The primary product was tentatively identified as 7,8-epoxide (**2**), which is hydrated by a hydrase to give a *vic*-glycol (**3**). The latter compound is then dehydrogenated by a DPN-linked dehydrogenase to form 7,8-dihydroxykynurenic acid (**4**). This reaction sequence is shown in Figure 3. Although all available data

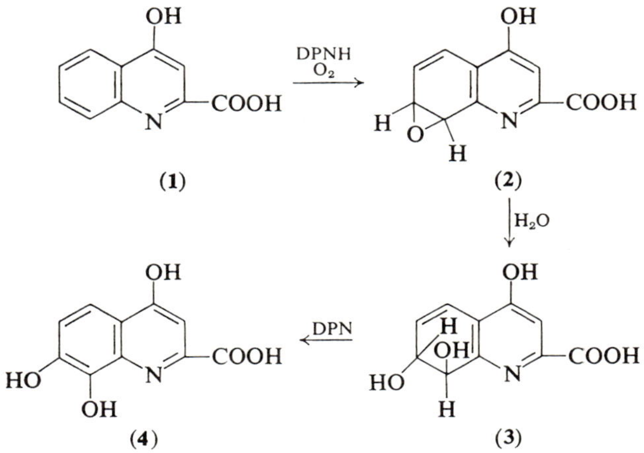

Figure 3

are in accord with this postulated scheme, experimental proof for the sequence could not be obtained by the use of ^{18}O and $H_2{}^{18}O$, presumably due to a rapid exchange reaction.

2. The Mechanism of Monooxygenase Reactions

Unfortunately, most monooxygenases are localized in the particulate fraction of cells or are unstable during purification. Crystalline enzymes have not been available until recently for use in studies on the mechanism of these reactions, except the lactic oxidative decarboxylase which was purified and crystallized from cells of *Mycobacterium phleii* by Sutton. Several monooxygenases, however, have recently been highly purified and were obtained in crystalline forms. *p*-Hydroxybenzoate hydroxylase crystallized from *Pseudomonas* by Hosokawa and Stanier, catalyzes the conversion of *p*-hydroxybenzoate to protocatechuate in the presence of oxygen and TPNH. Imidazoleacetate monooxygenase and L-lysine oxygenase were both obtained in crystalline form from *Pseudomonas* in our laboratory. The overall stoichiometry of monooxygenase reactions involves the incorporation of one atom of molecular oxygen into the substrate and the reduction of the other atom to water in the presence of a specific hydrogen donor. Until recently, TPNH was believed to be the most common reducing agent in such reactions, but recent evidence indicates that pyridine nucleotides act only as the ultimate electron donor in many cases, and only transfer electrons to the primary reductant, which may be FMN, FAD, pteridine derivatives, or the heme pigment, P_{450}.

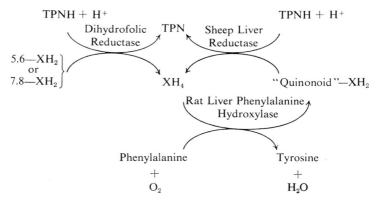

Figure 4

For phenylalanine hydroxylase of mammalian liver, certain reduced pteridine derivatives seemingly serve as the direct hydrogen donor in the formation of tyrosine from phenylalanine. The mechanism suggested by Kaufman and his co-workers for this reaction is presented in Figure 4, in which XH_2 stands for dihydropteridine and XH_4 for tetrahydropteridine. Tetrahydropteridine (cf. chapter by Huennekens) which is produced from 5,6- or 7,8-dihydropteridine by a reductase in the presence of TPNH, is utilized to reduce oxygen to H_2O during the formation of tyrosine. The dihydropteridine thus formed is not the original 5,6- or 7,8-XH_2 but rather a quinonoid-XH_2, the structure of which is shown in Figure 5.

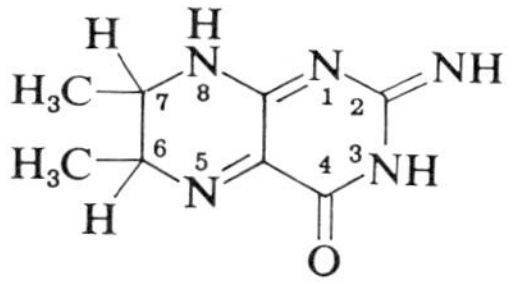

Figure 5

The lactonizing enzyme of 2,5-diketocamphane contains iron, which must be in the ferrous state for activity. DPNH acts by reducing FMN which in turn reduces the iron of the enzyme. The sequence of reactions is shown in Figure 6. E_1 is an FMN-coupled DPNH oxidase and E_2, containing non-heme iron, is the monooxygenase.

Since camphor can be converted to its lactone by the Baeyer-Villager reaction in which the oxygenating agent is peracetic acid, Gunsalus and his co-workers suggested that a reactive oxygen species for enzymic

$$DPNH + H^+ + FMN \xrightarrow{E_1} DPN^+ + FMNH_2$$

$$FMNH_2 + E_2 \cdot 2Fe^{3+} \longrightarrow FMN + 2H^+ + E_2 \cdot 2Fe^{2+}$$

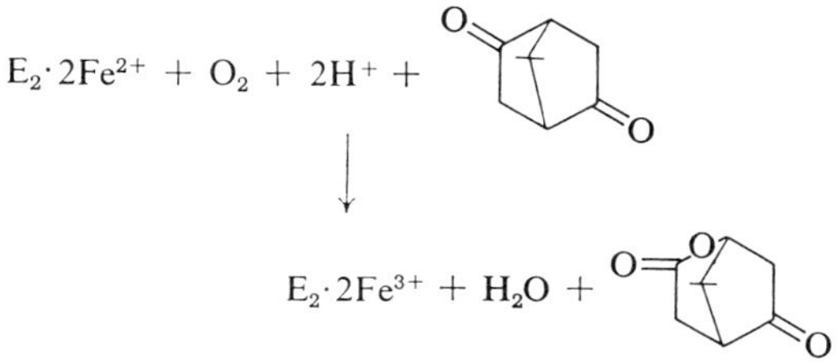

Figure 6

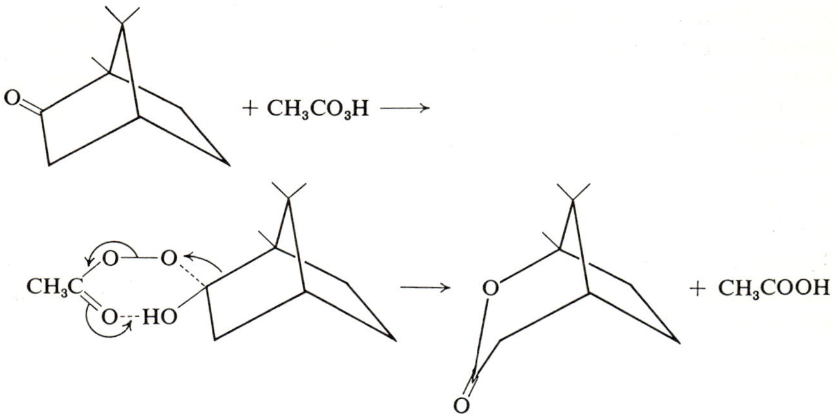

Figure 7

lactonization is similar to the Baeyer-Villager peracid as shown in Figure 7. Participation of free H_2O_2 in the enzymic reaction, however, was excluded.

In the above cases, the enzyme which reduces either pteridine or flavin in the presence of reduced pyridine nucleotide has been separated from the monooxygenase. On the other hand, salicylate hydroxylase and its associated reductase seem to be tightly bound. Salicylic hydroxylase catalyzes the conversion of salicylic acid to catechol, whereas the reductase catalyzes the reduction of FAD by DPNH. Although the monoxygenase has been highly purified and behaves as a single protein, no separation of reductase activity has been possible.

The involvement of heavy metals has been established in many monooxygenases, but not in all. The phenolase complex and dopamine β-hydroxylase both contain copper. For the latter enzyme a change in the valency state of copper during catalysis has been proposed by Kaufman. Interestingly, both these enzymes require ascorbic acid as a reducing agent.

There have been speculations concerning the nature of the activated oxygen in the monooxygenase reaction. In his recent review (6), Diner discussed various electronic aspects of the biochemical hydroxylation reaction. He infers that activated molecular oxygen is involved in metabolic hydroxylation of foreign organic compounds rather than free radicals of the hydroxyl or perhydroxyl type. If metals are involved in monooxygenase reactions, oxygen might well be activated by a mechanism similar to that involved in dioxygenase reactions. The resulting perferryl ion $Fe^{3+}O_2^-$ might then be the active species of oxygen. Although the participation of metals as electron-transferring and chelating agents is most attractive, an alternative mechanism involving a complex of oxygen, substrate, and reduced coenzyme might be suggested. In fact, in the case of highly purified salicylate hydroxylase, crystalline imidazoleacetate mono-oxygenase, and crystalline lysine oxygenase, neither iron, copper, nor any other metal has been detected in significant quantity. Inhibition experiments also failed to provide positive evidence for the participation of metals. It appears that the enzyme-bound FAD is first reduced either by DPNH, an exogenous reducing agent, or by the substrate; then in a ternary complex of oxygen, $FADH_2$, and the substrate, the oxygen molecule is reduced and the bond between the two oxygen atoms is cleaved, thereby forming a monooxygenated product and water.

3. The Mechanism of Double Hydroxylation

Since a great many catechol derivatives occur in nature, a comparison of various routes for their synthesis might be fruitful. Dihydroxy compounds may be formed, first, by two successive oxygenase reactions. Dihydroxyphenylalanine is synthesized in this way. Phenylalanine is oxidized to tyrosine by phenylalanine monooxygenase, and the resulting tyrosine is further hydroxylated at the *ortho* position by a monooxygenase, to form the product. The latter enzyme is located in the central nervous system, and requires a reduced pteridine derivative as an electron donor.

The formation of catecholic compounds by the combined action of a monooxygenase, a hydrase, and a dehydrogenase was previously mentioned. Kynurenic acid is converted to 7,8-dihydroxykynurenic acid by this mechanism by way of kynurenic acid-7,8-epoxide. Similarly, Boyland and his co-workers have reported that an epoxide of naphthalene is formed in the presence of TPNH, oxygen, and rat liver microsomes, and that the epoxide is further converted to a dihydrodiol of naphthalene. In addition, a *vic*-glycol, *trans*-5,6-dihydroxycyclohexadiene-1,3, is dehydrogenated to form catechol by a TPN-linked dehydrogenase from rabbit liver. Various other dihydrodiols and epoxides also occur in nature, although the mechanism of their formation is not understood.

Anthranilate hydroxylase, which catalyzes the formation of catechol from anthranilic acid, has been partially purified from cells of *Pseudomonas*. In the conversion of one mole of anthranilic acid to catechol, one mole each of oxygen and of DPNH is used, and one mole each of CO_2 and ammonia is produced. Both atoms of oxygen in catechol are exclusively derived from molecular oxygen. Since one mole each of oxygen and DPNH is utilized for the formation of one mole of catechol, the reaction apparently involves the direct incorporation of two atoms of molecular oxygen into the substrate rather than two successive single hydroxylation reactions. Two oxygen atoms, presumably in the same molecule, add to the double bond between carbons 1 and 2 with the concomitant release of ammonia and CO_2 (dioxygenase-type reaction), and simultaneously the cyclic peroxide intermediate is reductively cleaved as shown in Reaction (20) (hydroxylase reaction).

$$
\text{(structure: benzene ring with —COOH and —NH}_2\text{)} \longrightarrow \left[\text{(bicyclic peroxide intermediate with COOH, O, O, NH}_2\text{)}\right] \xrightarrow{\text{DPNH}_2} \text{(benzene ring with —OH and —OH)} + CO_2 + NH_3 \quad (20)
$$

VIII. Concluding Remarks

In conclusion, a few speculations might be made concerning the appearance and function of oxygenases in the evolution of living things. There is general agreement that life on earth arose in an anaerobic environment. It follows, therefore, that the minimum manifestations of life must be capable of occurring in the absence of molecular oxygen. Certainly the existence of many bacteria, which survive and grow solely under anaerobic conditions, is a present-day reminder that molecular oxygen is not essential for life. With the appearance of oxygen in the earth's atmosphere, however, advanced forms of life began to emerge. Perhaps the initial advantage of primitive oxygen-utilizing organisms was their ability to produce and utilize energy more efficiently in the presence of oxygen than in its absence. Thus, the role of oxygen as the terminal electron acceptor in respiration, which was early linked to the formation of energy donors such as ATP, became a prominent part of the metabolic apparatus of most living things.

With the subsequent development of extensive biosynthetic pathways for essential compounds, of intracellular structures with defined function, and of regulatory mechanisms for metabolism and growth, oxygenases probably appeared for the first time. Certainly, the oxidative formation of vitamins essential for coenzyme synthesis, of unsaturated fatty acids and sterols present in membrane structures, and of hormones and mediators of nerve impulses presently require, and may always have required, molecular oxygen. Since oxygen and heavy metals were a ubiquitous part of the environment of these primitive beings, why indeed should they not be utilized for these important reactions in preference to other more complex or less common oxidizing agents? In an evolutionary sense, therefore, oxidases preceded oxygenases, and the latter may be regarded as the most advanced class of oxidative enzymes.

Certainly the appearance of oxygen on earth not only affected the energy-producing apparatus of living things but also the nature of compounds which play indispensable roles in most forms of life. In contrast to the eons of time required for the development of oxygenase enzymes, we have known about them, oddly enough, for only a decade.

References

1. O. Hayaishi, in *Oxygenases*, O. Hayaishi, Ed., Academic Press, New York, 1962, p. 1.
2. D. Samuel, in *Oxygenases*, O. Hayaishi, Ed., Academic Press, New York, 1962, p. 32.
3. O. Hayaishi, *Proc. Plenary Sessions, Sixth Intern. Congr. Biochem. New York, 1964*, p. 31.
4. H. S. Mason, *Science*, **125**, 1185 (1957).
5. O. Hayaishi, *Ann. Rev. Biochem.*, **31**, 25 (1962).
6. S. Diner, in *Electronic Aspects of Biochemistry*, B. Pullman, Ed., Academic Press, New York, 1964, p. 237.

The Pyridine Nucleotide Coenzymes*

HORST SUND, *Chemisches Laboratorium der Universität, Freiburg im Breisgau, Germany*

I.   Introduction 603
II. The Structure of the Pyridine Nucleotide Coenzymes   .   . 604
     A. Nomenclature   604
     B. DPN and TPN   604
     C. Analogs 605
III. Chemical and Physical Properties of the Pyridine Nucleotide
     Coenzymes and the Model Compounds   609
     A. Reduction and Oxidation of Pyridine Nucleotide Coenzymes
         and Model Compounds 609
         1. Reduction of Pyridinium Compounds to Dihydropyridines 610
         2. Addition Reactions of Pyridinium Compounds   .   . 612
         3. Reduction to Tetrahydropyridines   616
         4. One-Electron Reduction   618
         5. Oxidation of Dihydropyridines   619
         6. Hydride Transfer or Free-Radical Mechanism in the
            Enzyme-Catalyzed Reaction?   621
         7. Pyridine Nucleotide Coenzymes and Oxidative Phosphoryl-
            ation   624
     B. Oxidation-Reduction Potentials of the Pyridine Nucleotide
         Coenzymes   624
     C. Absorption and Fluorescence Spectra of the Pyridine Nucleo-
         tide Coenzymes   626
         1. Ultraviolet Absorption Spectra   626
         2. Fluorescence Spectra   626
     D. Stereochemistry of the Pyridine Nucleotide Coenzymes   . 631
     Abbreviations   635
     References   636

I. Introduction

The naturally occurring pyridine nucleotides DPN and TPN are the coenzymes in more than a hundred enzyme-catalyzed reactions which differ chemically. A large number of enzymes catalyzing such reactions

* Dedicated to Professor Arthur Lüttringhaus on the occasion of his 60th birthday.

are known. Each of these enzymes uses DPN or TPN or both as the coenzyme.

The chemical principle connected with the biological oxidation by pyridine nucleotide coenzymes was discovered in 1935 by Warburg, Christian, and Griese (1). It consists of a reversible hydrogenation of the pyridinium salt to the dihydropyridine compound. This reversible hydrogenation is accompanied by a characteristic spectral change, the appearance and disappearance of the "*dihydro band*" at 340 mμ. This characteristic absorption band is the basis for the optical test due to Warburg (see refs. 2 and 3).

II. The Structure of the Pyridine Nucleotide Coenzymes

A. Nomenclature (*cf. ref. 4*)

No uniformity in the naming of the pyridine nucleotide coenzymes exists in the literature. Five different systems are in use:

1.	DPN	TPN
2.	NAD	NADP
3.	Cozymase	Phospho-cozymase
4.	Codehydrogenase I	Codehydrogenase II
	(Codehydrase I)	(Codehydrase II)
5.	Coenzyme I	Coenzyme II
	(Co I)	(Co II)

The last three systems are found only in the older literature, whereas the first two are used presently. DPN is the abbreviation for "*d*iphospho-*p*yridine *n*ucleotide," TPN for "*t*riphospho*p*yridine *n*ucleotide." The terms DPN and TPN, although widely used, fail to describe the chemical structure properly. The terms NAD (*n*icotinamide *a*denine *d*inucleotide) and NADP (*n*icotinamide *a*denine *d*inucleotide *p*hosphate) are in accord with the chemical structure but have not been universally accepted.

B. DPN and TPN

The formulas of the oxidized forms of DPN and TPN are given in Figure 1. The only difference between DPN and TPN is an additional phosphate group in the TPN molecule at position 2' of the ribose in the adenylic acid moiety. With respect to the nicotinamide moiety of the coenzyme molecule, the reactions and properties of DPN and TPN are identical. Therefore results obtained with DPN are also valid for TPN.

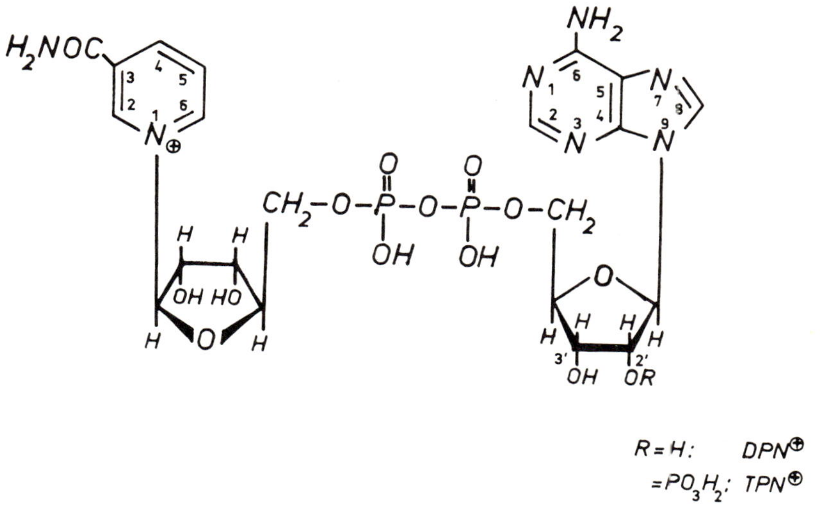

$$R = H: \quad DPN^{\oplus}$$
$$= PO_3H_2: \quad TPN^{\oplus}$$

Fig. 1. Structure of DPN$^+$ and TPN$^+$.

During the reactions catalyzed by pyridine nucleotide-dependent dehydrogenases, only the nicotinamide moiety of the coenzyme molecule is directly involved in the oxidation-reduction reaction. According to Reaction (1), the pyridinium salt is reversibly reduced to the 1,4-dihydro compound.

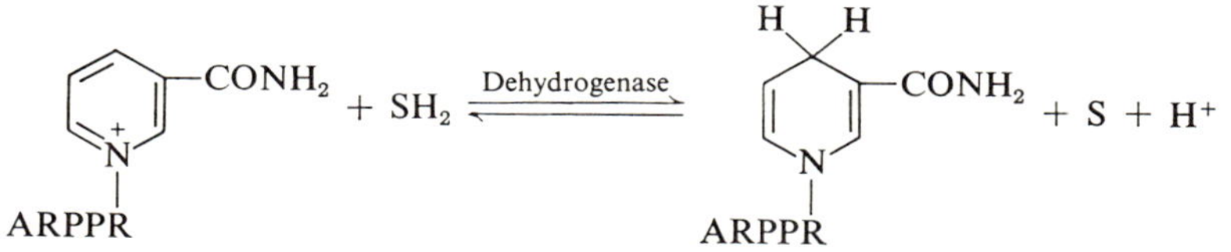

$$(\lambda_{max} = 340 \text{ m}\mu)$$

SH_2 = reduced substrate, e.g., CH_3—CH_2OH
S = oxidized substrate, e.g., CH_3—CHO

C. Analogs (5)

Besides the naturally occurring coenzymes, TPN and DPN, many "coenzyme analogs" have been synthesized chemically and enzymically (cf. Tables I and II). Numerous analogs have been prepared by substitution of the carbonamide group of the nicotinamide moiety or the amino group

TABLE I

Coenzyme Analog Specificity, Spectral Properties, and Oxidation-Reduction Potentials of DPN Analogs (3,6–10)

$$\left(DPN^+ \text{ analogs} = ARPPR - \overset{+}{N} \diagdown X \right)$$

X	Relative rate of $DPN^{\oplus}$ analog reduction[a]				Oxidation reduction potential (mV)	Characteristic absorption maxima (mμ)[a]	
	Yeast alcohol dehydrogenase	Horse liver alcohol dehydrogenase	Beef heart lactate dehydrogenase	Rabbit muscle glyceraldehyde-3-phosphate dehydrogenase		ARPPR (H H / X)	ARPPR (H CN / X)
$C(=O)NH_2$ (DPN)	100	100	100	100	−320	340	325
$C(=O)CH_3$	10	540	4	50	−258	363	343
$C(=O)H$	2	150	20	1	−262	358	327

Structure							
$O{=}C{-}CH(CH_3)_2$	52	792	37	1	−248	360	340
$O{=}C{-}NH{-}NH_2$	8	68	9	38	−344	335	320
$O{=}C{-}NHOH$	3	44	10	19	−320	340	325
$O{=}C{-}C_6H_5$	0	31	0	0	−247	365	360
$H{-}C{=}NHOH$	6	50	1	1	−347	330	320
$S{=}C{-}NH_2$	16	348	41	16	−285	398	359
NH_2	0	0	0	0			

[a] For experimental details see the references.

of the adenine by other groups. Different bases were introduced in place of adenine. The substitution of ribose by deoxyribose reduces the ability to act as a coenzyme (Table II). Substitution by glucose causes inactivity in dehydrogenase systems (17).

TABLE II

Coenzyme Analog Specificity of Some Dehydrogenases[a] (11–16)
(Coenzyme analogs = X—RPPRN)

X	Relative rate of DPN$^\oplus$ analog reduction			
	Yeast alcohol dehydrogenase	Horse liver alcohol dehydrogenase	Pig heart lactate dehydrogenase	Beef liver glutamate dehydrogenase
Adenine (DPN$^\oplus$)	100	100	100	100
Hypoxanthine	17	67	50	91
6-Mercaptopurine	14	56	38	71
6-(2-Hydroxyethylamino) purine	43	72	55[b]	72
Purine	36	73	106	
1-Desazapurine	7	61	88	
Adenine N^1-oxide	40		67[c]	
1-(2-Hydroxyethyl)adenine	0.8	35	11[b]	61
3-Isoadenine	6	140	60[c]	121
Cytosine	0.1	42	56	53
Uracil	2	75	71[b]	42
Benzimidazole	2	49	88	
Nicotinamide	0.01	0.3	0.1	3
H (NRPPR)	0.02	1.4	0.1	0.3
Deoxy-AMP instead of AMP	9	26	40[b]	24
Cyanoethyl ester—PPRN	0.2	5		

[a] For experimental details see the references.
[b] Beef heart lactate dehydrogenase.
[c] Rabbit muscle lactate dehydrogenase.

The relative rate of the reaction in the presence of the analogs (Tables I and II) depends on the concentrations of both the substrate and the coenzyme, as well as on the pH. In addition, the relative rate can change if one uses the reduced coenzyme instead of its oxidized form. In some cases the rate with the analogs is lower than with the natural coenzyme or no reaction is observed. In other cases the enzymic activity is higher. These results show that all parts of the coenzyme molecule (carbonamide group, adenine moiety, ribose or 2'-phosphoribose) play a role in the binding between enzyme and coenzyme.

Coenzyme analogs can be particularly useful in the study of enzyme–coenzyme binding and the mechanisms of the dehydrogenase reactions. They are also of value in the understanding of the chemistry and function of pyridine nucleotide coenzymes.

The α Isomer of DPN (18,19, see also 20).   The glycosidic linkage between the ribose and the nicotinamide in the enzymically active coenzyme is β-glycosidic ("β-DPN" or "β-TPN"). DPN preparations (and perhaps also TPN preparations) contain an additional compound with properties similar to those of DPN but with no activity in dehydrogenase systems. This compound was identified as the "α isomer" of DPN ("α-DPN"). In contrast to the β-DPN, the α-DPN contains an α-glycosidic linkage in the nicotinamide ribosyl moiety. The adenosine moiety possesses the β configuration in both the α-DPN and the β-DPN.

The 3' Isomer of TPN (18).   Treatment of TPN$^{\oplus}$ with acid causes a migration of the phosphate group of the adenosine ribose from the 2' position to the 3' position, via an intermediate cyclic 2',3'-phosphate. This "3' isomer" of TPN is inactive in those dehydrogenase systems which specifically require TPN. Dehydrogenases which are active with both DPN and TPN can also utilize the 3' isomer.

III. Chemical and Physical Properties of the Pyridine Nucleotide Coenzymes and the Model Compounds

A. Reduction and Oxidation of Pyridine Nucleotide Coenzymes and Model Compounds*

DPN$^{\oplus}$ and TPN$^{\oplus}$ are reversibly reduced to DPNH and TPNH, respectively (Eq. 1) in the reactions catalyzed by dehydrogenases. Four possible reduction mechanisms are to be discussed (21):

1. Transfer of one hydride ion:

$$\text{Py}^{\oplus} \xrightarrow{\text{H}^{\ominus}} \text{PyH} \tag{2a}$$

Hydride ion transfer means that the reducing agent gives a hydrogen nucleus with an electron pair to the oxidant. A hydride ion is never free during the process (22).

* For a review, see reference 21.

2. Simultaneous transfer of two electrons followed by a transfer of one proton:

$$Py^{\oplus} \xrightarrow{2e} Py^{\ominus} \xrightarrow{H^{\oplus}} PyH \tag{2b}$$

3. Successive transfer of two single electrons and one proton:

$$Py^{\oplus} \xrightarrow{e} Py \cdot \xrightarrow{e} Py^{\ominus} \xrightarrow{H^{\oplus}} PyH \tag{2c}$$

4. Separate transfer of one electron and one hydrogen atom:

$$Py^{\oplus} \xrightarrow{e} Py \cdot \xrightarrow{H \cdot} PyH \tag{2d}$$

Differentiation between these four mechanisms is only possible if the rates of the different steps in Reactions (2b)–(2d) can be separated.

To elucidate the problem of the mechanism of oxidation and reduction of pyridine nucleotide coenzymes, many studies on simpler model compounds have been performed. The only model compounds to be considered here are those pyridine compounds which possess substituents in positions 1 and 3. The chemical properties of pyridine compounds with substituents in other positions are so different from those of the nicotinamide moiety in the coenzyme molecule that one should compare only at least 1,3-disubstituted pyridine compounds with the coenzyme (21).

1. Reduction of Pyridinium Compounds to Dihydropyridines

The enzymic reduction of $DPN^{\oplus}$, as well as the chemical reduction of both $DPN^{\oplus}$ and other pyridinium salts by dithionite (hydrosulfite) or sodium borohydride, give the partially reduced dihydropyridine compounds. All three types of dihydronicotinamides [1,2 (I); 1,4 (II); 1,6 (III)] are known. They can be distinguished by their ultraviolet absorption spectra (Fig. 2) or nuclear magnetic resonance spectra (24–26).

It is evident that the single-peak spectra of 1,2- and 1,4-dihydropyridines above 230 mμ are attributable to a straight conjugated system of double bonds, while the double-peak spectrum of the 1,6-isomer is due to a crossed system. Moreover, the chain in the 1,2-isomer is longer by one

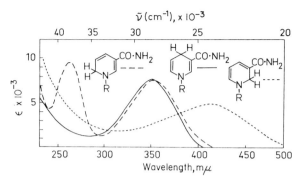

Fig. 2. Absorption spectra of the three isomeric dihydronicotinamides (R = 2,6-dichlorobenzyl) (23). ϵ in cm^2 mmole^{-1}. The spectrum of the 1,2-dihydronicotinamide was calculated from the difference in the spectra of the crude product and the pure 1,6-dihydronicotinamide.

double bond, which corresponds to a shift of 50–70 mμ towards longer wavelengths (27).

The enzymic reduction yields the 1,4-dihydro compound exclusively. The same is valid for the reduction of DPN$^{\oplus}$, TPN$^{\oplus}$, or PN$^{\oplus}$ model compounds (with the exception of N^1-[2,6-dichlorobenzyl]4,6-dimethylnicotinamide bromide) with dithionite, the most convenient chemical preparation method. The mechanism of the dithionite reduction was elucidated with N^1-[2,6-dichlorobenzyl]3,5-dicarbamoylpyridinium bromide. In this case it was possible to isolate the intermediate addition compound (IV) in crystalline form (28). The mechanism can be described as an addition of the sulfoxylate ion (generated from dithionite) in the 2-position by which the sulfinate (IV) is formed followed by a hydrolytic cleavage, analogous to the decarboxylation of β,γ-unsaturated carbonic acids (28) (Reaction (3)). According to this formulation the actual reduction step is the uptake of the electron pair of the C—S bond by the pyridine ring with the simultaneous elimination of SO_2.

Depending on the experimental conditions and the pyridinium compound used, the reduction by sodium borohydride yields, in varying amounts, all three dihydropyridines (29,30).

N-Methyl-3,4,5-tricyanopyridinium perchlorate, a pyridinium salt of extremely high electron affinity, forms very strong complexes with cyanide ions (logK_{Py-CN} = 14.9) and hydroxyl ions (log K_{Py-OH} = 14.5) (31,34). It acts as an oxidizing agent: fluorenol is oxidized to fluorenon. This is the first known model reaction for the enzymic dehydrogenation of an alcohol by a pyridinium salt.

$$(3)$$

2. Addition Reactions of Pyridinium Compounds

Analogous to the reduction according to Reaction (1) is the addition of anions to the pyridinium ring. In most cases, the addition of the nucleophilic agent gives 1,4-dihydropyridine derivatives whereas addition of the hydroxyl ion or, according to Reaction (3), the sulfoxylate anion leads to 1,6-dihydropyridine compounds (Reaction (4)). OH adducts could not be obtained in crystalline state until recently (34).

$$(4)$$

$$X = CN^-, RS^-, SO_3^{2-}, {}^-CH_2NO_2, {}^-CH_2COR, \quad \substack{\text{imidazolide}}^- , \quad {}^-NHOH, {}^-NHAr$$

Variation of the substituents at the ring nitrogen (series 1) and at C-3 (series 2) of the pyridine ring gave two series of compounds the properties of which are recorded in Table III and Figure 3.

The equilibrium constants of the addition compounds of $DPN^{\oplus}$ and various 1,3-disubstituted pyridinium compounds with cyanide (32) and sulfite (33) have been estimated (Table III). From the results one can conclude that substitution which decreases the electron density through an inductive (series 1) or mesomeric (series 2) effect, increases the affinity for the nucleophilic agents. Those pyridinium compounds which, like the coenzyme, contain an oxygen atom adjacent to the nitrogen bound carbon atom (e.g., X and XI in Fig. 3) form the most stable cyanide complexes.

TABLE III

Association Constants[a] of the Addition Compounds of Pyridinium Salts with Cyanide (32) and Sulfite (33)

	R	$\log K_{\text{Py-CN}}$	$\log K_{\text{Py-SO}_3}$
Series 1	CH_3		-1.2
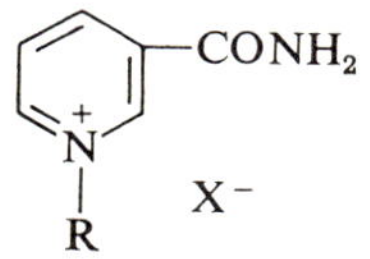	$CH_2—CH_2—CH_3$	-0.1	
	$CH_2—C_6H_5$	$+0.8$	-0.4
	$CH_2—C_6H_3Cl_2$	$+1.2$	$+0.1$
	RPPRA (DPN⊕)	$+2.4$	$+1.7$
	$CH_2—O—CH_2—C_6H_5$	$+2.9$	
	Tetraacetylglucose	$+4.1$	$+4.0$
Series 2		$(R' = DCB^b)$	$(R' = CH_3)$
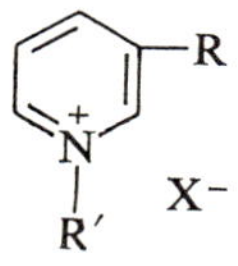	COOH	-0.7	-2.0
	$CON(CH_3)_2$	-0.2	-1.9
	$COOCH_3$	$+0.9$	-0.3
	$CONH_2$	$+1.2$	-1.2
	CN	$+3.3$	$+3.2$
	$COCH_3$	$+3.4$	$+2.0$

For Series 1 the structure shown is a pyridinium ring bearing a —CONH₂ substituent, with N—R and X⁻.

For Series 2 the structure shown is a pyridinium ring bearing —R, with N—R′ and X⁻.

[a] $K_{\text{Py-CN}} = [\text{Py—CN}]/[\text{Py}^\oplus][\text{CN}^\ominus]$, and analogous for the sulfite addition compound.

[b] DCB = dichlorobenzyl.

In some cases, with 3,5-disubstituted pyridinium salts, it has been shown (31,34,35) that in a fast reaction a 2-cyano-1,2-dihydropyridine derivative (VI) is formed upon cyanide addition to the pyridinium salt. This compound (VI) subsequently rearranges to give the thermodynamically more stable 4-cyano-1,4-dihydropyridine derivative.

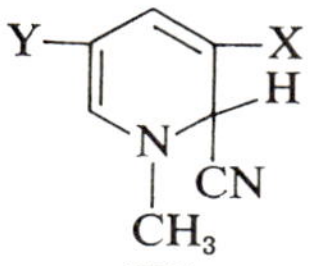

$$X = CN, \ COOC_2H_5$$
$$Y = CN, \ Br$$

(VI)

The ultraviolet absorption spectra of the addition compounds are similar to the spectra of the dihydro compounds. With regard to the dihydropyridine derivative, and depending on the nucleophilic agent, the maximum of absorption undergoes a shift in most cases to shorter wavelengths (Tables I and IV).

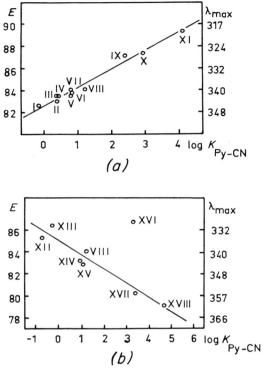

Fig. 3. Relationship between the complex stability (log K_{Py-CN}) and the maximum of the dihydro band (λ_{max} in mμ) of the cyanide addition compounds of pyridinium salts. E in kcal mole^{-1} (32).

(a) (Series 1):

	I	$R = CH_2 \cdot CH_2 \cdot CH_3$
	II	$CH_2 \cdot CH_2 \cdot C_6H_5$
	III	$CH_2 \cdot CH_2 \cdot CH_2 \cdot SO_3^{\ominus}$
	IV	$CH_2 \cdot CH{:}CH_2$
	V	$CH_2 \cdot C_6H_5$
	VI	$CH_2 \cdot CH_2 \cdot O \cdot C_6H_5$
	VII	$CH_2 \cdot CH(OH) \cdot CH_2 \cdot SO_3^{\ominus}$
	VIII	$CH_2 \cdot C_6H_3Cl_2(2.6)$
	IX	$RPPRA \ (= DPN^{\oplus})$
	X	$CH_2 \cdot O \cdot CH_2 \cdot C_6H_5$

TABLE IV

Ultraviolet Absorption of Addition Compounds of N^1-Dichloro-benzyl-3-carbamoylpyridinium Bromide in Dimethylformamide (38)
(Only the dihydro band is shown)

	X	λ_{max} (mμ)
	H	343
	SH	348
	CN	329
	CH_2NO_2	326
	$S-CH_2-CH_3$	330
	$SO_3^{\ominus\ominus}$	340

Substitution at the N-1 atom or at the C-3 atom in the pyridine ring will either favor or hinder the process described in Reaction (5) (cf. ref. 27).

$$\text{(5)}$$

It was found (32) that a relationship exists between the stability of the cyanide addition compounds and the maximum of the dihydro band: the wave number of the compounds is directly proportional to $\log K_{Py-CN}$ in the case of series 1, and inversely proportional in the case of series 2 (Fig. 3). From the equilibrium constants of the cyanide addition compounds of pyridinium salts it is possible to calculate the oxidation-reduction potential

(*b*) (Series (2):

	R_1	R_2
XII	$R_1=COOH$	$R_2=H$
XIII	$CON(CH_3)_2$	H
XIV	$COOCH_3$	H
XV	$COOC_2H_5$	H
XVI	CN	H
XVII	$COCH_3$	H
XVIII	$CONH_2$	$CONH_2$

of those compounds, whereas it is not possible to estimate the potential experimentally.

On binding to liver alcohol dehydrogenase (Fig. 4) or to lactate dehydrogenase, the dihydro band of DPNH is shifted by 15 mμ towards shorter wavelengths (9,21) and the anion affinity of DPN⊕ is increased (14,32). These effects may be considered as a "substituent effect" which is equivalent to a substitution at the ring nitrogen atom according to the compounds of series 1 listed in Table III and Figure 3 (27). These results show that the specific binding of the coenzyme to the enzyme not only creates favorable steric conditions for the enzyme reaction but also leads to a change in the reactivity of the coenzyme. (For the spectral properties of ternary enzyme–coenzyme–inhibitor complexes see ref. 36.)

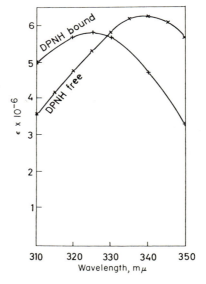

Fig. 4. The molar extinction coefficient of DPNH in the free state and bound to horse liver alcohol dehydrogenase (9).

3. Reduction to Tetrahydropyridines*

Tetrahydropyridines are obtained from dihydropyridines in three ways:

(1) Reduction of N^1-dichlorobenzyl-1,6-dihydronicotinamide, obtained by reaction of the pyridinium salt with hydroxyl ions, with Pd/BaSO₄ gives the 1,4,5,6-tetrahydronicotinamide compound (38).

* For the significance of the formation of a tetrahydropyridine derivative from DPNH in the presence of enzyme and oxidized substrate see references 36 and 37.

(2) Reduction of pyridinium salts with sodium borohydride gives, besides the dihydropyridines, Δ^3-tetrahydropyridines (39,40).

(3) Acid treatment of dihydropyridines causes disappearance of the dihydro band and appearance of a new band between 290 and 300 mμ which is characteristic for 1,4,5,6,-tetrahydropyridines. The acid instability of DPNH, TPNH, and reduced model compounds has been known for a long time (1,41).

The investigations of the reaction of weak acids with dihydropyridines, frequently a reversible reaction (37,42–44), the acid-catalyzed addition of the water molecule (45,46), or the base-catalyzed addition of phenylacetonitrile (47) to dihydropyridines, have shown that the electrophilic agents

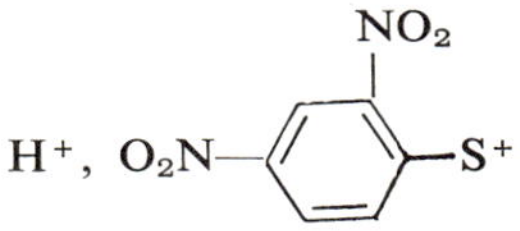

occupy the 5-position, and the nucleophilic agents the 6-position (in the case of 1,4-dihydropyridines, cf. ref. 26) or the 4-position (in the case of the 1,6-dihydropyridines) (see Reactions (6) and (7)).

1,4-Dihydropyridines:

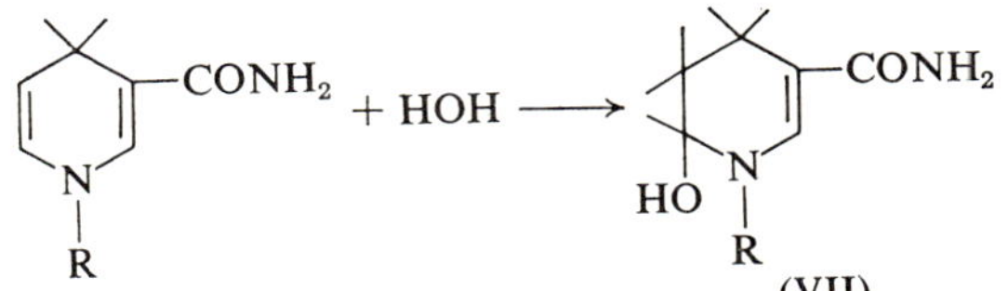

(6a)

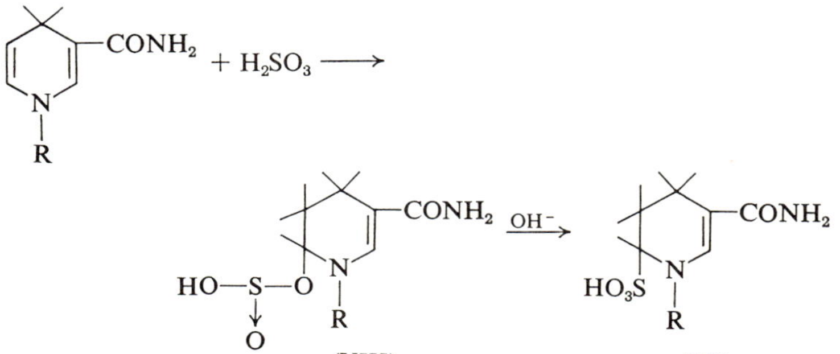

(6b)

1,6-Dihydropyridines:

(X)   (7a)

(XI)   (XII)   (7b)

Besides the tetrahydropyridine compound (VII), dimeric products are formed during the reaction of acid with the dihydro compound (45,48). The exact structures of these compounds are not known.* The "primary acid modification product" (VII) is unstable in strong acids; under these conditions the "secondary acid modification product" (perhaps a compound generated by ring opening of the 2,6-dihydroxy-piperidine derivative) is formed (18,43,46).

In the presence of glyceraldehyde-3-phosphate dehydrogenase, phosphate (or arsenate), and DPNH, a tetrahydropyridine derivative designated as DPNH-X is formed (49; see also 36,43,50). The structure of this compound is still unknown, but it is perhaps identical with those compounds which are formed by the reaction of 1,4-dihydropyridines with weak acids [Reaction (6a), VII with R = RPPRA].

4. One-Electron Reduction

Besides the dihydro- and tetrahydropyridines, a third reduction product is known (for review see ref. 21; cf. also ref. 51). The one-electron reduction of N^1-alkylnicotinamide derivatives with Zn/Cu, Mg, or Cr(II) acetate as well as the electrochemical reduction (and perhaps also the interaction with x-rays) give dimeric compounds (Reaction (8)).

The absorption spectra of the dimeric reduction products (XIII) are similar to the spectra of the 1,6-dihydro compounds. Hydrogen transfer from dimeric compounds to acceptors such as malachite green could not be

* The structure of the "dimeric acid product" has been elucidated very recently by x-ray crystallographic studies (H. L. Ammon and L. H. Jensen, *J. Am. Chem. Soc.* **88**, 613 (1966)).

$$(8)$$

observed. However, reduction of electron acceptors such as viologen occurred readily (23).

Evidence suggests that the final product of the electrochemical reduction of DPN$^{\oplus}$ is a 4,4′ dimer of DPN (52). This compound is electrochemically reoxidizable to DPN$^{\oplus}$ or reducible to DPNH.

5. Oxidation of Dihydropyridines*

Some reductions of pyridinium salts are known which have to be described as radical reactions (cf. Section III-A-4). Also, in the oxidation of dihydropyridines, some reactions follow a radical mechanism: oxidation of DPNH to DPN$^{\oplus}$ by spirocyclohexylporphyrexide, porphyrindine, and other radicals (54), oxidation of N^1-alkyl-1,4-dihydronicotinamide derivatives by diphenylpicryl hydrazyl (55) or the photochemical reaction between bromotrichloromethane and dihydropyridine compounds (56) (Reaction (9)). On the other hand, the majority of the investigated oxidations can be explained in terms of the direct transfer of a hydride ion from the dihydropyridine to the oxidizing agent (21).†

To the class of compounds which are reduced by hydride transfer from DPNH belong α-keto acids, ketones, thioketones (Reaction (10); ref. 58), α,β-unsaturated ketones (Reaction (11); ref. 59), quinones (Reaction (12);

$$(9a)$$

* For review see references 21 and 53.
† For a review of hydride transfer reactions in organic chemistry see reference 57.

$$\underset{\underset{CH_3}{|}}{CH_3O_2C}\text{-ring with H, CH}_3, CO_2CH_3, H_3C, CH_3 \quad + \ BrCCl_3 \longrightarrow \tag{9b}$$

$$CH_3O_2C\text{-ring}\text{-}CO_2CH_3,\ H_3C,\ CH_3,\ N^+,\ CH_3 \quad + \ Br^- + \cdot CCl_3$$

ref. 60), quinolines, indophenol dyes, malachite green, tetracyanoethylene as well as nicotinamide derivatives $DPN^{\oplus}$, $TPN^{\oplus}$, $DPN^{\oplus}$ analogs, N^1-benzylnicotinamide derivatives (Reaction (13); refs. 61–63), and flavin compounds (Reaction (14); ref. 64).

$$\underset{R}{\overset{H\ H}{\text{dihydropyridine}}}\text{-}CONH_2 \ + \ S{=}C\overset{C_6H_5}{\underset{C_6H_5}{}} \ + \ H^+ \longrightarrow \underset{R}{\text{pyridinium}}\text{-}CONH_2 \ + \ HS{-}\underset{H}{\overset{C_6H_5}{C}}{-}C_6H_5 \tag{10}$$

with a transition state (XIV):

(The dotted lines in XIV indicate bonds which are either broken or formed during the rate-controlling step.)

(XIV)

$$\underset{R}{\overset{H\ H}{CH_3O_2C\text{-ring}}}\text{-}CO_2CH_3,\ H_3C,\ CH_3 \ + \ C_6H_5{-}CO{-}CH{=}CH{-}CF_3 \longrightarrow \tag{11a}$$

$$CH_3O_2C\text{-ring}\text{-}CO_2CH_3,\ H_3C,\ N^+,\ CH_3 \ + \ C_6H_5{-}\underset{O^-}{\overset{|}{C}}{=}CH{-}CH_2{-}CF_3$$

$$C_6H_5{-}\underset{O^-}{\overset{|}{C}}{=}CH{-}CH_2{-}CF_3 \ + \ H^+ \longrightarrow C_6H_5{-}\underset{O}{\overset{\|}{C}}{-}CH_2{-}CH_2{-}CF_3 \tag{11b}$$

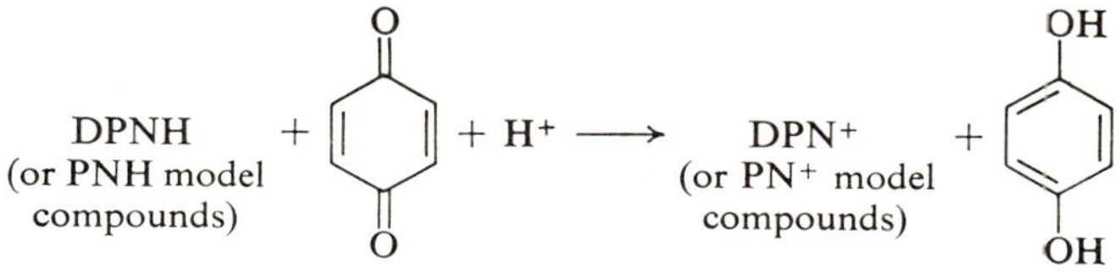

$$\text{DPNH} \; + \; \substack{\text{quinone}} \; + \; \text{H}^+ \longrightarrow \; \text{DPN}^+ \; + \; \substack{\text{hydroquinone}} \qquad (12)$$

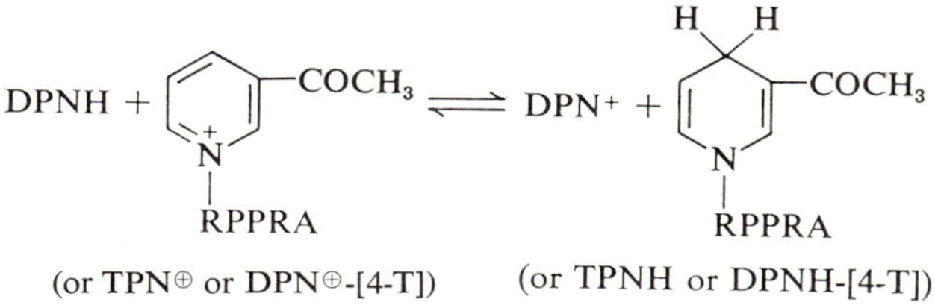

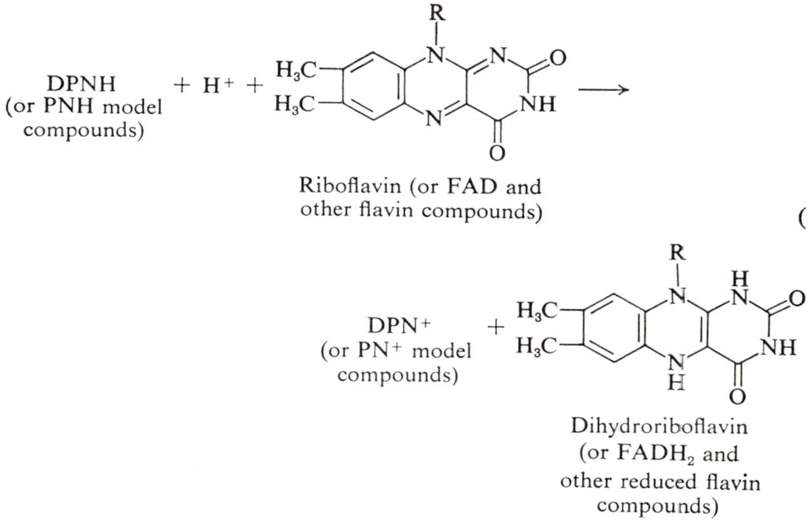

DPNH
(or PNH model
compounds)

DPN⁺
(or PN⁺ model
compounds)

$$\text{DPNH} \; + \; \substack{\text{pyridinium-COCH}_3} \; \rightleftharpoons \; \text{DPN}^+ \; + \; \substack{\text{dihydropyridine-COCH}_3} \qquad (13)$$

RPPRA

(or TPN⊕ or DPN⊕-[4-T])

RPPRA

(or TPNH or DPNH-[4-T])

DPNH   + H⁺ +
(or PNH model
compounds)

Riboflavin (or FAD and
other flavin compounds)   (14)

DPN⁺
(or PN⁺ model
compounds)

+

Dihydroriboflavin
(or FADH₂ and
other reduced flavin
compounds)

6. Hydride Transfer or Free-Radical Mechanism in the Enzyme-Catalyzed Reaction?

The model reactions in the absence of enzyme do not allow a definite answer to the question as to whether the enzymically catalyzed dehydrogenation according to Reaction (1) proceeds via a direct transfer of a hydride ion (i.e., a two-electron oxidation-reduction process) or whether two successive one-electron transfers take place with a free radical as an intermediate. Although ionic mechanisms seem to be favored in model reactions, both ionic and free-radical mechanisms have been observed. But in general, pyridinium salts behave as typical anion acceptors.

Hydrogen Transfer with DPNH Models.   A comparison of the reactions of the three isomeric dihydropyridines and the dimeric reduction product (XIII) with viologen (electron acceptor), dichlorophenolindophenol (both electron and hydride acceptor), or malachite green (hydride acceptor) may give some more information about the mechanism. The data in Table V show that while the hydride transfer from the 1,4-dihydropyridine is

TABLE V

Oxidation of Reduced Pyridine Compounds (23)

Reduced pyridine compound $R_1 = CH_2—C_6H_3Cl_2$ $R_2 = CH_2—C_6H_5$	Oxidizing agents		
		2,6-Dichlorophenol-indophenol	
	Viologen (Ethanol 78°C)	k_2, M^{-1} min^{-1} (pH 7.0, 25°C)	Malachite green (Ethanol, 25°C)
	no reaction	220	fast
	very slow	190	very slow
	slow	fast	very slow
	fast	20,000	no reaction

favored, no reaction is observed with the electron acceptor and therefore the tendency to form a radical intermediate is very poor. The opposite is observed for the dimeric compound.

Stability of the Cyanide Complexes and Rate of Hydrogen Transfer. The favored hydride transfer from 1,4-dihydropyridines to an acceptor is also deduced from the comparison of the K_{Py-CN} values and the rate constants of the reduction of dichlorophenolindophenol (Table VI). The stronger the

TABLE VI

Association Constants (K_{Py-CN}) of Cyanide Addition Compounds of Pyridinium Salts and Rate Constants (k_2) for the Oxidation of Dichlorophenolindophenol by the Corresponding Dihydropyridines (ref. 60)

R_1	$COOC_2H_5$	$CONH_2$	$COCH_3$	$CONH_2$
R_2	H	H	H	$CONH_2$
$K_{Py-CN}{}^a$	11	16	2300	50,000
$k_2{}^b$	420	220	36	2.6

^{a} $K_{Py-CN} = [Py-CN]/[Py^\oplus][CN^\ominus]$.
^{b} $[M^{-1} min^{-1}]$.

complex of a pyridinium salt with cyanide, the slower the reduction of dichlorophenolindophenol by the corresponding dihydropyridine compound (and vice versa). Reduction of dichlorophenolindophenol can proceed by an ionic mechanism or by a radical mechanism. From the results of Table VI one can conclude that the principle of hydrogen transfer from a dihydropyridine to an acceptor (Reaction (15a)) shows a strong resemblance to the cyanide addition reaction (Reaction (15b)). One can therefore assume that in the reduction of dichlorophenolindophenol an ionic mechanism is preferred.

$$Py^\oplus + H^\ominus \rightleftharpoons PyH \qquad (15a)$$

$$Py^\oplus + CN^\ominus \rightleftharpoons Py-CN \qquad (15b)$$

Another supporting fact is that a relationship is observed between the stability of the cyanide addition compounds of $DPN^\oplus$ and $DPN^\oplus$ analogs and the equilibrium constants of the enzyme-catalyzed reactions with DPN or the corresponding analogs (32).

Conclusion. All the results from the study of the reactivity of the coenzyme molecule and the model compounds described suggest that, in the reactions catalyzed by pyridine nucleotide-dependent dehydrogenases, a hydride ion is transferred reversibly from DPNH (or TPNH) to the substrate rather than that a free-radical reduction takes place. Such an ionic mechanism has been postulated for the reaction catalyzed by alcohol dehydrogenase (9,66).

7. Pyridine Nucleotide Coenzymes and Oxidative Phosphorylation

Many observations have shown that DPN derivatives, differing from $DPN^{\oplus}$ and DPNH, are formed during the process of oxidative phosphorylation. [This problem has been summarized very recently (67,68; see also 69).]

(*1*) The "extra-DPN" formed in mitochondria (70) is alkali-stable, does not act as coenzyme in the alcohol dehydrogenase system, and is converted to $DPN^{\oplus}$ on incubation with mitochondria in the presence of phosphate and ADP. This compound possesses the properties expected of "$DPN \sim I$," one of the postulated intermediates of oxidative phosphorylation (71). At present, it is not known whether this compound is a derivative of the oxidized or of the reduced form of DPN.

(*2*) The oxidation state of the high-energy derivative of DPN derived from DPNH on incubation with the phosphorylating particles from *Alcaligenes faecalis* (72–74) is also unknown. In the presence of phosphate and ADP this compound is converted into $DPN^{\oplus}$ and ATP is formed.

(*3*) A high-energy phosphorylated derivative of DPN is thought to be formed during incubation of $DPN^{\oplus}$ with heart mitochondria in the presence of phosphate and succinate (68). In the presence of submitochondrial particles, this compound transfers phosphate to ADP which gives ATP and mainly $DPN^{\oplus}$. But in the presence of an electron-transport inhibitor (e.g., antimycin A), DPNH and an unidentified compound are observed. The latter possesses an absorption maximum at 290–295 mμ and the available evidence indicates that it is a phosphorylated (tetrahydropyridine) derivative of DPNH possibly analogous to compound VII.

B. Oxidation-Reduction Potentials of the Pyridine Nucleotide Coenzymes

The $DPN^{\oplus}$–DPNH oxidation-reduction potential was determined by different methods to be -309 to -320 mV at $25°C$ at pH 7 (75,76). The temperature dependence was found to be −1.31 mV per degree (75). The oxidation-reduction potential for the TPN system is almost the same (77). The small differences reported may be due to varying purities of the coenzyme preparations (the potential of the α-DPN, for example, is different from that of the β-DPN) as well as the methods used. For practical purposes the potentials of both the DPN and TPN systems can be considered to be identical.

The potentials of the coenzyme analogs (Table I and ref. 18) are in most cases more positive than the potential of the DPN system. On the other hand, the potential of the α isomer of DPN is lower by about 20 mV.

The more positive potentials of the 3-acetylpyridine and the pyridine-3-aldehyde analogs favor the formation of the reduced coenzyme and therefore the formation of the oxidized substrate in the reactions catalyzed by dehydrogenases. This altered equilibrium of the enzyme reactions favors the quantitative estimation of ethanol, lactate, and other reduced substrates if the coenzyme analog is used instead of DPN.

The oxidation-reduction potential of the coenzyme can be altered on binding to an enzyme. In the case of alcohol dehydrogenase from horse liver (9) [and lactate dehydrogenase from beef heart (78)] the potential of the enzyme–coenzyme complex between pH 6 and 8 is 60–80 mV more positive than the potential of the free coenzyme. For this reason the equilibrium of the reaction catalyzed by alcohol dehydrogenase is changed if enzyme and coenzyme are present in equimolar amounts, instead of the presence of a large excess of coenzyme over enzyme.

The concentrations of DPN and TPN in the cells lie between 10^{-3} and $10^{-6} M$ (18), and the concentration of the pyridine nucleotide-dependent dehydrogenases at $10^{-6} M$ [in the case of yeast alcohol dehydrogenase $> 10^{-5} M$ (79)]. Assuming the presence of about 100 different pyridine nucleotide-dependent dehydrogenases in the cell, regardless of distribution in the different cellular compartments, comparison of the total amount of pyridine nucleotides and the total amount of coenzyme binding sites shows that the concentration of coenzyme binding sites is of the same order of magnitude as the concentration of pyridine nucleotides. Therefore, it appears that most of the pyridine nucleotides in the cell are probably bound to enzymes. This means that in these cases the enzyme-bound pyridine nucleotide system determines the kinetics and the thermodynamics of the enzyme reaction and not the free coenzyme system as normally investigated *in vitro* where only catalytic amounts of enzymes are present. It was observed (80,81) that the rates of different dehydrogenase reactions, measured in the presence of enzyme-bound pyridine nucleotides, differ to some extent from the rates measured with the free pyridine nucleotides. In addition, the complexes of DPN$^\oplus$ with anions are stronger with the enzyme-bound DPN$^\oplus$ than with the free coenzyme (14,18,32,82,83).

C. Absorption and Fluorescence Spectra of the Pyridine Nucleotide Coenzymes

1. Ultraviolet Absorption Spectra

The absorption spectrum of DPN$^\oplus$ above 230 mμ shows only one strong absorption maximum at 259 mμ due to both the adenine and the nicotinamide moieties of the coenzyme molecule (Fig. 5). The reduced coenzyme possesses a weaker band at 259 mμ due solely to the adenine moiety and a second characteristic absorption band, the "dihydro band," at 340 mμ due to the dihydronicotinamide moiety (Fig. 5). A value of 6.22 for the millimolar extinction coefficient of DPNH at 340 mμ is generally accepted (7,18). The ultraviolet absorption spectra of TPN$^\oplus$ and TPNH, as far as they are known, are the same as for DPN$^\oplus$ and DPNH, respectively.

Enzymic hydrolysis of the pyrophosphate linkages of DPN$^\oplus$ or of DPNH, causes an increase in the absorbancy at the absorption maximum of both the adenine and nicotinamide moieties (7). The difference in absorption between the intact and the hydrolyzed coenzyme molecule suggests intramolecular interactions between the two moieties in the intact coenzyme (cf. ref. 84 and the next section).

The spectra of the coenzyme analogs, as well as of the addition compounds, are similar to the spectrum of DPNH. The maximum of the dihydro band has shifted to shorter or longer wavelengths (Table I and Fig. 5; cf. also Section III-A-2). The millimolar extinction coefficients of the dihydro band of the pyridine-3-aldehyde-DPNH ($\epsilon = 9.3$) and the 3-acetylpyridine-DPNH ($\epsilon = 9.1$) are about 50% higher than the value for DPNH (7). On the other hand, the millimolar extinction coefficient of the cyanide addition compound of pyridine-3-aldehyde-DPN$^\oplus$ ($\epsilon = 4.6$, $\lambda_{max} = 327$ mμ) is less than that of the DPN$^\oplus$-cyanide addition compound ($\epsilon = 5.9$) (7).

2. Fluorescence Spectra*

DPNH fluoresces when irradiated by light in the 340 mμ region, but DPN$^\oplus$ does not (90). The excitation maxima of DPNH, corrected for

* For general discussions of the theoretical and practical basis of fluorescence measurements cf. references 85–89.

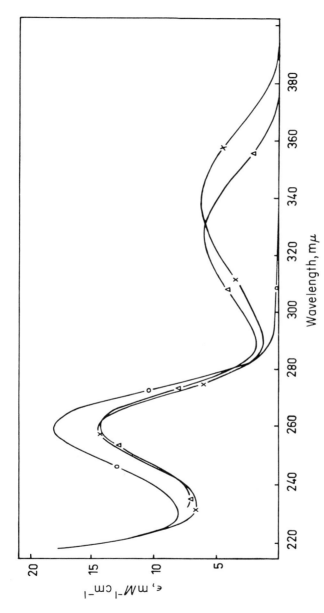

Fig. 5. Ultraviolet absorption spectra of DPN⊕ (○), DPNH (×), and the cyanide addition product of DPN⊕ (△) (7).

instrumental characteristics, were found to be 260 and 340 mμ, and the fluorescence maximum to be 457 mμ (in aqueous solution at pH 8.3 and 18°C) (84,91). The fluorescence intensity of DPNH is very strongly dependent upon the temperature and the solvent (92).

The strong excitation maximum of DPNH at 260 mμ, where more than 90% of the light is absorbed by the adenine moiety, represents an energy transfer from the excited adenine to the reduced nicotinamide, because the latter does not absorb at this wavelength. The lifetime of the adenine excited state is very short; adenine fluorescence is not observed in neutral aqueous solution at room temperature. Therefore, for the transfer to occur with the observed efficiency, the adenine and the dihydronicotinamide moieties of the coenzyme molecule must be in close juxtaposition to each other, thus permitting energy transfer to take place (84,91,92).*

In contrast to previous reports (18), it was recently shown that the α isomer of DPNH, like the β isomer, fluoresces upon irradiation at 260 mμ (19).

After hydrolysis of the pyrophosphate bond of DPNH, no fluorescence occurs on excitation at 260 mμ. The excitation maximum at 340 mμ is still observable (84,91). The 260 mμ excitation also disappears under conditions where the intramolecular complex dissociates, e.g., at higher temperatures or in solvents such as propylene glycol (84,92).

TPNH and the DPNH analogs, as well as most of the DPN⊕ addition compounds, fluoresce upon excitation of the absorption maximum of the dihydro band. The fluorescence intensity of 3-acetylpyridine-DPNH is only one-fourth of that shown by DPNH, although the extinction coefficient of the analog is about 50% higher than that of DPNH. The nicotinic acid analog of DPNH shows only one-tenth of the DPNH fluorescence intensity (18).

The excitation maximum at the purine absorption is also observed with TPNH, 3-acetylpyridine-DPNH, deoxyadenosine-DPNH, and 6-(2-hydroxyethylamino)purine-DPNH, but is only faintly discernible, if at all, with deamino-DPNH, 1-(2-hydroxyethyl)adenine-DPNH, 3-iso-adenine-DPNH, and the purine and uracil analogs of DPNH (5,15,19). This lack of energy transfer from the purine or pyrimidine to the dihydronicotinamide does not necessarily mean that the molecule does

---

* Solutions of DPNH or TPNH show a chemical relaxation. It was concluded that this chemical relaxation is due to a monomolecular reaction originating from an equilibrium between two forms of the reduced pyridine nucleotides, probably the open and the folded (closed) forms (65). NMR spectra of pyridine nucleotides indicate that the dinucleotides exist predominantly in the folded conformation (O. Jardetzky and N. G. Wade-Jardetzky, *J. Biol. Chem.*, **241**, 85 (1966).

not form a folded inner complex, since the electronic structure may simply be unfavorable for transfer by the coupled oscillator mechanism (15,84).

Intramolecular interactions are not restricted to reduced pyridine nucleotides. The ultraviolet absorption of DPN⊕ (cf. Section III-C-1), as well as fluorescence studies with 3-aminopyridine-DPN⊕, suggest that the purine and pyridine moieties are in close association with each other in the oxidized pyridine nucleotides. 3-Aminopyridine fluoresces, but if bound in the DPN⊕ analog the adenine moiety quenches the fluorescence of the 3-aminopyridine (5).

Fluorescence measurements offer a powerful tool for the study of the kinetics of enzyme-catalyzed reactions and the coenzyme–enzyme interaction and also demonstrate energy transfer from enzyme to coenzyme. Upon combination with dehydrogenases, the fluorescence maximum of DPNH shifts towards the ultraviolet and in most cases increases manyfold (Table VII, Fig. 6). The enhancement of fluorescence is due either to rigid orientation of the dihydronicotinamide moiety on the enzyme surface, which causes inhibition of an internal conversion operating in the free reduced coenzyme molecule, or to sensitized fluorescence (87).

The decrease of the fluorescence intensity of DPNH upon binding to D-glyceraldehyde-3-phosphate dehydrogenase may be the result of specific group interaction on the protein (92). That such a specific interaction

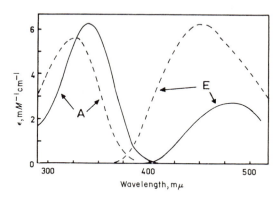

Fig. 6. Absorption, i.e., excitation, spectra (curves A) and emission spectra (curves E) of DPNH, both free (———) and bound to horse liver alcohol dehydrogenase (– – –). The absorption spectrum of the enzyme–coenzyme complex has been corrected for the absorption of alcohol dehydrogenase. The emission spectra have been corrected for blank fluorescence but not for the characteristics of the instrument components (emitted energy in arbitrary units) (87).

TABLE VII

Activation and Emission Maxima of DPNH, TPNH,
and 3-Acetylpyridine-DPNH, Free and Bound to Dehydrogenases
(The values have been corrected for instrumental characteristics.)

Compound (in parentheses, source of the dehydrogenase)	λ_{max} of absorption (activation) (mμ)	λ_{max} of emission (mμ)	Q^a	Ref.
DPNH, TPNH	340	462		11,93
3-Acetylpyridine-DPNH	365	478		11,94
Alcohol dehydrogenase-DPNH				
(horse liver), Fig. 6	325	440	13.5	11
(yeast)	340–345	443	ca. 3	11
Alcohol dehydrogenase-3-acetylpyridine-DPNH				
(horse liver)	350	440	7	11
Glucose-6-phosphate dehydrogenase-TPNH				
(rat mammary gland, human erythrocytes)	345	460	ca. 3	11
Glutamate dehydrogenase-DPNH or TPNH				
(beef liver)	343	452	ca. 2	11
Glycerol-1-phosphate dehydrogenase-DPNH				
(rabbit skeletal muscle)		454	7–10	11
Isocitrate dehydrogenase-DPNH				
(beef heart)		444	>2	95
Isocitrate dehydrogenase-TPNH				
(pig heart)	335	440	ca. 3	11
Lactate dehydrogenase-DPNH				
(beef brain)		440	>1	11
(beef heart, H_4)	325–335	440	8[b]	11,96,97
(beef muscle, M_4)			4[b]	96,97
(chicken heart, (H_4) and muscle, (M_4))			4–6[b]	96
(rabbit skeletal muscle)		445	3.8	11
(rat liver)	325	440	2.3	11,98
Lactate dehydrogenase-3-acetyl-pyridine-DPNH				
(beef heart, H_4)	350	440	26[b]	11,96,97
(beef muscle, M_4)			33[b]	96,97
(chicken heart (H_4) and muscle (M_4))			19–22[b]	96
(rat liver)		423	100	98

(*continued*)

TABLE VII (continued)

Compound (in parentheses, source of the dehydrogenase)	λ_{max} of absorption (activation) (mμ)	λ_{max} of emission (mμ)	Q^a	Ref.
Malate dehydrogenase-DPNH				
(beef brain)		450	>1	11
(pig heart)	350	454	2–3	11
(rat liver)			>1	11
Mannitol-1-phosphate-dehydrogenase-DPNH				
(*Aerobacter aerogenes*)		450	>1	11
Glyceraldehyde-3-phosphate dehydrogenase-DPNH				
(rabbit skeletal muscle)	340	462	<1	11

^{a} $Q = \dfrac{\text{molar fluorescence extinction coefficient of enzyme-bound PNH}}{\text{molar fluorescence extinction coefficient of free PNH}}$

^{b} Absolute quantum yield in % (3% for DPNH, 1.1% for 3-acetylpyridine-DPNH).

occurs is indicated by the broad absorption band at 365 mμ observed when DPN$^\oplus$ is bound to D-glyceraldehyde-3-phosphate dehydrogenase. This band may be due to a charge-transfer complex formation (99–101).*

D. Stereochemistry of the Pyridine Nucleotide Coenzymes†

$$\text{H} \quad \begin{array}{c} \text{CONH}_2 \\ \text{N}^+ \\ \text{R} \end{array} + SD_2 \rightleftharpoons \begin{array}{c} \text{H} \quad \text{D} \\ \text{CONH}_2 \\ \text{N} \\ \text{R} \end{array} + S + D^+ \qquad (16)$$

R = RPPRA
SD$_2$ = reduced substrate, e.g., CH$_3$—CD$_2$OD
S = oxidized substrate, e.g., CH$_3$—CDO

(XV)   (XVI)   (R=RPPRA)

* For charge-transfer complexes of DPN, TPN, and pyridinium compounds cf. refs. 28,102–107.

† For review see reference 108.

Carbon atom 4 becomes asymmetric when the nicotinamide moiety of $DPN^{\oplus}$ or $TPN^{\oplus}$ is reduced with deuterium- or tritium-labeled agents (Reaction (16)). Two stereoisomers (XV) and (XVI) can be formed. If $DPN^{\oplus}$ or $TPN^{\oplus}$ is reduced chemically with dithionite, a mixture of the two stereoisomers is obtained.* On the other hand, in the enzymic reduction, only one stereoisomer results.

The first experiment with an enzyme was conducted with alcohol dehydrogenase from yeast. In 1951 Westheimer, Fisher, Conn, and Vennesland showed (111) that the reduced DPN obtained in Reaction (17a) does not contain deuterium when the reaction is carried out in D_2O. In another experiment, 1,1-dideuteroethanol was oxidized in ordinary water (Reaction (17b)).

$$CH_3\!-\!CH_2OD + DPN^{\oplus} \longrightarrow CH_3\!-\!CHO + DPNH + D^{\oplus} \qquad (17a)$$

$$CH_3\!-\!CD_2OH + DPN^{\oplus} \longrightarrow CH_3\!-\!CDO + DPND + H^{\oplus} \qquad (17b)$$

In this case, the reduced coenzyme isolated from the reaction mixture contains one mole of deuterium per mole of reduced DPN. These results prove that the hydrogen is directly transferred from the α-carbon atom of the ethanol molecule and does not exchange with the protons of the medium.

When enzymically deuterated DPND, formed according to Reaction (17b), was reoxidized with acetaldehyde in the presence of alcohol dehydrogenase, all the deuterium in the coenzyme was lost (Reaction (18)). On the

$$CH_3\!-\!CHO + DPND + H^{\oplus} \underset{}{\overset{\substack{\text{alcohol}\\\text{dehydrogenase}}}{\rightleftharpoons}} CH_3\!\!-\!\!\overset{\overset{\textstyle D}{\vdots}}{\underset{\underset{\textstyle H}{\vdots}}{C}}\!\!\blacktriangleleft OH + DPN^{\oplus} \qquad (18)$$

other hand, only about half of the deuterium was lost when chemically deuterated DPND (prepared by the nonstereospecific reduction of $DPN^{\oplus}$ with dithionite in D_2O) was oxidized enzymically. The experiments described show that alcohol dehydrogenase catalyzes the reversible transfer of the hydride ion to only one side of the plane of the nicotinamide

* The two sides at the 4-position of the nicotinamide moiety of DPN behave differently in non-enzymic reactions. The chemical reduction of $DPN^{\oplus}$ (109) as well as the chemical oxidation of DPNH (110) causes a preferential attack on the A side. The probable explanation of this different behavior of the two sides is based upon the properties of the folded inner complex of the coenzyme molecule (92) (cf. Section III-C).

ring at carbon atom 4 in DPN. This side is called the "A side," and the DPND formed in reaction (18), "$DPND_A$."*

Previously the term α (or I) instead of A was used (and β or II instead of B) but the terms α and β should only be used to characterize the glycosidic linkage at the anomeric carbon atom.

From detailed investigation of numerous other pyridine nucleotide-dependent dehydrogenases, summarized in Table VIII, it is known that one group of enzymes shows the same *A-stereospecificity* as yeast alcohol dehydrogenase. The other group of enzymes uses the opposite side of the

TABLE VIII
The Coenzyme Stereospecificity of Hydrogen Transfer of
Pyridine Nucleotide-Dependent Dehydrogenases

Enzyme (Dehydrogenase = DH)	Source	Ref.
A-Stereospecificity		
Alcohol DH (with different substrates)	Horse liver, wheat germ, yeast, *Pseudomonas*	108
Acetaldehyde DH	Beef liver	108
Dicarbonylhexose reductase	*Gluconobacter cerinus*	114
Dihydrofolate reductase	*Streptococcus faecalis*	115
Dihydroorotate DH	*Zymobacterium oroticum*	108
DPNH-cytochrome b_5 reductase	Calf liver microsomes	116
Ferredoxin-TPNH reductase	Spinach chloroplasts	117, 118
Formate DH	Pea seeds	119
D-Glycerate DH	Parsley leaves	108
Glyoxylate reductase	Spinach leaves, pea seeds	119, 120
D_s-Isocitrate DH	Beef and pig heart muscle	108, 121
D-Lactate DH	*Lactobacillus arabinosus, Escherichia coli*	108, 120
L-Lactate DH	Beef heart muscle, rabbit skeletal muscle, potato tuber, *Lactobacillus arabinosus*	108
L-Malate DH	Pig heart muscle, wheat germ	108
Malic enzyme	Pigeon liver	119
5,10-Methylenetetrahydrofolate DH	Yeast	122
Mevaldate reductase	Pig and rat liver	123

(continued)

* The stereospecificity is not restricted to the coenzyme. The reaction occurs also stereospecifically with regard to the alcohol (36). With the aid of alcohol dehydrogenase, it is possible to synthesize both D- and L-monodeuteroethanol (112,113).

TABLE VIII (*continued*)

Enzyme (Dehydrogenase = DH)	Source	Ref.
B-Stereospecificity		
Dihydrolipoamide DH	Pig heart, spinach leaves	118, 124
DPNH-cytochrome *c* reductase	Rat liver mitochondria	108
D-Glucose DH	Beef liver	108
D-Glucose-6-phosphate DH	Yeast	108
L-Glutamate DH	Beef liver	108
Glutathione reductase	Yeast, *Escherichia coli*	108
D-Glyceraldehyde-3-phosphate DH	Rabbit skeletal muscle, yeast	108
Glycerolphosphate DH	Rabbit skeletal muscle	108
Hydrogen DH	*Hydrogenomonas ruhlandii*	125
L-β-Hydroxybutyryl-CoA DH	Pig heart muscle	108
β-Hydroxysteroid DH	*Pseudomonas testosteroni*	108
3α-Hydroxysteroid DH	*Pseudomonas testosteroni*	108
17β-Hydroxysteroid DH	Human placenta	108
Ketone reductase	*Curvularia falcata*, pig liver	126
6-Phospho-D-gluconate DH	Yeast	108
Tartronic semialdehyde reductase	*Escherichia coli*	127
Thioredoxin reductase	*Escherichia coli*	128
Transhydrogenase	*Pseudomonas fluorescens*	108
Uridine diphosphoglucose DH	Beef liver	119

4-position of the nicotinamide, therefore possessing *B-stereospecificity*. No nonstereospecific pyridine nucleotide-dependent dehydrogenase has been found up to now.

The absolute configuration of $DPND_A$ and $DPND_B$ has been established in the following manner (129). The nicotinamide moieties of $DPND_A$ and $DPND_B$ were converted into deuterated succinic acids of known

(19)

stereochemistry (see Reaction (19) for DPND$_A$). It follows that the hydrogen A (H$_A$) and the hydrogen B (H$_B$) in DPNH have the stereochemistry shown in formula (XVII)*: DPND$_A$ is (4R)-DPND, DPND$_B$ is (4S)-DPND.

(XVII)

Stereospecificity of the hydrogen transfer is the result of asymmetric binding of the coenzyme molecule to the active site of the enzyme. This could be demonstrated directly by measuring the optical rotatory dispersion (133): The binding of DPNH to alcohol dehydrogenase results in a characteristic Cotton effect.

Acknowledgments

The author wishes to express his thanks to the *Deutsche Forschungsgemeinschaft*, the *Fonds der Chemischen Industrie*, and the *Badische Anilin- & Soda-Fabrik* for financial support.

Abbreviations

ARPPR	Adenosine diphosphate ribose
DPN	Diphosphopyridine nucleotide (= ARPPRN)
DPN$\oplus$	Oxidized DPN
DPNH	Reduced DPN
FAD	Flavin adenine dinucleotide
PN$\oplus$	Oxidized pyridine nucleotide
PNH	Reduced pyridine nucleotide
Py$\oplus$	Pyridinium salt
Py$\ominus$	Pyridine ion
Py$\cdot$	Pyridine radical
PyH	Dihydropyridine
TPN	Triphosphopyridine nucleotide
TPN$\oplus$	Oxidized TPN
TPNH	Reduced TPN
K_{Py-CN}	$= \dfrac{[Py-CN]}{[Py\oplus][CN\ominus]}$

* The x-ray data confirm (130) the postulate (131) that 1,4-dihydronicotinamide is planar (cf. also ref. 132).

References

1. O. Warburg, W. Christian, and A. Griese, *Biochem. Z.*, **282**, 157 (1935).
2. O. H. Lowry, J. V. Passonneau, D. W. Schulz, and M. K. Rock, *J. Biol. Chem.*, **236**, 2746 (1961).
3. T. Bücher, W. Luh, and D. Pette, in *Hoppe-Seyler/Thierfelder, Handbuch der physiologisch- und pathologisch-chemischen Analyse*, 10th ed., Vol. VI/A, K. Lang, E. Lehnartz, O. Hoffmann-Ostenhof, and G. Siebert, Eds., Springer-Verlag, Berlin, 1964, p. 292.
4. *Report of the Commission on Enzymes* (I.U.B. Symposium Series, Vol. 20), Pergamon Press, New York, 1961, p. 16.
5. N. O. Kaplan, in *Molecular Basis of Enzyme Action and Inhibition*, P. A. E. Desnuelle, Ed. (Proc. 5th Intern. Congr. Biochem., Moscow, 1961, Vol. IV), Pergamon Press, New York, 1963, p. 295.
6. B. M. Anderson, C. J. Ciotti, and N. O. Kaplan, *J. Biol. Chem.*, **234**, 1219 (1959).
7. J. M. Siegel, G. A. Montgomery, and R. M. Bock, *Arch. Biochem. Biophys.*, **82**, 288 (1959).
8. B. M. Anderson and N. O. Kaplan, *J. Biol. Chem.*, **234**, 1226 (1959).
9. H. Sund and H. Theorell, in *The Enzymes*, 2nd ed., Vol. 7, P. D. Boyer, H. Lardy, and K. Myrbäck, Eds., Academic Press, New York, 1963, p. 25.
10. A. M. Stein, J. K. Lee, C. D. Anderson, and B. M. Anderson, *Biochemistry*, **2**, 1015 (1963).
11. H. G. Windmueller and N. O. Kaplan, *J. Biol. Chem.*, **236**, 2716 (1961).
12. C. P. Fawcett and N. O. Kaplan, *J. Biol. Chem.*, **237**, 1709 (1962).
13. I. Göhring, A. Wacker, and G. Pfleiderer, *Biochem. Z.*, **339**, 514 (1964).
14. G. Pfleiderer, in *Mechanismen enzymatischer Reaktionen* (14. Colloquium der Gesellschaft für physiologische Chemie), Springer-Verlag, Berlin, 1964, p. 300.
15. N. J. Leonard and R. A. Laursen, *Biochemistry*, **4**, 365 (1965).
16. C. Woenckhaus and G. Pfleiderer, *Biochem. Z.*, **341**, 495 (1965).
17. C. Woenckhaus, M. Volz, and G. Pfleiderer, *Z. Naturforsch.*, **19b**, 467 (1964).
18. N. O. Kaplan, in *The Enzymes*, 2nd ed., Vol. 3, P. D. Boyer, H. Lardy, and K. Myrbäck, Eds., Academic Press, New York, 1960, p. 105.
19. G. Pfleiderer, C. Woenckhaus, and M. Nelböck-Hochstetter, *Ann.*, **690**, 170 (1965).
20. S. Suzuki, K. Suzuki, T. Imai, N. Suzuki, and S. Okuda, *J. Biol. Chem.*, **240**, 554 (1965).
21. H. Sund, H. Diekmann, and K. Wallenfels, in *Advances in Enzymology*, Vol. 26, F. F. Nord, Ed., Interscience, New York, 1964, p. 115.
22. F. H. Westheimer, in G. E. W. Wolstenholme and C. M. O'Connor, Eds., *Steric Course of Microbiological Reactions* (Ciba Foundation Study Group No. 2), Churchill, London, 1959, p. 98.
23. K. Wallenfels and M. Gellrich, *Ber.*, **92**, 1406 (1959).
24. R. F. Hutton and F. H. Westheimer, *Tetrahedron*, **3**, 73 (1958).
25. H. E. Dubb, M. Saunders, and J. H. Wang, *J. Am. Chem. Soc.*, **80**, 1767 (1958).
26. H. Diekmann, G. Englert, and K. Wallenfels, *Tetrahedron*, **20**, 281 (1964).
27. K. Wallenfels, in *Steric Course of Microbiological Reactions* (Ciba Foundation Study Group No. 2), G. W. E. Wolstenholme and C. M. O'Connor, Eds., Churchill, London, 1959, p. 10.

28. K. Wallenfels and H. Schüly, *Ann.*, **621**, 178 (1959).
29. K. Wallenfels, H. Schüly, and D. Hofmann, *Ann.*, **621**, 106 (1959).
30. S. Chaykin and L. Meissner, *Biochem. Biophys. Res. Commun.*, **14**, 233 (1964).
31. K. Wallenfels and W. Hanstein, *Angew. Chem.*, **77**, 861 (1965), *Intern. Ed.*, **4**, 869 (1965).
32. K. Wallenfels and H. Diekmann, *Ann.*, **621**, 166 (1959).
33. G. Pfleiderer, E. Sann, and A. Stock, *Ber.*, **93**, 3083 (1960).
34. W. Hanstein, Dissertation, Freiburg, 1966.
35. R. E. Lyle and G. J. Gauthier, *Tetrahedron Letters*, **1965**, 4615.
36. H. Sund, see p. 641, this volume.
37. K. Wallenfels and D. Hofmann, *Tetrahedron Letters*, **1959**, No. 15, p. 10.
38. K. Wallenfels and H. Schüly, *Ann.*, **621**, 86 (1959).
39. J. J. Panouse, *Compt. Rend.*, **233**, 260 and 1200 (1951).
40. R. E. Lyle, D. A. Nelson, and P. S. Anderson, *Tetrahedron Letters*, **1962**, 553.
41. E. Haas, *Biochem. Z.*, **288**, 123 (1936).
42. K. Wallenfels, D. Hofmann, and H. Schüly, *Ann.*, **621**, 188 (1959).
43. A. Stock, E. Sann, and G. Pfleiderer, *Ann.*, **647**, 188 (1961).
44. H. Diekmann, D. Hofmann, and K. Wallenfels, *Ann.*, **674**, 79 (1964).
45. A. C. Anderson and G. Berkelhammer, *J. Am. Chem. Soc.*, **80**, 992 (1958).
46. C. C. Johnston, J. L. Gardner, C. H. Suelter, and D. E. Metzler, *Biochemistry*, **2**, 689 (1963).
47. K. Schenker and J. Druey, *Helv. Chim. Acta*, **42**, 2571 (1959).
48. H. Diekmann, Dissertation, Freiburg, 1962.
49. S. Chaykin, J. O. Meinhart, and E. G. Krebs, *J. Biol. Chem.*, **220**, 811 (see also p. 821) (1956).
50. G. Pfleiderer and A. Stock, *Biochem. Z.*, **336**, 56 (1962).
51. M. S. Spritzer, J. M. Costa, and P. J. Elving, *Anal. Chem.*, **37**, 211 (1965); J. N. Burnett and A. L. Underwood, *J. Org. Chem.*, **30**, 1154 (1965).
52. J. N. Burnett and A. L. Underwood, *Biochemistry*, **4**, 2060 (1965).
53. F. H. Westheimer, in *Advances in Enzymology*, Vol. 24, F. F. Nord, Ed., Interscience, New York, 1962, p. 441.
54. K. A. Schellenberg and L. Hellerman, *J. Biol. Chem.*, **231**, 547 (1958).
55. D. Mauzerall and F. H. Westheimer, *J. Am. Chem. Soc.*, **77**, 2261 (1955).
56. J. L. Kurz, R. Hutton, and F. H. Westheimer, *J. Am. Chem. Soc.*, **83**, 584 (1961).
57. N. C. Deno, H. J. Peterson, and G. S. Saines, *Chem. Rev.*, **60**, 7 (1960).
58. R. H. Abeles, R. F. Hutton, and F. H. Westheimer, *J. Am. Chem. Soc.*, **79**, 712 (1957).
59. B. E. Norcross, P. E. Klinedinst, and F. H. Westheimer, *J. Am. Chem. Soc.*, **84**, 797 (1962).
60. K. Wallenfels and M. Gellrich, *Ann.*, **621**, 149 (1959).
61. M. J. Spiegel and G. R. Drysdale, *J. Biol. Chem.*, **235**, 2498 (1960).
62. G. Cilento, *Arch. Biochem. Biophys.*, **88**, 352 (1960).
63. J. Ludowieg and A. Levy, *Biochemistry*, **3**, 373 (1964).
64. C. H. Suelter and D. E. Metzler, *Biochim. Biophys. Acta*, **44**, 23 (1960).
65. G. Czerlinski and F. Hommes, *Biochim. Biophys. Acta*, **79**, 46 (1964).
66. K. Wallenfels and H. Sund, *Biochem. Z.*, **329**, 59 (1957).
67. L. Ernster and C. P. Lee, *Ann. Rev. Biochem.*, **33**, 729 (1964).

68. D. E. Griffiths, in *Essays in Biochemistry*, Vol. I, P. N. Campbell and G. D. Greville, Eds., Academic Press, New York, 1965, p. 91.
69. P. D. Boyer, see p. 193, this volume.
70. J. L. Purvis, *Biochim. Biophys. Acta*, **38**, 435 (1960).
71. E. C. Slater and J. M. Tager, *Biochim. Biophys. Acta*, **77**, 276 (1963).
72. G. B. Pinchot, *Proc. Natl. Acad. Sci. U.S.*, **46**, 929 (1960).
73. G. B. Pinchot, *Federation Proc.*, **22**, 1076 (1963).
74. G. B. Pinchot and M. Hormanski, *Proc. Natl. Acad. Sci. U.S.*, **48**, 1970 (1962).
75. F. L. Rodkey, *J. Biol. Chem.*, **234**, 188 (1959).
76. W. M. Clark, *Oxidation-Reduction Potentials of Organic Systems*, Williams & Wilkins, Baltimore, 1960, p. 487.
77. F. L. Rodkey and J. A. Donovan, *J. Biol. Chem.*, **234**, 677 (1959).
78. J. B. Neilands, *J. Biol. Chem.*, **199**, 373 (1952).
79. Unpublished results.
80. A. P. Nygaard and W. J. Rutter, *Acta Chem. Scand.*, **10**, 37 (1956).
81. L. Astrachan, S. P. Colowick, and N. O. Kaplan, *Biochim. Biophys. Acta*, **24**, 141 (1957).
82. H. Terayama and C. S. Vestling, *Biochim. Biophys. Acta*, **20**, 586 (1956).
83. G. Pfleiderer, D. Jeckel, and T. Wieland, *Biochem. Z.*, **328**, 187 (1956), D. Gerlach, G. Pfleiderer, and J. J. Holbrook, *Biochem. Z.*, **343**, 354 (1965).
84. G. Weber, *J. Chim. Phys.*, **55**, 878 (1958).
85. T. Förster, *Fluoreszenz organischer Verbindungen*, Vandenhoeck & Ruprecht, Göttingen, 1951.
86. G. Weber, in M. L. Anson, K. Bailey, and J. T. Edsall, Eds., *Advances in Protein Chemistry*, Vol. 8, Academic Press, New York, 1953, p. 415; in *Light and Life*, W. D. McElroy and B. Glass, Eds., The Johns Hopkins Press, Baltimore, 1961, p. 82.
87. A. Ehrenberg and H. Theorell, in *Comprehensive Biochemistry*, Vol. 3, M. Florkin and E. H. Stotz, Eds., Elsevier, Amsterdam, 1962, p. 169.
88. C. S. Vestling, in *Methods of Biochemical Analysis*, Vol. 10, D. Glick, Ed., Interscience, New York, 1962, p. 137.
89. S. Udenfriend, *Fluorescence Assay in Biology and Medicine*, Academic Press, New York, 1962.
90. O. Warburg and W. Christian, *Biochem. Z.*, **287**, 291 (1936).
91. G. Weber, *Nature*, **180**, 1409 (1957).
92. S. F. Velick, in *Light and Life*, W. D. McElroy and B. Glass, Eds., The Johns Hopkins Press, Baltimore, 1961, p. 108.
93. L. N. M. Duysens and J. Amesz, *Biochim. Biophys. Acta*, **24**, 19 (1957).
94. S. Shifrin, N. O. Kaplan, and M. M. Ciotti, *J. Biol. Chem.*, **234**, 1555 (1959).
95. R. F. Chen and G. W. E. Plaut, *Biochemistry*, **2**, 1023 (1963).
96. R. H. McKay and N. O. Kaplan, *Biochim. Biophys. Acta*, **79**, 273 (1964).
97. S. R. Anderson and G. Weber, *Biochemistry*, **4**, 1948 (1965).
98. S. R. Anderson, E. D. Ihnen, and C. S. Vestling, *Federation Proc.*, **23**, 428 (1964).
99. E. M. Kosower, *J. Am. Chem. Soc.*, **78**, 3497 (1956).
100. S. F. Velick, *J. Biol. Chem.*, **233**, 1455 (1958).
101. P. Friedrich, *Biochim. Biophys. Acta*, **99**, 371 (1965).
102. E. M. Kosower, in *The Enzymes*, 2nd ed., Vol. 3, P. D. Boyer, H. Lardy, and K. Myrbäck, Eds., Academic Press, New York, 1960, p. 171.
103. I. Isenberg and A. Szent-Györgyi, *Proc. Natl. Acad. Sci. U.S.*, **45**, 1229 (1959).

104. S. G. A. Alivisatos, F. Ungar, A. Jibril, and G. A. Mourkides, *Biochim. Biophys. Acta*, **51**, 361 (1961).
105. G. Cilento and P. Tedeschi, *J. Biol. Chem.*, **236**, 907 (1961).
106. G. Cilento and S. Schreier, *Arch. Biochem. Biophys.*, **107**, 102 (1964).
107. S. Shifrin, *Biochim. Biophys. Acta*, **81**, 205 (1964); **96**, 173 (1965).
108. H. R. Levy, P. Talalay, and B. Vennesland, in *Progress in Stereochemistry*, Vol. 3, P. B. D. de la Mare and W. Klyne, Eds., Butterworths, London, 1962, p. 299.
109. H. F. Fisher, E. E. Conn, B. Vennesland, and F. H. Westheimer, *J. Biol. Chem.*, **202**, 687 (1953).
110. A. S. Pietro, N. O. Kaplan, and S. P. Colowick, *J. Biol. Chem.*, **212**, 941 (1955).
111. F. H. Westheimer, H. F. Fisher, E. E. Conn, and B. Vennesland, *J. Am. Chem. Soc.*, **73**, 2403 (1951).
112. F. A. Loewus, F. H. Westheimer, and B. Vennesland, *J. Am. Chem. Soc.*, **75**, 5018 (1953).
113. H. R. Levy, F. A. Loewus, and B. Vennesland, *J. Am. Chem. Soc.*, **79**, 2949 (1957).
114. S. Englard, G. Avigad, and L. Prosky, *J. Biol. Chem.*, **240**, 2302 (1965).
115. R. L. Blakley, B. V. Ramasastri, and B. M. McDougall, *J. Biol. Chem.*, **238**, 3075 (1963).
116. P. Strittmatter, *J. Biol. Chem.* **239**, 3043 (1964).
117. G. Krakow, R. N. Ammeraal, and B. Vennesland, *J. Biol. Chem.*, **240**, 1820 (1965).
118. R. N. Ammeraal, G. Krakow, and B. Vennesland, *J. Biol. Chem.*, **240**, 1824 (1965).
119. G. Krakow, J. Ludowieg, J. H. Mather, W. M. Normore, L. Tosi, S. Udaka, and B. Vennesland, *Biochemistry*, **2**, 1009 (1963).
120. G. Krakow and B. Vennesland, *Biochem. Z.*, **338**, 31 (1963).
121. R. F. Chen and G. W. E. Plaut, *Biochemistry*, **2**, 752 (1963).
122. B. V. Ramasastri and R. L. Blakley, *J. Biol. Chem.*, **239**, 112 (1964).
123. C. Donninger and G. Popják, *Biochem. J.*, **91**, 10P (1964).
124. M. M. Weber, N. O. Kaplan, A. S. Pietro, and F. E. Stolzenbach, *J. Biol. Chem.*, **227**, 27 (1957).
125. D. H. Bone, *Biochim. Biophys. Acta*, **67**, 589 (1963).
126. V. Prelog, in *Mechanismen enzymatischer Reaktionen*, Springer-Verlag, Berlin, 1964, p. 288.
127. G. Krakow, S. Udaka, and B. Vennesland, *Biochemistry*, **1**, 254 (1962).
128. A. Larsson and L. Thelander, *J. Biol. Chem.*, **240**, 2691 (1965).
129. J. W. Cornforth, G. Ryback, G. Popják, C. Donninger, and G. Schroepfer, *Biochem. Biophys. Res. Commun.*, **9**, 371 (1962).
130. I. L. Karle, *Acta Cryst.*, **14**, 497 (1961); H. Koyama, *Z. Krist.*, **118**, 51 (1963).
131. D. Hofmann, E. M. Kosower, and K. Wallenfels, *J. Am. Chem. Soc.*, **83**, 3314 (1961).
132. W. L. Meyer, H. R. Mahler, and R. H. Baker, *Biochim. Biophys. Acta*, **64**, 353 (1962).
133. D. D. Ulmer, T. K. Li, and B. L. Vallee, *Proc. Natl. Acad. Sci. U.S.*, **47**, 1155 (1961).

The Pyridine Nucleotide-Dependent Dehydrogenases

Horst Sund,* *Universität Konstanz, Konstanz, Germany*

I.	Introduction	642
II.	Chemical and Physical Properties of the PN Enzymes which Have Been Obtained in Crystalline or Homogenous Form	643
	A. Molecular Weight, Subunit Structure, and Enzymic Activity of the Dehydrogenases	643
	1. Molecular Weight and Subunit Structure . . .	643
	2. Relationship between Quaternary Structure and Activity for Glutamate Dehydrogenase from Beef Liver . .	658
	3. Lactate Dehydrogenase and Its Isoenzymes . . .	662
	B. Primary Structure of the Active Centers; Primary Structure and Evolution.	664
	C. Dehydrogenases: Zinc Enzymes?	667
	1. Alcohol Dehydrogenases	668
	2. Glyceraldehyde-3-phosphate Dehydrogenase . .	669
	3. Glutamate Dehydrogenase	670
III.	Enzymic Properties	670
	A. Types of Reactions Catalyzed by PN Enzymes . . .	670
	B. Substrate Specificity.	672
	1. Glutamate Dehydrogenase	672
	2. Aldehyde Dehydrogenases	673
	3. Polyol Dehydrogenases, Hemiacetal Dehydrogenases .	673
	4. Hydroxysteroid Dehydrogenases	675
	5. Alcohol Dehydrogenases	675
	6. Multiple Functions of Glyceraldehyde-3-phosphate Dehydrogenase	677
	C. Coenzyme Specificity and the Binding between Enzyme and Coenzyme	678
	1. Coenzyme Specificity	678
	2. Binary Enzyme–Coenzyme Complexes and Ternary Enzyme–Coenzyme–Substrate and Enzyme–Coenzyme–Inhibitor Complexes	679
	Abbreviations	694
	References	695

* The author wishes to express his thanks to the *Deutsche Forschungsgemeinschaft*, Badische Anilin- und Soda-Fabrik, and the *Fonds der Chemischen Industrie* for financial support.

I. Introduction

Great advances in the understanding of the mechanisms of reactions catalyzed by pyridine nucleotide-dependent dehydrogenases (PN enzymes) have been made in the last fifteen years. This is largely the result of the investigation of their stereochemistry utilizing labeled coenzymes and substrates, and the study of the coenzyme and substrate binding as well as of the number of active sites, using fluorescence, optical rotatory dispersion, and ultracentrifugation techniques. The synthesis of coenzyme model compounds and coenzyme analogs made possible an understanding of the chemistry of the reactions catalyzed by PN enzymes and the nature of the binding between enzyme and coenzyme. Investigations of the primary and quaternary structure were important in the elucidation of the molecular structure of various dehydrogenases and provided insight into the evolution of these enzymes and the existence of isoenzymes. In some cases the stepwise course of several reactions was clarified by kinetic investigations.

This chapter will deal with typical *simple* PN enzymes, i.e., those which require only the pyridine nucleotides as coenzymes. Simple PN enzymes are to be distinguished from the *complex* PN enzymes, which require a second coenzyme—for instance, a flavin coenzyme—for activity (for the pyridine nucleotide-dependent flavoproteins see refs. 1 and 2). In most cases the general reaction catalyzed by simple DPN-TPN enzymes is shown in Equation (1).

$$SH_2 + \underset{\substack{| \\ ARPPR}}{\overset{}{\bigcirc}} -CONH_2 \rightleftharpoons S + \underset{\substack{| \\ ARPPR}}{\overset{H \quad H}{\bigcirc}} -CONH_2 + H^+ \tag{1}$$

$$SH_2 = \text{reduced substrate, e.g., } CH_3 - CH_2OH$$
$$S = \text{oxidized substrate, e.g., } CH_3 - CHO$$

In some cases the reaction is more complicated. In addition to the coenzyme and the "virtual" substrate (SH, SH_2) a third reactant (HA) is involved (Reactions 2a and 2b):

$$SH + PN^+ + HA \rightleftharpoons S{-}A + PNH + H^+ \tag{2a}$$

$$SH_2 + 2PN^+ + HA \rightleftharpoons S{-}A + 2PNH + H^+ \tag{2b}$$

$$(HA = H_2O, H_3PO_4, CoASH)$$

It is not the author's intention to provide an extensive description and discussion of the properties of all the numerous individual dehydrogenases. For further information on aspects not dealt with here (especially kinetic aspects) the reader is referred to references 3–8.

II. Chemical and Physical Properties of the PN Enzymes which Have Been Obtained in Crystalline or Homogenous Form

A voluminous literature has developed in recent years dealing with the properties of a large number of PN enzymes from animal and plant tissues and microorganisms. The studies relating to these enzymes have been carried out at various levels of technical sophistication. Current interest in protein structure and its relation to the catalytic process has focused attention on homogenous enzyme preparation. The PN enzymes which have been obtained in crystalline or homogenous form are listed in Table I. Also included in this table are some of the physical characteristics of these enzymes and references which can be read for more extensive information on individual enzymes.

A. Molecular Weight, Subunit Structure, and Enzymic Activity of the Dehydrogenases

1. Molecular Weight and Subunit Structure

The molecular weights of the dehydrogenases investigated so far vary from 23,000 to two million (Table I). Most of the molecular weights are in the range 50,000 to 200,000. Dihydrofolate reductase and mannitol-1-phosphate dehydrogenase possess the smallest and the glutamate dehydrogenases the highest molecular weights.

The majority of the dehydrogenases are composed of more than one polypeptide chain. Alcohol dehydrogenase from horse liver, for instance, contains two polypeptide chains, yeast enzyme as well as lactate dehydrogenases and glyceraldehyde-3-phosphate dehydrogenases contain four polypeptide chains, and glutamate dehydrogenases up to 40. All the dehydrogenases consist of polypeptide chains with molecular weights between 14,000 and 50,000; higher values have not been detected.

It can be seen from Table I that the molecular weights of the polypeptide chains generally approximate the equivalent weights of the coenzyme binding sites. The one exception to this rule is the malate dehydrogenase from *Neurospora crassa* (267). In general, one active site is assigned to each coenzyme binding site. This raises the question of the smallest enzymically active unit of the dehydrogenases containing two or more polypeptide chains. It seems reasonable to assume that each polypeptide chain should

TABLE

Molecular Properties of Crystalline or Homogenous

Enzyme (DH = Dehydro-genase)	Reaction	Source	Molecular weight	Molecular weight of the polypeptide chain	Equivalent weight per coenzyme binding site
Alanine DH	L-Alanine + PN$^+$ + H$_2$O $\rightleftharpoons$ pyruvate + PNH + NH$_3$ + H$^+$	*Bacillus subtilis* (9)	228,000 (9)	38,000 (10)[b]	
Aldose DH	D-Aldopyranose + PN$^+$ $\rightleftharpoons$ aldono-lactone + PNH + H$^+$	*Pseudomonas* (12)	140,000 (13)	38,000 (13)	
Alcohol DH	Alcohol + PN$^+$ $\rightleftharpoons$ aldehyde (or ketone) + PNH + H$^+$	Horse liver (14,15)	84,000 (14)	42,000 (16–18)[b,c]	42,000 (14,17)
		Human liver (43)	87,000 (43)		41,000 (43)
		Rhesus monkey liver (45)	$\approx$ 85,000 (45)		
		Yeast (14)	157,000 (46)	36,000–37,000 (16,46–48)[b,c]	30,000–42,000 (14,85)
D-Arabinose DH	D-Arabinopyranose + PN$^+$ $\rightleftharpoons$ arabono lactone + PNH + H$^+$	Pseudomonas (12)	104,000 (13)	20,000 (13)	
Dihydrofolate reductase	Folate + 2PNH + 2H$^+$ $\rightleftharpoons$ tetrahydro-folate + 2PN$^+$	Chicken liver (63)	23,000 (63)		Probably 23,000 (63)
	Dihydrofolate + PNH + H$^+$ $\rightleftharpoons$ tetra-hydrofolate + PN$^+$	Calf thymus (68)	34,000 (68)		
Galactose DH	β-D-galactopyranose + PN$^+$ $\rightleftharpoons$ D-galactono-δ-lactone + PNH + H$^+$	*Pseudomonas saccharophila* (70)	103,000 (71)	25,000 (71)	
		Pseudomonas (12)	64,000 (13)	30,000 (13)	
Glucose-6-phosphate DH	D-Glucose 6-phos-phate + PN$^+$ $\rightleftharpoons$ D-glucono-δ-lactone 6-phos-phate + PNH + H$^+$	Bovine mammary gland (73)			
		Yeast (75)	128,000 (76)		

I

Pyridine Nucleotide-Dependent Dehydrogenases

n^{a}	$K_{E,R}$ (μM) $\left(= \dfrac{[E][R]}{[ER]}\right)$	$K_{E,O}$ (μM) $\left(= \dfrac{[E][O]}{[EO]}\right)$	References to kinetic work and enzymic properties	References to amino acid analysis and structure investigations	Remarks
6; ?			10, 11	9, 10	
3–4; ?			12		At low protein concentrations spontaneous dissociation into the polypeptide chains occurs (13)
2; 2	DPNH: 0.3 (pH = 7); 0.7–0.9 (pH = 9) (14)	DPN$^+$: 141–160 (pH = 7); 12–16 (pH = 9) (14)	4, 14, 15, 17, 19–39	15, 18, 40	Zinc content: 2 (41) to 4 (42) moles Zn^{++} per mole alcohol DH
?; 2			43, 44		Zinc content: 2.2 moles Zn^{++} per mole alcohol DH (43)
			45		
4; 4	DPNH: 10–23 (pH 7.2–7.9) (14)	DPN$^+$: 160–270 (pH 6.0–7.8) (14)	4, 14, 32, 33, 35, 39, 49–58	14, 59	Zinc content: 3.2–5.9 moles Zn^{++} per mole alcohol DH (41, 59, 60–62)
5; ?			12		At low protein concentrations dissociation into subunits with molecular weight 50,000 occurs (13)
?; 1			63–67		
			68, 69		
4; ?			72		
2; ?			12		At low protein concentrations spontaneous dissociation into the polypeptide chains occurs (13)
			74		
			74, 77		

(continued)

Enzyme (DH = Dehydrogenase)	Reaction	Source	Molecular weight	Molecular weight of the polypeptide chain	Equivalent weight per coenzyme binding site
Glutamate DH	Glutamate + PN$^+$ + H_2O ⇌ α-ketoglutarate + PNH + NH_3 + H$^+$	Beef liver	2,000,000 (4,78)	30,000–60,000 (79–82)[d] probably 50,000	30,000–50,000 (83–87)
		Chicken liver (120)	≈ 500,000 (121,122)		
		Frog liver (123)	500,000 (123)		
		Pig liver (126)	≈ 1,000,000 (126)	≈ 50,000 (126)	≈ 67,000 (126)
		Dogfish (128)			
		Neurospora crassa (129)	267,000 (129)	28,000 (129–131)[b]	
		Yeast (134)	≈ 300,000 (135)		

n^{a}	$K_{E,R}\ (\mu M)$ $\left(=\dfrac{[E][R]}{[ER]}\right)$	$K_{E,O}\ (\mu M)$ $\left(=\dfrac{[E][O]}{[EO]}\right)$	References to kinetic work and enzymic properties	References to amino acid analysis and structure investigations	Remarks
Probably 40 for each	TPNH: 26 (pH = 8.0, from kinetic experiments) (88); 0.5–2.0 (pH = 8,0 from direct estimation) (89); DPNH: 1.5–2.0 (pH = 7.5–7.7, from direct estimation) (83,84,86)	TPN⁺: 230 (pH = 8.0, from kinetic experiments) (88)	32, 87, 88–108	79, 87, 88, 94, 95, 99, 100, 109–115	At low protein concentrations (about < 6 mg/ml) reversible dissociation into subunits with molecular weights 1 million, ½ million, and ¼ million occurs (76,81,116–119)
			103,120,121		The enzyme undergoes association and dissociation (121,122)
			123–125		At low protein concentrations the enzyme dissociates into subunits with a molecular weight of 250,000; the equilibrium constant for this reaction is $4\mu M$ (123). Enzyme from premetamorphic tadpole liver differed from that of the adult frog liver in enzymic properties and molecular weight (125a)
20; ≈15	DPNH: ≈1.3 (126) (recalculated value)	DPN⁺: ≈30 (126) (recalculated value)	127		At protein concentrations below 2 mg/ml the enzyme molecule dissociates into 3 subunits with molecular weight of about 300,000; dioxane causes dissociation into subunits with molecular weights of 300,000 and 140,000 (126). The human liver enzyme possesses approximately the same properties as the pig liver enzyme (116, 126)
			128		
≈10; ?			129, 132, 133		Hybridization between two inactive mutant proteins causes an enzymically active complementation enzyme. From the experiment it was concluded that each mutant protein consists of four subunits each containing two identical polypeptide chains (131).
			136, 137		

Enzyme (DH = Dehydrogenase)	Reaction	Source	Molecular weight	Molecular weight of the polypeptide chain	Equivalent weight per coenzyme binding site
Glyceraldehyde-3-phosphate DH	D-Glyceraldehyde 3-phosphate + HPO_4^{2-} + PN^+ $\rightleftharpoons$ 1,3-diphospho-D-glycerate + PNH + H^+	Baboon muscle (138)			
		Beef skeletal muscle (139,140)	$(143)^c$		
		Cat skeletal muscle (139)			
		Chicken muscle (141)		$(143)^c$	
		Dog skeletal muscle (139)			
		Human erythrocytes (144)			
		Human heart muscle (141)		$(143)^c$	
		Pig skeletal muscle (139)	146,000 (145)	$37,000^{b,c}$ (146) (145)	40,000 (147)
		Rabbit skeletal muscle (152)	144,000 (145)	$35,000^c$ (145,153)	35,000–39,000 (85,153–155)
		Pheasant breast muscle (141)			
		Turkey muscle (141)			
		Crawfish muscle (163)			
		Halibut muscle (141)		$(143)^c$	
		Lobster tail muscle (141)	140,000 (165)	35,000 $(143,165)^c$	
		Sturgeon muscle (141)		$(143)^c$	
		Bacillus stearothermophilus (166)			
		Escherichia coli (141)			
		Yeast (167)	144,000 (146)	35,000 (146, $153^{b,c,d}$	37,000–40,000 (146,168)

n^a	$K_{E,R}$ (μM) $\left(= \dfrac{[E][R]}{[ER]}\right)$	$K_{E,O}$ (μM) $\left(= \dfrac{[E][O]}{[EO]}\right)$	References to kinetic work and enzymic properties	References to amino acid analysis and structure investigations	Remarks
			141	142	$s^0_{20,w} = 7.43S$ (141)
			139		
			141	141, 143	$s^0_{20,w} = 7.38S$ (141)
			139		
			141	141, 143	$s^0_{20,w} = 7.57S$ (141)
4; 4		DPN$^+$: 0.4 (pH = 8.5) (147)	148, 149	146	Zinc content: 2.4–2.8 moles Zn^{++} per mole glyceraldehyde-3-phosphate DH (150,151).
4; 4	DPNH: 0.24 (pH = 7.1) (156)	DPN$^+$: 0.06 (pH = 7.1) (156); AP-DPN$^+$: 28 (pH = 7.4) (157)	141, 148, 152, 157–159	153, 160–162	
			141	141	$s^0_{20,w} = 7.47S$ (141)
			141,159	141	$s^0_{20,w} = 7.38S$ (141)
					$s^0_{20,w} = 8.5S$ (164)
			141	141,143	$s^0_{20,w} = 7.59S$ (141)
4; ?			141, 159	141, 143, 165	$s^0_{20,w} = 7.45S$ (141)
			141, 159	141, 143	$s^0_{20,w} = 7.43S$ (141)
					Markedly resistant to thermal inactivation at 90°C (166)
			141	141	$s^0_{20,w} = 7.36S$ (141)
4; 4		DPN$^+$: 150 for the enzyme in the T state, 53 for the R state (pH = 8.5 at 40°C) (168); 7.8 (pH = 7.0 at 26°C) (170)	141, 167, 169	141, 146, 153, 171	$s^0_{20,w} = 7.47S$ (141). Zinc content: 2.3–2.7 (150, 151) and < 0.1 (61) respectively, moles Zn^{++} per mole enzyme.

(*continued*)

TABLE

Enzyme (DH = Dehydrogenase)	Reaction	Source	Molecular weight	Molecular weight of the polypeptide chain	Equivalent weight per coenzyme binding site
Glycerol-3-phosphate DH	L-Glycerol 3-phosphate + PN$^+$ ⇌ dihydroxyacetone phosphate + PNH + H$^+$	Rabbit muscle (172)	78,000 (173)	40,000 (174)	39,000 (173,175)
Glyoxylate reductase	Glycollate + PN$^+$ ⇌ glyoxylate + PNH + H$^+$	Tobacco leaves (178)			
Histidinol DH	L-Histidinol + PN$^+$ ⇌ L-histidinal + PNH + H$^+$ L-Histidinal + PN$^+$ + H$_2$O ⇌ L-histidine + PNH + H$^+$	*Salmonella typhimurium* (179)	≈ 75,000 (179)		
β-Hydroxyacyl-CoA DH	L-β-Hydroxyacyl-CoA + PN$^+$ ⇌ β-ketoacyl-CoA + PNH + H$^+$	Pig heart (180)			
3-α-Hydroxy-steroid DH	Androsterone + PN$^+$ ⇌ 5-α-androstane-3,17-dione + PNH + H$^+$ (also acts on other 3-α-hydroxysteroids)	*Pseudomonas testosteroni* (181)	47,000 (181)		
3 (or 17)-β-Hydroxy-steroid DH	Testosterone + PN$^+$ ⇌ Δ^4-androstene-3,17-dione + PNH + H$^+$ (also acts on other 3-β- or 17-β-hydroxysteroids)	*Pseudomonas testosteroni* (181)	100,000 (181)		
20-β-Hydroxy-steroid DH	20-β-Hydroxysteroid + PN$^+$ ⇌ 20-keto-steroid + PNH + H$^+$	*Streptomyces hydrogenans* (184)	118,000 (185)		
L-Iditol DH	Iditol + PN$^+$ ⇌ L-sorbose + PNH + H$^+$	Sheep liver (187)	115,000 (187)		
Isocitrate DH	Isocitrate + PN$^+$ ⇌ oxalsuccinate + PNH + H$^+$ Oxalsuccinate ⇌ α-ketoglutarate + CO$_2$	Pig heart (188, 189)	61,000–64,000 (188,189)		≈ 32,000 (190)

I (*continued*)

n^a	$K_{E,R}$ (μM) $\left(= \dfrac{[E][R]}{[ER]}\right)$	$K_{E,O}$ (μM) $\left(= \dfrac{[E][O]}{[EO]}\right)$	References to kinetic work and enzymic properties	References to amino acid analysis and structure investigations	Remarks
2; 2	DPNH: 1.8 (pH = 7.5, an additional binding of 31 moles of DPNH per mole of enzyme with $K_{E,R} = 600\mu M$ was found) (173)		172, 173, 176	177	
			178		
			179		
			180		
			182, 183	181	
			182, 183	181	
			182, 186, 186a		
			187		
?; 2	TPNH: ≈ 0.01 (pH = 7.1) (191)	TPN$^+$: ≈ 2 (pH = 7.1) (191)	190, 192–197		The enzyme possesses both dehydrogenase and carboxylase activity (188).

(*continued*)

Enzyme (DH = Dehydrogenase)	Reaction	Source	Molecular weight	Molecular weight of the polypeptide chain	Equivalent weight per coenzyme binding site
Lactate DH[e] (See Section II-A-3)	L-Lactate + PN[+] ⇌ pyruvate + PNH + H[+]	Beef (heart muscle, skeletal muscle) (198)	135,000– 144,000 (199–201)	34,000–35,000 (200,202)	35,000–40,000 (203,204)
		Chicken (heart muscle, skeletal muscle) (218)	146,000– 154,000 (218)	37,000 (218,219)	≈ 35,000 (203)
		Human (brain, erythrocytes, heart, liver, kidney) (222–224)	140,000 (222)		
		Pig (skeletal muscle, heart muscle) (227,228)	144,000 (229)	36,000 (214)	36,000–39,000 (85,229)
		Rabbit (skeletal muscle) (236)	132,000 (237)		33,000 (237)
		Rat (heart, liver, kidney, muscle) (244–246)	125,000		34,000 (204)
		Bacillus subtilis (250)	146,000 (250)	36,000 (10,250)[b]	

n^a	$K_{E,R}$ (μM) $\left(= \dfrac{[E][R]}{[ER]}\right)$	$K_{E,O}$ (μM) $\left(= \dfrac{[E][O]}{[EO]}\right)$	References to kinetic work and enzymic properties	References to amino acid analysis and structure investigations	Remarks
4; 4	DPNH: 0.39 for H$_4$-LDH (pH = 7.4) (204, cf. also 205); 0.35 (pH = 7.1) (156) AP-DPNH: 0.07 (H$_4$- and M$_4$-LDH) (pH = 7.0) (203)	DPN$^+$: 80 (pH = 8.15); AP-DPN$^+$: 9 (pH = 8.15) (205)	32, 199, 206–215, 230	201, 210, 216	In contrast to H$_4$-LDH, the binding sites in M$_4$-LDH or in the hybrid molecules (M$_3$H, M$_2$H$_2$, MH$_3$) do not act independently of one another (204). From the comparison of kinetic and sedimentation data it is assumed that at low protein concentrations a substrate-dependent equilibrium between an inactive dimer (H$_2$: molecular weight = 72,000) and an active tetramer (H$_4$: molecular weight = 144,000) exists (217, see also ref. 201).
4; 4	DPNH: 0.13 (H$_4$-LDH) and 1.0 (M$_4$-LDH) (pH = 7.0) (203); AP-DPNH: 0.2–0.3 (H$_4$- and M$_4$-LDH) (pH = 7.0) (203)		209–211, 216, 218–221	210, 216, 218	
			221–225	214, 226	
4; 4	DPNH: 0.5 (pH 7.2) (229)		231–232	214, 226, 233–235	
?; 4	DPNH: 3.2 (pH 6.9) (237)	DPN$^+$: 910 (pH = 6.9) (237)	49, 215, 225, 232, 238–243	214, 233, 234	
?; 4	DPNH: 0.4 for the two very tight-binding sites (pH = 8.6) (246,247)	DPN$^+$: 16–63 for the two very tight-binding sites (pH = 8.6) (246,247)	221, 238, 245, 248	233, 234, 249	
4; ?			10, 251	10, 250	At pH 8.8 dissociation into inactive subunits with molecular weight 72,000 occurs (250).

(*continued*)

Enzyme (DH = Dehydrogenase)	Reaction	Source	Molecular weight	Molecular weight of the polypeptide chain	Equivalent weight per coenzyme binding site
Malate DH	L-Malate + PN$^+$ ⇌ oxaloacetate + PNH + H$^+$	Beef heart mitochondria (252,253)	62,000–65,000 (252,253)		
		Beef heart supernatant fraction (257)	52,000 (257)		52,000 (258)
		Horse heart mitochondria (259)	≈ 70,000 (259,260)		34,000 (260)
		Pig heart mitochondria (259)	70,000–85,000 (85,260)	51,000^d (259)	35,000–42,000 (85,260)
		Rat liver (264)	50,000 (264)		50,000 (264)
		Bacillus subtilis (266)	148,000 (10,266)	37,000 (10,266)^b	
		Neurospora crassa (267)	54,000 (267)	13,500 (268)	50,000 (267)
Mannitol DH	D-Mannitol + PN$^+$ ⇌ D-fructose + PNH + H$^+$	*Lactobacillus brevis* (270)			
Mannitol-1-phosphate DH	D-Mannitol 1-phosphate + PN$^+$ ⇌ D-fructose 6-phosphate + PNH + H$^+$	*Aerobacter aerogenes* (271)	40,000 (271)		
		Escherichia coli (272)	25,000 (272)		
Phosphogluconate DH	6-Phospho-D-gluconate + PN$^+$ ⇌ D-ribulose 5-phosphate + PNH + CO$_2$ + H$^+$	*Candida utilis* (273)	≈ 100,000 (274)		

n^a	$K_{E,R}$ (μM) $\left(= \dfrac{[E][R]}{[ER]}\right)$	$K_{E,O}$ (μM) $\left(= \dfrac{[E][O]}{[EO]}\right)$	References to kinetic work and enzymic properties	References to amino acid analysis and structure investigations	Remarks
			252–254	255, 256	
?; 1	DPNH: 0.53 (pH = 8.5) (258)		253, 255, 257	255	
?; 2			259	259	$s_{20,w} = 4.36S$ (259)
$\approx$2; 2	DPNH: 1.0 (pH = 7.15) (261)	DPN$^+$: 280 (pH = 7.15) (261)	32, 176, 259, 262	259	For the supernatant malate DH see reference 263.
?; 1	DPNH: 0.1 (pH = 9) (264)		265		
4; ?			10, 266	10, 266	
4; 1			267, 269	267–269	Three of the four polypeptide chains are identical ("composition" $\alpha_3\beta$) (268). At pH = 2.8 reversible dissociation into inactive subunits occurs: $\alpha_3\beta \rightleftharpoons \alpha_3 + \beta$ (268). The enzyme also possesses aspartate aminotransferase activity (269a). Malate DH of two genetically nonlinked malate mutants with lower enzymic activity differed from those of the wild type in molecular weight, amino acid composition, and kinetic properties (269).
			270		
			271		
			272		
			74, 273–276	274	No separation of the two steps of the reaction (oxidation and decarboxylation) was detected (273)

(*continued*)

TABLE

Enzyme (DH = Dehydrogenase)	Reaction	Source	Molecular weight
Ribitol DH	Ribitol + PN$^+$ $\rightleftharpoons$ D-ribulose + PNH + H$^+$	*Aerobacter aerogenes* (277)	100,000– 140,000 (277)
Sorbitol-6-phosphate DH	D-Sorbitol 6-phosphate + PN$^+$ $\rightleftharpoons$ D-fructose 6-phosphate + PNH + H$^+$	*Aerobacter aerogenes* (271)	
Tartronate semialdehyde reductase	D-Glycerate + PN$^+$ $\rightleftharpoons$ tartronate semialdehyde + PNH + H$^+$	*Pseudomonas ovalis* (280)	91,000 (281)
Uridine diphosphate glucose DH	UDP-glucose + 2PN$^+$ + H$_2$O $\rightarrow$ UDP-glucuronic acid + 2PNH + 2H$^+$	Calf liver (282)	300,000 (282)

[a] n = molecular weight of the DH/molecular weight of the polypeptide chain; molecular weight of the DH/equivalent weight per coenzyme binding site.
[b] Fingerprint analysis indicates that the polypeptide chains are identical.

be enzymically active, provided that the single polypeptide chain can possess the same conformation as it does in the associated molecule.

Aldose dehydrogenase and galactose dehydrogenase (13) dissociate spontaneously at low protein concentrations into the polypeptide chains and are still active. Unfortunately, the number of coenzyme binding sites per molecule of these enzymes is not known. In contrast to aldose and galactose dehydrogenase, in the case of the other dehydrogenases investigated, one can observe in the native state only the associated molecules; it has thus far been impossible to cause the other dehydrogenases to dissociate into enzymically active polypeptide chains.*

* The folding of a polypeptide chain is necessarily controlled by the specific bonding properties of the constituents of the polypeptide chain. Many experimental

I (*continued*)

$K_{E,R}$ (μM) $\left(= \dfrac{[E][R]}{[ER]}\right)$	$K_{E,O}$ (μM) $\left(= \dfrac{[E][O]}{[EO]}\right)$	References to kinetic work and enzymic properties	Remarks
DPNH: 47 (pH = 8.0), calculated from the published kinetic data (279)	DPN$^+$: 908 (pH = 8.0), calculated from the published kinetic data (279)	32, 278, 279	
		271	$s_{20} = 6.0S$ at 3.2 mg/ml (271)
		280	
		283–286	A single enzyme catalyzed the two-step oxidation with a yet unidentified intermediate which appears to differ from UDP-glucose only in the uracil moiety (286).

^{c} The partial elucidation of the amino acid sequence indicates that the polypeptide chains are identical.
^{d} End-group analysis indicates that the polypeptide chains are identical.
^{e} For other crystalline lactate dehydrogenases see references 216 and 221.

---

results support the hypothesis (287, cf. also 119) that the conformation of a protein molecule is simply a function of its amino acid sequence, i.e., all "structural information" is contained in the primary structure. [Renaturation of dehydrogenases after inactivation under denaturing conditions has been achieved with glucose-6-phosphate dehydrogenase (288), glutamate dehydrogenase (111,112), glycerol-3-phosphate dehydrogenase (176,177), lactate dehydrogenase (176,289,290), and malate dehydrogenase (176).] The stabilization of the native conformation by noncovalent bonds is not restricted to the interaction between the side chains within one polypeptide chain but is also possible between the side chains of different polypeptide chains. If in the native state only the associated molecule is observed then it can be assumed that the self-assembly process to the associated molecule (i.e., a spontaneous equilibrium association) leads to the minimum of free energy and therefore the single polypeptide chain is not stable.

In other cases—e.g., with lactate dehydrogenase from beef heart (217) or *Bacillus subtilis* (250)—where dissociation into inactive subunits* is observed it must be assumed that the specific three-dimensional structure which is essential for the catalytic activity is stable only in the associated state. On the other hand, glutamate dehydrogenase from beef liver can dissociate into subunits which are still active but exhibit different specific activities (cf. Section II-A-2).

In principle the possibility also exists that several polypeptide chains participate in one active site. In this case only the associated molecule can manifest enzymic activity.

From the experimental results it was deduced that the coenzyme binding sites and the active sites can act independently of one another, for instance in liver and in yeast alcohol dehydrogenase (14). However, in glyceraldehyde-3-phosphate dehydrogenase from yeast, strong cooperativity between the polypeptide chains was observed (168, cf. also Section III-C-2-h). Furthermore in beef M_4-LDH or in the hybrid isozyme molecule (204) the binding sites for DPNH do not act independently of one another.

2. Relationship between Quaternary Structure and Activity for Glutamate Dehydrogenase from Beef Liver

Conformational changes including changes of the quaternary structure can alter the enzymic properties of an enzyme molecule. Thus, conformation change can function the regulation of enzyme activity, i.e., regulation of the synthesis of an end product of a metabolic pathway ("end-product or feedback inhibition"). Many such *regulatory* or *allosteric proteins* are known (291). One of the most intensively studied of all allosteric proteins is glutamate dehydrogenase from beef liver. Besides this enzyme, glucose-6-phosphate dehydrogenase (74, 292) and homoserine dehydrogenase (293) have been shown to fall into this category of enzymes.

In its native associated state, glutamate dehydrogenase from beef liver is a prolate particle with a molecular weight of 2×10^6 (4,78,111). It exists in a fully reversible association–dissociation equilibrium with its subunits, the molecular weights of which are 1×10^6, 0.5×10^6, and 0.25×10^6 (Fig. 1); and all of which are enzymically active (4,76,81,106, 116–118).

Nucleotides (Fig. 2a), phenanthroline, inorganic ions, steroids, thyroxine (Fig. 2b), sulfonylurea derivatives, and even DPN, the coenzyme

---

* A protein molecule consisting of more than two polypeptide chains can dissociate stepwise into the polypeptide chains. A dissociation product that still contains at least two polypeptide chains will be referred to as a *subunit* (cf. 119).

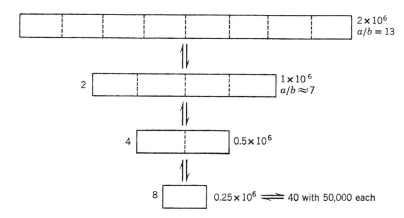

Fig. 1. Schematic representation of the association–dissociation equilibrium of glutamate dehydrogenase from beef liver (a/b = axial ratio). From Sund (4).

(Fig. 2a), can displace the association–dissociation equilibrium in one direction or the other. In this manner they act as inhibitors or activators of the enzyme reaction (4,88,94,95,100,109,294–296). The dissociation leads to an increase in the binding capacity for the coenzyme (83,84,86). From diffusion, sedimentation, and viscosity data it was concluded that the dissociation involves transverse cleavages (4,78, Fig. 1).

Some discrepancies exist with respect to the molecular weight of the smallest subunit formed during the spontaneous dissociation. For the native enzyme the molecular weight was estimated to be 250,000 (76,117,118), but after acetylation the molecular weight was found to be 400,000 (297). At the moment it is not clear whether the acetylated subunit is composed of two subunits with molecular weights of about 250,000 or whether the calculation of the molecular weight of the native enzyme from molecular sieve chromatography and zonal analytical centrifugation data are based on incorrect assumptions with respect to molecular shape, thereby leading to a molecular weight which is too low.

In the liver, glutamate dehydrogenase occurs almost entirely within the mitochondria. Here the enzyme concentration is very high (between 2 and 9 mg/ml), i.e., in a range in which the association–dissociation equilibrium and the influence on it of coenzymes, nucleotides, and the other compounds mentioned are significant (4,84,119).

Glutamate dehydrogenase is not completely specific for glutamic acid; it catalyzes, although at a much slower rate, the oxidative deamination of other amino acids. The oxidation of alanine has been particularly closely studied in connection with the quaternary structure (296). All compounds or experimental conditions that favor spontaneous dissociation

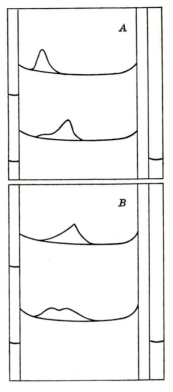

Fig. 2a. Sedimentation diagrams illustrating the effect of (A) ATP and (B) ADP on the dissociation of beef liver glutamate dehydrogenase by DPNH. In both experiments, two double sector cells were used; the lower pattern shows the effect of DPNH alone, while the upper pattern is the result of addition of either ADP or ATP in the presence of DPNH. Sedimentation at 42,040 rpm in 0.1M Tris-acetate buffer pH 8.0; enzyme concentration = 3.2 mg/ml. (A): ATP = 500μM, DPNH = 380μM, temperature 9°. (B): ADP 200μM, DPNH = 460μM, temperature 4°. From Frieden (379).

of the enzyme into subunits were found to cause an activation of the oxidation of alanine and an inhibition of the oxidation of glutamic acid. On the other hand, all compounds that favor spontaneous association activate the oxidation of glutamic acid and inhibit that of alanine. Contrary to earlier views (294,296,298) the basis of the relationship between the effects on quaternary structure and enzyme activity is the fact that the effector molecules react with the functional groups of the enzyme molecule and/or alter the conformation of the enzyme. Although both the

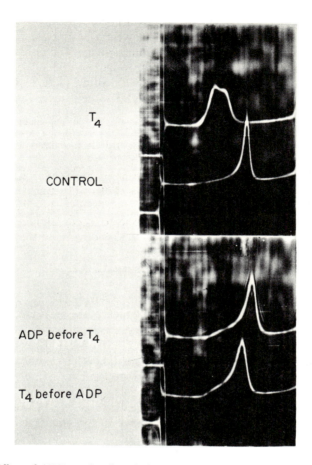

Fig. 2*b*. Effect of ADP on the dissociation of beef liver glutamate dehydrogenase produced by thyroxine (T_4). Sedimentation at 59,780 rpm in 0.05M Tris-HCl buffer pH 8.2, temperature 12.2° for the top two curves and 10.2° for the lower two curves. Thyroxine = 270μM, ADP = 560μM. From Wolff (109).

association–dissociation equilibrium and enzymic catalysis are affected, there is no direct connection between particle size and enzyme activity; both phenomena simply result from the same cause (4,84,98,100). In the very dilute solutions in which enzyme activity is usually assayed, glutamate dehydrogenase exists almost completely in the dissociated form. From this it follows directly that the subunit with the molecular weight of 250,000 is enzymically active.

In the presence of urea or long-chain alkyl sulfate, or at extreme pH values, the subunit having a molecular weight of 250,000 dissociates into polypeptide chains, with loss of enzymic activity. The polypeptide chains obtained in this way have a molecular weight of about 50,000, and end-group analysis indicates that they are all identical (4,79,80,82,109). It has been shown that this dissociation is also largely reversible, again with reactivation (111,112, Fig. 1).

3. Lactate Dehydrogenase and Its Isoenzymes

In 1952 it was observed that beef heart lactate dehydrogenase consists of two enzymically active components (199). Subsequently it could be shown (233,234) that various tissues can contain up to five lactate dehydrogenases and that it is very rare for one to occur alone. These lactate dehydrogenases can be separated from one another, and they are found to differ in their amino acid composition, their heat stability, and their enzymic and immunological properties (221,222,225,226,233,234,245, 299–301,304,305).

In the presence of $5M$ guanidine hydrochloride, lactate dehydrogenase dissociates into four subunits, each of which possesses a molecular weight of 34,000 (200). In general, every animal species contains two basic types of lactate dehydrogenase subunits (200,218,221,245,300,301). One of these occurs mainly in the muscle (M type), the other mainly in the heart (H type); the subunits are accordingly denoted by M and H. Thus the "formulas" of the lactate dehydrogenases containing four identical subunits are M_4 and H_4. Various combinations of the M and H subunits give (in vivo as well as in vitro) a maximum of five lactate dehydrogenases, or LDH isoenzymes: H_4, H_3M, H_2M_2, HM_3, and M_4. This is confirmed by fingerprint analysis (218,290,299). It was shown (302) that M_4-LDH from rat liver and H_4-LDH and H_3M-LDH from beef heart contain 6–8 moles of acetyl groups per mole of lactate dehydrogenase. Very probably these groups blocked the N-terminal groups, and therefore the large number of acetyl residues suggests the presence of more than one polypeptide chain per subunit of molecular weight of 34,000 (see refs. 214, 303).

The dissociation of lactate dehydrogenase into subunits can be largely reversed, with reactivation of the enzyme (176,289). Moreover, hybridization between lactate dehydrogenases from various sources, e.g., between bovine H_4-LDH and chicken H_4-LDH or frog M_4-LDH, can be carried out in vitro. On the other hand, hybridization between frog H_4-LDH and frog M_4-LDH has not been possible (221,290,304). Since hybridization occurs between lactate dehydrogenases from widely divergent species it must be assumed that during evolution there has been a considerable degree of conservation of the structural features which are required for tetramer formation. No information is available which would indicate that the subunits by themselves are enzymically active. The interaction between the four subunits seems to be essential for enzymic activity.

The enzymic and immunological properties of the hybrid molecules H_3M, H_2M_2, and HM_3 are roughly equivalent to an additive combination of the contributions from the H and M subunits (218,221,222,308, Fig. 3). For example, substrate inhibition is strong in the case of H_4-LDH and weak in the case of M_4-LDH; inhibition of the hybrid lactate dehydrogenases lies somewhere between these extremes (225,299,305). In agreement with this, the H type favors the aerobic oxidation of pyruvic acid and occurs mainly in "aerobic" tissue (heart muscle). The M type, on the other hand, favors the anaerobic pyridine nucleotide-dependent formation

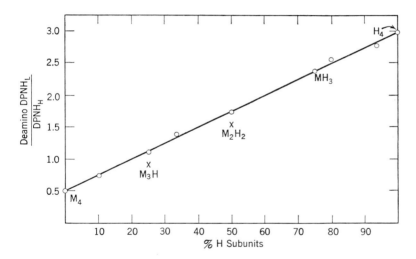

Fig. 3. Catalytic stoichiometry of hybrid LDH activity. Relationship of the ratio deamino $DPNH_L/DPNH_H$ for chicken LDH to the percentage of H subunits. (○) Values obtained when different proportions of crystalline H_4 and M_4 are added. (×) Values obtained with hybrid molecules isolated by electrophoresis from crude extracts. The data show that the subunits have the same catalytic action whether they exist separated in pure tetramers or together as hybrid tetramers. (Deamino $DPNH_L$ = activity with hypoxanthine analog of DPNH, measured at low pyruvate concentration; $DPNH_H$ = activity with DPNH, measured at high pyruvate concentration.) From Kaplan (221).

of lactic acid, and occurs mainly in "anaerobic" tissues (skeletal muscle) (299,305).

The lactate dehydrogenase isoenzyme pattern changes during ontogeny (299,306,307,309). In the embryonic state one type (H or M) predominates. In some species it is M, in others H ("embryonic LDH"). With increasing cell differentiation an isoenzyme pattern develops which is characteristic of the differentiated cell. It is assumed that the synthesis of H and M subunits is controlled by different genes (299).

B. Primary Structure of the Active Centers; Primary Structure and Evolution

In general, PN enzymes are inhibited by sulfhydryl reagents (14,182, 210,216,229,231,310–313) and therefore such enzymes have been termed "sulfhydryl" enzymes. In the sulfhydryl enzymes SH groups play, either directly or indirectly, an important role in the catalytic activity. However, in most cases the exact role of the SH groups is not known. In some enzymes SH groups have been shown to be located in the active center and/ or directly involved in the binding of the substrate; in these cases SH groups have a primary role in the catalytic process. On the other hand the SH groups may be involved only in maintaining that three-dimensional structure which is necessary for the catalytic properties of the enzyme. In each case chemical modification of the SH groups causes inactivation. For example, the interaction of yeast alcohol dehydrogenase with heavy metal ions causes its conformation to be altered in such a way that dissociation into subunits occurs (4,314–316).

As mentioned in Section II-A-1 (see also Table I) a great number of PN enzymes are built up of identical or nearly identical polypeptide chains and the molecular weights of the polypeptide chains approximate the equivalent weights of the coenzyme binding sites. The selective inhibition of alcohol dehydrogenases (16,317), lactate dehydrogenases (202,216,232), glutamate dehydrogenase (87), and glyceraldehyde-3-phosphate dehydrogenases (16,142,143,146,162,318) with [1-^{14}C] iodoacetic acid or with maleimide derivatives made it possible to label reactive cysteine residues in these enzymes. Each polypeptide chain contains only one unique reactive cysteine residue which probably belongs to an active center. After tryptic or peptic digestion the amino acid sequence around the reactive cysteine residues of these enzymes was elucidated (16).

The amino acid sequences around the reactive cysteine residues of glyceraldehyde-3-phosphate dehydrogenases from different sources which have been investigated are shown in Figure 4. Apart from differences in positions 1, 2, 3, and 16 the primary structure of all the octadecapeptides

Beef, pig, rabbit,
chicken, ostrich,
sturgeon, yeast
Lys-Ile-Val-Ser-Asn-Ala-Ser-Cys-Thr-Thr-Asn-Cys-Leu-Ala-Pro-Leu-Ala-Lys

Human Lys/Arg-Ile-Ile-Ser-Asn-Ala-Ser-Cys-Thr-Thr-Asn-Cys-Leu-Ala-Pro-Leu-Ala-Lys

Halibut Lys/Arg-Val-Val-Ser-Asn-Ala-Ser-Cys-Thr-Thr-Asn-Cys-Leu-Ala-Pro-Leu-Ala-Lys

Lobster Val-Met-Val-Ser-Asn-Ala-Ser-Cys-Thr-Thr-Asn-Cys-Leu-Ala-Pro-Val-Ala-Lys

       1   5   (8)   10   15   18

Fig. 4. Comparison of amino acid sequence around the reactive cysteine residues (marked by a circle) in glyceraldehyde-3-phosphate dehydrogenase from different sources (143). Identical residues are enclosed by lines.

Lys-····-Cys-⑧····-Lys-Val-Ile-His-Asp-His-Phe-Gly-Ile-Val-Glu-Gly-Leu-Met-Thr-(Thr, Val, His)-Ala-Ile-Thr-Ala-Thr-Gln-Lys-㊷
1   ⑧   18   20   30   40   ㊷

Thr-Val-Asp-Gly-Pro-Ser-Gly-Lys-Leu
50 51

Fig. 5. Amino acid sequence around the reactive cysteine and lysine residues (marked by circles) in glyceraldchyde-3 phosphate dehydrogenase from pig skeletal muscle (142). For the sequence 1–18 see Figure 4. An identical sequence was found in the enzyme from beef skeletal muscle, at least for the sequence 1–32.

is identical. In spite of considerable differences in the overall amino acid composition of the enzyme proteins each enzyme appears to be composed of four identical polypeptide chains with the same molecular weight (141,145). Strikingly, the primary structure in the neighborhood of the reactive cysteine residues has been conserved to a remarkable extent during the evolution of the different species. The differences appear less radical if one considers that valine and isoleucine or leucine are homologous pairs of amino acids ("conservative substitutions," see ref. 319).

In addition to catalyzing the oxidative phosphorylation of glyceraldehyde 3-phosphate, glyceraldehyde-3-phosphate dehydrogenase also catalyzes the hydrolysis of acetyl phosphate and p-nitrophenyl acetate. This reaction involves the formation of an intermediate acetyl–enzyme compound which contains ϵ-N-acetyl-lysine in addition to S-acetyl-cysteine (318,320). The specific acetylation of the ϵ-amino group of a unique lysine residue (lys-42 in Fig. 5) during reaction of NAD-free enzyme occurs by intramolecular migration of an acetyl group from S-acetyl-cysteine (cys-8 in Fig. 5) rather than by direct reaction of the lysine residues with the substrate (16,318).

The amino acid sequence around the unique acetylated lysine residue (Fig. 5) shows that the reactive cysteine residue and the reactive lysine residue in the primary structure are separated by 33 amino acid residues. It appears that the two reactive residues are brought into close juxtaposition by folding of the polypeptide chain thus permitting acyl migration to take place. At the moment it is not possible to decide whether the migration occurs between the reactive groups of the individual chains or between the groups of different chains.

The amino acid sequence around the reactive cysteine residues of different lactate dehydrogenases has also been conserved to a remarkable extent during evolution although this is not the case with respect to the overall primary structure (216,232).

In marked contrast to the results obtained with the different glyceraldehyde-3-phosphate dehydrogenases, the amino acid sequence around the reactive cysteine residues in alcohol dehydrogenases from yeast and horse liver are significantly different (16,317, Fig. 6). The same amino acids are

(a) $\begin{cases} \text{Arg} \\ \text{Lys-Tyr-Ser-Gly-Val-Cys-His-Thr-Asp-Leu-His-Ala-Try-His-Gly-Asp} \end{cases}$

1   5   ⑥   10   15

(b) Val-Ala-Thr-Gly-Ile-Cys-Arg-Ser-Asp-Asp-His-Val-Thr-Ser-Gly-Leu

Fig. 6. Comparison of amino acid sequences around the reactive cysteine residues (marked by a circle) in alcohol dehydrogenase from (a) yeast and (b) horse liver (16).

found only in positions 4, 9, 11, and 15 of the peptide; some of the other amino acids (positions 3, 5, and 8) are structurally related but half the amino acids are unrelated. In connection with this it is not surprising that the properties of yeast and liver alcohol dehydrogenase, although catalyzing the same general reaction, differ more than the properties of the different glyceraldehyde-3-phosphate dehydrogenases or lactate dehydrogenases. The yeast enzyme (molecular weight 157,000) is composed of four polypeptide chains whereas the liver enzyme (molecular weight 84,000) has only two. The former dissociates under denaturing conditions into the polypeptide chains while the latter enzyme does not appear to do so (47). Also the coenzyme specificities (39) and substrate specificities (14) of the two alcohol dehydrogenases are very different. The actual pattern of metabolic events in liver is quite different from the process of alcoholic fermentation in yeast. The difference in the biochemical function of the two alcohol dehydrogenases is marked and the extent of the biological adaptation appears to be paralleled by the great chemical changes which have occurred in alcohol dehydrogenases during the evolutionary development of the respective enzyme proteins (16).

C. Dehydrogenases: Zinc Enzymes?

Since the detection of metal ions, notably zinc ions (41,60), in dehydrogenases, there have been many speculations about the role of metal ions in the reactions catalyzed by PN enzymes (4,14,21,41,151,322–326,335). First it seemed that zinc ions play a general role in these reactions: various metal-complex-forming agents inhibit the PN enzymes; and zinc ions were found in various enzyme proteins (41,323,326).* But careful investigations showed that lactate dehydrogenases from different sources (327,328,334; see also 222) as well as glyceraldehyde-3-phosphate dehydrogenase from rabbit muscle (330) do not contain zinc ions in stoichiometric amounts relative to the enzyme protein. Glyceraldehyde-3-phosphate dehydrogenase from swine skeletal muscle, on the other hand, was shown to contain 2.8 moles of Zn^{++} per mole of enzyme. Furthermore, under proper conditions this zinc was shown to be exchangeable for radioactive labeled zinc ions (151). The results obtained with the yeast enzyme are still in controversy (61,150,329). In glutamate dehydrogenase from beef liver a very small amount of zinc was found, 0.015–0.032% (41,323). This is equivalent to 0.1–0.3 zinc ions per polypeptide chain of molecular weight 50,000.

* In some instances other metal ions were found (326).

1. Alcohol Dehydrogenases

Various amounts of zinc have been found in alcohol dehydrogenases from liver and yeast (see Table I).

The zinc content of yeast alcohol dehydrogenase depends on the activity of the preparation investigated. During recrystallization, alcohol dehydrogenase loses some of its activity and some of the enzyme-bound zinc (59). Extrapolation to the fully active molecule gives an amount of four zinc ions per enzyme molecule. Therefore, in yeast alcohol dehydrogenase the number of zinc ions per molecule is the same as the number of binding sites for the coenzyme molecule or the number of polypeptide chains.

The 1:1 relationship of metal ion to active site found in yeast alcohol dehydrogenase is not valid for horse liver enzyme. This enzyme is built up of two polypeptide chains and contains two active sites but the zinc content was found to be 4.1 zinc ions per enzyme molecule (42). Two of these zinc ions can be exchanged for $^{65}Zn^{++}$ (42; cf. also 331). Removal of any of the four zinc ions is accompanied by loss of activity. This result shows that all four zinc ions are essential components of the enzymically active molecule but the existence of enzyme–zinc complexes of different stabilities suggests the possibility of different functions for the zinc ions.

Alcohol dehydrogenases belong to the group *metalloenzymes* (41,323), because the zinc ions are bound firmly enough to remain associated with the protein throughout the isolation procedure. The identity of the ligands which bind the zinc ions to the protein molecule is not clear. Comparison of the absorption spectra of cadmium alcohol dehydrogenase (in which the zinc ions of horse liver alcohol dehydrogenase are replaced by cadmium ions) and metallothionein (a cadmium-containing protein in horse kidney cortex) suggests that the cadmium is bound to alcohol dehydrogenase by a mercaptide linkage (325). With the assumption that the cadmium ions are bound to the same sites it follows that the zinc ions are also bound to alcohol dehydrogenase by mercaptide linkages (325; cf. also 14). On the other hand, from studies of the ultraviolet absorption spectra of 1,10-phenanthroline bound to horse liver alcohol dehydrogenase and of various mixed complexes of zinc ions with 1,10-phenanthroline and other ligands, it was concluded that three or four imidazole ligands provide the best model for the environment of the zinc ions in liver alcohol dehydrogenase (332).

Zinc ions appear in only one stable oxidation state. Thus they are probably not involved directly in transferring electrons in the oxidation–reduction reaction. Therefore the function of the zinc ions might be based on (*1*) participation of the metal ion in the binding of substrate and/or coenzyme, (2) activation of the enzyme–coenzyme–substrate complex(es), and (3) maintenance of that conformation of the protein molecule which is necessary for the catalytic properties. In the first two cases the metal ion would have a primary role in the catalytic process; in the third, a secondary role.

From investigations of the optical rotatory properties, the spectral properties, and the inhibition of alcohol dehydrogenase by complex-forming agents it has been concluded that the coenzyme molecule is bound at or near the zinc binding site (see ref. 325 and Section III-C). The zinc ions seem to be involved in the binding of coenzyme to enzyme. In addition, zinc ions seem to be an important factor in the stabilization of the conformation of yeast alcohol dehydrogenase. On incubation with 1,10-phenanthroline this enzyme lost not only its activity and the enzyme-bound zinc ions, but also dissociated into its constituent polypeptide chains (47). A strong correlation between enzymatic activity, zinc content, and degree of dissociation was observed.

A third functional role of zinc ions was suggested (14,333). Hydrogen transfer in the alcohol dehydrogenase reaction is analogous to that postulated for the Meerwein-Ponndorf-Verley-Oppenauer reaction with zinc substituting for aluminum. In the enzyme the zinc ion probably produces the necessary polarization of C-1 of the substrate.

2. Glyceraldehyde-3-phosphate Dehydrogenase

The role of zinc ions in glyceraldehyde-3-phosphate oxidation is not well understood at the moment. The enzyme from swine skeletal muscle, like the other enzymes, is probably made up of four identical polypeptide chains (cf. Section II-B) each containing one coenzyme binding site and one active center. Therefore one would expect to find a zinc content of at least four zinc ions per enzyme molecule. Only 2.8 have been found (151). The reason this figure is low might be that the enzyme preparations which were investigated were not fully active (see the discussion of yeast alcohol dehydrogenase in Section II-C-1).

It has already been pointed out that the primary structure around the reactive cysteine residues is identical in the enzymes from beef, pig, rabbit, and yeast. (The same is probably true for the amino acid sequence around the reactive lysine residues; see Section II-B.) From this point of view it can be assumed that the mechanisms of the reactions catalyzed by these enzymes are also very similar, if not identical. However, if zinc ions play an important role in the catalytic process, the mechanisms of the reactions catalyzed by the different glyceraldehyde-3-phosphate dehydrogenases must be different. The swine enzyme contains zinc; in the rabbit enzyme zinc is absent. The question whether the yeast enzyme contains zinc is still open (61,150,329).

3. Glutamate Dehydrogenase

Glutamate dehydrogenase contains only 0.1–0.3 zinc ions per poly-peptide chain (41,323). There is obviously no relation between this number and the number of polypeptide chains and the coenzyme binding sites (83). A detailed investigation is necessary to decide whether the zinc ion has a functional role in glutamate dehydrogenase or whether the small zinc content which has been found is only an impurity.

III. Enzymic Properties

A. Types of Reactions Catalyzed by PN Enzymes

Several hundred PN enzymes, differing in their substrate specificity and their molecular properties, have been purified. Among this bewildering array, however, a relatively small number of reaction types occur (ref. 5 and Reactions 3a–3h).

$$R''\!-\!\underset{\underset{H}{|}}{\overset{\overset{R'}{|}}{C}}\!-\!OH + PN^+ \rightleftharpoons R''\!-\!\overset{\overset{R'}{|}}{C}\!=\!O + PNH + H^+ \qquad (3a)$$

$R' = H$ $R' = CR_3$ $R' = CR_2OH$ $R' = R$ $R' = COOH$
$R'' = H, CR_3,$ $R'' = CR_3,$ $R'' = CR_2OH$ $R'' = NHR$ $R'' = CR_2COOH$
     CR_2OH, C—R,   C—R,
               $\overset{||}{O}$ $\overset{||}{O}$
     COOH, CH_2SH COOH
$(R = alkyl\ or\ H)$

$$H\!-\!\underset{\underset{O}{|}}{\overset{\overset{OH}{|}}{C}}\!\rightharpoondown + PN^+ \rightleftharpoons \underset{\underset{O}{|}}{\overset{\overset{O}{||}}{C}}\!\rightharpoondown + PNH + H^+ \qquad (3b)$$

$$R'\!-\!\underset{\underset{S\!-\!R''}{|}}{\overset{\overset{H}{|}}{C}}\!-\!OH + PN^+ \rightleftharpoons R'\!-\!\underset{\underset{S\!-\!R''}{|}}{C}\!=\!O + PNH + H^+ \qquad (3c)$$

$R' = H,\ \underset{\underset{O}{||}}{C}\!-\!H,\ \underset{\underset{O}{||}}{C}\!-\!R$

$$\underset{H}{R-\overset{\displaystyle H}{\underset{|}{C}}=O} + PN^+ + HA \rightleftharpoons R-\overset{\displaystyle O}{\underset{A}{C}} + PNH + H^+ \qquad (3d)$$

$$A^- = OH, PO_4H_2^-, CoAS^-, R \text{ also } H$$

(If $R = \overset{\displaystyle CH_2}{\underset{COOH}{|}}$ and $A^- = CoAS^-$ in Reaction 3d, $CH_3-CO-S-CoA + CO_2$
are formed.)

$$R-\overset{\displaystyle H}{\underset{\displaystyle H}{\underset{|}{\overset{|}{C}}}}-OH + 2PN^+ + HA \rightleftharpoons R-\overset{\displaystyle O}{\underset{A}{C}} + 2PNH + 2H^+ \qquad (3e)$$

$$A^- = OH^-, CoAS^-$$

$$HCOOH + PN^+ \rightleftharpoons CO_2 + PNH + H^+ \qquad (3f)$$

$$R'-\overset{\displaystyle H}{\underset{\displaystyle R''}{\underset{|}{\overset{|}{C}}}}-\overset{\displaystyle H}{\underset{|}{N}}-R'' + PN^+ \rightleftharpoons R'-\overset{}{\underset{\displaystyle R''}{\underset{|}{C}}}=N-R'' + PNH + H^+ \qquad (3g)$$

$$R' = H, R, R$$
$$R'' = R, R, COOH$$
$$R''' = R, R, H$$

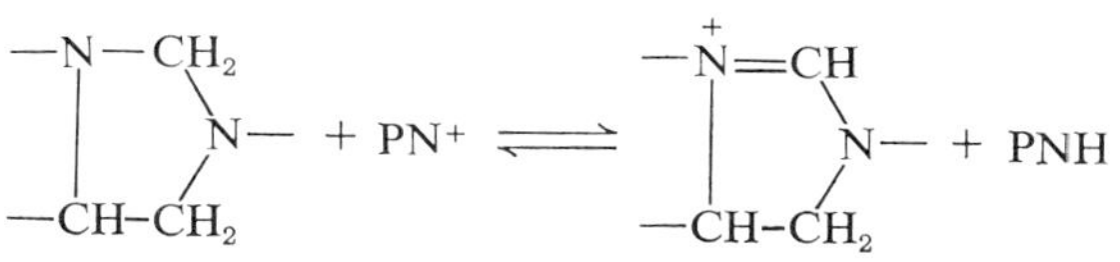

The majority of PN enzymes (e.g., alcohol dehydrogenases, polyol dehydrogenases, hydroxysteroid dehydrogenases, β-hydroxy acid dehydrogenases) catalyze reactions according to Equation (3a). In some cases in addition to the coenzyme and the "virtual" substrate, a third reactant is involved, e.g., water (Reactions 3d and 3e), phosphate (Reaction 3d), or coenzyme A (Reactions 3d and 3e).

The oxidations involving two coenzyme molecules (Reaction 3e) may be catalyzed by a single enzyme. Recently it was found (286) that in the reaction catalyzed by uridine diphosphate glucose dehydrogenase, an as yet unidentified intermediate is formed which appears to differ from UDP-glucose only in the uracil moiety.

Reactions (3g) and (3h) differ from the others in that the dehydrogenation occurs at a C—N function of the substrate molecule instead of a C—O function. Reaction-type (3g) includes those with folic acid derivatives and α-amino acids. The reversible oxidation of the latter probably proceeds in two steps. The oxidation of the α-amino acid to the α-imino acid (shown for glutamic acid in Reaction 4a) is followed by the hydrolysis of the imino acid to the α-keto acid (Reaction 4b). The free imino acid is not

$$HOOC—CH_2—CH_2—\underset{\underset{NH_2}{|}}{CH}—COOH + PN^+ \rightleftharpoons$$

$$HOOC—CH_2—CH_2—\underset{\underset{NH}{\|}}{C}—COOH + PNH + H^+ \quad (4a)$$

$$HOOC—CH_2—CH_2—\underset{\underset{NH}{\|}}{C}—COOH + H_2O \rightleftharpoons$$

$$HOOC—CH_2—CH_2—\underset{\underset{O}{\|}}{C}—COOH + NH_3 \quad (4b)$$

observed during the reaction (cf. refs. 336–338). It is possible that hydrolysis takes place while the imino acid is still bound to the enzyme.* According to another suggestion (340) an intermediate is formed in which an SH group of the enzyme is linked to the α-carbon of glutamic acid. Hydrogen transfer to the coenzyme followed by hydrolysis yields the hydrated imino acid which then decomposes to the α-keto acid. As yet no definite reaction pathway has been proposed which could be experimentally verified.

B. Substrate Specificity

Compared to the hydrolases and transferases, the substrate specificity of the PN enzymes has been neither systematically nor thoroughly investigated. In the cases which have been studied it was found that the PN enzymes generally do not possess a pronounced substrate specificity. In a few cases the enzymic activity is restricted to a few compounds.

From the many papers which deal with the substrate specificity of the individual enzymes only a few characteristic examples will be described. (For general information see references 3 and 6.)

1. Glutamate Dehydrogenase

First it was thought that glutamate dehydrogenase is exceedingly specific for glutamate or α-ketoglutarate, but later it was found that this enzyme can react also with other α-amino acids or α-keto acids (for review

---

* In the reaction catalyzed by D-amino acid oxidase free imino acid is formed during the course of the reaction. The rate-determining step is the dissociation of the imino acid from the enzyme (339).

see ref. 88). The activity with α-ketovalerate is about 25% that obtained with α-ketoglutarate. The rate with α-ketobutyrate or α-ketoisovalerate is only 2% compared to α-ketoglutarate under the conditions used. The relative activities of the different α-amino acids are pH dependent. At pH 9, relative to the oxidation of glutamate at pH 8, the rates were 2.3% for L-α-aminobutyrate, 1.7% for L-leucine, 1.6% for L-valine and D,L-norleucine, 0.95% for L-isoleucine, 0.82% for L-methionine, and 0.3% for L-alanine. Aspartate, α-aminoadipate, ornithine, and lysine are not active as substrates. From the results it is clear that either the incorrect positioning of the ω-carboxyl group or the absence of the second carboxyl group will cause a greatly diminished activity. Obviously the γ-carboxyl group of glutamate is not essential for activity although it plays an important role in the binding of the substrate to the enzyme. The oxidation of alanine has been intensively studied in connection with the conformation and the specific activity of the enzyme molecule (cf. Section II-A-2).

2. Aldehyde Dehydrogenases

Some aldehyde dehydrogenases have a pronounced substrate specificity while others are known to catalyze the oxidation of a wide variety of substrates (311,341–343). The oxidation of substituted benzaldehydes (342) and of aliphatic aldehydes (343) has been particularly well studied. The maximal rates of oxidation of aliphatic aldehydes have been found to correlate with the Taft sigma values (344) for the corresponding substituent groups. The higher the sigma value (i.e., the greater the electron withdrawal from the carbonyl carbon atom) the higher the maximal rate of oxidation (Fig. 7).* Therefore the rate seems to depend mainly on the electropositive nature of the carbonyl carbon atom and not on the steric properties of the substrate. Using substituted benzaldehydes (342) a biphasic curve is obtained which is apparently linear in each slope with a maximum at $\sigma^* = 0$ (if the logarithms of the maximal rates are plotted against the Hammet sigma values (345)).

3. Polyol Dehydrogenases, Hemiacetal Dehydrogenases

This class of enzymes shows a very broad substrate specificity (for review see ref. 347). For example, the aldose reductase from sheep seminal vesicles reduces D-glyceraldehyde with the same rate as D-glucoson. The maximal rates with pentoses (L-arabinose, D-ribose, D- or L-xylose) and

---

* The same relation was found for the reduction of substituted benzaldehydes by horse liver alcohol dehydrogenase (346).

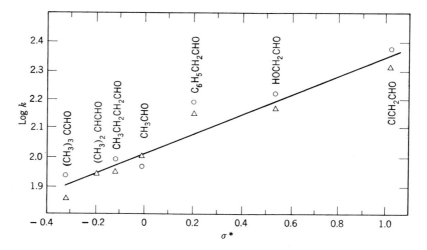

Fig. 7. Relation between the maximal rates (log *k*) of oxidation of various aliphatic aldehydes and the Taft *sigma* values (σ*). Rates are calculated as a percentage of the maximal rate obtained with propionaldehyde. (Δ) Aldehyde dehydrogenase purified from bovine brain mitochondria. (○) Aldehyde dehydrogenase purified from frozen-thawed bovine brain supernatant fluid. From Erwin and Deitrich (343).

hexoses (D-galactose, D-mannose, D-glucose) are about 50% that obtained with glyceraldehyde. Compared to D-glucoson the maximal rate in the presence of D-glyceroheptoses is 17–33%.

Detailed investigations of galactose dehydrogenase from *Pseudomonas saccharophila* have been published (72). This enzyme oxidizes galactopyranose to galactono-δ-lactone (Reaction 5), which rearranges at pH < 7 to

$$\text{β-D-Galactopyranose} + \text{DPN}^+ + \text{H}_2\text{O} \rightleftharpoons$$
$$\text{D-galactono-δ-lactone} + \text{DPNH} + \text{H}^+ \quad (5)$$

the γ-lactone whereas at pH > 7 it hydrolyzes to D-galactonate. Relative to the rate in the presence of D-galactose (100%) the rate with D-fucose is 100%, with 3-deoxy-D-galactose 40%, with 2-deoxy-D-galactose 10%, and with 4-deoxy-D-galactose, 3,6-dideoxy-D-galactose, D-galactosamine, L-arabinose, D-talose, and D-gulose between 10 and 1%. L-galactose, D-glucosamine, D-mannose, D-fructose, D-galacturonic acid, D-glucuronic acid, D-ribose, L-altrose, lactose, and maltose cannot serve as substrates for galactose dehydrogenase. From these results it follows that only those compounds can be oxidized which have the hydroxyl groups at carbon atoms 1, 2, and 3 as well as the hydroxymethyl group in equatorial positions; on the other hand, an axial disposition of the hydroxyl group at carbon atom 4 is necessary. With the exception of the hydroxyl group at carbon atom 1 the other hydroxyl groups are not essential. However, the reactivity is less if the hydroxyl groups are replaced by hydrogen.

4. Hydroxysteroid Dehydrogenases*

The 17β-hydroxysteroid dehydrogenase of human placenta (350,351) acts upon a large number of substituted derivatives of 17β-estradiol and estrone as substrates. It does not react with steroids bearing 3α-, 3β-, 11α-, 11β-, 16β-, 17α-, and 21-hydroxyl groups and is inert toward 3- and 7-ketones in the reduction reaction. This enzyme shows absolute steric specificity for the 17β-hydroxyl group; in the reduction reaction only the 17β-hydroxyl compound is formed. From a comparison of the reactivities of several estrone diastereomers it follows that the enzyme interacts with the entire steroid surface.

An enzyme preparation from *Pseudomonas testosteroni* catalyzes the oxidation of various 3β- and 17β-hydroxysteroids and only very slowly that of the 16β-hydroxysteroids (348). It is conceivable that the 3β-hydroxysteroid activity and the 17β-hydroxysteroid activity are due to the presence of distinct dehydrogenases which have not as yet been separated.

5. Alcohol Dehydrogenases (14,15,24,30,37,38,346,352–357)

The alcohol dehydrogenases from horse liver and yeast do not possess a pronounced substrate specificity. They can react with a great number of normal and branched-chain aliphatic and aromatic alcohols (both primary and secondary) and carbonyl compounds (14).

In aqueous solution an equilibrium between acetaldehyde and its hydrate exists. It was shown (51) that the real substrate of yeast alcohol dehydrogenase is the free and not the hydrated acetaldehyde.

With the exception of methanol in the straight-chain homologous series, the reactivity with the yeast enzyme decreases as the chain length increases. A straight-line relationship was obtained between the logarithm of the relative velocity and the number of carbon atoms, indicating the importance of the electronegativity and the dissociation ability of the carbinol group. Not only does the electronegativity influence the substrate specificity, but so does the substitution of the substrate. This effect is due to steric hindrance and orientation of the substrate on the enzyme surface. For instance, the oxidation of methyl-*n*-alkyl carbinols goes through a maximum at pentanol-2 (356).

* For review see references 186, 348, and 349.

With respect to the fact that the physiological function of horse liver alcohol dehydrogenase is still unknown, it is interesting that alcohol dehydrogenase preparations have been shown recently to exhibit substantial activity toward steroids (15,24,38). The only feature common to the 36 bile acids which were investigated was the presence of a 3β-hydroxyl or 3-keto group. It was established that liver alcohol dehydrogenase is stereospecific for 3β-hydroxy-5β-cholanic acid and that it is also capable of oxidizing primary alcohol groups in sterols. Kinetic studies indicate that the Theorell-Chance mechanism is not applicable to Reaction (6).

$$3\beta\text{-Hydroxy-5}\beta\text{-cholanic acid} + \text{NAD}^+ \rightleftharpoons 3\text{-Keto-5}\beta\text{-cholanic acid} + \text{NADH} + \text{H}^+ \tag{6}$$

Horse liver alcohol dehydrogenase exhibits multiple zones upon electrophoresis. Recently it was demonstrated (15,38) that a subfraction, which also was crystallized, possesses hydroxysteroid dehydrogenase activity ("LADH$_S$") in addition to alcohol dehydrogenase activity. The main fraction ("LADH$_E$") was inactive with hydroxysteroids. The only marked difference between the physicochemical properties of these two alcohol dehydrogenases was found to be a difference in their isoelectric points (6.8 for LADH$_E$ and 10.0 for LADH$_S$). It is thought that these dissimilar isoelectric points probably reflect differences in the amide group content (15). Studies of the inhibition of LADH$_S$ with 3α-hydroxy-5β-cholanic acid have shown this steroid to be a strongly competitive inhibitor in the reaction with 3β-hydroxy-5β-cholanic acid but practically without effect when ethanol was used as the substrate. These results have led to the hypothesis that LADH$_S$ has different binding sites for the two substrates, ethanol and 3β-hydroxy-5β-cholanic acid. The existence of two substrate binding sites would also explain why the Theorell-Chance mechanism is not valid if the hydroxysteroid is the substrate.

The transfer of hydrogen in the reactions catalyzed by alcohol dehydrogenases is stereospecific with respect to both the alcohol and the pyridine nucleotide (39). Enzymic reduction of monodeuteroacetaldehyde in the presence of DPNH (Eq. 7a) or of acetaldehyde in the presence of DPND$_A$ (Eq. 7b) gives the two enantiomeric monodeuteroethanols.

The steric course of the reduction of a great number of aldehydes and cyclic ketones of known absolute configuration has been determined

$$\text{D}-\overset{\displaystyle \|}{\underset{\displaystyle \text{O}}{\text{C}}}-\text{CH}_3 + \text{DPNH} + \text{H}^+ \rightleftharpoons \text{H}-\overset{\displaystyle \text{D}}{\underset{\displaystyle \text{CH}_3}{\text{C}}}-\text{OH} + \text{DPN}^+ \tag{7a}$$

$$(-)\text{-}(S)\text{-ethanol-1-}d$$

$$H\!-\!\underset{\underset{O}{\|}}{C}\!-\!CH_3 + DPND_A + H^+ \;\rightleftharpoons\; D\!\blacktriangleright\!\underset{\underset{CH_3}{|}}{\overset{\overset{H}{|}}{C}}\!\blacktriangleleft\!OH + DPN^+ \qquad (7b)$$

$(+)$-(R)-ethanol-1-d

(5,30,37,354). From studies with the ketone reductases from *Curvularia falcata* and pig liver (both having B stereospecificity with respect to the coenzyme ("B enzyme", cf. ref. 39), and horse liver alcohol dehydrogenase (having A stereospecificity, "A enzyme") a rule for the substrate specificity of the A and B enzymes was deduced which made it possible to specify allowed and forbidden positions of the substituents in the substrates. The carbon skeleton of the substrate molecule can be regarded as portions cut out of a diamond lattice. This section defines the space in the transition state of the reacting complex which is not occupied by the enzyme, coenzyme, or tightly bound solvent. Then the active site of the enzyme–coenzyme complex must in some way be the "negative" of the characteristic diamond lattice section. Predictions may be made, with caution, regarding the reactivity of hitherto unexamined substrates (30; for this problem see also refs. 24 and 37).

From the comparison of the rates of reactions with methylcyclohexanols and -hexanones and with α- and β-decalols and -decalones, it follows that the spatial structure of the enzyme protein is responsible for the specificity for substrates in which the transferred hydrogen has an equatorial or axial position of cyclohexane. Liver alcohol dehydrogenase and the ketone reductase from pig liver preferentially transfer the *equatorial* hydrogen ("*e* enzyme"); on the other hand, the ketone reductase from *Curvularia falcata* transfers preferentially the *axial* hydrogen of the substrate ("*a* enzyme").

Based on the substrate specificity and the coenzyme stereospecificity of the ketone reductases and liver alcohol dehydrogenase, the spatial arrangement of substrate and coenzyme in the transition state, shown in Figure 8, was postulated (5,30). Compounds in which R_2 is a carbon atom cannot be reduced by an enzyme having A stereospecificity; on the other hand, if R_{-2} is a carbon atom, these compounds cannot serve as substrates for enzymes with B stereospecificity.

6. Multiple Functions of Glyceraldehyde-3-phosphate Dehydrogenase*

The major physiological role of this enzyme is substrate level phosphorylation. However, in addition to this catalytic function, glyceraldehyde-3-phosphate dehydrogenase exhibits transacylase, hydrolase (hydrolysis of

* For a review, see reference 357.

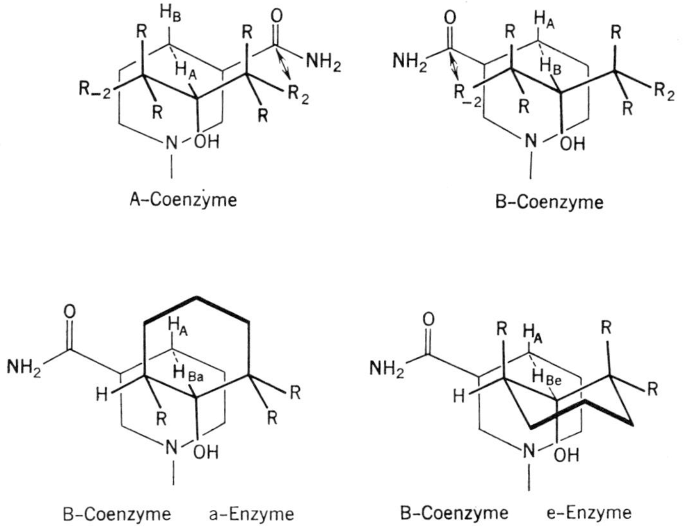

Fig. 8. Spatial arrangement of substrate and coenzyme in the transition state of the hydrogen transfer (for explanation see text). From Prelog (30).

acyl phosphates and of esters), hydratase (formation of DPNH-X), diaphorase, and transphosphorylase activities. The biological significance of these additional activities is only speculative, but the study of the multiple functions has been useful in the characterizations of the intermediate steps in the oxidation of glyceraldehyde-3-phosphate catalyzed by glyceraldehyde-3-phosphate dehydrogenase.

C. Coenzyme Specificity and the Binding between Enzyme and Coenzyme*

1. Coenzyme Specificity

In the cell the dehydrogenases use DPN or TPN as coenzymes. In some cases (e.g., glutamate dehydrogenase from beef liver (88) or horse liver alcohol dehydrogenase (29)) enzymes can react with both coenzymes. In other cases (e.g., L-leucine dehydrogenase from *Bacillus cereus* (358), isocitrate dehydrogenase from *Neurospora crassa* (359), or glycerol-3-phosphate dehydrogenase from rabbit muscle (173)), the dehydrogenase can work only with DPN, and TPN is not active as coenzyme. Glucose-6-phosphate dehydrogenase from yeast is an example of a TPN-specific

* For reviews, see references 5, 39, 321, and 357.

dehydrogenase (74). Some organisms contain distinct enzymes, one specific for DPN, the other specific for TPN (e.g., glutamate dehydrogenases from *Neurospora crassa* (360), *Fusarium oxysporum* (361), and yeast (88), or alcohol dehydrogenases from yeast (14)). In this connection it is interesting to note that the 3' isomer of TPN (phosphate group of the adenosine ribose in 3' position) cannot serve as coenzyme in those reactions which are catalyzed by DPN- or TPN-specific enzymes (39). In general the isomer will react only in systems which can function with either DPN or TPN. The 2'-monoester phosphate group appears to be essential for activity in enzymes which are absolutely specific for TPN.

Comparative studies on glutamate dehydrogenases from different sources have shown (362) that purine nucleotides strongly and specifically affect the enzymes from all animal sources investigated. On the other hand, glutamate dehydrogenases from non-animal sources, which in contrast to the enzymes from animals are specific either for DPN or TPN, are in general not influenced by purine nucleotides. It was suggested that in the more complex animal metabolism the role of purine nucleotides is to control the rate of utilization of one coenzyme (either DPN or TPN) relative to the other for the reaction catalyzed by glutamate dehydrogenase (362; cf. also 134, 363).

Besides the naturally occurring coenzymes, DPN and TPN, many "coenzyme analogs" have been synthesized *in vitro*. In the presence of these artificial coenzymes some of the PN enzymes exhibit a much higher activity compared to DPN or TPN; in other cases the activity is lower or no activity is observed (39). The coenzyme analogs are useful in the study of the enzyme–coenzyme interactions although there is no certain indication for the majority of these analogs that there exists a physiological role.

2. Binary Enzyme–Coenzyme Complexes and Ternary Enzyme–Coenzyme–Substrate and Enzyme–Coenzyme–Inhibitor Complexes

In the presence of DPN or TPN, PN enzymes form binary enzyme–coenzyme complexes and ternary enzyme–coenzyme–substrate and enzyme–coenzyme–inhibitor complexes which are important in the enzyme catalysis. These complexes can be demonstrated by physical methods: (*1*) spectrophotometric analysis, (*2*) fluorimetric analysis, (*3*) optical rotatory dispersion, (*4*) ultracentrifugation, and (*5*) equilibrium dialysis. The most frequently used method for the quantitative estimation of the number of binding sites and the dissociation constants is fluorimetric analysis. The last two methods are less precise than the others.

a. Ultraviolet Absorption Spectra of Binary Complexes. The maximum of the characteristic ultraviolet absorption band of DPNH is shifted from 340 mμ to 325 mμ in the presence of alcohol dehydrogenase from horse liver (14) or human liver (43) or lactate dehydrogenase from beef heart and rat liver (364). Similar shifts of 10–15 mμ toward shorter wavelengths appear in the presence of 3-AP-DPNH and Py-3-ald-DPNH (364). In the presence of malate dehydrogenase from pig heart the maximum of the enzyme–coenzyme complex is observed at 351 mμ compared to 340 mμ for the free coenzyme (365). The spectrophotometric analysis of the enzyme–coenzyme binding is restricted to these enzymes because the other PN enzymes which have been investigated do not cause a similar shift of the characteristic ultraviolet absorption of the reduced coenzyme. However, the characteristic absorption maximum of the addition compounds of DPN^+, 3-AP-DPN^+, or Py-3-ald-DPN^+ with cyanide, hydroxylamine, sulfite, and sulfhydryl compounds structurally related to the substrate is shifted to shorter wavelengths on binding to a number of PN enzymes which do not cause a spectral shift of the reduced coenzyme (14,364). These addition compounds can be regarded as dihydropyridine compounds and therefore as analogous to the reduced coenzyme molecule.

Changes in the ultraviolet absorption spectra occur also when horse liver alcohol dehydrogenase forms a binary complex with ARPPR (a coenzyme-competitive inhibitor with $K_{E,I} = 35\mu M$ at pH 7) and ternary complexes enzyme–DPN^+–caprate ($K_{EI,O} = 7.6\mu M$ between pH 6 and 9), enzyme–DPN^+–pyrazole, and enzyme–DPNH–isobutyramide (Fig. 9, Table II) (21,23).

In the complexes with ARPPR and DPN^+-caprate, an increase with its maximum at 281 mμ is followed by a decrease in the adenine-absorption region below 260 mμ (Figs. 9a and 9b). The difference spectrum from 290 mμ to 400 mμ (Fig. 9c) of the ternary enzyme–DPNH–isobutyramide complex reflects the shift of the dihydropyridine absorption from 340 mμ to 325 mμ, which is identical with the shift observed when DPNH is bound to liver alcohol dehydrogenase. The 281-mμ band is equal to that in the enzyme–ARPPR complex. It may be concluded that the positive 281-mμ band of the difference spectra (Fig. 9) and the negative band in the 260-mμ region result from the coupling of the ARPPR moiety with alcohol dehydrogenase. It is not clear if this effect is caused by a "red shift" of the adenine band or whether it is derived from a change in the absorption of the aromatic amino acid residues. However, the fact that these absorption changes were caused by the binding of the ARPPR moiety to liver alcohol dehydrogenase is emphasized by the finding that adenosine, which has

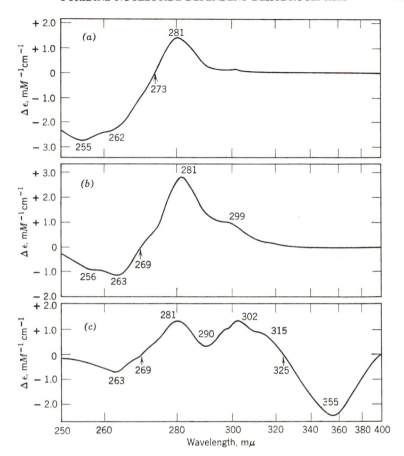

Fig. 9. Light absorption changes accompanying complex formation of horse liver alcohol dehydrogenase with coenzyme and inhibitor. Double difference spectra were recorded in sodium phosphate buffer, pH 7.0 (ionic strength 0.1) at the following concentrations of reactants: (a) 45.8μN alcohol dehydrogenase and 45.0μM ARPPR; (b) 46.3μN alcohol dehydrogenase, 50.0μM DPN$^+$, and 1000μM sodium caprate; (c) 80.0μN alcohol dehydrogenase, 80.0μM DPNH, and 0.1M isobutyramide. Scales at the ordinates are based upon the calculated values of different-extinction coefficients of these complexes. From Theorell and Yonetani (23).

practically no inhibitory effect, does not cause any increase in the 281-mμ region when mixed with liver alcohol dehydrogenase.

   b. *Fluorescence Spectra of Binary Complexes* (for reviews see 39, 366). The fluorescence intensity of the binary complex between glyceraldehyde-

TABLE II
Ternary Complexes between PN Enzymes, PNH, and
Reduced Substrates or Inhibitors

Enzyme (E)	Complex	λ_{max} of fluorescent emission[a] (mμ)	Remarks[b]	Ref.
Alcohol DH horse liver	E–DPNH– ethanol	≈ 430	$K_{ER,S} = 40mM$ ("abortive complex")	14, 368
	E–DPNH– imidazole	429	$K_{E,R} = 0.31\mu M$; pH $= 7$ $Q_{ER} = 13.5$ $K_{E,I} = 550\mu M$; $K_{ER,I} = 3500\mu M$; $Q_{ERI} = 47$ $K_{EI,R} = 2.03\mu M$	5, 14
	E–DPNH– acetamide	425	$K_{E,I} = 36000\mu M$; pH $= 7$ $K_{ER,I} = 5000\mu M$; $Q_{ERI} = 16.6$ $K_{EI,R} = 0.05\mu M$	5, 14
	E–DPNH– butyramide	425	$K_{E,I} = 1000\mu M$; pH $= 7$ $K_{ER,I} = 64\mu M$; $Q_{ERI} = 15.6$ $K_{EI,R} = 0.02\mu M$	5, 14
	E–DPNH- isobutyramide	425	$K_{E,I} = 9300\mu M$; pH $= 7$ $K_{ER,I} = 140\mu M$; $Q_{ERI} = 40$ $K_{EI,R} = 0.005\mu M$	5, 14
yeast	E–DPNH- ethanol	438		5
Dihydrofolate reductase murine lymphoma	E–TPNH– triampterene		$K_{E,R} = 0.05\mu M$; pH $= 7.5$ $K_{ER,I} = 0.005\mu M$ $K_{EI,R} = 0.002\mu M$	369

(*continued*)

TABLE II (*continued*)

Enzyme (E)	Complex	λ_{max} of fluorescent emission[a] (mμ)	Remarks[b]	Ref.
Glutamate DH beef liver	E–DPHN–L-glutamate	452	$Q \approx 2.5$; pH = 7.5	5
Lactate DH beef brain	E–DPNH–L-lactate	430	pH = 7.0	5
	E–DPNH-oxalate	420		
beef heart	E–DPNH–L-lactate	430 (pH = 6.88)	$K_{E,R} = 2\mu M$; $Q_{ER} = 7.6$ $K_{ER,S} = 40000\mu M$; $Q_{ERS} = 11.5$; pH = 7.1	370, 372
	E–DPNH–oxamate	440	$K_{ER,I} = 70\mu M$; $Q'_{ERI} < 1$ (pH = 6.0)	371
	E–DPNH–oxalate	≈ 425	$K_{ER,I} = 330\mu M$; pH = 7.3 and $K_{ER,I} = 170\mu M$; pH = 6.1 $Q'_{ERI} = 3.4$ pH = 6.01 (also formation of ternary complexes with other hydroxy acids)	5
rabbit skeletal muscle	E–DPNH–L-lactate	435	$K_{ER,S} = 38000\mu M$; $Q_{ERS} = 37^c$; pH = 7.6	237
	E–DPNH–oxalate	424	$Q_{ERI} \approx 3$; pH = 7.0	5

(*continued*)

TABLE II (*continued*)

Enzyme (E)	Complex	λ_{max} of fluorescent emission[a] (mμ)	Remarks[b]	Ref.
Malate DH beef brain	E–DPNH–L-malate	445	pH = 6.0	5
beef heart supernatant fraction	E–DPNH–D-malate	415–420	$K_{E,R} = 0.53\mu M$; pH = 8.5 $K_{ER,S} = 500\mu M$ $K_{ES,R} = 0.2\mu M$ (also formation of ternary complexes with other hydroxy acids)	258
pig heart	E–DPNH–D-malate	425	$K_{E,R} = 1.0\mu M$; $K_{E,I} = 15000\mu M$; $Q_{EI} = 2$–3 $K_{ER,I} = 6000\mu M$ $K_{EI,R} = 0.4\mu M$ $Q_{ERI} \approx 8$	261; cf. also 260

[a] Values corrected with respect to the characteristics of the instruments used in the investigations.

[b] Q = deflection ratio ER/R(Q_{ER}) and ERI/R(Q_{ERI}), respectively. Q' = deflection ratio ERI/ER(Q'_{ERI}) and ERS/ER(Q'_{ERS}), respectively. $K_{ER,I} = $[ER][I]/[ERI]; $K_{EI,R} = $[EI][R]/[ERI], etc.

[c] Calculated from the published data (237).

3-phosphate dehydrogenase from rabbit skeletal muscle and DPNH is lower than that of free DPNH (156). With this exception the fluorescence intensities of all other binary enzyme–PNH complexes are higher than those of PNH (Figs. 10 and 11) (5,39). On combination with reduced coenzymes the fluorescence of the enzyme proteins is quenched. This diminution in fluorescence, when excited at 280 mμ, is balanced by an appearance of PNH fluorescence ("sensitized fluorescence") indicating transfer of excitation energy from the aromatic amino acid residues of the protein to the bound reduced coenzyme. Either this or the quenching of the tryptophan fluorescence may be used to determine quantitatively the binding between enzyme and coenzyme. Even DPN$^+$, which is not itself

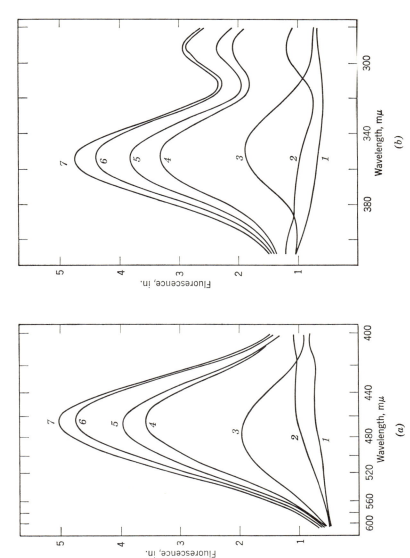

Fig. 10. Fluorescence spectra for beef liver glutamate dehyrogenase, DPNH, and glutamate in sodium phosphate buffer (ionic strength 0.1), pH 7.5, at 20°. The figure also shows the fluorescence spectra of the zinc complexes. (a) Emission spectra (activating wavelength 350mμ). (b) Activation spectra (measuring wavelength 468mμ). (1) Buffer; (2) 0.64 mg/ml glutamate dehydrogenase; (3) 0.90μM DPNH; (4) 0.64 mg/ml glutamate dehydrogenase, 0.90μM DPNH; (5) 0.64 mg/ml glutamate dehydrogenase, 0.90μM DPNH, 254μM glutamate; (6) 0.64 mg/ml glutamate dehydrogenase, 0.90μM DPNH, 47.7μM ZnSO₄; (7) 0.64 mg/ml glutamate dehydrogenase, 0.90μM DPNH, 254μM glutamate, 47.7μM ZnSO₄. From Sund (83).

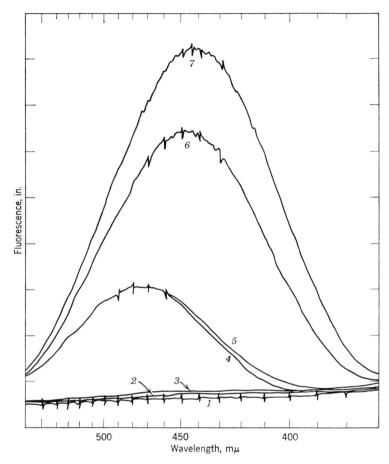

Fig. 11. Fluorescence emission spectra for horse liver alcohol dehydrogenase, DPNH, and imidazole in phosphate buffer (ionic strength 0.1), pH 7, at 23.5°. Excitation: 330 mμ. (1) Buffer; (2) 0.1M imidazole, (3) 2.2μN alcohol dehydrogenase; (4) 2.1μM DPNH; (5) 2.1μM DPNH, 0.1M imidazole; (6) 2.2μN alcohol dehydrogenase, 2.1μM DPNH; (7) 2.2μN alcohol dehydrogenase, 2.1μM DPNH, 0.1M imidazole. From Sund and Theorell (14).

fluorescent, can be titrated fluorimetrically since it quenches protein fluorescence.

The combination of lactate dehydrogenase from rabbit muscle with DPNH was investigated fluorimetrically with the temperature jump method (373). From the result it can be assumed that a fast bimolecular step ($k_1 \approx 10^9 M^{-1} \sec^{-1}$) is followed by a slow monomolecular inter-conversion of the ER complex ($k_3 = 1000 \sec^{-1}$) (Eq. 8). The dissociation constant, $K_{E,R}$, was found to be $6 \mu M$, the equilibrium constant, $K_{ER \to ER^*} \cdot (k_4/k_3)$, to be 0.3.

$$E + R \underset{k_2}{\overset{k_1}{\rightleftharpoons}} ER \underset{k_4}{\overset{k_3}{\rightleftharpoons}} ER^* \qquad (8)$$

   c. Binding Sites and Dissociation Constants. It has already been pointed out (Section II-A-1) that the molecular weight of the polypeptide chains generally approximate the equivalent weights of the coenzyme binding sites. There is very little doubt that the coenzyme binding site is indeed part of the active center of the PN enzymes because in many cases the physically determined dissociation constants of the enzyme–coenzyme complexes agree very well with the kinetically determined values (for instance, alcohol dehydrogenases (14), lactate dehydrogenases (212)). However, the kinetically determined dissociation constant does not agree with that obtained by fluorimetric analysis in the case of beef liver gluta-mate dehydrogenase (88). The reason for this discrepancy is not clear.

   Again, with the exception of glyceraldehyde-3-phosphate dehydrogenase from rabbit muscle (156), a generalization can be made with regard to the dissociation constants. The binding between enzyme and reduced coenzyme is stronger than the binding between enzyme and oxidized coenzyme (cf. Table I).

   From kinetic experiments and direct estimation of enzyme–coenzyme binding it may be concluded for the majority of PN enzymes that the binding sites act equally and independently of one another in those enzymes which contain at least two binding sites. Therefore it may be further con-cluded (367) that each coenzyme molecule quenches only the fluorescence of the tryptophan residues within its own domain and that there is neglig-ible overlap between domains. The enzyme proteins behave as though they possess n-fold symmetry corresponding to n binding sites with n equivalent subunits. However, this is not true of lactate dehydrogenase from beef skeletal muscle (M_4 and the hybrid molecules; cf. Section III-C-2-g). Of particular interest are two other PN enzymes. Glyceralde-hyde-3-phosphate dehydrogenase from yeast binds DPN^+ in two different protein conformations (R and T states) (cf. Section III-C-2-h). The coenzyme binding capacity of beef liver glutamate dehydrogenase is higher in the dissociated state than in the associated state (cf. Section II-A-2).

   d. Ternary Complexes. Fluorimetric methods demonstrate ternary complexes between the PN enzymes and (*1*) reduced coenzyme and reduced substrate (Fig. 10 and Table II), (2) oxidized coenzyme and oxidized substrate, and (3) oxidized or reduced coenzyme and inhibitor (Fig. 11 and Table II). For example, the formation of ternary complexes by the addition of imidazole (Fig. 11) or fatty acid amides to the binary liver alcohol dehydrogenase–DPNH complex causes further change in the fluorescence emission (14). The fluorescence of the binary enzyme–DPNH complex ($Q = 13.5$) increases after addition of imidazole ($Q = 47$), acetamide ($Q = 16.6$), propionamide ($Q = 20.3$), butyramide ($Q = 15.6$), and particularly isobutyramide ($Q = 40$). On the other hand, the fluorescence is quenched by formamide ($Q = 11.7$), valeramide ($Q = 8.9$), and hexamide ($Q = 11.1$). The maximum of fluorescence emission of the binary enzyme–DPNH complex is shifted toward shorter wavelengths upon formation of ternary complexes (Table II).

Amides and imidazole also form binary complexes with liver alcohol dehydrogenase, but amides do not form ternary complexes with the binary enzyme–DPN$^+$ complex. Imidazole also forms a ternary complex with enzyme–DPN$^+$. Furthermore, fatty acids do not form ternary complexes with enzyme–DPNH but do with the binary enzyme–DPN$^+$ complex and also form binary complexes with the enzyme. It was found (14) that all the fatty acids studied compete with the alcohol-binding site in the enzyme but not with the aldehyde-binding site, whereas all the amides studied compete with the aldehyde-binding site and not with the alcohol-binding site. The estimation of the dissociation constants describing the effect of the inhibitors, fatty acids or amides, on liver alcohol dehydrogenase ($K_{E,I}$) and on the binary enzyme–coenzyme complexes ($K_{ER,I}$ and $K_{EO,I}$), and of the coenzyme on the binary enzyme–inhibitor complexes ($K_{EI,R}$ and $K_{EI,O}$) shows that the inhibitor stabilizes the binding between liver alcohol dehydrogenase and DPN ($K_{EI,R} < K_{E,R}$ and $K_{EI,O} < K_{E,O}$) and likewise the coenzyme stabilizes the binding of the inhibitor to the enzyme ($K_{ER,I} < K_{E,I}$ and $K_{EO,I} < K_{E,I}$). The opposite was found for imidazole and therefore this compound can act as an activator of the reaction because it labilizes the enzyme–coenzyme binding and favors the dissociation of the coenzyme from the enzyme, the rate-determining step of the reaction.

The mutual stabilization of binding when the PN enzymes form ternary complexes with the coenzyme and either substrate or inhibitor was also observed with other enzymes, e.g., dihydrofolate reductase and malate dehydrogenases (Table II).

In this connection it is interesting to note that D,L-isocitrate blocks competitively the binding of TPNH to isocitrate dehydrogenase from pig heart (191). This result is a reflection of the fact that complexes of the type enzyme–reduced coenzyme–reduced substrate are not formed by this enzyme, and emphasizes the role of coenzyme in determining the substrate-binding properties of isocitrate dehydrogenase.

e. Conformational Changes following Formation of Binary and Ternary Complexes. Protein conformations are not rigid structures. They are susceptible to critical rearrangements in response to variation of the environment. Considerable attention has been given in recent years to the possible role of conformational changes in enzymes during the course of their action. Conformational changes probably play an important role in the catalytic process.

By various methods it has been possible to demonstrate that the interaction with coenzyme or substrate or the formation of enzyme–coenzyme–inhibitor complexes induces conformational changes in PN enzymes.

Yeast alcohol dehydrogenase. The deuterium exchange of yeast alcohol dehydrogenase was studied in the presence and absence of DPN$^+$ and DPNH (374). About 4% of all peptide hydrogen atoms of the enzyme protein was protected completely by DPN$^+$ and partially by DPNH from exchanging with the solvent. It appears most likely that this result reflects specific changes in the conformation of yeast alcohol dehydrogenase when the enzyme combines with the coenzyme. Analogous results were obtained with lactate dehydrogenase from chicken heart (375).

Horse liver alcohol dehydrogenase. X-ray investigations (18) and measurements of optical rotatory dispersion (376) of horse liver alcohol dehydrogenase and its binary complexes with DPNH have shown that both exhibit an orthorhombic symmetry with identical dimensions and only slightly different values of the Moffitt-Yang parameter b_0 (-100 for the enzyme and -119 for the enzyme–DPNH complex). Addition of isobutyramide to enzyme–DPNH (or pyrazole to enzyme–DPN$^+$) giving a ternary enzyme–coenzyme–inhibitor complex induces conformational changes. The Moffitt-Yang parameter b_0 decreases drastically to -185 and a monoclinic symmetry is observed instead of orthorhombic symmetry.

The general conclusion from the experiments is that the formation of the binary complex does not significantly change the conformation of the enzyme as measured by optical rotatory dispersion or x-ray diffraction. The small competitive inhibitor molecule when combining with the enzyme–coenzyme complex causes a considerable conformational change.

Glyceraldehyde-3-phosphate dehydrogenase from yeast. This enzyme also undergoes conformational changes accompanying its interaction with substrate or coenzyme (377). At pH 7.7 the b_0 parameter increases from -131 (apoprotein) to -98 upon reaction with the substrate, glyceraldehyde-3-phosphate, and decreases to -158 upon addition of DPN$^+$. The apparent helical content is 21% for the apoenzyme, 25% for the enzyme–DPN$^+$ complex, and 16% for the enzyme–substrate compound (glyceraldehyde-3-phosphate-hemimercaptal-enzyme) if one assumes that a b_0 value of −630 reflects a helical content of 100%.

f. The Cotton Effect of the Alcohol Dehydrogenase–DPNH Complex (for review see ref. 17).   Upon binding of DPNH the optical rotatory dispersion of alcohol dehydrogenase becomes anomalous due to appearance of a pronounced, single, negative Cotton effect (Fig. 12). The point of inflection of the Cotton effect of horse liver alcohol dehydrogenase, at 327 mµ, corresponds closely to the absorption maximum of the enzyme–DPNH complex. Upon binding of 1,10-phenanthroline, the optical rotatory dispersion of alcohol dehydrogenase also becomes anomalous as a result of a Cotton effect at 297 mµ, the wavelength of the absorption maximum of the complex between the enzyme-bound zinc ion and 1,10-phenanthroline.

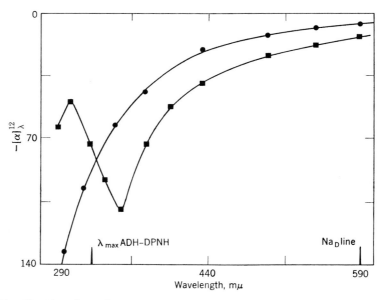

Fig. 12. The effect of DPNH on the optical rotatory dispersion of horse liver alcohol dehydrogenase (ADH) in 0.1M phosphate buffer, pH 7.5, at 12°. (●)ADH, 2 × 10^{-5}M. (■)ADH, 2 × 10^{-5}M + DPNH, 4 × 10^{-5}M. In the presence of DPNH, the rotatory dispersion of the enzyme is anomalous due to a pronounced negative Cotton effect centered about the absorption maximum of the enzyme–DPNH complex at 327 mµ. From Ulmer and Vallee (17).

In kinetic experiments 1,10-phenanthroline competes with both DPN^+ and DPNH. In agreement with the inhibition studies 1,10-phenanthroline diminishes the amplitude of the Cotton effect of the enzyme–coenzyme complex and vice versa. From all the experiments it is very probable that 1,10-phenanthroline is bound to the enzyme-bound zinc and therefore the coenzyme is also bound at or near the zinc ion, which is assumed to be located in the active center of alcohol dehydrogenase.

The *extrinsic* Cotton effect is the result of the formation of an optically active complex between enzyme and coenzyme (or 1,10-phenanthroline). The asymmetric binding of the chromophoric molecule to the native enzyme protein induces the optically active absorption band of the chromophore. This asymmetric binding is the basis for the stereospecificity with regard to the transferred hydrogen.

Rotatory dispersion titrations have shown (17) that ADP, ARPPR, and AMP reversibly reduce the magnitude of the Cotton effect of the horse liver alcohol dehydrogenase-DPNH complex and compete with the coenzyme-binding site, whereas nicotinamide mononucleotide and N-methyl nicotinamide neither inhibit the enzyme nor displace DPNH. In addition, it was shown spectropolarimetrically that AMP displaces 1,10-phenanthroline from the binding of liver alcohol dehydrogenase. From these results it was concluded (17) that AMP interacts with the zinc ion and that the coenzyme binds to the zinc ion of alcohol dehydrogenase through the AMP moiety. In contrast to these findings it was shown by double difference spectrophotometry that ARPPR, a potent inhibitor of liver alcohol dehydrogenase, competes strictly with DPN^+ and DPNH, but the enzyme–ARPPR interaction is independent of 1,10-phenanthroline (31, cf. also 21). Therefore, the interaction between the enzyme–bound zinc ion and the adenine moiety of the coenzyme in the enzyme–coenzyme complex seems unlikely.

From kinetic experiments at pH 7.0 the dissociation constant of the complex between horse liver alcohol dehydrogenase and 3-AP-DPNH was calculated to $4.8 \mu M$, $K_{E,R}$ for DPNH is lower about ten times ($0.3 \mu M$) (378). On the other hand, $K_{E,O}$ was found to be about the same for both DPN^+ and 3-AP-DPN^+. From these results it was concluded that the amide group of the nicotinamide moiety of DPN plays an important role in the binding of DPNH but not in the binding of DPN^+.

g. The DPNH Binding by Beef Lactate Dehydrogenase. The binding of DPNH by the beef lactate dehydrogenase isoenzymes was studied by fluorescence methods (204). At pH 7.4 each enzyme molecule binds four coenzyme molecules. Beef heart lactate dehydrogenase (H_4) was found to have four independent, equal binding sites with a dissociation constant $K_{E,R} = 0.39\mu M$.

The titrations of lactate dehydrogenase from beef muscle (M_4) as well as of the hybrid molecules HM_3, H_2M_2, and H_3M showed marked changes in the order of binding reaction (0.43–1.5). The most pronounced change was observed upon the binding of the first DPNH molecule. The binding of DPNH by these enzyme molecules cannot be described by an equation of the Adair type (204). The most probable cause of the phenomenon may be the existence of relaxation effects in the protein molecule.

h. The DPN$^+$ Binding by Yeast D-Glyceraldehyde-3-Phosphate Dehydrogenase. Detailed kinetic studies of the binding DPN$^+$ to yeast D-glyceraldehyde-3-phosphate dehydrogenase with the aid of temperature-jump techniques using the absorption of the EO complex at 360 mμ have shown that this enzyme exhibits cooperative ligand binding (168). Each enzyme molecule binds four coenzyme molecules. On raising the temperature from 20° to 50° at pH 8.5 the saturation curve for the DPN$^+$ binding is shifted from hyperbolic to weakly sigmoidal (Fig. 13). Under conditions leading to sigmoidal titration curves (e.g., pH 8.5 and 40°) the equilibration of a solution perturbed by a temperature jump proceeds in three discrete steps. The relaxation spectrum jump proceeds in three discrete

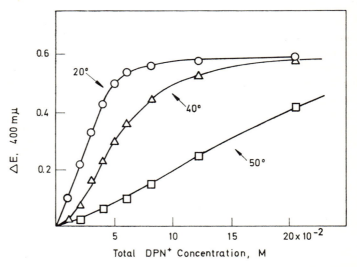

Fig. 13. Effect of temperature on the spectrophotometric titration at pH 8.5 of yeast D-glyceraldehyde-3-phosphate dehydrogenase with DPN$^+$. From Kirschner, Eigen, Bittman, and Voigt (168).

steps. The relaxation spectrum consists of three relaxation processes which are well separated from each other on the time axis with the time constants $1/\tau_1 = 7000$ sec^{-1}, $1/\tau_2 = 690$ sec^{-1}, and $1/\tau_3 = 0.2$ sec^{-1}. The three processes depend on the coenzyme concentration. With increasing DPN$^+$ concentrations the rates of the two rapid processes increase whereas the rate of the slow process drops steeply to a plateau value. From these results the rapid processes can be tentatively identified with two discrete, bimolecular reactions involving the binding of DPN$^+$, probably the two different kinds of coenzyme binding sites. The slow process is independent of enzyme concentration; this indicates an isomerization process of the enzyme.

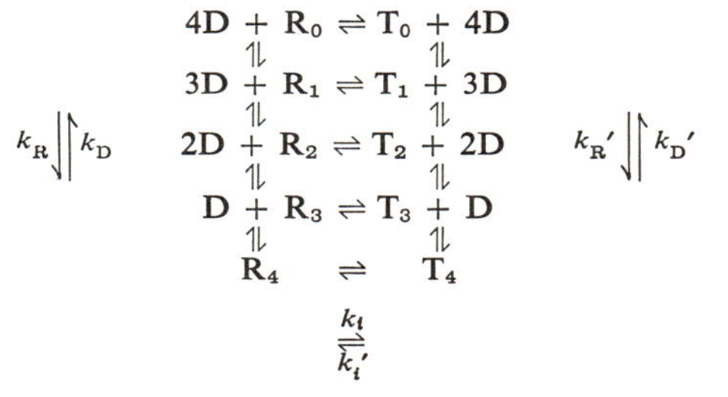

$$
\begin{array}{ccc}
& 4D + R_0 \rightleftharpoons T_0 + 4D & \\
& \Updownarrow \qquad\qquad \Updownarrow & \\
& 3D + R_1 \rightleftharpoons T_1 + 3D & \\
& \Updownarrow \qquad\qquad \Updownarrow & \\
k_R \Big\Updownarrow k_D \quad & 2D + R_2 \rightleftharpoons T_2 + 2D & \quad k_R' \Big\Updownarrow k_D' \\
& \Updownarrow \qquad\qquad \Updownarrow & \\
& D + R_3 \rightleftharpoons T_3 + D & \\
& \Updownarrow \qquad\qquad \Updownarrow & \\
& R_4 \quad \rightleftharpoons \quad T_4 & \\
& \overset{k_t}{\underset{k_i'}{\rightleftharpoons}} &
\end{array}
$$

Fig. 14. Scheme of the different equilibria between yeast D-glyceraldehyde-3-phosphate dehydrogenase and DPN$^+$. From Kirschner, Eigen, Bittman, and Voigt (168). R_0 (= E): enzyme molecule in the R conformation with four free coenzyme binding sites; R_1 (= ED): one binding site of the enzyme molecule in the R conformation occupied by DPN$^+$, . . . ; R_4 (= ED$_4$): all four binding sites of the enzyme molecule in the R conformation occupied by DPN$^+$, and analogous for the enzyme in the T conformation. k_R, k_D, k_R', and k_D' are "intrinsic" rate constants; $i = 0, 1, 2,$ 3, and 4.

From the results it was concluded (168) that the coenzyme–free enzyme molecule exists in two conformations (R and T) only. The dissociation constants $K_{E,0}$ of the various binding sites on each isomer are identical and independent on each other. Furthermore, the cooperativity between the polypeptide chains appears to be so high that if one polypeptide chain changes conformation, all the other polypeptide chains in the tetrameric molecule change conformation also. Intermediates such as R_3T or R_2T_2 were not detectable. The scheme describing these phenomena is shown in Figure 14. The equilibrium and rate constants, shown in Table III, are defined as follows:

$$ L_0 = [T_0]/[R_0] = k_0/k_0'; \quad L_i = [T_i]/[R_i] = c^i \cdot L_0 = k_i/k_i' \qquad (9a) $$

$$K = 4[R_0] \cdot [D]/[R_1] = 3[R_1] \cdot [D]/2[R_2]$$
$$= 2[R_2] \cdot [D]/3[R_3] = [R_3] \cdot [D]/4[R_4] = k_D/k_R \tag{9b}$$

$$K' = K/c = 4[T_0] \cdot [D]/[T_1] = 3[T_1] \cdot [D]/2[T_2]$$
$$= 2[T_2] \cdot [D]/3[T_3] = [T_3] \cdot [D]/4[T_4] = k_D'/k_R' \tag{9c}$$

$K_{E,O}$ for the enzyme in the T conformation (with the lower affinity for the ligand) is three times greater than $K_{E,O}$ in the R conformation (with the greater affinity for the ligand).

The extreme simplicity of the relaxation spectrum can be explained by identity of the intrinsic affinities of each of the two conformations and by strong cooperativity between the polypeptide chains (168).

TABLE III

Rate and Equilibrium Constants of the Reaction of Yeast D-Glyceraldehyde-3-phosphate Dehydrogenase with DPN$^+$ Obtained from Kinetic Data at pH 8.5 and 40° (cf. also Fig. 14).

Rate constants	Equilibrium constants
$k_R = 1.9 \times 10^7 \ M^{-1} \sec^{-1}$	$K_{E,O} = 53 \mu M$
$k_D = 1 \times 10^3 \sec^{-1}$	$K'_{E,O} = 150 \mu M$
$k_R' = 1.37 \times 10^6 \ M^{-1} \sec^{-1}$	$c = K_{E,O}/K'_{E,O} = 0.35$
$k_D' = 210 \sec^{-1}$	$L_0 = 30.5$
$k_0 = 5.5 \sec^{-1}$	
$k_0' = k_1' = k_2' = k_3' = k_4' = 0.18 \sec^{-1}$	
$k_1 = c \times k_0 = 2.0 \sec^{-1}$	
$k_2 = c^2 \times k_0 = 0.7 \sec^{-1}$	
$k_3 = c^3 \times k_0 = 0.24 \sec^{-1}$	
$k_4 = c^4 \times k_0 = 0.08 \sec^{-1}$	

Abbreviations

3-AP-DPN	3-Acetylpyridine analog of DPN
ARPPR	Adenosine diphosphate ribose
DH	Dehydrogenase
DPN$^+$ and DPNH	Oxidized and reduced forms of diphosphopyridine nucleotide
I	Inhibitor
LDH	Lactate dehydrogenase
PN	Pyridine nucleotide (PN$^+$ or O, oxidized PN; PNH or R, reduced PN)
PN enzymes	Pyridine nucleotide-dependent dehydrogenases
Py-3-ald-DPN	Pyridine-3-aldehyde analog of DPN
TPN$^+$ and TPNH	Oxidized and reduced forms of triphosphopyridine nucleotide

$$K_{E,R} = \frac{[E][R]}{[ER]}; \quad K_{ER,I} = \frac{[ER][I]}{[ERI]}; \quad K_{EI,R} = \frac{[EI][R]}{[ERI]}; \quad \text{etc.}$$

References

1. V. Massey and C. Veeger, *Ann. Rev. Biochem.*, **32**, 579 (1963).
2. G. Palmer and V. Massey, see p. 263, this volume.
3. P. D. Boyer, H. Lardy, and K. Myrbäck, Eds., *The Enzymes*, 2nd ed., Vol. 7, Academic Press, New York, 1963.
4. H. Sund, in *Mechanismen enzymatischer Reaktionen* (14. Colloquium der Gesellschaft für Physiologische Chemie, 1963), Springer-Verlag, Berlin, 1964, p. 318.
5. H. Sund, H. Diekmann, and K. Wallenfels, in *Advances in Enzymology*, Vol. 26, F. F. Nord, Ed., Interscience, New York, 1964, p. 115.
6. *Hoppe-Seyler/Thierfelder: Handbuch der physiologisch- und pathologisch-chemischen Analyse*, 10th ed., Vol. VI/A, K. Lang, E. Lehnarz, O. Hoffmann-Ostenhof, and G. Siebert, Eds., Springer-Verlag, Berlin, 1964.
7. M. Dixon and E. C. Webb, *Enzymes*, 2nd ed., Academic Press, New York, 1964, p. 672.
8. P. Strittmatter, *Ann. Rev. Biochem.*, **35**, (I), 125 (1966).
9. A. Yoshida and E. Freese, *Biochim. Biophys. Acta*, **92**, 33 (1964).
10. A. Yoshida, *Biochim. Biophys. Acta*, **105**, 70 (1965).
11. A. Yoshida and E. Freese, *Biochim. Biophys. Acta*, **96**, 248 (1965).
12. A. L. Cline and A. S. L. Hu, *J. Biol. Chem.*, **240**, 4488 and 4493 (1965).
13. A. L. Cline and A. S. L. Hu, *J. Biol. Chem.*, **240**, 4498 (1965).
14. H. Sund and H. Theorell, in reference 3, p. 25.
15. H. Theorell, S. Taniguchi, Å. Åkeson, and L. Skursky, *Biochem. Biophys. Res. Commun.*, **24**, 603 (1966).
16. J. I. Harris, in *Structure and Activity of Enzymes*, T. W. Goodwin, J. I. Harris, and B. S. Hartley, Eds., Academic Press, New York, 1964, p. 97.
17. D. D. Ulmer and B. L. Vallee, in *Advances in Enzymology*, Vol. 27, F. F. Nord, Ed., Interscience, New York, 1965, p. 37.
18. C.-I. Brändén, *Arch. Biochem. Biophys.*, **112**, 215 (1965).
19. C. L. Woronick, *Acta Chem. Scand.*, **15**, 2062 (1961); **17**, 1789, 1791 (1963).
20. H. Theorell and T. Yonetani, *Arch. Biochem. Biophys.*, **Suppl. 1**, 209 (1962).
21. H. Theorell and T. Yonetani, *Biochem. Z.*, **338**, 537 (1963).
22. T. Yonetani and H. Theorell, *Arch. Biochem. Biophys.*, **106**, 243 (1964).
23. H. Theorell and T. Yonetani, *Arch. Biochem. Biophys.*, **106**, 252 (1964).
24. G. Waller, H. Theorell, and J. Sjövall, *Arch. Biochem. Biophys.*, **111**, 671 (1965).
25. H. Theorell, B. Chance, and T. Yonetani, *J. Mol. Biol.*, **17**, 513 (1966).
26. K. Dalziel, *Biochem. J.*, **84**, 244 (1962).
27. K. Dalziel, *J. Biol. Chem.*, **238**, 2850 (1963).
28. K. Dalziel, *Acta Chem. Scand.*, **17**, S 27 (1963).
29. K. Dalziel and F. M. Dickinson, *Biochem. J.*, **95**, 311 (1965).
30. V. Prelog, in *Mechanismen enzymatischer Reaktionen* (14. Colloquium der Gesellschaft für Physiologische Chemie, 1963), Springer-Verlag, Berlin, 1964, p. 288; *Pure Appl. Chem.*, **9**, 119 (1964).
31. T. Yonetani, *Acta Chem. Scand.*, **17**, S96 (1963); *Biochem. Z.*, **338**, 300 (1963).
32. V. Bloomfield, L. Peller, and R. A. Alberty, *J. Am. Chem. Soc.*, **84**, 4375 (1962).
33. C. C. Wratten and W. W. Cleland, *Biochemistry*, **2**, 935 (1963).
34. C. C. Wratten and W. W. Cleland, *Biochemistry*, **4**, 2442 (1965).
35. E. Silverstein and P. D. Boyer, *J. Biol. Chem.*, **239**, 3908 (1964).

36. J. S. McKinley-McKee, *Progress in Biophysics and Molecular Biology*, Vol. 14, J. A. V. Butler and H. E. Huxley, Eds., Pergamon Press, Oxford, 1964, p. 223.
37. J. M. H. Graves, A. Clark, and H. J. Ringold, *Biochemistry*, 4, 2655 (1965).
38. R. Pietruszko, A. Clark, J. M. H. Graves, and H. J. Ringold, *Biochem. Biophys. Res. Commun.*, 23, 526 (1966).
39. H. Sund, see p. 603, this volume.
40. H. Jörnvall, *Acta Chem. Scand.*, 19, 1483 (1965).
41. B. L. Vallee, in *The Enzymes*, 2nd ed., Vol. 3, P. D. Boyer, H. Lardy, and K. Myrbäck, Eds., Academic Press, New York, 1960, p. 225.
42. Å. Åkeson, *Biochem. Biophys. Res. Commun.*, 17, 211 (1964).
43. J. P. von Wartburg, J. L. Bethune, and B. L. Vallee, *Biochemistry*, 3, 1775 (1964).
44. A. H. Blair and B. L. Vallee, *Biochemistry*, 5, 2026 (1966).
45. J. Papenberg, J. P. von Wartburg, and H. Aebi, *Biochem. Z.*, 342, 95 (1965).
46. M. Bühner and H. Sund, unpublished result.
47. J. H. R. Kägi and B. L. Vallee, *J. Biol. Chem.*, 235, 3188 (1960).
48. R. T. Hersh, *Biochim. Biophys. Acta*, 58, 353 (1962).
49. J. F. Thomson, *Biochemistry*, 2, 224 (1963).
50. B. Müller-Hill and K. Wallenfels, *Biochem. Z.*, 339, 338 (1964).
51. B. Müller-Hill and K. Wallenfels, *Biochem. Z.*, 339, 349 (1964).
52. K. A. Schellenberg, *J. Biol. Chem.*, 240, 1165 (1965).
53. K. A. Schellenberg, *J. Biol. Chem.*, 241, 2446 (1966).
54. D. Palm, *Biochem. Biophys. Res. Commun.*, 22, 151 (1966).
55. D. Palm, *Z. Naturforsch.*, 21b, 540 (1966).
56. D. Palm, *Z. Naturforsch.*, 21b, 547 (1966).
57. B. M. Anderson, M. L. Reynolds, and C. D. Anderson, *Biochim. Biophys. Acta*, 99, 46 (1965); *Arch. Biochem. Biophys.*, 111, 202, (1965).
58. B. M. Anderson and M. L. Reynolds, *Biochim. Biophys. Acta*, 96, 45 (1965); *Arch. Biochem. Biophys.*, 111, 1 (1965).
59. A. Arens, H. Sund, and K. Wallenfels, *Biochem. Z.*, 337, 1 (1963).
60. B. L. Vallee and F. L. Hoch, *Proc. Natl. Acad. Sci. U.S.*, 41, 327 (1955).
61. B. S. Vanderheiden, J. O. Meinhart, R. G. Dodson, and E. G. Krebs, *J. Biol. Chem.*, 237, 2095 (1962).
62. H. Sund, *Z. Naturwiss.-Mediz. Grundlagenforsch.*, 3, 284 (1965).
63. B. T. Kaufman and R. C. Gardiner, *J. Biol. Chem.*, 241, 1319 (1966).
64. S. F. Zakrzewski and C. A. Nichol, *J. Biol. Chem.*, 235, 2984 (1960).
65. S. F. Zakrzewski, *J. Biol. Chem.*, 235, 1780 (1960); 238, 1485 and 4002 (1963); 241, 2962 (1966).
66. C. K. Mathews and F. M. Huennekens, *J. Biol. Chem.*, 238, 3436 (1963).
67. B. T. Kaufman, *J. Biol. Chem.*, 239, 669 (1964); *Proc. Natl. Acad. Sci. U.S.*, 56, 695 (1966).
68. D. M. Greenberg, B. D. Tam, E. Jenny, and B. Payes, *Biochim. Biophys. Acta*, 122, 423 (1966).
69. R. Nath and D. M. Greenberg, *Biochemistry*, 1, 435 (1962).
70. K. Wallenfels and G. Kurz, in *Methods in Enzymology*, Vol. 9, S. P. Colowick and N. O. Kaplan, Eds., Academic Press, New York, 1966, p. 112.
71. H. Sund and G. Kurz, unpublished results.
72. K. Wallenfels and G. Kurz, *Biochem. Z.*, 335, 559 (1962); G. Kurz, Dissertation, Freiburg, 1967.

73. G. R. Julian, R. G. Wolfe, and F. J. Reithel, *J. Biol. Chem.*, **236**, 754 (1961).
74. E. A. Noltmann and S. A. Kuby, in reference 3, p. 223.
75. E. A. Noltmann, C. J. Gubler, and S. A. Kuby, *J. Biol. Chem.*, **236**, 1225 (1961).
76. P. Andrews, *Biochem. J.*, **96**, 595 (1965).
77. I. Eger-Neufeldt, A. Teinzer, L. Weiss, and O. Wieland, *Biochem. Biophys. Res. Commun.*, **19**, 43 (1965).
78. H. Sund, *Acta Chem. Scand.*, **17**, S102 (1963).
79. B. Jirgensons, *J. Am. Chem. Soc.*, **83**, 3161 (1961).
80. H. F. Fisher, L. L. McGregor, and D. G. Cross, *Biochim. Biophys. Acta*, **65**, 175 (1962).
81. C. Frieden, *J. Biol. Chem.*, **237**, 2396 (1962).
82. E. Marler and C. Tanford, *J. Biol. Chem.*, **239**, 4217 (1964).
83. H. Sund, *Acta Chem. Scand.*, **15**, 940 (1961).
84. H. Sund, Habilitationsschrift, Freiburg, 1964.
85. G. Pfleiderer and F. Auricchio, *Biochem. Biophys. Res. Commun.*, **16**, 53 (1964).
86. P. M. Bayley and G. K. Radda, *Biochem. J.*, **98**, 105 (1966).
87. G. Pfleiderer, J. J. Holbrook, L. Nowicki, and R. Jeckel, *Biochem. Z.*, **346**, 297 (1966).
88. C. Frieden, in reference 3, p. 3.
89. C. Frieden, *J. Biol. Chem.*, **238**, 146 (1963).
90. H. F. Fisher and L. L. McGregor, *Biochem. Biophys. Res. Commun.*, **3**, 629 (1960).
91. H. F. Fisher and L. L. McGregor, *J. Biol. Chem.*, **236**, 791 (1961).
92. H. F. Fisher and D. G. Cross, *Biochem. Biophys. Res. Commun.*, **20**, 120 (1965).
93. G. di Prisco, S. M. Arfin, and H. J. Strecker, *J. Biol. Chem.*, **240**, 1611 (1963).
94. G. M. Tomkins and K. L. Yielding, *Cold Spring Harbor Symp. Quant. Biol.*, **26**, 331 (1961).
95. K. L. Yielding and G. M. Tomkins, *Biochim. Biophys. Acta*, **62**, 327 (1962).
96. K. L. Yielding, G. M. Tomkins, and D. S. Trundle, *Biochim. Biophys. Acta*, **85**, 342 (1964).
97. M. W. Bitensky, K. L. Yielding, and G. M. Tomkins, *J. Biol. Chem.*, **240**, 663 and 668 (1965).
98. G. M. Tomkins, K. L. Yielding, J. F. Curran, M. R. Summers, and M. W. Bitensky, *J. Biol. Chem.*, **240**, 3793 (1965).
99. J. Wolff, *J. Biol. Chem.*, **237**, 236 (1962).
100. C. Frieden, *J. Biol. Chem.*, **238**, 3286 (1963); *Biochem. Biophys. Res. Commun.*, **10**, 410 (1963).
101. R. F. Colman and C. Frieden, *Biochem. Biophys. Res. Commun.*, **22**, 100 (1966),
102. A. Dobry-Duclaux, *Biochim. Biophys. Acta*, **89**, 1 (1964).
103. L. Corman and N. O. Kaplan, *Biochemistry*, **4**, 2175 (1965).
104. H. Sund, *Acta Chem. Scand.*, **19**, 390 (1965).
105. E. Kun and B. Achmatowicz, *J. Biol. Chem.*, **240**, 2619 (1965).
106. J. E. Churchich and F. Wold, *Biochemistry*, **2**, 781 (1963).
107. B. M. Anderson and M. L. Reynolds, *J. Biol. Chem.*, **241**, 1688 (1966).
108. K. Taketa and B. M. Pogell, *J. Biol. Chem.*, **241**, 720 (1966).
109. J. Wolff, *J. Biol. Chem.*, **237**, 230 (1962).
110. H. Sund and Å. Åkeson, *Biochem. Z.*, **340**, 421 (1964).
111. H. Sund, *Angew. Chem.*, **76**, 954 (1964); *Intern. Ed.*, **3**, 802 (1964).
112. M. Minssen and H. Sund, unpublished result.

113. N. Talal, G. M. Tomkins, J. F. Mushinski, and K. L. Yielding, *J. Mol. Biol.*, **8**, 46 (1964).
114. B. Jirgensons, *J. Biol. Chem.*, **240**, 1064 (1965).
115. D. G. Cross and H. F. Fisher, *Biochemistry*, **5**, 880 (1966).
116. H. Kubo, T. Yamano, M. Iwatsubo, H. Watari, T. Soyama, J. Shiraishi, S. Sawada, N. Kawashima, S. Mitani, and K. Ito, *Bull. Soc. Chim. Biol.*, **40**, 431 (1958).
117. R. Cohen, *9th Meeting, Biophys. Soc., San Francisco 1965*, p. 140.
118. K. S. Rogers, L. Hellerman, and T. E. Thompson, *J. Biol. Chem.*, **240**, 198, (1965).
119. H. Sund and K. Weber, *Angew. Chem.*, **78**, 217 (1966); *Intern. Ed.*, **5**, 231 (1966).
120. J. E. Snoke, *J. Biol. Chem.*, **223**, 271 (1956).
121. C. Frieden, *Biochim. Biophys. Acta*, **62**, 421 (1962).
122. K. S. Rogers, P. J. Geiger, T. E. Thompson, and L. Hellerman, *J. Biol. Chem.*, **238**, 481 (1963).
123. L. A. Fahien, B. O. Wiggert, and P. P. Cohen, *J. Biol. Chem.*, **240**, 1083 (1965).
124. L. A. Fahien, B. O. Wiggert, and P. P. Cohen, *J. Biol. Chem.*, **240**, 1091 (1965).
125. B. O. Wiggert, and P. P. Cohen, *J. Biol. Chem.*, **240**, 4790 (1965).
125a. B. O. Wiggert and P. P. Cohen, *J. Biol. Chem.*, **241**, 210 (1966).
126. H. Kubo, M. Iwatsubo, H. Watari, and T. Soyama, *J. Biochem. (Tokyo)*, **46**, 1171 (1959).
127. M. Inagaki, *J. Biochem. (Tokyo)*, **46**, 893 and 1001 (1959).
128. L. Corman and N. O. Kaplan, *Federation Proc.*, **25**, 746 (1966).
129. R. W. Barratt and W. N. Strickland, *Arch. Biochem. Biophys.*, **102**, 66 (1963).
130. J. R. S. Fincham and A. Coddington, *J. Mol. Biol.*, **6**, 361 (1963).
131. A. Coddington, J. R. S. Fincham, and T. K. Sundaram, *J. Mol. Biol.*, **17**, 503 (1966).
132. J. R. S. Fincham, in *Advances in Enzymology*, Vol. 22, F. F. Nord, Ed., Interscience, New York, 1960, p. 1; *J. Mol. Biol.*, **4**, 257 (1962).
133. J. R. S. Fincham and P. A. Bond, *Biochem. J.*, **77**, 96 (1960).
134. N. O. Kaplan, *Bacteriol. Rev.*, **27**, 155 (1963).
135. W. Bernhardt, M. Zink, and H. Holzer, *Biochim. Biophys. Acta*, **118**, 549 (1966).
136. H. Holzer and S. Schneider, *Biochem. Z.*, **329**, 361 (1957).
137. S. Grisolia, C. L. Quijada, and M. Fernandez, *Biochim. Biophys. Acta*, **81**, 61 (1964).
138. B. Szörényi, *Acta Physiol. Acad. Sci. Hung.*, **17**, 195 (1960).
139. P. Elödi and E. Szörényi, *Acta Physiol. Acad. Sci. Hung.*, **9**, 339 (1956).
140. T. Dévényi, Á. Pusztai, M. Sajgó, and B. Szörényi, *Acta Physiol. Acad. Sci. Hung.*, **13**, 95 (1958).
141. W. S. Allison and N. O. Kaplan, *J. Biol. Chem.*, **239**, 2140 (1964).
142. R. N. Perham, *Biochem. J.*, **99**, 14C (1966).
143. W. S. Allison and J. I. Harris, *Abstr., 2nd Meeting, Fed. European Biochem. Soc., Vienna, 1965*, p. 140.
144. M. Oguchi, T. Hashimoto, S. Minakami, and H. Yoshikawa, *J. Biochem. (Tokyo)*, **60**, 99 (1966).
145. W. F. Harrington and G. M. Karr, *J. Mol. Biol.*, **13**, 885 (1965).
146. J. I. Harris and R. N. Perham, *J. Mol. Biol.*, **13**, 876 (1965).
147. P. Friedrich, *Biochim. Biophys. Acta*, **99**, 371 (1965).
148. P. Friedrich, L. Polgár, and G. Szabolcsi, *Nature*, **202**, 1214 (1964).

149. L. Polgár, *Biochim. Biophys. Acta*, **118**, 276 (1966).
150. T. Keleti, S. Györgyi, M. Telegdi, and H. Zaluska, *Acta Physiol. Acad. Sci. Hung.*, **22**, 11 (1962).
151. T. Keleti, *Biochim. Biophys. Acta*, **89**, 422 (1964); *Biochem. Biophys. Res. Commun.*, **22**, 640 (1966).
152. S. F. Velick, in *Methods in Enzymology*, Vol. 1, S. P. Colowick and N. O. Kaplan, Eds., Academic Press, New York, 1955, p. 401.
153. R. N. Perham and J. I. Harris, *J. Mol. Biol.*, **7**, 316 (1963).
154. A. L. Murdock and O. J. Koeppe, *J. Biol. Chem.*, **239**, 1983 (1964).
155. M. E. Kirtley and D. E. Koshland, *Biochem. Biophys. Res. Commun.*, **23**, 810 (1966).
156. S. F. Velick, *J. Biol. Chem.*, **233**, 1455 (1958).
157. C. S. Furfine and S. F. Velick, *J. Biol. Chem.*, **240**, 844 (1965).
158. A. G. Hilvers, K. van Dam, and E. C. Slater, *Biochim. Biophys. Acta*, **85**, 206 (1964).
159. W. S. Allison and N. O. Kaplan, *Biochemistry*, **3**, 1792 (1964).
160. L. Cunningham and A. M. Schepman, *Biochim. Biophys. Acta*, **73**, 406 (1963).
161. I. Krimsky and E. Racker, *Biochemistry*, **2**, 512 (1963).
162. A. H. Gold and H. L. Segal, *Biochemistry*, **3**, 778 (1964).
163. E. Szörényi, P. Elödi, and T. Dévényi, *Acta Physiol. Acad. Sci. Hung.*, **9**, 351 (1956).
164. P. Elödi, *Acta Physiol. Acad. Sci. Hung.*, **13**, 199 (1958).
165. H. C. Watson and L. J. Banaszak, *Nature*, **204**, 918 (1964).
166. R. E. Amelunxen, *Biochim. Biophys. Acta*, **122**, 175 (1966).
167. E. G. Krebs, in *Methods in Enzymology*, Vol. 1, S. P. Colowick and N. O. Kaplan, Eds., Academic Press, New York, 1955, p. 407.
168. K. Kirschner, M. Eigen, R. Bittman, and B. Voigt, *Proc. Natl. Acad. Sci. U.S.*, **56**, 1661 (1966).
169. E. L. Taylor, B. P. Meriwether, and J. H. Park, *J. Biol. Chem.*, **238**, 734 (1963).
170. A. Stockell, *J. Biol. Chem.*, **234**, 1286 (1959).
171. B. H. Havsteen, *Acta Chem. Scand.*, **19**, 1643 (1965).
172. T. Baranowski, in reference 3, p. 85.
173. J. van Eys, B. J. Nuenke, and M. K. Patterson, *J. Biol. Chem.*, **234**, 2308 (1959).
174. W. C. Deal and W. H. Holleman, *Federation Proc.*, **23**, 264 (1964).
175. G. Pfleiderer and F. Auricchio, *Biochem. Biophys. Res. Commun.*, **16**, 53 (1964).
176. O. P. Chilson, G. B. Kitto, J. Pudles, and N. O. Kaplan, *J. Biol. Chem.*, **241**, 2431 (1966).
177. I. van Eys, J. Judd, J. Ford, and W. B. Womack, *Biochemistry*, **3**, 1755 (1964).
178. I. Zelitch, *J. Biol. Chem.*, **216**, 553 (1955).
179. J. C. Loper and E. Adams, *J. Biol. Chem.*, **240**, 788 (1965).
180. J. R. Stern, *Biochim. Biophys. Acta*, **26**, 448 (1957).
181. P. G. Squire, S. Delin, and J. Porath, *Biochim. Biophys. Acta*, **89**, 409 (1964).
182. P. Talalay, in reference 3, p. 177.
183. P. Talalay, in *Methods in Enzymology*, Vol. 5, S. P. Colowick and N. O. Kaplan, Eds., Academic Press, New York, 1962, p. 512.
184. H. J. Hübener and F. G. Sahrholz, *Biochem. Z.*, **333**, 95 (1960).
185. M. Gehatia, *Z. Naturforsch.*, **17b**, 432 (1962).

186. P. Talalay, in *Methods in Enzymology*, Vol. 5, S. P. Colowick and N. O. Kaplan, Eds., Academic Press, New York, 1962, p. 512.

186a. J. Schmidt-Thomé, G. Nesemann, H. J. Hübener, and I. Alester, *Biochem. Z.*, **336**, 322 (1962).

187. M. G. Smith, *Biochem. J.*, **83**, 135 (1962).

188. J. Moyle and M. Dixon, *Biochem. J.*, **63**, 548 (1956).

189. G. Siebert, J. Dubuc, R. C. Warner, and G. W. E. Plaut, *J. Biol. Chem.*, **226**, 965 (1957).

190. G. W. E. Plaut, in reference 3, p. 105.

191. T. A. Langan, *Acta Chem. Scand.*, **14**, 936 (1960).

192. J. Moyle, *Biochem. J.*, **63**, 552 (1956).

193. G. Siebert, M. Carsiotis, and G. W. E. Plaut, *J. Biol. Chem.*, **226**, 977 (1957).

194. Z. B. Rose, *J. Biol. Chem.*, **235**, 928 (1960).

195. S. Englard and I. Listowsky, *Biochem. Biophys. Res. Commun.*, **12**, 356 (1963).

196. G. E. Lienhard and I. A. Rose, *Biochemistry*, **3**, 185 (1964).

197. H. Zalkin and D. B. Sprinson, *J. Biol. Chem.*, **241**, 1067 (1966).

198. G. W. Schwert, D. B. S. Millar, and Y. Takenaka, *J. Biol. Chem.*, **237**, 2131 (1962).

199. J. B. Neilands, *J. Biol. Chem.*, **199**, 373 (1952); **208**, 225 (1954).

200. E. Appella and C. L. Markert, *Biochem. Biophys. Res. Commun.*, **6**, 171 (1961).

201. D. B. S. Millar, *J. Biol. Chem.*, **237**, 2135 (1962).

202. A. H. Gold and H. L. Segal, *Biochemistry*, **4**, 1506 (1965).

203. R. H. McKay and N. O. Kaplan, *Biochim. Biophys. Acta*, **79**, 273 (1964).

204. S. R. Anderson and G. Weber, *Biochemistry*, **4**, 1948 (1965).

205. A. D. Winer, *Acta Chem. Scand.*, **17**, S203 (1963).

206. J. F. Thomson and S. L. Nance, *Biochim. Biophys. Acta*, **99**, 369 (1965).

207. A. P. Nygaard, *Acta Chem. Scand.*, **10**, 397 (1956).

208. A. P. Nygaard, *Acta Chem. Scand.*, **10**, 408 (1956).

209. L. Brand, J. Everse, and N. O. Kaplan, *Biochemistry*, **1**, 423 (1962).

210. G. Di Sabato and N. O. Kaplan, *Biochemistry*, **2**, 776 (1963).

211. D. Robinson, D. Stollar, S. White, and N. O. Kaplan, *Biochemistry*, **2**, 486 (1963).

212. G. W. Schwert and A. D. Winer, in reference 3, p. 127.

213. D. B. S. Millar and G. W. Schwert, *J. Biol. Chem.*, **238**, 3249 (1963).

214. R. Jaenicke, *Biochim. Biophys. Acta*, **85**, 186 (1964).

215. E. Silverstein and P. D. Boyer, *J. Biol. Chem.*, **239**, 3901 (1964).

216. T. P. Fondy, J. Everse, G. A. Driscoll, F. Castillo, F. E. Stolzenbach, and N. O. Kaplan, *J. Biol. Chem.*, **240**, 4219 (1965).

217. G. Hathaway and R. S. Criddle, *Proc. Natl. Acad. Sci. U.S.*, **56**, 680 (1966).

218. T. P. Fondy, A. Pesce, I. Freedberg, F. Stolzenbach, and N. O. Kaplan, *Biochemistry*, **3**, 522 (1964).

219. G. Di Sabato and N. O. Kaplan, *J. Biol. Chem.*, **239**, 438 (1964).

220. O. P. Chilson, G. B. Kitto, J. Pudles, and N. O. Kaplan, *J. Biol. Chem.*, **241**, 2431 (1966).

221. N. O. Kaplan, *Brookhaven Symp. Biol.*, **17**, 131 (1964).

222. J. S. Nisselbaum and O. Bodansky, *J. Biol. Chem.*, **236**, 323 (1961); **238**, 969 (1963); *Ann. New York Acad. Sci.*, **103**, 930 (1963).

223. J. S. Nisselbaum, D. E. Packer, and O. Bodansky, *J. Biol. Chem.*, **239**, 2830 (1964).

224. E. D. Wachsmuth and G. Pfleiderer, *Biochem. Z.*, **336**, 545 (1963).
225. P. G. W. Plagemann, K. F. Gregory, and F. Wróblewski, *J. Biol. Chem.*, **235**, 2282 and 2288 (1960); *Biochem. Z.*, **334**, 37 (1961).
226. E. D. Wachsmuth, G. Pfleiderer, and T. Wieland, *Biochem. Z.*, **340**, 80 (1964).
227. G. Jécsai, *Acta Physiol. Acad. Sci. Hung.*, **20**, 339 (1961).
228. W. J. Reeves and G. M. Fimognari, *J. Biol. Chem.*, **238**, 3853 (1963).
229. J. J. Holbrook, *Biochem. Z.*, **344**, 141 (1966).
230. E. H. Eisman, H. A. Lee and A. D. Winer, *Biochemistry*, **4**, 606 (1965).
231. W. Gruber, K. Warzecha, G. Pfleiderer, and T. Wieland, *Biochem. Z.*, **336**, 107 (1962).
232. J. J. Holbrook, G. Pfleiderer, J. Schnetger, and S. Diemair, *Biochem. Z.*, **344**, 1 (1966).
233. G. Pfleiderer and D. Jeckel, *Biochem. Z.*, **329**, 370 (1957).
234. T. Wieland and G. Pfleiderer, *Angew. Chem.*, **74**, 261 (1962), *Intern. Ed.*, **1**, 169 (1962).
235. B. Pickles, B. A. Jeffery, and M. G. Rossmann, *J. Mol. Biol.*, **9**, 598 (1964).
236. E. Racker, *J. Biol. Chem.*, **196**, 347 (1952).
237. H. J. Fromm, *J. Biol. Chem.*, **238**, 2938 (1963).
238. J. van Eys, F. E. Stolzenbach, L. Sherwood, and N. O. Kaplan, *Biochim. Biophys. Acta*, **27**, 63 (1958).
239. J. F. Thomson and J. J. Darling, *Biochem. Biophys. Res. Commun.*, **9**, 334 (1962).
240. J. F. Thomson, J. J. Darling, and L. F. Bordner, *Biochim. Biophys. Acta*, **85**, 177 (1964).
241. H. J. Fromm, *Biochim. Biophys. Acta*, **99**, 540 (1965).
242. V. Zewe and H. J. Fromm, *J. Biol. Chem.*, **237**, 1668 (1962); *Biochemistry*, **4**, 782 (1965).
243. R. Stambaugh and D. Post, *J. Biol. Chem.*, **241**, 1462 (1966).
244. F. Kubowitz and P. Ott, *Biochem. Z.*, **314**, 94 (1943).
245. T. Wieland, G. Pfleiderer, and F. Ortanderl, *Biochem. Z.*, **331**, 103 (1959).
246. A. C. Maehly, Ed., *Biochemical Preparations*, Vol. 11, Wiley, New York, 1966, p. 69.
247. C. S. Vestling, in *Methods of Biochemical Analysis*, Vol. 10, D. Glick, Ed., Interscience, New York, 1962, p. 137.
248. S. R. Anderson, J. R. Florini, and C. S. Vestling, *J. Biol. Chem.*, **239**, 2991 (1964).
249. D. M. Gibson, E. O. Davisson, B. K. Bachhawat, B. R. Ray, and C. S. Vestling, *J. Biol. Chem.*, **203**, 397 (1953).
250. A. Yoshida and E. Freese, *Biochim. Biophys. Acta*, **99**, 56 (1965).
251. A. Yoshida, *Biochim. Biophys. Acta*, **99**, 66 (1965).
252. L. Siegel and S. Englard, *Biochim. Biophys. Acta*, **54**, 67 (1961).
253. F. C. Grimm and D. G. Doherty, *J. Biol. Chem.*, **236**, 1980 (1961).
254. D. D. Davies, and E. Kun, *Biochem. J.*, **66**, 307 (1957).
255. L. Siegel and S. Englard, *Biochim. Biophys. Acta*, **64**, 101 (1962).
256. F. C. Grimm and D. G. Doherty, *Biochim. Biophys. Acta*, **57**, 381 (1962).
257. S. Englard and H. H. Breiger, *Biochim. Biophys. Acta*, **56**, 571 (1962).
258. M. Cassman and S. Englard, *J. Biol. Chem.*, **241**, 787 (1966).
259. C. J. R. Thorne, *Biochim. Biophys. Acta*, **59**, 624 (1962).
260. C. J. R. Thorne and N. O. Kaplan, *J. Biol. Chem.*, **238**, 1861 (1963).
261. H. Theorell and T. A. Langan, *Acta Chem. Scand.*, **14**, 933 (1960).

262. D. N. Raval, and R. G. Wolfe, *Biochemistry*, **1**, 263, 1112, and 1118 (1962); **2**, 220 (1963).
263. C. J. R. Thorne and P. M. Cooper, *Biochim. Biophys. Acta*, **81**, 397 (1963).
264. M. J. Coon, *Biochemical Preparations*, Vol. 9, Wiley, New York, 1962, p. 102.
265. E. Kun and P. Volfin, *Biochem. Biophys. Res. Commun.*, **22**, 187 (1966).
266. A. Yoshida, *J. Biol. Chem.*, **240**, 1113 and 1118 (1965).
267. K. D. Munkres and F. M. Richards, *Arch. Biochem. Biophys.*, **109**, 466 (1965).
268. K. D. Munkres, *Biochemistry*, **4**, 2180 and 2186 (1965).
269. K. D. Munkres and F. M. Richards, *Arch. Biochem. Biophys.*, **109**, 457 (1965).
269a. K. D. Munkres, *Arch. Biochem. Biophys.*, **112**, 340 and 347 (1965).
270. G. Martinez, H. A. Barker, and B. L. Horecker, *J. Biol. Chem.*, **238**, 1598 (1963).
271. M. Liss, S. B. Horwitz, and N. O. Kaplan, *J. Biol. Chem.*, **237**, 1342 (1962).
272. L. Klungsøyr, *Biochim. Biophys. Acta*, **128**, 55 (1966).
273. S. Pontremoli, A. de Flora, E. Grazi, G. Mangiarotti, A. Bonsignore, and B. L. Horecker, *J. Biol. Chem.*, **236**, 2975 (1961).
274. E. Grazi, M. Rippa, and S. Pontremoli, *J. Biol. Chem.*, **240**, 234 (1965).
275. M. Rippa, E. Grazi, and S. Pontremoli, *J. Biol. Chem.*, **241**, 1632 (1966).
276. G. E. Lienhard and I. A. Rose, *Biochemistry*, **3**, 190 (1964).
277. H. J. Fromm and J. A. Bietz, *Arch. Biochem. Biophys.*, **115**, 510 (1966).
278. R. C. Nordlie and H. J. Fromm, *J. Biol. Chem.*, **234**, 2523 (1959).
279. H. J. Fromm and D. R. Nelson, *J. Biol. Chem.*, **237**, 215 (1962).
280. A. M. Gotto and H. L. Kornberg, *Biochem. J.*, **81**, 273 (1961).
281. A. Rodgers, *Biochem. J.*, **81**, 285 (1961).
282. D. Wilson, *Anal. Biochem.*, **10**, 472 (1965).
283. J. L. Strominger, E. S. Maxwell, J. Axelrod, and H. M. Kalckar, *J. Biol. Chem.* **224**, 79 (1957).
284. N. D. Goldberg, J. L. Dahl, and R. E. Parks, *J. Biol. Chem.*, **238**, 3109 (1963).
285. G. Salitis and I. T. Oliver, *Biochim. Biophys. Acta*, **81**, 55 (1964).
286. P. C. Simonart, W. L. Salo, and S. Kirkwood, *Biochem. Biophys. Res. Commun.*, **24**, 120 (1966).
287. R. Lumry and H. Eyring, *J. Phys. Chem.*, **58**, 110 (1954).
288. B. H. Nevaldine and H. R. Levy, *Biochem. Biophys. Res. Commun.*, **21**, 28 (1965).
289. C. J. Epstein, M. M. Carter, and R. F. Goldberger, *Biochim. Biophys. Acta*, **92**, 391 (1964).
290. C. L. Markert and E. J. Massaro, *Arch. Biochem. Biophys.*, **115**, 417 (1966).
291. J. Monod, J. P. Changeux, and F. Jacob, *J. Mol. Biol.*, **6**, 306 (1963). J. Monod, J. Wyman, and J. P. Changeux, *J. Mol. Biol.*, **12**, 88 (1965).
292. H. N. Kirkman and E. M. Hendrickson, *J. Biol. Chem.*, **237**, 2371 (1962); E. A. Tsutsui and P. A. Marks, *Biochem. Biophys. Res. Commun.*, **8**, 338 (1962).
293. P. Datta, H. Gest, and H. L. Segal, *Proc. Natl. Acad. Sci. U.S.*, **51**, 125 (1964); P. Datta and H. Gest, *J. Biol. Chem.*, **240**, 3023 (1965).
294. C. Frieden, *J. Biol. Chem.*, **234**, 809 and 815 (1959).
295. H. Sund, in *Proc. 4ᵉ Congrès de la Fédération Intern. du Diabète*, Vol. 1, M. Demole, Ed., Éditions Médecine et Hygiène, Geneva, 1961, p. 726.
296. G. M. Tomkins and E. S. Maxwell, *Ann. Rev. Biochem.*, **32**, 677 (1963); G. M. Tomkins, K. L. Yielding, and J. Curran, *Proc. Natl. Acad. Sci. U.S.*, **47**, 270 (1961).
297. R. F. Colman and C. Frieden, *J. Biol. Chem.*, **241**, 3661 (1966).

298. A. White, P. Handler, and E. L. Smith, *Principles of Biochemistry*, 3rd ed., McGraw-Hill, New York, 1964, pp. 343, 864.
299. C. L. Markert and H. Ursprung, *Develop. Biol.* **5**, 363 (1962); C. L. Markert, in *Cytodifferentiation and Macromolecular Synthesis*, M. Locke, Ed., Academic Press, New York, 1963, p. 65.
300. A. L. Koen and C. R. Shaw, *Biochim. Biophys. Acta*, **96**, 231 (1965).
301. C. L. Markert, *Science*, **140**, 1329 (1963); *Abstr., 6th Intern. Congr. Biochem.*, *New York 1964*, p. 320.
302. L. D. Stegink and C. S. Vestling, *J. Biol. Chem.*, **241**, 4923 (1966).
303. E. Appella, in reference 221, p. 151.
304. S. N. Salthe, O. P. Chilson, and N. O. Kaplan, *Nature*, **207**, 723 (1965).
305. A. C. Wilson, R. D. Cahn, and N. O. Kaplan, *Nature*, **197**, 331 (1963).
306. D. T. Lindsay, *J. Exptl. Zool.*, **152**, 75 (1963).
307. I. H. Fine, N. O. Kaplan, and D. Kuftinec, *Biochemistry*, **2**, 116 (1963).
308. K. Rajewsky, S. Avrameas, P. Grabar, G. Pfleiderer, and E. D. Wachsmuth, *Biochim. Biophys. Acta*, **92**, 248 (1964).
309. G. Pfleiderer and E. D. Wachsmuth, *Biochem. Z.*, **334**, 185 (1961).
310. P. D. Boyer, in *The Enzymes*, 2nd ed., Vol. 1, P. D. Boyer, H. Lardy, and K. Myrbäck, Eds., Academic Press, New York, 1959, p. 511.
311. W. B. Jakoby, in reference 3, p. 203.
312. S. F. Velick and C. Furfine, in reference 3, p. 243.
313. G. Pfleiderer, J. J. Holbrook, L. Nowicki, and R. Jeckel, *Biochem. Z.*, **346**, 297 (1966).
314. H. Sund, *Biochem. Z.*, **333**, 205 (1960).
315. P. J. Snodgrass, B. L. Vallee, and F. L. Hoch, *J. Biol. Chem.*, **235**, 504 (1960).
316. K. Wallenfels and B. Müller-Hill, *Biochem. Z.*, **339**, 352 (1964).
317. T. K. Li and B. L. Vallee, *Biochemistry*, **3**, 869 (1964).
318. J. I. Harris and L. Polgár, *J. Mol. Biol.*, **14**, 630 (1965).
319. E. Margoliash and E. L. Smith, in *Evolving Genes and Proteins*, V. Bryson and H. J. Vogel, Eds., Academic Press, New York, 1965, p. 221.
320. E. Mathew, C. F. Agnello, and J. H. Park, *J. Biol. Chem.*, **240**, 3232 (1965).
321. S. Shifrin and N. O. Kaplan, in *Advances in Enzymology*, Vol. 22, F. F. Nord, Ed., Interscience, New York, 1960, p. 337.
322. E. M. Kosower, *Biochim. Biophys. Acta*, **56**, 474 (1962).
323. B. L. Vallee and J. E. Coleman, in *Comprehensive Biochemistry*, Vol. 12, M. Florkin and E. H. Stotz, Eds., Elsevier, Amsterdam, 1964, p. 165.
324. H. Sund., *Z. Naturwiss.-Medizin. Grundlagenforsch.*, **3**, 284 (1965).
325. D. D. Ulmer and B. L. Vallee, in *Advances in Enzymology*, Vol. 27, F. F. Nord, Ed., Interscience, New York, 1965, p. 37.
326. B. L. Vallee, F. L. Hoch, S. J. Adelstein, and W. E. C. Wacker, *J. Am. Chem. Soc.*, **78**, 5879 (1956).
327. G. Pfleiderer, D. Jeckel, and T. Wieland, *Biochem. Z.*, **330**, 296 (1958).
328. C. S. Vestling, W. T. Hsieh, H. Terayama, and J. N. Baptist, *Acta Chem. Scand.*, **17**, S23 (1963).
329. T. Keleti, *Biochem. Biophys. Res. Commun.*, **22**, 640 (1966).
330. W. Ferdinand, *Biochem. J.*, **92**, 578 (1964).
331. R. Druyan and B. L. Vallee, *Biochemistry*, **3**, 944 (1964).
332. R. A. Plane and T. V. Long, *Acta Chem. Scand.*, **17**, S91 (1963).
333. K. Wallenfels and H. Sund, *Biochem. Z.*, **329**, 59 (1957).

334. A. Pesce, R. H. McKay, F. Stolzenbach, R. D. Cahn, and N. O. Kaplan, *J. Biol. Chem.*, **239**, 1753 (1964).
335. H. R. Mahler, in *Mineral Metabolism*, Vol. 1, C. L. Comar and F. Bronner, Eds., Academic Press, New York, 1961, p. 743.
336. J. A. Olson and C. B. Anfinsen, *J. Biol. Chem.*, **202**, 841 (1953).
337. H. J. Strecker, *Arch. Biochem. Biophys.*, **46**, 128 (1953).
338. D. S. Goldman, *Biochim. Biophys. Acta*, **34**, 527 (1959).
339. V. Massey and Q. H. Gibson, *Federation Proc.*, **23**, 18 (1964).
340. L. Hellerman, K. A. Schellenberg, and O. K. Reiss, *J. Biol. Chem.*, **233**, 1468 (1958).
341. W. Lamprecht and F. Heinz, in reference 6, p. 556.
342. R. A. Deitrich, L. Hellerman, and J. Wein, *J. Biol. Chem.*, **237**, 560 (1962).
343. V. G. Erwin and R. A. Deitrich, *J. Biol. Chem.*, **241**, 3533 (1966).
344. R. W. Taft, in *Steric Effects in Organic Chemistry*, M. S. Newman, Ed., Wiley, New York, 1956, p. 556.
345. L. P. Hammett, *Physical Organic Chemistry*, McGraw-Hill, New York, 1940, p. 184.
346. C. H. Blomquist, *Acta Chem. Scand.*, **20**, 1747 (1966).
347. S. Hollmann, in reference 6, p. 704.
348. P. Talalay, in reference 3, p. 177.
349. P. Talalay. in *Methods of Biochemical Analysis*, Vol. 8, D. Glick, Ed., Interscience, New York, 1960, p. 119.
350. L. J. Langer, J. A. Alexander, and L. L. Engel, *J. Biol. Chem.*, **234**, 2609 (1959).
351. J. A. Adams, J. Jarabak, and P. Talalay, *J. Biol. Chem.*, **237**, 3069 (1962).
352. K. Dalziel and F. M. Dickinson, *Biochem. J.*, **160** 34 (1966).
353. K. Dalziel and F. M. Dickinson, *Biochem. J.*, **100**, 491 (1966).
354. G. J. Karabatsos, J. S. Fleming, N. Hsi, and R. H. Abeles, *J. Am. Chem. Soc.*, **88**, 849 (1966).
355. J. D. Shore and H. Thoerell, *Arch. Biochem. Biophys.*, **117**, 375 (1966).
356. J. van Eys and N. O. Kaplan, *J. Am. Chem. Soc.*, **79**, 2782 (1957).
357. S. P. Colowick, J. van Eys, and J. H. Park, in *Comprehensive Biochemistry*, Vol. 14, M. Florkin and E. H. Stotz, Eds., Elsevier, Amsterdam, 1966, p. 1.
358. B. D. Sanwal and M. W. Zink, *Arch. Biochem. Biophys.*, **94**, 430 (1961).
359. B. D. Sanwal, C. S. Stachow, and R. A. Cook, *Biochemistry*, **4**, 410 (1965).
360. B. D. Sanwal and M. Lata, *Arch. Biochem. Biophys.*, **98**, 420 (1962).
361. B. D. Sanwal, *Arch. Biochem. Biophys.*, **93**, 377 (1961).
362. C. Frieden, *J. Biol. Chem.*, **240**, 2028 (1965).
363. G. Hierholzer and H. Holzer, *Biochem. Z.*, **339**, 175 (1963).
364. J. van Eys, F. E. Stolzenbach, L. Sherwood, and N. O. Kaplan, *Biochim. Biophys. Acta*, **27**, 63 (1958).
365. G. Pfleiderer and E. Hohnholz, *Biochem. Z.*, **331**, 245 (1959).
366. S. Udenfriend, *Fluorescence Assay in Biology and Medicine*, Academic Press, New York, 1962.
367. S. F. Velick, in *Light and Life*, W. D. McElroy and B. Glass, Eds., The Johns Hopkins Press, Baltimore, 1961, p. 108.
368. A. D. Winer and G. W. Schwert, *Biochim. Biophys. Acta*, **29**, 424 (1958).
369. J. P. Perkins and J. R. Bertino, *Biochemistry*, **5**, 1005 (1966).
370. A. D. Winer, *Biochim. Biophys. Acta*, **59**, 219 (1962).
371. A. D. Winer and G. W. Schwert, *J. Biol. Chem.*, **234**, 1155 (1959).

372. A. D. Winer and G. W. Schwert, *Science*, **128**, 660 (1958).
373. G. H. Czerlinski and G. Schreck, *J. Biol. Chem.*, **239**, 913 (1964).
374. A. Hvidt and J. H. R. Kägi, *Compt. Rend. Trav. Lab. Carlsberg*, **33**, 497 (1963).
375. G. di Sabato and M. Ottesen, *Biochemistry*, **4**, 422 (1965).
376. A. Rosenberg, H. Theorell, and T. Yonetani, *Arch. Biochem. Biophys.*, **110**, 413 (1965).
377. B. H. Havsteen, *Acta Chem. Scand.*, **19**, 1643 (1965).
378. J. D. Shore and K. Woo, *Federation Proc.*, **26**, 841 (1967).
379. C. Frieden, *J. Biol. Chem.*, **234**, 815 (1959).

Subject Index

A

Acid reductases, of bacteria, 95
"Activated carbon dioxide," hypothesis of CO_2 fixation, 126–127
Activation, in phosphorylation, nature of, 195–196
Active and controlled state, computer representation of, 35
  difference spectrum, of liver mitochondria, 33
  DPN reduction in, 34–36
  inhibition of respiration in, 31–32
  redox pattern of respiratory chain in, 33
  redox states of DPN and ubiquinone in, 36–37
  transitions from, 30–31
"Active formaldehyde," 475–478
  and "active methyl," interconversion of, 461–463
  glyoxylate analog of, 477–478
  nature, 476
"Active formate," in histidine synthesis, 472–473
  in protein synthesis, 473–474
  in purine synthesis, 471–472
  and "active formaldehyde," interconversion of, 460–461
Acyl phosphate intermediates, in P_i or ADP activation, 196
Adenine nucleotides, compartmentation of, in mitochondria, 204
Adenosine-5-phosphosulfate (APS). See APS reductase.
Adenosyl-B_{12}, structure of, 483
  synthesizing system, properties of, 488–489
ADP, effect on respiration, 28–30
ADP level, as metabolic control on oxygen uptake, 215

Alcohol dehydrogenase, amino acid sequences around reactive cysteine residues in, 666
DPNH effect on optical rotatory dispersion of, 690
  fluorescence emission spectra of, 686
  of horse liver, 689
  light absorption changes accompanying complex formation of, 681
  spatial arrangement of substrate and coenzyme in transition state of, in hydrogen transfer, 678
  steady state kinetics of, 288
  of yeast, 689
Alcohol dehydrogenases, 668–669, 675–677
Aldehyde dehydrogenase, maximal oxidation rates of, and Taft sigma values, 674
Aldehyde dehydrogenases, 673
Aldehyde oxidase, catalytic activities of, 311
  cyanide inactivation of, 318
  development of EPR signals in, 325
  electron transport in, 315
  electron transport sequence of, 322–325
  EPR signals during reoxidation of, 322–323
  inhibition of, 314–315
  metal–flavin ratio of, 307–308
  role of lipid in structure of, 307
  specificity of, 308–309
  spectrophotometric changes in, 323–324
  substrate reduction of, 323–325
Aldehyde and xanthine oxidases, anaerobic titrations of, 323–327
  EPR spectra of, 317
  EPR studies on, 320–327
  methanol inhibition of, 315–316

role of electron carriers in catalysis by, 312–320
substrate activation and hydroxylation by, 315
Alkaline photolysis, of flavins, 241
Alloxane, 241
Alloxazines, 241
D-Amino acid oxidase, fluorescence of, 283–284
  reaction mechanism for, 293–295
  shifts in flavin spectra of, 266–267
  spectra of transient intermediates of, 276
  steady state kinetics of, 289
  substrate-induced spectral changes in, 279
L-Amino acid oxidase, of snake venom, reaction mechanism for, 291–292
Aminopterin derivatives, structures of, 456
1,2- and 1,3-Aminothiols, possible role in phosphorylation, 211
Amytal, effect on DPNH oxidation, in bacteria, 97
  inhibition by, 20–21
  role of, in reversed electron transfer, 41–42
Antimycin, inhibition by, 22–23
  respiratory inhibition by, 207–208
Antimycin A, as inhibitor of cyclic photophosphorylation, 148
  role in reversed electron transfer, 41–42
APS reductase, 95
Arsenate, uncoupling by, 209
Artificial electron acceptors, and function of quinones, 23–24
Ascorbic acid oxidase, 428–431
  mechanism of action of, 430–431
  reaction catalyzed, 428
  reaction inactivation, 428–429
  state of copper in, 429–430
ATP, formation in photosynthesis. See Photophosphorylation.
  secondary uncoupling of, 209
ATPase, of mitochondria, 219–220
ATP-driven reduction, characteristics of, 221

ATP formation, increased ATP hydrolysis in, 210
  in noncyclic photophosphorylation, 136
  nucleotide transfer or interconversion block of, 210
  other inhibitor actions on, 210–212
  phosphorylation block of, 209–210
  primary uncoupling of, 208–209
  respiratory inhibition of, 207–208
"ATP-jump," 223
"ATP synthetases," as coupling factors, 214
Atractyloside, inhibition by, 210
Azide, inhibition of ATP formation by, 210–211

B

Bacteria, membrane structures of, 57–60
  structure of, 57–60
Bacterial membranes, composition and properties of, 68–72
  protein content of, 60
Bacterial respiratory chain, cytochrome requirement of, 73
B_{12} coenzymes, 482–502
  (See also Vitamin B_{12}).
  routes for synthesis of, 491
  structure, nomenclature, and chemistry of, 482–487
B_{12} enzymes, diol dehydrase, 495–497
  (see also Diol dehydrase)
  glutamate and methylmalonyl CoA mutase, 492–494
  methionine synthetase, 498–502
  ribonucleotide reductase, 497–498
B_{12} reductase, 489–491
B_{12a} reductase, mechanism for, 491
Benzoquinones. See Quinones.
Broken-cell extracts, bacterial, subfractions of, 72
  oxidative activities of, from bacteria, 64–65
  preparation of, 60–62

C

Calcium ion, uncoupling by, 209
Carbohydrate, in membrane fractions of bacteria, 69

Carbon assimilation, ATP and $NADPH_2$ requirement of, 159
  energy requirements for, 125–128
  by isolated chloroplasts, 128–130
  and photophosphorylation. *See* Photophosphorylation.
Carbon monoxide, inhibition of bacterial oxidases, 80–82
Carrier–inhibitor intermediates, in oxidative phosphorylation, 226
Catalytic efficiency, of mitochondria, 205
Ceruloplasmin. *See* Laccase.
C_1 group, intramolecular transfer of, 468–471
  transfer of, to water or nitrogen, 471–475
C_1 groups, attachment sites to tetrahydrofolate, 441
C_1 units, tetrahydrofolate bound, mechanism for reduction of, 461
  oxidoreduction of, 460–463
Charge separation, as energized state in oxidative phosphorylation, 203
Charge-transfer absorptions, and flavin semiquinone absorption, 280
Chemiluminescence, 313–314
Chlorophyll, in cyclic photophosphorylation, 135
  in noncyclic photophosphorylation, 136–137
Chloroplast fluorescence, quenching of, by ferredoxin in cyclic photophosphorylation, 163
  by NADP in noncyclic photophosphorylation, 164
Chloroplasts, early reactions of photosynthesis in, 131–132
  photosynthesis in, separation of light and dark phases of, 130–131
  photosynthetic capacity of, 128–131
CO binding pigment, of microsomes, 176–177
Coenzyme Q, of chromatophores of *Rhodospirillum*, 566
  function in coenzyme Q reductases, 553–555 (*see also* Quinones)
  function in intact tissues, 552–553
  function in phosphorylating systems, 551–552

function in reduced coenzyme Q_{10}-cytochrome c reductase, 555–556
  lipid enclosed pathway for, in cytochrome oxidase, 575
  mitochondrial function of, 543–556
  oxidation-reduction of, 549–551
  redox state in heart and mitochondria, 554
  solvent extraction studies of function of, 544–548
Coenzyme Q function, site for, in electron transport, 543
CO_2 fixation. *See* Carbon assimilation.
Contractile processes, and oxidative phosphorylation, 227
Controlled state. *See* Active and controlled state.
Control of respiration, by coupling, 28–29
Copper enzymes, discovery of, 415–417
Copper protein, in photosynthetic apparatus, 165
Coupling, of phosphorylation in bacterial extracts, 107
"Coupling factor" proteins, 213–214
Coupling factors, 214
Crossover theorem, and site of energy transfer of respiratory chain, 32–34
Cyanide, xanthine oxidase inhibition by, 316–317
Cyclic photophosphorylation, 134–136
  in chloroplasts scheme for, 157
  ferredoxin in, 147–148
  inhibition of, 148
  participation of cytochromes in, 156
  sensitivity to inhibitors, 156
Cyclohydrolase, reaction mechanism for, 470–471
Cytochrome a, rate of oxidation of, 47
Cytochromes a and a_3, difference spectra of, 388–391
  ratios of, 392–394
Cytochrome b, 383–384
  of bacteria, 86
  function in phosphorylating and non-phosphorylating preparations, 25–26
  kinetics of oxidation of, 48–49
  in reversed electron transfer, 27
  role of, 25–27

Cytochrome b_1, of bacteria, relation to quinones, 92

Cytochrome b_2, 384

Cytochrome b_5, 174–176, 384
absorption spectra of, 175
apoenzyme, recombination with heme, 176

Cytochrome b_5 reductase, absorption spectra of, 178
apoprotein preparations of, 178–179
difference spectra of complexes of, 185
flavin reduction of, by DPNH, 185
oxidation and reduction rates of, 186
reaction sequence in, 186–188
substrate specificity of, 184

Cytochrome b_{555}, as intermediate in oxidative phosphorylation, 226

Cytochromes b and c, of bacteria, 84–86

Cytochrome c, 380–383
comparative studies of, 382
comparison of bacterial and mammalian types, 84–85
DPNH reducible, relation to viable count and turbidity, 77
oxidation of, by cytochrome oxidase, 395–397
oxidation-reduction by bacterial particles, 98
reduction of, 313
structure, 380–382

Cytochrome c_1, 384–386
electron acceptor for, 385
removal of, effect on respiratory rate in bacteria, 101

Cytochrome c reductase, titration of, 182

Cytochrome chain, as energy-producing reaction in photosynthetic bacteria, 112

Cytochrome components, of bacteria, 73

Cytochrome content, of bacteria, effect of age on, 76–77
effect of growth medium on, 75
effect of oxygen tension on, 75–76
of particle fractions of bacteria, 80

Cytochrome f, 151
redox potential of, 152

Cytochrome o, 81

Cytochrome oxidase, copper in, 394–395
history of, 386–388

reaction kinetics of, 395–397
reaction mechanism of, 402–404
reaction stoichiometry of, 394–395
reaction with oxygen, 400
reaction with oxygen and ligands, 397–402
spectrophotometric appearance of, 388–391

Cytochromes, bacterial, attachment to membrane fragments of, 67
effect of growth conditions on, 75–79
maximum levels attained, 79
of bacterial respiratory chain, 72–86
interaction between chains, 18
kinetics of oxidation-reduction in bacteria, 96
in reversed electron transfer. See Reversed electron transfer in anaerobiosis.
role in two light reactions of photophosphorylation, 152–153

D

Dark respiratory chain, of photosynthetic bacteria, 112–113

Dehydrogenase content, of bacteria, effect of age of culture on, 77

Dehydrogenases, bacterial, 88 (see also Respiratory dehydrogenases)
of bacterial preparations, 89
cytoplasmic, of bacteria, 86
electron acceptors in assay of, 344–347
firmness of attachment to bacterial membranes, 66–67
interaction with cytochromes, of bacteria, 98–100
membrane linked, of bacteria, 87–88
pyridine nucleotide-dependent coenzyme stereospecificity of, 633–634
as zinc enzymes, 667–670

Deoxycytidylate hydroxymethylase, 478–481

Desaspidin, inhibition, in two light reactions of photophosphorylation, 155

2,6-Dichlorophenol indophenol, as carrier in noncyclic photophosphorylation, 154–155

Difference spectra, of mammalian and bacterial cytochromes, 74

Dihydrofolate, tautomeric forms of, 449–452 (see also Folate, Folic acid, and Tetrahydrofolate)

Dihydrofolate reductase, 452–458
  aminopterin titration of, 457
  enhanced activity of, 454–455
  inhibitors of, 456–458
  properties of, 453

Dihydroorotic dehydrogenase, 327–332
  absorption spectrum of, 304
  allosteric effect of DPN on, 330–331
  catalytic activities of, 311–312
  effect of substrates on absorbancy of, 330
  electron transport in, 329
  EPR study of kinetics of, 332
  flavin semiquinone in, 329
  metal–flavin ratio of, 308
  role of iron in, 327–328
  role of sulfhydryl in, 327–329
  spectra during anaerobic titration of, 330
  spectra of orotate binding to, 328
  substrate specificity of, 312
  transport sequence of, 331–332

Dihydropteridine, structure of, 597

7,8-Dihydropteridine, proton magnetic resonance spectra of, 451

Dinitrophenol, uncoupling of phosphorylation by, in bacteria, 107–108

2,4-Dinitrophenol, uncoupling by, 208–209, 212

Diol dehydrase, reaction mechanism for, 496 (see also B_{12} enzymes)

Dioxygenases, 585–588 (see also Pyrocatechase, Metapyrocatechase, Monooxygenases, and Oxygenases)
  heme-containing, tryptophan pyrrolase, 593–594
  inorganic iron containing, mechanism of, 591–593
  subgroups of, 586–587
  tryptophan pyrrolase, mechanism for, 594

DPN, α-isomer of, 609
  hydrogen transport function of, 17

DPN analogs, specificity, spectra, and oxidation-reduction potentials of, 606–607

DPN and TPN, 604–605 (see also NAD and Pyridinium compounds)
  structure of, 605

DPN and TPN analogs, 605–609

DPNH, absorption, excitation, and emission spectra of free and bound forms of, 631
  free and bound, molar extinction coefficient of, 616
  kinetics of oxidation of, 43–49

DPNH analogs, in mechanism of cytochrome b_5 reductase, 185–186

DPNH-CoQ reductase, 356–361
  in extracts of particles, 358

DPNH-cytochrome b_5 reductase, 177, 184–188
  reaction mechanism for, 293

DPNH-cytochrome c reductase, of microsomes, 173–174

DPNH dehydrogenase, 347–356
  acceptor specificity of, 348
  of bacteria, 86–87
  comparison with DPNH-CoQ reductase, 349–350
  conversion to DPNH-CoQ reductase, 360
  CoQ_1 reductase activity on degradation of, 351
  difference spectrum of, 348
  effect of p-chloromercurisulfonate on, 354
  fragmentation of, 350–356
  as multienzyme complex, 355–356
  physical and chemical properties of, 347–348
  substrate-induced transformation of, 353
  thermal inactivation of, 351

DPNH models, hydrogen transfer with, 622
  stability of cyanide complexes and rate of hydrogen transfer by, 623–624

DPNH oxidation, characteristics of, 352

E

Electron acceptors, artificial, criteria for, 344–345
  specificity of, 345–346
  of cyclic photophosphorylation, 135

Electron flow, in cyclic photophosphorylation, 135

Electron paramagnetic resonance, of flavoproteins, 284–286

Electron transfer, inhibitors of, 20–23
    temperature dependence of, 19

Electron transport, sequence of, in bacteria, 96–104

Electron transport and phosphorylation, stoichiometry of, 203

"Elementary particle," identity of, 17–18

Energized states, in oxidative phosphorylation, 202–203

Energy-linked reduction, independence of, from adenine nucleotides, 221
    relation to oxidative phosphorylation, 220–221

Enzymes of carbon assimilation, separation of, from chloroplasts, 130

EPR signals, of aldehyde and xanthine oxidases, 320–327

ETP, of bacteria, 97–98

Exchange reactions, of purified enzymes, 216–217

F

FAD, hydrolysis of, 240
    internal complex formation of, 253–254

Far-red light, in demonstrating cyclic photophosphorylation, 147–148
    as inhibitor of oxygen production by chloroplasts, 147,148

Fermentation, relation to photophosphorylation, 133

Ferredoxin, absorption spectra of, 518
    (see also Non-heme iron proteins)
    bacterial and spinach, properties of, 143
    bacterial type, spectra of, 140
    catalytic activity of, 521–526
    chemical properties of, 142–143
    chloroplast type, spectrum of, 141
    of C. pasteurianum, 525–526
    comparison of properties of, 519
    in cytochrome c reduction, 523–524
    definition of, 138–139
    historical background of, 137–138
    iron of, 142–143
    iron and labile sulfur of, 528–529
    and NADP reduction, 143–145

    in nitrite reduction, 524–525
    photoreduction of and oxygen production, stoichiometry, 145–146
    photoreduction reactions of, 521
    properties of, 518–521
    proposed structure of, 526–527
    redox potential of, 138
    in TPN reduction, 521–523

Ferredoxin-NADP reductase, 143–144

Ferredoxins, of bacteria and green plants, 137–138
    "labile sulfide" of, 142–143
    oxidation-reduction potentials of, 142
    spectral characteristics of, 139–142

Ferricyanide, as electron acceptor, 24

Flavin, fixation of, to apoprotein, 245–247
    valence state and chemical and physical properties of, 244

Flavin adenine dinucleotide. See FAD.

Flavin and flavoproteins, semiquinoid oxidation state in, spectra of, 267–274

Flavin-copper chelate, scheme of charge transfer in, 255

Flavin–metal complex, charge transfer in, 260
    effects on oxidation-reduction, 256–258
    spectra of, 258

Flavin–metal interaction, 254–261

Flavin–metal systems, electron flow in, 257

Flavin mononucleotide. See FMN.

"Flavin nucleotides," nomenclature of, 240

Flavin peptides, fluorescence curves of, 364

Flavin radicals, ESR signal saturation of, 260 (see also Flavosemiquinone)

Flavins, emission spectra, effect of solvent on, 282
    environmental influences on properties of, 247–248
    fluorescence of, pH effects on, 281
    solvent effects on, 281
    mechanism of photolysis of, 241
    molecular complexes of, 253–254
    redox mechanisms and kinetics, 261–262
    spectra of, solvent effects on, 265
    synthesis, biosynthesis, and degradation of, 242

Flavins and flavoproteins, charge-transfer absorptions of, 277–280
  charge-transfer complexes of, 279
  oxidized, spectra of, 264–267
  reduced, oxygen reactivity of, 286–287
  spectra of, 280
  substrate-produced long-wavelength absorption of, 274–277
Flavin side chain, nature of, 240–241
Flavin solutions, partially reduced, equilibria of, 250
Flavin species, interrelationships, at different pH and redox states, 243
  visible absorption of, 259
Flavocoenzymes, nomenclature of, 239–244
  structure and cleavage loci of, 240
Flavohydroquinone, 248–250
  absorption spectra of, 249
  alkylation of, 248
  O- and N-alkylation of, 246
  ionization states of, 248–250
  lack of chelation by, 254
  planarity of, 249–250
Flavoprotein mechanisms, kinetic parameters for, 288
Flavoproteins, 239–240
  fluorescence of, 281–284
  fluorescence quenching of, 282–283
  lack of EPR signal, mechanism for, 285
  of microsomes, mechanisms of, 181–188
  photodenaturation of, by flavin, 241
  properties of, 268–273
  reaction mechanisms of, 289–296
    of fully oxidized and fully reduced forms, 290–291
    fully reduced flavin forms, 292–293
    number of flavin molecules at active center, 290
    semiquinoid forms involving additional redox groups, 295–296
    substrate complexed flavin semiquinone forms, 293–295
    true semiquinoid forms, 291–292
  respiratorychain-linked, concentration in cells, 374
  redox state of, 374–375

Flavoprotein systems, steady-state kinetics of, 287–289
Flavoquinone, 244–248
  alkylation of, 245
  copper complex of, 255–256
  lack of metal chelation by, 254
  protonation of, 244–245
  reactivity of 8-methyl group of, 247
Flavosemiquinone, 250–253 (see also Flavin radicals)
  absorption maxima of, 251
  ESR spectra of, 251–253
  paramagnetism of, 251–253
  stabilization of by metal chelation, 261
Fluorescence, of chloroplasts, quenching of, by photophosphorylation, 163–164
FMN free radicals, ESR spectra, 252
Folate, oxidoreduction of pyrazine ring of, 446–449 (see also Dihydrofolate, Folic acid, and Tetrahydrofolic acid)
  routes for interconversion of, 446
Folate coenzymes, 440–482
  absorption spectra of, 443–444
  fluorescence spectra of, 444
  involvement in oxidative reactions, 446–463
  lability of, 443
  metabolic reactions of, 442
  non-oxidative reactions of, 463–482
  nucleophilic adducts of, 448
  optical rotation of, 444–445
  structure, nomenclature, and chemistry of, 440–446
Folate and dihydrofolate, mechanisms for reduction of, 447
Folic acid, structure of, 440 (see also Dihydrofolate, Folate, and Tetrahydrofolic acid)
Folinic isomerase, 468–470
  reaction mechanisms for, 469
Formate-activating enzyme, 463–468
  Michaelis constants for, 464
  reaction mechanisms for, 467
Formate activation, mechanism for, 464–468
Formyl- and formimino-transferring enzymes, 468–475

Free energy changes, in oxidations, 206
Fumarate reductase, 368–369

G

Glucose oxidase, reaction mechanism for, 290–291
  spectra of semiquinoid species of, 274
Glutamate dehydrogenase, 670, 672–673
  association–dissociation equilibrium of, schematic representation, 659
  dissociation of by DPNH, sedimentation diagrams of, 660
  fluorescence spectra of, 685
  quaternary structure and activity of, 658–662
  thyroxine dissociation of, effect of ADP on, 661
Glutamate mutase. *See* B_{12} enzymes.
Glutathione reductase, half-reduced form of, 274–275
  reaction mechanism for, 295–296
Glyceraldehyde-3-phosphate dehydrogenase, 669–670
  amino acid sequence around reactive cysteine and lysine residues in, 665
  multiple functions of, 677–678
  reaction of yeast and DPN, rate and equilibrium constants of, 694
  scheme for equilibria between yeast and DPN, 694
  temperature effect on spectrophotometric titration of, 693
  of yeast, 689–690
    DPN binding by, 692–694
Glycerol, effect on electron transport, 19
Glycine, oxidative decarboxylation pathway of, 480
Glycine dehydrogenase, 480–481
Gram-negative bacteria, membrane structure of, 59–60
Gram-positive bacteria, membrane structure of, 59
Group translocation hypothesis, of oxidative phosphorylation in bacteria, 110–111
Guanidines, inhibition of phosphorylation by, 211

H

Heme protein. *See* Cytochromes.
Hemiacetal dehydrogenases, 673–675
Hemophilus parainfluenzae, 57–58
  broken cells of, 63
  membrane bound respiratory system of, 102
Hill reaction, 128
Hydrogenase, and ferredoxin in obligate anaerobes, 137–138
C_{21}-Hydroxylase, 179–180
Hyroxylation reactions, CO-binding pigment in, 180–181
  mechanism of, in microsomes, 179–181
  TPNH-specific reductase in, 180–181
Hydroxysteroid dehydrogenases, 675

I

Inhibitors, of oxidase, in sequence studies in bacteria, 96
  of phosphorylation, complex action of, 212
Inorganic phosphate, compartmentation of, in mitochondria, 204
  oxygen loss of, 196–200
Inorganic substrates, oxidation by bacteria, 89
Intrinsic factor, B_{12} binding by, 487
Iron flavoproteins, absorption characteristics of, 306
  absorption spectra of, 302–306
  oxygen reduction by, 314
Iron proteins, of dihydroorotic dehydrogenase, absorption spectra of, 305
  of xanthine oxidase, absorption spectra of, 305
Isoalloxazines, 241

K

Keilin-Hartree preparation, nature of, 213

L

Labile sulfur, nature of, in ferredoxin, 529
Laccase, 431–435
  ESR spectra of, 432
  mechanism of, 434–435
  specificity of, 433–434
  state of copper in, 431–433

Lactate dehydrogenase, of beef, DPNH binding by, 692
  catalytic stoichiometry of hybrid forms of, 663
  and isoenzymes of, 662–664
Lipid, in membrane fraction of bacteria, 69
Lipoprotein membrane fragments, and electron transport, 172
Lipoyl dehydrogenase, charge transfer complexes of, 277–278
  fluorescence of, 283
  half-reduced form of, 274–275
  reaction mechanism for, 295–296
  steady state kinetics of, 287
Lumichrome, formation of, 241–242
Lumiflavin, formation of, 241–242

M

Membrane-bound enzymes, interaction of soluble enzymes with, in bacteria, 104
  role in bacterial transport, 105–106
Menadione reductase, of bacteria, 91
Metal-free flavoproteins, magnetic resonance studies on, 284–286
Metalloflavoproteins, catalytic activities of, 308–312
  EPR studies of, 320–327
  nature, 301–302
  physical properties and chemical composition of, 306–308
  role of labile sulfur in, 333–334
  xanthine and aldehyde oxidases, 312–327
Metaphosphate, formation, in mechanism of phosphorylation, 197
Metapyrocatechase, mechanism for, 592 (see also Dioxygenases and Pyrocatechase)
Methane formation, B_{12} coenzymes in, 502
Methionine synthetase, 481–482 (see also B_{12} enzymes)
  mechanisms proposed for, 500
  multiple forms of, 499
5,10-Methylene tetrahydrofolate dehydrogenase, 460–461
Methylmalonyl CoA mutase. See B_{12} enzymes.

Microsomal electron transport, overall function of, 188–189
Microsomes, carbon monoxide binding pigment of, 176–177
  oxidative components and chemical properties of, 174–179
  oxidative reactions of, 173–174
  physical and chemical properties of, 172
Mitchell hypothesis, 228
Mitochondria, as carriers of respiratory chain, 15
  cytochrome content of, 8–9, 11
  effect of aging on, 204–205, 209
  fractions of, phosphorylation by, 212–214
  metal ions in, 204–205
  molar composition of respiratory chain of, 11
  oxidative phosphorylation in, inhibition and uncoupling of, 207–212
  P:O ratios and phosphorylation sites of, 205–207
  respiratory control in, 214–215
  surface packing of respiratory components, 16
Mitochondrial lipids, role in electron transfer, 19
Mixed function oxidases, 171
  hydroxylases, 173 (see also Monooxygenases)
Molybdenum, in xanthine dehydrogenase, 319
Monooxygenase reactions, mechanism of, 596–599
Monooxygenases, 585–588 (see also Mixed function oxidase and Oxygenase)
  camphane lactonizing enzyme, reaction sequence for, 597
  general mechanism of, 594–600
  mechanism for double hydroxylation by, 599–600
  oxidation of kyneurenic acid, 595
  peracid mechanism of camphor lactonization, 598
  phenylalanine hydroxylase, mechanism for, 596
  subgroups of, 587–588

Multicomponent fragments, of respiratory
    chain. *See* Multicomponent
    particles.
Multicomponent particles, 12–14
Multiple oxidases of bacteria. *See*
    Oxidases.
Myosin ATPase, 227

N

NADP, oxidized form of, in pseudo-
    cyclic photophosphorylation,
    149–150 (*see also* DPN, TPN,
    and Pyridinium compounds)
NADP reduction, by chloroplasts,
    mechanism, 143–145
Naphthoquinones. *See* Quinones.
Net phosphorylation capacity, of mito-
    chondria, 205–215
Nitrate, cytochrome reduction by, in
    bacteria, 93
Nitrate reductase, 92–94
    competitive inhibition by oxygen, 94
Noncyclic photophosphorylation, 136–137
    ATP formation in, 157–158
    in chloroplasts, scheme for, 158
    energy balance of, 160
    energy output of, 144–145
    equation for, 147
    with ferredoxin, 146–147
    oxygen evolution in, 153–155
    sensitivity to inhibitors, 156
    as sum of two light reactions, 152–153
    two light reactions in, scheme of, 154
Non-heme iron, absorption spectra of,
    303–305
    of bacterial respiratory chain, 95–96
    in mechanism of phosphorylation, 197
Non-heme iron complex, 332–334
Non-heme iron proteins, classification,
    516–518 (*see also* Ferredoxin)
    optical rotatory dispersion of, 333
Nonphosphorylated intermediates,
    possible nature of, 201–202
Nucleic acids, of bacterial membranes, 69
Nucleophiles, in phosphorylation, 196–
    197

O

Oligomycin, inhibition of phosphorylation
    by, 209–210

One-carbon metabolism. *See* Folate
    coenzymes.
Osmotic pressure, effect on respiration of
    bacterial membranes, 71–72
Oxidases, bacterial, 80–84 (*see also*
    Oxygenases)
    relation to nitrite reductase, 83
    copper-containing, biological function
        of, 419
    classification and nomenclature of,
        418
    properties of, 417
Oxidative phosphorylation, ATPase in,
    219–220
    basic mechanistic considerations, 195–
        203
    common chemistry for phosphoryla-
        tions of, 202
    covalent bond formation by ADP in,
        scheme for, 225
    dynamic reversibility of, 215
    energy-linked reductions of, 220–221
    exchange and partial reactions of, 215–
        220
    experimental approaches to, 203–229
    inhibitors of, effects on purified enzyme
        systems, 212
    intermediates of, 222–227
        from ADP, 225–226
        derived from P_i, 223–224
        existence and amount of, 222–223
    ion transport, swelling and contraction
        in, 221–222
    mitochondria in relation to, 204–205
    nonphosphorylated intermediates in,
        226–227
    participation of adenine ring in, 225
    P_i or ADP activation in, 195–196
    relation to other energy-linked systems,
        227–228
    schemes for water formation from P_i,
        200
    studies with model systems, 228–229
    substrate interconversion and dissocia-
        tion steps of, scheme for, 216–217
Oxygen, production of, by noncyclic
    photophosphorylation, 145–147

Oxygenases, classification of, 585–588
(*see also* Dioxygenases, Monooxygenases, *and* Oxidases)
based on cofactors, 586
demonstration of action of, 584–585
distribution and relative activities of,
588–589
general mechanism of, 591
general properties and mechanism of
action, 591–600
historical development of, 582–584
oxidation and oxygenation, 581–582
physiological significance of, 589–591
role in evolution, 600–601
Oxygen-dependent photophosphorylation.
See Pseudocyclic photophosphorylation.
Oxygen exchange, ADP requirement of,
218
between ATP and water, 215–220
differential rates of, explanations for,
216–217
in phosphorylation, 196–200
between P_i and ATP, 215–220
between P_i and water, 215–220
Oxygen transferases. *See* Dioxygenases.

P

Particulate-bound DPN, as coupling
factors, 214
Peroxidase, 404–405
kinetics of, 406–410
mechanism of action of, 406–410
nature of kinetic intermediates, 409–410
properties of, 405
prosthetic group of, 405
reaction pathways of, 407–408
spectra of, 406
spectroscopic observations on, 406
Phenylalanine hydroxylase, 460
p-Phenylenediamine, oxidation by bacterial preparations, 80
Phosphate esters, in mechanism of phosphorylation, 198
Phosphohistidine, as intermediate in oxidative phosphorylation, 223–224
Phospholipids, of bacterial membranes,
68–69
function in mitochondria, 204

Phosphorylated "coupling factor," as intermediate in oxidative phosphorylation, 223–224
Phosphorylated intermediate(s), possible
nature of, 200–201
Phosphorylated oxidation-reduction components, 223
Phosphorylating systems, of bacteria,
similarity to mitochondria, 107
Phosphorylation, in bacteria, 106–111
energy requirement of, 206
stimulation by soluble factors of bacterial cells, 108–109
Phosphorylation sites, in cyclic photophosphorylation, 156
Phosphoserine proteins, as intermediates
in oxidative phosphorylation, 224
Photodecomposition, of flavin side chain,
241
Photophosphorylation, 127–128
and carbon assimilation, 134
cyclic and noncyclic, mechanisms of,
156–159
discovery of, 132–134
hypothesis of, 159–162
and oxidative phosphorylation, 227–228
relation to respiration, 132–133
Photosynthesis, two light reactions in, of
whole cells, 150–151
Photosynthetic bacteria, 112–113
Photosynthetic carbon cycle. *See* Reductive pentose phosphate cycle.
Photosynthetic phosphorylation. *See*
Photophosphorylation.
Plastocyanin, 165
Plastoquinones, in Hill reaction, 563 (*see
also* Quinones)
in TPN and ferricyanide reduction,
563–564
Polyol dehydrogenases, 673–675
P:O ratios, of cells and cell-free extracts
of bacteria, 106
Protein, in membrane fractions of bac-
teria, 69
Proteins, conformation change of, in oxi-
dative phosphorylation, 202–203
Protoplasts, 59–60
Pseudocyclic photophosphorylation,
148–150

Pteridines. *See* Dihydropteridines *and* Folate.

Pyridine coenzyme analogs, specificity of dehydrogenases for, 608

Pyridine compounds, reduced, oxidation of, 622

Pyridine nucleotide coenzymes, absorption and fluorescence spectra of, 626–631 (*see also* DPN, TPN, NAD, *and* Pyridinium compounds)

activation and emission maxima of free and bound forms of, 629–630

chemical and physical properties of, 609–635

fluorescence spectra of, 626–631

nomenclature of, 604

oxidation-reduction potentials of, 624–626

oxidative phosphorylation and, 624

reduction and oxidation of, 609–624

stereochemistry of, 631–635

structure of, 604–609

ultraviolet absorption spectra of, 626, 627

Pyridine nucleotide dehydrogenases, binary and ternary enzyme–coenzyme–substrate and –inhibitor complexes of, 679–694

coenzyme specificity of, 678–679

coenzyme specificity and binding between enzyme and coenzyme of, 678–694

crystalline or homogenous, chemical and physical properties of, 643–670

molecular properties of, 644–657

enzymic activity of, 659–664

enzymic properties of, 670–694

molecular weight and subunit structure of, 643–658

primary structure of active centers of, 664–667

substrate specificity of, 672–678

types of reactions catalyzed by, 670–672

Pyridine nucleotide reductase, photosynthetic, catalytic activity of, 521–525

Pyridine nucleotides, binding sites and dissociation constants of, 687

conformational changes following formation of binary and ternary complexes of, 689–690

Cotton effect of alcohol dehydrogenase–DPNH complex of, 690–691

fluorescence spectra of binary complexes of, 681

reduction of, by chloroplasts, 143–144

ternary complexes of, 687–689

between enzymes, and reduced substrates or inhibitors, 682–684

ultraviolet absorption spectra of binary complexes of, 680–681

Pyridinium addition compounds, association constants of salts of, 613

relation between cyanide complex stability and absorption maxima of, 614

Pyridinium compounds, addition reactions of, 612–616 (*see also* DPN, TPN, *and* NAD)

cyanide addition complexes, association and DCIP oxidation rate constants of, 623

isomeric, absorption spectra of, 611

mechanism in enzyme-catalyzed reaction of, 621–624

one-electron reduction of, 618–619

oxidation of dihydropyridines, 619

reduction to dihydropyridines, 610–612

reduction to tetrahydropyridines, 616–618

ultraviolet absorption of, 615

Pyrocatechase, mechanism for, 593 (*see also* Dioxygenases *and* Metapyrocatechase)

Q

Quantum efficiency, of *Chlorella* cells, at two light wavelengths, 150–151

Quinone changes, growth conditions and, 540–543

Quinone content, effect of growth conditions on, 542

of spinach chloroplasts, 562

Quinone function, effect of other lipids on, 568–569 (*see also* Coenzyme Q, Plastoquinone, Vitamin K, *and* Ubiquinone)

ESR studies on, 568
genetic studies of, 575–576
study of, techniques for, 540
Quinone reductases, 556–557
Quinones, association of non-heme iron
    with, in phosphorylation, 571
bacterial, reduction by substrate, 90–91
of bacterial respiratory chain, 89–92
in chloroplasts, 165, 561–565
distribution of, 538–540
of *E. coli*, electron transport path in
    ETP of, 560
effect of addition to bacterial extracts,
    90
electron transport pathway of *Hemo-
    philus*, 561
as link between flavoproteins and cyto-
    chromes in bacteria, 100
mechanisms of quinol phosphate for-
    mation, 570
mitochondrial, in split-pathway of elec-
    tron transport, 548–549
of non-mitochondrial electron trans-
    port, 567–568
phosphate derivatives in oxidative
    phosphorylation, 228
and phosphorylation mechanisms,
    569–575
in the photosynthetic apparatus, 165
relation to soluble phosphorylation
    factors of bacteria, 109–110
in restoration of electron transport and
    phosphorylation, 559
restoration of succinoxidase activity by,
    548
role in bacterial electron transport,
    89–91
sites for, in chloroplast electron trans-
    port, 564
succinic cytochrome *c* reductase and
    succinic oxidase, restoration by, 545
types of, 533–538
in various photosynthetic systems,
    565–567

R

Reconstitution, 13–14
of succinate dehydrogenase, with
    "*b*-*c*₁ particle," 14

Redox changes, of respiratory com-
    ponents, 31
Redox equilibria, anaerobic, redox ratios
    of DPN and cytochrome *c* in,
    44
Redox pattern, competitive effect of ATP
    and ADP on, 38–40
Reduced cytochrome *c*, reaction with bac-
    terial oxidases, 82–83
Reductive pentose phosphate cycle, 125–
    126, 131
energy requirement for, 127
Regulatory mechanism(s), in photophos-
    phorylation, 159
Respiration, control of, 28–29
control mechanisms in, 34–36
Respiration rates, of bacteria, indepen-
    dence of cytochrome content, 101–
    102
of bacteria and aged particles, 99
comparison of in cytochrome mutants
    of bacteria, 103
Respiratory carriers, kinetics of oxidation
    and reduction of, 46–49
Respiratory chain, artificial electron
    donors and acceptors of, 23–25
ATP and ADP effect on redox pattern
    of, 40
of bacteria, properties of components
    of, 72–96
    scheme of, 56
based on redox potentials, 7
carriers of, 6–8
components of, 6
construction of, from complexes, 14
definition, 3–4
electron and hydrogen pathways in, 5
energy-linked redox equilibrium in, 37,
    39
external donors and acceptors of, 24
inhibition sites in, 20
mechanism of interaction, 18–19
multicomponent fragments of, 13
principle of, 4–6
redox patterns of, 45
redox potentials of carriers of, 7–8
steady states of, 27–36
steady state redox behavior of, 29–36
structural and molecular organization
    of, 15–18

structure of, in bacterial extracts, 60–72
submitochondrial preparations, fragments, and reconstitution of, 12–14
Respiratory chain enzymes, of bacteria, location and nature of linkages to membranes, 62–67
  localization of, in bacterial membranes, 61–62
  loss of, in broken-cell extracts, 66
Respiratory chain preparations, composition of, 8–12
  cytochrome content of, 8–10
Respiratory components, rates of oxidation of, by oxygen, 47
  steady state reduction of, 30
Respiratory control, 29
Respiratory dehydrogenases, electron acceptor specificity of, 342 (see also Dehydrogenases)
  extraction and lability of, 341–342
  general properties, 340–342
  linkage to respiratory chain, 342–344
  as metalloflavoproteins, 340–341
  phospholipid linkage to transport particles, 342–343
  reaction with natural electron acceptors, 346–347
  salt linkage to transport particles, 343
Respiratory enzyme activities, comparison of, between ETP and bacterial particle, 70
Respiratory pigments, of bacterial membranes, mechanism of action of, 96–104
  lack of fixed stoichiometry in bacteria, 101–103
Reverse electron flow, in mitochondria, 220
Reversed electron transfer, 36–45
  under aerobic conditions, 36–39
  in anaerobiosis, 42–45
  ascorbate and TMPD in, 44–45
  DPN reduction in, 36–37
  external acceptors and donors in, 39–40
  pathways of, 41–42
  rates of, 49
  redox pattern of, 43
Reversed hydrogen and electron transfer, pathways of, 41

Riboflavin, 3-methyl tetraacetyl-, spectra of, 265
Ribonucleotide reductase. See B_{12} enzymes.
Rotenone, effect on DPNH oxidation, in bacteria, 97
  inhibition by, 21–22

S

Serine hydroxymethylase, 478–481
  reaction mechanism for, 479
Serum albumin, protective effect of, on phosphorylation, 211
Soluble ATPase, as coupling factor, 213
Soluble DPNH-CoQ reductase, rotenone titration of, 359
Soluble factors, of oxidative phosphorylation in bacteria, 108–109
Soluble redox systems, addition of, to bacterial fractions, 98
Spectra, substrate influence on, 274–277
Spin density, localization of, in flavosemiquinone, 253
Steroid and drug hydroxylation reactions. See Hydroxylation reactions.
Structural changes, in bacterial membranes, 67–68
Submitochondrial preparations, prosthetic group content of, 9, 12
"Substrate-level" phosphorylation, 194
Succinate dehydrogenase, 361–373
  activation of, 365–366
  activation effect on succinate-PMS and fumarate-$FMNH_2$ reactions of, 366
  of aerobic cells, 367–368
  of anaerobic cells, 368
  assay and chemical determination, 362
  catalytic activities of alkali-treated and reactivated preparations of, 370
  evolutionary development of, 367–369
  iron of, 367
  nature and composition of preparations, 363
  properties of, 363–364
Succinate oxidase, function of cytochrome b in, 27
  restoration by quinones, 545–548
Succinate reductase, reconstitution of, 369–373

Succinoxidase, alkali inactivation and reactivation of, 372
  catalytic activities of alkali-treated and reactivated preparations of, 370
  reconstituted, functioning of, 371
Sulfhydryl, in aldehyde and xanthine oxidases, 319
Sulfhydryl group reagents, inhibition of phosphorylation by, 211
Sulfite, oxidation of, 313

T

Terminal reductases, of bacteria, 92–96
Tetrahydrofolate, formaldehyde interaction with, mechanism for, 476 (*see also* Dihydrofolate, Folate, *and* Folic acid)
  interaction of aldehydes with, 475–478
  pK_a values for groups in, 445
  transamidation, mechanism for, 475
Tetrahydrofolic acid, structure of, 440
Tetramethylparaphenylenediamine, as mediator of electrons from ascorbate, 25
Thymidylate synthetase, 458–460
  reaction mechanism for, 459
TMPD. *See* Tetramethylparaphenylenediamine.
Tocopherylquinones. *See* Quinones.
TPN, 3′-isomer of, 609
TPNH-cytochrome *c* reductase, 177
  mechanism for, 183
  reaction mechanism for, 292–293
  reduction-oxidation of, 182–183
TPNH requirement, in microsomes, 173
"TPN-reducing factor." *See* Ferredoxin.
Transamidation. *See* Tetrahydrofolate.
Transport, into bacterial cells, 104–106
  energy source for, in bacteria, 105
Tryptophane pyrrolase. *See* Dioxygenases.
Two-electron transfer, of flavoproteins, linkage to one-electron transfer cytochromes, 289–290, 296–297
Tyrosinase, 419–428
  active sites of, 422–423
  detailed reaction mechanism of, 426–428
  double reaction specificity of, 421–423
  induction period of, 421

mechanisms for, 427
reaction mechanism of, 425–428
reactions catalyzed by, 420
sources of, 420
state of copper in, 423–425
substrates and products for catecholase activity of, 425
substrate and products for cresolase activity of, 425–426

U

Ubiquinone, hydrogen transport function of, 17–19 (*see also* Quinones)

V

Vitamin B_{12}, coenzyme form of, 483–484 (*see also* B_{12} coenzymes)
  lytic reactions of, 486–487
  mechanism for non-enzymic reduction of, 490
  reduction of, 485–486
    by dithiols, 489–490
  spectra of, 484
  structure of, 483
Vitamin B_{12} coenzymes, chemical and enzymic synthesis of, 487–492
  spectra of, 485
Vitamin K, in bacterial electron transport, 558–561 (*see also* Quinones)
  in electron transport, 557–561
  electron transport path in bacteria, 559

W

Water formation, from P_i with ADP, schemes for, 199–200
Water uptake and swelling, by mitochondria, 222
Wavelength of light, relation to shifts from noncyclic to cyclic photophosphorylation, 160–162

X

Xanthine oxidase, absorption spectra of, 304
  anaerobic titration of, 325–326
  catalytic activities of, 308–310
  chemiluminescence induction by, 313–314

cyanide inhibition of, 316–318
cytochrome *c* reduction by, 313
electron transport sequence of, 321–322
kinetics of, 310
metal–flavin ratios of, 306–307
methanol inactivation of, 316
oxygen effect on, 313

oxygen radical formation by, 313–314
pH dependence of, 309
reaction mechanism of, 310
"single turnover" EPR experiment
with, 321
specificity of, 308–310
sulfite oxidation by, 313

BIOLOGICAL OXIDATIONS